T0252351

Multivariate Time Series Analysis and Applications

WILEY SERIES IN PROBABILITY AND STATISTICS

Established by *Walter A. Shewhart and Samuel S. Wilks*

Editors: *David J. Balding, Noel A. C. Cressie, Garrett M. Fitzmaurice, Geof H. Givens, Harvey Goldstein, Geert Molenberghs, David W. Scott, Adrian F. M. Smith, Ruey S. Tsay*

Editors Emeriti: *J. Stuart Hunter, Iain M. Johnstone, Joseph B. Kadane, Jozef L. Teugels*

The ***Wiley Series in Probability and Statistics*** is well established and authoritative. It covers many topics of current research interest in both pure and applied statistics and probability theory. Written by leading statisticians and institutions, the titles span both state-of-the-art developments in the field and classical methods.

Reflecting the wide range of current research in statistics, the series encompasses applied, methodological and theoretical statistics, ranging from applications and new techniques made possible by advances in computerized practice to rigorous treatment of theoretical approaches. This series provides essential and invaluable reading for all statisticians, whether in academia, industry, government, or research.

A complete list of titles in this series can be found at
http://www.wiley.com/go/wsps

Multivariate Time Series Analysis and Applications

William W.S. Wei
Department of Statistical Science
Temple University, Philadelphia, PA, USA

This edition first published 2019
© 2019 John Wiley & Sons Ltd

All rights reserved. No part of this publication may be reproduced, stored in a retrieval system, or transmitted, in any form or by any means, electronic, mechanical, photocopying, recording or otherwise, except as permitted by law. Advice on how to obtain permission to reuse material from this title is available at http://www.wiley.com/go/permissions.

The right of William W.S. Wei to be identified as the author of this work has been asserted in accordance with law.

Registered Offices
John Wiley & Sons, Inc., 111 River Street, Hoboken, NJ 07030, USA
John Wiley & Sons Ltd, The Atrium, Southern Gate, Chichester, West Sussex, PO19 8SQ, UK

Editorial Office
9600 Garsington Road, Oxford, OX4 2DQ, UK

For details of our global editorial offices, customer services, and more information about Wiley products visit us at www.wiley.com.

Wiley also publishes its books in a variety of electronic formats and by print-on-demand. Some content that appears in standard print versions of this book may not be available in other formats.

Limit of Liability/Disclaimer of Warranty
MATLAB® is a trademark of The MathWorks, Inc. and is used with permission. The MathWorks does not warrant the accuracy of the text or exercises in this book. This work's use or discussion of MATLAB® software or related products does not constitute endorsement or sponsorship by The MathWorks of a particular pedagogical approach or particular use of the MATLAB® software. While the publisher and authors have used their best efforts in preparing this work, they make no representations or warranties with respect to the accuracy or completeness of the contents of this work and specifically disclaim all warranties, including without limitation any implied warranties of merchantability or fitness for a particular purpose. No warranty may be created or extended by sales representatives, written sales materials or promotional statements for this work. The fact that an organization, website, or product is referred to in this work as a citation and/or potential source of further information does not mean that the publisher and authors endorse the information or services the organization, website, or product may provide or recommendations it may make. This work is sold with the understanding that the publisher is not engaged in rendering professional services. The advice and strategies contained herein may not be suitable for your situation. You should consult with a specialist where appropriate. Further, readers should be aware that websites listed in this work may have changed or disappeared between when this work was written and when it is read. Neither the publisher nor authors shall be liable for any loss of profit or any other commercial damages, including but not limited to special, incidental, consequential, or other damages.

Library of Congress Cataloging-in-Publication data applied for

Hardback: 9781119502852

Cover design: Wiley
Cover image: Courtesy of William W.S. Wei

Set in 10/12pt Times by SPi Global, Pondicherry, India

Printed in the UK

To my Grandpa, Parents, Wife, Sons,
and Daughter

Contents

About the author

William W.S. Wei is a Professor of Statistical Science at Temple University in Philadelphia, Pennsylvania, United States of America. He earned his BA in Economics from the National Taiwan University (1966), BA in Mathematics from the University of Oregon (1969), and M.S. (1972) and Ph.D. (1974) in Statistics from the University of Wisconsin, Madison. From 1982 to 1987, he was the Chair of the Department of Statistics at Temple University. He has been a Visiting Professor at many universities including the Nankai University in China, National University of Colombia in Colombia, Korea University in Korea, National Sun Yat-Sen University, National Chiao Tung University, and National Taiwan University in Taiwan, and Middle East Technical University in Turkey. His research interests include time series analysis, forecasting methods, high dimensional problems, statistical modeling, and their applications. He has developed new methodologies in seasonal adjustment, aggregation and disaggregation, outlier detection, robust estimation, and vector time series analysis. Some of his most significant contributions include extensive research on the effects of aggregation, methods of measuring information loss due to aggregation, new stochastic procedures of performing data disaggregation, model-free outlier detection techniques, robust methods of estimating autocorrelations, statistics for analyzing multivariate time series, and dimension reduction for high-dimensional time series. His first book, *Time Series Analysis – Univariate and Multivariate Methods*, the first edition published in 1990 and the second edition published in 2006, has been translated into several languages and heavily cited by researchers worldwide. He has just completed his second book, *Multivariate Time Series Analysis and Applications*. He is an active educator and researcher. He has successfully supervised many Ph.D. students, who hold teaching positions at universities or leadership positions in government and industry throughout the world. He is a Fellow of the American Statistical Association, a Fellow of the Royal Statistical Society, and an Elected Member of the International Statistical Institute. He was the 2002 President of ICSA (International Chinese Statistical Association). He is currently an Associate Editor of the *Journal of Forecasting* and the *Journal of Applied Statistical Science*. In addition to teaching and research, he is also active in community service. He served on the educational advisory committee of his local school district, as the chair of the selection committee for a community high school scholarship program, and as the president of several community organizations, including the Taiwanese Hakka Associations of America. Among the many awards he has received are the 2014 Lifetime Achievement Award and the 2016 Musser Award for Excellence in Research from the Temple University Fox School of Business.

Preface

My main research area is time series analysis and forecasting. I have written a book, *Time Series Analysis – Univariate and Multivariate Methods*. Since the first edition was published in 1990 and the second edition in 2006, the book has been used by many researchers and universities worldwide, and I have received many encouraging letters and emails from researchers, instructors, and students about the usefulness of the book in their research and studies. It has been translated into several languages including Chinese, Spanish, and Portuguese. With the development of computers and the internet, we have had a data explosion, and many new theories and methods have been developed in high-dimensional time series analysis. Many publishers have contacted me asking for a new edition. Because of the development of so much new material, it would be impractical to include it all in a new edition of the book. Therefore, I decided to write a new book and call it *Multivariate Time Series Analysis and Applications*. Due to the enthusiasm of multiple publishing editors, I had a difficult time deciding which publisher to choose. After consulting with Dr. Sarkar, our department chair, I decided to choose Wiley. After four stages of rigorous reviews, the Wiley directors unanimously approved the publication of this book.

Many research studies involve multivariate time series. For example, a study of monthly cancer rates in the United States during the past 10 years can involve 50, many hundreds, or even thousands of time series depending on whether we investigate the cancer rates for states, cities, or counties, and a study of the quarterly sales performance of one company's different branches in the country or the world from 2010 to 2018 may involve many hundreds or thousands of time series, depending on the number of products and the number of branches within the company. Multivariate time series analysis methods are needed to properly analyze the data in these studies, which are different from standard statistical theory and methods based on random samples that assume independence. Dependence is the fundamental nature of time series. The use of highly correlated high-dimensional time series data introduces many complications and challenges. The methods and theory to solve these issues will make up the contents of this book. After introducing the fundamental concepts and reviewing the standard vector time series models, we will explore many important issues, including multivariate time series regression, dimension reduction and model simplification, multivariate GARCH (generalized autoregressive conditional heteroskedasticity) models, repeated measurement phenomenon, space–time series modeling, multivariate spectral analysis for both stationary and nonstationary vector time series, and the high-dimension problem in multivariate time series analysis.

In this book, I follow the same fundamental themes of my research with a balanced emphasis on both theory and applications. Methodologies are introduced with proper theoretical justifications and illustrated with simulated and empirical data sets. It should be pointed out that this book is designed for a research-oriented second time series analysis course and therefore standard exercises normally found in the first course will not be provided. Instead, research-oriented projects will be suggested.

I am grateful to the Department of Statistical Science, Fox School of Business, and Temple University for granting me a sabbatical research leave to finish this book. Fox School is celebrating its 100-year anniversary in 2018, and this book is my way of joining in its centennial celebration. I am also grateful to my wife, Susanna, a professor at Saint Joseph's University, sons and daughter, Stephen, Stanley, and Jessica, who are all married physicians, for their help in proofreading my manuscripts even as they are so busy with their own careers and families. My very sincere thanks go to Zeda Li, Andrew Gehman, Nandi Shinjini, and Kaijun Wang, who are either my Ph.D. dissertation students, research/teaching assistants, or both, and Kevin Liu, who was one of our excellent Masters students in the department, for their outstanding assistance in collecting data sets and developing software programs used in the book. Finally, I would like to thank Ms. Alison Oliver, Publications Manager, and Ms. Jemima Kingsly, Project Editor; Ms. Francesca McMahon, Editorial Assistant, Ms. Blesy Regulas, Project Editor; Mr. Mustaq Ahamed Noorullah, Production Editor, all at John Wiley & Sons, and Ms. Lynette Woodward, Editorial & Academic Services, UK, who have been invaluable in completing this important book project.

William W.S. Wei
March 2018

About the companion website

This book is accompanied by a companion website:

www.wiley.com/go/wei/datasets

The website includes:

- Data Appendix (Bookdata)

Scan this QR code to visit the companion website.

1

Fundamental concepts and issues in multivariate time series analysis

With the development of computers and the internet, we have had a data explosion. For example, a study of monthly cancer rates in the United States during the past 10 years can involve 50 or many hundreds or thousands of time series depending on whether we investigate the cancer rates for states, cities, or counties. Multivariate time series analysis methods are needed to properly analyze these data in a study, and these are different from standard statistical theory and methods based on random samples that assume independence. Dependence is the fundamental nature of the time series. The use of highly correlated high-dimensional time series data introduces many complications and challenges. The methods and theory to solve these issues will make up the content of this book.

1.1 Introduction

In studying a phenomenon, we often encounter many variables, $Z_{i,t}$, where $i = 1, 2, \ldots, m$, and the observations are taken according to the order of time, t. For convenience we use a vector, $\mathbf{Z}_t = [Z_{1,t}, Z_{2,t}, \ldots, Z_{m,t}]'$, to denote the set of these variables, where $Z_{i,t}$ is the ith component variable at time t and it is a random variable for each i and t. The time t in $\mathbf{Z}_t$ can be continuous and any value in an interval, such as the time series of electric signals and voltages, or discrete and be a specific time point, such as the daily closing price of various stocks or the total monthly sales of various products at the end of each month. In practice, even for a continuous time series, we take observations only at digitized discrete time points for analysis. Hence, we will consider only discrete time series in this book, and with no loss of generalizability, we will consider $Z_{i,t}$, for $i = 1, 2, \ldots, m$, $t = 0, \pm 1, \pm 2, \ldots$, and hence $\mathbf{Z}_t = [Z_{1,t}, Z_{2,t}, \ldots, Z_{m,t}]'$, $t = 0, \pm 1, \pm 2, \ldots$.

We call $\mathbf{Z}_t = [Z_{1,t}, Z_{2,t}, \ldots, Z_{m,t}]'$ a multivariate time series or a vector time series, where the first subscript refers to a component and the second subscript refers to the time. The fundamental characteristic of a multivariate time series is that its observations depend not only on component i but also time t. The observations between $Z_{i,s}$ and $Z_{j,t}$ can be correlated when $i \neq j$,

Multivariate Time Series Analysis and Applications, First Edition. William W.S. Wei.
© 2019 John Wiley & Sons Ltd. Published 2019 by John Wiley & Sons Ltd.
Companion website: www.wiley.com/go/wei/datasets

regardless of whether the times s and t are the same or not. They are vector-valued random variables. Most standard statistical theory and methods based on random samples are not applicable, and different methods are clearly needed. The body of statistical theory and methods for analyzing these multivariate or vector time series is referred to as multivariate time series analysis.

Many issues are involved in multivariate time series analysis. They are different from standard statistical theory and methods based on a random sample that assumes independence and constant variance. In multivariate time series, $\mathbf{Z}_t = [Z_{1,t}, Z_{2,t}, \ldots, Z_{m,t}]'$, a fundamental phenomenon is that dependence exists not only in terms of i but also in terms of t. In addition, we have the following important issues to consider:

1. **Fundamental concepts and representations related to dependence.**
   We will introduce the variance–covariance and correlation matrix functions, vector white noise processes, vector autoregressive and vector moving average representations, vector autoregressive models, vector moving average models, and vector autoregressive moving average models.

2. **Relationship between several multivariate time series.**
   A multiple regression is known to be a useful statistical model that describes a relationship between a response variable and several predictor variables. The error term in the model is normally assumed to be uncorrelated noise with zero mean and constant variance. We will extend the results to a multivariate time series regression model where both response variables, and predictor variables are vectors. More importantly, not only are all components in the multivariate regression equation time series variables, but also the error term follows a correlated time series process.

3. **Dimension reduction and model simplification.**
   Without losing information, we will introduce useful methods of dimension reduction and representation including principal components and factor analysis in time series.

4. **Representations of time variant variance–covariance structure.**
   Unlike most classical linear methods, where the variance of error term has been assumed to be constant, in time series analysis a non-constant variance often occurs and generalized autoregressive conditional heteroscedasticity (GARCH) models are been introduced. The literature and theory of GARCH models for univariate time series was introduced in chapter 15 of Wei (2006). In this book, we will extend the results to the multivariate GARCH models.

5. **Repeated measurement phenomenon.**
   Many fields of study, including medicine, biology, social science, and education, involve time series measurements of treatments for different subjects. They are multivariate time series but often relatively short, and the applications of standard time series methods are difficult, if not impossible. We will introduce some methods and various models that are useful for analyzing repeated measures data. Empirical examples will be used as illustrations.

6. **Space and time series modeling.**
   In many multivariate time series applications, the components i in $Z_{i,t}$ refer to regions or locations. For example, in a crime study, the observations can be the crime rates of

different counties in a state, and in a market study, one could look at the price of a certain product in different regions. Thus, the analysis will involve both regions and time, and we will construct space and time series models.

7. **Multivariate spectral analysis.**
   Similar to univariate time series analysis where one can study a time series through its autocovariance/autocorrelation functions and lag relationships or through its spectrum properties, we can study a multivariate time series through a time domain or a frequency domain approach. In the time domain approach we use covariance/correlation matrices, and in the frequency domain approach we will use spectral matrices. We will introduce spectral analysis for both multivariate stationary and nonstationary time series.

8. **High dimension problem in multivariate time series.**
   Because of high-speed internet and the power and speed of the new generation of computers, a researcher now faces some very challenging phenomena. First, he/she must deal with an ever-increasing amount of data. To find useful information and hidden patterns underlying the data, a researcher may use various data-mining methods and techniques. Adding a time dimension to these large databases certainly introduces new aspects and challenges. In multivariate time series analysis, a very natural issue is the high dimension problem where the number of parameters may exceed the length of the time series. For example, a simple second order vector autoregressive VAR(2) model for the 50 states in the USA will involve more than 5000 parameters, and the length of the time series may be much shorter. For example, the length of the monthly observations for 20 years is only 240. Traditional time series methods are not designed to deal with these kinds of high-dimensional variables. Even with today's computer power and speed, there are many difficult problems that remain unsolved. As most statistical methods are developed for a random sample, the use of highly correlated time series data certainly introduces a new set of complications and challenges, especially for a high-dimensional data set.

The methods and theory to solve these issues will be the focus of this book. Examples and applications will be carefully chosen and presented.

1.2 Fundamental concepts

The m-dimensional vector time series process, $\mathbf{Z}_t = [Z_{1,t}, Z_{2,t}, \ldots, Z_{m,t}]'$, is a stationary process if each of its component series is a univariate stationary process and its first two moments are time-invariant. Just as a univariate stationary process or model is characterized by its moments such as mean, autocorrelation function, and partial autocorrelation function, a stationary vector time series process or model is characterized by its mean vector, correlation matrix function, and partial correlation matrix function.

1.2.1 Correlation and partial correlation matrix functions

Let $\mathbf{Z}_t = [Z_{1,t}, Z_{2,t}, \ldots, Z_{m,t}]'$, $t = 0, \pm 1, \pm 2, \ldots$ be a m-dimensional stationary real-valued vector process so that $E(Z_{i,t}) = \mu_i$ is constant for each $i = 1, 2, \ldots, m$ and the cross-covariance between $Z_{i,t}$ and $Z_{j,s}$, for all $i = 1, 2, \ldots, m$ and $j = 1, 2, \ldots, m$, are functions only of the time difference $(s - t)$. Hence, we have the mean vector

$$E(\mathbf{Z}_t) = \boldsymbol{\mu} = \begin{bmatrix} \mu_1 \\ \mu_2 \\ \vdots \\ \mu_m \end{bmatrix}, \tag{1.1}$$

and the lag k covariance matrix

$$\begin{aligned}
\boldsymbol{\Gamma}(k) &= \text{Cov}\{\mathbf{Z}_t, \mathbf{Z}_{t+k}\} = E\left[(\mathbf{Z}_t - \boldsymbol{\mu})(\mathbf{Z}_{t+k} - \boldsymbol{\mu})'\right] \\
&= E\begin{bmatrix} Z_{1,t} - \mu_1 \\ Z_{2,t} - \mu_2 \\ \vdots \\ Z_{m,t} - \mu_m \end{bmatrix} [Z_{1,t+k} - \mu_1, Z_{2,t+k} - \mu_2, \ldots, Z_{m,t+k} - \mu_{\text{m}}] \\
&= \begin{bmatrix} \gamma_{1,1}(k) & \gamma_{1,2}(k) & \cdots & \gamma_{1,m}(k) \\ \gamma_{2,1}(k) & \gamma_{2,2}(k) & \cdots & \gamma_{2,m}(k) \\ \vdots & \vdots & \vdots & \vdots \\ \gamma_{m,1}(k) & \gamma_{m,2}(k) & \cdots & \gamma_{m,m}(k) \end{bmatrix},
\end{aligned} \tag{1.2}$$

where

$$\gamma_{i,j}(k) = E(Z_{i,t} - \mu_i)(Z_{j,t+k} - \mu_j)$$

for $k = 0, \pm 1, \pm 2, \ldots, i = 1, 2, \ldots, m$, and $j = 1, 2, \ldots, m$. As a function of k, $\boldsymbol{\Gamma}(k)$ is called the covariance matrix function for the vector process $\mathbf{Z}_t$. Also, $i = j$, $\gamma_{i,i}(k)$ is the autocovariance function for the ith component process, $Z_{i,t}$; and $i \neq j$, $\gamma_{i,j}(k)$ is the cross-covariance function between component series $Z_{i,t}$ and $Z_{j,t}$. The matrix $\boldsymbol{\Gamma}(0)$ can be easily seen to be the contemporaneous variance–covariance matrix of the process.

The covariance matrix function has the following properties:

1. $\boldsymbol{\Gamma}(k) = \boldsymbol{\Gamma}'(-k)$. This follows because

$$\gamma_{i,j}(k) = E\left[(Z_{i,t} - \mu_i)(Z_{j,t+k} - \mu_j)\right] = E\left[(Z_{j,t+k} - \mu_j)(Z_{i,t} - \mu_i)\right] = \gamma_{j,i}(-k).$$

2. $|\gamma_{i,j}(k)| \leq [\gamma_{i,i}(0)\gamma_{j,j}(0)]^{1/2}$, for all i, $j = 1, \ldots, m$, because of the Cauchy–Schwarz inequality.
3. The covariance matrix function is positive semidefinite in the sense that

$$\sum_{i=1}^{n}\sum_{j=1}^{n} \boldsymbol{\alpha}_i' \boldsymbol{\Gamma}(t_i - t_j)\boldsymbol{\alpha}_j \geq 0, \tag{1.3}$$

for any set of time points $t_1, t_2, \ldots, t_n$ and any set of real vectors $\boldsymbol{\alpha}_1, \boldsymbol{\alpha}_2, \ldots, \boldsymbol{\alpha}_n$. The result follows immediately from $\mathrm{Var}\left[\sum_{i=1}^{n} \boldsymbol{\alpha}_i' \mathbf{Z}_{t_i}\right] \geq 0$.

The correlation matrix function for the vector process is defined by

$$\boldsymbol{\rho}(k) = \mathbf{D}^{-1/2}\boldsymbol{\Gamma}(k)\mathbf{D}^{-1/2} = \left[\rho_{i,j}(k)\right] \tag{1.4}$$

for $i = 1, 2, \ldots, m$, and $j = 1, 2, \ldots, m$, where $\mathbf{D}$ is the diagonal matrix in which the ith diagonal element is the variance of the ith process; that is, $\mathbf{D} = diag\,[\gamma_{1,1}(0), \gamma_{2,2}(0), \ldots, \gamma_{m,m}(0)]$. Thus, the ith diagonal element of $\boldsymbol{\rho}(k)$ is the autocorrelation function for the ith component series $Z_{i,t}$ whereas the (i, j)th off-diagonal element of $\boldsymbol{\rho}(k)$ is the cross-correlation function between component series $Z_{i,t}$ and $Z_{j,t}$.

Similarly, the correlation matrix functions have the following properties:

1. $\boldsymbol{\rho}(k) = \boldsymbol{\rho}'(-k)$.
2. The correlation matrix function is positive semidefinite so that

$$\sum_{i=1}^{n}\sum_{j=1}^{n} \boldsymbol{\alpha}_i' \boldsymbol{\rho}\left(t_i - t_j\right)\boldsymbol{\alpha}_j \geq 0, \tag{1.5}$$

for any set of time points $t_1, t_2, \ldots, t_n$ and any set of real vectors $\boldsymbol{\alpha}_1, \boldsymbol{\alpha}_2, \ldots, \boldsymbol{\alpha}_n$.

Other than the correlation matrix function, another useful function for describing a vector time series process is the partial correlation matrix function. The concept of a partial correlation matrix function was introduced much later, and there are different versions.

Heyse and Wei (1985) extended the definition of univariate partial autocorrelation to vector time series and derived the correlation matrix between $\mathbf{Z}_t$ and $\mathbf{Z}_{t+s}$ after removing the linear dependence of each on the intervening vectors $\mathbf{Z}_{t+1}, \ldots, \mathbf{Z}_{t+s-1}$. This partial correlation matrix is defined as the correlation between the residual vectors

$$\begin{aligned}\mathbf{U}_{s-1,t+s} &= \mathbf{Z}_{t+s} - \boldsymbol{\alpha}_{s-1,1}\mathbf{Z}_{t+s-1} - \cdots - \boldsymbol{\alpha}_{s-1,s-1}\mathbf{Z}_{t+1} \\ &= \begin{cases} \mathbf{Z}_{t+s} - \sum_{j=1}^{s-1} \boldsymbol{\alpha}_{s-1,j}\mathbf{Z}_{t+s-j}, & s \geq 2, \\ \mathbf{Z}_{t+1}, & s = 1, \end{cases}\end{aligned} \tag{1.6}$$

and

$$\begin{aligned}\mathbf{V}_{s-1,t} &= \mathbf{Z}_t - \boldsymbol{\beta}_{s-1,1}\mathbf{Z}_{t+1} - \cdots - \boldsymbol{\beta}_{s-1,s-1}\mathbf{Z}_{t+s-1} \\ &= \begin{cases} \mathbf{Z}_t - \sum_{j=1}^{s-1} \boldsymbol{\beta}_{s-1,j}\mathbf{Z}_{t+j}, & s \geq 2, \\ \mathbf{Z}_{t+1}, & s = 1. \end{cases}\end{aligned} \tag{1.7}$$

Let $\mathbf{C}_{\mathbf{VU}}(s)$ be the covariance between $\mathbf{V}_{s-1,t}$ and $\mathbf{U}_{s-1,t+s}$, i.e. $\mathbf{C}_{\mathbf{VU}}(s) = \mathrm{Cov}(\mathbf{V}_{s-1,t}, \mathbf{U}_{s-1,t+s})$, Heyse and Wei (1985) showed that

$$\mathbf{C_{VU}}(s)=\boldsymbol{\Gamma}(s)-[\boldsymbol{\Gamma}(s-1),\boldsymbol{\Gamma}(s-2),\ldots,\boldsymbol{\Gamma}(1)]\begin{bmatrix}\boldsymbol{\Gamma}(0) & \boldsymbol{\Gamma}'(1) & \cdots & \boldsymbol{\Gamma}'(s-2)\\ \boldsymbol{\Gamma}(1) & \boldsymbol{\Gamma}(0) & \cdots & \boldsymbol{\Gamma}'(s-3)\\ \vdots & \vdots & \vdots & \vdots\\ \boldsymbol{\Gamma}(s-2) & \boldsymbol{\Gamma}(s-3) & \cdots & \boldsymbol{\Gamma}(0)\end{bmatrix}^{-1}\begin{bmatrix}\boldsymbol{\Gamma}(1)\\ \boldsymbol{\Gamma}(2)\\ \vdots\\ \boldsymbol{\Gamma}(s-1)\end{bmatrix}, \tag{1.8}$$

$$\begin{aligned}\mathbf{C_{UU}}(0)&=\mathrm{Var}(\mathbf{U}_{s-1,t+s})=E\left(\mathbf{U}_{s-1,t+s}\mathbf{U}'_{s-1,t+s}\right)\\ &=\boldsymbol{\Gamma}(0)-[\boldsymbol{\Gamma}'(1),\boldsymbol{\Gamma}'(2),\ldots,\boldsymbol{\Gamma}'(s-1)]\begin{bmatrix}\boldsymbol{\Gamma}(0) & \boldsymbol{\Gamma}'(1) & \cdots & \boldsymbol{\Gamma}'(s-2)\\ \boldsymbol{\Gamma}(1) & \boldsymbol{\Gamma}(0) & \cdots & \boldsymbol{\Gamma}'(s-3)\\ \vdots & \vdots & \vdots & \vdots\\ \boldsymbol{\Gamma}(s-2) & \boldsymbol{\Gamma}(s-3) & \cdots & \boldsymbol{\Gamma}(0)\end{bmatrix}^{-1}\begin{bmatrix}\boldsymbol{\Gamma}(1)\\ \boldsymbol{\Gamma}(2)\\ \vdots\\ \boldsymbol{\Gamma}(s-1)\end{bmatrix},\end{aligned} \tag{1.9}$$

and

$$\begin{aligned}\mathbf{C_{VV}}(0)&=\mathrm{Var}(\mathbf{V}_{s-1,t})=E\left(\mathbf{V}_{s-1,t}\mathbf{V}'_{s-1,t}\right)\\ &=\boldsymbol{\Gamma}(0)-[\boldsymbol{\Gamma}(s-1),\boldsymbol{\Gamma}(s-2),\ldots,\boldsymbol{\Gamma}(1)]\begin{bmatrix}\boldsymbol{\Gamma}(0) & \boldsymbol{\Gamma}'(1) & \cdots & \boldsymbol{\Gamma}'(s-2)\\ \boldsymbol{\Gamma}(1) & \boldsymbol{\Gamma}(0) & \cdots & \boldsymbol{\Gamma}'(s-3)\\ \vdots & \vdots & \vdots & \vdots\\ \boldsymbol{\Gamma}(s-2) & \boldsymbol{\Gamma}(s-3) & \cdots & \boldsymbol{\Gamma}(0)\end{bmatrix}^{-1}\begin{bmatrix}\boldsymbol{\Gamma}'(s-1)\\ \boldsymbol{\Gamma}'(s-2)\\ \vdots\\ \boldsymbol{\Gamma}'(1)\end{bmatrix},\end{aligned} \tag{1.10}$$

where $\boldsymbol{\Gamma}(k)=\mathrm{Cov}\{\mathbf{Z}_t,\mathbf{Z}_{t+k}\}=E\left(\mathbf{Z}_t\mathbf{Z}'_{t+k}\right)$. Thus, the partial autocorrelation matrix function is

$$\mathbf{P}(s)=[\mathbf{D_V}(s)]^{-1}\mathbf{C_{VU}}(s)[\mathbf{D_U}(s)]^{-1}, \tag{1.11}$$

where $\mathbf{D_V}(s)$ is the diagonal matrix in which the ith diagonal element is the square root of the ith diagonal element of $\mathbf{C_{VV}}(0)$ and $\mathbf{D_U}(s)$ is similarly defined for $\mathbf{C_{UU}}(0)$.

Tiao and Box (1981) defined the partial autoregression matrix at lag s, denoted by $\boldsymbol{\Phi}_{s,s}$, to be the last matrix coefficient when the data is fitted to a vector autoregressive process of order s. It can be shown that

$$\boldsymbol{\Phi}_{s,s}=\mathbf{C}'_{\mathbf{VU}}(s)[\mathbf{D_V}(s)]^{-1}. \tag{1.12}$$

Ansley and Newbold (1979) defined the multivariate partial autocorrelation matrix at lag s to be

$$\mathbf{Q}(s)=[\mathbf{W_U}(s)]^{-1}\mathbf{C}'_{\mathbf{VU}}(s)[\mathbf{W_V}(s)]^{-1}, \tag{1.13}$$

where $\mathbf{W}_{\mathbf{U}}(s)$ and $\mathbf{W}_{\mathbf{V}}(s)$ are the symmetric square roots of $\mathbf{C}_{\mathbf{UU}}(0)$ and $\mathbf{C}_{\mathbf{VV}}(0)$, respectively, defined such that $|\mathbf{W}_{\mathbf{U}}(s)|^2 = \mathbf{C}_{\mathbf{UU}}(0)$ and $|\mathbf{W}_{\mathbf{V}}(s)|^2 = \mathbf{C}_{\mathbf{VV}}(0)$. However, it should be noted that although $\mathbf{P}(s)$, $\mathbf{\Phi}_{s,s}$ and $\mathbf{Q}(s)$ all share the same cut-off property for vector VAR(s) models, the elements of $\mathbf{P}(s)$ are proper correlation coefficient but those of $\mathbf{\Phi}_{s,s}$ and $\mathbf{Q}(s)$ are not correlation coefficient except when $m = 1$; that is, except in the univariate case in which $\mathbf{P}(s) = \mathbf{\Phi}_{s,s} = \mathbf{Q}(s)$. For more details, we refer readers to Wei (2006, chapter 16).

1.2.2 Vector white noise process

The m-dimensional vector process, $\mathbf{a}_t$, is said to be a vector white noise process with mean vector $\mathbf{0}$ and covariance matrix function $\mathbf{\Sigma}$ if

$$E\left[\mathbf{a}_t\mathbf{a}'_{t+k}\right] = \begin{cases} \mathbf{\Sigma}, \text{if } k = 0, \\ \mathbf{0}, \text{if } k \neq 0, \end{cases} \tag{1.14}$$

where $\mathbf{\Sigma}$ is a $m \times m$ symmetric positive definite matrix. Note that although the components of the white noise process are uncorrelated at different times, they may be contemporaneously correlated. It is a Gaussian white noise process if $\mathbf{a}_t$ also follows a multivariate normal distribution. Unless mentioned otherwise, $\mathbf{a}_t$ will be used to denote a Gaussian vector white noise process with mean vector $\mathbf{0}$ and covariance matrix function $\mathbf{\Sigma}$, **VWN**($\mathbf{0}$, $\mathbf{\Sigma}$), in this book.

1.2.3 Moving average and autoregressive representations of vector processes

A m-dimensional stationary vector time series process $\mathbf{Z}_t$ that is purely nondeterministic can always be written as a linear combination of a sequence of vector white noises, that is

$$\begin{aligned} \mathbf{Z}_t &= \boldsymbol{\mu} + \mathbf{a}_t + \mathbf{\Psi}_1\mathbf{a}_{t-1} + \mathbf{\Psi}_2\mathbf{a}_{t-2} + \cdots \\ &= \boldsymbol{\mu} + \sum_{\ell=0}^{\infty}\mathbf{\Psi}_\ell\mathbf{a}_{t-\ell} = \boldsymbol{\mu} + \mathbf{\Psi}(B)\mathbf{a}_t, \end{aligned} \tag{1.15}$$

where $\mathbf{\Psi}(B) = \sum_{\ell=0}^{\infty}\mathbf{\Psi}_\ell B^\ell, \mathbf{\Psi}_0 = \mathbf{I}$, and the sequence of $m \times m$ coefficient matrices $\mathbf{\Psi}_\ell$ is square summable, $\sum_{\ell=0}^{\infty}\|\mathbf{\Psi}_\ell\|^2 < \infty$, in the sense that if we write $\mathbf{\Psi}(B) = \sum_{\ell=0}^{\infty}\mathbf{\Psi}_\ell B^\ell = \left[\psi_{i,j}(B)\right]$, and $\psi_{i,j}(B) = \sum_{\ell=0}^{\infty}\psi_{i,j,\ell}B^\ell$, we have $\sum_{\ell=0}^{\infty}\psi_{i,j,\ell}^2 < \infty$, for all $i = 1, 2, \ldots, m$ and $j = 1, 2, \ldots, m$. The B is the backshift operator such that $B^j\mathbf{a}_t = \mathbf{a}_{t-j}$.

The infinite sum of random variables is defined as the limit in quadratic mean of the finite partial sums. Thus, $\mathbf{Z}_t$ in Eq. (1.15) is defined such that

$$E\left[\left(\mathbf{Z}_t - \boldsymbol{\mu} - \sum_{\ell=0}^{n}\mathbf{\Psi}_\ell\mathbf{a}_{t-\ell}\right)'\left(\mathbf{Z}_t - \boldsymbol{\mu} - \sum_{\ell=0}^{n}\mathbf{\Psi}_\ell\mathbf{a}_{t-\ell}\right)\right] \to 0 \text{ as } n \to \infty. \tag{1.16}$$

The Eq. (1.15) is known as the vector moving average (MA) representation.

For a given sequence of covariance matrices, $\mathbf{\Gamma}(k)$, $k = 0, \pm 1, \pm 2, \ldots$, the covariance matrix generating function is defined as

$$\mathbf{\Gamma}(B) = \sum_{k=-\infty}^{\infty} \mathbf{\Gamma}(k) B^k, \tag{1.17}$$

where the covariance matrix of lag k is the coefficient of B^k and B^{-k}. For a stationary vector process given in Eq. (1.15), it can be easily seen that

$$\mathbf{\Gamma}(B) = \sum_{k=-\infty}^{\infty} \mathbf{\Gamma}(k) B^k = \mathbf{\Psi}(B)\mathbf{\Sigma}\mathbf{\Psi}'\left(B^{-1}\right), \tag{1.18}$$

where $\mathbf{\Psi}'(B^{-1}) = \sum_{j=0}^{\infty}\mathbf{\Psi}_j' B^{-j}$.

A vector time series process $\mathbf{Z}_t$ is said to be invertible if it can be written as a vector autoregressive (AR) representation

$$\begin{aligned}\mathbf{Z}_t &= \mathbf{\theta}_0 + \mathbf{\Pi}_1 \mathbf{Z}_{t-1} + \mathbf{\Pi}_2 \mathbf{Z}_{t-2} + \cdots + \mathbf{a}_t \\ &= \mathbf{\theta}_0 + \sum_{\ell=1}^{\infty} \mathbf{\Pi}_\ell \mathbf{Z}_{t-\ell} + \mathbf{a}_t,\end{aligned} \tag{1.19}$$

or equivalently,

$$\mathbf{\Pi}(B)\mathbf{Z}_t = \mathbf{\theta}_0 + \mathbf{a}_t, \tag{1.20}$$

where $\mathbf{\Pi}(B) = \sum_{\ell=0}^{\infty}\mathbf{\Pi}_\ell B^\ell$, $\mathbf{\Pi}_0 = \mathbf{I}$, so that the sequence of $m \times m$ autoregressive coefficient matrices $\mathbf{\Pi}_\ell$ is absolutely summable in the sense that if we write $\mathbf{\Pi}(B) = \sum_{\ell=0}^{\infty}\mathbf{\Pi}_\ell B^\ell = \left[\pi_{i,j}(B)\right]$, and $\pi_{i,j}(B) = \sum_{\ell=0}^{\infty}\pi_{i,j,\ell}B^\ell$, we have $\sum_{\ell=0}^{\infty}\left|\pi_{i,j,\ell}\right| < \infty$, for all i = 1, 2, …, m and j = 1, 2, …, m.

Remarks

1. Statistical software is needed to perform time series data analysis. We will provide the associated software code for the empirical examples used in the book at the end of chapters. Since R is a free software package supported by researchers through the R Foundation for Statistical Computing, we will use R (2018 version, Rx64 3.4.2) most of the time. Some procedures are not available or difficult to implement in R, so then SAS (2015) or MATLAB (matrix laboratory) will be used. Readers should know that even in the same software, available functions may vary from version to version. For various analyses of empirical examples in the book, readers can simply copy the software code and paste into the relevant software on your laptop to get the presented outputs. When you run into difficulty in using the code, it could be because required functions are missing in the software version you are using.
2. This book is designed for a research-oriented advanced time series analysis course. Problem sets normally found in an introductory course will not be provided. Instead, research-oriented projects will be suggested.

3. To illustrate the use of multivariate time series analysis methods introduced in the book, we have used very extensive empirical examples with data sets from "Bookdata" installed on the C drive of my laptop computer as shown in the software codes. The printed data sets are provided in the book appendix. To help readers, I have uploaded these data sets as "**Data Set 2**" on my website (http://astro.temple.edu/~wwei/). Sincerely thank to Wiley, you can also access the data sets through its provided website: www.wiley.com/go/wei/datasets.

Projects

1. Find a univariate time series book, for example, Wei (2006, ch. 1–13), to review your background.
2. Write a detailed study plan on multivariate time series analysis and applications so that you can evaluate your accomplishments at the end of the semester if you use this book in your course.

References

Ansley, C.F. and Newbold, P. (1979). Multivariate partial autocorrelations. *American Statistical Association Proceedings of Business and Economic Statistics Section*, pp. 349–353.

Heyse, J.F. and Wei, W.W.S. (1985). Inverse and partial lag autocorrelation for vector time series. *American Statistical Association Proceedings of Business and Economic Statistics Section*, pp. 233–237.

MATLAB (matrix laboratory). A proprietary programming language developed by MathWorks. www.mathworks.com.

R Foundation. R: a free programming language and software environment for statistical computing and graphics, supported by the R Foundation for Statistical Computing. https://www.r-project.org.

SAS Institute, Inc. (2015). SAS for Windows, 9.4. Cary, NC: SAS Institute, Inc.

Tiao, G.C. and Box, G.E.P. (1981). Modeling multiple time series with applications. *Journal of the American Statistical Association* **76**: 802–816.

Wei, W.W.S. (2006). *Time Series Analysis – Univariate and Multivariate Methods*, 2e. Boston, MA: Pearson Addison-Wesley.

2

Vector time series models

Although as shown in Chapter 1, a stationary vector time series process $\mathbf{Z}_t$ that is purely nondeterministic can always be written as a moving average representation, and an invertible vector time series process can always be written as an autoregressive representation; they involve an infinite number of coefficient matrices in the representation. In practice, with a finite number of observations, we will construct a time series model with a finite number of coefficient matrices. Specifically, we will present VMA(q), VAR(p), VARMA(p, q), and VARMA(p, q) × (P, Q)$_s$ models in this chapter. After introducing the properties of these models, we will discuss their parameter estimation and forecasting. We will also introduce the concept of multivariate time series (MTS) outliers and their detections. Detailed empirical examples will be given.

2.1 Vector moving average processes

The m-dimensional vector moving average process or model in the order of q, shortened to VMA(q), is given by

$$\begin{aligned} \mathbf{Z}_t &= \boldsymbol{\mu} + \mathbf{a}_t - \boldsymbol{\Theta}_1 \mathbf{a}_{t-1} - \cdots - \boldsymbol{\Theta}_q \mathbf{a}_{t-q} \\ &= \boldsymbol{\mu} + \boldsymbol{\Theta}_q(B)\mathbf{a}_t, \end{aligned} \tag{2.1}$$

where $\boldsymbol{\Theta}_q(B) = \mathbf{I} - \boldsymbol{\Theta}_1 B - \cdots - \boldsymbol{\Theta}_q B^q$, $\mathbf{a}_t$ is a sequence of the m-dimensional vector white noise process, VWN($\mathbf{0}$, $\boldsymbol{\Sigma}$), with mean vector, $\mathbf{0}$, and covariance matrix function

$$E\left(\mathbf{a}_t \mathbf{a}'_{t+k}\right) = \begin{cases} \boldsymbol{\Sigma}, \text{if } k = 0, \\ \mathbf{0}, \text{if } k \neq 0, \end{cases} \tag{2.2}$$

Multivariate Time Series Analysis and Applications, First Edition. William W.S. Wei.
© 2019 John Wiley & Sons Ltd. Published 2019 by John Wiley & Sons Ltd.
Companion website: www.wiley.com/go/wei/datasets

and $\mathbf{\Sigma}$ is a $m \times m$ symmetric positive-definite matrix. The VMA(q) model is clearly stationary with the mean vector,

$$E(\mathbf{Z}_t) = \boldsymbol{\mu}, \tag{2.3}$$

and covariance matrix function,

$$\begin{aligned} \mathbf{\Gamma}(k) &= E\left[(\mathbf{Z}_t - \boldsymbol{\mu})(\mathbf{Z}_{t+k} - \boldsymbol{\mu})'\right] \\ &= E\left[\left(\mathbf{a}_t - \mathbf{\Theta}_1 \mathbf{a}_{t-1} - \cdots - \mathbf{\Theta}_q \mathbf{a}_{t-q}\right)\left(\mathbf{a}_{t+k} - \mathbf{\Theta}_1 \mathbf{a}_{t+k-1} - \cdots - \mathbf{\Theta}_q \mathbf{a}_{t+k-q}\right)'\right] \\ &= \begin{cases} \sum_{j=0}^{q-k} \mathbf{\Theta}_j \mathbf{\Sigma} \mathbf{\Theta}'_{j+k}, & \text{for } k = 0, 1, \ldots, q, \\ \mathrm{O}, & k > q, \end{cases} \end{aligned} \tag{2.4}$$

where $\mathbf{\Theta}_0 = \mathbf{I}$ and $\mathbf{\Gamma}(-k) = \mathbf{\Gamma}'(k)$. Thus, $\mathbf{\Gamma}(k)$ cuts off after lag q.

Let $\dot{\boldsymbol{Z}}_t = \boldsymbol{Z}_t - \boldsymbol{\mu}$. The VMA($q$) model is invertible if we can write it as the autoregressive representation

$$\mathbf{\Pi}(B)\dot{\mathbf{Z}}_t = \mathbf{a}_t, \tag{2.5}$$

with

$$\mathbf{\Pi}(B) = \left(\mathbf{I} - \mathbf{\Pi}_1 B - \mathbf{\Pi}_2 B^2 - \cdots\right) = \left[\mathbf{\Theta}_q(B)\right]^{-1}$$

so that the sequence of $m \times m$ autoregressive coefficient matrices $\mathbf{\Pi}_\ell$ is absolutely summable in the sense that if we write $\mathbf{\Pi}(\boldsymbol{B}) = \sum_{\ell=0}^{\infty} \mathbf{\Pi}_\ell \boldsymbol{B}^\ell = \left[\pi_{i,j}(B)\right]$, with $\pi_{i,j}(B) = \sum_{\ell=0}^{\infty} \mathbf{\Pi}_{i,j,\ell} \boldsymbol{B}^\ell$, we have $\sum_{\ell=0}^{\infty} \left|\pi_{i,j,\ell}\right| < \infty$, for all i = 1, 2, …, m and j = 1, 2, …, m. Since

$$\mathbf{\Pi}(B) = \left[\mathbf{\Theta}_q(B)\right]^{-1} = \frac{1}{\left|\mathbf{\Theta}_q(B)\right|} \mathbf{\Theta}_q^+(B), \tag{2.6}$$

and the element of the adjoint matrix $\mathbf{\Theta}_q^+(B)$ are polynomials in B of maximum order of $(m-1)q$, the model will be invertible if all zeros of $|\mathbf{\Theta}_q(B)|$ are outside of the unit circle.

Because of

$$\left[\mathbf{I} - \mathbf{\Pi}_1 B - \mathbf{\Pi}_2 B^2 - \cdots\right]\left\{\mathbf{I} - \mathbf{\Theta}_1 B - \cdots - \mathbf{\Theta}_q B^q\right] = \mathbf{I}, \tag{2.7}$$

we have

$$\begin{aligned} &\mathbf{I} - \mathbf{\Pi}_1 B - \mathbf{\Pi}_2 B^2 - \mathbf{\Pi}_3 B^3 - \mathbf{\Pi}_4 B^4 - \cdots \\ &- \mathbf{\Theta}_1 B - \mathbf{\Pi}_1 \mathbf{\Theta}_1 B^2 - \mathbf{\Pi}_2 \mathbf{\Theta}_1 B^3 - \mathbf{\Pi}_3 \mathbf{\Theta}_1 B^4 - \cdots \\ &\quad - \mathbf{\Theta}_2 B^2 - \mathbf{\Pi}_1 \mathbf{\Theta}_2 B^3 - \mathbf{\Pi}_2 \mathbf{\Theta}_2 B^4 - \mathbf{\Pi}_3 \mathbf{\Theta}_2 B^5 - \cdots \\ &\quad \cdots \qquad\qquad\qquad\qquad = \mathbf{I}. \end{aligned} \tag{2.8}$$

The $\mathbf{\Pi}_\ell$ can be calculated from the $\mathbf{\Theta}_j$ by equating the coefficient of B^ℓ on both sides of Eq. (2.8) as follows

$$\begin{aligned}
\mathbf{\Pi}_1 &= -\mathbf{\Theta}_1 \\
\mathbf{\Pi}_2 &= -\mathbf{\Pi}_1\mathbf{\Theta}_1 - \mathbf{\Theta}_2 \\
\mathbf{\Pi}_3 &= -\mathbf{\Pi}_2\mathbf{\Theta}_1 - \mathbf{\Pi}_1\mathbf{\Theta}_2 - \mathbf{\Theta}_3 \\
&\vdots
\end{aligned} \tag{2.9}$$

and in general,

$$\mathbf{\Pi}_\ell = -\mathbf{\Pi}_{\ell-1}\mathbf{\Theta}_1 - \mathbf{\Pi}_{\ell-2}\mathbf{\Theta}_2 - \cdots - \mathbf{\Pi}_{\ell-k}\mathbf{\Theta}_k, \text{for } \ell = 1, 2, \ldots \tag{2.10}$$

where $\mathbf{\Pi}_0 = \mathbf{I}$, $\mathbf{\Pi}_j = \mathbf{O}$, for $j < 0$, and $\mathbf{\Theta}_k = \mathbf{O}$, for $k > q$.

Example 2.1 Consider the m-dimensional VMA(1) model,

$$\mathbf{Z}_t = \boldsymbol{\mu} + \mathbf{a}_t - \mathbf{\Theta}_1\mathbf{a}_{t-1}. \tag{2.11}$$

Its covariance matrix function is given by

$$\begin{aligned}
\mathbf{\Gamma}(0) &= E\left[(\mathbf{Z}_t - \boldsymbol{\mu})(\mathbf{Z}_t - \boldsymbol{\mu})'\right] \\
&= E\left[(\mathbf{a}_t - \mathbf{\Theta}_1\mathbf{a}_{t-1})(\mathbf{a}_t - \mathbf{\Theta}_1\mathbf{a}_{t-1})'\right] \\
&= E\left[(\mathbf{a}_t - \mathbf{\Theta}_1\mathbf{a}_{t-1})\left(\mathbf{a}_t' - \mathbf{a}_{t-1}'\mathbf{\Theta}_1'\right)\right] \\
&= \mathbf{\Sigma} + \mathbf{\Theta}_1\mathbf{\Sigma}\mathbf{\Theta}_1',
\end{aligned} \tag{2.12}$$

$$\begin{aligned}
\mathbf{\Gamma}(1) &= E\left[(\mathbf{Z}_t - \boldsymbol{\mu})(\mathbf{Z}_{t+1} - \boldsymbol{\mu})'\right] \\
&= E\left[(\mathbf{a}_t - \mathbf{\Theta}_1\mathbf{a}_{t-1})(\mathbf{a}_{t+1} - \mathbf{\Theta}_1\mathbf{a}_t)'\right] \\
&= E\left[(\mathbf{a}_t - \mathbf{\Theta}_1\mathbf{a}_{t-1})\left(\mathbf{a}_{t+1}' - \mathbf{a}_t'\mathbf{\Theta}_1'\right)\right] \\
&= \mathbf{\Sigma}\mathbf{\Theta}_1',
\end{aligned} \tag{2.13}$$

$$\begin{aligned}
\mathbf{\Gamma}(k) &= E\left[(\mathbf{Z}_t - \boldsymbol{\mu})(\mathbf{Z}_{t+k} - \boldsymbol{\mu})'\right] \\
&= E\left[(\mathbf{a}_t - \mathbf{\Theta}_1\mathbf{a}_{t-1})(\mathbf{a}_{t+k} - \mathbf{\Theta}_1\mathbf{a}_{t+k-1})'\right] \\
&= E\left[(\mathbf{a}_t - \mathbf{\Theta}_1\mathbf{a}_{t-1})\left(\mathbf{a}_{t+k}' - \mathbf{a}_{t+k-1}'\mathbf{\Theta}_1'\right)\right] \\
&= \mathrm{O}, \text{for } k > 1,
\end{aligned} \tag{2.14}$$

which can be easily seen as a special case of Eq. (2.4) with $q = 1$. The model is invertible if the zeros of $|\mathbf{I} - \mathbf{\Theta}_1 B|$ are outside of the unit circle or equivalently when the eigenvalues of $\mathbf{\Theta}_1$ are all inside the unit circle. In such a case, since

$$\mathbf{\Pi}(B) = (\mathbf{I} - \mathbf{\Theta}_1 B)^{-1} = \left(\mathbf{I} + \mathbf{\Theta}_1 B + \mathbf{\Theta}_1^2 B^2 + \cdots\right), \tag{2.15}$$

we can write it as

$$\dot{\mathbf{Z}}_t + \sum_{j=1}^{\infty} \mathbf{\Theta}_1^j \dot{\mathbf{Z}}_{t-j} = \mathbf{a}_t, \tag{2.16}$$

where $\dot{\mathbf{Z}}_t = \mathbf{Z}_t - \boldsymbol{\mu}$, and the coefficient matrix $\mathbf{\Theta}_1^j$ decreases toward the $\mathbf{O}$ matrix as j increases.

The fundamental property of a VMA(q) model is that its covariance matrix function cuts off after lag q. This property will be used to identify the underlying model when we try to build a vector time series model from a series of vector time series observations.

2.2 Vector autoregressive processes

The m-dimensional vector autoregressive process or model of order p, shortened to VAR(p), is given by

$$\mathbf{Z}_t = \boldsymbol{\theta}_0 + \mathbf{\Phi}_1 \mathbf{Z}_{t-1} + \cdots + \mathbf{\Phi}_p \mathbf{Z}_{t-p} + \mathbf{a}_t, \tag{2.17}$$

or

$$\mathbf{\Phi}_p(B)\mathbf{Z}_t = \boldsymbol{\theta}_0 + \mathbf{a}_t, \tag{2.18}$$

where $\mathbf{a}_t$ is a sequence of m-dimensional vector white noise process, VWN($\mathbf{0}$, $\mathbf{\Sigma}$), and

$$\mathbf{\Phi}_p(B) = \mathbf{I} - \mathbf{\Phi}_1 B - \cdots - \mathbf{\Phi}_p B^p. \tag{2.19}$$

The model is clearly invertible. It will be stationary if the zeros of $|\mathbf{I} - \mathbf{\Phi}_1 B - \cdots - \mathbf{\Phi}_p B^p|$ lie outside of the unit circle or equivalently, the roots of

$$\left|\lambda^p \mathbf{I} - \lambda^{p-1} \mathbf{\Phi}_1 - \cdots - \mathbf{\Phi}_p\right| = 0 \tag{2.20}$$

are all inside the unit circle. In this case, its mean is a constant vector, $E(\mathbf{Z}_t) = \boldsymbol{\mu}$, which can be found by noting that

$$\begin{aligned}\boldsymbol{\mu} = E(\mathbf{Z}_t) &= E\left(\boldsymbol{\theta}_0 + \mathbf{\Phi}_1 \mathbf{Z}_{t-1} + \cdots + \mathbf{\Phi}_p \mathbf{Z}_{t-p} + \mathbf{a}_t\right) \\ &= \boldsymbol{\theta}_0 + \mathbf{\Phi}_1 \boldsymbol{\mu} + \cdots + \mathbf{\Phi}_p \boldsymbol{\mu}\end{aligned}$$

and hence

$$\boldsymbol{\mu} = \left(\mathbf{I} - \mathbf{\Phi}_1 - \cdots - \mathbf{\Phi}_p\right)^{-1} \boldsymbol{\theta}_0. \tag{2.21}$$

Since

$$\boldsymbol{\theta}_0 = \left(\mathbf{I} - \mathbf{\Phi}_1 - \cdots - \mathbf{\Phi}_p\right)\boldsymbol{\mu} = \left(\mathbf{I} - \mathbf{\Phi}_1 B - \cdots - \mathbf{\Phi}_p B^p\right)\boldsymbol{\mu},$$

Equation (2.18) can be written as

$$\boldsymbol{\Phi}_p(B)\dot{\mathbf{Z}}_t = \mathbf{a}_t, \tag{2.22}$$

or

$$\dot{\mathbf{Z}}_t = \boldsymbol{\Phi}_1\dot{\mathbf{Z}}_{t-1} + \cdots + \boldsymbol{\Phi}_p\dot{\mathbf{Z}}_{t-p} + \mathbf{a}_t, \tag{2.23}$$

where $\dot{\mathbf{Z}}_t = \mathbf{Z}_t - \boldsymbol{\mu}$.

The stationary VAR(p) model in Eq. (2.17) or equivalently Eq. (2.22) can be written as a moving average representation,

$$\begin{aligned}\dot{\mathbf{Z}}_t &= [\boldsymbol{\Phi}_p(B)]^{-1}\mathbf{a}_t \\ &= \boldsymbol{\Psi}(B)\mathbf{a}_t,\end{aligned} \tag{2.24}$$

where $\boldsymbol{\Psi}(B) = \sum_{k=0}^{\infty}\boldsymbol{\Psi}_k B^k = \mathbf{I} + \boldsymbol{\Psi}_1 B + \boldsymbol{\Psi}_2 B^2 + \boldsymbol{\Psi}_3 B^3 + \cdots = [\psi_{i,j}(B)]$ with $\psi_{i,j}(B) = \sum_{k=0}^{\infty}\psi_{i,j,k}B^k$, such that the sequence of $m \times m$ coefficient matrices $\boldsymbol{\Psi}_k = [\psi_{i,j,k}]$ is square summable, that is, $\sum_{k=0}^{\infty}\psi_{i,j,k}^2 < \infty$, for all $i = 1, 2, \ldots, m$ and $j = 1, 2, \ldots, m$.

The $\boldsymbol{\Psi}_k$ weight matrices in Eq. (2.24) can be found from the VAR(p) coefficient matrices $\boldsymbol{\Phi}_j$*'s* by equating the coefficient of B^i in the following equation

$$(\mathbf{I} - \boldsymbol{\Phi}_1 B - \cdots - \boldsymbol{\Phi}_p B^p)(\mathbf{I} + \boldsymbol{\Psi}_1 B + \boldsymbol{\Psi}_2 B^2 + \boldsymbol{\Psi}_3 B^3 + \cdots) = \mathbf{I}. \tag{2.25}$$

More specifically,

$$\begin{aligned}&\mathbf{I} + \boldsymbol{\Psi}_1 B + \boldsymbol{\Psi}_2 B^2 + \boldsymbol{\Psi}_3 B^3 + \cdots \\ &-\boldsymbol{\Phi}_1 B - \boldsymbol{\Phi}_1\boldsymbol{\Psi}_1 B^2 - \boldsymbol{\Phi}_1\boldsymbol{\Psi}_2 B^3 - \cdots \\ &\qquad -\boldsymbol{\Phi}_2 B^2 - \boldsymbol{\Phi}_2\boldsymbol{\Psi}_1 B^3 - \boldsymbol{\Phi}_2\boldsymbol{\Psi}_2 B^4 - \cdots = \mathbf{I}\end{aligned}$$

so

$$\begin{aligned}&\boldsymbol{\Psi}_1 = \boldsymbol{\Phi}_1 \\ &\boldsymbol{\Psi}_2 = \boldsymbol{\Phi}_1\boldsymbol{\Psi}_1 + \boldsymbol{\Phi}_2 \\ &\boldsymbol{\Psi}_3 = \boldsymbol{\Phi}_1\boldsymbol{\Psi}_2 + \boldsymbol{\Phi}_2\boldsymbol{\Psi}_1 + \boldsymbol{\Phi}_3 \\ &\vdots \\ &\boldsymbol{\Psi}_p = \boldsymbol{\Phi}_1\boldsymbol{\Psi}_{p-1} + \boldsymbol{\Phi}_2\boldsymbol{\Psi}_{p-2} + \cdots + \boldsymbol{\Phi}_p \\ &\boldsymbol{\Psi}_k = \boldsymbol{\Phi}_1\boldsymbol{\Psi}_{k-1} + \boldsymbol{\Phi}_2\boldsymbol{\Psi}_{k-2} + \cdots + \boldsymbol{\Phi}_p\boldsymbol{\Psi}_{k-p}, k \ge p,\end{aligned} \tag{2.26}$$

where $\boldsymbol{\Psi}_0 = \mathbf{I}$, and $\boldsymbol{\Psi}_j = \mathbf{O}$, for $j < 0$.

The covariance matrix function is computed as

$$\begin{aligned}
\boldsymbol{\Gamma}(k) &= E(\dot{\mathbf{Z}}_t)(\dot{\mathbf{Z}}_{t+k})' \\
&= E(\dot{\mathbf{Z}}_t)(\boldsymbol{\Phi}_1\dot{\mathbf{Z}}_{t+k-1} + \boldsymbol{\Phi}_2\dot{\mathbf{Z}}_{t+k-2} + \cdots + \boldsymbol{\Phi}_p\dot{\mathbf{Z}}_{t+k-p} + \mathbf{a}_{t+k})' \\
&= E(\dot{\mathbf{Z}}_t)\left(\dot{\mathbf{Z}}'_{t+k-1}\boldsymbol{\Phi}'_1 + \dot{\mathbf{Z}}'_{t+k-2}\boldsymbol{\Phi}'_2 + \cdots + \dot{\mathbf{Z}}'_{t+k-p}\boldsymbol{\Phi}'_p + \mathbf{a}'_{t+k}\right) \\
&= \begin{cases} \boldsymbol{\Gamma}(-1)\boldsymbol{\Phi}'_1 + \boldsymbol{\Gamma}(-2)\boldsymbol{\Phi}'_2 + \cdots + \boldsymbol{\Gamma}(-p)\boldsymbol{\Phi}'_p + \boldsymbol{\Sigma}, \text{for } k=0, \\ \boldsymbol{\Gamma}(k-1)\boldsymbol{\Phi}'_1 + \boldsymbol{\Gamma}(k-2)\boldsymbol{\Phi}'_2 + \cdots + \boldsymbol{\Gamma}(k-p)\boldsymbol{\Phi}'_p, \text{for } k>0. \end{cases}
\end{aligned} \tag{2.27}$$

Thus, we have

$$\begin{bmatrix} \boldsymbol{\Gamma}(0) & \boldsymbol{\Gamma}(-1) & \boldsymbol{\Gamma}(-2) & \cdots & \boldsymbol{\Gamma}(1-p) \\ \boldsymbol{\Gamma}(1) & \boldsymbol{\Gamma}(0) & \boldsymbol{\Gamma}(-1) & \cdots & \boldsymbol{\Gamma}(2-p) \\ \vdots & \vdots & \vdots & \vdots & \vdots \\ \boldsymbol{\Gamma}(p-1) & \boldsymbol{\Gamma}(p-2) & \boldsymbol{\Gamma}(p-3) & \cdots & \boldsymbol{\Gamma}(0) \end{bmatrix} \begin{bmatrix} \boldsymbol{\Phi}'_1 \\ \boldsymbol{\Phi}'_2 \\ \vdots \\ \boldsymbol{\Phi}'_p \end{bmatrix} = \begin{bmatrix} \boldsymbol{\Gamma}(1) \\ \boldsymbol{\Gamma}(2) \\ \vdots \\ \boldsymbol{\Gamma}(p) \end{bmatrix},$$

which is the system of generalized Yule–Walker matrix equations and can be used to find the coefficient matrices $\boldsymbol{\Phi}_j^{'}s$ from $\boldsymbol{\Gamma}(j)'s$, that is,

$$\begin{bmatrix} \boldsymbol{\Phi}'_1 \\ \boldsymbol{\Phi}'_2 \\ \vdots \\ \boldsymbol{\Phi}'_p \end{bmatrix} = \begin{bmatrix} \boldsymbol{\Gamma}(0) & \boldsymbol{\Gamma}(-1) & \boldsymbol{\Gamma}(-2) & \cdots & \boldsymbol{\Gamma}(1-p) \\ \boldsymbol{\Gamma}(1) & \boldsymbol{\Gamma}(0) & \boldsymbol{\Gamma}(-1) & \cdots & \boldsymbol{\Gamma}(2-p) \\ \vdots & \vdots & \vdots & \vdots & \vdots \\ \boldsymbol{\Gamma}(p-1) & \boldsymbol{\Gamma}(p-2) & \boldsymbol{\Gamma}(p-3) & \cdots & \boldsymbol{\Gamma}(0) \end{bmatrix}^{-1} \begin{bmatrix} \boldsymbol{\Gamma}(1) \\ \boldsymbol{\Gamma}(2) \\ \vdots \\ \boldsymbol{\Gamma}(p) \end{bmatrix}. \tag{2.28}$$

Once $\boldsymbol{\Phi}_j^{'}s$ are found, we have

$$\boldsymbol{\Sigma} = \boldsymbol{\Gamma}(0) - \boldsymbol{\Gamma}(-1)\boldsymbol{\Phi}'_1 + \boldsymbol{\Gamma}(-2)\boldsymbol{\Phi}'_2 + \cdots + \boldsymbol{\Gamma}(-p)\boldsymbol{\Phi}'_p. \tag{2.29}$$

Example 2.2 The m-dimensional VAR[1] model is given by

$$\mathbf{Z}_t = \boldsymbol{\theta}_0 + \boldsymbol{\Phi}_1\mathbf{Z}_{t-1} + \mathbf{a}_t, \tag{2.30}$$

or

$$(\mathbf{I} - \boldsymbol{\Phi}_1 B)\dot{\mathbf{Z}}_t = \mathbf{a}_t, \tag{2.31}$$

where $\boldsymbol{\theta}_0 = (\mathbf{I} - \boldsymbol{\Phi}_1 B)\boldsymbol{\mu}$. The model is invertible. It is stationary if the zeros of $|\mathbf{I} - \boldsymbol{\Phi}_1 B|$ lie outside of the unit circle. Let $\lambda = B^{-1}$. Since

$$|\mathbf{I} - \boldsymbol{\Phi}_1 B| = 0 \Leftrightarrow |\boldsymbol{\lambda}\mathbf{I} - \boldsymbol{\Phi}_1| = 0, \tag{2.32}$$

the stationarity condition can also be checked whether all the eigenvalues of $\mathbf{\Phi}_1$ are inside the unit circle.

For a stationary **VAR**(1) model, its vector moving average representation is

$$\begin{aligned}
\dot{\mathbf{Z}}_t &= (\mathbf{I}-\mathbf{\Phi}_1 B)^{-1}\mathbf{a}_t \\
&= \left(\mathbf{I}+\mathbf{\Phi}_1 B+\mathbf{\Phi}_1^2 B^2+\cdots\right)\mathbf{a}_t \\
&= \mathbf{a}_t+\mathbf{\Phi}_1\mathbf{a}_{t-1}+\mathbf{\Phi}_1^2\mathbf{a}_{t-2}+\mathbf{\Phi}_1^3\mathbf{a}_{t-3}+\cdots \\
&= \sum_{j=0}^{\infty}\mathbf{\Psi}_j\mathbf{a}_{t-j},
\end{aligned} \tag{2.33}$$

where $\mathbf{\Psi}_0 = \mathbf{I}$, and $\mathbf{\Psi}_j=\mathbf{\Phi}_1^j$ that agrees with Eq. (2.26) with $p = 1$. The coefficient matrix $\mathbf{\Psi}_j=\mathbf{\Phi}_1^j$ decreases toward a zero matrix as j increases. The covariance matrix function is given by

$$\begin{aligned}
\mathbf{\Gamma}(k) &= E\left(\dot{\mathbf{Z}}_t\right)\left(\dot{\mathbf{Z}}_{t+k}\right)' \\
&= E\left(\dot{\mathbf{Z}}_t\right)\left(\mathbf{\Phi}_1\dot{\mathbf{Z}}_{t+k-1}+\mathbf{a}_{t+k}\right)' \\
&= E\left(\dot{\mathbf{Z}}_t\right)\left(\dot{\mathbf{Z}}'_{t+k-1}\mathbf{\Phi}'_1+\mathbf{a}'_{t+k}\right) \\
&= \begin{cases} \mathbf{\Gamma}(-1)\mathbf{\Phi}'_1+\mathbf{\Sigma}, & k=0, \\ \mathbf{\Gamma}(k-1)\mathbf{\Phi}'_1=\mathbf{\Gamma}(0)\left(\mathbf{\Phi}'_1\right)^k, k\geq 1. \end{cases}
\end{aligned} \tag{2.34}$$

More specifically,

$$\begin{aligned}
\mathbf{\Gamma}(0) &= E\left(\dot{\mathbf{Z}}_t\right)\left(\dot{\mathbf{Z}}_t\right)' \\
&= E\left[\mathbf{a}_t+\mathbf{\Phi}_1\mathbf{a}_{t-1}+\mathbf{\Phi}_1^2\mathbf{a}_{t-2}+\mathbf{\Phi}_1^3\mathbf{a}_{t-3}+\cdots\right]\left[\mathbf{a}'_t+\mathbf{a}'_{t-1}\mathbf{\Phi}'_1+\mathbf{a}'_{t-2}\left(\mathbf{\Phi}_1^2\right)'+\mathbf{a}'_{t-3}\left(\mathbf{\Phi}_1^3\right)'+\cdots\right] \\
&= \sum_{j=0}^{\infty}\mathbf{\Phi}_1^j\mathbf{\Sigma}\left(\mathbf{\Phi}_1^j\right)',
\end{aligned} \tag{2.35}$$

and

$$\mathbf{\Gamma}(k) = \left[\sum_{j=0}^{\infty}\mathbf{\Phi}_1^j\mathbf{\Sigma}\left(\mathbf{\Phi}_1^j\right)'\right]\left(\mathbf{\Phi}'_1\right)^k, \tag{2.36}$$

where $\mathbf{\Phi}_0 = \mathbf{I}$.

Using Eq. (2.34), we can express the parameters as functions of covariance matrices. Specifically,

$$\mathbf{\Phi}_1 = \mathbf{\Gamma}'(1)\mathbf{\Gamma}^{-1}(0), \tag{2.37}$$

and

$$\begin{aligned}\mathbf{\Sigma} &= \mathbf{\Gamma}(0) - \mathbf{\Gamma}(-1)\mathbf{\Gamma}^{-1}(0)\mathbf{\Gamma}(1) \\ &= \mathbf{\Gamma}(0) - \mathbf{\Gamma}'(1)\mathbf{\Gamma}^{-1}(0)\mathbf{\Gamma}(1).\end{aligned} \tag{2.38}$$

2.2.1 Granger causality

One of the interesting problems in studying a vector time series is that we often want to know whether there are any causal effects among these variables. Specifically, in a VAR(p) model

$$\mathbf{\Phi}_p(B)\mathbf{Z}_t = \mathbf{\theta}_0 + \mathbf{a}_t, \tag{2.39}$$

we can partition the vector $\mathbf{Z}_t$ into two components, $\mathbf{Z}_t = \left[\mathbf{Z}'_{1,t}, \ \mathbf{Z}'_{2,t}\right]'$ so that

$$\begin{bmatrix} \mathbf{\Phi}_{11}(B) & \mathbf{\Phi}_{12}(B) \\ \mathbf{\Phi}_{21}(B) & \mathbf{\Phi}_{22}(B) \end{bmatrix} \begin{bmatrix} \mathbf{Z}_{1,t} \\ \mathbf{Z}_{2,t} \end{bmatrix} = \begin{bmatrix} \mathbf{\theta}_1 \\ \mathbf{\theta}_2 \end{bmatrix} + \begin{bmatrix} \mathbf{a}_{1,t} \\ \mathbf{a}_{2,t} \end{bmatrix}. \tag{2.40}$$

When $\mathbf{\Phi}_{12}(B) = \mathbf{0}$, Eq. (2.40) becomes

$$\begin{cases} \mathbf{\Phi}_{11}(B)\mathbf{Z}_{1,t} = \mathbf{\theta}_1 + \mathbf{a}_{1,t}, \\ \mathbf{\Phi}_{22}(B)\mathbf{Z}_{2,t} = \mathbf{\theta}_2 + \mathbf{\Phi}_{21}(B)\mathbf{Z}_{1,t} + \mathbf{a}_{2,t}. \end{cases} \tag{2.41}$$

The future values of $\mathbf{Z}_{2,\,t}$ are influenced not only by its own past but also by the past of $\mathbf{Z}_{1,\,t}$, while the future values of $\mathbf{Z}_{1,\,t}$ are influenced only by its own past. In other words, we say that variables in $\mathbf{Z}_{1,\,t}$ cause $\mathbf{Z}_{2,\,t}$, but variables in $\mathbf{Z}_{2,\,t}$ do not cause $\mathbf{Z}_{1,\,t}$. This concept is often known as the Granger causality, because it is thought to have been Granger who first introduced the notion in 1969. For more discussion about causality and its tests, we refer readers to Granger (1969), Hawkes (1971a, b), Pierce and Haugh (1977), Eichler et al. (2017), and Zhang and Yang (2017), among others.

2.3 Vector autoregressive moving average processes

The m-dimensional vector autoregressive moving average (VARMA) process or model of orders p and q, shortened to VARMA(p,q) is given by

$$\mathbf{Z}_t = \mathbf{\theta}_0 + \mathbf{\Phi}_1\mathbf{Z}_{t-1} + \cdots + \mathbf{\Phi}_p\mathbf{Z}_{t-p} + \mathbf{a}_t - \mathbf{\Theta}_1\mathbf{a}_{t-1} - \cdots - \mathbf{\Theta}_q\mathbf{a}_{t-q}, \tag{2.42}$$

or

$$\boldsymbol{\Phi}_p(B)\mathbf{Z}_t = \boldsymbol{\theta}_0 + \boldsymbol{\Theta}_q(B)\mathbf{a}_t, \tag{2.43}$$

where $\mathbf{a}_t$ is a sequence of m-dimensional vector white noise process, VWN($\mathbf{0}$, $\boldsymbol{\Sigma}$), and

$$\begin{aligned} \boldsymbol{\Phi}_p(B) &= \mathbf{I} - \boldsymbol{\Phi}_1 B - \cdots - \boldsymbol{\Phi}_p B^p, \\ \boldsymbol{\Theta}_q(B) &= \mathbf{I} - \boldsymbol{\Theta}_1 B - \cdots - \boldsymbol{\Theta}_q B^q. \end{aligned} \tag{2.44}$$

The model is stationary if the zeros of the determinant polynomial $|\boldsymbol{\Phi}_p(B)|$ are all outside of the unit circle so that

$$\begin{aligned} \dot{\mathbf{Z}}_t &= \left[\boldsymbol{\Phi}_p(B)\right]^{-1}\boldsymbol{\Theta}_q(B)\mathbf{a}_t \\ &= \sum_{j=0}^{\infty} \boldsymbol{\Psi}_j \mathbf{a}_{t-j}. \end{aligned} \tag{2.45}$$

The model is invertible if zeros of the determinant polynomial $|\boldsymbol{\Theta}_q(B)|$ are all outside of the unit circle so that

$$\boldsymbol{\Pi}(B)\dot{\mathbf{Z}}_t = \mathbf{a}_t, \tag{2.46}$$

where

$$\boldsymbol{\Pi}(B) = \left[\boldsymbol{\Theta}_q(B)\right]^{-1}\boldsymbol{\Phi}_p(B) = \mathbf{I} - \sum_{j=0}^{\infty} \boldsymbol{\Pi}_j B^j. \tag{2.47}$$

Example 2.3 The m-dimensional VARMA(1,1) model is given by

$$\mathbf{Z}_t = \boldsymbol{\theta}_0 + \boldsymbol{\Phi}_1 \mathbf{Z}_{t-1} + \mathbf{a}_t - \boldsymbol{\Theta}_1 \mathbf{a}_{t-1}, \tag{2.48}$$

or

$$(\mathbf{I} - \boldsymbol{\Phi}_1 B)\dot{\mathbf{Z}}_t = (\mathbf{I} - \boldsymbol{\Theta}_1 B)\mathbf{a}_t. \tag{2.49}$$

The model is stationary if the zeros of $|\mathbf{I} - \boldsymbol{\Phi}_1 B|$ lie outside of the unit circle so that

$$\begin{aligned} \dot{\mathbf{Z}}_t &= (\mathbf{I} - \boldsymbol{\Phi}_1 B)^{-1}(\mathbf{I} - \boldsymbol{\Theta}_1 B)\mathbf{a}_t \\ &= \sum_{j=0}^{\infty} \boldsymbol{\Psi}_j \mathbf{a}_{t-j}, \end{aligned} \tag{2.50}$$

where $\boldsymbol{\Psi}_0 = \mathbf{I}$, and

$$\boldsymbol{\Psi}_j = \boldsymbol{\Phi}_1^{j-1}(\boldsymbol{\Phi}_1 - \boldsymbol{\Theta}_1), j \geq 1, \tag{2.51}$$

which decreases to a zero matrix as j increases just like $\boldsymbol{\Phi}_1^j$.

The model is invertible if the zeros of $|\mathbf{I} - \mathbf{\Theta}_1 B|$ lie outside of the unit circle, and in such a case we can write it as

$$\mathbf{\Pi}(B)\dot{\mathbf{Z}}_t = \mathbf{a}_t, \tag{2.52}$$

or

$$\dot{\mathbf{Z}}_t = \mathbf{\Pi}_1 \dot{\mathbf{Z}}_{t-1} + \mathbf{\Pi}_2 \dot{\mathbf{Z}}_{t-2} + \cdots + \mathbf{a}_t, \tag{2.53}$$

where

$$\mathbf{\Pi}(B) = (\mathbf{I} - \mathbf{\Theta}_1 B)^{-1}(\mathbf{I} - \mathbf{\Phi}_1 B) = \mathbf{I} - \mathbf{\Pi}_1 B - \mathbf{\Pi}_2 B^2 - \cdots$$

and

$$\mathbf{\Pi}_j = \mathbf{\Theta}_1^{j-1}(\mathbf{\Phi}_1 - \mathbf{\Theta}_1), \tag{2.54}$$

which follows the pattern of $\mathbf{\Theta}_1^j$ decreasing to a zero matrix as j increases.

Since

$$E\left[\dot{\mathbf{Z}}_t\left(\dot{\mathbf{Z}}'_{t+k} - \dot{\mathbf{Z}}'_{t+k-1}\mathbf{\Phi}'_1\right)\right] = E\left[\dot{\mathbf{Z}}_t\left(\mathbf{a}'_{t+k} - \mathbf{a}'_{t+k-1}\mathbf{\Theta}'_1\right)\right], \tag{2.55}$$

the covariance matrix function of the **VARMA**(1, 1) model can be obtained as follows

$$\begin{aligned}
&k = 0, \mathbf{\Gamma}(0) - \mathbf{\Gamma}(-1)\mathbf{\Phi}'_1 = \mathbf{\Sigma} - E\left[\dot{\mathbf{Z}}_t\left(\mathbf{a}'_{t-1}\mathbf{\Theta}'_1\right)\right] \\
&\qquad\qquad = \mathbf{\Sigma} - E\left[\mathbf{\Phi}_1\dot{\mathbf{Z}}_{t-1} + \mathbf{a}_t - \mathbf{\Theta}_1\mathbf{a}_{t-1}\right)\left(\mathbf{a}'_{t-1}\mathbf{\Theta}'_1\right)\right] = \mathbf{\Sigma} - (\mathbf{\Phi}_1 - \mathbf{\Theta}_1)\mathbf{\Sigma}\mathbf{\Theta}_1, \\
&k = 1, \mathbf{\Gamma}(1) - \mathbf{\Gamma}(0)\mathbf{\Phi}'_1 = -\mathbf{\Sigma}\mathbf{\Theta}'_1, \\
&k \ge 2, \mathbf{\Gamma}(k) - \mathbf{\Gamma}(k-1)\mathbf{\Phi}'_1 = \mathrm{O}.
\end{aligned} \tag{2.56}$$

More specifically, we can express $\mathbf{\Gamma}(k)$ in terms of model parameters $\mathbf{\Phi}_1$, $\mathbf{\Theta}_1$, and $\mathbf{\Sigma}_1$ as follows,

$$\begin{aligned}
\mathbf{\Gamma}(0) &= E\left(\dot{\mathbf{Z}}_t\dot{\mathbf{Z}}'_t\right) = \sum_{j=0}^{\infty}\mathbf{\Psi}_j\mathbf{\Sigma}(\mathbf{\Psi}_j)' = \mathbf{\Sigma} + \sum_{j=1}^{\infty}\mathbf{\Phi}_1^{j-1}(\mathbf{\Phi}_1 - \mathbf{\Theta}_1)\mathbf{\Sigma}(\mathbf{\Phi}'_1 - \mathbf{\Theta}'_1)\left(\mathbf{\Phi}_1^{j-1}\right)', \\
\mathbf{\Gamma}(1) &= \left[\mathbf{\Sigma} + \sum_{j=1}^{\infty}\mathbf{\Phi}_1^{j-1}(\mathbf{\Phi}_1 - \mathbf{\Theta}_1)\mathbf{\Sigma}(\mathbf{\Phi}'_1 - \mathbf{\Theta}'_1)\left(\mathbf{\Phi}_1^{j-1}\right)'\right]\mathbf{\Phi}'_1 - \mathbf{\Sigma}\mathbf{\Theta}'_1, \\
\mathbf{\Gamma}(k) &= \left\{\left[\mathbf{\Sigma} + \sum_{j=1}^{\infty}\mathbf{\Phi}_1^{j-1}(\mathbf{\Phi}_1 - \mathbf{\Theta}_1)\mathbf{\Sigma}(\mathbf{\Phi}'_1 - \mathbf{\Theta}'_1)\left(\mathbf{\Phi}_1^{j-1}\right)'\right]\mathbf{\Phi}'_1 - \mathbf{\Sigma}\mathbf{\Theta}'_1\right\}(\mathbf{\Phi}'_1)^{k-1}, k \ge 1,
\end{aligned} \tag{2.57}$$

where $\mathbf{\Phi}_1^0 = \mathbf{I}$.

We can also write the parameters $\mathbf{\Phi}_1$, $\mathbf{\Theta}_1$, and $\mathbf{\Sigma}_1$ as functions of $\mathbf{\Gamma}(k)$'s. For details, we refer readers to Wei (2006, chapter 16).

2.4 Nonstationary vector autoregressive moving average processes

In univariate time series analysis, a nonstationary time series is reduced to a stationary time series by proper power transformations and differencing. These can still be used in vector time series analysis. However, it should be noted that these transformations should be applied to a component series individually because not all component series can be reduced to stationary by exactly the same power transformation and the same number of differencing. To be more flexible, after using proper power transformations to a component series, we will use the following presentation for a nonstationary vector time series model

$$\mathbf{\Phi}_p(B)\mathbf{D}(B)\mathbf{Z}_t = \mathbf{\Theta}_q(B)\mathbf{a}_t, \tag{2.58}$$

where

$$\mathbf{D}(B) = \begin{bmatrix} (1-B)^{d_1} & 0 & . & \cdots & 0 & 0 \\ 0 & (1-B)^{d_2} & 0 & \cdots & . & 0 \\ . & 0 & . & \cdots & . & . \\ \vdots & \vdots & \vdots & \ddots & \vdots & \vdots \\ 0 & \cdots & . & \cdots & . & 0 \\ 0 & 0 & . & \cdots & 0 & (1-B)^{d_m} \end{bmatrix}, \tag{2.59}$$

and the zeros of $|\mathbf{\Phi}_p(B)|$ and $|\mathbf{\Theta}_q(B)|$ are outside of the unit circle. The unit root test introduced by Dickey and Fuller (1979) can be used to determine the order of d_i. For more recent references on a unit root, see Teles et al. (2008), Chambers (2015), Cavaliere et al. (2015), and Hosseinkouchack and Hassler (2016), among others.

2.5 Vector time series model building

2.5.1 Identification of vector time series models

In constructing a vector time series model, just like in univariate time series model building, the first step is to plot the vector time series. By plotting all the component series in one graph, we obtain a good idea of the movements of different components and the general pattern of their relationships. In principle, the vector time series model building procedure is similar to the univariate time series model building procedure. We identify an underlying model from its correlation and partial correlation matrix functions. Table 2.1 gives a useful summary.

Given an observed vector time series $\mathbf{Z}_1, \ldots, \mathbf{Z}_n$, we compute its sample correlation and partial correlation matrices after proper transformations are applied to reduce a nonstationary series to a stationary series.

Table 2.1 Characteristics of stationary vector time series models.

Process	Correlation matrix function	Partial correlation matrix function
VAR(p)	Non-zero matrix with diminishing elements	Zero matrix after lag p
VMA(q)	Zero matrix after lag q	Non-zero matrix with diminishing elements
VARMA(p,q)	Non-zero matrix with diminishing elements	Non-zero matrix with diminishing elements

2.5.2 Sample moments of a vector time series

2.5.2.1 Sample mean and sample covariance matrices

For a stationary m-dimensional vector time series process, given the n observations, $\mathbf{Z}_1, \dots, \mathbf{Z}_n$, the mean vector, $\boldsymbol{\mu}$, is estimated by its sample mean vector,

$$\bar{\mathbf{Z}} = \frac{1}{n}\sum_{t=1}^{n}\mathbf{Z}_t = \frac{1}{n}\begin{bmatrix}\sum_{t=1}^{n} Z_{1,t} \\ \sum_{t=1}^{n} Z_{2,t} \\ \vdots \\ \sum_{t=1}^{n} Z_{m,t}\end{bmatrix} = \begin{bmatrix}\bar{Z}_1 \\ \bar{Z}_2 \\ \vdots \\ \bar{Z}_m\end{bmatrix}. \tag{2.60}$$

It can be easily seen that $E(\bar{\mathbf{Z}}) = \boldsymbol{\mu}$, and $\bar{\mathbf{Z}}$ is an unbiased estimator of $\boldsymbol{\mu}$. Since

$$E(\bar{\mathbf{Z}}-\boldsymbol{\mu})(\bar{\mathbf{Z}}-\boldsymbol{\mu})' = \frac{1}{n^2}\sum_{i=1}^{n}\sum_{j=1}^{n}\boldsymbol{\Gamma}(i-j) = \frac{1}{n}\sum_{k=-(n-1)}^{n-1}\left(1-\frac{k}{n}\right)\boldsymbol{\Gamma}(k), \tag{2.61}$$

for a stationary vector process with $\rho_{i,i}(k) \to 0$ as $k \to 0$, $\bar{\mathbf{Z}}$ is actually a consistent estimator of $\boldsymbol{\mu}$. So,

$$E(\bar{\mathbf{Z}}-\boldsymbol{\mu})(\bar{\mathbf{Z}}-\boldsymbol{\mu})' \to 0 \text{ as } n \to \infty, \tag{2.62}$$

and

$$\lim_{n\to\infty}\left\{nE\left[(\bar{\mathbf{Z}}-\boldsymbol{\mu})(\bar{\mathbf{Z}}-\boldsymbol{\mu})'\right]\right\} = \sum_{k=-\infty}^{\infty}\boldsymbol{\Gamma}(k). \tag{2.63}$$

It follows that, in terms of the infinite vector moving average representation in Eq. (1.15), the large sample covariance matrix of $\bar{\mathbf{Z}}$ is given by

$$E(\bar{\mathbf{Z}}-\boldsymbol{\mu})(\bar{\mathbf{Z}}-\boldsymbol{\mu})' \approx \frac{1}{n}\sum_{k=-\infty}^{\infty}\boldsymbol{\Gamma}(k) = \frac{1}{n}\left(\sum_{j=0}^{\infty}\boldsymbol{\Psi}_j\right)\boldsymbol{\Sigma}\left(\sum_{j=0}^{\infty}\boldsymbol{\Psi}_j'\right). \tag{2.64}$$

For the covariance matrix function, it is naturally estimated by its sample covariance matrix function given by

$$\begin{aligned}\hat{\boldsymbol{\Gamma}}(k) &= \frac{1}{n}\sum_{t=1}^{n-k}(\mathbf{Z}_t-\bar{\mathbf{Z}})(\mathbf{Z}_{t+k}-\bar{\mathbf{Z}})', \text{ if } 0 \le k \le (n-1)\\ &= \left[\hat{\gamma}_{i,j}(k)\right]\end{aligned} \tag{2.65}$$

where

$$\hat{\gamma}_{i,j} = \frac{1}{n}\sum_{t=1}^{n-k}\left(Z_{i,t}-\bar{Z}_i\right)\left(Z_{j,t+k}-\bar{Z}_j\right). \tag{2.66}$$

2.5.2.2 *Sample correlation matrix function*

The correlation matrix function, $\boldsymbol{\rho}(k) = [\rho_{i,j}(k)]$ is estimated by the sample correlation matrix function given by

$$\hat{\boldsymbol{\rho}}(k) = \left[\hat{\rho}_{i,j}(k)\right] \tag{2.67}$$

where the $\hat{\rho}_{i,j}(k)$ represents the sample cross-correlations between the ith and jth component series at lag k, that is,

$$\hat{\rho}_{i,j}(k) = \frac{\hat{\gamma}_{i,j}(k)}{\left[\hat{\gamma}_{i,i}(0)\right]^{1/2}\left[\hat{\gamma}_{j,j}(0)\right]^{1/2}} = \frac{\sum_{t=1}^{n-k}\left(Z_{i,t}-\bar{Z}_i\right)\left(Z_{j,t+k}-\bar{Z}_j\right)}{\left[\sum_{t=1}^{n}\left(Z_{i,t}-\bar{Z}_i\right)^2\sum_{t=1}^{n}\left(Z_{j,t}-\bar{Z}_j\right)^2\right]^{1/2}}. \tag{2.68}$$

Although $\hat{\boldsymbol{\Gamma}}(k)$ and $\hat{\boldsymbol{\rho}}(k)$ are not unbiased estimators of $\boldsymbol{\Gamma}(k)$ and $\boldsymbol{\rho}(k)$ for a stationary vector process, they are consistent estimators, and the $\hat{\rho}_{i,j}(k)$ are asymptotically normally distributed with means $\rho_{i,j}(k)$ and the variance given by Bartlett (1955) shown next

$$\operatorname{Var}\left[\hat{\rho}_{i,j}(k)\right] \approx \frac{1}{n-k}\sum_{h=-\infty}^{\infty}\left\{\begin{aligned}&\rho_{i,i}(h)\rho_{j,j}(h)+\rho_{i,j}(k+h)\rho_{i,j}(k-h)\\ &+\rho_{i,j}^2(k)\left[\rho_{i,j}^2(h)+\frac{1}{2}\rho_{i,i}^2(h)+\frac{1}{2}\rho_{j,j}^2(h)\right]\\ &-2\rho_{i,j}(k)\left[\rho_{i,i}(h)\rho_{i,j}(h+k)+\rho_{j,j}(h)\rho_{i,j}(h-k)\right]\end{aligned}\right\}. \tag{2.69}$$

If $\rho_{i,j}(k) = 0$ for $|k| > q$ for some q, then

$$\mathrm{Var}\left[\hat{\rho}_{i,j}(k)\right] \approx \frac{1}{n-k}\left[1 + 2\sum_{h=1}^{q}\{\rho_{i,i}(h)\rho_{j,j}(h)\},\ \text{for } |k| > q.\right. \tag{2.70}$$

When the $\mathbf{Z}_t$ series are vector white noise, we have

$$\mathrm{Var}\left[\hat{\rho}_{i,j}(k)\right] \approx \frac{1}{n-k}. \tag{2.71}$$

For large samples, $(n-k)$ is often replaced by n in these expressions. The other related reference is Hannan (1970, p. 228).

2.5.2.3 *Sample partial correlation matrix function and extended cross-correlation matrices*

As shown in Eqs. (1.8)–(1.11), the partial correlation function, $\mathbf{P}(s)$, is a function of $\boldsymbol{\Gamma}(j)$. Thus, the sample partial correlation matrices, denoted by $\hat{\mathbf{P}}(s)$, is obtained by using $\hat{\boldsymbol{\Gamma}}(j)$ in place of $\boldsymbol{\Gamma}(j)$ for $j = 0, 1, \ldots, (s-1)$ in Eqs. (1.8)–(1.10). Because $\hat{\mathbf{P}}(s)$ is a proper correlation matrix, the results of sample correlation matrices can be used for its inference. Specifically, the elements of $\hat{\mathbf{P}}(s)$ denoted by $\hat{p}_{i,j}(s)$, are independent and asymptotically normally distributed with mean 0 and variance $1/n$. Thus,

$$X(s) = n\sum_{i=1}^{m}\sum_{j=1}^{m}\left[\hat{p}_{i,j}(s)\right]^2 \tag{2.72}$$

is asymptotically distributed as a χ^2 with m^2 degrees of freedom.

One other useful identification tool is the extended cross-correlation matrix function, which is the extension of the extended autocorrelation function (EACF), introduced by Tsay (2014, chapter 3), and has been implemented as a MTS function, Eccm, in R. We will illustrate its use in the empirical examples later.

2.5.3 Parameter estimation, diagnostic checking, and forecasting

Given a vector time series of n observations, $\mathbf{Z}_t, t = 1, 2, \ldots, n$, once a tentative VARMA$(p,q)$ model is identified,

$$\mathbf{Z}_t - \boldsymbol{\Phi}_1\mathbf{Z}_{t-1} - \cdots - \boldsymbol{\Phi}_p\mathbf{Z}_{t-p} = \boldsymbol{\theta}_0 + \mathbf{a}_t - \boldsymbol{\Theta}_1\mathbf{a}_{t-1} - \cdots - \boldsymbol{\Theta}_q\mathbf{a}_{t-q}. \tag{2.73}$$

The efficient estimates of the parameters, $\boldsymbol{\Phi} = (\boldsymbol{\Phi}_1, \ldots, \boldsymbol{\Phi}_p)$, $\boldsymbol{\theta}_0$, $\boldsymbol{\Theta} = (\boldsymbol{\Theta}_1, \ldots, \boldsymbol{\Theta}_q)$, and $\boldsymbol{\Sigma}$, are obtained by using a maximum likelihood method. Specifically, assuming the vector white noise is Gaussian, the log-likelihood function is given by

$$\begin{aligned} \ln L(\boldsymbol{\Phi}, \boldsymbol{\theta}_0, \boldsymbol{\Theta}, \boldsymbol{\Sigma}|\mathbf{Z}) &= -\frac{nm}{2}\ln 2\pi - \frac{n}{2}|\boldsymbol{\Sigma}| - \frac{1}{2}\sum_{t=1}^{n}\mathbf{a}_t'\boldsymbol{\Sigma}^{-1}\mathbf{a}_t \\ &= -\frac{nm}{2}\ln 2\pi - \frac{n}{2}|\boldsymbol{\Sigma}| - \frac{1}{2}\mathrm{tr}\,\boldsymbol{\Sigma}^{-1}S(\boldsymbol{\Phi}, \boldsymbol{\theta}_0, \boldsymbol{\Theta}), \end{aligned} \tag{2.74}$$

where

$$\mathbf{a}_t = \mathbf{Z}_t - \mathbf{\Phi}_1\mathbf{Z}_{t-1} - \cdots - \mathbf{\Phi}_p\mathbf{Z}_{t-p} - \mathbf{\theta}_0 + \mathbf{\Theta}_1\mathbf{a}_{t-1} + \cdots + \mathbf{\Theta}_q\mathbf{a}_{t-q},$$

and

$$S(\mathbf{\Phi}, \mathbf{\theta}_0, \mathbf{\Theta}) = \sum_{t=1}^{n} \mathbf{a}_t\mathbf{a}_t'.$$

The computation method is available in many statistical packages such as EViews (2016), MATLAB (2017), R (2009), SAS (2015), SCA (2008), and SPSS (2009).

After the parameters are estimated, it is important to check the adequacy of the model through a careful analysis of the residuals

$$\hat{\mathbf{a}}_t = \mathbf{Z}_t - \hat{\mathbf{\Phi}}_1\mathbf{Z}_{t-1} - \cdots - \hat{\mathbf{\Phi}}_p\mathbf{Z}_{t-p} - \hat{\mathbf{\theta}}_0 + \hat{\mathbf{\Theta}}_1\hat{\mathbf{a}}_{t-1} + \cdots + \hat{\mathbf{\Theta}}_q\hat{\mathbf{a}}_{t-q}. \tag{2.75}$$

For an adequate model, the sequence of residual vectors should behave as a vector white noise process.

After residual analysis, if the model is adequate, then it can be used for forecasting future values. For the general model in Eq. (2.73), the $\ell-$ step ahead forecast at time n is given by

$$\hat{\mathbf{Z}}_n(\ell) = \hat{\mathbf{\Phi}}_1\hat{\mathbf{Z}}_n(\ell-1) + \cdots + \hat{\mathbf{\Phi}}_p\hat{\mathbf{Z}}_n(\ell-p) + \hat{\mathbf{\theta}}_0 + \hat{\mathbf{a}}_n(\ell) - \hat{\mathbf{\Theta}}_1\hat{\mathbf{a}}_n(\ell-1) - \cdots - \hat{\mathbf{\Theta}}_q\hat{\mathbf{a}}_n(\ell-q), \tag{2.76}$$

where $\hat{\mathbf{Z}}_n(j) = \mathbf{Z}_{n+j}$ for $j \le 0$, $\hat{\mathbf{a}}_{n+j} = \mathbf{0}$ for $j > 0$, and $\hat{\mathbf{a}}_{n+j} = \mathbf{a}_{n+j}$ when $j \le 0$. It can also be used for inference and control using the estimates of parameters and the relationship presented in the vector model.

For more references on model building, we refer readers to papers by Akaike (1974), Box and Cox (1964), Parzen (1977), Ljung and Box (1978), Schwartz (1978), Dickey and Fuller (1979), Dickey et al. (1984), Tsay and Tiao (1984), and Reinsel and Ahn (1992). One can also study vector time series through state space representation, as shown in Libert et al. (1993), Jong and Lin (1994), Guo (2003), Wei (2006, chapter 18), Casals et al. (2016), and Zhang and Yang (2017), among others.

2.5.4 Cointegration in vector time series

It is well known that any stationary VARMA model can be approximated by a VAR model. Moreover, because of its easier interpretation, a VAR model is often used in practice. It should be noted that for a given vector time series $\mathbf{Z}_t$, it could occur that each component series $Z_{i,t}$ is nonstationary but its linear combination $\mathbf{Y}_t = \mathbf{\beta}'\mathbf{Z}_t$ is stationary for some $\mathbf{\beta}'$. In such a case, one should use its error-correction representation

$$\Delta\mathbf{Z}_t = \mathbf{\theta}_0 + \mathbf{\gamma}\mathbf{Z}_{t-1} + \mathbf{\alpha}_1\Delta\mathbf{Z}_{t-1} + \cdots + \mathbf{\alpha}_{p-1}\Delta\mathbf{Z}_{t-p+1} + \mathbf{a}_t, \tag{2.77}$$

where $\mathbf{\gamma}$ is related to $\mathbf{\beta}'$ and $\mathbf{\gamma}\mathbf{Z}_{t-1}$ is a stationary error-correction term. For more details, we refer readers to Engle and Granger (1987), Granger (1986), Wei (2006, chapter 17), Ghysels and Miller (2015), Miller and Wang (2016), and Wagner and Wied (2017).

2.6 Seasonal vector time series model

The VARMA(p,q) model in Eq. (2.42) can be extended to a seasonal vector model that contains both seasonal and non-seasonal AR and MA polynomials as follows,

$$\boldsymbol{\alpha}_P(B^s)\boldsymbol{\Phi}_p(B)\mathbf{Z}_t = \boldsymbol{\theta}_0 + \boldsymbol{\beta}_Q(B^s)\boldsymbol{\Theta}_q(B)\mathbf{a}_t, \tag{2.78}$$

where

$$\boldsymbol{\alpha}_P(B^s) = \mathbf{I} - \boldsymbol{\alpha}_1 B^s - \cdots - \boldsymbol{\alpha}_P B^{Ps} = \mathbf{I} - \sum_{k=1}^{P} \boldsymbol{\alpha}_k B^{ks},$$

$$\boldsymbol{\beta}_Q(B^s) = \mathbf{I} - \boldsymbol{\beta}_1 B^s - \cdots - \boldsymbol{\beta}_Q B^{Qs} = \mathbf{I} - \sum_{k=1}^{Q} \boldsymbol{\beta}_k B^{ks},$$

and s is a seasonal period. $\boldsymbol{\alpha}_k$ and $\boldsymbol{\beta}_k$ are seasonal AR and seasonal MA parameters, respectively. P is the seasonal AR order, Q is the seasonal MA order. The zeros of $|\boldsymbol{\alpha}_P(B)|$ and $|\boldsymbol{\beta}_Q(B)|$ are outside of the unit circle. For simplicity, we will denote the seasonal vector time series model in Eq. (2.78) with a seasonal period s as VARMA(p, q) × (P, Q)$_s$.

It is important to point out that for a univariate time series, the polynomials in Eq. (2.78) are scalar and they are commutative. For example, for a univariate seasonal AR model, the two representations,

$$\alpha_P(B^s)\phi_p(B)Z_t = a_t, \text{ and } \phi_p(B)\alpha_P(B^s)Z_t = a_t,$$

are exactly the same. However, for a vector seasonal VAR model, the two representations,

$$\boldsymbol{\alpha}_P(B^s)\boldsymbol{\Phi}_p(B)\mathbf{Z}_t = \mathbf{a}_t, \text{ and } \boldsymbol{\Phi}_p(B)\boldsymbol{\alpha}_P(B^s)\mathbf{Z}_t = \mathbf{a}_t,$$

are not the same, because the vector multiplications are not commutative. This leads to many complications in terms of model identification, parameter estimation, and forecasting. We refer readers to an interesting and excellent paper by Yozgatligil and Wei (2009) for more details and examples.

For nonstationary seasonal vector time series, we can further extend model in Eq. (2.58) as follows,

$$\boldsymbol{\alpha}_P(B^s)\boldsymbol{\Phi}_p(B)\mathbf{D}(B^s)\mathbf{D}(B)\mathbf{Z}_t = \boldsymbol{\beta}_Q(B^s)\boldsymbol{\Theta}_q(B)\mathbf{a}_t, \tag{2.79}$$

where

$$\mathbf{D}(B^s) = \begin{bmatrix} (1-B^s)^{d_1} & 0 & . & \cdots & 0 & 0 \\ 0 & (1-B^s)^{d_2} & 0 & \cdots & . & 0 \\ . & 0 & . & \cdots & . & . \\ \vdots & \vdots & \vdots & \ddots & \vdots & \vdots \\ 0 & . & . & \cdots & . & 0 \\ 0 & 0 & . & \cdots & 0 & (1-B^s)^{d_m} \end{bmatrix}. \tag{2.80}$$

The identification, parameter estimation, and forecasting for seasonal vector time series are similar to VARMA models, except we need to examine their correlation matrix and partial correlation matrix functions at the multiple of the seasonal period s. Again, we remind readers of the remarks in the last paragraph on the different representations of seasonal vector time models, regardless of whether they are stationary or nonstationary.

It should be noted that in applications, we may find many vector time series are not only nonstationary but also seasonal. However, in vector time series modeling, we often build a vector time series model based on differenced and seasonally adjusted data, especially because of the representation issues related to seasonal vector AR and MA models.

2.7 Multivariate time series outliers

In time series analysis, it is important to examine the possible outliers and do some proper adjustments because outliers can lead to inaccurate parameter estimation, model misspecification, and poor forecasts. Outlier detection has been studied extensively for univariate time series, including Fox (1972), Abraham and Box (1979), Martin (1980), Hillmer et al. (1983), Chang et al. (1988), Tsay (1986, 1988), Chen and Liu (1993), Lee and Wei (1995), Wang et al. (1995), Sanchez and Pena (2003), and many others. We normally classify outliers in four categories, additive outliers, innovational outliers, level shifts, and temporal changes. For MTS, a natural approach is first to use univariate techniques to the individual component and remove outliers, then treat the adjusted series as outlier–free and model them jointly. However, there are several disadvantages of this approach. First, in MTS an outlier of its univariate component may be induced by an outlier from other component within the multivariate series. Overlooking this situation may lead to overspecification of the number of outliers. Second, an outlier impacting all the components may not be detected by using the univariate outlier detection methods because they do not use the joint information from all time series components in the system at the same time.

To overcome the difficulties, Tsay et al. (2000) extended four types of outliers for univariate time series to MTS and their detections, and Galeano et al. (2006) further proposed a detection method based on projection pursuit, which sometime is more powerful than testing the multivariate series directly. Other references on MTS outliers include Helbing and Cleroux (2009), Martinez-Alvarez et al. (2011), and Cucina et al. (2014), among others.

2.7.1 Types of multivariate time series outliers and detections

Suppose we have m-dimensional time series $\mathbf{X}_t = (X_{1t}, X_{2t}, \ldots, X_{mt})'$ that follows a VARMA model

$$\mathbf{\Phi}(B)\mathbf{X}_t = \mathbf{\Theta}_0 + \mathbf{\Theta}(B)\mathbf{a}_t, \tag{2.81}$$

where $\mathbf{a}_t = (a_{1,t}, a_{2,t}, \ldots, a_{m,t})'$ is m-dimensional Gaussian white noise with mean $\mathbf{0}$ and positive-definite covariance matrix $\mathbf{\Sigma}$. Recall that if $\mathbf{X}_t$ is invertible, the VARMA model in Eq. (2.81) has its corresponding AR representation,

$$\mathbf{\Pi}(B)\mathbf{X}_t = \mathbf{C}_1 + \mathbf{a}_t, \tag{2.82}$$

where $\mathbf{\Pi}(B) = \{\mathbf{\Theta}(B)\}^{-1}\mathbf{\Phi}(B)$ and $\mathbf{C}_1 = \{\mathbf{\Theta}(1)\}^{-1}\mathbf{\Theta}_0$. If $\mathbf{X}_t$ is stationary, we can also have the MA representation,

$$\mathbf{X}_t = \mathbf{C}_2 + \mathbf{\Psi}(B)\mathbf{a}_t, \tag{2.83}$$

where $\mathbf{\Psi}(B) = \{\mathbf{\Phi}(B)\}^{-1}\mathbf{\Theta}(B)$ and $\mathbf{C}_2 = \{\mathbf{\Phi}(1)\}^{-1}\mathbf{\Theta}_0$.

Let $I_t^{(h)}$ be the indicator variable such that $I_h^{(h)} = 1$ and $I_t^{(h)} = 0$ if $t \neq h$. Suppose we observe time series $\mathbf{Z}_t = (Z_{1t}, Z_{2t}, \ldots, Z_{mt})'$, and $\boldsymbol{\omega} = (\omega_1, \omega_2, \ldots, \omega_m)'$ is the size of the outlier. The four types of outlier in univariate time series can be generalized by following formulation

$$\mathbf{Z}_t = \mathbf{X}_t + \boldsymbol{\alpha}(B)\boldsymbol{\omega} I_t^{(h)}. \tag{2.84}$$

The outlier type is defined by the matrix polynomial $\boldsymbol{\alpha}(B)$. Specifically,

- if $\boldsymbol{\alpha}(B) = \mathbf{\Psi}(B)$, we have a multivariate innovational outlier (MIO);
- if $\boldsymbol{\alpha}(B) = \mathbf{I}$, we have a multivariate additive outlier (MAO);
- if $\boldsymbol{\alpha}(B) = (1 - B)^{-1}\mathbf{I}$, then the outlier is a multivariate level shift (MLS);
- if $\boldsymbol{\alpha}(B) = (\mathbf{I} - \delta\mathbf{I}B)^{-1}$, with a constant $0 < \delta < 1$, we have a multivariate temporary change (MTC).

A MIO is an outlier in the innovations that can be interpreted as an internal change in the structure of the time series. It will affect several consecutive observations. A MAO is typically due to external causes, such as typos or measurement errors, and only affect single observation. MLS changes the mean level of the series and its effect is permanent. MTC causes an initial impact, and its effect decreases at a fixed rate in the subsequent observations. Different components of the time series may have different outlier effects. Tsay et al. (2000) showed that when the parameters of the VARMA model are known, the series of innovations is $\boldsymbol{\varepsilon}_t = \mathbf{\Pi}(B)\mathbf{Z}_t - \mathbf{C}_1$, where $\mathbf{Z}_t = \mathbf{X}_t$ and $\boldsymbol{\varepsilon}_t = \mathbf{a}_t$ for $t < h$. The relationship between the computed innovations, $\boldsymbol{\varepsilon}_t$, and the true white noise innovations is actually

$$\boldsymbol{\varepsilon}_t = \mathbf{\Pi}(B)\boldsymbol{\alpha}(B)\boldsymbol{\omega} I_t^{(h)} + \mathbf{a}_t. \tag{2.85}$$

Let $\mathbf{\Pi}(B)\boldsymbol{\alpha}(B) = \mathbf{I} - \sum_{j=1}^{\infty}\mathbf{\Gamma}_j B^j$. They showed that when the model is known, the estimation of the size of a multivariate outlier of type i at time h is given by

$$\boldsymbol{\omega}_{i,h} = -\left(\sum_{j=0}^{n-h}\mathbf{\Gamma}_j'\mathbf{\Sigma}^{-1}\mathbf{\Gamma}_j\right)^{-1}\left(\sum_{j=0}^{n-h}\mathbf{\Gamma}_j'\mathbf{\Sigma}^{-1}\boldsymbol{\varepsilon}_{h+j}\right), \quad i = I, A, L, T, \tag{2.86}$$

where $\mathbf{\Gamma}_0 = -\mathbf{I}$, and we denote MIO as I, MAO as A, MLS as L, and MTC as T for the subscripts. The covariance of the estimate in Eq. (2.86) is $\mathbf{\Sigma}_{i,h} = \left(\sum_{j=0}^{n-h}\mathbf{\Gamma}_j'\mathbf{\Sigma}^{-1}\mathbf{\Gamma}_j\right)^{-1}$, and the test statistic used to detect multivariate outliers is

$$J_{i,h} = \boldsymbol{\omega}_{i,h}'\mathbf{\Sigma}_{i,h}^{-1}\boldsymbol{\omega}_{i,h}, \quad i = I, A, L, T, \tag{2.87}$$

which follows a noncentral Chi-square $\chi_m^2(\eta_i)$ with noncentral parameters $\eta_i = \boldsymbol{\omega}'\boldsymbol{\Sigma}_{i,h}^{-1}\boldsymbol{\omega}$. Under the null hypothesis $\boldsymbol{\omega} = \mathbf{0}$, the distribution of $J_{i,h}$ will be chi-squared with m degrees of freedom. To detect outliers in practice, the parameter matrices are substituted by their estimates, and the following overall test statistic is used:

$$J_{\max}(i, h_i) = \max_{1 \le h \le n} J_{i,h}, \tag{2.88}$$

where h_i is the time indices at which the maximum of the test statistic occurs. After an outlier is identified, its impact on the underlying time series is removed. The resulting time series is then treated as a new time series and the detection procedure is iterated until no significant outlier can be found. Then, a joint time series model with outliers is fitted.

2.7.2 Outlier detection through projection pursuit

Galeano et al. (2006) proposed a method that first projects the MTS into a univariate time series, and then detects outliers within the derived or projected univariate time series. They also suggested an algorithm that finds optimal projection directions by using kurtosis coefficients. This approach has two main advantages. First, it is simple because no multivariate model has to be prespecified. Second, an appropriate projection direction can lead to a test statistic that is more powerful.

To illustrate the method, we use projection of the VARMA model as an example. A linear combination of m multiple time series that follow a VARMA model is a univariate ARMA model. Specifically, let $\mathbf{A}'$ be a vector contains the weights of the linear combination and assume $\mathbf{X}_t$ is a m-dimensional VARMA(p,q) process. Then, Lütkepohl (1984, 1987) showed that $x_t = \mathbf{A}'\mathbf{X}_t$ follows a ARMA(p^*, q^*) process with $p^* \le mp$ and $q^* \le (m-1)p + q$. Thus, x_t has the following representation

$$\phi(B)x_t = c + \theta(B)e_t, \tag{2.89}$$

where $\phi(B) = |\boldsymbol{\Phi}(B)|$, $c = \mathbf{A}'\boldsymbol{\Phi}(1)^*\boldsymbol{\Theta}_0$, and $\mathbf{A}'\boldsymbol{\Phi}(B)^*\boldsymbol{\Theta}(B)\mathbf{a}_t = \theta(B)e_t$, $\boldsymbol{\Phi}(B)^*$ is the adjoint matrix of $\boldsymbol{\Phi}(B)$, and e_t is a univariate white noise process with mean 0 and constant variance σ^2.

Suppose we observe a time series $\mathbf{Z}_t$ that is affected by an outlier as shown in Eq. (2.84), the projected time series $z_t = \mathbf{A}'\mathbf{Z}_t$ can be represented as

$$z_t = x_t + \mathbf{A}'\boldsymbol{\alpha}(B)\boldsymbol{\omega}I_t^{(h)}. \tag{2.90}$$

Particularly, if $\mathbf{Z}_t$ is affected by MAO, then the projected time series is

$$z_t = x_t + \boldsymbol{\beta}I_t^{(h)}, \tag{2.91}$$

so that it has an additive outlier of size $\boldsymbol{\beta} = \mathbf{A}'\boldsymbol{\omega}$ at $t = h$. Similarly, if $\mathbf{Z}_t$ is affected by an MLS, then its corresponding projected time series will have a level shift with size $\boldsymbol{\beta} = \mathbf{A}'\boldsymbol{\omega}$ at $t = h$. The same argument can also be made in the case when we have MTC. Thus, we can formulate the following hypothesis

$$H_0: \boldsymbol{\beta} = \mathbf{0} \ \ v.s. \ \ H_A: \boldsymbol{\beta} \ne \mathbf{0}, \tag{2.92}$$

which is the equivalent to test

$$H_0^* : \boldsymbol{\omega} = \mathbf{0} \ \ v.s. \ \ H_A^* : \boldsymbol{\omega} \neq \mathbf{0}. \tag{2.93}$$

A MIO has more complicated effect. It may lead to a patch of consecutive outliers with sizes $\mathbf{A}'\boldsymbol{\omega}$, $\mathbf{A}'\boldsymbol{\Psi}_1\boldsymbol{\omega}$, …, $\mathbf{A}'\boldsymbol{\Psi}_{n-h}\boldsymbol{\omega}$, starting with time index $t = h$. Suppose h is not close to n and because $\boldsymbol{\Psi}_j \to \mathbf{0}$, the size of the outlier tends to 0. In addition, the univariate series z_t obtained by the projection can be affected by an additive outlier, a patch of outliers, or a level shift. For more details, please see Galeano et al. (2006).

To obtain the direction of outliers, Galeano et al. (2006) proved that the direction of the outlier can be found by maximizing or minimizing the kurtosis coefficient of the projected series. Specifically,

- For a MAO, the kurtosis coefficient of z_t is maximized when $\mathbf{A}'$ is proportional to $\boldsymbol{\omega}$ and is minimized when $\mathbf{A}'$ is orthogonal to $\boldsymbol{\omega}$.
- For a MTC, the kurtosis coefficient of z_t is maximized or minimized when $\mathbf{A}'$ is proportional to $\boldsymbol{\omega}$, and is minimized or maximized when $\mathbf{A}'$ is orthogonal to $\boldsymbol{\omega}$.
- For a MLS, the kurtosis coefficient of z_t is minimized when $\mathbf{A}'$ is proportional to $\boldsymbol{\omega}$ and is maximized when $\mathbf{A}'$ is orthogonal to $\boldsymbol{\omega}$ if

$$h \in \left[1 + \frac{1}{2}\left(1 - \frac{1}{\sqrt{3}}\right)n, 1 + \frac{1}{2}\left(1 + \frac{1}{\sqrt{3}}\right)n\right].$$

Otherwise, the kurtosis coefficient of z_t is maximized when $\mathbf{A}'$ is proportional to $\boldsymbol{\omega}$ and is minimized when $\mathbf{A}'$ is orthogonal to $\boldsymbol{\omega}$.

There are two implications based on these results. First, for an MAO, MLS, or MTC, one of the directions obtained by maximizing or minimizing the kurtosis coefficient is the direction of the outlier. Second, these directions are obtained without the information of the time index at which the outlier occurs.

When there are multiple outliers, it is insufficient if we only consider the projections that maximize or minimize the kurtosis coefficient because of the masking effects. To overcome the difficulty, they suggest constructing the following $2m$ projection directions:

1. Let $k = 1$ and $\mathbf{Y}_t^{(k)} = \mathbf{Z}_t$
2. Let $\boldsymbol{\Sigma}_{\mathbf{Y}_t^{(k)}} = \frac{1}{n}\sum_{t=1}^{n}\left(\mathbf{Y}_t^{(k)} - \bar{\mathbf{Y}}_t^{(k)}\right)\left(\mathbf{Y}_t^{(k)} - \bar{\mathbf{Y}}_t^{(k)}\right)'$, and find $\mathbf{A}_k$ such that
$$\mathbf{A}_k = \underset{\mathbf{A}_k'\boldsymbol{\Sigma}_{Y_t^{(k)}}\mathbf{A}_k = 1}{\arg\max} \ \frac{1}{n}\sum_{t=1}^{n}\left[\mathbf{A}_k'\left(\mathbf{Y}_t^{(k)} - \bar{\mathbf{Y}}_t^{(k)}\right)\right]^4. \tag{2.94}$$
3. If $k = m$, then stop; otherwise, define $\mathbf{Y}_t^{(k+1)} = \left(\mathbf{I} - \mathbf{A}_k\mathbf{A}_k'\boldsymbol{\Sigma}_{\mathbf{Y}_t^{(k)}}\right)\mathbf{Y}_t^{(k)}$, which means that $\mathbf{Y}_t^{(k+1)}$ is the projection of the observations in an orthogonal direction to $\mathbf{A}_k$. Let $k = k + 1$ and go to step 2.
4. Repeat the same procedure to minimize the objective function in Eq. (2.94), to obtain extra m directions, that is, $\mathbf{A}_{m+1}$, …, $\mathbf{A}_{2m}$.

They also propose identifying the level-shift outliers before checking for other types of outlier. For this they further suggest using the likelihood ratio test (LRT) of the iterative univariate outlier detection procedure proposed by Chang et al. (1988) to identify level shifts, and then use the level shift adjusted series $\hat{\mathbf{Z}}_t$ to fit the model,

$$\left(\mathbf{I}-\mathbf{\Phi}_1 B-\cdots-\mathbf{\Phi}_{\hat{p}} B^{\hat{p}}\right)\hat{\mathbf{Z}}_t=\hat{\boldsymbol{\varepsilon}}_t, \tag{2.95}$$

and its parameter estimates to detect additive outliers, temporal changes, and innovative outliers through the $2m$ projected univariate series $\hat{z}_{i,t}-\mathbf{A}'_i\hat{\mathbf{Z}}_t$ and their associated innovational series $\hat{e}_{i,t}=\mathbf{A}'_i\hat{\boldsymbol{\varepsilon}}_t$ for $i=1, \ldots, 2m$, where the LRT is used again. We refer readers to the numerous illustrated examples of using LRT given in Wei (2006, chapter 10). The steps are:

1. Model each projected time series $\hat{z}_{i,t}$ using AR(p) model with the order selected by AIC assuming there are no outliers. Then, compute LRT statistics $\lambda^i_{A,t}$ and $\lambda^i_{T,t}$. In addition, compute $\lambda^i_{I,t}$ based on the innovational series $\hat{e}_{i,t}$. Then, obtain the following statistics:

$$\Lambda_A=\max_{1\le i\le 2m}\max_{1\le t\le n}|\lambda^i_{A,t}|,$$

$$\Lambda_T=\max_{1\le i\le 2m}\max_{1\le t\le n}|\lambda^i_{T,t}|,$$

and

$$\Lambda_I=\max_{1\le i\le 2m}\max_{1\le t\le n}|\lambda^i_{I,t}|.$$

2. Let $C_{A,\alpha}$, $C_{T,\alpha}$, and $C_{I,\alpha}$ be the critical values at significance level α. (i) If $\Lambda_i<C_{i,\alpha}$ for $i=I, A, T$, then no outliers are found, and the procedure stops. (ii) If $\Lambda_i>C_{i,\alpha}$ for only one i, then identify an outlier of type i and remove its effect using the multivariate parameter estimates. (iii) If $\Lambda_i>C_{i,\alpha}$ for more than one i, then identify the outlier based on the most significant test statistics and remove its effect using the multivariate parameter estimates. For the recommended critical values of the test statistics, please refer to table 2 in Galeano et al. (2006).

3. Repeat steps 1 and 2 until no more outlier is detected.

4. Let $\left\{h^A_1,\ldots,h^A_{i_A}\right\}$, $\left\{h^T_1,\ldots,h^T_{i_T}\right\}$, and $\left\{h^I_1,\ldots,h^I_{i_I}\right\}$ be the time indexes of the detected additive outliers, temporal changes, and innovational outliers. We estimate jointly the model and the detected outlier as

$$\begin{aligned}\widetilde{\mathbf{Z}}_t&=\hat{\mathbf{Z}}_t-\sum_{i_A=1}^{r_A}\boldsymbol{\omega}_{i_A}I_t^{\left(h^A_{i_A}\right)}-\sum_{i_T=1}^{r_T}\frac{\boldsymbol{\omega}_{i_T}}{1-\delta B}I_t^{\left(h^T_{i_T}\right)},\\ \widetilde{\boldsymbol{\varepsilon}}_t&=\hat{\boldsymbol{\varepsilon}}_t-\sum_{i_I=1}^{r_I}\boldsymbol{\omega}_{i_I}I_t^{\left(h^I_{i_I}\right)}.\end{aligned} \tag{2.96}$$

If some of the outliers become insignificant, then remove the least significant outlier and re-estimate the model. Repeat until all outliers are significant.

Finally, if needed, let $\left\{h_1^L,\ldots,h_{r_L}^L\right\}$, $\left\{h_1^A,\ldots,h_{r_A}^A\right\}$, $\left\{h_1^T,\ldots,h_{r_T}^T\right\}$, and $\left\{h_1^I,\ldots,h_{r_I}^I\right\}$ be the time indexes of the detected level shifts, additive outliers, temporal changes, and innovational outliers. We can jointly estimate the model and the detected outliers for the observed series $\mathbf{Y}_t$,

$$\left(\mathbf{I}-\mathbf{\Phi}_1 B-\cdots-\mathbf{\Phi}_{\hat{p}} B^{\hat{p}}\right)\mathbf{Y}_t=\xi_t, \tag{2.97}$$

where

$$\mathbf{Y}_t=\mathbf{Z}_t-\sum_{i_L=1}^{r_L}\boldsymbol{\omega}_{i_L} I_t^{\left(h_{i_L}^L\right)}-\sum_{i_A=1}^{r_A}\boldsymbol{\omega}_{i_A} I_t^{\left(h_{i_A}^A\right)}-\sum_{i_T=1}^{r_T}\frac{\boldsymbol{\omega}_{i_T}}{1-\delta B} I_t^{\left(h_{i_T}^T\right)},$$

and

$$\xi_t=\varepsilon_t-\sum_{i_I=1}^{\gamma_I}\boldsymbol{\omega}_{i_I} I_t^{\left(h_{i_I}^I\right)},$$

and $\mathbf{Y}_t$ is the outlier adjusted series.

We will stop the chapter's major topics here. Before going into detailed empirical examples, we would like to suggest a few good time series analysis books that cover both univariate and vector time series to our readers. They include Mills (1991), Murteira et al. (1993), Chatfield (2004), Lütkepohl (2007), Priestley (1982), Hamilton (1994), Wei (2006), Tsay (2014), and Box et al. (2015), among others.

2.8 Empirical examples

2.8.1 First model of US monthly retail sales revenue

Example 2.4 In this example, we try to build a multivariate vector time series model for the US monthly retail sales revenue (in millions of dollars) including AUT (automobile), BUM (building materials), CLO (clothing), BWL (beer wine and liquor), FUR (furniture), GEM (general merchandize), GRO (grocery), and HOA (household appliances), from June 2009 to November 2016, with $n = 90$. The data set is listed as series WW2a in the Data Appendix and given in Table 2.2.

There are eight series in total, with their plots shown in Figure 2.1. There is an upward trend in each of them. The p-values of the unit root test are shown in Table 2.3.

After taking difference $(1 - B)\mathbf{Z}_t$, where $\mathbf{Z}_t = (Z_{1,t}, Z_{2,t}, \ldots, Z_{8,t})$, we show their sample correlation matrix and partial correlation matrix functions in Figure 2.2.

Since the sample partial correlation matrix function cuts off after lag 13, it suggests a VAR (13) model,

$$\mathbf{Z}_t=\boldsymbol{\theta}_0+\mathbf{\Phi}_1\mathbf{Z}_{t-1}+\cdots+\mathbf{\Phi}_{13}\mathbf{Z}_{t-13}+\mathbf{a}_t. \tag{2.98}$$

Unfortunately, there are $8 + 13(8 \times 8) = 840$ parameters to be estimated and the limited available observations make the estimation impossible.

Table 2.2 US monthly retail sales revenue in millions of dollars from 6/2009 to 11/2016.

Period	AUT	BUM	CLO	BWL	FUR	GEM	GRO	HOA
June 09	48078	26350	11533	3306	3703	47089	42146	1280
July 09	51093	24003	12042	3556	3804	47097	43838	1302
August 09	55677	21522	12963	3399	3939	48826	42994	1308
September 09	42299	21304	11757	3263	3827	44202	41537	1203
October 09	45268	21263	13020	3425	3723	48509	43020	1257
November 09	41448	19915	13837	3356	3848	53489	42508	1398
December 09	46936	19182	18727	4626	3940	69970	45395	1489
January 10	42143	15349	9898	2877	3683	43854	42986	1112
February 10	43434	15468	10565	2916	3833	44756	40061	1129
March 10	56482	21692	13122	3214	4090	48643	43433	1223
April 10	52987	26682	12896	3310	3675	47751	42236	1245
May 10	54178	26231	13119	3466	3965	50476	44604	1363
June 10	53049	25446	12160	3438	3748	48290	42877	1342
July 10	55854	23100	12674	3657	3964	48620	44558	1355
August 10	55120	21698	13161	3455	3997	49314	43431	1291
September 10	51556	21261	12318	3365	3932	45678	42642	1226
October 10	51463	21882	13165	3497	3714	49037	43535	1192
November 10	49393	21458	15100	3524	4009	55587	43723	1461
December 10	55231	20277	20049	4683	4051	71770	46656	1514
January 11	49692	16006	10157	2888	3576	44323	44280	1077
February 11	53590	16003	11366	2985	3871	45805	41229	1142
March 11	63583	21964	13711	3249	4305	49677	45069	1216
April 11	58199	24377	13855	3363	3902	50482	45586	1182
May 11	57518	27709	13642	3471	3862	51437	46319	1276
June 11	57722	27009	13265	3551	3793	50907	45883	1316
July 11	57538	23530	13410	3740	4004	50661	47170	1336
August 11	59341	23733	13880	3576	4133	51267	46449	1362
September 11	56231	22501	13425	3517	4053	48238	44876	1384
October 11	55574	23077	13788	3515	3742	50877	45725	1279
November 11	54186	22371	15809	3646	4128	57198	46045	1613
December 11	61128	21201	21721	4892	4303	73907	48840	1639
January 12	52989	17981	10677	2995	3898	45642	45460	1170
February 12	60345	18502	12608	3202	4327	49635	44172	1266
March 12	68799	24156	15195	3549	4442	53114	47406	1325
April 12	61761	25987	13995	3409	3800	50601	45753	1202
May 12	67063	29374	14586	3786	4193	53091	48418	1400
June 12	63336	25772	14030	3815	4075	52279	47215	1412
July 12	63364	23921	13716	3733	4145	50733	47472	1374
August 12	68423	24170	15158	3752	4371	53679	48046	1410
September 12	60831	22327	13719	3503	4127	49326	46048	1305
October 12	61295	24659	14256	3626	3790	51398	46978	1298
November 12	59772	23512	16575	3869	4234	59013	47314	1640
December 12	64800	21172	21616	5126	4350	73802	49363	1613
January 13	59912	19323	11325	3158	3966	46616	47082	1239
February 13	63076	18702	12276	3231	3985	48503	43837	1288

(continued overleaf)

Table 2.2 (*continued*)

Period	AUT	BUM	CLO	BWL	FUR	GEM	GRO	HOA
March 13	71529	23216	15410	3625	4475	54650	48869	1377
April 13	68820	28906	14139	3465	3975	49955	45316	1306
May 13	72952	32849	15197	3973	4213	54774	49549	1477
June 13	69509	27891	14169	3817	4083	53179	47765	1402
July 13	73058	27611	14350	4010	4238	51958	48783	1492
August 13	75544	26242	15677	4078	4573	55227	49288	1469
September 13	64308	24424	13505	3643	4269	49910	46499	1402
October 13	67070	26255	14852	3799	4274	52923	48283	1441
November 13	64894	23974	16829	4043	4589	60375	48904	1596
December 3	68348	22399	21756	5235	4231	73798	50374	1635
January 14	60664	19900	11088	3373	3912	47319	49142	1251
February 14	64398	18989	12101	3353	4078	48886	45091	1264
March 14	77683	24343	14864	3679	4598	53967	48997	1392
April 14	74600	30447	15009	3699	4164	52789	48690	1281
May 14	79113	33758	15645	4187	4616	56900	51742	1430
June 14	73221	30409	14037	4086	4187	53872	49434	1430
July 14	77685	29003	14678	4240	4438	53646	51362	1462
August 14	79966	26798	15888	4216	4704	57050	51309	1449
September 14	71294	26574	13894	3856	4586	50663	48861	1435
October 14	72328	27748	15237	4087	4470	54625	51057	1449
November 14	68450	25248	17845	4133	4721	61843	51042	1594
December 14	76041	24498	22721	5606	4798	75313	53011	1625
January 15	68158	21161	11649	3576	4410	49436	51536	1271
February 15	68559	19923	12646	3517	4240	49100	47088	1252
March 15	82234	26526	15576	3881	4741	54686	51167	1419
April 15	79070	31512	15202	3864	4357	52424	50091	1342
May 15	82410	33282	16279	4369	4867	57441	53201	1519
June 15	79265	31709	14673	4241	4632	54135	50950	1531
July 15	82841	30676	15315	4524	4788	54829	53116	1543
August 15	83872	27916	16247	4248	4919	56674	52120	1433
September 15	77506	27377	14365	4091	4999	51588	50244	1464
October 15	77134	28546	15618	4291	4860	55544	51987	1394
November 15	72505	26889	17485	4241	4887	61397	51355	1543
December 15	80868	27044	22884	5834	5076	76391	54146	1592
January 16	68984	22100	11560	3559	4464	49050	52167	1226
February 16	76652	23290	13073	3718	4638	50661	49193	1321
March 16	84611	30027	15828	3986	5013	55303	52741	1378
April 16	81446	32496	14917	4043	4611	52960	50958	1296
May 16	82428	34787	15546	4322	4814	55722	53502	1415
June 16	81439	34371	14826	4388	4848	54481	52745	1487
July 16	84085	30347	15134	4655	4821	54710	53756	1435
August 16	88224	30257	15950	4386	4924	54859	52769	1433
September 16	81027	29210	14619	4336	5080	50670	51621	1434
October 16	78994	29010	15310	4337	4731	54221	52769	1308
November 16	77389	29095	17344	4571	5077	60520	52793	1532

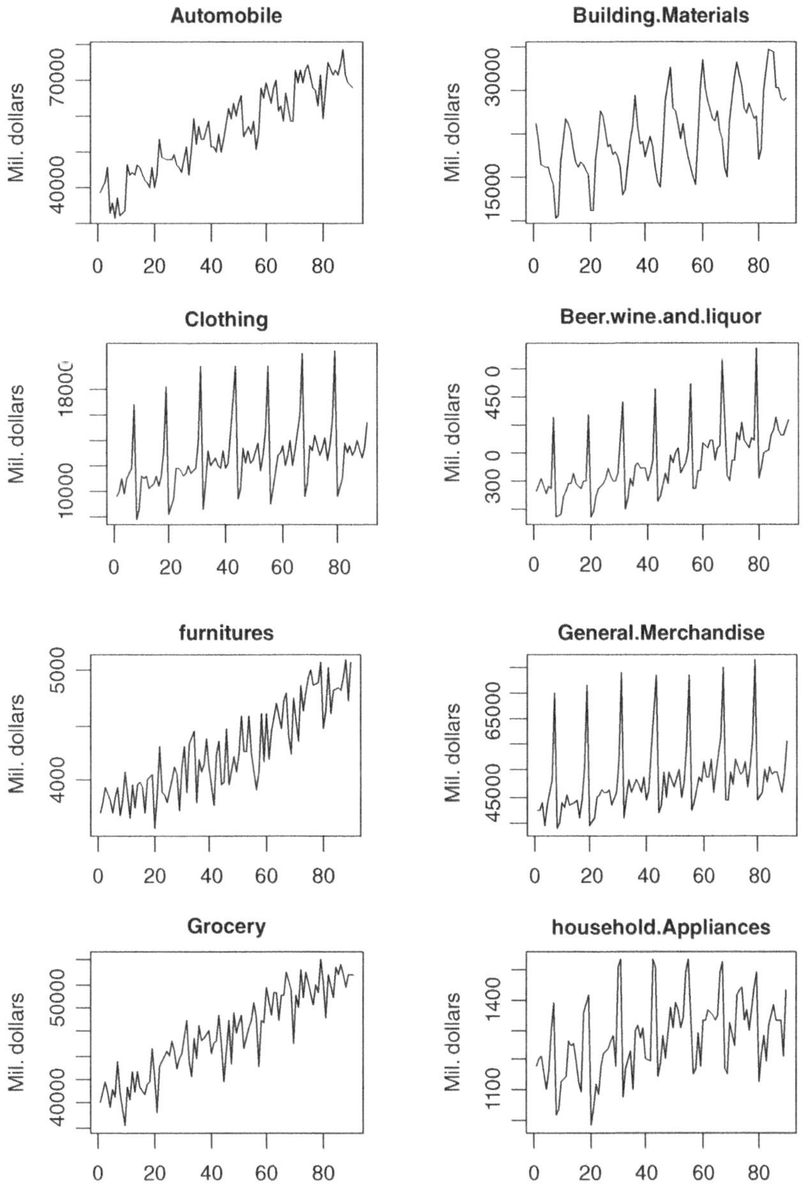

Figure 2.1 US monthly retail sales revenue from June 2009 to November 2016.

Table 2.3 Unit root test *P*-values for eight series.

	Original	First Difference	Series Name
$Z_{1,t}$	0.75	<0.01	Automobile
$Z_{2,t}$	0.45	<0.01	Building Materials
$Z_{3,t}$	0.42	<0.01	Clothing
$Z_{4,t}$	0.53	<0.01	Beer, Wine, Liquor
$Z_{5,t}$	0.76	<0.01	Furniture
$Z_{6,t}$	0.48	<0.01	General Merchandise
$Z_{7,t}$	0.83	<0.01	Grocery
$Z_{8,t}$	0.53	<0.01	Household Appliances

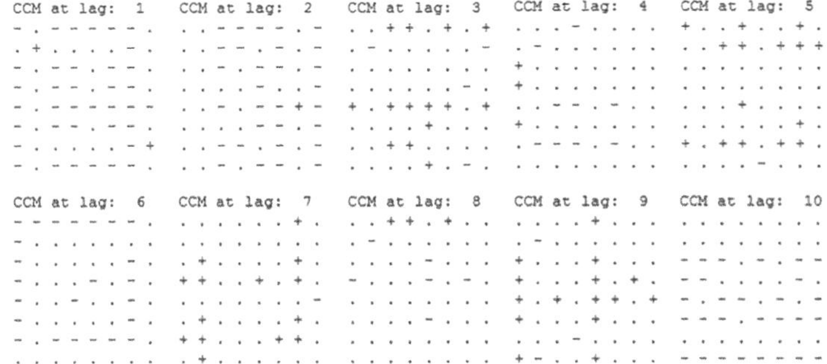

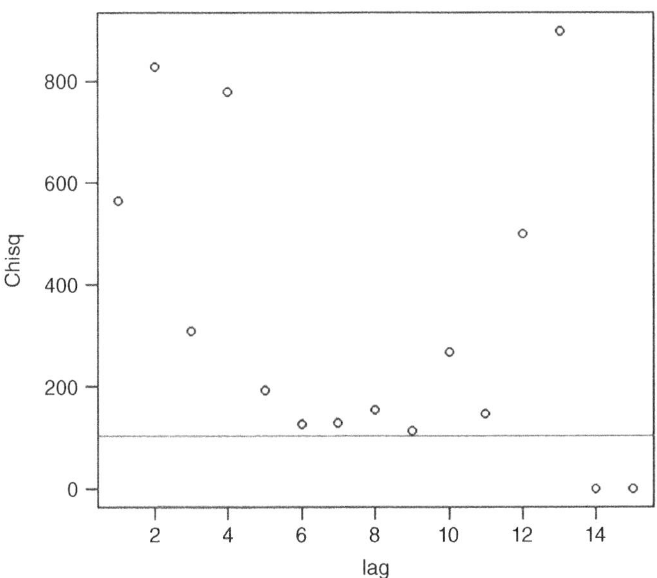

Figure 2.2 Sample correlation matrix and partial correlation matrix functions.

To solve this problem, we decide to reduce the number of data dimensions by combining clothing ($Z_{3,t}$), beer, wine, liquor ($Z_{4,t}$), furniture ($Z_{5,t}$), and household appliances ($Z_{8,t}$) into one sales group and name it as "consumer materials" (COM). The new classified data set is referred to as the WW2b series shown in Table 2.4, and their plots are given in Figure 2.3, which include individual and combined plots so that readers can get a better feel about their related patterns.

Based on the earlier results of the unit root test, we again take differencing and the differenced new five series are given in Figure 2.4.

Now, let the new five-dimensional vector variable be $\mathbf{W}_t = (1 - B)\mathbf{Z}_t$. To identify a possible model, we calculate its sample correlation matrix function shown in Figure 2.5. Some of them

Table 2.4 Sales revenue in millions of dollars for the five new US monthly retail groups.

Period	AUT	BUM	GEM	GRO	COM
June 09	48078	26350	47089	42146	19822
July 09	51093	24003	47097	43838	20704
August 09	55677	21522	48826	42994	21609
September 09	42299	21304	44202	41537	20050
October 09	45268	21263	48509	43020	21425
November 09	41448	19915	53489	42508	22439
December 09	46936	19182	69970	45395	28782
January 10	42143	15349	43854	42986	17570
February 10	43434	15468	44756	40061	18443
March 10	56482	21692	48643	43433	21649
April 10	52987	26682	47751	42236	21126
May 10	54178	26231	50476	44604	21913
June 10	53049	25446	48290	42877	20688
July 10	55854	23100	48620	44558	21650
August 10	88120	21698	49314	43431	21904
September 10	51886	21261	45678	42642	20841
October 10	51463	21882	49037	43535	21568
November 10	49393	21458	55587	43723	24094
December 10	55231	20277	71770	46656	30297
January 11	49692	16006	44323	44280	17698
February 11	53590	16003	45808	41229	19364
March 11	63583	21964	49677	45069	22481
April 11	58199	24377	50482	45586	22302
May 11	57518	27709	51437	46319	22251
June 11	57722	27009	50907	45883	21925
July 11	57538	23530	80661	47170	22490
August 11	59341	23733	51267	46449	22951
September 11	56231	22501	48238	44876	22379
October 11	55574	23077	80877	45725	22324
November 11	54,I86	22371	57198	46045	25196
December 11	61128	21201	73907	48840	32555
January 12	52989	17981	45642	45460	18740
Febuary 12	60345	18502	49635	44172	21403
March 12	68799	24156	53114	47406	24511
April 12	61761	25987	80601	45753	22406
May 12	67063	29374	53091	48418	23965
June 12	63336	25772	52279	47215	23332
July 12	63364	23921	50733	47472	22968
August 12	68423	24170	53679	48046	24691
September 12	60831	22327	49326	46048	22654
October 12	61295	24659	51398	46978	22970
November 12	59772	23512	59013	47314	26318
December 12	64800	21172	73802	49363	32705
January 13	59912	19323	46616	47082	19688

(continued overleaf)

Table 2.4 (*continued*)

Period	AUT	BUM	GEM	GRO	COM
February 13	63076	18702	48503	43837	20780
March 13	71529	23216	54650	48869	24887
April 13	68820	28906	49955	45316	22885
May 13	72952	32849	54774	49549	24860
June 13	69509	27891	53179	47765	23471
July 13	73058	27611	51958	48783	24090
August 13	75544	26242	55227	49288	25797
September 13	64308	24424	49910	46499	22819
October 13	67070	26255	52923	48283	24366
November 13	64894	23974	60375	48904	27057
December 13	68348	22399	73798	50374	32857
January 14	60664	19900	47319	49142	19624
February 14	64398	18989	48886	45091	20796
March 14	77683	24343	53967	48997	24533
April 14	74600	30447	52789	48690	24153
May 14	79113	33758	56900	51742	25878
June 14	73221	30409	53872	49434	23740
July 14	77685	29003	53646	51362	24818
August 14	79966	26798	57050	51309	26257
September 14	71294	26574	80663	48861	23771
October 14	72328	27748	54625	51057	25243
November 14	68450	25248	61843	51042	28293
December 14	76041	24498	75313	53011	34750
January 15	68158	21161	49436	51536	20906
February 15	68559	19923	49100	47088	21655
March 15	82234	26526	54686	51167	25617
April 15	79070	31512	52424	50091	24765
May 15	82410	33282	57441	53201	27034
June 15	79265	31709	54135	50950	25077
July 15	82841	30676	54829	53116	26170
August 15	83872	27916	56674	52120	26847
September 15	77506	27377	51588	50244	24919
October 15	77134	28546	55544	51987	26163
November 15	72505	26889	61397	51355	28156
December 15	80868	27044	76391	54146	35386
January 16	68984	22100	49050	52167	20809
February 16	76652	23290	80661	49193	22750
March 16	84611	30027	55303	52741	26205
April 16	81,446	32496	52960	50958	24867
May 16	82428	34787	55722	53502	26097
June 16	81439	34371	54481	52745	25549
July 16	84085	30347	54710	53756	26045
August 16	88224	30257	54859	52769	26693
September 16	81,027	29210	80670	51621	25469
October 16	78994	29010	54221	52769	25686
November 16	77389	29095	60520	52793	28524

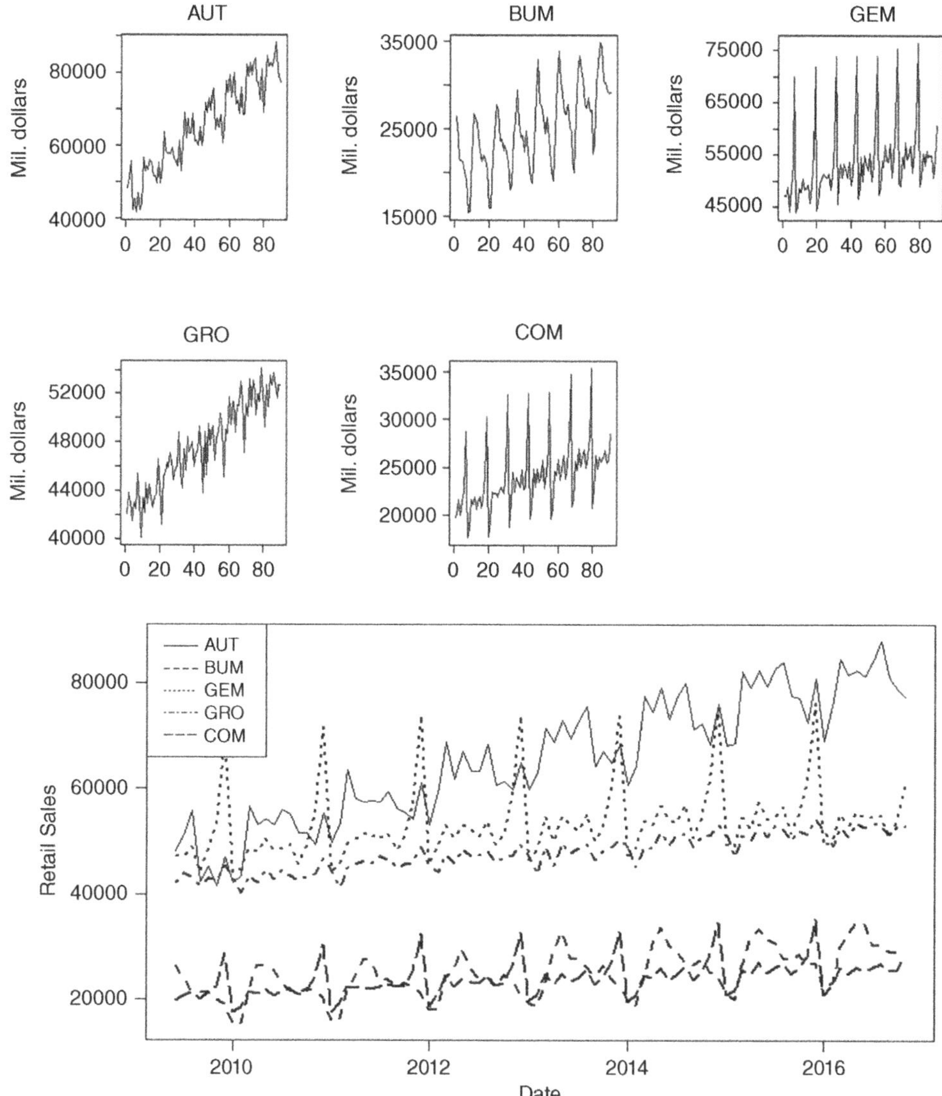

Figure 2.3 Individual and combined plots for the five new US monthly retail group sales revenue from June 2009 to November 2016, in million dollars.

are significant at relatively high lags. To help find a best initial choice, we will now try the sample extended cross-correlation matrix function, and its result is given in Table 2.5.

From Table 2.5, it is clear that we should try a VAR(4) model.

$$\left(\mathbf{I}-\mathbf{\Phi}_1 B-\mathbf{\Phi}_2 B^2-\mathbf{\Phi}_3 B^3-\mathbf{\Phi}_4 B^4\right)\mathbf{W}_t=\mathbf{\theta}_0+\mathbf{a}_t. \tag{2.99}$$

The estimation result is given in Equation (2.100).

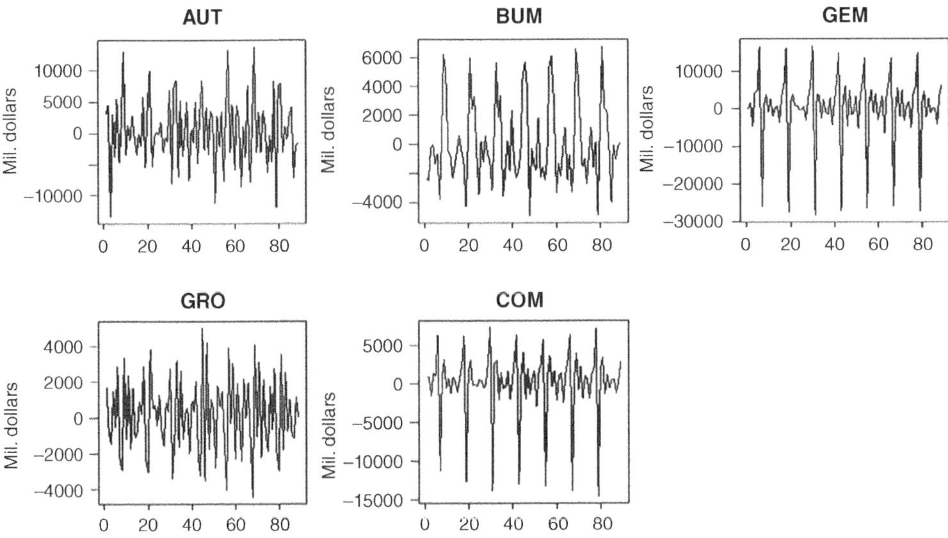

Figure 2.4 Five new differenced US monthly retail group sales revenue from July 2009 to November 2016.

$$\hat{\mathbf{W}}_t = \hat{\boldsymbol{\theta}}_0 + \hat{\boldsymbol{\Phi}}_1 \mathbf{W}_{t-1} + \hat{\boldsymbol{\Phi}}_2 \mathbf{W}_{t-2} + \hat{\boldsymbol{\Phi}}_3 \mathbf{W}_{t-3} + \hat{\boldsymbol{\Phi}}_4 \mathbf{W}_{t-4}, \tag{2.100}$$

where

$$\hat{\boldsymbol{\theta}}_0 = \begin{bmatrix} \underset{(260.57)}{1753.66} & \underset{(140.36)}{138.92} & \underset{(442.73)}{2363.07} & \underset{(87.26)}{594.85} & \underset{(210.07)}{1072.01} \end{bmatrix}',$$

$$\hat{\boldsymbol{\Phi}}_1 = \begin{bmatrix} \underset{(0.13)}{-0.41} & \underset{(0.20)}{0.24} & \underset{(0.29)}{0.67} & \underset{(0.41)}{-1.24} & \underset{(0.62)}{-2.04} \\ \underset{(0.07)}{0.16} & \underset{(0.11)}{-0.23} & \underset{(0.16)}{0.19} & \underset{(0.22)}{-0.34} & \underset{(0.34)}{-0.81} \\ \underset{(0.23)}{-1.13} & \underset{(0.34)}{0.26} & \underset{(0.49)}{-1.07} & \underset{(0.70)}{0.46} & \underset{(1.05)}{0.70} \\ \underset{(0.04)}{-0.15} & \underset{(0.07)}{0.24} & \underset{(0.10)}{0.08} & \underset{(0.14)}{-1.18} & \underset{(0.21)}{0.06} \\ \underset{(0.11)}{-0.50} & \underset{(0.16)}{0.09} & \underset{(0.23)}{-0.21} & \underset{(0.33)}{0.23} & \underset{(0.05)}{-0.37} \end{bmatrix},$$

$$\hat{\boldsymbol{\Phi}}_2 = \begin{bmatrix} \underset{(0.13)}{-0.2} & \underset{(0.17)}{0.56} & \underset{(0.28)}{0.89} & \underset{(0.61)}{-1.55} & \underset{(0.64)}{-2.86} \\ \underset{(0.07)}{0.17} & \underset{(0.09)}{-0.14} & \underset{(0.15)}{0.1} & \underset{(0.33)}{-0.10} & \underset{(0.35)}{-0.94} \\ \underset{(0.22)}{-1.10} & \underset{(0.29)}{0.52} & \underset{(0.48)}{-0.92} & \underset{(1.03)}{-2.03} & \underset{(1.09)}{0.28} \\ \underset{(0.04)}{-0.08} & \underset{(0.06)}{0.22} & \underset{(0.10)}{0.19} & \underset{(0.2)}{-0.99} & \underset{(0.22)}{-0.47} \\ \underset{(0.1)}{-0.43} & \underset{(0.14)}{0.18} & \underset{(0.23)}{-0.28} & \underset{(0.48)}{-0.82} & \underset{(0.52)}{-0.40} \end{bmatrix},$$

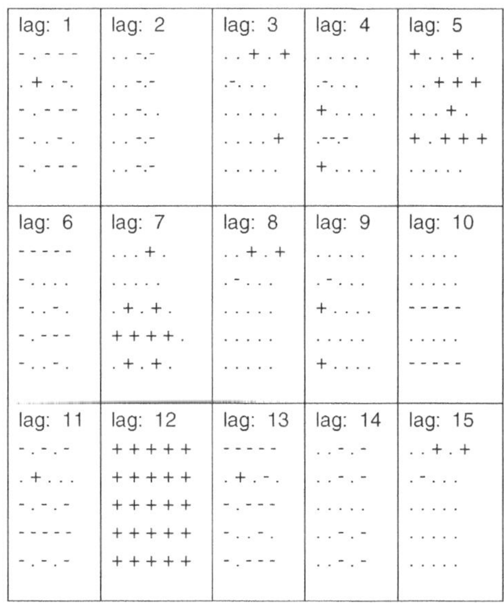

lag: 1
- . - - -
. + . - .
- . - - -
- . . - .
- . - - -
lag: 2
. . - . -
. . - . -
. . - . .
. . - . -
. . - . -
lag: 3
. . + . +
. - . . .
. . . . .
. . . . +
. . . . .
lag: 4
. . . . .
. - . . .
+ . . . .
. - - . -
+ . . . .
lag: 5
+ . . + .
. . + + +
. . . + .
+ . + + +
. . . . .
lag: 6
- - - - -
- . . . .
- . . - .
- . - - -
- . . - .
lag: 7
. . . + .
. . . . .
. + . + .
+ + + + .
. + . + .
lag: 8
. . + . +
. - . . .
. . . . .
. . . . .
. . . . .
lag: 9
. . . . .
. - . . .
+ . . . .
. . . . .
+ . . . .
lag: 10
. . . . .
. . . . .
- - - - -
. . . . .
- - - - -
lag: 11
- . - . -
. + . . .
- . - . -
- - - - -
- . - . -
lag: 12
+ + + + +
+ + + + +
+ + + + +
+ + + + +
+ + + + +
lag: 13
- - - - -
. + . - .
- . - - -
- . . - .
- . - - -
lag: 14
. . - . -
. . - . -
. . . . .
. . - . -
. . - . -
lag: 15
. . + . +
. - . . .
. . . . .
. . . . .
. . . . .

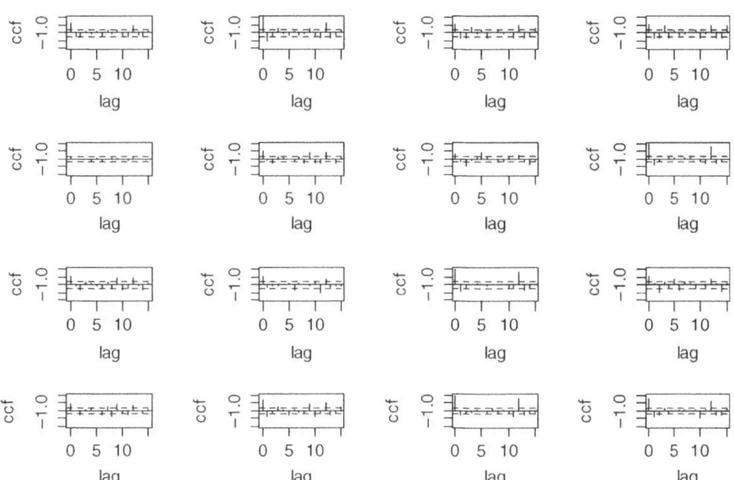

Figure 2.5 Sample correlation matrix function for the five differenced series.

Table 2.5 *P*-value table of sample extended cross-correlation matrices.

Column: MA order							
Row: AR order							
	0	1	2	3	4	5	6
0	0.0000	0.0000	0.0000	0.0000	0.0000	0.0000	0.0000
1	0.0000	0.0000	0.0000	0.0000	0.0000	0.0001	0.0095
2	0.0000	0.0000	0.0000	0.0008	0.0458	0.0080	0.3032
3	0.0136	0.8083	0.3415	0.8067	0.7359	0.7964	0.9866
4	0.4227	0.1624	0.3317	0.0321	0.2480	0.7221	0.7815
5	0.3028	0.2942	0.9156	0.4218	0.3694	0.3813	0.8819

$$\hat{\mathbf{\Phi}}_3 = \begin{bmatrix} \underset{(0.14)}{-0.32} & \underset{(0.17)}{0.89} & \underset{(0.28)}{0.91} & \underset{(0.61)}{-2.24} & \underset{(0.62)}{-2.23} \\ \underset{(0.07)}{0.21} & \underset{(0.09)}{0.07} & \underset{(0.15)}{0.22} & \underset{(0.33)}{-0.09} & \underset{(0.33)}{-1.23} \\ \underset{(0.24)}{-1.10} & \underset{(0.29)}{0.37} & \underset{(0.47)}{-1.61} & \underset{(1.03)}{-4.82} & \underset{(1.05)}{3.50} \\ \underset{(0.05)}{-0.16} & \underset{(0.06)}{0.26} & \underset{(0.09)}{-0.05} & \underset{(0.2)}{-0.68} & \underset{(0.21)}{0.11} \\ \underset{(0.11)}{-0.42} & \underset{(0.14)}{0.25} & \underset{(0.22)}{-0.59} & \underset{(0.49)}{-2.21} & \underset{(0.50)}{1.13} \end{bmatrix},$$

$$\hat{\mathbf{\Phi}}_4 = \begin{bmatrix} \underset{(0.15)}{-0.09} & \underset{(0.19)}{0.70} & \underset{(0.29)}{0.99} & \underset{(039)}{-2.02} & \underset{(0.68)}{-2.02} \\ \underset{(0.08)}{0.1} & \underset{(0.1)}{-0.31} & \underset{(0.16)}{0.18} & \underset{(0.21)}{-0.33} & \underset{(0.37)}{-0.65} \\ \underset{(0.25)}{0.30} & \underset{(0.32)}{1.33} & \underset{(0.49)}{0.77} & \underset{(0.65)}{-3.88} & \underset{(1.15)}{-1.57} \\ \underset{(0.05)}{-0.00} & \underset{(0.06)}{0.2} & \underset{(0.10)}{0.15} & \underset{(0.13)}{-0.33} & \underset{(0.23)}{-0.44} \\ \underset{(0.12)}{-0.42} & \underset{(0.15)}{0.66} & \underset{(0.23)}{0.49} & \underset{(0.31)}{-1.73} & \underset{(0.55)}{-1.07} \end{bmatrix},$$

$$\hat{\mathbf{\Sigma}} = \begin{bmatrix} 2\,867\,199.1 & 566\,391.03 & 958\,958.0 & 398\,422.33 & 592\,479.6 \\ 566\,391.0 & 831\,937.33 & -696\,103.54 & 43\,853.66 & -290\,911.7 \\ 958\,958.0 & -696\,103.54 & 8\,277\,476.9 & 707\,274.74 & 3\,808\,155.6 \\ 398\,422.3 & 43\,853.66 & 707\,274.7 & 321\,570.65 & 298\,087.8 \\ 592\,479.6 & -290\,911.69 & 3\,808\,155.6 & 298\,087.78 & 1\,863\,635.2 \end{bmatrix}.$$

and AIC = 70.118 29, BIC = 72.914 51, respectively.

To check the model adequacy, we can compute the sample extended correlation matrices for the residuals from the fitted model in Eq. (2.100) and the test result is shown in Table 2.6. The suggested model for the residuals is VARMA(0,0), so the model is adequate.

Table 2.6 *P*-values table of sample extended cross-correlation matrices for residuals.

	Column: MA order						
	Row: AR order						
	0	1	2	3	4	5	6
0	0.3104	0.3295	0.2868	0.2546	0.5524	0.5227	0.4075
1	0.9499	0.4955	0.6249	0.1380	0.9396	0.7601	0.7417
2	0.9690	1.0000	0.8985	0.9981	0.9931	0.9136	0.9840
3	0.9903	1.0000	1.0000	0.9999	0.9990	0.9818	0.9640
4	1.0000	1.0000	1.0000	1.0000	1.0000	0.9906	0.9776
5	1.0000	0.9992	0.9994	1.0000	1.0000	0.9991	0.9732

Table 2.7 One-step ahead forecasts for five US monthly retail group sales revenue using R.

Variable $W_{i,t}$	One-step ahead forecast (Ste) $\hat{W}_{i,t}(1)$	95% forecast limit	Adjusted forecast for $\hat{Z}_{i,t}(1)$
$W_{1,t}$	5115 (1693.3)	(1795.85, 8433.51)	82504
$W_{2,t}$	−1785 (912.1)	(−3573.14, 2.31)	27310
$W_{3,t}$	15625 (2877.1)	(9986.08, 21264.16)	76145
$W_{4,t}$	2779 (567.1)	(1667.18, 3890.10)	55572
$W_{5,t}$	6155 (1365.2)	(3779.17, 9130.56)	34979

The one-step ahead forecast for the differenced five monthly US retail sales using Eq. (2.100) is given in Table 2.7. Since

$$\hat{W}_{i,t}(1) = (1-B)\hat{Z}_{i,t}(1) = \hat{Z}_{i,t}(1) - Z_t, \ i = 1,2,3,4,5, \tag{2.101}$$

with the available sales for November 2016 from Table 2.4, we can compute the adjusted forecast in terms of original variables as

$$\hat{Z}_{i,t}(1) = \hat{W}_{i,t}(1) + Z_t, \tag{2.102}$$

and they are given in the last column of Table 2.7.

2.8.2 Second model of US monthly retail sales revenue

Example 2.5 In Example 2.4, we constructed a VAR(4) model in Eq. (2.99) for the differenced new five US monthly retail group sales revenue from July 2009 to November 2016, which is denoted as the five-dimensional vector series $\mathbf{W}_t$. The model is shown to be adequate based on the residual analysis output from the MTS program in R. However, a careful reader may notice that as shown in Figure 2.3 for the original series and Figure 2.4 for the differenced series, each of the five series clearly contains seasonal phenomenon. How can the VAR(4) model ignoring seasonality in Eq. (2.99) be adequate? This is a valid concern. The current version of MTS in R does not handle the seasonal vector models. So, we will now illustrate the use of seasonal vector time series modeling on the same data set using SAS.

To avoid repetition, we will skip the identification portion, and simply state that the identification results suggest a VAR(2) model for $(1 - B)(1 - B^{12})\mathbf{Z}_t$,

$$\left(\mathbf{I} - \mathbf{\Phi}_1 B - \mathbf{\Phi}_2 B^2\right)(1-B)\left(1-B^{12}\right)\mathbf{Z}_t = \mathbf{a}_t. \tag{2.103}$$

The fitted result is

$$\left(\mathbf{I} - \hat{\mathbf{\Phi}}_1 B - \hat{\mathbf{\Phi}}_2 B^2\right)(1-B)\left(1-B^{12}\right)\hat{\mathbf{Z}}_t = \mathbf{a}_t, \tag{2.104}$$

where

$$\hat{\boldsymbol{\Phi}}_1 = \begin{bmatrix} \underset{(0.13)}{-0.55} & \underset{(0.23)}{-0.11} & \underset{(0.53)}{0.13} & \underset{(0.61)}{-0.61} & \underset{(0.74)}{-0.20} \\ \underset{(0.06)}{0.01} & \underset{(0.11)}{-0.46} & \underset{(0.24)}{0.62} & \underset{(0.28)}{-0.21} & \underset{(0.33)}{-0.49} \\ \underset{(0.06)}{-0.01} & \underset{(0.1)}{0.28} & \underset{(0.23)}{-0.71} & \underset{(0.27)}{-0.06} & \underset{(0.32)}{-0.00} \\ \underset{(0.04)}{-0.03} & \underset{(0.07)}{0.1} & \underset{(0.17)}{-0.17} & \underset{(0.19)}{-0.52} & \underset{(0.23)}{-0.16} \\ \underset{(0.03)}{-0.00} & \underset{(0.06)}{0.05} & \underset{(0.15)}{-0.04} & \underset{(0.17)}{-0.10} & \underset{(0.20)}{-0.59} \end{bmatrix}$$

$$\hat{\boldsymbol{\Phi}}_2 = \begin{bmatrix} \underset{(0.13)}{-0.28} & \underset{(0.25)}{-0.14} & \underset{(0.53)}{1.04} & \underset{(0.58)}{-1.42} & \underset{(0.76)}{-0.19} \\ \underset{(0.06)}{-0.07} & \underset{(0.12)}{-0.44} & \underset{(0.24)}{0.69} & \underset{(0.26)}{-0.2} & \underset{(0.34)}{-0.20} \\ \underset{(0.06)}{0.03} & \underset{(0.11)}{0.8} & \underset{(0.23)}{0.05} & \underset{(0.25)}{-0.52} & \underset{(0.33)}{-0.07} \\ \underset{(0.04)}{0.02} & \underset{(0.08)}{0.03} & \underset{(0.16)}{0.25} & \underset{(0.18)}{-0.64} & \underset{(0.23)}{-0.03} \\ \underset{(0.04)}{0.34} & \underset{(0.07)}{-0.08} & \underset{(0.14)}{0.21} & \underset{(0.16)}{-0.29} & \underset{(0.2)}{-0.47} \end{bmatrix}$$

$$\hat{\boldsymbol{\Sigma}} = \begin{bmatrix} 4673030.19 & 712513.46 & 580689.52 & 224320.08 & 666104.96 \\ 712513.46 & 959436.64 & 75567.01 & 30145.5 & 128138.15 \\ 580689.52 & 75567.01 & 882037.51 & 491109.19 & 429157.93 \\ 224320.08 & 30145.5 & 491109.19 & 441654.65 & 248705.3 \\ 666104.96 & 128138.15 & 429157.93 & 248705.3 & 341606.89 \end{bmatrix}$$

The one-step ahead forecasts for the five monthly US retail group sales revenue using Eq. (2.104) are given in Table 2.8. They are directly outputted from SAS, and we do not need to make any adjustments like those in the last column of Table 2.7 based on R.

Although MTS in R does not carry the unit root test for a seasonal unit root, it can certainly perform both regular and seasonal differences. For comparison, we will use Eccm in R MTS to

Table 2.8 One-step ahead forecast for five US monthly retail group sales revenue using SAS.

Forecasts					
Variable	Obs	Forecast	Standard Error	95% Confidence Limits	
AUT	91	84 666.85153	2 161.71927	80 429.95963	88 903.74344
BUM	91	28 963.43163	979.50837	27 043.63051	30 883.23275
GEM	91	75 815.81267	939.16852	73 975.07619	77 656.54915
GRO	91	55 532.68196	664.57103	54 230.14669	56 835.21724
COM	91	35 877.58782	584.47147	34 732.04480	37 023.13084

get the sample extended cross-correlation matrix function for $\mathbf{Y}_t = (1 - B)(1 - B^{12})\mathbf{Z}_t$ shown in Table 2.9.

From Table 2.9, a VAR(2) model is clearly suggested, and we obtain the following fitted model,

$$\left(\mathbf{I}-\hat{\boldsymbol{\Phi}}_1 B-\hat{\boldsymbol{\Phi}}_2 B^2\right)\mathbf{Y}_t=\mathbf{a}_t, \tag{2.105}$$

where

$$\hat{\boldsymbol{\Phi}}_1 = \begin{bmatrix} \underset{(0.13)}{-0.55} & \underset{(0.23)}{-0.11} & \underset{(0.53)}{0.13} & \underset{(0.61)}{-0.61} & \underset{(0.74)}{-0.20} \\ \underset{(0.06)}{0.01} & \underset{(0.11)}{-0.46} & \underset{(0.24)}{0.62} & \underset{(0.28)}{-0.21} & \underset{(0.33)}{-0.49} \\ \underset{(0.06)}{-0.01} & \underset{(0.1)}{0.28} & \underset{(0.23)}{-0.71} & \underset{(0.27)}{-0.06} & \underset{(0.32)}{-0.00} \\ \underset{(0.04)}{-0.03} & \underset{(0.07)}{0.1} & \underset{(0.17)}{-0.17} & \underset{(0.19)}{-0.52} & \underset{(0.23)}{-0.16} \\ \underset{(0.03)}{-0.00} & \underset{(0.06)}{0.05} & \underset{(0.15)}{-0.04} & \underset{(0.17)}{-0.10} & \underset{(0.20)}{-0.59} \end{bmatrix}$$

$$\hat{\boldsymbol{\Phi}}_2 = \begin{bmatrix} \underset{(0.13)}{-0.28} & \underset{(0.25)}{-0.14} & \underset{(0.53)}{1.04} & \underset{(0.58)}{-1.42} & \underset{(0.76)}{-0.19} \\ \underset{(0.06)}{-0.07} & \underset{(0.12)}{-0.44} & \underset{(0.24)}{0.69} & \underset{(0.26)}{-0.2} & \underset{(0.34)}{-0.20} \\ \underset{(0.06)}{0.03} & \underset{(0.11)}{0.8} & \underset{(0.23)}{0.05} & \underset{(0.25)}{-0.52} & \underset{(0.33)}{-0.07} \\ \underset{(0.04)}{0.02} & \underset{(0.08)}{0.03} & \underset{(0.16)}{0.25} & \underset{(0.18)}{-0.64} & \underset{(0.23)}{-0.03} \\ \underset{(0.04)}{0.34} & \underset{(0.07)}{-0.08} & \underset{(0.14)}{0.21} & \underset{(0.16)}{-0.29} & \underset{(0.2)}{-0.47} \end{bmatrix}$$

Table 2.9 *P*-value table of sample extended cross-correlation matrices.

	Column: MA order						
Row: AR order							
	0	1	2	3	4	5	6
0	0.0000	0.0000	0.0000	0.0065	0.0417	0.0697	0.2795
1	0.0400	0.1303	0.4739	0.9133	0.9636	0.6779	0.8458
2	0.9753	0.9530	0.5771	0.9976	0.9940	0.9865	0.9505
3	0.9998	1.0000	0.9998	0.9995	0.9978	0.9894	0.9954
4	1.0000	1.0000	0.9999	1.0000	1.0000	0.9945	0.9917
5	1.0000	1.0000	1.0000	1.0000	1.0000	1.0000	0.9154

and

$$\hat{\boldsymbol{\Sigma}} = \begin{bmatrix} 4049959.5 & 617511.66 & 503264.25 & 194410.7 & 577291.0 \\ 617511.66 & 831511.76 & 65491.41 & 26126.10 & 111053.06 \\ 503264.25 & 65491.41 & 764432.51 & 425628.97 & 371936.87 \\ 194410.7 & 26216.10 & 425627.97 & 382767.4 & 215544.6 \\ 577291.0 & 111053.06 & 371936.87 & 215544.6 & 296059.3 \end{bmatrix}$$

The one-step ahead forecast for the five monthly US retail group sales revenue using Eq. (2.105) from R are given in columns 2 and 3 in Table 2.10. However, the column 2 is the forecast of $\mathbf{Y}_{t+1}$, that is, $\hat{\mathbf{Y}}_t(1)$. To obtain the forecast for the original variable $\mathbf{Z}_t$, we note that

$$\hat{\mathbf{Y}}_t(1) = (1-B)\left(1-B^{12}\right)\hat{\mathbf{Z}}_t(1) = \hat{\mathbf{Z}}_t(1) - \mathbf{Z}_t - \mathbf{Z}_{t-11} + \mathbf{Z}_{t-12}.$$

This implies that

$$\hat{\mathbf{Z}}_t(1) = \hat{\mathbf{Y}}_t(1) + \mathbf{Z}_t + \mathbf{Z}_{t-11} - \mathbf{Z}_{t-12}. \tag{2.106}$$

Thus, we can compute the forecast for original sales revenue as follows:

$$\hat{Z}_{i,t}(1) = \hat{Y}_{i,t}(1) + Z_{i,t} + Z_{i,t-11} - Z_{i,t-12}, i = 1,2,3,4,5. \tag{2.107}$$

Specifically, using the information from Tables 2.4 and 2.10, we have

$$\begin{aligned} \hat{Z}_{1,t}(1) &= -1085.15 + Z_{1,t} + Z_{1,t-11} - Z_{1,t-12} \\ &= -1085.15 + 77\,389 + 80\,868 - 72\,505 = 84\,666.85 \end{aligned}$$

Table 2.10 One-step ahead forecasts for five US monthly retail group sales revenue using R.

Variable $Y_{i,t}$	1-step ahead forecast (Ste) $\hat{Y}_{i,t}(1)$	95% forecast limit	Adjusted forecast for $\hat{Z}_{i,t}(1)$
$Y_{1,t}$	−1 085.15 (2 012.5)	(−5 029.55, 2 859.26)	84 666.85
$Y_{2,t}$	−250.57 (911.9)	(−2 037.89, 1 536.75)	28 963.43
$Y_{3,t}$	301.81 (874.3)	(−1 411.85, 2 015.48)	75 815.81
$Y_{4,t}$	−51.32 (618.7)	(−1 263.94, 1 161.30)	55 532.68
$Y_{5,t}$	123.59 (544.1)	(−942.87, 1 190.05)	35 877.59

$$\hat{Z}_{2,t}(1) = -286.57 + Z_{2,t} + Z_{2,t-11} - Z_{2,t-12}$$

$$= -250.57 + 29\,059 + 27\,044 - 26\,889 = 28\,963.43$$

$$\hat{Z}_{3,t}(1) = 301.81 + Z_{3,t} + Z_{3,t-11} - Z_{3,t-12}$$

$$= 301.81 + 60\,520 + 76\,391 - 61\,397 = 75\,815.81$$

$$\hat{Z}_{4,t}(1) = -51.32 + Z_{4,t} + Z_{4,t-11} - Z_{4,t-12}$$

$$= -51.32 + 52\,793 + 54\,146 - 51\,355 = 55\,532.68$$

$$\hat{Z}_{5,t}(1) = 123.59 + Z_{5,t} + Z_{5,t-11} - Z_{5,t-12}$$

$$= 123.59 + 28\,524 + 35\,386 - 28\,156 = 35\,877.59$$

and they are given in the last column of Table 2.10.

They are exactly the same as the SAS output given in Table 2.8. In fact, the VAR(2) parameter estimates in Eq. (2.105) from R are also the same as those in Eq. (2.104) from SAS except the estimate of $\hat{\mathbf{\Sigma}}$, which must be due to the use of different constants in their divisions. Based on the same software R for the model in Eq. (2.105), the AIC = 66.68, and BIC = 68.20, and they are smaller than the AIC = 70.118 29 and BIC = 72.914 51 for the model in Eq. (2.104).

In general, to evaluate the performance of ℓ– step ahead forecasts between different vector time series models, we normally use $(n-\ell)$ observations for model fitting and leave the last ℓ observations for forecast comparisons. We leave this for readers to try.

2.8.3 US macroeconomic indicators

Example 2.6 In this example, we will try a macroeconomic time series data set. It is a collection of time series of US macroeconomic indicators, which is a subset of that used in Stock and Watson (2009) and Koop (2013). The full data list contains 168 quarterly macroeconomic variables from Quarter 1 of 1959 to Quarter 4 of 2007, representing information about many aspects of the US economy. We retrieve 61 time series from the full dataset, leading to a dataset with $m = 61$ and $n = 196$. Variables are originally at a monthly frequency, and they are transformed to quarterly by taking the average of three months in a quarter. Seasonal adjustments are made if necessary. All variables are transformed to stationary using the suggestion of Stock and Watson (2009) with $n = 195$ remained. Table 2.11 contains a brief description of each variable along with a transformation code, where 1 = first differencing of log of variables and 2 = second differencing of log of variables. The numerical values of the data set are listed as WW10 in the Data Appendix at the end of the book.

The purpose of this example is mainly to show the problems of vector time series analysis when the number of dimensions is high. In the first case of Example 2.4, the vector time series only contains $m = 8$ variables. So, we can show its plot, its sample correlation, and partial correlation matrix functions. However, we cannot estimate the model parameters when a VAR(13) model was suggested. In this example, with $m = 61$, we can still show its plot in Figure 2.6 with time ranging from 0 to 200, although not very clearly.

Table 2.11 Information of variables used in the Example 2.6.

Variable	Description	Code	Group
GDP252	Real personal consumption exp: quantity index	1	GDP
GDP253	Real personal consumption exp: durable goods	1	GDP
GDP254	Real personal consumption exp: nondurable goods	1	GDP
GDP255	Real personal consumption exp: services	1	GDP
GDP256	Real gross private domestic inv: quantity index	1	GDP
GDP257	Real gross private domestic inv: xed inv	1	GDP
GDP258	Real gross private domestic inv: nonresidential	1	GDP
GDP259	Real gross private domestic inv: nonres structure	1	GDP
GDP260	Real gross private domestic inv: nonres equipment	1	GDP
GDP261	Real gross private domestic inv: residential	1	GDP
GDP266	Real gov consumption exp, gross inv: federal	2	GDP
GDP267	Real gov consumption exp, gross inv: state and local	2	GDP
GDP268	Real final sales of domestic product	2	GDP
GDP269	Real gross domestic purchases	2	GDP
GDP271	Real gross national product	2	GDP
GDP272	Gross domestic product: price index	2	GDP
GDP274	Personal cons exp: durable goods, price index	2	GDP
GDP275	Personal cons exp: nondurable goods, price index	2	GDP
GDP276	Personal cons exp: services, price index	2	GDP
GDP277	Gross private domestic investment, price index	2	GDP
GDP278	Gross priv dom inv: fixed inv, price index	2	GDP
GDP279	Gross priv dom inv: nonresidential, price index	2	GDP
GDP280	Gross priv dom inv: nonres structures, price index	2	GDP
GDP281	Gross priv dom inv: nonres equipment, price index	2	GDP
GDP282	Gross priv dom inv: residential, price index	2	GDP
GDP284	Exports, price index	2	GDP
GDP285	Imports, price index	2	GDP
GDP286	Government cons exp and gross inv, price index	2	GDP
GDP287	Gov cons exp and gross inv: federal, price index	2	GDP
GDP288	Gov cons exp and gross inv: state and local, price index	2	GDP
GDP289	Final sales of domestic product, price index	2	GDP
GDP290	Gross domestic purchases, price index	2	GDP
GDP291	Final sales to domestic purchasers, price index	2	GDP
GDP292	Gross national products, price index	2	GDP
IPS11	Industrial production index: products total	1	IPS
IPS299	Industrial production index: final products	1	IPS
IPS12	Industrial production index: consumer goods	1	IPS
IPS13	Industrial production index: consumer durable	1	IPS
IPS18	Industrial production index: consumer nondurable	1	IPS
IPS25	Industrial production index: business equipment	1	IPS
IPS32	Industrial production index: materials	1	IPS
IPS34	Industrial production index: durable goods materials	1	IPS
IPS38	Industrial production index: nondurable goods material	1	IPS
IPS43	Industrial production index: manufacturing	1	IPS
IPS307	Industrial production index: residential utilities	1	IPS
IPS306	Industrial production index: consumer fuels	1	IPS
CES275	Avg hrly earnings, prod wrkrs, nonfarm-goods prod	2	CES

Table 2.11 (*continued*)

Variable	Description	Code	Group
CES277	Avg hrly earnings, prod wrkrs, nonfarm-construction	2	CES
CES278	Avg hrly earnings, prod wrkrs, nonfarm-manufacturing	2	CES
CES003	Employees, nonfarm: goods-producing	1	CES
CES006	Employees, nonfarm: mining	1	CES
CES011	Employees, nonfarm: construction	1	CES
CES015	Employees, nonfarm: manufacturing	1	CES
CES017	Employees, nonfarm: durable goods	1	CES
CES033	Employees, nonfarm: nondurable goods	1	CES
CES046	Employees, nonfarm: service providing	1	CES
CES048	Employees, nonfarm: trade, transportation, and utilities	1	CES
CES049	Employees, nonfarm: wholesale trade	1	CES
CES053	Employees, nonfarm: retail trade	1	CES
CES088	Employees, nonfarm: financial activities	1	CES
CES140	Employees, nonfarm: government	1	CES

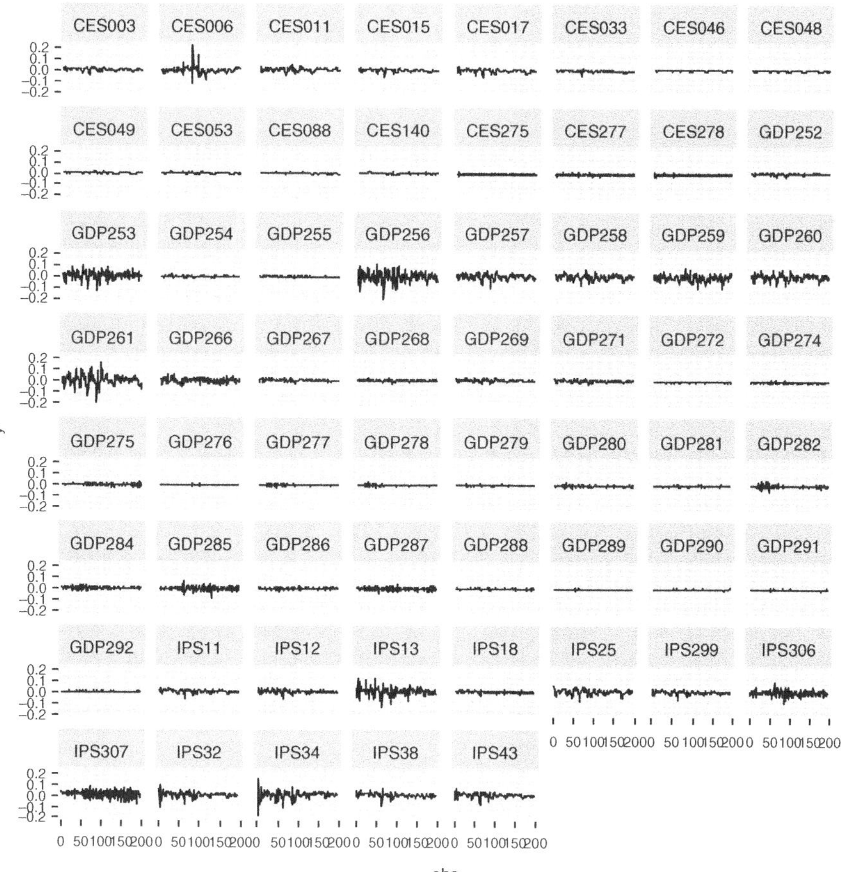

Figure 2.6 Plot of 61 transformed quarterly macroeconomic variables from Quarter 1 of 1959 to Quarter 4 of 2007.

However, we have difficulty getting its sample correlation and partial correlation matrix functions. In fact, error messages with the system being computationally singular were returned from R software all the time, so we decide to simply fit a VAR(1) model,

$$\mathbf{Z}_t = \boldsymbol{\Phi}_1 \mathbf{Z}_{t-1} + \mathbf{a}_t, \tag{2.108}$$

and see what happens.

The following is the estimate $\hat{\boldsymbol{\Phi}}_1$ from the output:

$$\hat{\boldsymbol{\Phi}}_1 = \left[\hat{\phi}_{i,j}\right], i = 1, \ldots, 31, \text{ and } j = 1, \ldots, 14$$

	1	2	3	4	5	6	7	8	9	10	11	12	13	14
1	1.11	7.51	1.32	−0.32	7.16	1.61	2.7	5.21	2.42	−1.63	1.17	−1.88	1.25	1.64
2	−0.21	−1.37	−0.24	0.08	−0.39	−0.39	−0.43	−0.57	−0.46	−0.18	−0.53	0.26	−0.29	−0.2
3	−0.41	−2.25	−0.73	0.18	−0.46	−0.9	−1.24	−2.48	−0.85	0.1	−1.71	0.59	−0.74	−0.4
4	−0.27	−3.1	−0.61	0.63	0.13	−0.28	−0.93	−1.97	−0.59	1.81	−1.65	0.98	−0.62	−0.2
5	0	0.05	−0.12	0.11	−0.02	−0.49	−0.44	−0.59	−0.28	−0.5	−0.36	−0.07	−0.13	−0
6	0.22	−0.07	0.05	0.33	0.02	0.37	−0.13	−1.29	0.16	−0.68	0.15	0.13	0.19	0.16
7	−0.65	−3.11	−0.01	−0.38	−2.22	−1.78	−0.68	0.9	−1.37	−2.7	−1.62	0.08	−1	−0.9
8	0.18	0.99	0.04	0.05	0.59	0.47	0.36	0.28	0.42	0.72	0.43	0.01	0.29	0.25
9	0.34	2.23	−0.01	0.09	2.19	1.36	0.85	0.68	1	2.41	0.96	−0.06	0.65	0.67
10	−0.03	0.24	0	−0.11	0.56	0.16	0.25	0.6	0.16	0.61	−0.16	−0.05	−0.01	0.05
11	0	0.08	−0.08	0.05	0.56	−0.16	−0.07	0.01	−0.04	−0.36	−0.39	−0.06	−0.08	0.04
12	0.03	0.19	0	0.01	−0.18	−0.46	−0.27	−0.17	−0.26	−0.89	0.29	0.3	0.01	0.04
13	0.12	−0.38	0.02	0.29	−3.72	−1.58	−2.2	−4.52	−1.12	0.42	1.76	−0.52	−0.02	−0.3
14	0.27	0.17	0.96	−0.4	−6.14	1.13	0.45	0.59	−0.25	2.58	4.4	0.5	0.94	−0.3
15	−0.09	−0.06	0	−0.12	3.62	2.21	2.54	3.56	2.15	1.06	−2.21	0.04	0.06	0.25
16	−0.33	1.91	−0.28	−1.06	31.93	9.71	4.19	11.6	0.64	21.6	−5.99	5.49	0.06	5.07
17	−0.28	−1.13	−0.17	−0.18	1.25	−0.68	−0.23	−0.28	−0.2	−1.87	0.5	−0.46	−0.48	0
18	−0.13	−0.07	−0.14	−0.23	4.18	0.89	1.71	1.22	1.96	−1.12	−1.72	−1.45	−0.44	0.26
19	0.3	1.66	0.4	−0.28	7.16	2.19	3.11	2.24	3.6	−0.02	−3.09	−2.16	−0.17	0.81
20	0.29	−1.34	0.44	0.59	13.88	6.29	3.95	0.48	6.46	11.8	−1.98	−2.22	0.52	2
21	−0.38	−12.1	2.24	0.1	−27.4	−12.9	−11.9	−5.44	−16.3	−18.2	0.21	−2.32	−3.34	−4.8
22	2.45	22.69	−0.31	−0.52	23.27	11.5	11.44	8.65	14.6	13.6	2.69	2.81	4.07	5.74
23	−0.86	−4.28	−0.65	−0.07	−2.58	−1.31	−0.8	−0.98	−1.14	−2.43	−1.05	−0.08	−0.78	−1.1
24	−1.68	−9.16	−1.05	−0.09	−8.85	−4.92	−3.85	−2.99	−5.25	−7.44	−2.32	−0.64	−1.7	−2.7
25	0.19	4.73	−0.64	−0.24	5.06	2.17	2.65	1.86	3.17	1.92	0.45	1.05	0.91	1.05
26	−0.12	−0.42	−0.19	0.02	0.46	−0.29	0.02	−0.32	0.25	−1.09	0.5	−0.09	−0.02	0.03
27	0.11	0.36	0.2	−0.05	−0.58	0.03	−0.43	−0.07	−0.69	1.18	−0.51	0.14	0	−0.1
28	−0.28	1.52	−1.01	−0.38	3.85	0.75	2.17	0.73	2.32	−3.73	−1.28	−1.85	−0.61	0.15
29	0.14	−0.28	0.44	0.07	0.02	−0.01	−0.38	0	−0.28	1.29	0.02	0.31	0.12	0.12
30	0.37	0.37	0.77	0.06	0.55	−0.04	−0.46	0.33	−0.42	1.79	0.45	0.44	0.26	0.35
31	−2.64	−15.3	−1.36	0.24	−6.87	−9.35	−13.4	−21.8	−9.05	−1.08	1.27	−1.87	−3.06	−3

$$\hat{\Phi}_1 = \left[\hat{\phi}_{i,j}\right], i = 32, \ldots, 61, \text{ and } j = 1, \ldots, 14$$

	1	2	3	4	5	6	7	8	9	10	11	12	13	14
32	−4.35	−28	−3.88	2.23	−21	−24.4	−17.7	−15.6	−19.1	−41.8	−2.33	1.74	−5.76	−6.2
33	3.02	21.35	1.76	−0.85	3.95	19.5	13.25	10.7	15.5	35.5	13.9	3.5	7.4	4.24
34	3.96	15.94	3.51	0.49	−27.4	0.83	6.87	10.8	3.85	−10.6	0.89	−2.64	2.95	−2
35	−0.04	0.33	−0.12	−0.09	2.9	2.23	1.72	3.41	0.96	3.17	2.71	0.73	0.81	0.83
36	−0.45	−0.39	−0.55	−0.41	2	−0.63	−0.77	0.81	−1.57	−0.94	1.66	0.04	0.03	0.21
37	1.27	1.28	0.72	1.7	1.22	3.17	3.03	−0.33	5.15	5.99	−1.14	−1.47	1.28	0.83
38	−0.36	−0.63	−0.15	−0.46	−0.88	−1.04	−0.94	−0.15	−1.48	−1.94	−0.14	0.4	−0.45	−0.4
39	−0.89	−1.33	−0.52	−1.09	−3.06	−2.37	−1.98	0.73	−3.7	−4.77	−0.57	0.83	−1.04	−1.1
40	0.05	−0.41	0.07	0.12	−1.12	−0.36	0.01	−0.45	0.2	−1.05	−0.82	−0.16	−0.08	−0.3
41	−0.21	−0.75	−0.14	−0.13	0.67	−0.04	0.43	1.36	0	−1.21	−0.16	0.06	−0.1	−0
42	−0.01	−0.01	−0.03	−0.02	0.55	0.49	0.27	0.85	−0.07	0.84	0.44	0.13	0.29	0.16
43	0	0.13	−0.04	0	−0.56	−0.39	−0.31	−0.1	−0.44	−0.73	0.81	0.13	0.15	0.01
44	0.57	1.65	0.53	0.4	−1.43	−0.93	−1.04	−5.43	1.23	0.04	−2.12	−0.75	−0.8	−0.2
45	0.04	0.17	0.06	0	0.02	−0.02	−0.06	−0.35	0.07	0.1	−0.17	0.02	−0.02	0.02
46	0.01	−0.01	0	0.01	−0.15	−0.02	−0.1	−0.25	−0.03	0.17	−0.03	−0.02	0	−0
47	0.52	3.5	0.02	0.11	−1.03	0.71	−0.37	−2.77	0.8	3.52	−1.27	0.85	0.52	0.1
48	−0.28	−1.27	−0.15	−0.12	−0.65	0.08	0.11	0.77	−0.29	−0.1	0.41	−0.31	−0.12	−0.3
49	−0.31	−2.34	−0.02	0.02	1.6	−0.04	0.2	1.86	−0.68	−0.84	0.07	−0.64	−0.28	0.03
50	1.61	4.31	2.71	0.23	1.86	1.08	4.26	2.77	4.12	−6.31	3.43	−0.3	1.45	1.4
51	−0.09	−0.3	−0.09	−0.04	0.01	−0.16	−0.27	−0.12	−0.32	0.06	−0.14	−0.05	−0.12	−0.1
52	−0.31	−0.43	−0.53	−0.12	0.3	0.12	−0.7	−0.47	−0.64	2	−0.93	0.02	−0.29	−0.2
53	−9.48	−52	−5.5	−2.32	72.88	−7.28	−6.13	12.5	−16.9	−29.2	−25.8	3.89	−10.1	4.19
54	4.85	29.45	1.95	1.13	−47.5	2.9	1.13	−9.44	7.77	18.7	14.3	−2.14	5.26	−3.6
55	3.16	18.39	1.4	0.82	−25.6	3.81	2.01	−4.61	6.07	15.6	7.82	−1.66	3.48	−1.7
56	0.57	2.42	0.15	0.38	−1.98	−1.47	−0.69	−1.91	−0.32	−3.93	−1.04	1.17	0.3	0.11
57	−1	−4.62	−0.9	−0.13	−2.13	−0.63	−0.67	−0.19	−0.92	0.34	4.03	−0.48	−0.43	−0.6
58	0.37	1.42	0.36	0.14	3.03	2.19	2.28	2.68	2.05	1.92	−1.44	−0.03	0.5	0.58
59	0.64	2.69	0.63	0.06	1.27	0.05	0.23	0.31	0.31	−0.86	−1.76	−0.03	0.2	0.37
60	−0.13	−1.01	−0.03	0.06	−1.54	−0.28	−0.33	0.17	−0.59	−0.16	0.3	−0.05	−0.07	−0.3
61	0.01	−0.78	0.1	0.19	0.51	0.27	0.08	0.05	0.15	0.87	−0.4	0.16	0.05	0.05

$$\hat{\Phi}_1 = \left[\hat{\phi}_{i,j}\right], i = 1, \ldots, 31, \text{and} j = 15, \ldots, 29$$

	15	16	17	18	19	20	21	22	23	24	25	26	27	28	29
1	1.78	−0.25	0.2	1.2	−0.1	−0.19	−0.1	0.11	−0.01	0.15	−0.49	−0.36	1.75	−1.6	−3.49
2	−0.28	0.05	−0.1	−0.13	0.06	0.08	0.07	0.06	0.02	0.1	0.08	0.01	−0.11	0.24	0.45
3	−0.63	0.17	−0.2	−0.18	0.16	0.24	0.21	0.11	0.04	0.19	0.41	0.15	−0.15	0.77	1.36
4	−0.52	0.22	−0.3	−0.28	0.21	0.39	0.34	0.15	0.15	0.22	0.74	0.29	−0.32	0.93	1.59
5	−0.03	0.02	−0	−0.01	0.07	0.1	0.11	0.11	0	0.18	0.11	−0.07	−0.09	−0.1	−0.19
6	0.04	−0.11	−0.4	0.05	−0.06	0.1	0.11	0.05	0.44	−0.09	0.18	−0.51	0.57	0.13	−0.05
7	−0.99	0.04	0.27	0	−0.03	0.23	0.2	0.25	−0.17	0.43	0.14	0.69	0.5	−0	0.19
8	0.29	0	−0	−0.01	0.02	−0.09	−0.08	−0.08	−0.04	−0.11	−0.08	−0.14	−0.26	−0	−0.07
9	0.74	0.04	0	−0.02	0.06	−0.19	−0.18	−0.18	−0.1	−0.22	−0.18	−0.18	−0.57	0.02	−0.04
10	0.06	0.04	0.1	0.01	0.02	0.02	0.01	0.01	−0.06	0.02	0.05	0.14	−0.06	−0	0
11	−0.01	0.01	−0	0.03	0.06	0.08	0.07	0.08	0	0.12	0.06	−0.02	0.08	−0	−0.14
12	−0.02	0	0.05	−0.03	0.01	0.11	0.11	0.14	0.01	0.22	0.06	−0.07	0.07	0.04	−0.08
13	0.14	−0.05	0.32	−0.46	−0.16	−0.12	−0.08	−0.09	−0.2	−0.06	−0.08	−0.09	−1.95	−0.5	−0.64
14	0.15	−0.42	0.22	−0.9	−0.62	−0.66	−0.69	−0.73	−0.26	−1.08	−0.62	−0.02	−0.89	−0.5	0.18

15	−0.2	0.25	−0.1	0.86	0.24	0.03	0.02	0.02	0.33	−0.12	0.02	0.15	1.4	0.7	0.95
16	3.54	−1.79	0.3	−1.4	−0.69	−1.91	−2.92	−0.99	−1.16	−0.84	−6.51	−4.64	−2.41	−2.7	−3.35
17	−0.23	0.06	−0.1	−0.07	−0.12	−0.04	−0.06	−0.05	−0.18	0.05	−0.1	0.14	−1.11	−0.1	−0.15
18	0	−0.09	−0.2	−0.68	−0.24	0.03	−0.01	−0.16	−0.13	−0.15	0.37	0.32	−0.35	0.09	−0.2
19	0.56	−0.32	−0.5	−1.02	−0.89	0.02	−0.06	−0.37	−0.48	−0.3	0.7	0.09	−1.68	−0.3	−0.77
20	1.67	0.34	−0.4	1.11	0.11	−0.43	0	−0.58	0.17	−0.89	1.19	0.81	−0.75	0.42	−0.23
21	−5.95	−0.66	1.99	−2.28	−1.05	−0.72	−0.7	0.69	0.62	1.05	−2.6	−2.16	−7.44	−1.5	−1.86
22	6.17	−0.58	−1.4	0.63	0.36	−1.49	−2.06	−1.72	−4.05	−0.39	−4.2	1.88	6.98	−1	−1.02
23	−1.02	0.28	0.3	−0.01	0.05	0.69	0.74	0.49	0.92	0.14	1.54	−0.05	−0.62	0.46	0.66
24	−2.3	0.53	0.03	−0.23	0.13	1.72	1.91	1.06	2.52	0.03	4.22	−0.57	−1.12	1.36	2.06
25	1.48	0.05	−0.7	0.25	0.21	0.12	−0.01	−0.21	−0.43	−0.2	0.07	0.35	2.22	0.43	0.71
26	0.1	−0.02	−0	0.02	0	−0.11	−0.07	−0.09	−0.03	0.11	0.06	0.38	−0.23	−0.1	−0.29
27	−0.08	−0.02	−0	0.02	−0.01	0.08	0.03	0.08	0.07	0.09	−0.07	−0.12	0.16	0.06	0.11
28	−0.23	−0.08	−0.6	−0.59	−0.36	0.16	0.04	−0.16	0.44	−0.4	0.71	0.13	−1.12	−0.6	−0.11
29	0.16	−0.02	0.24	0.19	0.08	−0.1	−0.08	−0.06	−0.33	0.06	−0.21	0.17	0.46	0.01	−0.59
30	0.36	−0.12	0.21	0.08	0.05	−0.21	−0.19	−0.1	−0.7	0.17	−0.49	−0.41	0.42	−0.1	−0.39
31	−3.21	0.77	−2.2	−1.6	2.14	2.06	2.52	−0.44	3.28	−2.06	8.65	6.59	−2.14	−0.8	−2.04

$$\hat{\Phi}_1 = \left[\hat{\phi}_{i,j}\right], i = 32, \ldots, 61, \text{and} j - 15, \ldots, 29$$

	15	16	17	18	19	20	21	22	23	24	25	26	27	28	29
32	−4.98	−0.13	3.51	−2.21	0.82	1.06	1.72	0.82	2.79	−0.04	3.35	3.63	−4.24	−3.5	−3.7
33	5.01	0.97	−2.1	3.61	0.79	−1.61	−1.67	−0.67	−2.2	−0.16	−3.87	−2.83	2.17	2.78	3.37
34	−1	0.52	1.49	3.35	−1.7	0.82	1.01	2.69	−1.09	4.26	−2.82	−3.18	9.65	4.45	7.36
35	1.09	−0.2	0.53	−0.69	−0.09	−0.46	−0.49	−0.08	−0.81	0.2	−1.57	−1.03	−1.74	−0.2	0.51
36	0.44	−0.04	−0.5	−0.1	0.02	−0.02	0.03	−0.16	0.06	−0.24	0.53	0.08	−0.38	−0.3	−0.62
37	1.14	0.07	−0.7	−1.08	−0.33	−0.33	−0.26	−0.79	−0.94	−0.77	1.03	2.66	−4.58	−0.3	−0.52
38	−0.48	−0.03	0.16	0.35	0.09	0.08	0.05	0.2	0.24	0.19	−0.3	−0.7	1.37	0.05	0.09
39	−1.31	−0.02	0.56	0.78	0.27	0.27	0.2	0.56	0.71	0.53	−0.63	−1.6	3.16	0.06	0.06
40	−0.22	0	−0	−0.03	0.02	0.02	0.01	0.01	−0.03	0.03	0.01	0.23	0.1	−0.1	−0.18
41	0.02	−0.13	0.07	−0.04	0.03	−0.11	−0.13	−0.17	−0.23	−0.15	0.06	−0.02	0.34	−0.4	−0.73
42	0.35	−0.04	−0.1	−0.28	−0.05	−0.06	−0.06	0.02	−0.06	0.05	−0.28	−0.14	−0.87	−0.1	0.04
43	0.16	−0.02	−0	−0.15	−0.01	−0.03	−0.02	0.01	−0.04	0.03	−0.12	−0.18	−0.48	−0	0.06
44	−1.08	0.41	0.15	1.15	0.09	0.48	0.5	0.35	0.8	0.2	0.92	0.85	2.93	1.15	1.15
45	−0.02	0.02	0.01	0.06	0.02	0.03	0.03	0.03	0.06	0.01	0.03	0.04	0.16	0.07	0.09
46	−0.01	0	−0	0.01	−0.01	0.01	0.01	0	0	0	0.03	0	0.03	0.04	0.04
47	0.23	0.04	−0.2	0.14	−0.05	0.11	0.08	0.03	0.07	0.02	0.18	0.29	0.86	0.29	0.46
48	−0.21	0.06	0.01	0.18	0.02	−0.05	−0.04	0	−0.04	0.03	−0.15	−0.22	−0.14	0.15	0.16
49	−0.02	0.04	0.25	−0.29	−0.01	−0.01	0.01	−0.01	0.11	−0.07	0.06	0.06	−1.27	−0.1	−0.08
50	1.59	0.25	0.53	1.21	−0.22	0.69	0.87	0.68	1.46	0.18	1.32	2.8	3.17	−0	−0.54
51	−0.1	0	−0	−0.06	0.01	0	0	0	−0.01	0.01	−0.03	−0.14	−0.14	0.04	0.1
52	−0.27	−0.02	−0.1	−0.18	0.07	−0.16	−0.2	−0.17	−0.26	−0.1	−0.27	−0.39	−0.53	0.09	0.26
53	2.97	−2.34	2.91	−2.94	0.32	−0.08	−1.3	0.64	3.49	−0.9	−5.42	−1.63	9.54	−3.2	−7.8
54	−2.99	1.33	−2	1.46	−0.09	−0.19	0.46	−0.67	−2.68	0.46	2.83	−0.65	−7.22	1.93	4.89
55	−1.33	0.9	−1.3	0.65	−0.11	−0.06	0.39	−0.39	−1.53	0.24	2.06	0.59	−4.4	1.52	3.6
56	0.3	0.09	−0.4	−0.04	0.09	0.04	0.04	−0.17	0.47	−0.52	0.56	−1	−0.65	0.53	1.28
57	−0.78	0.16	0.2	0.08	0.23	0.41	0.5	0.3	−0.16	0.55	0.99	0.43	−0.46	−0.6	−1.26
58	0.66	−0.05	0.15	0.1	−0.09	−0.12	−0.15	−0.01	−0.01	−0.05	−0.4	−0.22	0.46	0.27	0.37
59	0.44	−0.25	−0.1	−0.22	−0.18	−0.39	−0.46	−0.26	−0.24	−0.24	−0.95	−0.04	0.13	−0.3	−0.39
60	−0.27	0.01	0.05	−0.01	−0.01	0.02	0.03	0.09	−0.03	0.13	−0.1	0.18	0.11	−0.1	−0.19
61	0.08	−0.05	0.05	−0.02	−0.03	0.04	0.03	0.11	−0.02	0.17	−0.16	0.32	0.47	−0.1	−0.32

$$\hat{\Phi}_1 = \left[\hat{\phi}_{i,j}\right], i = 1, \ldots, 31, \text{and } j = 30, \ldots, 43$$

	30	31	32	33	34	35	36	37	38	39	40	41	42	43
1	0.06	−0.28	−0.04	−0.06	−0.32	1.93	1.75	1.32	5.83	−0.44	2.57	7.06	11.74	3.41
2	0.04	0.05	0.03	0.04	0.05	−0.26	−0.24	−0.14	−0.58	0.05	−0.5	−0.84	−1.3	−0.58
3	0.24	0.17	0.15	0.16	0.19	−0.52	−0.52	−0.33	−1.17	0.08	−0.97	−1.63	−2.66	−1.09
4	0.25	0.22	0.18	0.19	0.24	−0.39	−0.33	0.05	−1.64	0.78	−1.21	−2.64	−4.71	−1.46
5	0.02	0.01	0.01	0.01	0.01	−0.15	−0.22	−0.12	0.09	−0.13	−0.5	0.21	0.48	−0.15
6	0.34	−0.12	−0.03	−0.03	0.12	0.01	−0.06	0.24	−1.77	0.93	−0.28	−1.28	−2.9	−1.16
7	−0.22	0.04	0.06	0.05	0.06	−0.72	−0.54	−0.56	−0.33	−0.67	−1.02	−1.16	−0.76	0.13
8	−0.01	0	−0.02	−0.02	0	0.23	0.19	0.13	0.46	0.01	0.42	0.59	0.76	0.17
9	0.03	0.05	−0.01	0	0.03	0.65	0.56	0.36	1.41	0	1.19	1.72	2.4	0.49
10	−0.07	0.04	0.02	0.02	0.04	0.1	0.12	0.02	0.86	−0.28	0.24	0.47	1.03	0.37
11	0.06	0.01	0.03	0.03	0.01	0.01	0	−0.02	0.2	−0.07	−0.05	0.13	0.28	−0.09
12	0.1	0	0.01	0.02	0	−0.21	−0.25	−0.2	−0.29	−0.12	−0.46	−0.12	−0.13	−0.16
13	−0.39	−0.05	−0.26	−0.26	−0.05	−0.77	−0.92	−0.71	−1.3	−0.39	−1.35	−0.8	−1.01	−0.03
14	−0.79	−0.38	−0.5	−0.52	−0.38	−0.56	−0.3	−0.33	−2.69	0.26	−0.19	−3.29	−5.82	−0.24
15	0.52	0.25	0.4	0.41	0.25	1.29	1.38	0.99	1.88	0.59	2.57	1.79	2.5	0.82
16	−2.82	−1.18	−1.3	−1.34	−1.13	3.43	0.76	0.04	7.33	−3.32	6.48	16.14	26.49	8.25
17	−0.15	0.04	−0.05	−0.05	0.04	−0.15	−0.16	−0.06	0.34	−0.23	−0.22	0.03	0.54	−0.58
18	0	−0.1	−0.12	−0.11	−0.11	0.8	0.65	0.39	3.38	−0.77	2.16	1.8	4.21	−0.52
19	−0.32	−0.36	−0.45	−0.45	−0.35	1.32	1.09	0.71	6.09	−1.38	3.39	3.53	7.42	−0.38
20	1.03	0.36	0.36	0.33	0.38	4.35	4.35	3.37	9.2	1.16	5.42	5.38	8.28	2.49
21	−1.2	−0.7	−1.16	−1.08	−0.73	−6.82	−7.13	−5.11	−16.9	−1.09	−10.8	−10.5	−12.4	−9.13
22	−0.97	−0.62	−0.25	−0.23	−0.62	3.88	4.2	3.24	15.1	−1.18	6.21	9.01	10.31	7.09
23	0.29	0.29	0.24	0.22	0.29	−0.14	−0.25	−0.34	−1.65	0.26	0.16	−0.53	−0.4	−0.12
24	0.69	0.56	0.53	0.48	0.56	−1.43	−1.53	−1.53	−5.61	0.15	−0.96	−3.45	−3.8	−2.58
25	0.11	0.05	0.19	0.17	0.05	0.86	0.97	0.66	2.93	−0.09	1.87	1.52	1.68	1.49
26	−0.01	−0.02	−0.05	−0.03	−0.03	−0.09	−0.02	−0.19	−0.5	−0.07	0.14	−0.1	−0.18	−0.1
27	0.01	−0.04	0.03	0.02	−0.02	−0.14	−0.26	−0.14	−0.2	−0.12	−0.48	0.16	0.31	0.06
28	−0.47	−0.07	−0.21	−0.23	−0.1	0.16	0.15	0.24	3.76	−1.23	1.78	1.67	4.97	−2.64
29	0.16	−0.03	0.04	0.05	−0.02	0.16	0.1	0.01	−0.56	0.28	−0.17	0.03	−0.51	0.95
30	−0.35	−0.15	−0.05	−0.04	−0.12	0.27	0.17	0.21	0.44	0.2	−0.37	−0.09	−0.75	1.83
31	1.34	0.17	0.27	0.34	0.72	−2.17	−2.55	−3.11	−9.52	−0.17	−2.79	−1.18	−2	−2.53

$$\hat{\Phi}_1 = \left[\hat{\phi}_{i,j}\right], i = 32, \ldots, 61, \text{and } j = 30, \ldots, 43$$

	30	31	32	33	34	35	36	37	38	39	40	41	42	43
32	−2.41	−0.1	−1.16	−0.31	−0.21	−4.66	−3.66	0.06	−14.7	5.27	−10.6	−12.4	−19.4	−3.43
33	2.82	1.14	1.35	0.53	1.15	2.01	2.78	−1.17	−2.05	−0.46	5.01	2.25	−3.71	4.85
34	1.62	0.42	1.38	1.3	−0.08	−1.93	0.1	2.18	3.45	1.92	−6.05	−13.4	−20.7	−5.58
35	−0.79	−0.19	−0.25	−0.22	−0.18	0.92	1.11	0.42	2.91	−0.5	2.65	2.73	2.95	2.75
36	0.03	−0.05	−0.1	−0.12	−0.05	−0.28	−0.16	−0.5	−0.85	−0.42	−1.74	−0.45	−0.67	−1.53
37	−0.33	0.18	−0.61	−0.58	0.13	2.18	1.96	1.99	7	0.01	3.91	4.59	5.55	3.92
38	0.07	−0.06	0.17	0.16	−0.05	−0.79	−0.76	−0.66	−2.42	0.02	−1.24	−1.57	−2.05	−0.96
39	0.16	−0.1	0.43	0.4	−0.07	−2	−1.86	−1.67	−5.79	−0.06	−2.82	−4.35	−5.51	−3.09
40	−0.03	0	−0.02	−0.02	0	−0.29	−0.28	−0.18	−0.74	0.03	0.06	−0.63	−0.93	−0.26
41	−0.04	−0.13	−0.09	−0.09	−0.13	0.27	0.4	0.4	1.07	0.16	0.22	0.29	0.51	0.6
42	−0.25	−0.04	−0.11	−0.11	−0.03	−0.2	−0.13	−0.26	−0.06	−0.37	0.18	−0.03	−0.57	−0.04
43	−0.14	−0.02	−0.05	−0.05	−0.02	−0.28	−0.24	−0.24	−0.61	−0.13	−0.22	−0.36	−0.99	0.09
44	1.22	0.41	0.6	0.6	0.39	0.69	0.23	0.71	−0.41	1.21	−0.28	0.12	2.21	−1.11

45	0.06	0.02	0.04	0.04	0.02	0.05	0.02	0.04	0.06	0.04	0	0.09	0.22	0
46	0.04	0	0.01	0.01	0	0.01	0.01	0.03	0.1	0.01	0.01	0.1	0.1	0.12
47	0.16	0.02	0.05	0.07	0.02	−0.32	−0.51	−0.68	−1.21	−0.53	−0.5	0.25	0.29	−0.46
48	0.09	0.07	0.08	0.07	0.07	−0.17	−0.14	−0.2	−0.72	0	0.1	−0.42	−0.59	−0.26
49	−0.04	0.04	−0.04	−0.06	0.05	0.49	0.57	0.66	1.56	0.33	0.74	0.63	0.86	0.65
50	0.38	0.26	0.32	0.34	0.22	2.2	2.07	1.82	2.24	1.61	3.88	3.88	7.54	1.26
51	−0.01	0	0	0	0	−0.14	−0.13	−0.12	−0.26	−0.07	−0.21	−0.29	−0.44	−0.18
52	−0.03	−0.02	−0.03	−0.03	−0.01	−0.25	−0.24	−0.21	0.03	−0.29	−0.53	−0.37	−0.85	−0.01
53	2.03	−2.52	−1.17	−1.44	−2.22	−3.39	−5.83	−6.58	−20	−0.21	−4.03	31.26	55.86	−5.5
54	−1.4	1.43	0.62	0.77	1.27	0.66	2.24	2.74	9.56	−0.62	0.19	−21.9	−39.1	2.59
55	−0.78	0.97	0.38	0.48	0.86	1.51	2.48	2.87	9.19	−0.02	1.49	−12.1	−22.2	2.4
56	−0.1	0.09	0.1	0.1	0.08	0.15	0.21	−0.08	0.97	−0.4	0.97	0.23	2.04	0.07
57	−0.04	0.17	0.06	0.07	0.14	−2.46	−2.44	−2.26	−5.92	−0.95	−3.32	−4.94	−8.2	−4.32
58	0.21	−0.05	0.04	0.03	−0.04	1.3	1.17	0.92	2.77	0.25	1.96	2.55	3.54	1.89
59	−0.25	−0.26	−0.24	−0.25	−0.25	1.02	1.05	1.15	2.64	0.6	1.02	2.33	3.9	1.88
60	−0.01	0.01	−0.01	−0.01	0.01	−0.27	−0.29	−0.22	−1.24	0.14	−0.5	−0.81	−1.84	0.16
61	0.06	−0.05	−0.02	−0.03	−0.04	0.18	0.16	0.36	0.49	0.3	−0.15	0.54	0.31	0.16

$$\hat{\Phi}_1 = \left[\hat{\phi}_{i,j}\right], i = 1, \ldots, 31, \text{ and } j = 44, \ldots, 57$$

	44	45	46	47	48	49	50	51	52	53	54	55	56	57
1	4.75	−9.83	−5.89	1.6	1.32	1.62	1.84	−2.1	1.11	1.97	2.91	0.64	0.11	0.33
2	−0.6	0.97	0.58	−0.17	−0.09	−0.19	−0.26	0.15	−0.1	−0.29	−0.4	−0.12	−0.02	−0.02
3	−1.22	2.83	2	−0.44	−0.32	−0.43	−0.48	0.39	−0.19	−0.54	−0.73	−0.26	−0.02	0.02
4	−1.66	3.99	3.58	−0.59	−0.44	−0.73	−0.89	2.41	−0.48	−1.06	−1.4	−0.58	0.05	−0.08
5	−0.03	−0.42	−0.8	0.02	0.07	−0.02	−0.1	−0.63	−0.12	−0.08	−0.02	−0.15	−0.03	−0.04
6	−0.97	2.61	1.88	−0.05	−0.08	−0.13	−0.47	2.26	0.1	−0.7	−0.96	−0.29	0	−0.07
7	−0.55	−1.89	−1.74	−0.11	0.52	−0.09	−0.33	−2.96	−0.62	−0.22	−0.33	−0.1	−0.14	−0.28
8	0.35	0.03	0.24	0.06	−0.16	0.07	0.2	0.68	0.19	0.2	0.28	0.08	0.03	0.1
9	1.08	−0.05	0.31	0.17	−0.32	0.21	0.56	1.3	0.57	0.57	0.8	0.24	0.12	0.26
10	0.41	−1.02	−0.38	0.05	0.03	0.08	0.22	−0.42	0.13	0.26	0.36	0.12	0.02	0.06
11	0.07	−0.55	−0.24	0.05	0.06	0.05	0	−0.05	0	−0.01	0.02	−0.05	−0.01	0.01
12	−0.18	−0.98	−0.24	−0.03	−0.01	−0.05	−0.17	−0.29	−0.09	−0.19	−0.24	−0.1	0	−0.01
13	−0.91	2.49	−2.56	−0.4	−0.23	−0.47	−0.55	−3.32	−1.38	−0.2	−0.16	−0.29	−0.16	−0.38
14	−1.8	3.78	1.72	−0.79	−0.73	−0.76	−0.48	0.15	−0.35	−0.44	−0.99	0.31	0.06	−0.17
15	1.66	−0.34	2.2	0.55	0.35	0.63	0.86	3.01	1.1	0.68	0.82	0.49	0.14	0.28
16	10.22	−23.2	9.55	−0.02	−5.16	0.96	2.92	10.12	11.21	0.95	1.9	−0.63	−0.1	0.75
17	−0.07	−0.25	−0.23	0.22	0.09	0.25	−0.11	1.26	−0.57	−0.04	0.04	−0.15	0.03	0.1
18	1.42	−0.47	−0.39	0.45	−0.08	0.64	0.44	2.73	−0.58	0.61	1.08	−0.09	0.04	0.35
19	2.63	−3.23	−1.11	0.66	−0.15	0.86	0.9	5.63	−0.46	1.04	1.8	−0.08	0.26	0.74
20	5.19	5.75	0.83	1.08	1.36	1.22	2.13	2.04	−0.38	2.72	3.71	1.22	0.34	0.78
21	−9.44	8.35	−15.4	−1.55	0.16	−2.8	−4.63	−12.4	−2.61	−4.72	−6.41	−2.22	−0.01	−0.64
22	6.54	−11.5	9.6	1.67	−0.13	2.73	3.64	10.11	3.96	3.3	4.46	1.53	0.38	1.36
23	−0.22	−0.26	0.42	−0.33	−0.42	−0.33	−0.19	−0.18	−0.38	−0.15	−0.18	−0.11	−0.21	−0.35
24	−2.29	1.82	−0.11	−0.67	−0.41	−0.86	−1.32	−1.53	−1.69	−1.26	−1.49	−0.87	−0.45	−0.95
25	1.38	−3.79	4.01	0.27	−0.35	0.61	0.79	3.73	0.59	0.7	1.01	0.25	−0.08	0.03
26	−0.11	−0.26	−0.2	−0.01	0.3	−0.05	−0.02	0.35	−0.31	0.03	0.06	0	0.02	0.03
27	−0.04	0.21	0.28	−0.04	−0.33	−0.02	0.02	−0.12	0.39	−0.06	−0.07	−0.05	−0.04	−0.03
28	1.13	−1.31	1.33	0.31	−1.22	0.63	−0.53	2.56	−2.28	−0.23	0.18	−0.82	−0.11	0.23
29	0.01	0.65	−0.76	0	0.48	−0.06	0.37	−0.26	0.63	0.34	0.35	0.3	0.08	0.06
30	0.16	−0.52	−0.73	−0.1	0.26	−0.14	0.51	−0.55	1.33	0.35	0.35	0.36	0.12	0.2
31	−2.51	23.98	−9.74	2.94	7.37	0.72		−14.4	−5.15	0.27	0.91	−0.57	−0.09	−0.35

$$\hat{\Phi}_1 = [\hat{\phi}_{i,j}], i = 32, \ldots, 61, \text{and} j = 44, \ldots, 57$$

	44	45	46	47	48	49	50	51	52	53	54	55	56	57
32	−8.23	10.27	−15.3	−2.88	−0.94	−4.25	−4.56	−26.1	−7.08	−3.09	−4.93	−0.35	−0.71	−2.44
33	1.55	−8.99	12.49	1.4	4.85	1.7	2.32	14.35	6.03	0.89	0.43	1.56	0.77	0.73
34	−7.56	1.59	4.55	−3.24	−5.7	−1.65	−1.56	2.97	−2.22	−1.74	−3.18	0.43	−0.27	−0.76
35	1.95	−4.66	−0.7	−0.42	0.22	−0.59	0.84	3.53	0.52	0.78	0.84	0.7	0.05	−0.01
36	−0.54	−0.91	−1.4	0.13	−0.4	0.15	0.1	3.9	0.5	−0.16	−0.05	−0.3	0.09	0.08
37	3.45	2.12	−5.98	0	0.62	0.08	1.15	0.13	1.43	1.12	1.73	0.19	−0.19	0.2
38	−1.2	−0.47	1.81	−0.08	−0.19	−0.12	−0.42	−0.46	−0.38	−0.42	−0.68	−0.04	0.03	−0.05
39	−3.2	−0.93	3.73	0.02	−0.34	−0.04	−1.12	−0.56	−1.01	−1.16	−1.79	−0.22	0.08	−0.16
40	−0.49	0.44	−0.25	−0.13	−0.05	−0.12	−0.22	−0.3	−0.14	−0.22	−0.34	−0.04	−0.01	−0.01
41	0.41	−0.55	−1.5	−0.04	−0.07	−0.02	0.27	−1.34	0.39	0.3	0.29	0.33	0.09	0.08
42	−0.25	−1.02	0.1	−0.1	−0.08	−0.15	0.01	3.46	0.17	−0.17	−0.23	−0.07	−0.05	−0.01
43	−0.41	−0.26	0.33	−0.18	−0.07	−0.24	−0.12	1.93	0.02	−0.24	−0.34	−0.08	−0.05	−0.06
44	0.68	5.5	2.99	0.8	0.36	1.07	−0.43	−10.3	−1.55	0.23	0.63	−0.38	−0.02	−0.06
45	0.08	−0.02	0.24	0.03	0.01	0.03	0.02	−0.38	0	0.04	0.06	0	0	0
46	0.05	−0.02	−0.05	0.04	0.04	0.05	0	0.05	−0.08	0.02	0.04	0	0	0
47	−0.29	2.14	0.13	−0.57	0.02	−0.18	0.06	−0.4	0.2	0.07	0.22	−0.15	−0.02	0.03
48	−0.28	−0.37	−0.21	0.09	−0.36	0.06	−0.16	0.06	−0.19	−0.17	−0.26	−0.04	−0.02	−0.08
49	0.67	−1.27	0.14	0.25	0.01	−0.15	0.17	1.78	0.21	0.09	0.01	0.21	0.16	0.11
50	3.45	−3.65	4.66	0.29	−0.98	0.07	1.46	6.62	2.96	0.05	0.67	−0.93	0.13	−0.22
51	−0.22	0	0.15	−0.02	0.02	0	−0.09	−0.12	−0.19	−0.04	−0.07	0	−0.02	0
52	−0.34	0.49	−0.83	0.03	0.28	0.11	−0.09	−1.34	−0.07	0.1	0.05	0.19	0.01	0.11
53	9.5	0.88	8.6	3.32	−0.08	3.08	7.27	7.89	3.33	7.88	10.7	2.98	1.05	−0.22
54	−8.1	1.44	−7.13	−2.36	0.62	−2.19	−5.07	−8.3	−3.44	−4.76	−6.73	−1.36	−0.77	0.19
55	−3.62	0.79	−4.37	−1.08	0.28	−0.86	−2.67	−4.52	−2.02	−2.34	−3.57	−0.19	−0.36	0.29
56	0.52	−1.53	0.99	−0.08	0.12	0.2	−0.06	−4.26	0.49	−0.05	−0.1	0.04	0.71	0.54
57	−4.05	0.99	−0.69	−0.06	0.58	−0.69	−1.3	3.82	−0.28	−1.71	−2.06	−1.2	−0.29	−0.31
58	1.97	−0.9	0.36	0.29	0.14	0.33	0.86	1.18	0.42	0.96	1.25	0.52	0.25	0.51
59	1.82	−0.56	−0.33	−0.22	−0.79	0.14	0.68	−1.41	0.08	0.89	1.06	0.64	0.09	0.17
60	−0.64	0.73	−0.23	−0.01	−0.02	−0.07	−0.23	0.25	0.09	−0.31	−0.48	−0.06	0.01	−0.04
61	0.2	1.43	0.15	0.21	0.13	0.23	0.15	0.63	−0.14	0.21	0.32	0.04	−0.02	−0.06

$$\hat{\Phi}_1 = [\hat{\phi}_{i,j}], i = 1, \ldots, 31, \text{ and } j = 58, \ldots, 61$$

	58	59	60	61
1	−0.3	0.13	−0.39	−0.11
2	−0.01	0.02	0.05	0
3	0.07	0.12	0.17	−0.09
4	0.01	0.17	0.27	0.14
5	−0.1	−0.06	0	0
6	−0.16	−0.05	0.02	0.02
7	−0.1	−0.3	−0.08	0.16
8	0.04	0.12	0.01	−0.07
9	0.18	0.29	0.05	−0.16
10	0.06	0.07	0.01	−0.03

11	−0.04	0.01	0	−0.06
12	−0.1	0.02	0	0.04
13	−0.03	−0.64	−0.03	0.35
14	0.33	−0.16	0	0.46
15	0.13	0.4	0.01	−0.38
16	−1.36	1.63	0.03	−3.47
17	−0.04	0.11	−0.05	−0.12
18	0.18	0.32	−0.08	−0.53
19	0.5	0.68	−0.11	−0.88
20	0.85	0.59	−0.11	−0.47
21	−0.87	0.06	0.49	1.32
22	1.38	1.19	−0.69	−1.7
23	−0.36	−0.44	0.11	0.16
24	−0.9	−1.08	0.18	0.54
25	0.01	−0.11	−0.12	−0.32
26	0.04	0.01	−0.03	0.05
27	−0.04	−0.03	0.03	−0.09
28	−0.19	0.23	−0.38	−1.03
29	0.16	0.05	0.13	0.19
30	0.25	0.24	0.21	0.14
31	−0.03	−0.29	−1.63	2.95

$\hat{\Phi}_1 = \left[\hat{\phi}_{i,j}\right], i = 32, \ldots, 61$, and $j = 58, \ldots, 61$

	58	59	60	61
32	−2.55	−1.31	−1.87	2.43
33	1.83	−0.26	2.04	1.3
34	1.04	−1.78	1.74	−0.26
35	0.33	−0.06	0.04	−0.11
36	0	0.11	0.03	0.01
37	0.22	0.37	−0.06	−0.98
38	−0.04	−0.1	−0.01	0.24
39	−0.17	−0.3	−0.03	0.69
40	0	−0.03	−0.03	−0.01
41	−0.01	0.02	−0.01	0.26
42	0.13	0.01	−0.03	−0.24
43	0.08	−0.06	−0.04	−0.03
44	−0.49	−0.01	0.14	0.17
45	−0.02	0	0	−0.01
46	0	0	−0.01	0
47	−0.11	−0.02	0.03	−0.29

48	0	−0.07	0.03	0.05
49	0.2	0.13	0	0.39
50	0.09	−0.89	−0.36	0.62
51	0.01	0.03	0.02	−0.06
52	0.04	0.24	0.06	−0.1
53	−2.69	−4.02	2.01	0.37
54	1.59	2.75	−1.06	−0.39
55	1.1	2	−0.65	−0.38
56	0.1	0.42	0.11	0.11
57	−0.17	−0.35	−0.05	−0.12
58	0.74	0.53	−0.04	0.08
59	0.13	0.24	0.07	0.01
60	0.02	0	0.82	0
61	0.04	0.01	0	0.45

This is clearly impractical. We will skip the output of the estimate of residual variance–covariance matrix $\hat{\boldsymbol{\Sigma}}$, and try to make the condensed form of the (61×61) matrix $\hat{\boldsymbol{\Phi}}_1$ from this output shown next.

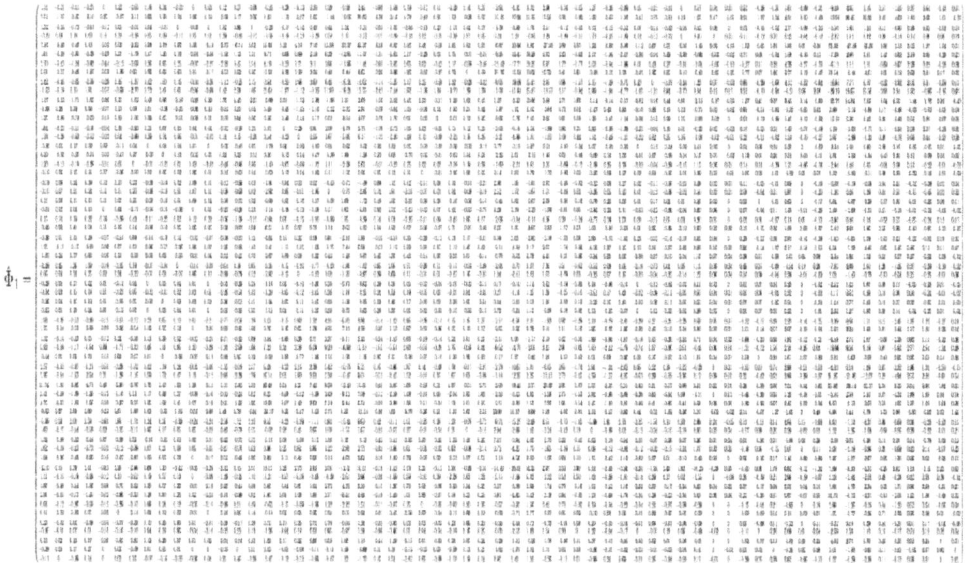

Readers may not be able to make out the numbers clearly. One parameter estimate takes 11 pages to list its elements, and this is only for a dimension of 61. We print out the estimate simply to illustrate the issues related to the high dimensional vector time series analysis. We will end the example here and try to introduce some solutions in the later chapters. ◁

2.8.4 Unemployment rates with outliers

Example 2.7 In this example, we apply the projection pursuit outlier detection method to US unemployment data. The data contains the monthly unemployment rates of the 10 US states, including Alabama, Arkansas, Florida, Georgia, Louisiana, Mississippi, Oklahoma, South Carolina, Tennessee, and Texas, from January 1976 to August 2010 for 416 observations. The data information is obtained from the website of US Bureau of Labor Statistics, listed as WW2c in the Data Appendix, and plotted in Figure 2.7.

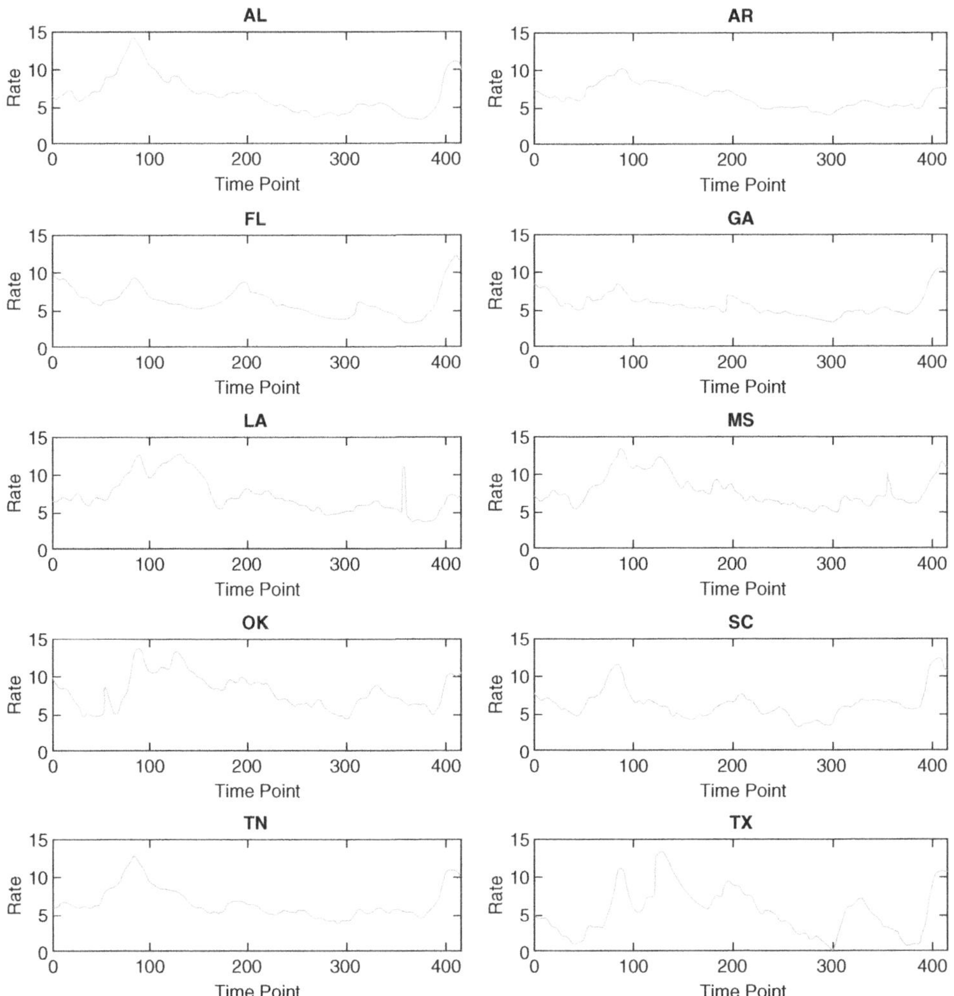

Figure 2.7 The unemployment rates of 10 US states, including AL, AR, FL, GA, LA, MS, OK, SC, TN, and TX.

◁

Based on BIC, we fitted VAR(3) model, $\mathbf{Z}_t = \sum_{j=1}^{3} \mathbf{\Phi}_j \mathbf{Z}_{t-1} + \varepsilon_t$, to the data set and got the following estimates of coefficient matrices:

$\hat{\mathbf{\Phi}}_1 =$									
1.52	0.11	0.01	0.06	0.00	−0.03	0.00	0.13	0.22	0.05
0.01	1.29	0.07	0.06	−0.01	0.02	0.00	0.10	0.14	0.04
0.08	0.10	1.15	0.06	−0.05	0.09	0.02	0.14	0.14	0.12
0.12	0.25	0.19	1.07	0.00	0.02	0.01	0.21	0.21	0.14
−0.05	0.26	−0.27	−0.04	1.24	−0.46	0.12	0.25	0.39	0.38
0.19	0.13	−0.09	0.00	−0.12	1.21	0.05	0.07	0.34	0.21
0.24	0.20	−0.04	0.04	−0.01	0.04	1.16	0.06	0.22	0.08
0.10	0.03	−0.02	0.07	−0.01	0.01	0.03	1.64	0.11	0.08
0.27	0.19	−0.03	0.05	0.01	0.01	0.00	0.08	1.39	0.04
0.23	0.18	0.06	0.01	0.00	0.02	0.07	0.12	0.11	1.08

$\hat{\mathbf{\Phi}}_2 =$									
−0.33	−0.19	0.00	−0.08	0.00	0.03	−0.06	−0.16	−0.37	−0.05
−0.10	−0.07	−0.09	−0.06	0.01	−0.02	−0.02	−0.12	−0.21	−0.02
−0.04	−0.23	−0.07	0.00	0.08	−0.16	−0.09	−0.17	−0.18	−0.06
−0.27	−0.33	−0.14	−0.01	0.01	−0.04	−0.08	−0.38	−0.32	−0.10
−0.06	−0.40	0.04	0.12	−0.48	0.98	−0.14	−0.34	−0.80	0.12
−0.35	−0.23	0.03	0.07	0.03	0.07	−0.06	−0.04	−0.53	0.02
−0.30	−0.10	−0.01	−0.09	0.02	−0.06	−0.06	−0.11	−0.38	0.09
−0.23	−0.15	0.01	−0.04	0.05	−0.13	−0.11	−0.54	−0.09	−0.03
−0.40	−0.30	0.06	−0.06	0.00	−0.03	−0.03	−0.05	−0.21	−0.09
−0.35	−0.33	−0.05	−0.05	0.00	−0.01	−0.05	−0.14	−0.14	0.07

$\hat{\mathbf{\Phi}}_3 =$									
−0.22	0.12	−0.01	0.02	0.01	−0.01	0.05	0.03	0.17	0.00
0.10	−0.23	0.02	0.01	0.00	0.01	0.01	0.02	0.07	−0.03
−0.05	0.14	−0.10	−0.05	−0.06	0.08	0.06	0.02	0.06	−0.04
0.13	0.10	−0.04	−0.11	−0.01	0.02	0.05	0.17	0.15	−0.02
0.13	0.22	0.18	−0.04	0.12	−0.62	0.09	0.01	0.55	−0.42
0.18	0.13	0.02	−0.04	0.07	−0.37	0.07	−0.04	0.25	−0.22
0.08	−0.10	0.05	0.04	−0.01	0.04	−0.15	0.05	0.15	−0.13
0.13	0.12	−0.01	−0.01	−0.05	0.12	0.06	−0.13	0.00	−0.03
0.14	0.12	−0.03	0.02	−0.01	0.03	0.03	−0.02	−0.21	0.04
0.12	0.13	0.02	0.01	0.00	0.01	−0.01	0.03	0.01	−0.19

We applied the projection pursuit outlier detection method to this time series and found an interesting outlier. For the critical values, we choose 4.3 for level-shift outliers and 5.8 for other types of outliers as suggested by table 2 in Galeano et al. (2006). During the first iteration, no level-shift outliers were found. To find other types of outliers, we calculated the following maximized LRT statistics at $t = 357$: $\Lambda_A = 22.16$, $\Lambda_T = 29.37$, $\Lambda_I = 17.85$. Since Λ_T is the largest of all and its value is greater than the critical value 5.8, we identify that there is a temporal change at $t = 357$. The detection procedure stops at the 15th iteration as no statistics are greater than the critical value.

iteration	(Λ_I, h_I)	(Λ_A, h_A)	(Λ_T, h_T)	Time	Type
1	(17.86, 357)	(22.17, 357)	(29.37, 357)	357, Sep 2005	TC
2	(16.29, 122)	(13.73, 54)	(17.67, 54)	54, Jun 1980	TC
3	(16.30, 122)	(12.86, 194)	(16.68, 194)	194, Feb 1992	TC
4	(16.30, 122)	(10.42, 359)	(13.16, 360)	122, Feb 1986	IO
5	(15.42, 357)	(10.74, 360)	(13.55, 360)	357, Sep 2005	IO
6	(14.23, 360)	(9.47, 360)	(13.16, 360)	360, Dec 2005	IO
7	(11.06, 310)	(7.69, 310)	(9.5, 310)	310, Oct 2001	IO
8	(12.97, 202)	(6.44, 201)	(8.79, 202)	202, Oct 1992	IO
9	(8.67, 361)	(4.75, 360)	(5.94, 416)	361, Jan 2006	IO
10	(7.75, 412)	(5.35, 357)	(6.37, 416)	412, Apr 2010	IO
11	(7.48, 358)	(5.51, 362)	(6.32, 416)	358, Oct 2005	IO
12	(6.95, 363)	(5.40, 362)	(5.72, 362)	363, Jun 2006	IO
13	(5.97, 409)	(4.45, 408)	(4.31, 408)	409, Dec 2009	IO
14	(6.04, 415)	(4.35, 246)	(4.60, 362)	415, Jul 2010	IO
15	(5.51, 362)	(4.64, 246)	(4.37, 362)		

Then, we examined the significance of the detected temporal change outlier by using step 4, such that

$$\begin{aligned} \widetilde{\mathbf{Z}}_t &= \hat{\mathbf{Z}}_t - \sum_{i_T=1}^{r_T} \frac{\hat{\boldsymbol{\omega}}_{i_T}}{1-\delta B} I_t^{\left(h_{i_T}^T\right)}, r_T = 3, h_{i_T}^T = 357, 54, 194, \\ \widetilde{\boldsymbol{\varepsilon}}_t &= \hat{\boldsymbol{\varepsilon}}_t - \sum_{i_I=1}^{r_I} \hat{\boldsymbol{\omega}}_{i_I} I_t^{\left(h_{i_I}^I\right)}, \quad r_I = 11, h_{i_I}^I = 122, 357, 360, \ldots, 415. \end{aligned} \tag{2.109}$$

The temporal change outlier effects and their corresponding standard errors are shown in the following table:

States	$\hat{\omega}_{1_T}$	$\hat{\omega}_{2_T}$	$\hat{\omega}_{3_T}$
AL	0.022	0.028	−0.097
	(0.049)	(0.050)	(0.070)
AR	−0.016	−0.010	−0.080
	(0.048)	(0.049)	(0.061)
FL	0.063	0.044	−0.016
	(0.094)	(0.094)	(0.108)
GA	0.074	0.052	−0.080
	(0.118)	(0.118)	(0.078)
LA	2.776	−0.071	−2.375
	(0.276)	(0.262)	(0.176)
MS	1.444	−0.094	−0.187
	(0.145)	(0.112)	(0.095)

OK	0.102	1.827	−0.061
	(0.127)	(0.126)	(0.092)
SC	0.033	0.045	−0.094
	(0.056)	(0.060)	(0.086)
TN	0.030	−0.001	−0.081
	(0.049)	(0.051)	(0.067)
TX	0.041	0.003	−0.004
	(0.085)	(0.086)	(0.058)

and the innovational change outlier effects and their corresponding standard errors are shown in the following table:

States	$\hat{\omega}_{1_I}$	$\hat{\omega}_{2_I}$	$\hat{\omega}_{3_I}$	$\hat{\omega}_{4_I}$	$\hat{\omega}_{5_I}$	$\hat{\omega}_{6_I}$	$\hat{\omega}_{7_I}$
AL	0.066	−0.047	−0.097	0.163	−0.024	0.001	0.155
	(0.068)	(0.068)	(0.070)	(0.070)	(0.069)	(0.069)	(0.069)
AR	0.101	−0.097	−0.080	0.013	−0.015	0.121	0.003
	(0.061)	(0.060)	(0.061)	(0.061)	(0.060)	(0.060)	(0.060)
FL	−0.027	−0.131	−0.016	1.123	−1.034	0.094	−0.290
	(0.109)	(0.109)	(0.108)	(0.108)	(0.090)	(0.071)	(0.071)
GA	0.018	0.025	−0.080	−0.058	0.048	−0.013	−0.155
	(0.078)	(0.077)	(0.078)	(0.078)	(0.078)	(0.077)	(0.077)
LA	0.193	3.244	−2.375	−0.052	−0.195	0.920	0.077
	(0.260)	(0.258)	(0.095)	(0.116)	(0.116)	(0.115)	(0.102)
MS	0.085	1.676	−0.187	0.007	−0.028	0.005	−0.094
	(0.132)	(0.132)	(0.092)	(0.094)	(0.094)	(0.094)	(0.094)
OK	0.369	−0.171	−0.061	0.099	0.077	0.044	0.082
	(0.095)	(0.092)	(0.086)	(0.092)	(0.092)	(0.091)	(0.086)
SC	0.035	−0.054	−0.094	0.016	0.123	0.077	−0.398
	(0.084)	(0.084)	(0.067)	(0.086)	(0.086)	(0.086)	(0.086)
TN	0.053	−0.132	−0.081	0.147	0.073	0.039	0.050
	(0.068)	(0.067)	(0.058)	(0.067)	(0.067)	(0.066)	(0.066)
TX	1.510	−0.010	−0.004	0.027	0.034	0.060	0.157
	(0.100)	(0.057)	(0.058)	(0.058)	(0.058)	(0.057)	(0.057)

States	ω_{8_I}	ω_{9_I}	ω_{10_I}	ω_{11_I}	ω_{12_I}
AL	−0.018	−0.065	−0.049	0.203	−0.108
	(0.068)	(0.068)	(0.068)	(0.068)	(0.067)
AR	−0.034	0.074	−0.035	−0.050	0.013
	(0.060	(0.060	(0.060)	(0.060)	(0.060)
FL	−0.147	0.128	−0.025	0.172	0.256
	(0.069)	(0.068)	(0.068)	(0.068)	(0.060)
GA	−0.077	0.081	−0.056	0.071	0.150
	(0.077)	(0.077)	(0.076)	(0.076)	(0.076)
LA	−0.401	0.112	−0.658	0.080	0.003
	(0.102)	(0.100)	(0.100)	(0.094)	(0.094)
MS	−0.624	−0.446	−0.182	0.420	0.227
	(0.093)	(0.087)	(0.083)	(0.083)	(0.080)

OK	−0.049	−0.017	−0.072	−0.091	0.097
	(0.091)	(0.091)	(0.091)	(0.091)	(0.091)
SC	−0.076	0.233	−0.060	0.035	0.336
	(0.083)	(0.082)	(0.081)	(0.081)	(0.081)
TN	−0.095	0.149	0.015	0.069	0.031
	(0.066)	(0.066)	(0.065)	(0.065	(0.065)
TX	−0.094	0.088	−0.036	−0.030	0.098
	(0.057)	(0.056)	(0.056)	(0.056)	(0.056)

The dotted lines shown in Figure 2.8 represent the outlier removed time series

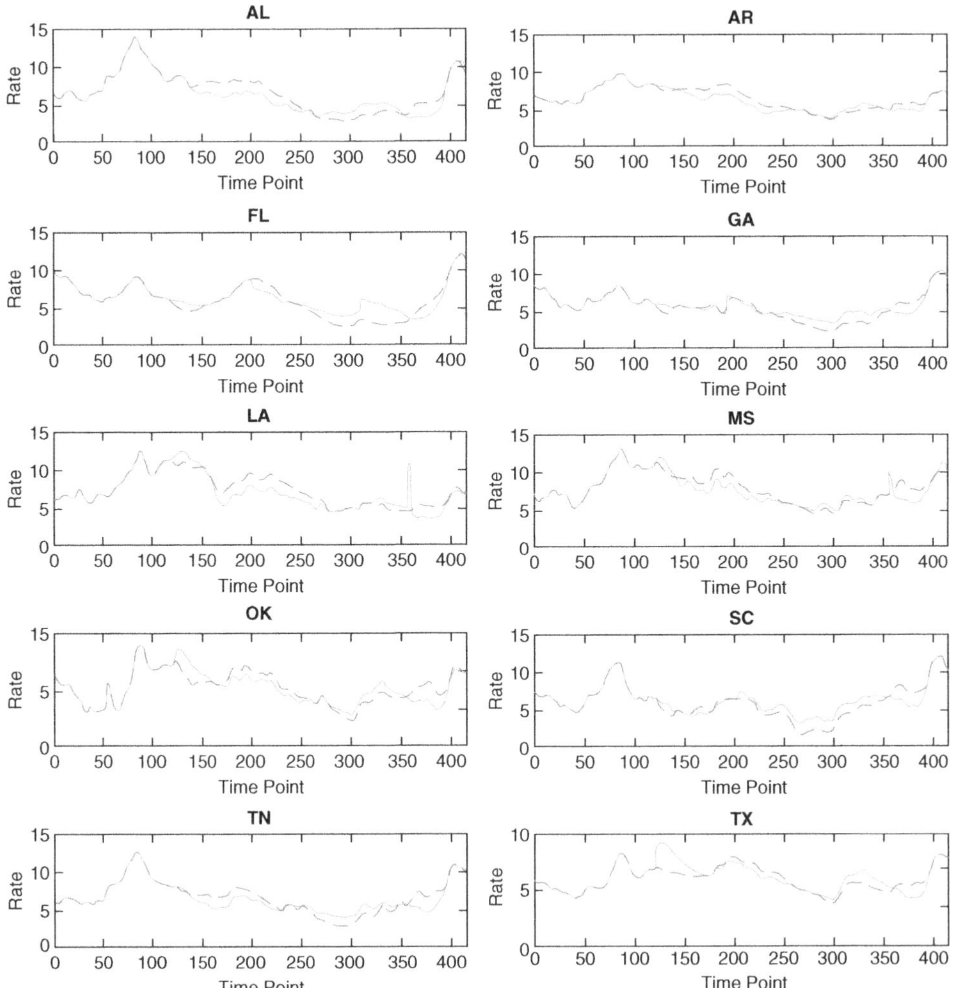

Figure 2.8 The original and outlier adjusted unemployment rates of 10 US states.

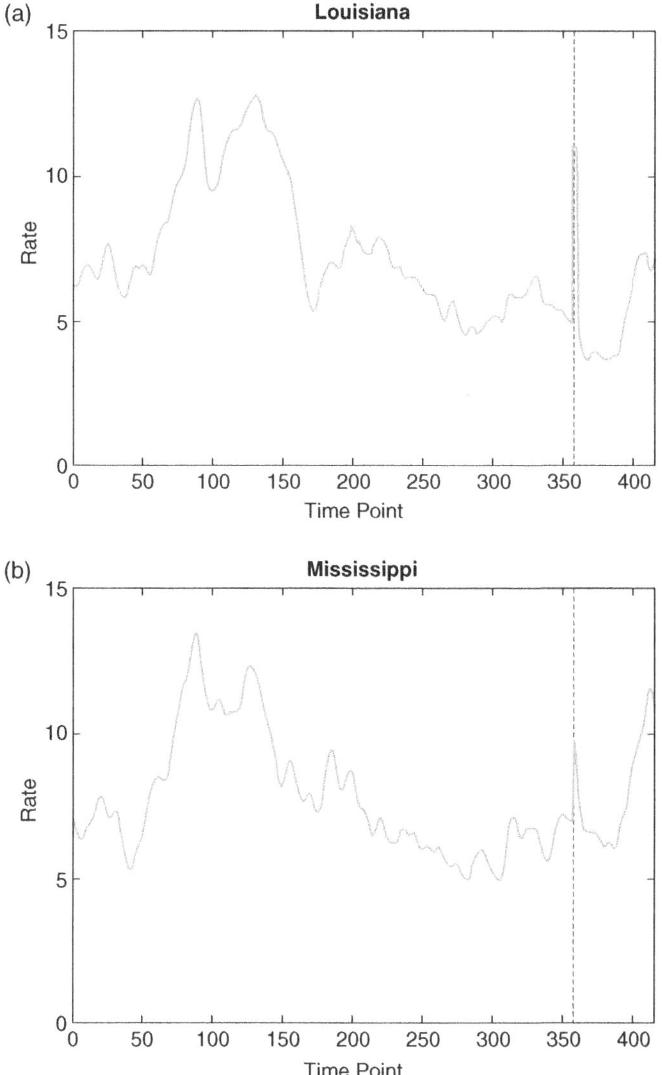

Figure 2.9 The unemployment rates of (a) Louisiana and (b) Mississippi, where the dotted red line indicates the observation at 357.

Let us carefully examine the outliers found at $t = 357, 358, 360, 361$, which correspond to September 2005, October 2005, December 2005, and January 2006. This was the period when Hurricane Katrina hit the United States. More specifically, the states that were severely impacted by Hurricane Katrina are Louisiana and Mississippi. This is consistent with the estimated outlier effects where the coefficients corresponding to LA and MS are most significant. The following time series plots of LA and MS in Figure 2.9 clearly show their significantly high unemployment rates at $t = 357$.

The other interesting findings here are: (i) the outlier detected at observation 54, which corresponds to June 1980, is possibly due to early 1980s recession in the United States; and (ii) the outlier detected at observation 194, which corresponds to February 1992, is likely the result of

Federal Reserve's Tax Reform Act that intended to reduce inflation but also limited economic expansion.

Finally, we re-fit the following VAR(3) model,

$$\mathbf{Y}_t = \sum_{j=1}^{3} \mathbf{\Phi}_j \mathbf{Y}_{t-j} + \xi_t \tag{2.110}$$

to the outliers' adjusted series, and obtain the following coefficient matrices:

$\hat{\mathbf{\Phi}}_1 =$

1.51	0.09	0.11	0.09	0.05	−0.02	0.00	0.08	0.23	0.15
0.03	1.23	0.10	0.16	0.05	0.04	−0.02	0.04	0.15	0.11
0.09	0.05	1.20	0.13	0.02	0.05	0.02	0.07	0.09	0.26
0.14	0.10	0.18	1.33	−0.02	0.10	−0.10	0.08	0.16	0.13
−0.04	0.00	−0.16	−0.04	1.66	0.03	0.07	0.06	0.19	0.20
0.06	0.01	0.00	0.07	0.01	1.64	−0.01	0.06	0.21	0.17
0.15	0.07	0.08	0.02	0.09	0.09	1.35	0.05	0.13	0.08
0.14	0.03	0.04	0.13	0.00	0.14	0.00	1.57	0.11	0.08
0.27	0.14	0.03	0.13	0.05	0.09	−0.01	0.01	1.37	0.14
0.13	0.11	0.11	0.10	0.09	0.07	0.06	0.04	0.15	1.06

$\hat{\mathbf{\Phi}}_2 =$

−0.35	−0.16	−0.11	−0.19	−0.09	−0.03	−0.05	−0.10	−0.37	−0.21
−0.14	−0.01	−0.17	−0.29	−0.06	−0.09	0.02	−0.01	−0.25	−0.07
−0.10	−0.21	0.05	−0.10	−0.04	−0.11	−0.11	−0.11	−0.14	−0.22
−0.27	−0.23	−0.18	−0.08	0.06	−0.21	0.06	−0.18	−0.29	−0.08
0.02	−0.13	0.20	0.04	−0.64	−0.08	−0.06	−0.12	−0.30	−0.07
−0.10	−0.10	−0.02	−0.11	−0.01	−0.62	0.05	−0.11	−0.27	−0.14
−0.26	−0.12	−0.12	−0.03	−0.17	−0.20	−0.14	−0.05	−0.25	0.03
−0.27	−0.10	0.02	−0.13	0.05	−0.33	−0.11	−0.43	−0.18	−0.08
−0.39	−0.22	−0.10	−0.19	−0.08	−0.21	−0.03	0.08	−0.20	−0.22
−0.28	−0.27	−0.13	−0.17	−0.11	−0.14	−0.05	0.02	−0.20	0.26

$\hat{\mathbf{\Phi}}_3 =$

−0.19	0.11	0.01	0.10	0.06	0.04	0.04	0.02	0.16	0.04
0.12	−0.25	0.07	0.14	0.01	0.05	−0.02	−0.03	0.10	−0.04
0.00	0.17	−0.27	−0.01	−0.01	0.07	0.08	0.02	0.07	−0.02
0.13	0.14	0.00	−0.28	−0.04	0.10	0.04	0.11	0.15	−0.06
0.03	0.14	−0.03	0.00	−0.05	0.06	−0.01	0.05	0.12	−0.13
0.05	0.11	0.00	0.04	0.01	−0.08	0.00	0.05	0.09	−0.05
0.12	0.05	0.03	0.01	0.08	0.13	−0.24	0.01	0.10	−0.09
0.13	0.07	−0.06	0.02	−0.06	0.20	0.10	−0.16	0.08	0.00
0.13	0.09	0.07	0.07	0.03	0.13	0.03	−0.07	−0.20	0.06
0.15	0.15	0.03	0.06	0.02	0.09	0.00	−0.06	0.04	−0.35

In concluding the chapter, we would like to point out that model building in multivariate vector time series is far more difficult when there are interventions and non-linear phenomenon involved. This will not be a topic discussed in this book. We refer readers to Box and Tiao (1975), Chen and Tsay (1993), Wei (2006, chapter 19), Lütkepohl (2007), Little (2013), Wei (2015), Chan et al. (2015), Li et al. (2015), Cai et al. (2017), Chan et al. (2017), Garthoff and Schmid (2017), Giurcanu (2017), Plakandaras and Gogas (2017), Wong et al. (2017), Nyberg (2018), Wang and Samworth (2018), among others.

Software code

R code for Example 2.4

```
if("tseries" %in% rownames(installed.packages()) == FALSE)
{install.packages("tseries")}
if("MTS" %in% rownames(installed.packages()) == FALSE)
{install.packages("MTS")}
if("ppcor" %in% rownames(installed.packages()) == FALSE)
{install.packages("ppcor")}
library(tseries)
library(MTS)
library(ppcor)
library(TSA)
library(forecast)
library(ggplot2)

## Read in original data set
ms <- as.data.frame(read.csv("C:/Bookdata/WW2a.csv")[,-1])
rownames(ms) <- seq(as.Date("2009/6/1"), by="month",
length=90)
x.orig <- ms

##Plot original vector time series
par(mfrow=c(3,3))
   for(i in 1:8){
   plot.ts(x.orig[,i],main=colnames(x.orig)[i],xlab=NULL,
ylab='Mil. dollars')}

## Unit root test
adf.test(x.orig[,1])
adf.test(x.orig[,2])
adf.test(x.orig[,3])
adf.test(x.orig[,4])
adf.test(x.orig[,5])
adf.test(x.orig[,6])
```

```
adf.test(x.orig[,7])
adf.test(x.orig[,8])

## Model building
## Take difference
zt=diffM(x.orig)

## Correlation and partial correlation matrix functions
## CCM with default lag=12
zt.ccm<-ccm(zt, lag=15)

## Modeling on a revised vector series
ms <- as.data.frame(read.csv("C:/Data/Bookdata/WW2b.csv")
[,-1])
rownames(ms) <- seq(as.Date("2009/6/1"), by="month",
length=90)
x.orig <- ms

## Plot the new five series
par(mfrow=c(2,3))
   for(i in 1:5){
   plot.ts(x.orig[,i],main=colnames(x.orig)[i],xlab=NULL,
ylab='Mil. dollars')}

plot(seq(as.Date("2009/6/1"), by="month", length=90),x.orig
[,1],type='l',ylab="Retail Sales",xlab="Date",ylim=c(min(x.
orig),max(x.orig)))
for(i in 2:5){
      lines(seq(as.Date("2009/6/1"), by="month", length=90),
x.orig[,i],type='l',lty=i,lwd=2)
}
legend("topleft",legend=colnames(zt),,lty=1:5)

## Differencing the five series
ms5<-ms[,1:5]
zt=diffM(ms5)

## Plot the five differenced series
par(mfrow=c(2,3))
   for(i in 1:5){
   plot.ts(zt[,i],main=colnames(x.orig)[i],xlab=NULL,
ylab='Mil. dollars')}

# Model identication

## CCM with default lag=12
zt.ccm<-ccm(zt, lag=15)
```

```
zt.pacf<-pacf(zt, 15)
eccmz=Eccm(zt)

## Vector time series model fit
m4=VAR(zt, p=4, include.mean=TRUE)

## Diagnostic checking
m4$residuals
eccmres=Eccm(m4$residuals)

## Vector time series model forecasting
ff <- VARpred(m4)
ff$pred

# Upper 95%
upper<-ff$pred+1.96*ff$se
upper
# Lower 95%
lower<-ff$pred-1.96*ff$se
lower
q()
```

SAS Code for Example 2.5

```
proc import datafile='C:/Bookdata/WW2b.csv'
out=ms dbms=csv replace;
getnames=yes;
run;
proc varmax data=ms;
model AUT BUM GEM GRO COM / p=2 dify=(1,12) noint print=
(corry parcoef pcorr
pcancorr roots estimates diagnose);
output out=for lead=1;
run;
```

R Code for Example 2.5

```
## Read in original data set
ms <- as.data.frame(read.csv("C:/Bookdata/WW2b.csv")[,-1])
rownames(ms) <- seq(as.Date("2009/6/1"), by="month",
length=90)

## Model building
## Take seasonal difference first and then regular
difference
dszt=diffM(ms,d=12)
drszt=diffM(dszt)
```

```
## Plot the five differenced series
par(mfrow=c(2,3))
   for(i in 1:5){
   plot.ts(drszt[,i],main=colnames(ms)[i],xlab=NULL,
ylab='Differenced series')}

## Correlation and partial correlation matrix functions
## CCM with default lag=12
drszt.ccm<-ccm(drszt, lag=15)
drszt.pacf<-pacf(drszt, 15)
eccmz=Eccm(drszt)

## Model fitting
## Vector time series model fit
m=VAR(drszt, p=2, include.mean=FALSE)

## Vector time series model forecasting
ff <- VARpred(m)
ff$pred

# upper 95%
upper<-ff$pred+1.96*ff$se
upper

# lower 95%
lower<-ff$pred-1.96*ff$se
lower
q()
```

R Code for Example 2.6

```
library(MTS)
setwd("C:/Bookdata/")
koopact <- read.csv("WW10.csv",head=TRUE)
attach(koopact)
dat <- as.matrix(cbind(GDP252, GDP253, GDP254, GDP255, GDP256,
              GDP257, GDP258, GDP259, GDP260, GDP261,
              GDP266, GDP267, GDP268, GDP269, GDP271,
              GDP272, GDP274, GDP275, GDP276, GDP277,
              GDP278, GDP279, GDP280, GDP281, GDP282,
              GDP284, GDP285, GDP286, GDP287, GDP288,
              GDP289, GDP290, GDP291, GDP292, IPS11,
              IPS299, IPS12, IPS13, IPS18, IPS25,
              IPS32, IPS34, IPS38, IPS43, IPS307, IPS306,
```

```
              CES275, CES277, CES278, CES003, CES006,
              CES011, CES015, CES017, CES033, CES046,
              CES048, CES049, CES053, CES088, CES140))

#==================================================
# Plot
#==================================================

dat2 <- matrix(dat,195*61)
dat3 <- data.frame(y=dat2, group=rep(colnames(dat),
each=195), obs=rep(1:195,61) )
ggplot(data = dat3, aes(x = obs, y = y, group = group)) +
    geom_line() +
    facet_wrap(~ group)

#==================================================
# VAR
#==================================================

m1 <- VAR(dat[1:(131+64-1),],p=1,output=FALSE,include.
mean=FALSE)

# estimation
round(m1$coef,2) # estimator
round(m1$secoef,2) # standard error
q()
```

SAS Code for Example 2.6

```
proc import datafile='C:/Bookdata/WW10.csv'
out=ms dbms=csv replace;
getnames=yes;
run;
proc varmax data=ms;
model GDP252 GDP253 GDP254 GDP255 GDP256 GDP257 GDP258
      GDP259 GDP260 GDP261 GDP266 GDP267 GDP268 GDP269
      GDP271 GDP272 GDP274 GDP275 GDP276 GDP277 GDP278
      GDP279 GDP280 GDP281 GDP282 GDP284 GDP285 GDP286
      GDP287 GDP288 GDP289 GDP290 GDP291 GDP292 IPS11
      IPS299 IPS12 IPS13 IPS18 IPS25 IPS32 IPS34 IPS38 IPS43
IPS307
      IPS306 CES275 CES277 CES278 CES003 CES006 CES011 CES015
      CES017 CES033 CES046 CES048 CES049 CES053 CES088
CES140/ p=1 noint
print=(corry estimates diagnose);
output out=for lead=1;
run;
```

MATLAB Code for Multivariate Outlier Detection
Multivariate Outlier Detection – Example 2.7

```
zt = csvread('C:/Bookdata/WW2c.csv',1,1);
names = readtable('C:/Bookdata/WW2c.csv');
states = names.Properties.VariableNames(:);
%states
 figure
for j=1:10
subplot(5,2,j)
plot(zt(j,:))
title(states{j+1})
xlabel("Time Point")
ylabel("Rate")
xlim([0 416])
end

zt=zt';
vc=[4.3 5.8];
[YL,detecto,tipos]=algoritmoestacionario(zt,vc);

figure
subplot(1,2,1)
plot(zt(18,:))
line([357 357], [0 15],'Color','red','LineStyle',
'-','LineWidth',.1)
title("Louisiana")
xlabel("Time Point")
ylabel("Rate")
xlim([0 416])
subplot(1,2,2)
plot(zt(24,:))
line([357 357], [0 15],'Color','red','LineStyle',
'-','LineWidth',.1)
title("Mississippi")
xlabel("Time Point")
ylabel("Rate")
xlim([0 416])


%%%%%%%%%%%%%%%%%%%%%%%%%

function [YL,detecto,tipos] = algoritmoestacionario(Y,vc);

nc = vc(1);
vcio = vc(2);
vcao = vc(2);
vctc = vc(2);
```

```
[k n] = size(Y);
[lst,amcu] = cambiosdenivel(Y,nc);

les = length(lst);
YL = Y;

if les > 0
   lsts = zeros(1,les);
   westls = zeros(k*les,1);
   sdwestls = zeros(k*les,1);
   for i = 1 : les
       [YL,lsts(i),westls(((i-1)*k+1):(i*k),:),sdwestls
(((i-1)*k+1):(i*k),:)] = limpiacambiosdenivel(YL,lst(i));
   end
end

[YL,momtot,tiptot,westot,sdwestot] = restoatipicos(YL,vcio,
vcao,vctc);

if les > 0
   detecto = [lst momtot];
   tipos = [3 * ones(1,les) tiptot];
else
   detecto = momtot;
   tipos = tiptot;
end

%%%%%%%%%%%%%%%%%%%%%%%%%

function [lst,amcu] = cambiosdenivel(Y,nc);

warning off;
[n k] = size(Y');


X = Y';
mm = mean(X);
S = cov(X);
X = X - ones(n,1) * mm;
Rr = chol(S);
X = ((Rr')\(X'))';
V = kur_nwa(X,0);

for i = 1 : (2 * k)
    y(1:n,i) = (X * V(:,i));
end
```

```
hay = 1;
t1 = 1;
t2 = n;
mom = [];
momfin = [];
amcu = [];
iter = 0;

while (hay == 1) & (iter<(n/10))

    iter = iter + 1
    cu = zeros(n,2*k);

    kg = floor(sqrt(t2-t1+1));
    for i = 1 : (2*k)
        for h = 1 : (kg + 1)
            aux = cov(y(t1:(t2-h+1),i),y((t1+h-1):t2,i));
            covg(h) = aux(1,2);
        end
        ma(i) = covg(1);
        for h = 1 : kg
            ma(i) = ma(i) + 2 * (1 - h/kg) * covg(h+1);
        end
        ma(i) = sqrt(ma(i));
    end

    for i = 1 : (2*k)
        for h = t1 : t2
            no = (t2 - t1 + 1);
            cu(h,i) = (h - t1 + 1) * sqrt(1/no) * abs(mean(y
 (t1:h,i)) - mean(y(t1:t2,i))) / ma(i);
        end
    end

    [md hd] = max(max(abs(cu),[],2));
    if (md > nc)
       tm = hd;
    else
       hay = 0;
    end

    if hay == 1

       cambio1 = 1;
       t22 = tm;
       while cambio1 == 1
```

```
        kg = floor(sqrt(t22-t1+1));
        for i = 1 : (2*k)
            for h = 1 : (kg + 1)
                aux = cov(y(t1:(t22-h+1),i),y((t1+h-1):
t22,i));
                covg(h) = aux(1,2);
            end
            ma(i) = covg(1);
            for h = 1 : kg
               ma(i) = ma(i) + 2 * (1 - h/kg) * covg(h+1);
            end
            ma(i) = sqrt(ma(i));
        end

        cu = zeros(n,2*k);
        for i = 1 : (2*k)
            for h = t1 : t22
                no = (t22 - t1 + 1);
                cu(h,i) = (h - t1 + 1) * sqrt(1/no) * abs
(mean(y(t1:h,i)) - mean(y(t1:t22,i))) / ma(i);
            end
        end
        [md hd] = max(max(abs(cu),[],2));

        if (md > nc)
           if hd == t22
              cambio1 = 0;
              hfirst = t22;
           end
           if hd < t22
              if abs(hd-t1) < 3
                 cambio1 = 0;
                 hfirst = t22;
              else
                 t22 = hd;
              end
           end
        else
           cambio1 = 0;
           hfirst = t22;
        end
    end

    cambio2 = 1;
    t11 = tm + 1;
    while cambio2 == 1
```

```
            kg = floor(sqrt(t2-t11+1));
            for i = 1 : (2*k)
                for h = 1 : (kg + 1)
                    aux = cov(y(t11:(t2-h+1),i),y
((t11+h-1):t2,i));
                    covg(h) = aux(1,2);
                end
                ma(i) = covg(1);
                for h = 1 : kg
                 ma(i) = ma(i) + 2 * (1 - h/kg) * covg(h+1);
                end
                ma(i) = sqrt(ma(i));
            end

            cu = zeros(n,2*k);
            for i = 1 : (2*k)
                for h = t11 : t2
                    no = (t2 - t11 + 1);
                  cu(h,i) = (h - t11 + 1) * sqrt(1/no) * abs
(mean(y(t11:h,i)) - mean(y(t11:t2,i))) / ma(i);
                end
            end

            [md hd] = max(max(abs(cu),[],2));
            if (md > nc)
               if hd == t11
                  cambio2 = 0;
                  hlast = t11;
               end
               if hd > t11
                  if abs(hd-t2) < 3
                     cambio2 = 0;
                     hlast = t11;
                  else
                     t11 = hd;
                  end
               end
            else
               cambio2 = 0;
               hlast = t11;
            end
      end

      if abs(hfirst - hlast) < (k + 10)
```

```
            hay = 0;
            mom = [mom floor(median([hfirst hlast]))];
         else
            t1 = hfirst + 1;
            t2 = hlast;
            if sum(mom==hfirst)==0
                mom = [mom hfirst];
            end
            if sum(mom==hlast)==0
                mom = [mom hlast];
            end

         end

     end

end

if iter == 1
    lst =[];
end

% Aqui empieza el paso 3

hay = 0;
mom = sort(mom);
lo = length(mom);
if lo > 0
    hay = 1;
    mom = [1 mom n];
end

while (hay == 1)

    lo = length(mom) - 2;
    if lo > 0
       j = 2;
       eliminar = 1;
       while eliminar == 1
             if length(mom) > 3
                t1 = mom(j-1);
                t2 = mom(j+1);
             else
                t1 = 1;
                t2 = n;
             end
```

```
            kg = floor(sqrt(t2-t1+1));
            for i = 1 : (2*k)
                for h = 1 : (kg + 1)
                  aux = cov(y(t1:(t2-h+1),i),
y((t1+h-1):t2,i));
                  covg(h) = aux(1,2);
                end
                ma(i) = covg(1);
                for h = 1 : kg
                 ma(i) = ma(i) + 2 * (1 - h/kg) * covg(h+1);
                end
                ma(i) = sqrt(ma(i));
            end
            cu = zeros(n,2*k);
            for i = 1 : (2*k)
                for h = t1 : t2
                   no = (t2 - t1 + 1);
                   cu(h,i) = (h - t1 + 1) * sqrt(1/no) * abs
(mean(y(t1:h,i)) - mean(y(t1:t2,i))) / ma(i);
                end
            end
            [md hd] = max(max(abs(cu),[],2));
            if (md > nc)
               amcu = [amcu md];
               mom(j) = hd;
               j = j + 1;
            else
               amcu = [];
              mom = [mom(1:(j-1)) mom((j+1):length(mom))];
            end
            if length(mom) == 2
                eliminar = 0;
            end
            if j == length(mom)
                eliminar = 0;
            end
      end
   end
   momf = mom;

   lof = length(momf ) - 2;
   if lof == 0
      hay = 0;
      momfin = [];
   else
```

```
        if lo == lof
           dife = max(abs(mom-momf ));
           if dife < 2
              lfin = length(momf-2);
              momfin = momf(2:(lfin-1));
              hay = 0;
           else
              mom = momf;
           end
        else
           mom = momf;
        end
     end

end

if length(mom) > 2
   lst = momf(2:(length(mom)-1));
else
   lst = [];
end

llst = length(lst);
if llst > 0
    lst = lst + ones(1,llst);
else
    lst = [];
end

%%%%%%%%%%%%%%%%%%%%%%%%%%

function [cnst,stdcst,phi,stdphi,resi] = estima(z,ip,ist,
idx,idxc);

[nob k] = size(z);
lag = 1:ip;
dnob = nob - ist + 1;
stdcst = zeros(1,k);

for ii = 1 : k


    icnt = 0;
    if idxc(ii) == 0
       icnt = icnt + 1;
       for it = ist : nob
```

```
            x(it,icnt) = 1;
        end
    end
    for i = 1 : ip
        m = lag(i);
        for j = 1 : k
            if idx(ii,j,i) == 0
                icnt = icnt + 1;
                for it = ist : nob
                    x(it,icnt) = z(it-m,j);
                end
            end
        end
    end


    for i = 1 : icnt
        tmp = 0;
        for it = ist : nob
            tmp = tmp + x(it,i) * z(it,ii);
        end
        xpy(i) = tmp;
        for j = 1 : i
            tmp = 0;
            for it = ist : nob
                tmp = tmp + x(it,i) * x(it,j);
            end
            xpx(i,j) = tmp;
            xpx(j,i) = tmp;
        end
    end

    xpxinv = inv(xpx);


    for i = 1 : icnt
        tmp = 0;
        for j = 1 : icnt
            tmp = tmp + xpxinv(i,j) * xpy(j);
        end
        beta(i) = tmp;
    end

    for it = ist : nob
        tmp = 0;
```

```
        for i = 1 : icnt
            tmp = tmp + beta(i) * x(it,i);
        end
        resi(it,ii) = z(it,ii) - tmp;
    end


    tmp = 0;
    for it = ist : nob
        tmp = tmp + resi(it,ii) * resi(it,ii);
    end
    rvar = tmp / (nob-ist+1-icnt);


    cnst(ii) = 0;
    m = 0;
    if idxc(ii) == 0
       cnst(ii) = beta(1);
       m = 1;
       stdcst(ii) = sqrt(rvar * xpxinv(1,1));
    end
    for i = 1 : ip
        for j = 1 : k
            phi(ii,j,i) = 0;
            if idx(ii,j,i) == 0
               m = m + 1;
               phi(ii,j,i) = beta(m);
               stdphi(ii,j,i) = sqrt(rvar*xpxinv(m,m));
            end
        end
    end

end

if idxc == ones(1,k)
   cnst = zeros(1,k);
   stdcst(1,1:k) = zeros(1,k);
end

%%%%%%%%%%%%%%%%%%%%%%%%%

function Vv = kur_nwa(x,mode)

if nargin < 2,
  mode = 0;
end
```

```
% Initializations

%% Parameters (tolerances)

maxit = 100;

tol = 1.0e-5;
tol1 = 1.0e-6;
beta = 1.0e-4;
rho0 = 0.1;

[n,p0] = size(x);

%% Initialization of vectors

Vv = [];
ti = 1;

% Computing directions

%% Choice of minimization/maximization

cff = 1;
if mode < 0,
  cffmax = 2;
else
  cffmax = 3;
end

%% Main loop to compute 2p directions

while (cff
  xx = x;
  p = p0;
  pin = p - 1;
  M = eye(p0);

  for i = 1:pin,
    if cff == 1,
      a = max_kur(xx);
    else
      a = min_kur(xx);
    end
    la = length(a);
    za = zeros(la,1); za(1) = 1;
```

```
    w = a - za; nw = w'*a;
    if abs(nw) > eps,
      Q = eye(la) - w*w'/nw;
    else
      Q = eye(la);
    end

%% Compute projected values

    Vv = [ Vv (M*a) ];

    Qp = Q(:,2:p);
    M = M*Qp;
    ti = ti + 1;

%% Reduce dimension

    Tt = xx*Q;
    xx = Tt(:,(2:p));

    p = p - 1;

  end

%% Compute last projection

  Vv = [ Vv M ];
  ti = ti + 1;

%% Proceed to minimization

  cff = cff + 1;

end

%%%%%%%%%%%%%%%%%%%%%%%%%

function [XL,lst,westls,sdwestls] = limpiacambiosdenivel
(Y,lst);

[k n] = size(Y);
Y = Y';
hay = 1;
pmax= floor(log(n));

while hay == 1
```

```
    lst = sort(lst);
    minlst = min(lst);
    r = length(lst);
    lstb = lst - (minlst - 1) * ones(1,r);

    for j = 1 : pmax
        [cnst,stdcst,phiest,stdphi,resid,covresi] = varfit
(Y,j,zeros(k,k,j),zeros(1,k));
        bic(j) = n * log(det(covresi)) + log(n) * j * k^2;
    end
    [aux MP] = min(bic);
    [cnst,stdcst,phiest,stdphi,resid,covresi] = varfit(Y,MP,
zeros(k,k,MP),zeros(1,k));

    A = [];
    for i = minlst : n
        A = [A;resid(i,:)'];
    end

    M = zeros(k*(n-minlst+1),k*r);
    for i = 1 : r
        M((k*(lstb(i)-1)+1):(k*lstb(i)),(k*(i-1)+1):(i*k)) =
eye(k);
        if lstb(i) < n
            for j = 1 : (n - (lst(i)))
                if j < (MP + 1)
                 M((k*(lstb(i)+j-1)+1):(k*(lstb(i)+j)),(k*(i-
1)+1):(i*k)) = eye(k)-sum(phiest(:,:,1:j),3);
                else
                 M((k*(lstb(i)+j-1)+1):(k*(lstb(i)+j)),(k*(i-
1)+1):(i*k)) = eye(k)-sum(phiest(:,:,1:MP),3);
                end
            end
        end
    end

    C = zeros(n - minlst + 1,n - minlst + 1);
    for i = 1 : (n - minlst + 1)
        C((k*(i-1)+1):(k*i),(k*(i-1)+1):(k*i)) = covresi;
    end

    west = inv(M' * inv(C) * M) * (M' * inv(C) * A);
    varwest = inv(M' * inv(C) * M);

    testat = west ./ sqrt(diag(varwest));
```

```
    wess = zeros(r,k);
    for i = 1 : r
        wess(i,:) = west((k*(i-1)+1):(k*i))';
        mtestat(i) = max(abs(testat((k*(i-1)+1):(k*i),:)));
    end
    [tmin rmin] = min(mtestat);

    if tmin < 0
       if r > 1
          if rmin == 1
              lst = lst(2:r);
          elseif rmin == r
              lst = lst(1:(r-1));
          else
              lst = [lst(1:(rmin-1)) lst((rmin+1):r)];
          end
       else
          XL = Y;
          lst = [];
          hay = 0;
          west = [];
          sdwestls = [];
       end
    else
        hay = 0;
        XL = Y;
        for i = 1 : r
            XL(lst(i):n,:) = XL(lst(i):n,:) - ones(n-lst(i)
+1,1) * wess(i,:);
        end
        sdwestls = sqrt(diag(varwest));
    end

end

XL = XL';
westls = west;

%%%%%%%%%%%%%%%%%%%%%%%%%

function [v,f] = max_kur(x,d0,maxit,show)


% DP/FJP 23/5/00

% Initialization
```

```
maxitdefault = 30;

if nargin < 4,
  show = 0;
end
if nargin < 3,
  maxit = maxitdefault;
end
if maxit <= 0,
  maxit = maxitdefault;
end
if nargin < 2,
  d0 = [];
end

%% Tolerances

maxitini = 1;
tol = 1.0e-4;
tol1 = 1.0e-7;
tol2 = 1.0e-2;
beta = 1.0e-4;
rho0 = 0.1;

[n,p] = size(x);

%% Initial estimate of the direction

if length(d0) == 0,

  uv = sum((x.*x)')';
  uw = 1../(eps + sqrt(uv));
  uu = x.*(uw*ones(1,p));

  Su = cov(uu);
  [V,D] = eig(Su);

  r = [];
  for i = 1:p,
    r = [ r val_kur(x,V(:,i)) ];
  end
  [v,ik] = max(r);
  a = V(:,ik(1));

  itini = 1;
```

```
  difa = 1;
  while (itini <= maxitini)&(difa > tol2),
    z = x*a;
    zaux = z.^2;
    xaux = x'.*(ones(p,1)*zaux');
    H = 12*xaux*x;
    [V,E] = eig(H);
    [vv,iv] = max(diag(E));
    aa = V(:,iv(1));
    difa = norm(a - aa);
    a = aa;
    itini = itini + 1;
  end
else
  a = d0/norm(d0);
end

%% Values at iteration 0 for the optimization algorithm

z = x*a;
sk = sum(z.^4);
lam = 2*sk;
f = sk;
g = (4*(z.^3)'*x)';
zaux = z.^2;
xaux = x'.*(ones(p,1)*zaux');
H = 12*xaux*x;

al = 0;
it = 0;
diff = 1;
rho = rho0;
clkr = 0;

c = 0;

% Newton method starting from the initial direction

if show,
  disp(' It. Obj.F. | g | c alpha rho');
end

while 1,

%% Check termination conditions
```

```
  gl = g - 2*lam*a;

  A = 2*a';
  [Q,W] = qr(a);
  Z = Q(:,(2:p));

  if show,
    aa = sprintf('%3.0f  %12.5e %13.4e',it,f,norm(gl));
    bb = sprintf(' %13.4e %8.3f %11.2e',abs(a'*a-1),al,rho);
    disp([ aa bb ]);
  end

  crit = norm(gl) + abs(c);
  if (crit <= tol)|(it >= maxit),
    break
  end

%% Compute search direction

  Hl = H - 2*lam*eye(p,p);
  Hr = Z'*Hl*Z;
  [V,E] = eig(Hr);
  Es = min(-abs(E),-1.0e-4);
  Hs = V*Es*V';

  py = - c/(A*A');
  rhs = Z'*(g + H*A'*py);
  pz = - Hs\rhs;
  pp = Z*pz + py*A';

  dlam = (2*a)\(gl + H*pp);

%% Adjust penalty parameter

  f0d = gl'*pp - 2*rho*c*a'*pp - dlam*c;
  crit1 = beta*norm(pp)^2;

  if f0d < crit1,
    rho1 = 2*(crit1 - f0d)/(eps + c^2);
    rho = max([2*rho1 1.5*rho rho0]);
    f = sk - lam*c - 0.5*rho*c^2;
    f0d = gl'*pp - 2*rho*c*a'*pp - dlam*c;
    clkr = 0;
  elseif (f0d > 1000*crit1)&(rho > rho0),
    rho1 = 2*(crit1 - gl'*pp + dlam*c)/(eps + c^2);
```

```
    if (clkr == 4)&(rho > 2*rho1),
      rho = 0.5*rho;
      f = sk - lam*c - 0.5*rho*c^2;
      f0d = gl'*pp - 2*rho*c*a'*pp - dlam*c;
      clkr = 0;
    else
      clkr = clkr + 1;
    end
  end
  if (abs(f0d/(norm(g-2*rho*a*c)+norm(c))) < tol1),
    break
  end

%% Line search

  al = 1;
  itbl = 0;
  while itbl < 20,
    aa = a + al*pp;
    lama = lam + al*dlam;
    zz = x*aa;
    cc = aa'*aa - 1;
    sk = sum(zz.^4);
    ff = sk - lama*cc - 0.5*rho*cc^2;

    if ff > f + 0.0001*al*f0d,
      break
    end
    al = al/2;
    itbl = itbl + 1;
  end
  if itbl >= 20,
    if show,
      disp('Error in the line search');
    end
    break
  end

%% Update values for the next iteration

  a = aa;
  lam = lama;
  z = zz;

  nmd2 = a'*a;
  c = nmd2 - 1;
```

```
  f = sk - lam*c - 0.5*rho*c^2;
  g = (4*(z.^3)'*x)';
  zaux = z.^2;
  xaux = x'.*(ones(p,1)*zaux');
  H = 12*xaux*x;

  it = it + 1;

end

% Values to be returned

v = a/norm(a);
xa = x*v;
f = sum(xa.^4)/n;

%%%%%%%%%%%%%%%%%%%%%%%%%

function [v,f] = min_kur(x,d0,maxit,show)

% Initialization

maxitdefault = 30;

if nargin < 4,
  show = 0;
end
if nargin < 3,
  maxit = maxitdefault;
end
if maxit <= 0,
  maxit = maxitdefault;
end
if nargin < 2,
  d0 = [];
end

%% Tolerances

maxitini = 1;
tol = 1.0e-4;
tol1 = 1.0e-7;
tol2 = 1.0e-2;
beta = 1.0e-4;
rho0 = 0.1;
```

```
[n,p] = size(x);

%% Initial estimate of the direction

if length(d0) == 0,

  uv = sum((x.*x)')';
  uw = 1../(eps + sqrt(uv));
  uu = x.*(uw*ones(1,p));

  Su = cov(uu);
  [V,D] = eig(Su);

  r = [];
  for i = 1:p,
    r = [ r val_kur(x,V(:,i)) ];
  end
  [v,ik] = min(r);
  a = V(:,ik(1));

  itini = 1;
  difa = 1;
  while (itini <= maxitini)&(difa > tol2),
    z = x*a;
    zaux = z.^2;
    xaux = x'.*(ones(p,1)*zaux');
    H = 12*xaux*x;
    [V,E] = eig(H);
    [vv,iv] = min(diag(E));
    aa = V(:,iv(1));
    difa = norm(a - aa);
    a = aa;
    itini = itini + 1;
  end
else
  a = d0/norm(d0);
end

%% Values at iteration 0 for the optimization algorithm

z = x*a;
sk = sum(z.^4);
lam = 2*sk;
f = sk;
```

```
g = (4*(z.^3)'*x)';
zaux = z.^2;
xaux = x'.*(ones(p,1)*zaux');
H = 12*xaux*x;

al = 0;
it = 0;
diff = 1;
rho = rho0;
clkr = 0;

c = 0;

% Newton method starting from the initial direction

if show,
  disp(' It. F.obj. | g | c alfa rho');
end

while 1,

%% Check termination conditions

  gl = g - 2*lam*a;

  A = 2*a';
  [Q,W] = qr(a);
  Z = Q(:,(2:p));

  if show,
    aa = sprintf('%3.0f  %12.5f %13.4e',it,f,norm(gl));
    bb = sprintf(' %13.4e %8.3f %11.2e',abs(a'*a-1),al,rho);
    disp([ aa bb ]);
  end

  crit = norm(gl) + abs(c);
  if (crit = maxit),
    break
  end

%% Compute search direction

  Hl = H - 2*lam*eye(p,p);
  Hr = Z'*Hl*Z;
```

```
  [V,E] = eig(Hr);
  Es = max(abs(E),1.0e-4);
  Hs = V*Es*V';

  py = - c/(A*A');
  rhs = Z'*(g + H*A'*py);
  pz = - Hs\rhs;
  pp = Z*pz + py*A';

  dlam = (2*a)\(gl + H*pp);

%% Adjust penalty parameter

  f0d = gl'*pp + 2*rho*c*a'*pp - dlam*c;
  crit1 = beta*norm(pp)^2;

  if f0d > -crit1,
    rho1 = 2*(crit1 + f0d)/(eps + c^2);
    rho = max([2*rho1 1.5*rho rho0]);
    f = sk - lam*c + 0.5*rho*c^2;
    f0d = gl'*pp + 2*rho*c*a'*pp - dlam*c;
    clkr = 0;
  elseif (f0d < -1000*crit1)&(rho > rho0),
    rho1 = 2*(crit1 - gl'*pp + dlam*c)/(eps + c^2);
    if (clkr == 4)&(rho > 2*rho1),
      rho = 0.5*rho;
      f = sk - lam*c + 0.5*rho*c^2;
      f0d = gl'*pp + 2*rho*c*a'*pp - dlam*c;
      clkr = 0;
    else
      clkr = clkr + 1;
    end
  end
  if (abs(f0d) < tol1),
    break
  end

%% Line search

  al = 1;
  itbl = 0;
  while itbl < 20,
    aa = a + al*pp;
    lama = lam + al*dlam;
    zz = x*aa;
```

```
    cc = aa'*aa - 1;
    sk = sum(zz.^4);
    ff = sk - lama*cc + 0.5*rho*cc^2;

    if ff < f + 0.0001*al*f0d,
      break
    end
    al = al/2;
    itbl = itbl + 1;
  end
  if itbl >= 20,
    if show,
      disp('Error in the line search');
    end
    break
  end

%% Update values for the next iteration

  a = aa;
  lam = lama;
  z = zz;

  nmd2 = a'*a;
  c = nmd2 - 1;
  f = sk - lam*c + 0.5*rho*c^2;
  g = (4*(z.^3)'*x)';
  zaux = z.^2;
  xaux = x'.*(ones(p,1)*zaux');
  H = 12*xaux*x;

  it = it + 1;

end

% Values to be returned

v = a/norm(a);
xa = x*v;
f = sum(xa.^4)/n;

%%%%%%%%%%%%%%%%%%%%%%%%%%

function [fi,theta,s2,C,FPE,AIC]=mlest(x,p,q)

x=x(:); %column
n=length(x);
```

```
model=armax(x,[p q]);

fi=get(model,'a');
theta=get(model,'c');
s2=get(model,'noisevar');
C=get(model,'cov');

info=get(model,'estim');
FPE=fpe(model);
AIC=aic(model);

si = length(fi);
if si > 1
    fi(2:si) = - fi(2:si);
end

%%%%%%%%%%%%%%%%%%%%%%%%%

function mawgt = parametrosma(phi,mxp);

if length(size(phi)) == 3
   [k k ip] = size(phi);
else
   [k1 k1] = size(phi);
   if k1 == 0
      ip = 0;
   else
      k = k1;
      ip = 1;
   end
end

mawgt(:,:,1) = phi(:,:,1);

if ip > 1
   for j1 = 2 : ip
       mawgt(:,:,j1) = zeros(k,k);
       for j2 = 1 : (j1 - 1)
          mawgt(:,:,j1) = mawgt(:,:,j1) + phi(:,:,j2) * mawgt
(:,:,j1 - j2);
       end
       mawgt(:,:,j1) = mawgt(:,:,j1) + phi(:,:,j1);
   end
end
```

```
for j1 = (ip + 1) : mxp
    mawgt(:,:,j1) = zeros(k,k);
    for j2 = 1 : ip
        mawgt(:,:,j1) = mawgt(:,:,j1) + phi(:,:,j2) * mawgt
(:,:,j1 - j2);
    end
end

%%%%%%%%%%%%%%%%%%%%%%%%%

function [YL,momtot,tiptot,westot,sdwestot] = restoatipicos
(Y,vcio,vcao,vctc);

warning off;

[k n] = size(Y);
PMAX = floor(log(n));
pmax = floor(sqrt(n));
hay = 1;
YL = Y';
de = 0.7;
tiptot = [];
momtot = [];
westot = [];
sdwestot = [];
iter = 0;

while hay == 1

    iter = iter + 1

    YT = YL;
    mm = mean(YT);
    S = cov(YT);
    YT = YT - ones(n,1) * mm;
    Rr = chol(S);
    YT = ((Rr')\(YT'))';
    V = kur_nwa(YT,0);
    hao = [];
    y = zeros(n,2*k);
    for c = 1 : (2*k)
        y(:,c) = YT * V(:,c);
    end
```

```
% Para los residuos del caso MIO

    for j = 1 : PMAX
        [cnst,stdcst,phiest,stdphi,resid,covresi] = varfit
(YL,j,zeros(k,k,j),zeros(1,k));
        aic(j) = n * log(det(covresi)) + 2 * j * k^2;
    end
    [aux MP] = min(aic);
    [cnst,stdcst,phiest,stdphi,resid,covresi] = varfit
(YL,MP,zeros(k,k,MP),zeros(1,k));

    A = resid;
    mm = mean(A);
    S = cov(A);
    A = A - ones(n,1) * mm;
    Rr = chol(S);
    A = ((Rr')\(A'))';
    VA = kur_nwa(A,0);
    for c = 1 : (2*k)
        a(:,c) = A * VA(:,c);
    end

% Comienza el algoritmo

    io = zeros(n,2*k);
    ao = zeros(n,2*k);
    tc = zeros(n,2*k);

    for c = 1 : (2*k)

        for j = 1 : pmax
            [fi,thet,s2,C,FPE,AIC] = mlest(y(:,c),j,0);
            bic(j) = n * log(s2) + log(n) * j;
        end
        [aux pap(c)] = min(bic);

        [fi,thet,s2,C,FPE,AIC] = mlest(y(:,c),pap(c),0);
        m = zeros(n,pap(c));
        for r = 1 : pap(c)
            m((r+1),:) = [y(r:-1:1,c)' zeros(1,pap(c)-r)];
        end
        for r = (pap(c) + 1) : n
            m(r,:) = y((r-1):-1:(r-pap(c)),c)';
        end
        res = y(:,c) - m * fi(2:(pap(c)+1))';
```

```
        res(1:pap(c)) = zeros(1,pap(c));

        s2 = sqrt(var(a(:,c)));
        io(:,c) = abs(a(:,c)./s2);

        Xa = [1;-fi(2:(pap(c)+1))'];
        for j = (pmax + 1) : (n - pap(c))
            Ya = res(j:(j+pap(c)));
            west = inv(Xa' * Xa) * Xa' * Ya;
            s2 = sqrt(var(res([1:(j-1) (j+1):n])));
            vwest = s2^2 * inv(Xa' * Xa);
            ao(j,c) = abs(west / sqrt(vwest));
        end

        fitc(1) = 1;
        fitc(2) = de - fi(2);
        if pap(c) > 1
           for j = 3 : (pap(c) + 1)
               fitc(j) = de * fitc(j - 1) - fi(j);
           end
        end
        for j = (pap(c) + 2) : n
           fitc(j) = de^(j - pap(c) - 1) * fitc(pap(c) + 1);
        end
        for j = 1 : n
            Yt = res(j:n);
            Xt = fitc(1:(n - j + 1))';
            west = inv(Xt' * Xt) * Xt' * Yt;
            s2 = sqrt(var(res([1:(j-1) (j+1):n])));
            vwest = s2^2 * inv(Xt' * Xt);
            tc(j,c) = abs(west / sqrt(vwest));
        end

    end

    [mio iio] = max(io);
    [mmio miio] = max(mio);
    hi = iio(miio);

    [mao iao] = max(ao);
    [mmao miao] = max(mao);
    ha = iao(miao);

    [mtc itc] = max(tc);
    [mmtc mitc] = max(mtc);
```

```
ht = itc(mitc);

[mmio hi mmao ha mmtc ht]

msig = [];
asig = [];
hsig = [];

if mmio > vcio
    asig = [asig 1];
    hsig = [hsig hi];
    msig = [msig mmio];
end

if mmao > vcao
    asig = [asig 2];
    hsig = [hsig ha];
    msig = [msig mmao];
end

if mmtc > vctc
    asig = [asig 4];
    hsig = [hsig ht];
    msig = [msig mmtc];
end

lhsig = length(hsig);

if lhsig == 0
   hay = 0;
end

if lhsig > 0

   [mmsig,imsig] = max(msig);
   tiposig = asig(imsig);
   momsig = hsig(imsig);

   if (tiposig == 1)

      tiptot = [tiptot 1];
      momtot = [momtot momsig];
      tamano = resid(momsig,:);
      westot = [westot;tamano];
      mawgt = parametrosma(phiest,n-momsig);
```

```
          YL(momsig,:) = YL(momsig,:) - tamano;
          if momsig < n
            for j = (momsig + 1) : n
                YL(j,:) = YL(j,:) - tamano * mawgt(:,:,j -
momsig)';
            end
          end
          sdwestot = [sdwestot;sqrt(diag(covresi)')];

       end

       if (tiposig == 2)

          tiptot = [tiptot 2];
          momtot = [momtot momsig];
          VAO = inv(covresi);
          estAO = inv(covresi) * resid(momsig,:)';
          if (momsig + MP) < (n + 1)
             for j = 1 : MP
               VAO = VAO + phiest(:,:,j)' * inv(covresi) *
phiest(:,:,j);
                estAO = estAO - phiest(:,:,j)' * inv(covresi)
* resid(momsig + j,:)';
             end
          else
             aux = n - momsig;
             for j = 1 : aux
                 VAO = VAO + phiest(:,:,j)' * inv(covresi) *
phiest(:,:,j);
                 estAO = estAO - phiest(:,:,j)' * inv(covresi)
* resid(momsig + j,:)';
             end
          end
          tamano = (inv(VAO) * estAO)';
          westot = [westot;tamano];
          YL(momsig,:) = YL(momsig,:) - tamano;
          sdwestot = [sdwestot;sqrt(diag(inv(VAO)))'];

       end

       if (tiposig == 4)

          tiptot = [tiptot 4];
          momtot = [momtot momsig];

          phiMTC(:,:,1) = phiest(:,:,1) - de * eye(k);
          if MP > 1
```

```
            for j = 2 : MP
               phiMTC(:,:,j) = phiest(:,:,j) + de * phiMTC
(:,:,j-1);
            end
         end
          for j = (MP + 1) : (n - MP)
             phiMTC(:,:,j) = de ^ (j - MP) * phiMTC(:,:,MP);
          end
          VTC = inv(covresi);
          estTC = inv(covresi) * resid(momsig,:)';
          if (momsig + MP) < (n + 1)
             for j = 1 : MP
             VTC = VTC + phiMTC(:,:,j)' * inv(covresi) *
phiMTC(:,:,j);
             estTC = estTC - phiMTC(:,:,j)' * inv(covresi) *
resid(momsig + j,:)';
             end
          else
             aux = n - momsig;
             for j = 1 : aux
               VTC = VTC + phiMTC(:,:,j)' * inv(covresi) *
phiMTC(:,:,j);
               estTC = estTC - phiMTC(:,:,j)' * inv(covresi)
* resid(momsig + j,:)';
             end
          end
          tamano = (inv(VTC) * estTC)';
          westot = [westot;tamano];
          YL(momsig,:) = YL(momsig,:) - tamano;
          if momsig < n
             for j = (momsig + 1) : n
              YL(j,:) = YL(j,:) - de ^ (j - momsig) * tamano;
             end
          end
          sdwestot = [sdwestot;sqrt(diag(inv(VTC)))'];

        end

     end

     [momtot' tiptot']

end

YL = YL';
```

```
%%%%%%%%%%%%%%%%%%%%%%%%%

function mc = val_kur(x,d,km)


if nargin < 3,
  km = 4;
end

[n,p] = size(x);
[p1,p2] = size(d);
if p ~= p1,
  disp('Data dimensions are not correct');
  return
end

t = x*d;
tm = mean(t);
tt = abs(t - tm);
vr = sum(tt.^2)/(n-1);
kr = sum(tt.^km)/n;
mc = kr/vr^(km/2);

%%%%%%%%%%%%%%%%%%%%%%%%%

function [cnst,stdcst,phi,stdphi,resi,covresi] = varfit(z,p,
idx,idxc);

[nob k] = size(z);
ip = p;
ist = p + 1;

[cnst,stdcst,phi,stdphi,resi] = estima(z,ip,ist,idx,idxc);
covresi = (nob/(nob-(ip*k+1)))*cov(resi);
```

Projects

1. For the data set, WW2b, used in Examples 2.4 and 2.5, fit a VAR(2) model to the five-dimensional vector $(1 - B^{12})\mathbf{Z}_t$ with $(n - 3)$ observations. Use the fitted model to do three-step ahead forecasts and evaluate the forecasts. Complete your analysis with a written report with your software code attached.
2. For the data set, WW2b, fit the models in Eqs. (2.99) and (2.103) with $(n - 3)$ observations. Use the fitted models to do three-step ahead forecasts and compare the forecast performance

of the two models and the model from Project 1. Complete your analysis with a written report including the software code used.

3. Perform a cointegration analysis on the variables given in WW2b and complete your analysis with a written report including the software code used.

4. Find a m-dimensional social science-related time series data set of your interest with m no less than 10. Complete your detailed analysis including model identification, estimation, and forecast with a written report and the software code for your analysis.

5. Find a m-dimensional time series data set of your interest. Perform multivariate outlier detections. Complete your detailed analysis with a written report and the software code for your analysis.

References

Abraham, B. and Box, G.E.P. (1979). Bayesian analysis of some outlier problems in time series. *Biometrika* **66**: 229–236.

Akaike, H. (1974). A new look at the statistical identification. *IEEE Transactions on Automatic Control* **AC-19**: 716–723.

Bartlett, M.S. (1955). *An Introduction to Stochastic Processes*. Cambridge University Press.

Box, G.E.P. and Cox, D.R. (1964). An analysis of transformations. *Journal of Royal Statistical Society, Series B* **26**: 211–252.

Box, G.E.P., Jenkins, G.M., Reinsel, G.C., and Ljung, G.M. (2015). *Time Series Analysis: Forecasting and Control*, 5e. Wiley.

Box, G.E.P. and Tiao, G.C. (1975). Intervention analysis with applications to economic and environmental problems. *Journal of American Statistical Association* **70**: 70–79.

Cai, B., Gao, J., and Tjostheim, D. (2017). A new class of bivariate threshold cointegration models. *Journal of Business & Economic Statistics* **35**: 288–305.

Casals, J., Garcia-Hiernaux, A., Jerez, M., Sotoca, S. and Trindade, A.A. (2016). *State–Space Methods for Time Series Analysis: Theory, Applications and Software*. CRC Press.

Cavaliere, G., Harvey, D., Leybourne, S.J., and Taylor, A.M.R. (2015). Testing for unit roots under multiple trend breaks and non-stationary volatility using bootstrap minimum Dickey–Fuller statistics. *Journal of Time Series Analysis* **36**: 603–629.

Chambers, M. (2015). Testing for a unit root in a near-integrated model with skip-sampled data. *Journal of Time Series Analysis* **36**: 630–649.

Chan, W.S., Cheung, S.H., Chow, W.K., and Zhang, L.X. (2015). A robust test for threshold-type nonlinearity in multivariate time-series analysis. *Journal of Forecasting* **34**: 441–454.

Chan, N.H., Ing, C.K., Li, Y., and Yau, C.Y. (2017). Threshold estimation via group orthogonal greedy algorithm. *Journal of Business and Economic Statistics* **35**: 334–345.

Chang, I., Tiao, G.C., and Chen, C. (1988). Estimation of time series parameters in the presence of outliers. *Technometrics* **30** (2): 193–204.

Chatfield, C. (2004). *The Analysis of Time Series, an Introduction*, 6e. Chapman & Hall/CRC.

Chen, R. and Tsay, R.S. (1993). Functional-coefficient autoregressive models. *Journal of the American Statistical Association* **88**: 298–308.

Chen, C. and Liu, L.M. (1993). Forecasting time series with outliers. *Journal of Forecasting* **12**: 13–35.

Cucina, D., Salvatore, A., and Protopapas, M.K. (2014). Outliers detection in multivariate time series using genetic algorithms. *Chemometrics and Intelligent Laboratory Systems* **132**: 103–110.

Dickey, D.A. and Fuller, W.A. (1979). Distribution of the estimates for autoregressive time series with a unit root. *Journal of American Statistical Association* **74**: 427–431.

Dickey, D.A., Hasza, D.P., and Fuller, W.A. (1984). Testing for unit roots in seasonal time series. *Journal of American Statistical Association* **79**: 355–367.

Eichler, M., Dahlhaus, R., and Dueck, J. (2017). Graphical modeling for multivariate Hawkes processes with nonparametric link functions. *Journal of Time Series Analysis* **38**: 225–242.

Engle, R.F. and Granger, C.W.J. (1987). Co-integration and error correction: representation, estimation, and testing. *Econometrica* **55**: 251–276.

EViews Enterprise Edition. June 2016.

Fox, A.J. (1972). Outliers in time series. *Journal of Royal Statistical Society, Series B* **43**: 350–363.

Galeano, P., Pena, D., and Tsay, R. (2006). Outlier detection in multivariate time series by projection pursuit. *Journal of the American Statistical Association* **101**: 654–669.

Garthoff, R. and Schmid, W. (2017). Monitoring means and covariances of multivariate nonlinear time series with heavy tails. *Communication of Statistics, Theory and Methods* **46**: 10394–10415.

Ghysels, E. and Miller, J.I. (2015). Testing for cointegration with temporally aggregated and mixed-frequency time series. *Journal of Time Series Analysis* **36**: 797–816.

Giurcanu, M.C. (2017). Oracle M-estimation for time series models. *Journal of Time Series Analysis* **38**: 479–504.

Granger, C.W.J. (1969). Investigating causal relations by econometric models and cross-spectral methods. *Econometrika* **37**: 424–438.

Granger, C.W.J. (1986). Developments in the study of co-integrated economic variables. *Oxford Bulletin of Economics and Statistics* **48**: 213–228.

Guo, W. (2003). Dynamic state-space models. *Journal of Time Series Analysis* **24**: 149–158.

Hamilton, J.D. (1994). *Time Series Analysis*. Princeton University Press.

Hannan, E.J. (1970). *Multiple Time Series*. Wiley.

Hawkes, A. (1971a). Spectra of some self-exciting and mutually exciting point processes. *Biometrika* **58**: 83–90.

Hawkes, A. (1971b). Point spectra of some mutually exciting point processes. *Journal of the Royal Statistical Society, Series B* **33**: 438–443.

Helbing, J. and Cleroux, R. (2009). On outlier detection in multivariate time series. *ACTA Mathematica Vietnamica* **34**: 19–26.

Hillmer, S.C., Bell, W.R., and Tiao, G.C. (1983). Modelling considerations in the seasonal adjustment of economic time series. In: *Applied Time Series Analysis of Economic Data* (ed. A. Zellner), 74–100. Oxford: U.S. Bureau of the Census.

Hosseinkouchack, M. and Hassler, U. (2016). Powerful unit root tests free of nuisance parameters. *Journal of Time Series Analysis* **37**: 533–554.

Jong, P. and Lin, S. (1994). Stationary and nonstationary state space models. *Journal of Time Series Analysis* **15**: 151–166.

Koop, G.M. (2013). Forecasting with medium and large Bayesian VARS. *Journal of Applied Econometrics* **28**: 177–203.

Lee, J.H. and Wei, W.W.S. (1995). A model-independent outlier detection procedure. *Journal of Applied Statistical Science* **2**: 345–359.

Li, J.S.H., Ng, A.C.Y., and Chan, W.S. (2015). Managing financial risk in Chinese stock markets: option pricing and modeling under a multivariate threshold autoregression. *International Review of Economics and Finance* **40**: 217–230.

Libert, G., Wang, L., and Liu, B. (1993). An innovation state space approach for time series forecasting. *Journal of Time Series Analysis* **14**: 589–601.

Little, T.D. (2013). *The Oxford Handbook of Quantitative Methods*. Oxford University Press.

Ljung, G.M. and Box, G.E.P. (1978). On a measure of lack of fit in time series models. *Biometrika* **65**: 297–303.

Lütkepohl, H. (1984). *Forecasting Contemporaneously Aggregated Vector ARMA Processes*. Springer-Verlag.

Lütkepohl, H. (1987). *Forecasting Aggregated Vector ARMA Processes*. Springer.
Lütkepohl, H. (2007). *New Introduction to Multiple Time Series Analysis*. Springer.
Martin, R.D. (1980). Robust estimation of autoregressive models. In: *Directions in Time Series* (ed. D.R. Brillinger and G.C. Tiao), 254–288. Hagwood, CA: Institute of Mathematical Statistics.
Martinez-Alvarez, F., Troncoso, A., Riquelme, J.C., and Aguilar-Ruiz, J.S. (2011). Discovery of motifs to forecast outlier occurrence in time series. *Pattern Recognition Letters* **32**: 1652–1665.
MATLAB Documentation. MathWorks. 2017.
Miller, J.I. and Wang, X. (2016). Implementing residual-based KPSS tests for cointegration with data subject to temporal aggregation and mixed sampling frequencies. *Journal of Time Series Analysis* **37**: 810–824.
Mills, T.C. (1991). *Time Series Techniques for Economists*. Cambridge University Press.
Murteira, B.J.F., Muller, D.A., and Turkman, K.F. (1993). *Analise de Sucessoes Cronologicas*. McGraw-Hill.
Nyberg, H. (2018). Forecasting US interest rates and business cycle with a nonlinear regime switching VAR model. *Journal of Forecasting* **37**: 1–15.
Parzen, E. (1977). Multiple time series modeling: determining the order of approximating autoregressive schemes. In: *Multivariate Analysis IV* (ed. P. Krishnaiah), 283–295. Amsterdam: North-Holland.
Pierce, D.A. and Haugh, L.D. (1977). Causality in temporal systems: characterizations and a survey. *Journal of Econometrics* **5**: 265–293.
Plakandaras, V. and Gogas, P. (2017). The information content of the term spread in forecasting the US inflation rate: a nonlinear approach. *Journal of Forecasting* **36**: 109–121.
Priestley, M.B. (1982). *Spectral Analysis and Time Series*. Academic Press.
R Foundation for Statistical Computing (2009). *R: A Language and Environment for Statistical Computing, Reference Index Version 2.10.1*. Vienna, Austria: R Foundation for Statistical Computing http://www.R-project.org.
Reinsel, G.C. and Ahn, S.K. (1992). Vector autoregressive models with unit roots and reduced rank structure: estimation, likelihood ratio test, and forecasting. *Journal of Time Series Analysis* **13**: 133–145.
Sanchez, M.J. and Pena, D. (2003). The identification of multiple outliers in ARIMA models. *Communications in Statistics, Theory and Methods* **32**: 1265–1287.
SAS Institute, Inc. (2015). *SAS for Windows, 9.4*. Cary, NC: SAS Institute, Inc.
SCA Corp. (2008). *SCA WorkBench User's Guide, Release 5.4*. Villa Park, IL: SCA Corp.
Schwartz, G. (1978). Estimating the dimension of a model. *Annals of Statistics* **6**: 461–464.
SPSS, Inc. (2009). *SPSS 15.0 for Windows*. Chicago, IL: SPSS, Inc.
Stock, J.H. and Watson, M.W. (2009). Forecasting in dynamic factor models subject to structural instability. In: *The Methodology and Practice of Econometrics* (ed. N. Shephard and J. Castle), 173–206. Oxford: Oxford University Press.
Teles, P., Hodgess, E., and Wei, W.W.S. (2008). Testing a unit root based on aggregate time series. *Communications in Statistics – Theory and Methods* **37**: 565–590.
Tsay, R.S. (1986). Time series model specification in the presence of outliers. *Journal of American Statistical Association* **81**: 132–141.
Tsay, R.S. (1988). Outliers, level shifts, and variance changes in time series. *Journal of Forecasting* **7**: 1–22.
Tsay, R.S. (2014). *Multivariate Time Series Analysis with R and Financial Applications*. Wiley.
Tsay, R.S., Pena, D., and Pankratz, A. (2000). Outliers in multivariate time series. *Biometrika* **87**: 789–804.
Tsay, R.S. and Tiao, G.C. (1984). Consistent estimates of autoregressive parameters and extended sample autocorrelation function for stationary and non-stationary ARIMA models. *Journal of American Statistical Association* **79**: 84–96.
Wagner, M. and Wied, D. (2017). Consistent monitoring of cointegrating relationships: the US housing market and the subprime crisis. *Journal of Time Series Analysis* **38**: 960–980.

Wang, W., Chow, S.C., and Wei, W.W.S. (1995). On likelihood distance for outlier detection. *Journal of Biopharmaceutical Statistics* **5**: 307–322.

Wang, T. and Samworth, R. (2018). High dimensional change point estimation via sparse projection. *Journal of the Royal Statistical Society, Series B* **80**: 57–83.

Wei, W.W.S. (2006). *Time Series Analysis – Univariate and Multivariate Methods*, 2e. Boston, MA: Pearson Addison-Wesley.

Wei, W.W.S. (2013). Time series analysis. In: *The Oxford Handbook of Quantitative Methods* (ed. T.D. Little), 458–485. Oxford University Press.

Wei, W.W.S. (2015). Issues related to the use of time series in model building and analysis: review article. *Communications for Statistical Applications and Methods* **22**: 209–222.

Wong, S.F., Tong, H., Siu, T.K., and Lu, Z. (2017). A new multivariate nonlinear time series model for portfolio risk measurement: the threshold copula-based TAR approach. *Journal of Time Series Analysis* **38**: 243–265.

Yozgatligil, C. and Wei, W.W.S. (2009). Representation of multiplicative seasonal vector autoregressive moving average models. *The American Statistician* **63** (4): 328–334.

Zhang, T. and Yang, B. (2017). Box–Cox transformation in big data. *Technometrics* **59**: 189–201.

3

Multivariate time series regression models

Regression analysis is one of the most commonly used statistical methods. It is covered in most undergraduate and graduate statistical courses. However, the method discussed in these courses is the standard multiple regression model with one response variable. In this chapter, we will introduce multivariate time series regression models with several response variables. We will illustrate this method using many examples.

3.1 Introduction

In this chapter, we will discuss several different formulations of multivariate time series regression models. The multiple regression is one of the most commonly used statistical models, so we will start with its multivariate representation in the next section. Other extensions and representations will be introduced in Sections 3.3 and 3.4. They include the representation adapted from the vector autoregressive models, which will be referred to as vector time series regression models. The VARX model is another extension. We will discuss the similarities and differences among these extensions and presentations.

3.2 Multivariate multiple time series regression models

3.2.1 The classical multiple regression model

In a multiple regression model, a response variable Y is related to k predictor variables, $X_1, X_2, \ldots, X_k$, as follows,

$$Y = \beta_0 + \beta_1 X_1 + \cdots + \beta_k X_k + \xi, \tag{3.1}$$

Multivariate Time Series Analysis and Applications, First Edition. William W.S. Wei.
© 2019 John Wiley & Sons Ltd. Published 2019 by John Wiley & Sons Ltd.
Companion website: www.wiley.com/go/wei/datasets

where ξ is assumed to be uncorrelated white noise, often as i.i.d. $N(0, \sigma^2)$. When time series data are used to fit a multiple regression model, we often write Eq. (3.1) as

$$\begin{aligned} Y_t &= \beta_0 + \beta_1 X_{1,t} + \cdots + \beta_k X_{k,t} + \xi_t \\ &= \mathbf{X}_t' \boldsymbol{\beta} + \xi_t \end{aligned} \tag{3.2}$$

where t refers to time,

$$\mathbf{X}_t' = [1, X_{1,t}, X_{2,t}, \ldots, X_{k,t}]$$
$$\boldsymbol{\beta} = [\beta_0, \beta_1, \beta_2, \ldots, \beta_k]'$$

and in time series regression ξ_t is normally assumed to follow a time series model such as AR(p).

When we have time series data from time $t = 1$ to $t = n$, we can present Eq. (3.2) in the matrix form,

$$\underset{n \times 1}{\mathbf{Y}} = \underset{(n \times (k+1))}{\mathbf{X}} \underset{(k+1) \times 1}{\boldsymbol{\beta}} + \underset{(n \times 1)}{\boldsymbol{\xi}}, \tag{3.3}$$

where

$$\mathbf{Y} = \begin{bmatrix} Y_1 \\ Y_2 \\ \vdots \\ Y_n \end{bmatrix}, \quad \mathbf{X} = \begin{bmatrix} \mathbf{X}_1' \\ \mathbf{X}_2' \\ \vdots \\ \mathbf{X}_n' \end{bmatrix} = \begin{bmatrix} 1 & X_{1,1} & X_{2,1} & \cdots & X_{k,1} \\ 1 & X_{1,2} & X_{2,2} & \cdots & X_{k,2} \\ \vdots & \vdots & \vdots & \ddots & \vdots \\ 1 & X_{1,n} & X_{2,n} & \cdots & X_{k,n} \end{bmatrix}, \quad \boldsymbol{\beta} = \begin{bmatrix} \beta_0 \\ \beta_1 \\ \vdots \\ \beta_k \end{bmatrix}, \quad \boldsymbol{\xi} = \begin{bmatrix} \xi_1 \\ \xi_2 \\ \vdots \\ \xi_n \end{bmatrix},$$

and $\boldsymbol{\xi}$ follows a n-dimensional multivariate normal distribution $N(\mathbf{0}, \boldsymbol{\Sigma})$. Given $\boldsymbol{\Sigma}$, the generalized least squares estimator (GLS)

$$\hat{\boldsymbol{\beta}} = \left(\mathbf{X}' \boldsymbol{\Sigma}^{-1} \mathbf{X}\right)^{-1} \mathbf{X}' \boldsymbol{\Sigma}^{-1} \mathbf{Y} \tag{3.4}$$

is known to be the best unbiased estimator in the sense that for any constant vector $\mathbf{c}$, the estimator

$$\mathbf{c}' \hat{\boldsymbol{\beta}} = c_0 \hat{\beta}_0 + c_1 \hat{\beta}_1 + \cdots + c_k \hat{\beta}_k$$

has the smallest possible variance among all other unbiased estimators $\tilde{\boldsymbol{\beta}}$ of $\boldsymbol{\beta}$ in the form

$$\mathbf{c}' \tilde{\boldsymbol{\beta}} = c_0 \tilde{\beta}_0 + c_1 \tilde{\beta}_1 + \cdots + c_k \tilde{\beta}_k.$$

3.2.2 Multivariate multiple regression model

Now, suppose that instead of one response variable in Eq. (3.2), we have m response time series variables related to these k predictor time series variables, that is,

$$
\begin{aligned}
Y_{1,t} &= \beta_{1,0} + \beta_{1,1}X_{1,t} + \cdots \beta_{1,k}X_{k,t} + \varepsilon_{1,t} = \mathbf{X}_t'\boldsymbol{\beta}_{(1)} + \xi_{1,t} \\
Y_{2,t} &= \beta_{2,0} + \beta_{2,1}X_{1,t} + \cdots \beta_{2,k}X_{k,t} + \varepsilon_{2,t} = \mathbf{X}_t'\boldsymbol{\beta}_{(2)} + \xi_{2,t} \\
&\vdots \\
Y_{m,t} &= \beta_{m,0} + \beta_{m,1}X_{1,t} + \cdots \beta_{m,k}X_{k,t} + \varepsilon_{m,t} = \mathbf{X}_t'\boldsymbol{\beta}_{(m)} + \xi_{m,t},
\end{aligned}
$$

or

$$\mathbf{Y}_t' = \mathbf{X}_t'\left[\boldsymbol{\beta}_{(1)}, \boldsymbol{\beta}_{(2)}, \ldots, \boldsymbol{\beta}_{(m)}\right] + \boldsymbol{\xi}_t', \tag{3.5}$$

where

$$
\begin{aligned}
\mathbf{Y}_t' &= [Y_{1,t}, Y_{2,t}, \ldots, Y_{m,t}], \\
\mathbf{X}_t' &= [1, X_{1,t}, X_{2,t}, \ldots, X_{m,t}], \\
\boldsymbol{\beta}_{(i)} &= \left[\beta_{i,0}, \beta_{i,1}, \ldots \beta_{i,k}\right]', i = 1, 2, \ldots, m,
\end{aligned}
$$

and

$$\boldsymbol{\xi}_{\mathbf{t}}' = \left[\xi_{1,t}, \xi_{2,t}, \ldots, \xi_{m,t}\right].$$

For $i = 1, 2, \ldots, m$ and time $t = 1$ to $t = n$, let

$$
\mathbf{Y}_{(i)} = \begin{bmatrix} Y_{i,1} \\ Y_{i,2} \\ \vdots \\ Y_{i,n} \end{bmatrix}, \quad \mathbf{X} = \begin{bmatrix} \mathbf{X}_1' \\ \mathbf{X}_2' \\ \vdots \\ \mathbf{X}_n' \end{bmatrix} = \begin{bmatrix} 1 & X_{1,1} & X_{2,1} & \cdots & X_{k,1} \\ 1 & X_{1,2} & X_{2,2} & \cdots & X_{k,2} \\ \vdots & \vdots & \vdots & \ddots & \vdots \\ 1 & X_{1,n} & X_{2,n} & \cdots & X_{k,n} \end{bmatrix}, \text{and } \boldsymbol{\xi}_{(i)} = \begin{bmatrix} \xi_{i,1} \\ \xi_{i,2} \\ \vdots \\ \xi_{i,n} \end{bmatrix}.
$$

The matrix form of the multiple regression for the ith response variable of $\mathbf{Y}_t'$ is

$$\mathbf{Y}_{(i)} = \mathbf{X}\boldsymbol{\beta}_{(i)} + \boldsymbol{\xi}_{(i)}, \tag{3.6}$$

which, as expected, is exactly the same as Eq. (3.3). Putting all the multiple regressions for the m response variables together from $t = 1$ to $t = n$, we have

$$\underset{n\times m}{\mathbf{Y}} = \underset{(n\times(k+1))}{\mathbf{X}} \; \underset{(k+1)\times m}{\boldsymbol{\beta}} + \underset{(n\times m)}{\boldsymbol{\xi}}, \tag{3.7}$$

where

$$
\mathbf{Y} = \begin{bmatrix} \mathbf{Y}_1' \\ \mathbf{Y}_2' \\ \vdots \\ \mathbf{Y}_n' \end{bmatrix} = \begin{bmatrix} Y_{1,1} & Y_{2,1} & \cdots & Y_{m,1} \\ Y_{1,2} & Y_{2,2} & \cdots & Y_{m,2} \\ \vdots & \vdots & \ddots & \vdots \\ Y_{1,n} & Y_{2,n} & \cdots & Y_{m,n} \end{bmatrix} = \left[\mathbf{Y}_{(1)}, \mathbf{Y}_{(2)}, \ldots, \mathbf{Y}_{(m)}\right],
$$

$$\boldsymbol{\beta} = \begin{bmatrix} \beta_{1,0} & \beta_{2,0} & \cdots & \beta_{m,0} \\ \beta_{1,1} & \beta_{2,1} & \cdots & \beta_{m,1} \\ \vdots & \vdots & \ddots & \vdots \\ \beta_{1,k} & \beta_{2,k} & \cdots & \beta_{m,k} \end{bmatrix} = \left[\boldsymbol{\beta}_{(1)}, \boldsymbol{\beta}_{(2)}, \ldots, \boldsymbol{\beta}_{(m)}\right],$$

and

$$\boldsymbol{\xi} = \begin{bmatrix} \boldsymbol{\xi}_1' \\ \boldsymbol{\xi}_2' \\ \vdots \\ \boldsymbol{\xi}_n' \end{bmatrix} = \begin{bmatrix} \xi_{1,1} & \xi_{2,1} & \cdots & \xi_{m,1} \\ \xi_{1,2} & \xi_{2,2} & \cdots & \xi_{m,2} \\ \vdots & \vdots & \ddots & \vdots \\ \xi_{1,n} & \xi_{2,n} & \cdots & \xi_{m,n} \end{bmatrix} = \left[\boldsymbol{\xi}_{(1)}, \boldsymbol{\xi}_{(2)}, \ldots, \boldsymbol{\xi}_{(m)}\right].$$

Each $\boldsymbol{\xi}_{(i)}$ follows a n-dimensional multivariate normal distribution $N(\mathbf{0}, \boldsymbol{\Sigma}_{(i)})$, $i = 1, \ldots, m$, and $\boldsymbol{\xi}_{(i)}$ and $\boldsymbol{\xi}_{(j)}$ are uncorrelated for $i \neq j$. We will call the model given in Eq. (3.7) the multivariate multiple time series regression model.

3.3 Estimation of the multivariate multiple time series regression model

3.3.1 The Generalized Least Squares (GLS) estimation

As noted in Eq. (3.6), the ith response $\mathbf{Y}_{(i)}$ actually follows the general multiple time series regression model

$$Y_{i,t} = \beta_{i,0} + \beta_{i,1}X_{1,t} + \cdots + \beta_{i,k}X_{k,t} + \varepsilon_{i,t} = \mathbf{X}_t'\boldsymbol{\beta}_{(i)} + \xi_{i,t}, t = 1, \ldots n, \tag{3.8}$$

or

$$\mathbf{Y}_{(i)} = \mathbf{X}\boldsymbol{\beta}_{(i)} + \boldsymbol{\xi}_{(i)}, \tag{3.9}$$

where $\boldsymbol{\xi}_{(i)} = [\xi_{i,1}, \xi_{i,2}, \ldots, \xi_{i,n}]'$ follows a n-dimensional multivariate normal distribution $N(\mathbf{0}, \boldsymbol{\Sigma}_{(i)})$. In the time series regression, $\xi_{i,t}$ is often assumed to follow a time series model such as AR(p). From the results of the multiple regression, we know that when $\boldsymbol{\Sigma}_{(i)}$ is known, the GLS estimator

$$\hat{\boldsymbol{\beta}}_{(i)} = \left(\mathbf{X}'\boldsymbol{\Sigma}_{(i)}^{-1}\mathbf{X}\right)^{-1}\mathbf{X}'\boldsymbol{\Sigma}_{(i)}^{-1}\mathbf{Y}_{(i)} \tag{3.10}$$

is the best unbiased estimator.

Normally, we will not know the variance–covariance matrix $\boldsymbol{\Sigma}_{(i)}$ of $\boldsymbol{\xi}_{(i)}$. Even if $\xi_{i,t}$ follows a time series model such as AR(p) or ARMA(p, q), the $\boldsymbol{\Sigma}_{(i)}$ structure is not known because the related time series model parameters are usually unknown. In this case, we use the following GLS procedure suggested in Wei (2006, Chapter 15):

Step 1: Calculate the ordinary least squares (OLS) residuals $\hat{\xi}_{i,t}$ from the OLS fitting of the model in (3.8).

Step 2: Estimate the parameters of the assumed time series model based on the OLS residuals $\hat{\xi}_{i,t}$.

Step 3: Compute $\hat{\boldsymbol{\Sigma}}_{(i)}$ from the estimated model for $\xi_{i,t}$ obtained in step 2.

Step 4: Compute the GLS estimator, $\hat{\boldsymbol{\beta}}_{(i)} = \left(\mathbf{X}'\hat{\boldsymbol{\Sigma}}_{(i)}^{-1}\mathbf{X}\right)^{-1}\mathbf{X}'\hat{\boldsymbol{\Sigma}}_{(i)}^{-1}\mathbf{Y}_{(i)}$, using the $\hat{\boldsymbol{\Sigma}}_{(i)}$ obtained in step 3.

Compute the residuals $\hat{\xi}_{i,t}$ from the GLS model fitting in step 4, and repeat step 1 through step 4 until a convergence criterion (such as the maximum absolute value change in the estimates between iterations becomes less than a specified quantity) is reached.

Combining $\hat{\boldsymbol{\beta}}_{(i)}$ for $i = 1, \ldots, m$, we get

$$\underset{n\times m}{\hat{\mathbf{Y}}} = \underset{(n\times(k+1))}{\mathbf{X}}\ \underset{(k+1)\times m}{\hat{\boldsymbol{\beta}}}, \tag{3.11}$$

where

$$\hat{\boldsymbol{\beta}} = \left[\hat{\boldsymbol{\beta}}_{(1)}, \hat{\boldsymbol{\beta}}_{(2)}, \ldots, \hat{\boldsymbol{\beta}}_{(m)}\right] = \begin{bmatrix} \hat{\beta}_{1,0} & \hat{\beta}_{2,0} & \cdots & \hat{\beta}_{m,0} \\ \hat{\beta}_{1,1} & \hat{\beta}_{2,1} & \cdots & \hat{\beta}_{m,1} \\ \vdots & \vdots & \ddots & \vdots \\ \hat{\beta}_{1,k} & \hat{\beta}_{2,k} & \cdots & \hat{\beta}_{m,k} \end{bmatrix},$$

and the estimate of the variance–covariance matrix $\boldsymbol{\Sigma}_{(i)}$ of $\boldsymbol{\xi}_{(i)}$ is given by step 3 in the last GLS iteration.

It should be pointed out that although the error term can be autocorrelated in the time series regression model, it should be stationary. A nonstationary error structure could produce a spurious regression where a significant regression can be achieved for totally unrelated series as pointed out by Abraham and Ledolter (2006), Chatterjee, Hadi, and Price (2006), Draper and Smith (1998), Granger and Newbold (1986), and Phillips (1986).

3.3.2 Empirical Example I – U.S. retail sales and some national indicators

Example 3.1 In this example, we will first examine the OLS regression of U.S. retail sales and some important national indicator variables, including *GDP* (gross domestic product), *DJIA* (Dow Jones Industrial Average), and average *CPI* (consumer price index) from 1992 to 2015. The data set is the WW3a series listed in the Data Appendix and given in Table 3.1 with the plot given in Figure 3.1.

After examining the figure, we will first try the following regression equations

Table 3.1 Multivariate regression data set.

Year t	$Y_{1,t}$	$Y_{2,t}$	$Y_{3,t}$	$Y_{4,t}$	GDP_t	GDP_{t-1}	GDP_{t-2}	$DJIA_{t-1}$	CPI_t
1992	418 393	370 513	89 705	247 876	6.539	6.174	5.98	3 168.8	140.300
1993	472 916	374 516	92 594	265 996	6.879	6.539	6.174	3 301.1	144.500
1994	541 141	384 340	96 363	285 190	7.309	6.879	6.539	3 754.1	148.200
1995	579 715	390 386	101 635	300 498	7.664	7.309	6.879	3 834.4	152.400
1996	627 507	401 073	109 557	315 305	8.100	7.664	7.309	5 117.1	156.900
1997	653 817	409 373	118 672	331 363	8.609	8.100	7.664	6 448.3	160.500
1998	688 415	416 525	129 583	351 081	9.089	8.609	8.100	7 908.3	163.000
1999	764 204	433 699	142 699	380 179	9.661	9.089	8.609	9 181.4	166.600
2000	796 210	444 764	155 234	404 228	10.285	9.661	9.089	11 497.1	172.200
2001	815 579	462 429	166 533	427 468	10.622	10.285	9.661	10 786.9	177.100
2002	818 811	464 856	179 983	446 520	10.978	10.622	10.285	10 021.5	179.900
2003	841 588	474 385	192 426	468 771	11.511	10.978	10.622	8 341.6	184.000
2004	866 372	490 380	199 290	497 382	12.275	11.511	10.978	10 453.9	188.900
2005	888 307	508 484	210 085	528 385	13.094	12.275	11.511	10 783	195.300
2006	899 997	525 232	223 336	554 256	13.856	13.094	12.275	10 717.5	201.600
2007	910 139	547 837	237 164	578 582	14.478	13.856	13.094	12 463.2	207.342
2008	785 865	569 276	246 573	595 041	14.719	14.478	13.856	13 264.8	215.303
2009	671 772	568 418	252 794	588 918	14.419	14.719	14.478	8 776.4	214.537
2010	742 913	580 530	260 435	603 757	14.964	14.419	14.719	10 428.1	218.056
2011	812 938	609 137	271 612	624 766	15.518	14.964	14.419	11 577.5	224.939
2012	886 494	628 205	274 000	642 313	16.155	15.518	14.964	12 217.6	229.594
2013	959 294	640 847	281 840	651 874	16.692	16.155	15.518	13 104.1	232.957
2014	1 020 851	669 165	299 263	667 163	17.393	16.692	16.155	16 576.7	236.736
2015	1 095 412	685 568	315 257	674 928	18.037	17.393	16.692	17 823.1	237.017

Data sources: sales data (in millions) from United States Census of Bureau; *GDP* (in trillions) from World Bank national accounts data and OECD Nation Accounts data files; *CPI* ([1982–1984] = 100) from the U.S. Department of Labor, Bureau of Labor Statistics.

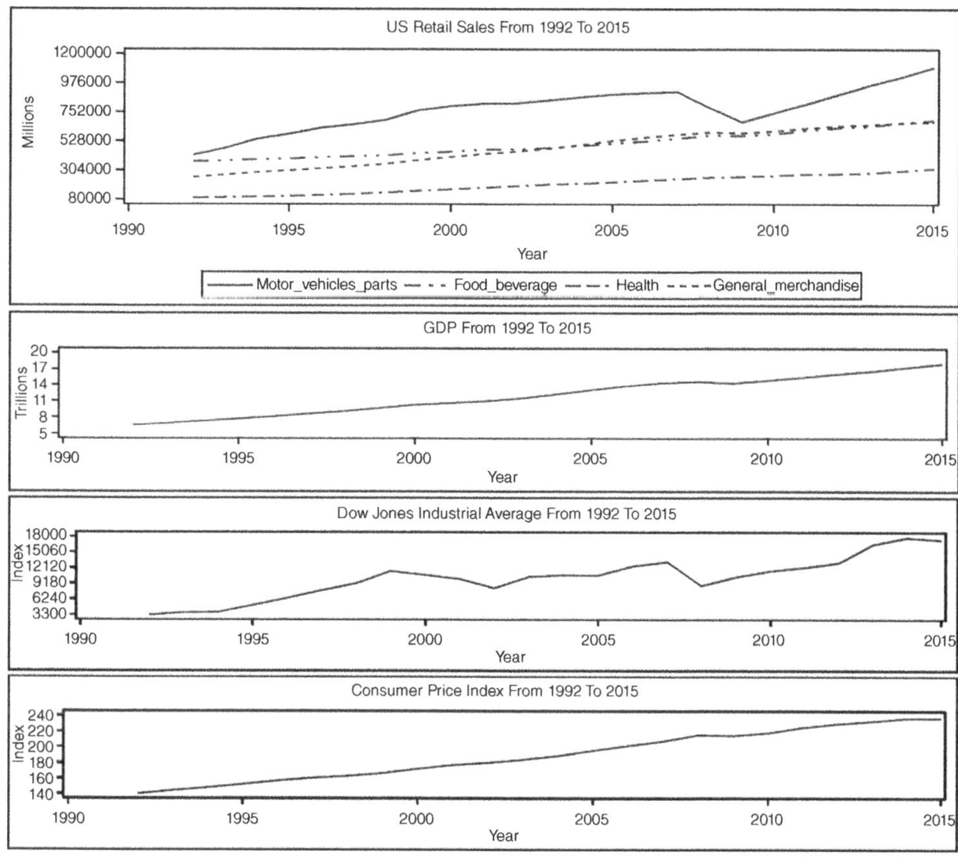

Figure 3.1 U.S. retail sales, *GDP*, *DJIA*, and *CPI*.

$$Y_{i,t} = \alpha_i + \beta_{i,1}GDP_t + \beta_{i,2}GDP_{t-1} + \beta_{i,3}GDP_{t-2} + \beta_{i,4}DJIA_{t-1} + \beta_{i,5}CPI_t + \xi_{i,t}, \tag{3.12}$$

where

$Y_{1,t}$ = sales of motor vehicle and parts dealers

$Y_{2,t}$ = sales of food and beverage stores

$Y_{3,t}$ = sales of health care and personal care stores

$Y_{4,t}$ = sales of general merchandise stores

GDP_t = GDP (trillion) at year t

$DJIA_{t-1}$ = Dow Jones Industrial Average at beginning of the year

CPI_t = average CPI of the year (1982–1984 = 100)

and the $\xi_{i,t}$ follows independent $N\left(0, \sigma_i^2\right)$. In terms of the OLS regression, Eq. (3.12) becomes

$$Y_{i,t} = \mathbf{X}_t \boldsymbol{\beta}_{(i)} + \xi_{i,t}, \tag{3.13}$$

where

$$\mathbf{X}_t = [1\ GDP_t\ GDP_{t-1}\ GDP_{t-2}\ DJIA_{t-1}\ CPI_t], \boldsymbol{\beta}_{(i)} = \begin{bmatrix} \alpha_i \\ \beta_{i,1} \\ \beta_{i,2} \\ \beta_{i,3} \\ \beta_{i,4} \\ \beta_{i,5} \end{bmatrix},$$

and $\xi_{i,t}$ follows $i.i.d.N(0, \sigma_i^2)$. The estimation results are given in Table 3.2.

The estimate of the variance–covariance matrix of the residual vectors is

$$\hat{\boldsymbol{\Sigma}} = \begin{bmatrix} 2547210725.5 & -214607937.5 & 20407059.72 & 176611585.98 \\ -214607937.5 & 189615009.67 & -1531483.06 & -167931270.7 \\ 20407059.72 & -1531483.06 & 18190562.48 & 19143159.42 \\ 176611585.98 & -167931270.7 & 19143159.42 & 182639666.60 \end{bmatrix}.$$

The models fit very well with very large R-squares. All sales are positively correlated with current year GDP. However, surprisingly most parameter estimates are not significant. Removing these predictor variables with p-value more than 0.2, we get the following estimation results in Table 3.3.

Table 3.2 OLS estimation results.

	$Y_{1,t}$	$Y_{2,t}$	$Y_{3,t}$	$Y_{4,t}$
$\hat{\alpha}_i$	777 627	−27 085	−5201.66	−54 006
(SE)	(467 638)	(127 589)	(39 519)	(125 220)
$\hat{\beta}_{i,1}(GDP_t)$	175 401	4173.50	13 975	29 242
(SE)	(61 485)	(16 775)	(5195.88)	(16 464)
$\hat{\beta}_{i,2}(GDP_{t-1})$	−36 457	−7546.19	−4211.73	9150.61
(SE)	(84 872)	(23 156)	(7172.27)	(22 726)
$\hat{\beta}_{i,3}(GDP_{t-2})$	−75 016	7829.35	14 931	−2403.84
(SE)	(56 716)	(15 474)	(4792.86)	(15187)
$\hat{\beta}_{i,4}(DJIA_{t-1})$	23.907	1.248	−0.0428	−1.868
(SE)	(7.967)	(2.174)	(0.6733)	(2.133)
$\hat{\beta}_{i,5}(CPI_t)$	−5779.45	2466.3	−451.62	619.56
(SE)	(5541.27)	(1511.87)	(468.27)	(1483.80)
R^2	0.929	0.985	0.997	0.993
$F(5,18)$	46.77	232.47	1324.91	503.67
$(p-value)$	(<0.0001)	(<0.0001)	(<0.0001)	(<0.0001)

Table 3.3 OLS estimation results with nonsignificant predictor variables removed.

	$Y_{1,t}$	$Y_{2,t}$	$Y_{3,t}$	$Y_{4,t}$
$\hat{\alpha}_i$	386 816	−83 242	−43 672	317.61
(SE)	(33 107)	(16 145)	(3282.28)	(9665.49)
$\hat{\beta}_{i,1}\,(GDP_t)$			9776.48	39 551
(SE)			(2268.01)	(771.18)
$\hat{\beta}_{i,2}\,(GDP_{t-1})$				
(SE)				
$\hat{\beta}_{i,3}\,(GDP_{t-2})$			10 810	
(SE)			(2333.61)	
$\hat{\beta}_{i,4}\,(DJIA_{t-1})$	39.86			
(SE)	(3.188)			
$\hat{\beta}_{i,5}\,(CPI_t)$		3088.86		
(SE)		(84.08)		
R^2	0.877	0.984	0.997	0.992
$F(5,18)$	156.3	1349.55	3568.57	2630.29
(p−value)	(<0.0001)	(<0.0001)	(<0.0001)	(<0.0001)

The results are quite surprising. All nonsignificant constant terms become significant except for $Y_{4,t}$. All sales are positively related to only one predictor variable with fitted R^2 values almost equal to those fitted with five predictor variables that were used in Table 3.2. The sales of motor vehicle and parts dealers ($Y_{1,t}$) are highly related to Dow Jones Industrial Average at the beginning of the year ($DJIA_{t-1}$). The sales of food and beverage stores ($Y_{2,t}$) is positively related to average *CPI* of the year (CPI_t) with R^2 value equal to 0.984. The sales of health care and personal care stores ($Y_{3,t}$) at year t are positively related to *GDP* at year t (GDP_t) and year $t - 2$ (GDP_{t-2}) with R^2 value almost 1, and their sales are not related to the previous year's *GDP*. Lastly, the sales of general merchandise stores ($Y_{4,t}$) are highly and positively related to the current year *GDP* (GDP_t) with fitted R^2 value almost 1.

Example 3.2 The residuals from the OLS regression in Example 3.1 are not quite white noise. In fact, they can all be approximated by an AR(1) process. We will improve the OLS estimation with GLS described in Section 3.3.1 on the model

$$\mathbf{Y}_{(i)} = \mathbf{X}\boldsymbol{\beta}_{(i)} + \boldsymbol{\xi}_{(i)}, \quad i = 1,\ 2,\ 3,\ 4 \tag{3.14}$$

where $\boldsymbol{\xi}_{(i)} = [\xi_{i,1}, \xi_{i,2}, \ldots, \xi_{i,n}]'$ follows a n-dimensional multivariate normal distribution $N(\mathbf{0}, \boldsymbol{\Sigma}_{(i)})$, and $\boldsymbol{\Sigma}_{(i)}$ is a variance–covariance matrix from AR(1) process, $\xi_{i,t} = \phi^{(i)}\xi_{i,t-1} + a_{i,t}$. The GLS estimate of $\boldsymbol{\beta}_{(i)}$ is

$$\hat{\boldsymbol{\beta}}_{(i)} = \left(\mathbf{X}'\hat{\boldsymbol{\Sigma}}_{(i)}^{-1}\mathbf{X}\right)^{-1}\mathbf{X}'\hat{\boldsymbol{\Sigma}}_{(i)}^{-1}\mathbf{Y}_{(i)}, \tag{3.15}$$

where $\hat{\boldsymbol{\Sigma}}_{(i)}^{-1}$ is the estimated variance–covariance matrix of the residuals from the fitted multiple regression of $\hat{\mathbf{Y}}_{(i)}$ in Example 3.1. The GLS results from SAS AUTOREG procedure are given in Table 3.4.

Table 3.4 Generalized least squares estimation results.

	$Y_{1,t}$	$Y_{2,t}$	$Y_{3,t}$	$Y_{4,t}$
$\hat{\alpha}_i$	863 639	81 461	−25 915	−174 716
(*SE*)	(374 348)	(68 777)	(40 269)	(59 045)
$\hat{\beta}_{i,1}(GDP_t)$	199 868	4 895	10 639	23 050
(*SE*)	(43 092)	(7 493)	(5 268)	(6 399)
$\hat{\beta}_{i,2}(GDP_{t-1})$	−13 260	9519	2 977	−2 276
(*SE*)	(41 573)	(7 200)	(5 465)	(6 147)
$\hat{\beta}_{i,3}(GDP_{t-2})$	−66 740	60.3340	8 829	333.9967
(*SE*)	(31 237)	(5 423)	(4 013)	(4631)
$\hat{\beta}_{i,4}(DJIA_{t-1})$	1.4737	1.4579	−0.1895	−1.2119
(*SE*)	(6.2360)	(1.0875)	(0.7394)	(0.9288)
$\hat{\beta}_{i,5}(CPI_t)$	−8 657	1 320	−202.9449	2 097
(*SE*)	(4 035)	(707.1509)	(470.5135)	(604.1294)
R^2	0.971	0.997	0.998	0.999
$\hat{\phi}^{(i)}$ used	−0.8917	−0.9448	−0.5545	−0.9485

Compared to the results in Table 3.2, we can see that the R^2 values using the GLS are much higher. More importantly, a few more parameter estimates become significant and meaningful.

3.4 Vector time series regression models

3.4.1 Extension of a VAR model to VARX models

Recall from Chapter 2 that the m-dimensional vector autoregressive model, VAR(p), is given by

$$\mathbf{Z}_t = \boldsymbol{\theta}_0 + \boldsymbol{\Phi}_1 \mathbf{Z}_{t-1} + \cdots + \boldsymbol{\Phi}_p \mathbf{Z}_{t-p} + \mathbf{a}_t, \tag{3.16}$$

where $\boldsymbol{\theta}_0$ is a $m \times 1$ constant vector, $\boldsymbol{\Phi}_i$ are $m \times m$ parameter coefficient matrices, $\mathbf{a}_t$ is a sequence of m-dimensional vector white noise process, $\mathbf{VWN}(\mathbf{0}, \boldsymbol{\Sigma})$. Eq. (3.16) can be extended to the following

$$\mathbf{Y}_t = \boldsymbol{\theta}_0 + \boldsymbol{\Phi}_1 \mathbf{X}_{1,t} + \cdots + \boldsymbol{\Phi}_p \mathbf{X}_{k,t} + \boldsymbol{\xi}_t, \tag{3.17}$$

where a response vector $\mathbf{Y}_t$ is related to k predictor vectors, $\mathbf{X}_{1,t}, \ldots, \mathbf{X}_{k,t}$, and the error vector, $\boldsymbol{\xi}_t$, is a m-dimensional Gaussian vector white noise process, $\mathbf{VWN}(\mathbf{0}, \boldsymbol{\Sigma})$. To make the model in Eq. (3.17) more general, some or all of the predictor vectors do not need to have the same dimension as the response vector $\mathbf{Y}_t$. For example, instead of the dimension m, $\mathbf{X}_{i,t}$ can have a dimension r. In such a case, the dimension of the associated parameter coefficient matrix $\boldsymbol{\Phi}_i$ will be $m \times r$, which will no longer be a square matrix like those in the VAR(p) model.

For the multivariate time series regression, some software packages, such as **MATLAB** (2017), use the following model,

$$\mathbf{Y}_t = \boldsymbol{\theta}_0 + \mathbf{X}_t \boldsymbol{\beta} + \boldsymbol{\Phi}_1 \mathbf{Y}_{t-1} + \cdots + \boldsymbol{\Phi}_p \mathbf{Y}_{t-p} + \boldsymbol{\xi}_t, \tag{3.18}$$

where $\mathbf{X}_t$ is a $m \times r$ design matrix for r exogenous variables. Since the model involves the VAR structure for $\mathbf{Y}_t$ and predictor vector $\mathbf{X}_t$, it is known as a VARX model. However, it should be noted that in Eq. (3.18) the associated regression coefficients $\boldsymbol{\beta}$ corresponding to the r exogenous variables is a $r \times 1$ vector, which implies that the column entries of $\mathbf{X}_t$ share a common regression coefficient for all t, and this is relatively restrictive. In this formulation, the VARX model in Eq. (3.18) without lagged response vector variables $\mathbf{Y}_{t-j}$ does not reduce to the multivariate multiple regression model given in Eq. (3.5).

Another representation of the VARX model is given by

$$\mathbf{Y}_t - \boldsymbol{\theta}_0 + \boldsymbol{\Phi}_1 \mathbf{Y}_{t-1} + \cdots + \boldsymbol{\Phi}_p \mathbf{Y}_{t-p} + \boldsymbol{\Theta}_0 \mathbf{X}_t + \boldsymbol{\Theta}_1 \mathbf{X}_{t-1} + \cdots + \boldsymbol{\Theta}_s \mathbf{X}_{t-s} + \boldsymbol{\xi}_t, \tag{3.19}$$

where $\boldsymbol{\Phi}_i$ is a $m \times m$ parameter matrix for $\mathbf{Y}_{t-i}$, $\mathbf{X}_t$ is a r-dimensional time series vector for the r exogenous variables, and $\boldsymbol{\Theta}_i$ is a $m \times r$ parameter matrix for $\mathbf{X}_{t-i}$. This representation is used by some other software such as SAS and is called the VARX(p,s) model, and it is the form that we recommend to use. The parameter estimation of vector time series regression models is achieved through either the least squares (LS) or the maximum likelihood (ML), similar to those of vector time series models introduced in Chapter 2. Once the model is fitted, it can be used to forecast $\mathbf{Y}_{t+\ell}$ as follows

$$\hat{\mathbf{Y}}_t(\ell) = \hat{\boldsymbol{\theta}}_0 + \hat{\boldsymbol{\Phi}}_1 \hat{\mathbf{Y}}_t(\ell-1) + \cdots + \hat{\boldsymbol{\Phi}}_p \hat{\mathbf{Y}}_t(\ell-p) + \hat{\boldsymbol{\Theta}}_0 \hat{\mathbf{X}}_t(\ell) + \hat{\boldsymbol{\Theta}}_1 \hat{\mathbf{X}}_t(\ell-1) + \cdots + \hat{\boldsymbol{\Theta}}_s \hat{\mathbf{X}}_t(\ell-s), \tag{3.20}$$

where a separate vector time series model of $\mathbf{X}_t$ may need to be constructed for $\hat{\mathbf{X}}_t(\ell-j)$, $j \geq 0$.

The forecasting procedures are exactly the same as those discussed in Chapter 2. Rather than repeating them, we will look at some useful empirical examples instead.

3.4.2 Empirical Example II – VARX models for U.S. retail sales and some national indicators

Example 3.3 In this example, we will use the vector time series regression model introduced in Eq. (3.17) to study the relationship for the variables presented in Table 3.1. However, to examine the efficiency of the model for forecasting, we will use the data from 1992 to 2014 for model fitting and leave the 2015 values for comparison. Specifically, we will examine the following model:

$$\underset{4\times 1}{\mathbf{Y}_t} = \underset{4\times 1}{\boldsymbol{\theta}_0} + \underset{4\times 5}{\boldsymbol{\Theta}} \underset{5\times 1}{\mathbf{X}_t} + \underset{4\times 1}{\boldsymbol{\xi}_t}, \tag{3.21}$$

where t from 1 (1992) to 23 (2014),

$$\mathbf{Y}_t = \begin{bmatrix} Y_{1,t} \\ Y_{2,t} \\ Y_{3,t} \\ Y_{4,t} \end{bmatrix}, \ \boldsymbol{\theta}_0 = \begin{bmatrix} \alpha_1 \\ \alpha_2 \\ \alpha_3 \\ \alpha_4 \end{bmatrix}, \boldsymbol{\Theta} = \begin{bmatrix} \beta_{1,1} & \beta_{1,2} & \beta_{1,3} & \beta_{1,4} & \beta_{1,5} \\ \beta_{2,1} & \beta_{2,2} & \beta_{2,3} & \beta_{2,4} & \beta_{2,5} \\ \beta_{3,1} & \beta_{3,2} & \beta_{3,3} & \beta_{3,4} & \beta_{3,5} \\ \beta_{4,1} & \beta_{4,2} & \beta_{4,3} & \beta_{4,4} & \beta_{4,5} \end{bmatrix}, \ \mathbf{X}_t = \begin{bmatrix} GDP_t \\ GDP_{t-1} \\ GDP_{t-2} \\ DJIA_{t-1} \\ CPI_t \end{bmatrix}, \boldsymbol{\xi}_t = \begin{bmatrix} \xi_{1,t} \\ \xi_{2,t} \\ \xi_{3,t} \\ \xi_{4,t} \end{bmatrix},$$

and $\boldsymbol{\xi}_t$ is a four-dimensional Gaussian vector white noise process, $\mathbf{VWN}(\mathbf{0}, \boldsymbol{\Sigma})$.

The least squares estimation result using SAS VARMAX procedure is given by

$$\begin{bmatrix} \hat{Y}_{1,t} \\ \hat{Y}_{2,t} \\ \hat{Y}_{3,t} \\ \hat{Y}_{4,t} \end{bmatrix} = \begin{bmatrix} 877552.87 \\ -222208.00 \\ -6585.67 \\ 140865.26 \end{bmatrix} + \begin{bmatrix} 1800005.14 & -33572.75 & -71354.50 & 24.37 & -7001.30 \\ -4816.33 & -13175.90 & 679.69 & 0.53 & 4852.21 \\ 32911.15 & -4251.66 & 14880.47 & -0.05 & -434.70 \\ 38220.73 & 9829.43 & 4736.60 & -1.15 & -1763.25 \end{bmatrix} \begin{bmatrix} GDP_t \\ GDP_{t-1} \\ GDP_{t-2} \\ DJIA_{t-1} \\ CPI_t \end{bmatrix}, \tag{3.22}$$

and the estimate of the variance–covariance matrix of the residual vectors is

$$\hat{\boldsymbol{\Sigma}} = \begin{bmatrix} 2671603096.9 & -17748743.5 & 21959875.95 & 137381372.32 \\ -177548743.5 & 103753320.82 & -2309697.58 & -80919156.35 \\ 21959875.95 & -2309697.58 & 19255714.70 & 20956467.65 \\ 137381372.32 & -80919156.35 & 20956467.65 & 96617679.23 \end{bmatrix}.$$

We now use Eq. (3.22) to forecast the 2015 sales and the results are given in Table 3.5.

Example 3.4 In this example, we will use the VARX(p,s) model introduced in Eq. (3.19) to study the relationship of U.S. retail sales and some important national indicator variables discussed in Examples 3.1 and 3.2 by using the full data set WW3a, which includes more related values of predictor variables, as shown in Table 3.6.

We recall that

$Y_{1,t}$ = sales of motor vehicle and parts dealers of the year t

$Y_{2,t}$ = sales of food and beverage stores of the year t

$Y_{3,t}$ = sales of health care and personal care stores of the year t

$Y_{4,t}$ = sales of general merchandise stores of the year t

and let

$X_{1,t}$ = GDP (trillion) of the year t

$X_{2,t}$ = Dow Jones Industrial Average of the year t

$X_{3,t}$ = average CPI of the year t

Table 3.5 Forecasts for 2015 sales from the model in Eq. (3.22).

Variable	Forecast	Standard Error	95% Forecast Limit	Actual Value
$Y_{1,t}$	1 122 531.94	51 687.55	(1 021 226.20, 1 223 837.68)	1 095 412
$Y_{2,t}$	632 611.36	11 185.94	(612 647.29, 625 275.43)	685 568
$Y_{3,t}$	314 881.38	4 388.13	(306 280.80, 323 481.96)	315 257
$Y_{4,t}$	727 816.36	9 829.43	(708 551.03, 747 081.69)	674 928

Table 3.6 Vector time series regression data set.

Year (t)	$Y_{1,t}$	$Y_{2,t}$	$Y_{3,t}$	$Y_{4,t}$	$X_{1,t}$	$X_{1,t-1}$	$X_{1,t-2}$	$X_{2,t}$	$X_{2,t-1}$	$X_{2,t-2}$	$X_{3,t}$	$X_{3,t-1}$	$X_{3,t-2}$
1992	418 393	370 513	89 705	247 876	6.539	6.174	5.98	3 301.1	3 168.83	2 633.66	140.3	136.2	130.7
1993	472 916	374 516	92 594	265 996	6.879	6.539	6.174	3 754.1	3 301.1	3 168.83	144.5	140.3	136.2
1994	541 141	384 340	96 363	285 190	7.309	6.879	6.539	3 834.4	3 754.1	3 301.1	148.2	144.5	140.3
1995	579 715	390 386	101 635	300 498	7.664	7.309	6.879	5 117.1	3 834.4	3 754.1	152.4	148.2	144.5
1996	627 507	401 073	109 557	315 305	8.1	7.664	7.309	6 448.3	5 117.1	3 834.4	156.9	152.4	148.2
1997	653 817	409 373	118 672	331 363	8.609	8.1	7.664	7 908.3	6 448.3	5 117.1	150.5	156.9	152.4
1998	688 415	416 525	129 583	351 081	9.089	8.609	8.1	9 181.4	7 908.3	6 448.3	163	160.5	156.9
1999	764 204	433 699	142 699	380 179	9.661	9.089	8.609	11 497.1	9 181.4	7 908.3	166.6	163	160.5
2000	796 210	444 764	155 234	404 228	10.285	9.661	9.089	10 786.9	11 497.1	9 181.4	172.2	166.6	163
2001	815 579	462 429	166 533	427 468	10.622	10.285	9.661	10 021.5	10 786.9	11 497.1	177.1	172.2	166.6
2002	818 811	464 856	179 983	446 520	10.978	10.622	10.285	8 341.6	10 021.5	10 786.9	179.9	177.1	172.2
2003	841 588	474 385	192 426	468 771	11.511	10.978	10.622	10 453.9	8 341.6	10 021.5	184	179.9	177.1
2004	866 372	490 380	199 290	497 382	12.275	11.511	10.978	10 783	10 453.9	8 341.6	188.9	184	179.9
2005	888 307	508 484	210 085	528 385	13.094	12.275	11.511	10 717.5	10 783	10 453.9	195.3	188.9	184
2006	899 997	525 232	223 336	554 256	13.856	13.094	12.275	12 463.2	10 717.5	10 783	201.6	195.3	188.9
2007	910 139	547 837	237 164	578 582	14.478	13.856	13.094	13 264.8	12 463.2	10 717.5	207.342	201.6	195.3
2008	785 865	569 276	246 573	595 041	14.719	14.478	13.856	8 776.4	13 264.8	12 463.2	215.303	207.342	201.6
2009	671 772	568 418	252 794	588 918	14.419	14.719	14.478	10 428.1	8 776.4	13 264.8	214.537	215.303	207.342
2010	742 913	580 530	260 435	603 757	14.964	14.419	14.719	11 577.5	10 428.1	8 776.4	218.056	214.537	215.303
2011	812 938	609 137	271 612	624 766	15.518	14.964	14.419	12 217.6	11 577.5	10 428.1	224.939	218.056	214.537
2012	886 494	628 205	274 000	642 313	16.155	15.518	14.964	13 104.1	12 217.6	11 577.5	229.594	224.939	218.056
2013	959 294	640 847	281 840	651 874	16.692	16.155	15.518	16 576.7	13 104.1	12 217.6	232.957	229.594	224.939
2014	1 020 851	669 165	299 263	667 163	17.393	16.692	16.155	17 823.07	16 576.7	13 104.1	236.736	232.957	229.594
2015	1 095 412	685 568	315 257	674 928	18.037	17.393	16.692	17 425.03	17 823.07	16 576.7	237.017	236.736	232.957

Specifically, we will examine the following VARX(1,2) model

$$\underset{4\times 1}{\mathbf{Y}_t} = \underset{4\times 1}{\boldsymbol{\theta}_0} + \underset{4\times 4}{\boldsymbol{\Phi}_1}\underset{4\times 1}{\mathbf{Y}_{t-1}} + \underset{4\times 3}{\boldsymbol{\Theta}_0}\underset{3\times 1}{\mathbf{X}_t} + \underset{4\times 3}{\boldsymbol{\Theta}_1}\underset{3\times 1}{\mathbf{X}_{t-1}} + \underset{4\times 3}{\boldsymbol{\Theta}_2}\underset{3\times 1}{\mathbf{X}_{t-2}} + \underset{4\times 1}{\boldsymbol{\xi}_t}, \tag{3.23}$$

where

$$\boldsymbol{\theta}_0 = \begin{bmatrix} \alpha_1 \\ \alpha_2 \\ \alpha_3 \\ \alpha_4 \end{bmatrix}, \boldsymbol{\Theta}_0 = \begin{bmatrix} \beta_{0,1,1} & \beta_{0,1,2} & \beta_{0,1,3} \\ \beta_{0,2,1} & \beta_{0,2,2} & \beta_{0,2,3} \\ \beta_{0,3,1} & \beta_{0,3,2} & \beta_{0,3,3} \\ \beta_{0,4,1} & \beta_{0,4,2} & \beta_{0,4,3} \end{bmatrix}, \ \boldsymbol{\Theta}_1 = \begin{bmatrix} \beta_{1,1,1} & \beta_{1,1,2} & \beta_{1,1,3} \\ \beta_{1,2,1} & \beta_{1,2,2} & \beta_{1,2,3} \\ \beta_{1,3,1} & \beta_{1,3,2} & \beta_{1,3,3} \\ \beta_{1,4,1} & \beta_{1,4,2} & \beta_{1,4,3} \end{bmatrix},$$
$$\boldsymbol{\Theta}_2 = \begin{bmatrix} \beta_{2,1,1} & \beta_{2,1,2} & \beta_{2,1,3} \\ \beta_{2,2,1} & \beta_{2,2,2} & \beta_{2,2,3} \\ \beta_{2,3,1} & \beta_{2,3,2} & \beta_{2,3,3} \\ \beta_{2,4,1} & \beta_{2,4,2} & \beta_{2,4,3} \end{bmatrix}, \mathbf{X}_t = \begin{bmatrix} X_{1,t} \\ X_{2,t} \\ X_{3,t} \end{bmatrix} = \begin{bmatrix} GDP_t \\ DJIA_t \\ CPI_t \end{bmatrix}, \tag{3.24}$$

and $\boldsymbol{\xi}_t$ is a four-dimensional Gaussian vector white noise process, $\mathbf{VWN}(\mathbf{0}, \boldsymbol{\Sigma})$.

The least square estimation result using SAS VARMAX procedure is given by

$$\begin{bmatrix} \hat{Y}_{1,t} \\ \hat{Y}_{2,t} \\ \hat{Y}_{3,t} \\ \hat{Y}_{4,t} \end{bmatrix} = \begin{bmatrix} 246291.33 \\ 20797.39 \\ -16601.31 \\ 51273.02 \end{bmatrix} + \begin{bmatrix} 0.60 & 1.76 & 1.84 & -0.21 \\ 0.00 & 0.49 & 0.89 & -1.20 \\ 0.01 & -0.13 & 0.73 & -0.15 \\ -0.02 & -0.39 & 0.37 & 0.82 \end{bmatrix} \begin{bmatrix} Y_{1,t-1} \\ Y_{2,t-1} \\ Y_{3,t-1} \\ Y_{4,t-1} \end{bmatrix}$$
$$+ \begin{bmatrix} 146505.35 & 14.74 & 2900.88 \\ 1130.01 & 1.48 & 4042.87 \\ 5036.70 & 0.09 & 148.13 \\ 21142.74 & 0.14 & 1038.33 \end{bmatrix} \begin{bmatrix} X_{1,t} \\ X_{2,t} \\ X_{3,t} \end{bmatrix} + \begin{bmatrix} -206911.78 & -4.24 & -5151.50 \\ 20853.16 & -1.20 & -1394.34 \\ 3848.29 & -0.53 & -786.20 \\ -16451.41 & 0.43 & -355.72 \end{bmatrix} \begin{bmatrix} X_{1,t-1} \\ X_{2,t-1} \\ X_{3,t-1} \end{bmatrix}$$
$$+ \begin{bmatrix} 12981.72 & 9.35 & -1814.22 \\ 14235.09 & -0.36 & -1620.60 \\ 2600.72 & 1.14 & 983.06 \\ -2811.35 & 1.29 & 18.66 \end{bmatrix} \begin{bmatrix} X_{1,t-2} \\ X_{2,t-2} \\ X_{3,t-2} \end{bmatrix}, \tag{3.25}$$

and

$$\hat{\boldsymbol{\Sigma}} = \begin{bmatrix} 4242276632.59 & 2555671.66 & 16950478.87 & 25348523.89 \\ 25554671.66 & 7875069.96 & 3792478.13 & 3855028.40 \\ 16950478.87 & 3792478.13 & 7856479.98 & -43175.39 \\ 25348523.89 & 3855028.40 & -43175.39 & 6824708.87 \end{bmatrix}. \tag{3.26}$$

Most parameter estimates in Eq. (3.25) are not significant, so we need to re-estimate the model by setting the parameters with insignificant estimates to zero. However, to examine the efficiency of this VARX model in forecasting, in the re-estimation, we will only use the data from 1992 to 2014 and leave the 2015 data for forecast comparison. From this re-estimation, we obtain the following simplified equation:

$$\begin{bmatrix} \hat{Y}_{1,t} \\ \hat{Y}_{2,t} \\ \hat{Y}_{3,t} \\ \hat{Y}_{4,t} \end{bmatrix} = \begin{bmatrix} 0.5207 & 0 & 0 & 0 \\ 0 & 1.02625 & 0.41863 & -0.16745 \\ 0 & 0 & 1.04812 & 0 \\ 0 & -0.02316 & 0 & 0.65136 \end{bmatrix} \begin{bmatrix} Y_{1,t-1} \\ Y_{2,t-1} \\ Y_{3,t-1} \\ Y_{4,t-1} \end{bmatrix} + \begin{bmatrix} 50174.73125 & 18.00883 & 0 \\ 0 & 0 & 0 \\ 0 & 0 & 0 \\ 15724.42719 & 0 & 0 \end{bmatrix} \begin{bmatrix} X_{1,t} \\ X_{2,t} \\ X_{3,t} \end{bmatrix}$$
$$+ \begin{bmatrix} -36281.40682 & 0 & 0 \\ 0 & 0 & 0 \\ 0 & 0 & 0 \\ 0 & 0 & 0 \end{bmatrix} \begin{bmatrix} X_{1,t-1} \\ X_{2,t-1} \\ X_{3,t-1} \end{bmatrix}, \tag{3.27}$$

and

$$\hat{\boldsymbol{\Sigma}} = \begin{bmatrix} 10699358614 & 715994057.43 & 438950329.55 & 995560416.12 \\ 715994057.43 & 173917231.03 & 62412049.314 & 95035472.176 \\ 438950329.55 & 62412049.314 & 61739028.580 & 48801287.342 \\ 995560416.12 & 95035472.176 & 48801287.342 & 166981942.52 \end{bmatrix}. \tag{3.28}$$

The result becomes much simpler, and we can conclude the following:

1. The sales of motor vehicle and parts dealers at year t depends mainly on the sales of its previous year, the current and previous year *GDP*, and the current year Dow Jones Industrial Average.
2. The sales of food and beverage stores at year t depends on the sales of the previous year. They are positively related to the previous year sales of health care and personal care stores but negatively related to the previous year sales of general merchandise stores.

Table 3.7 Forecasts for 2015 sales from the model in Eq. (3.27).

Variable	Forecast	Standard Error	95% Forecast Limit	Actual Value
$Y_{1,t}$	1 119 316.06	103 437.70	(916 581.89, 1 322 050.24)	1 095 412
$Y_{2,t}$	700 295.38	13 187.77	(674 447.83, 726 142.93)	685 568
$Y_{3,t}$	313 664.52	7 857.42	(298 264.26, 329 064.78)	315 257
$Y_{4,t}$	702 687.10	12 922.15	(677 360.16, 728 014.05)	674 928

Table 3.8 Forecast errors for 2015 sales from models in Eqs. (3.22) and (3.27).

Sales Variable	Forecast Error From (3.22)	Forecast Error From (3.27)
$Y_{1,t}$	27 119.94	23 904.06
$Y_{2,t}$	−52 956.64	14 727.38
$Y_{3,t}$	−2 115.62	−1 592.48
$Y_{4,t}$	72 816.37	27 749.10
Average of Absolute Errors	38 752.14	16 633.25

3. The sales of health care and personal care stores of the year t primarily depend on the sales performance of its previous year and very little on other variables.
4. The sales of general merchandise at year t depends on the sales of the previous year, negatively related to the previous year sales of food and beverage stores, and positively related to the current year *GDP*.

We now use Eq. (3.27) to forecast the 2015 sales and the results are given in Table 3.7.

By comparing Tables 3.5 and 3.7, we see that the forecast intervals based on the model in Eq. (3.22) are much shorter than those based on the model in Eq. (3.27), and these are true for all four of the response variables. However, Table 3.8 gives the exact comparison between forecasts and actual values of the 2015 sales. Based on the average of absolute forecast errors, the model in Eq. (3.27) outperforms the model in Eq. (3.22).

3.5 Empirical Example III – Total mortality and air pollution in California

Example 3.5 Air pollution is by far the largest contributor to early death. According to a WHO report, one in eight total global deaths was the result of air pollution exposure. Since California is one of the most air polluted states in the U.S., to further illustrate the VARX (p,s) model, we will examine the relationship between total mortality and air pollution in California. As we know, because of air pollution problems, Los Angeles established an Air Pollution Control Agency in the early 1950s. Since problems with air pollution remained, LA residents started to complain about the effectiveness of the agency. So, the agency decided to offer a research grant for scholars Box and Tiao to study the effectiveness of its various control schemes. It was through this study that Box and Tiao (1975) introduced the well-known time series intervention method.

We believe that a suitable time unit used to study the relationship between death and air pollution is yearly. Unfortunately, we cannot find any annual air pollution data sets. The only data sets that we can find are the daily mortality and air pollution, which contains a very large number of variables and observations, but there are many missing values that cannot be easily replaced by any available estimation method. Currently available software for multivariate regression either requires no missing values or known models for these missing values. As a result, we will consider the daily mortality and pollution levels for two air pollutants, ozone and carbon monoxide, from five Californian cities between April 13 and September 14, 1999. Thus, we will use only a portion of the original dataset, which includes a vector of five dependent variables and two vectors of independent variables with five elements each. The data set of these 15 time series is listed as WW3b in the Data Appendix. The detailed description of notations used is given in Table 3.9. The plot of the numbers of daily death in these five California cities is given in Figure 3.2.

To investigate the relationship between the daily death and the air pollutant O_3 and CO, we first examine the correlation matrix of these 15 variables as shown in Table 3.10.

With $\sqrt{Var(\hat{\rho}_{i,j})} \approx \sqrt{1/n} = \sqrt{1/155} = 0.08$, we can see that the air pollution levels among the five cities are highly correlated, but their correlations with death are not strong because the time unit used is daily and not yearly as we indicated earlier. Simply for an illustration of multivariate time series regression model building, we try the following VARX model,

$$\underset{5\times 1}{\mathbf{Y}_t} = \underset{5\times 1}{\boldsymbol{\theta}_0} + \underset{5\times 5}{\boldsymbol{\Theta}_0}\,\underset{5\times 1}{\mathbf{X}_{1,t}} + \underset{5\times 5}{\boldsymbol{\Theta}_1}\,\underset{5\times 1}{\mathbf{X}_{2,t}} + \underset{5\times 1}{\boldsymbol{\xi}_t}, \tag{3.29}$$

where

$$\mathbf{Y}_t = \begin{bmatrix} Y_{1,t} \\ Y_{2,t} \\ Y_{3,t} \\ Y_{4,t} \\ Y_{5,t} \end{bmatrix} = \begin{bmatrix} LA\_D \\ RS\_D \\ SD\_D \\ SJ\_D \\ SA\_D \end{bmatrix}, \mathbf{X}_{1,t} = \begin{bmatrix} LA\_O3 \\ RS\_O3 \\ SD\_O3 \\ SJ\_O3 \\ SA\_O3 \end{bmatrix}, \text{and } \mathbf{X}_{2,t} = \begin{bmatrix} LA\_CO \\ RS\_CO \\ SD\_CO \\ SJ\_CO \\ SA\_CO \end{bmatrix}.$$

Table 3.9 Description of notations used in Example 3.5.

Variable	Description
D	daily mortality counts from all causes
O3	daily ozone time series, measured in parts per billion (ppb)
CO	carbon monoxide, measured in μg m^{-3}
LA	Los Angeles, CA
RS	Riverside, CA
SD	San Diego, CA
SJ	San Jose, CA
SA	Santa Ana/Anaheim, CA

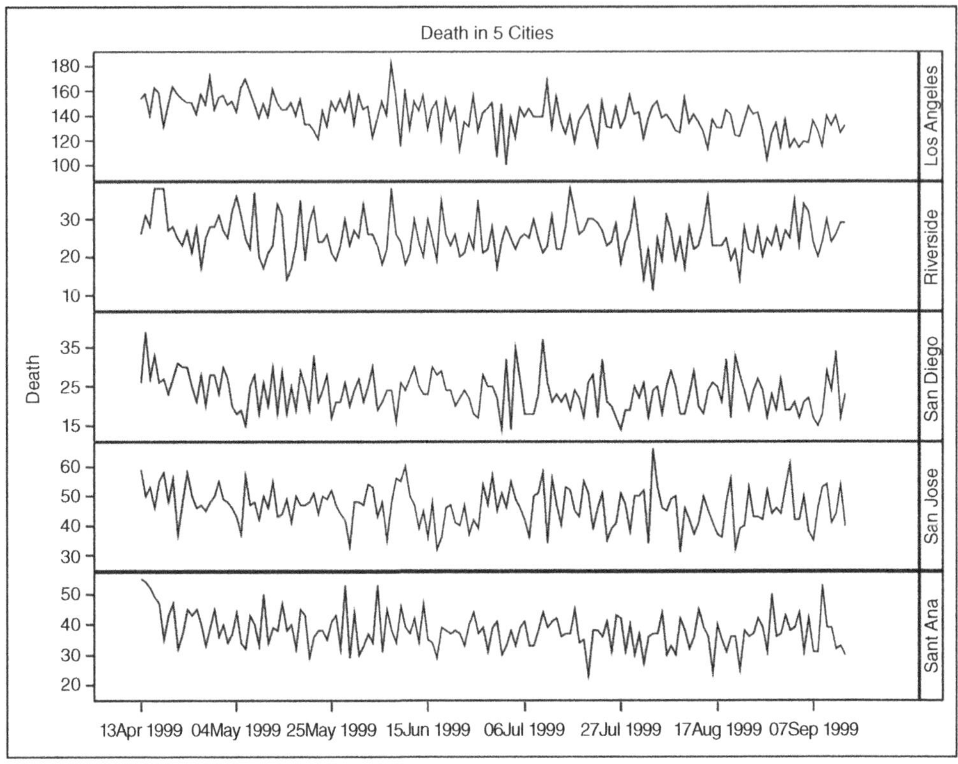

Figure 3.2 The daily deaths in five California cities.

We use $(n - 3)$ observations to fit the model and keep the last three observations for forecast comparison. The least squares estimation result using SAS VARMAX procedure is given next.

$$\begin{bmatrix} \hat{Y}_{1,t} \\ \hat{Y}_{2,t} \\ \hat{Y}_{3,t} \\ \hat{Y}_{4,t} \\ \hat{Y}_{5,t} \end{bmatrix} = \begin{bmatrix} 121.91 \\ 29.11 \\ 20.15 \\ 53.38 \\ 36.60 \end{bmatrix} + \begin{bmatrix} -0.37 & 0.60 & -0.41 & 0.06 & 0.36 \\ 0.04 & -0.15 & 0.05 & -0.03 & 0.02 \\ 0.02 & 0.17 & -0.14 & -0.13 & 0.14 \\ -0.24 & -0.11 & 0.16 & -0.30 & 0.33 \\ -0.22 & 0.10 & -0.04 & -0.01 & 0.18 \end{bmatrix} \begin{bmatrix} O3_{1,t} \\ O3_{2,t} \\ O3_{3,t} \\ O3_{4,t} \\ O3_{5,t} \end{bmatrix}$$
$$+ \begin{bmatrix} -0.01 & 0.00 & -0.00 & 0.00 & 0.03 \\ -0.01 & 0.01 & -0.01 & 0.01 & 0.01 \\ 0.00 & -0.01 & 0.00 & -0.00 & 0.01 \\ -0.02 & 0.00 & -0.01 & 0.00 & 0.02 \\ -0.00 & 0.00 & -0.01 & 0.00 & 0.01 \end{bmatrix} \begin{bmatrix} CO_{1,t} \\ CO_{2,t} \\ CO_{3,t} \\ CO_{4,t} \\ CO_{5,t} \end{bmatrix}, \quad (3.30)$$

Table 3.10 Correlation matrix for the 15 variables.

	Pearson Correlation Coefficients														
	LA_D	LA_o3	LA_co	RS_D	RS_o3	RS_co	SD_D	SD_o3	SD_co	SJ_D	SJ_o3	SJ_co	SA_D	SA_o3	SA_co
LA_D	1.00	0.10	0.10	0.04	0.12	0.10	0.12	0.04	0.04	0.13	0.24	0.15	0.10	0.14	0.31
LA_o3	0.10	1.00	−0.27	−0.07	0.70	−0.14	0.09	0.64	−0.12	−0.08	0.74	−0.26	0.00	0.87	−0.12
LA_co	0.10	−0.27	1.00	0.09	−0.12	0.67	0.11	0.06	0.62	−0.01	−0.34	0.82	0.14	−0.35	0.77
RS_D	0.04	−0.07	0.09	1.00	−0.17	0.12	0.10	−0.15	−0.02	0.02	−0.02	0.19	0.10	−0.06	0.20
RS_o3	0.12	0.70	−0.12	−0.17	1.00	0.01	0.07	0.88	0.05	−0.11	0.56	−0.22	0.04	0.63	−0.10
RS_co	0.10	−0.14	0.67	0.12	0.01	1.00	−0.05	0.10	0.67	0.02	−0.08	0.50	0.11	−0.17	0.55
SD_D	0.12	0.09	0.11	0.10	0.07	−0.05	1.00	0.03	0.00	0.07	0.09	0.13	0.10	0.09	0.23
SD_o3	0.04	0.64	0.06	−0.15	0.88	0.10	0.03	1.00	0.14	−0.08	0.39	−0.10	0.03	0.53	−0.03
SD_co	0.04	−0.12	0.62	−0.02	0.05	0.67	0.00	0.14	1.00	−0.09	−0.13	0.45	−0.02	−0.17	0.42
SJ_D	0.13	−0.08	−0.01	0.02	−0.11	0.02	0.07	−0.08	−0.09	1.00	−0.02	0.08	0.12	−0.02	0.14
SJ_o3	0.24	0.74	−0.34	−0.02	0.56	−0.08	0.09	0.39	−0.13	−0.02	1.00	−0.29	0.08	0.83	0.01
SJ_co	0.15	−0.26	0.82	0.19	−0.22	0.50	0.13	−0.10	0.45	0.08	−0.29	1.00	0.18	−0.33	0.78
SA_D	0.10	0.00	0.14	0.10	0.04	0.11	0.10	0.03	−0.02	0.12	0.08	0.18	1.00	0.04	0.26
SA_o3	0.14	0.87	−0.35	−0.06	0.63	−0.17	0.09	0.53	−0.17	−0.02	0.83	−0.33	0.04	1.00	−0.17
SA_co	0.31	−0.12	0.77	0.20	−0.10	0.55	0.23	−0.03	0.42	0.14	0.01	0.78	0.26	−0.17	1.00

and the estimate of the variance–covariance matrix of the residual vectors is

$$\hat{\boldsymbol{\Sigma}} = \begin{bmatrix} 187.45 & -1.82 & -0.61 & 7.65 & -1.86 \\ -1.82 & 27.91 & 2.17 & -2.31 & 1.54 \\ -0.61 & 2.17 & 22.38 & 1.75 & 1.15 \\ 7.65 & -2.31 & 1.75 & 42.33 & 2.21 \\ -1.86 & 1.54 & 1.15 & 2.21 & 34.88 \end{bmatrix}.$$

The next three period forecasts and their related information are given in Table 3.11. The result is not quite satisfactory. In fact, the real death value for San Diego on September 12, 1999 is outside of the 95% forecast limits. Next, let us try to fit the following VARX model

$$\underset{5\times 1}{\mathbf{Y}_t} = \underset{5\times 1}{\boldsymbol{\theta}_0} + \underset{5\times 5}{\boldsymbol{\Phi}_1}\,\underset{5\times 1}{\mathbf{Y}_{t-1}} + \underset{5\times 5}{\boldsymbol{\Theta}_0}\,\underset{5\times 1}{\mathbf{X}_{1,t}} + \underset{5\times 5}{\boldsymbol{\Theta}_1}\,\underset{5\times 1}{\mathbf{X}_{2,t}} + \underset{5\times 1}{\boldsymbol{\xi}_t}. \tag{3.31}$$

The estimation result is

$$\begin{bmatrix} \hat{Y}_{1,t} \\ \hat{Y}_{2,t} \\ \hat{Y}_{3,t} \\ \hat{Y}_{4t} \\ \hat{Y}_{5,t} \end{bmatrix} = \begin{bmatrix} 115.74 \\ 25.09 \\ 15.67 \\ 50.90 \\ 32.27 \end{bmatrix} + \begin{bmatrix} 0.04 & -0.21 & 0.24 & -0.22 & 0.37 \\ 0.03 & 0.04 & -0.09 & 0.03 & -0.01 \\ 0.02 & -0.01 & -0.12 & 0.17 & -0.06 \\ 0.00 & 0.08 & -0.17 & 0.01 & 0.08 \\ 0.01 & 0.02 & 0.01 & 0.13 & -0.13 \end{bmatrix} \begin{bmatrix} Y_{1,t-1} \\ Y_{2,t-1} \\ Y_{3,t-1} \\ Y_{4,t-1} \\ Y_{5,t-1} \end{bmatrix}$$
$$+ \begin{bmatrix} -0.39 & 0.68 & -0.47 & -0.08 & 0.43 \\ 0.06 & -0.17 & 0.06 & -0.01 & -0.00 \\ 0.10 & 0.10 & -0.09 & -0.05 & 0.04 \\ -0.22 & -0.11 & 0.18 & -0.26 & 0.26 \\ -0.16 & 0.03 & 0.05 & 0.10 & 0.06 \end{bmatrix} \begin{bmatrix} O3_{1,t} \\ O3_{2,t} \\ O3_{3,t} \\ O3_{4,t} \\ O3_{5,t} \end{bmatrix} + \begin{bmatrix} -0.01 & 0.00 & -0.00 & 0.00 & 0.03 \\ -0.01 & 0.00 & -0.00 & 0.01 & 0.01 \\ 0.00 & -0.01 & 0.00 & -0.00 & 0.01 \\ -0.02 & 0.00 & -0.01 & 0.00 & 0.01 \\ 0.00 & 0.00 & -0.01 & 0.00 & 0.00 \end{bmatrix} \begin{bmatrix} CO_{1,t} \\ CO_{2,t} \\ CO_{3,t} \\ CO_{4,t} \\ CO_{5,t} \end{bmatrix}, \tag{3.32}$$

and the estimate of the variance–covariance matrix of the residual vectors is

$$\hat{\boldsymbol{\Sigma}} = \begin{bmatrix} 186.23 & -1.03 & 1.88 & 7.70 & -0.38 \\ -1.03 & 28.74 & 1.75 & -2.78 & 1.45 \\ 1.88 & 1.75 & 21.68 & 1.50 & -.02 \\ 7.70 & -2.78 & 1.50 & 42.35 & 1.38 \\ -0.38 & 1.45 & -0.02 & 1.38 & 33.19 \end{bmatrix}.$$

The next three period forecasts and their related information from the model in Eq. (3.32) are given in Table 3.12. Again, the forecast result is not satisfactory. The real death value for San Diego on September 12, 1999, is also outside of the 95% forecast limits.

Table 3.11 The detailed information for the three period forecasts from the model in (3.30).

Variable	Period	Date	Forecast	Real Value	Error	Standard Error	95% Confidence Limits	
LA_death	153	September 12, 1999	133.69	140	−6.31	13.69134	06.8589	160.528
	154	September 13, 1999	138.30	126	12.30	13.69134	11.4681	165.1372
	155	September 14, 1999	129.71	132	−2.29	13.69134	02.8743	156.5434
RS_death	153	September 12, 1999	25.59	26	−0.41	5.28314	15.23418	35.9437
	154	September 13, 1999	27.65	29	−1.35	5.28314	17.29514	38.00466
	155	September 14, 1999	22.48	29	−6.52	5.28314	12.12868	32.8382
SD_death	153	September 12, 1999	21.50	34	−12.50	4.73102	12.23077	30.77602
	154	September 13, 1999	20.93	17	3.93	4.73102	11.65738	30.20263
	155	September 14, 1999	21.84	23	−1.16	4.73102	12.56833	31.11358
SJ_death	153	September 12, 1999	43.69	44	−0.31	6.50609	30.94079	56.4442
	154	September 13, 1999	45.87	54	−8.13	6.50609	33.1181	58.62151
	155	September 14, 1999	40.76	40	0.76	6.50609	28.00574	53.50915
SA_death	153	September 12, 1999	36.67	32	4.67	5.90621	25.09582	48.24773
	154	September 13, 1999	38.91	33	5.91	5.90621	27.32965	50.48155
	155	September 14, 1999	36.53	30	6.53	5.90621	24.95158	48.10348

Table 3.12 The detailed information for the three period forecasts from the model in (3.32).

Variable	Period	Date	Forecast	Real Value	Error	Standard Error	95% Confidence Limits	
LA_death	153	September 12, 1999	135.26	140	−4.74	13.64644	108.5122	162.0053
	154	September 13, 1999	137.45	126	11.45	13.94102	110.1261	164.7739
	155	September 14, 1999	129.59	132	−2.41	13.95659	102.2356	156.9444
RS_death	153	September 12, 1999	25.33	26	−0.64	5.36105	14.82444	35.83936
	154	September 13, 1999	27.79	29	−1.21	5.39594	17.20954	38.36124
	155	September 14, 1999	23.09	29	−5.91	5.39883	12.50404	33.66707
SD_death	153	September 12, 1999	20.67	34	−13.33	4.65587	11.54558	29.79624
	154	September 13, 1999	20.90	17	3.90	4.83337	11.42951	30.37596
	155	September 14, 1999	21.61	23	−1.39	4.84264	12.12102	31.10381
SJ_death	153	September 12, 1999	43.46	44	−0.54	6.50748	30.70738	56.21626
	154	September 13, 1999	45.79	54	−8.21	6.58074	32.88833	58.68437
	155	September 14, 1999	41.23	40	1.27	6.58203	28.32596	54.12705
SA_death	153	September 12, 1999	35.96	32	3.96	5.76135	24.6661	47.25018
	154	September 13, 1999	38.80	33	5.80	5.86926	27.29824	50.30531
	155	September 14, 1999	35.67	30	5.67	5.87366	24.16234	47.18665

Now, let us try the following VARX model with more parameters,

$$\underset{5\times1}{\mathbf{Y}_t} = \underset{5\times1}{\boldsymbol{\theta}_0} + \underset{5\times5}{\boldsymbol{\Theta}_0}\underset{5\times1}{\mathbf{X}_{1,t}} + \underset{5\times5}{\boldsymbol{\Theta}_1}\underset{5\times1}{\mathbf{X}_{2,t}} + \underset{5\times5}{\boldsymbol{\Theta}_2}\underset{5\times1}{\mathbf{X}_{1,t-1}} + \underset{5\times5}{\boldsymbol{\Theta}_3}\underset{5\times1}{\mathbf{X}_{2,t-1}} + \underset{5\times1}{\boldsymbol{\xi}_t}. \quad (3.33)$$

The estimation result is

$$\begin{bmatrix} \hat{Y}_{1,t} \\ \hat{Y}_{2,t} \\ \hat{Y}_{3,t} \\ \hat{Y}_{4,t} \\ \hat{Y}_{5,t} \end{bmatrix} = \begin{bmatrix} 131.10 \\ 31.24 \\ 21.20 \\ 56.48 \\ 37.04 \end{bmatrix} + \begin{bmatrix} -0.27 & -0.38 & -0.24 & -0.07 & 0.26 \\ -0.04 & -0.14 & -0.00 & -0.03 & 0.14 \\ 0.08 & 0.25 & -0.18 & -0.04 & 0.05 \\ -0.25 & -0.05 & 0.14 & -0.28 & 0.31 \\ -0.18 & 0.14 & -0.07 & 0.00 & 0.18 \end{bmatrix} \begin{bmatrix} O3_{1,t} \\ O3_{2,t} \\ O3_{3,t} \\ O3_{4,t} \\ O3_{5,t} \end{bmatrix}$$

$$+ \begin{bmatrix} -0.02 & 0.00 & -0.00 & 0.01 & 0.02 \\ -0.00 & 0.00 & -0.00 & 0.00 & 0.00 \\ 0.00 & -0.01 & 0.00 & -0.00 & 0.01 \\ -0.01 & -0.00 & 0.00 & 0.00 & 0.01 \\ 0.00 & 0.00 & -0.02 & 0.00 & 0.00 \end{bmatrix} \begin{bmatrix} CO_{1,t} \\ CO_{2,t} \\ CO_{3,t} \\ CO_{4,t} \\ CO_{5,t} \end{bmatrix}$$

$$+ \begin{bmatrix} 0.24 & 0.51 & -0.56 & 0.23 & -0.21 \\ 0.22 & -0.14 & 0.14 & -0.01 & -0.23 \\ -0.23 & -0.19 & 0.14 & -0.09 & 0.20 \\ 0.12 & -0.32 & 0.18 & 0.14 & -0.17 \\ 0.07 & -0.02 & 0.04 & 0.15 & -0.28 \end{bmatrix} \begin{bmatrix} O3_{1,t-1} \\ O3_{2,t-1} \\ O3_{3,t-1} \\ O3_{4,t-1} \\ O3_{5,t-1} \end{bmatrix}$$

$$+ \begin{bmatrix} 0.02 & -0.00 & -0.01 & -0.02 & 0.02 \\ -0.01 & 0.01 & -0.00 & 0.00 & 0.00 \\ -0.00 & 0.00 & -0.00 & -0.00 & 0.00 \\ -0.00 & 0.01 & -0.01 & 0.00 & 0.00 \\ -0.00 & -0.00 & 0.01 & 0.00 & 0.01 \end{bmatrix} \begin{bmatrix} CO_{1,t-1} \\ CO_{2,t-1} \\ CO_{3,t-1} \\ CO_{4,t-1} \\ CO_{5,t-1} \end{bmatrix}, \quad (3.34)$$

and the estimate of the variance–covariance matrix of the residual vectors is

$$\hat{\boldsymbol{\Sigma}} = \begin{bmatrix} 188.02 & -1.25 & 1.37 & 7.01 & -5.82 \\ -1.25 & 28.57 & 2.13 & -4.26 & 0.92 \\ 1.37 & 2.13 & 22.89 & 0.96 & 1.49 \\ 7.01 & -4.26 & 0.96 & 41.86 & -0.34 \\ -5.82 & 0.92 & 1.49 & -0.34 & 33.00 \end{bmatrix}.$$

The next three period forecasts and their related information from the model in Eq. (3.34) are given in Table 3.13.

Table 3.13 The detailed information for the three period forecasts from the model in Eq. (3.34).

Variable	Period	Date	Forecast	Real	Error	Standard Error	95% Confidence Limits	
LA_death	153	September 12, 1999	130.13	140	−9.87	13.71186	103.25687	157.00638
	154	September 13, 1999	136.35	126	10.35	13.71186	109.48000	163.22951
	155	September 14, 1999	127.92	132	−4.08	13.71186	101.04084	154.79036
RS_death	153	September 12, 1999	25.89	26	−0.11	5.34510	15.41636	36.36878
	154	September 13, 1999	27.03	29	−1.94	5.34510	16.55793	37.51034
	155	September 14, 1999	25.57	29	−3.43	5.34510	15.09440	36.04681
SD_death	153	September 12, 1999	22.66	34	−11.34	4.78436	13.28403	32.03839
	154	September 13, 1999	20.77	17	3.77	4.78436	11.38971	30.14407
	155	September 14, 1999	23.29	23	0.29	4.78436	13.91430	32.66867
SJ_death	153	September 12, 1999	44.16	44	0.16	6.46995	31.48270	56.84444
	154	September 13, 1999	45.21	54	−8.79	6.46995	32.52533	57.88708
	155	September 14, 1999	46.03	40	6.03	6.46995	33.34482	58.70657
SA_death	153	September 12, 1999	36.92	32	4.92	5.74493	25.66091	48.18061
	154	September 13, 1999	38.98	33	5.98	5.74493	27.71603	50.23574
	155	September 14, 1999	37.75	30	7.75	5.74493	26.49270	49.01241

The forecast result is again not satisfactory with the real death value for San Diego on September 12, 1999, falling very much outside of the 95% forecast limits.

The model in Eq. (3.33) has more parameters than those models in Eqs. (3.29) and (3.31). However, its sum of absolute forecast errors is much larger than the other two simpler models. This example shows that a more complicated model does not necessarily produce a better result.

There are not many references on multivariate time series regression models. In closing this chapter, we recommend some references to our readers, including Pankratz (1991), Rao (2002), Lütkepohl (2007), SAS Institute, Inc. (2015), and Islam (2017).

Software code

R code for Examples 3.1–3.3

```
##data import##
df1 <- read.csv(file = "c:/Bookdata/WW3a.csv")
colnames(df1)[2:5] <- c("Y1t","Y2t","Y3t","Y4t")

##Example 3.1##
#Table 3.2#
reg1 <- lm(cbind(df1$Y1t,df1$Y2t,df1$Y3t,df1$Y4t) ~ df1$GDP +
        df1$GDP.1 + df1$GDP.2 + df1$DJIA.1 + df1$CPI)
summary(reg1)

#Table 3.3#
lm1 <- lm(Y1t ~ DJIA.1,data = df1)
summary(lm1)

lm2 <- lm(Y2t ~ CPI,data = df1)
summary(lm2)

lm3 <- lm(Y3t ~ GDP + GDP.2, data = df1)
summary(lm3)

lm4 <- lm(Y4t ~ GDP, data = df1)
summary(lm4)

##End of Example 3.1##

##Example 3.2##
#Table 3.4#
library(nlme)

lm1 <- gls(Y1t ~ GDP + GDP.1 + GDP.2 + DJIA.1 + CPI,
    data = df1, method = "ML", correlation = corARMA
    (p = 1,form = ~ Year),
    verbose = TRUE)
summary(lm1)
```

```
lm2 <- gls(Y2t ~ GDP + GDP.1 + GDP.2 + DJIA.1 + CPI,
       data = df1, method = "ML", correlation = corARMA
        (p = 1,form = ~ Year),
        verbose = TRUE)
summary(lm2)

lm3 <- gls(Y3t ~ GDP + GDP.1 + GDP.2 + DJIA.1 + CPI,
       data = df1, method = "ML", correlation = corARMA
        (p = 1,form = ~ Year),
        verbose = TRUE)
summary(lm3)

lm4 <- gls(Y4t ~ GDP + GDP.1 + GDP.2 + DJIA.1 + CPI,
       data = df1, method = "ML", correlation = corARMA
        (p = 1,form = ~ Year),
        verbose = TRUE)
summary(lm4)
##End of Example 3.2##

##Example 3.3##
df2 <- df1[df1$Year != 2015,]
reg2 <- lm(cbind(Y1t,Y2t,Y3t,Y4t) ~ GDP +
        GDP.1 + GDP.2 + DJIA.1 + CPI,data = df2)
summary(reg2)
#Table 3.5
predict(reg2, newdata = df1[df1$Year==2015,])
##End of Example 3.3##
```

SAS Code

The output from R and SAS for Examples 3.1–3.3 are equivalent. However, the output from R and SAS for Examples 3.4 and 3.5 are not quite the same. We believe the SAS results are correct, so only SAS code will be provided for the last two examples. To help readers using SAS, we also provide SAS code for Examples 3.1–3.3 in the following.

SAS Code for Example 3.1

```
/*Example 3.1*/
proc import file="C:/Bookdata/WW3a.csv" out=df1 Replace;
run;

data df1;
set df1;
Y1t = Motor_vehicles_parts;
Y2t = Food_beverage;
Y3t = Health;
```

```
Y4t = General_merchandise;
run;

proc reg;
model Y1t = GDP GDP_1 GDP_2 DJIA_1 CPI;
model Y2t = GDP GDP_1 GDP_2 DJIA_1 CPI;
model Y3t = GDP GDP_1 GDP_2 DJIA_1 CPI;
model Y4t = GDP GDP_1 GDP_2 DJIA_1 CPI;
run;
quit;

/*Variance Covariance Matrix under Table 3.2*/
proc varmax data = df1;
model Y1t Y2t Y3t Y4t = gdp gdp_1 gdp_2 djia_1 cpi/ method = ls;
run;

proc reg data = df1;
model Y1t = DJIA_1;
model Y2t = CPI;
model Y3t = GDP GDP_2;
model Y4t = GDP;
run;
quit;

/*Figure 3.1*/
proc import file = "c:\bookdata\ww3a.csv" out = df1 dbms =
csv replace;
run;

data df1;
set df1;
motor_vehicles_parts = motor_vehicles___parts ;
food_beverage = food___beverage;
keep motor_vehicles_parts food_beverage health
general_merchandise gdp djia cpi year;
run;

ods rtf file = "C:\Bookdata\figure3.1.rtf" style = JOURNAL
startpage = no;
ods graphics / reset height = 2in width = 6.5in;
proc sgplot data = df1;
       series x = year y = motor_vehicles_parts;
       series x = year y = food_beverage/lineattrs = (pattern
       = dashdotdot);
       series x = year y = health/lineattrs = (pattern =
       dashdashdot);
```

```
        series x = year y = general_merchandise/lineattrs =
        (pattern = shortdash);
        xaxis values = (1990 to 2015 by 5);
        yaxis label = "millions" values = (80000 to 1200000 by
        112000);
        title "Us Retail Sales from 1992 to 2015";

run;
ods graphics / reset height = 1.2in width = 6.5in;
proc sgplot data = df1;
        series x = year y = gdp;
        xaxis values = (1990 to 2015 by 5);
        yaxis label = "trillions" values = (5 to 20 by 3);
        title "GDP from 1992 to 2015";
run;
proc sgplot data = df1;
        series x = year y = djia;
        xaxis values = (1990 to 2015 by 5);
        yaxis label = "index" values = (3300 to 18000 by 2940);
        title "Dow Jones Industrial Average from 1992 to 2015";
run;
proc sgplot data = df1;
        series x = year y = cpi;
        xaxis values = (1990 to 2015 by 5);
        yaxis label = "index" values = (140 to 240 by 20);
        title "Consumer Price Index From 1992 to 2015";
run;
ods rtf close;
```

SAS Code for Example 3.2

```
/*Example 3.2*/
proc import file = "C:\Bookdata\WW3a.csv" out = df2 replace;
run;

data df2;
set df2;
Y1t = Motor vehicles_parts;
Y2t = Food_beverage;
Y3t = Health;
Y4t = General_merchandise;
run;

proc autoreg data = df2;
model Y1t = GDP GDP_1 GDP_2 DJIA_1 CPI/NORMAL NLAG = 1 METHOD = ML;
output out = data1 residual = R1;
```

```
title "GLS Regression for Y1t";
run;
proc autoreg data = df2;
model Y2t = GDP GDP_1 GDP_2 DJIA_1 CPI/NORMAL NLAG = 1 METHOD = ML;
title "GLS Regression for Y2t";
output out = data2 residual = R2;
run;
proc autoreg data = df2;
model Y3t = GDP GDP_1 GDP_2 DJIA_1 CPI/NORMAL NLAG = 1 METHOD = ML;
title "GLS Regression for Y3t";
output out= data3 residual = R3;
run;
proc autoreg data = df2;
model Y4t = GDP GDP_1 GDP_2 DJIA_1 CPI/NORMAL NLAG = 1 METHOD = ML;
title "GLS Regression for Y4t";
output out = data4 residual = R4;
nloptions maxiter = 10000000 maxtime = 3000;
run;
```

SAS Code for Example 3.3

```
/*Example 3.3*/
proc import file = "c:\bookdata\ww3a.csv" out = df3 replace;
run;

data df3;
set df3;
Y1t = Motor_vehicles_parts;
Y2t = Food_beverage;
Y3t = Health;
Y4t = General_merchandise;
run;

data work.df3;
set work.df3;
array y(4) y1t y2t y3t y4t;
do i = 1 to 4;
if year = 2015 then y(i)=.;
end;
run;

proc varmax data= work.df3;
model y1t y2t y3t y4t = gdp gdp_1 gdp_2 djia_1 cpi/ method=ls;
output lead=1;
run;
```

SAS Code for Example 3.4

```
/*Example 3.4*/
proc import file = "c:\bookdata\ww3a.csv" out = df4 replace;
run;

data df4;
set df4;
Y1t = Motor_vehicles_parts;
Y2t = Food__beverage;
Y3t = Health;
Y4t = General_merchandise;
run;

/*model before restriction*/
proc varmax data = work.df4;
model y1t y2t y3t y4t = gdp djia cpi/
p = 1 xlag = 2 method = ls;
output lead=1;
run;

/*drop data in 2015*/
data work.df4;
set work.df4;
array y(4) y1t y2t y3t y4t;
do i = 1 to 4;
if year = 2015 then y(i)=.;
end;
run;

/*model after restriction*/
proc varmax data = work.df4;
model y1t y2t y3t y4t = gdp djia cpi/
p = 1 xlag = 2 method = ls;
restrict
const(1) = 0,xl(0,1,3) = 0,xl(1,1,2) = 0,
xl(1,1,3) = 0,xl(2,1,1) = 0,xl(2,1,2) = 0,
xl(2,1,3) = 0,ar(1,1,2) = 0,ar(1,1,3) = 0,
ar(1,1,4) = 0,const(2) = 0,xl(0,2,1) = 0,
xl(0,2,2) = 0,xl(0,2,3) = 0,xl(1,2,1) = 0,
xl(1,2,2) = 0,xl(1,2,3) = 0,xl(2,2,1) = 0,
xl(2,2,2) = 0,xl(2,2,3) = 0,ar(1,2,1) = 0,
const(3) = 0,xl(0,3,1) = 0,xl(0,3,2) = 0,
xl(0,3,3) = 0,xl(1,3,1) = 0,xl(1,3,2) = 0,
xl(1,3,3) = 0,xl(2,3,1) = 0,xl(2,3,2) = 0,
xl(2,3,3) = 0,ar(1,3,1) = 0,ar(1,3,2) = 0,
```

```
ar(1,3,4) = 0,const(4) = 0,xl(0,4,2) = 0,
xl(0,4,3) = 0,xl(1,4,1) = 0,xl(1,4,2) = 0,
xl(1,4,3) = 0,xl(2,4,1) = 0,xl(2,4,2) = 0,
xl(2,4,3) = 0,ar(1,4,1) = 0,ar(1,4,3) = 0
;
output lead=1;
run;
```

SAS Code for Example 3.5

```
/*Example 3.5*/
proc import out = df5 file = "c:\bookdata\ww3b.csv" dbms =
csv replace;
run;

/*delete last 3 death*/
proc sql;
create table df5 as
select *
from df5
order by date;
create table df5 as
select monotonic() as rownum,*
from df5;
select max(rownum) -3 into: filter
from df5;
quit;

data df5;
set df5;
array death(*) la_death rs_death sd_death sj_death sa_death;
do i = 1 to dim(death);
        if rownum > &filter. then death{i} = .;
end;
drop i;
run;

/*model 1*/
proc varmax data = df5;
id date interval = day;
model la_death rs_death sd_death sj_death sa_death = la_co
rs_co sd_co sj_co sa_co la_o3 rs_o3 sd_o3 sj_o3 sa_o3/xlag =
0 print = (diagnose);
output lead = 3;
run;
```

```
/*model 2*/
proc varmax data = df5;
id date interval = day;
model la_death rs_death sd_death sj_death sa_death = la_co
rs_co sd_co sj_co sa_co la_o3 rs_o3 sd_o3 sj_o3 sa_o3 /p = 1
xlag = 0 print = (diagnose);
output lead = 3;
run;

/*model 3*/
proc varmax data = df5;
id date interval = day;
model la_death rs_death sd_death sj_death sa_death = la_co
rs_co sd_co sj_co sa_co la_o3 rs_o3 sd_o3 sj_o3 sa_o3 /xlag =
1 print = (diagnose);
output lead = 3;
run;


/*Figure 3.2*/
proc import file = "c:\bookdata\ww3b.csv" out = df1 dbms =
csv replace;
run;

data df2;
set df1;
los_angeles = la_death;
riverside = rs_death;
san_diego = sd_death;
san_jose = sj_death;
sant_ana = sa_death;
run;

data df2;
set df2;
keep date los_angeles riverside san_diego san_jose sant_ana;
run;

data df2;
set df2;
array vars{*} los_angeles riverside san_diego san_jose
sant_ana;
do i = 1 to dim(vars);
death = vars{i};
city = vname(vars{i});
output;
end;
```

```
drop i los_angeles riverside san_diego san_jose sant_ana;
run;

ods graphics / reset height = 5in width = 6.5in;
ods rtf file = "C:\Bookdata\figure3.2.rtf" style = JOURNAL
startpage = no;
proc sgpanel data = df2;
panelby city/onepanel layout=rowlattice uniscale = column
novarname;
format date date9.;
series x =date y = death;
colaxis values = ('13apr99'd to '20sep99'd) display =
(nolabel);
title 'Death in 5 cities';
run;
ods rtf close;
```

Projects

1. Find m response and k predictor time series variables with $m \geq 3$ and $k \geq 5$ of your interest. Build a multivariate multiple time series regression model with a written report and associated software code.

2. Build a vector time series regression model on the response and predictor variables from Project 1 with a written report and associated software code.

3. Build a vector time series VARX(p,s) model on the response and predictor variables from Project 1 with a written report and associated software code.

4. Use $(n - 3)$ observations to estimate the models obtained from Projects 2 and 3. Forecast the next three periods of the response variables and compare the forecast results from the two models.

5. Find a social science or natural science related time series data set, which includes multivariate responses and predictors, construct a multivariate multiple time series regression analysis, and complete with a written report and analysis software code.

References

Abraham, B. and Ledolter, J. (2006). *Introduction to Regression Modeling*. Thompson Brooks/Cole.

Box, G.E.P. and Tiao, G.C. (1975). Intervention analysis with applications to economic and environmental problems. *Journal of the American Statistical Association* **70**: 70–79.

Chatterjee, S., Hadi, A.S., and Price, B. (2006). *Regression Analysis by Examples*, 4e. Wiley.

Draper, N.R. and Smith, H. (1998). *Applied Regression Analysis*, 3e. Wiley.

Granger, C.W.J. and Newbold, P. (1986). *Forecast Economic Time Series*, 2e. Academic Press.

Islam, M.Q. (2017). Estimation and hypothesis testing in multivariate linear regression models under non-normality. *Communication of Statistics, Theory and Methods* **46**: 8521–8543.

Lütkepohl, H. (2007). *New Introduction to Multiple Time Series Analysis*. Springer.

MATLAB (2017). https://www.mathworks.com/help/matlab

Pankratz, A. (1991). *Forecasting with Dynamic Regression Models*. Wiley.

Phillips, P.C.B. (1986). Understanding spurious regressions in econometrics. *Journal of Econometrics* **33**: 311–340.

Rao, C.R. (2002). *Linear Statistical Inference and its Applications*, 2e. Wiley.

SAS Institute, Inc. (2015). *SAS/ETS User's Guide*. Cary, NC: SAS Institute, Inc.

Wei, W.W.S. (2006). *Time Series Analysis – Univariate and Multivariate Methods*, 2e. Pearson Addison-Wesley.

4

Principle component analysis of multivariate time series

Principal component analysis (PCA) is a statistical technique used for explaining the variance–covariance matrix of a set of m-dimensional variables through a few linear combinations of these variables. In this chapter, we will illustrate the method to show that a large m-dimensional process can often be sufficiently explained by smaller k principal components and thus reduce a higher dimension problem to one with fewer dimensions.

4.1 Introduction

Because of the advances of computing technology, the dimension, m, used in data analysis has become larger and larger. PCA is a statistical method that converts a set of correlated variables into a set of uncorrelated variables through an orthogonal transformation. Hopefully, a small subset of the uncorrelated variables carries sufficient information of the original large set of correlated variables.

The PCA concept to achieve parsimony was first introduced by Karl Pearson (1901). However, it was Hotelling (1933) who developed the method of stochastic variables and officially introduced the term of principal components in 1933. The techniques can be used on general variables or standardized variables and hence either the covariance matrix or correlation matrix. To many people, these techniques seem related, but in reality, they could be quite different. The goal of the method is to represent a large m-dimensional process with much smaller k principal components and hence reduce a higher dimension problem to one with fewer dimensions.

We will begin with the population PCA, discuss its properties and interpretations, and then extend it to a sample PCA. We will also discuss the large sample properties of the sample principal components.

Multivariate Time Series Analysis and Applications, First Edition. William W.S. Wei.
© 2019 John Wiley & Sons Ltd. Published 2019 by John Wiley & Sons Ltd.
Companion website: www.wiley.com/go/wei/datasets

4.2 Population PCA

Given a m-dimensional random vector $\mathbf{Z} = [Z_1, \ \ldots, \ Z_m]'$, let $\boldsymbol{\Sigma}$ be the covariance matrix,

$$\boldsymbol{\Gamma} = E\left[(\mathbf{Z}-\boldsymbol{\mu})(\mathbf{Z}-\boldsymbol{\mu})'\right] = \left[\gamma_{i,j}\right],$$

where $\boldsymbol{\mu} = E(\mathbf{Z}_t)$. We will choose a vector $\boldsymbol{\alpha} = [\alpha_1, \ \ldots, \ \alpha_m]'$ such that $Y_1 = \boldsymbol{\alpha}'\mathbf{Z}$ has the maximum variance. Moreover, to obtain a unique solution, we also require that $\boldsymbol{\alpha}'\boldsymbol{\alpha} = 1$. That is, we will choose $\boldsymbol{\alpha} = [\alpha_1, \ \ldots, \ \alpha_m]'$ such that

$$\text{Var}(Y_1) = \max_{\boldsymbol{\alpha}} \left[\boldsymbol{\alpha}'\boldsymbol{\Gamma}\boldsymbol{\alpha}\right] \text{ subject to } \boldsymbol{\alpha}'\boldsymbol{\alpha} = 1. \tag{4.1}$$

Putting them together, we obtain the solution using the method of the Lagrange multiplier. That is, let

$$V = \boldsymbol{\alpha}'\boldsymbol{\Gamma}\boldsymbol{\alpha} - \lambda(\boldsymbol{\alpha}'\boldsymbol{\alpha} - 1), \tag{4.2}$$

where λ is a Lagrange multiplier, and we maximize V with the constraint. Thus,

$$\frac{\partial V}{\partial \boldsymbol{\alpha}} = 2\boldsymbol{\Gamma}\boldsymbol{\alpha} - 2\lambda\boldsymbol{\alpha} = \mathbf{0},$$

or

$$[\boldsymbol{\Gamma} - \lambda]\boldsymbol{\alpha} = \mathbf{0}.$$

Since $\boldsymbol{\alpha} \neq \mathbf{0}$, we have

$$|\boldsymbol{\Gamma} - \lambda\mathbf{I}| = 0.$$

That is, λ is an eigenvalue and $\boldsymbol{\alpha}$ is the corresponding eigenvector of $\boldsymbol{\Gamma}$, that is

$$\boldsymbol{\Gamma}\boldsymbol{\alpha} = \lambda\boldsymbol{\alpha}. \tag{4.3}$$

In fact, because

$$\boldsymbol{\alpha}'\boldsymbol{\Gamma}\boldsymbol{\alpha} = \lambda\boldsymbol{\alpha}'\boldsymbol{\alpha} = \lambda = \text{Var}(Y_1),$$

this λ is actually the largest eigenvalue of $\boldsymbol{\Gamma}$. We will call it λ_1 and its corresponding eigenvector as $\boldsymbol{\alpha}_1 = [\alpha_{1,1}, \ \ldots, \ \alpha_{1,m}]'$.

Because $\boldsymbol{\Gamma}$ is $m \times m$, there will be m such eigenvalues, $\lambda_1 \geq \lambda_2 \geq \cdots \geq \lambda_m \geq 0$, and we denote the corresponding normalized eigenvectors as $\boldsymbol{\alpha}_1, \boldsymbol{\alpha}_2, \ldots, \boldsymbol{\alpha}_m$, where $\boldsymbol{\alpha}_i = [\alpha_{i,1}, \ \ldots, \ \alpha_{i,m}]'$. Thus, we obtain the following linear combinations

$$\begin{aligned} Y_1 &= \alpha_1'\mathbf{Z} = \alpha_{1,1}Z_1 + \alpha_{1,2}Z_2 + \cdots + \alpha_{1,m}Z_m \\ Y_2 &= \alpha_2'\mathbf{Z} = \alpha_{2,1}Z_1 + \alpha_{2,2}Z_2 + \cdots + \alpha_{2,m}Z_m \\ &\vdots \\ Y_m &= \alpha_m'\mathbf{Z} = \alpha_{m,1}Z_1 + \alpha_{m,2}Z_2 + \cdots + \alpha_{m,m}Z_m \end{aligned} \tag{4.4}$$

such that

$$\begin{aligned}\operatorname{Var}(Y_i) &= \boldsymbol{\alpha}_i'\boldsymbol{\Gamma}\boldsymbol{\alpha}_i = \lambda_i,\ i=1,2,\ldots,m,\text{and}\\ \operatorname{Cov}\left(Y_i, Y_j\right) &= \boldsymbol{\alpha}_i'\boldsymbol{\Gamma}\boldsymbol{\alpha}_j = \lambda_j\boldsymbol{\alpha}_i'\boldsymbol{\alpha}_j = 0,\ i,j=1,2,\ldots,m.\end{aligned} \tag{4.5}$$

We will call $Y_1 = \boldsymbol{\alpha}_1'\mathbf{Z}$ the first principal component, $Y_2 = \boldsymbol{\alpha}_2'\mathbf{Z}$ the second principal component, and so on.

4.3 Implications of PCA

Let $\boldsymbol{\Lambda}$ be the diagonal matrix of eigenvalues and $\mathbf{P} = [\boldsymbol{\alpha}_1,\ \boldsymbol{\alpha}_2,\ \ldots,\ \boldsymbol{\alpha}_m]$ be the matrix formed from the corresponding normalized eigenvectors in the PCA. Since $\mathbf{P}'\mathbf{P} = \mathbf{P}\mathbf{P}' = \mathbf{I}$, Eq. (4.3) implies that

$$\boldsymbol{\Gamma}\mathbf{P} = \mathbf{P}\boldsymbol{\Lambda}, \tag{4.6}$$

and

$$\boldsymbol{\Gamma} = \mathbf{P}\boldsymbol{\Lambda}\mathbf{P}'. \tag{4.7}$$

Since $\gamma_{i,i} = \operatorname{Var}(Z_i)$, $i = 1, 2, \ldots, m$, we note that

$$\begin{aligned}\sum_{i=1}^{m}\operatorname{Var}(Z_i) &= \gamma_{1,1} + \gamma_{2,2} + \cdots + \gamma_{m,m} = tr(\boldsymbol{\Sigma}) = tr(\mathbf{P}\boldsymbol{\Lambda}\mathbf{P}')\\ &= \operatorname{tr}(\boldsymbol{\Lambda}) = \lambda_1 + \lambda_2 + \cdots + \lambda_m = \sum_{i=1}^{m}\operatorname{Var}(Y_i).\end{aligned} \tag{4.8}$$

The proportion of the total variance of $\mathbf{Z}_t$ explained by the ith principal component Y_i is given by

$$\frac{\lambda_i}{\lambda_1 + \lambda_2 + \cdots + \lambda_m},\ i = 1,2,\ldots,m. \tag{4.9}$$

In many applications, the sum of the variances of the first few principal components may account for more than 85 or 90% of the total variance. This implies that the study of a given high m-dimensional process $\mathbf{Z}$ can be accomplished through the careful study of a set containing a small number of principal components without losing much information.

Let $\mathbf{e}_i$ be the ith m-dimensional unit vector with its ith element being 1 and 0 otherwise, for example, $\mathbf{e}_1 = (1,\ 0,\ \ldots,\ 0)'$ and $\mathbf{e}_2 = (0,\ 1,\ 0,\ \ldots,\ 0)'$. In many applications, we may want to find the relationship between a principal component Y_i and variable Z_j in $\mathbf{Z}$. It is interesting to note that

$$\operatorname{Cov}\left(Y_i, Z_j\right) = \operatorname{Cov}\left(\boldsymbol{\alpha}_i'\mathbf{Z}, \mathbf{e}_j'\mathbf{Z}\right) = \boldsymbol{\alpha}_i'\boldsymbol{\Sigma}\mathbf{e}_j = \boldsymbol{\alpha}_i'\lambda_i\mathbf{e}_j = \lambda_i\alpha_{i,j}.$$

It follows that

$$\rho_{Y_i,Z_j} = \frac{\mathrm{Cov}(Y_i,Z_j)}{\sqrt{\mathrm{Var}(Y_i)}\sqrt{\mathrm{Var}(Z_j)}} = \frac{\lambda_i \alpha_{i,j}}{\sqrt{\lambda_i}\sqrt{\gamma_{j,j}}} = \frac{\alpha_{i,j}\sqrt{\lambda_i}}{\sqrt{\gamma_{j,j}}}, \tag{4.10}$$

for $i, j = 1, 2, \ldots, m$. Thus, one can use the relative magnitude of the correlations or the coefficients $\alpha_{i,j}$ associated with variable Z_j to interpret the principal components. These correlations and coefficients together with their positive or negative signs can also be used to measure the importance of the variables in $\mathbf{Z}$ to a given principal component. They sometimes lead to different rankings, but they often reach similar results.

In these discussions, we obtain principal components $Y_j = \boldsymbol{\alpha}_j'\mathbf{Z}$ for a given process $\mathbf{Z}$ through its variance–covariance matrix $\boldsymbol{\Gamma}$. Clearly, we can also construct these principal components through its correlation matrix,

$$\boldsymbol{\rho} = \mathbf{D}^{-1/2}\boldsymbol{\Gamma}\mathbf{D}^{-1/2}, \tag{4.11}$$

where $\mathbf{D}$ is the diagonal matrix in which the ith diagonal element is the variance of the ith process, that is

$$\mathbf{D} = \mathrm{diag}\left[\gamma_{1,1}, \gamma_{2,2}, \ldots, \gamma_{m,m}\right]. \tag{4.12}$$

In other words, we will construct the principal components for the standardized variables,

$$\mathbf{U} = [U_1, \ \ldots, \ U_m]' = \left[\frac{Z_1 - \mu_1}{\sqrt{\gamma_{1,1}}}, \ \ldots, \ \frac{Z_m - \mu_m}{\sqrt{\gamma_{m,m}}}\right]' = \mathbf{D}^{-1/2}(\mathbf{Z} - \boldsymbol{\mu}) \tag{4.13}$$

and $\boldsymbol{\rho}$ is the covariance matrix of $\mathbf{U}$ because $\mathrm{Cov}(\mathbf{U}) = \mathbf{D}^{-1/2}\boldsymbol{\Gamma}\mathbf{D}^{-1/2} = \boldsymbol{\rho}$. Hence, the procedure of constructing the principal components from these standardized variables is exactly the same as before. The first natural question to ask is, why do we use standardized variables? One obvious answer is that in many applications the unit used in variables may not be the same. For example, if values of one variable are mostly small numbers between 0 and 1 like percentages but values of one other variable are mostly numbers in the millions or billions like imports and exports, then to avoid the unnecessary impact of the different units used in these variables, we will naturally consider using standardized variables. The second question to ask is, are the two sets of principal components constructed from $\boldsymbol{\Gamma}$ and $\boldsymbol{\rho}$ the same? The unfortunate answer is decidedly no. In fact, they could be very different. The choice depends on applications and could be a challenge in applied research.

4.4 Sample principle components

In practice, we may not know the parameter value $\boldsymbol{\Gamma}$ of $\mathbf{Z}$. So, given a m-dimensional stationary vector time series $\mathbf{Z}_t$, we will simply replace the unknown population variance–covariance matrix, $\boldsymbol{\Gamma}$, by the sample variance–covariance matrix, $\hat{\boldsymbol{\Gamma}}$, computed from $\mathbf{Z}_t$, $t = 1, 2, \ldots, n$,

$$\hat{\mathbf{\Gamma}} = \frac{1}{n}\sum_{t=1}^{n}(\mathbf{Z}_t-\bar{\mathbf{Z}})(\mathbf{Z}_t-\bar{\mathbf{Z}})' = \begin{bmatrix} \hat{\gamma}_{1,1} & \hat{\gamma}_{1,2} & \cdots & \hat{\gamma}_{1,m} \\ \hat{\gamma}_{2,1} & \hat{\gamma}_{2,2} & \cdots & \hat{\gamma}_{2,m} \\ \vdots & \vdots & \vdots & \vdots \\ \hat{\gamma}_{m,1} & \hat{\gamma}_{m,2} & \cdots & \hat{\gamma}_{m,m} \end{bmatrix}, \tag{4.14}$$

where

$\bar{\mathbf{Z}} = (\bar{Z}_1, \bar{Z}_2, \ldots, \bar{Z}_m)'$ is the sample mean vector,

$\hat{\gamma}_{i,i} = \frac{1}{n}\sum_{t=1}^{n}(Z_{i,t}-\bar{Z}_i)^2$ is the sample variance of the Z_i,

$\hat{\gamma}_{i,j} = \frac{1}{n}\sum_{t=1}^{n}(Z_{i,t}-\bar{Z}_i)(Z_{j,t}-\bar{Z}_j)$ is the sample covariance between Z_i and Z_j.

In exactly the same approach, we will now choose a vector $\hat{\boldsymbol{\alpha}} = [\hat{\alpha}_1, \ \ldots, \ \hat{\alpha}_m]'$ such that $\hat{Y} = \hat{\boldsymbol{\alpha}}'\mathbf{Z}_t$ has the maximum sample variance, where we denote the vector and its resulting linear combination with a hat for distinction between the population and sample principal components. Thus, the first sample principal component will be obtained by

$$\hat{Y}_1 = \hat{\boldsymbol{\alpha}}_1'\mathbf{Z}_t \text{ that maximizes the sample variance of } \hat{\boldsymbol{\alpha}}_1'\mathbf{Z}_t \text{ subject to } \hat{\boldsymbol{\alpha}}'_1\hat{\boldsymbol{\alpha}}_1 = 1.$$

The second sample principal component will be obtained by

$$\hat{Y}_2 = \hat{\boldsymbol{\alpha}}'_2\mathbf{Z}_t \text{ that maximizes the sample variance of } \hat{\boldsymbol{\alpha}}_2'\mathbf{Z} \text{ subject to } \hat{\boldsymbol{\alpha}}_2'\hat{\boldsymbol{\alpha}}_2 = 1$$

$$\text{and zero sample covariance between } \hat{\boldsymbol{\alpha}}_1'\mathbf{Z} \text{ and } \hat{\boldsymbol{\alpha}}_2'\mathbf{Z}_t.$$

Continuing, the ith sample principal component will be obtained by

$$\hat{Y}_i = \hat{\boldsymbol{\alpha}}'_i\mathbf{Z}_t \text{ that maximizes the sample variance of } \hat{\boldsymbol{\alpha}}_i'\mathbf{Z}_t \text{ subject to } \hat{\boldsymbol{\alpha}}_i'\hat{\boldsymbol{\alpha}}_i = 1, \text{where}$$
$$\hat{\boldsymbol{\alpha}}_i = [\hat{\alpha}_{i,1}, \ \ldots, \ \hat{\alpha}_{i,m}]', \text{and zero sample covariance between the pairs } \left(\hat{\boldsymbol{\alpha}}_i'\mathbf{Z}_t, \hat{\boldsymbol{\alpha}}_j'\mathbf{Z}_t\right) \text{ for all } i > j.$$

Let $(\hat{\lambda}_1, \hat{\boldsymbol{\alpha}}_1), (\hat{\lambda}_2, \hat{\boldsymbol{\alpha}}_2), \ldots, (\hat{\lambda}_m, \hat{\boldsymbol{\alpha}}_m)$ be the eigenvalue–eigenvector pairs of the sample variance–covariance matrix, $\hat{\mathbf{\Gamma}}$. The ith sample principal component is in fact given by

$$\hat{Y}_i = \hat{\boldsymbol{\alpha}}'_i\mathbf{Z}_t = \hat{\alpha}_{i,1}Z_{1,t} + \hat{\alpha}_{i,2}Z_{2,t} + \cdots + \hat{\alpha}_{i,m}Z_{m,t}, \ i = 1, 2, \ldots, m,$$

where $\hat{\lambda}_1 \geq \hat{\lambda}_2 \geq \cdots \geq \hat{\lambda}_m \geq 0$. From the process and the results from Sections 4.2 and 4.3, we have

1. $\hat{\lambda}_i =$ sample variance $\hat{Y}_i$, $i = 1, 2, \ldots, m$.
2. Sample covariance between $\hat{Y}_i$ and $\hat{Y}_j$ is zero for $i \neq j$.
3. $\sum_{i=1}^{m}\hat{\gamma}_{i,i} = \hat{\lambda}_1 + \hat{\lambda}_2 + \cdots + \hat{\lambda}_m =$ total sample variance.
4. Sample covariance between $\hat{Y}_i$ and $Z_j = \hat{\lambda}_i\hat{\alpha}_{i,j}$, $i, j = 1, 2, \ldots, m$.
5. Sample correlation between $\hat{Y}_i$ and $Z_j = \sqrt{\hat{\lambda}_i}\hat{\alpha}_{i,j}/\sqrt{\hat{\gamma}_{j,j}}$, $i, j = 1, 2, \ldots, m$.

Similar to the population principal components, we can obtain the sample principal components through the sample variance–covariance matrix $\hat{\mathbf{\Gamma}}$ or the sample correlation matrix,

$$\hat{\boldsymbol{\rho}} = \hat{\mathbf{D}}^{-1/2}\hat{\mathbf{\Gamma}}\hat{\mathbf{D}}^{-1/2}, \tag{4.15}$$

where $\hat{\mathbf{D}}$ is the diagonal matrix in which the ith diagonal element is the sample variance of $Z_{i,\,t}$, that is

$$\hat{\mathbf{D}} = \text{diag}\left[\hat{\gamma}_{1,1}, \hat{\gamma}_{2,2}, \ldots, \hat{\gamma}_{m,m}\right]. \tag{4.16}$$

In other words, we will construct the sample principal components for the standardized sample variables,

$$\hat{\mathbf{U}} = \left[\hat{U}_1, \ \ldots, \ \hat{U}_m\right]' = \left[\frac{Z_1 - \bar{Z}_1}{\sqrt{\hat{\gamma}_{1,1}}}, \ \ldots, \ \frac{Z_m - \bar{Z}_m}{\sqrt{\hat{\gamma}_{m,m}}}\right]' = \hat{\mathbf{D}}^{-1/2}(\mathbf{Z}_t - \bar{\mathbf{Z}}) \tag{4.17}$$

and $\hat{\boldsymbol{\rho}}$ is the sample covariance matrix of $\hat{\mathbf{U}}$. The procedure is exactly the same as before. For simplicity, we will use the same notations, $\hat{Y}_i, \hat{\lambda}_i, \hat{\boldsymbol{\alpha}}_i$ for the sample principal components and their associated eigenvalues and eigenvectors irrespective of whether they are constructed from $\hat{\mathbf{\Gamma}}$ or $\hat{\boldsymbol{\rho}}$. Whether from $\hat{\mathbf{\Gamma}}$ or $\hat{\boldsymbol{\rho}}$ should be clear from the applications. Again, the two sets of sample principal components constructed from $\hat{\mathbf{\Gamma}}$ and $\hat{\boldsymbol{\rho}}$ will in general be different. The choice depends on applications.

Once the sample principal components are obtained, the next natural step is to investigate their properties through $\hat{\lambda}_i$ and $\hat{\boldsymbol{\alpha}}_i$. Their large sample properties are summarized in the following theorem, and we refer readers to Anderson (1963) for its proof. We also refer readers to reference books by Johnson and Wichern (2002), and Rao (2002).

Theorem 4.1 Let $\hat{\mathbf{\Gamma}}$ be the sample variance–covariance matrix from a multivariate Gaussian process $\mathbf{N}(\boldsymbol{\mu}, \mathbf{\Gamma})$, $\boldsymbol{\lambda} = [\lambda_1, \ \lambda_2, \ \ldots, \ \lambda_m]$, and $\hat{\boldsymbol{\lambda}} = \left[\hat{\lambda}_1, \ \hat{\lambda}_2, \ \ldots, \ \hat{\lambda}_m\right]$, where λ_i and $\hat{\lambda}_i$ are the eigenvalues of $\mathbf{\Gamma}$ and $\hat{\mathbf{\Gamma}}$, respectively. Let $\mathbf{\Lambda}$ = *diag* $[\lambda_1, \lambda_2, \ldots, \lambda_m]$ and assume that $\lambda_1 > \ \lambda_2 > \ \cdots > \ \lambda_m > 0$. Then,

1. $\lim_{n\to\infty} \sqrt{n}\left(\hat{\boldsymbol{\lambda}} - \boldsymbol{\lambda}\right) = \mathbf{N}\left(\mathbf{0}, 2\mathbf{\Lambda}^2\right)$.
2. Let $\boldsymbol{\alpha}_1, \ \boldsymbol{\alpha}_2, \ \ldots, \ \boldsymbol{\alpha}_m$ be the corresponding eigenvectors for the eigenvalues $\lambda_1, \ \lambda_2, \ \ldots, \ \lambda_m$ from $\mathbf{\Gamma}, \hat{\boldsymbol{\alpha}}_1, \hat{\boldsymbol{\alpha}}_2, \ldots, \hat{\boldsymbol{\alpha}}_m$ be the corresponding eigenvectors for the eigenvalues $\hat{\lambda}_1, \ \hat{\lambda}_2, \ \ldots, \ \hat{\lambda}_m$ from $\hat{\mathbf{\Gamma}}$, and

$$\mathbf{\Omega}_i = \lambda_i \sum_{\substack{i=1 \\ j \neq i}}^{m} \frac{\lambda_i}{\left(\lambda_i - \lambda_j\right)^2} \boldsymbol{\alpha}_i \boldsymbol{\alpha}_i'. \tag{4.18}$$

Then $\lim_{n\to\infty} \sqrt{n}(\hat{\boldsymbol{\alpha}}_i - \boldsymbol{\alpha}_i) = \mathbf{N}(\mathbf{0}, \mathbf{\Omega}_i)$.

3. Each eigenvalue $\hat{\lambda}_i$ is distributed independently of the elements $\hat{\alpha}_{i,j}, j = 1, \ldots, m$ of the associated eigenvector $\hat{\boldsymbol{\alpha}}_i$.

The result in (1) implies that the $\hat{\lambda}_i$ values are independently normally distributed as $N(\lambda_i, 2\lambda_i/n)$. Hence, the $100(1-\alpha)\%$ confidence interval for λ_i can be obtained as follows:

$$\frac{\hat{\lambda}_i}{1 + N_{\alpha/2}\sqrt{2/n}} \le \lambda_i \le \frac{\hat{\lambda}_i}{1 - N_{\alpha/2}\sqrt{2/n}}, \tag{4.19}$$

where $N_{\alpha/2}$ is the upper $100(\alpha/2)$th percentile of the standard normal distribution.

From these discussions, it is clear that PCA is a statistical method used to find k linear combinations of m original statistical variables through the analysis of the covariance or correlation matrix of these m variables with k being much less than m, so that the study of a given high m-dimensional process can be accomplished through the study of a much smaller number of principal components without losing much information. Obviously, this makes sense only when the covariance and correlation is stable and constant. Hence, for a time series process, it has to be stationary. For a nonstationary series, it needs to be reduced to stationary by using some transformations such as power transformation and differencing. Since the residual vector, $\mathbf{a}_t$, is actually a function of $\mathbf{Z}_t$, one can also analyze the covariance or correlation matrix of residuals after a VAR model fitting. Also, when using PCA, we normally work on a mean adjusted data set.

4.5 Empirical examples

4.5.1 Daily stock returns from the first set of 10 stocks

Example 4.1 Let us consider the data set, listed as WW4a in the Data Appendix, of the daily stock returns from 10 different stocks, including CVX (Chevron), XOM (Exxon), AAPL (Apple), FB (Facebook), MSFT (Microsoft), MRK (Merck), PFE (Pfizer), BAC (Bank of America), JPM (JP Morgan), and WFC (Wells Fargo & Co.) that were traded on the New York Stock Exchange from August 2, 2016 to December 30, 2016. The data are plotted in Figure 4.1.

In this example, the sample $\mathbf{Z}_t = [Z_{1,t}, Z_{2,t}, Z_{3,t}, Z_{4,t}, Z_{5,t}, Z_{6,t}, Z_{7,t}, Z_{8,t}, Z_{9,t}, Z_{10,t}]$, $t = 1, 2, \ldots, n$, is the daily stock returns for Chevron ($Z_{1,t}$), Exxon ($Z_{2,t}$), Apple ($Z_{3,t}$), Facebook ($Z_{4,t}$), Microsoft ($Z_{5,t}$), Merck ($Z_{6,t}$), Pfizer ($Z_{7,t}$), Bank of America ($Z_{8,t}$), JP Morgan ($Z_{9,t}$), and Wells Faro & Co ($Z_{10,t}$) that were traded on the New York Stock Exchange from August 2, 2016 to December 30, 2016 and we have $m = 10$ and $n = 106$. We have the sample mean vector,

$$\begin{aligned}\bar{\mathbf{Z}} &= (\bar{Z}_1, \bar{Z}_2, \ldots, \bar{Z}_{10})' \\ &= (0.001\,68, 0.000\,52, 0.000\,89, -0.000\,66, 0.000\,94, 0.000\,16, -0.001\,23, \\ &\quad 0.000\,42, 0.002\,911, 0.001\,41)',\end{aligned}$$

with sample covariance matrix, $\hat{\boldsymbol{\Gamma}}$,

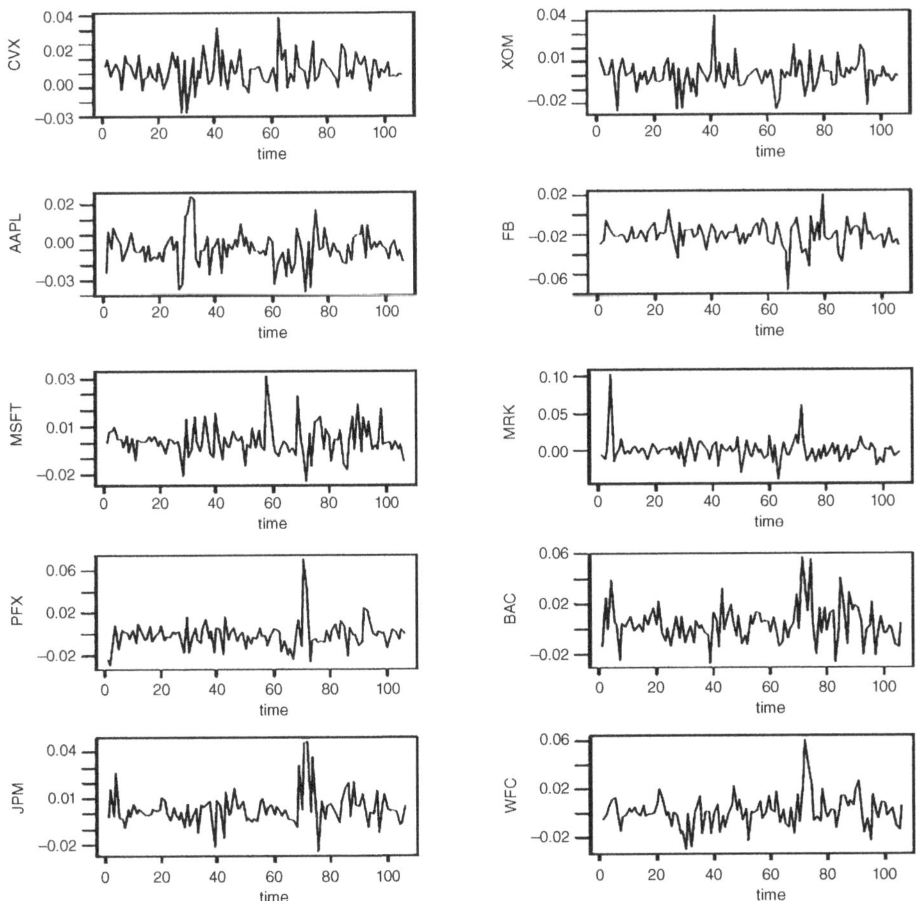

Figure 4.1 Daily stock returns for 10 stocks between August 2 and December 30, 2016.

	CVX	XOM	AAPL	FB	MSFT
CVX	1.087154e−04	6.983585e−05	8.576443e−06	2.841400e−05	2.349692e−05
XOM	6.983585e−05	1.063956e−04	1.159600e−05	2.475253e−05	2.188846e−05
AAPL	8.576443e−06	1.159600e−05	1.231327e−04	6.237891e−05	5.277169e−05
FB	2.841400e−05	2.475253e−05	6.237891e−05	1.487562e−04	6.697871e−05
MSFT	2.349692e−05	2.188846e−05	5.277169e−05	6.697871e−05	1.012891e−04
MRK	2.015158e−05	5.188890e−05	3.017592e−05	3.534955e−05	2.720992e−0
PFE	1.959854e−05	4.650749e−05	1.080636e−05	3.061439e−05	1.225974e−05
BAC	4.039745e−05	5.224031e−05	−1.339308e−05	−2.972687e−06	7.338406e−06
JPM	3.425218e−05	3.701334e−05	−1.168483e−05	6.079249e−07	6.054941e−06
WFC	1.889755e−05	2.852743e−05	−2.163606e−05	−1.736215e−05	−1.199474e−05

	MRK	PFE	BAC	JPM	WFC
CVX	2.015158e−05	1.959854e−05	4.039745e−05	3.425218e−05	1.889755e−05
XOM	5.188890e−05	4.650749e−05	5.224031e−05	3.701334e−05	2.852743e−05
AAPL	3.017592e−05	1.080636e−05	−1.339308e−05	−1.168483e−05	−2.163606e−05
FB	3.534955e−05	3.061439e−05	−2.972687e−06	6.079249e−07	−1.736215e−05
MSFT	2.720992e−05	1.225974e−05	7.338406e−06	6.054941e−06	−1.199474
MRK	2.522852e−04	1.180967e−04	9.456417e−05	7.451309e−05	2.881565e−05
PFE	1.180967e−04	1.659857e−04	7.508658e−05	6.907373e−05	3.707669e−05
BAC	9.456417e−05	7.508658e−05	2.364619e−04	1.434413e−04	1.192228e−04
JPM	7.451309e−05	6.907373e−05	1.434413e−04	1.184657e−04	8.792855e−05
WFC	2.881565e−05	3.707669e−05	1.192228e−04	8.792855e−05	1.720248e−04

and the sample correlation matrix, $\hat{\boldsymbol{\rho}}$,

	CVX	XOM	AAPL	FB	MSFT
CVX	1.00000000	0.6493384	0.07412677	0.223434058	0.22391548
XOM	0.64933841	1.0000000	0.10131175	0.196752448	0.21084926
AAPL	0.07412677	0.1013118	1.00000000	0.460907181	0.47253396
FB	0.22343406	0.1967524	0.46090718	1.000000000	0.54565465
MSFT	0.22391548	0.2108493	0.47253396	0.545654650	1.00000000
MRK	0.12167962	0.3167136	0.17120954	0.182473725	0.17021586
PFE	0.14589582	0.3499658	0.07558875	0.194828564	0.09455067
BAC	0.25195788	0.3293542	−0.07848980	−0.015850063	0.04741763
JPM	0.30181896	0.3296857	−0.09674747	0.004579479	0.05527545
WFC	0.13818618	0.2108654	−0.14866064	−0.108535145	−0.09086861

	MRK	PFE	BAC	JPM	WFC
CVX	0.1216796	0.14589582	0.25195788	0.301818963	0.13818618
XOM	0.3167136	0.34996582	0.32935425	0.329685727	0.21086537
AAPL	0.1712095	0.0755887	−0.07848980	−0.096747468	−0.14866064
FB	0.1824737	0.19482856	−0.01585006	0.004579479	−0.10853514
MSFT	0.1702159	0.09455067	0.04741763	0.055275452	−0.09086861
MRK	1.0000000	0.57710707	0.38716859	0.431012985	0.13832064
PFE	0.5771071	1.00000000	0.37900626	0.492585061	0.21941693
BAC	0.3871686	0.37900626	1.00000000	0.857032493	0.59113031
JPM	0.4310130	0.49258506	0.85703249	1.000000000	0.61593958
WFC	0.1383206	0.21941693	0.59113031	0.615939576	1.00000000

4.5.1.1 The PCA based on the sample covariance matrix

The eigenvalues and eigenvectors, which are often known as variances and component loadings, of the sample variance–covariance matrix, are given in Table 4.1.

Table 4.1 Sample PCA results for the 10 stock returns based on the sample covariance matrix.

	Comp. 1 $\hat{\alpha}_1$	Comp. 2 $\hat{\alpha}_2$	Comp. 3 $\hat{\alpha}_3$	Comp. 4 $\hat{\alpha}_4$	Comp. 5 $\hat{\alpha}_5$	Comp. 6 $\hat{\alpha}_6$	Comp. 7 $\hat{\alpha}_7$	Comp. 8 $\hat{\alpha}_8$	Comp. 9 $\hat{\alpha}_9$	Comp. 10 $\hat{\alpha}_{10}$
CVX	−0.166	−0.104	−0.407	0.576	0.120				0.622	−0.227
XOM	−0.230	−0.126	−0.243	0.552		0.203			−0.686	0.215
AAPL		−0.431	−0.173	−0.354	0.180	0.377	−0.639	−0.264		
FB		−0.513	−0.319	−0.231	−0.324	−0.213	0.505	−0.399		
MSFT		−0.362	−0.286	−0.197	0.127	−0.116		0.841		
MRK	−0.470	−0.329	0.596		0.396	0.223	0.320			
PFE	−0.369	−0.167	0.316	0.126	−0.725	−0.118	−0.358	0.136		−0.165
BAC	−0.539	0.289	−0.191	−0.222	0.279	−0.438	−0.164	−0.155	−0.217	−0.416
JPM	−0.391	0.175				−0.206			0.265	0.827
WFC	−0.324	0.379	−0.251	−0.265	−0.266	0.678	0.255			−0.110
Variance $(\hat{\lambda}_i)$	0.00058	0.00030	0.00018	0.00011	0.00009	0.00008	0.00007	0.00005	0.00003	0.00002
Percentage of Total Variance	0.384	0.199	0.119	0.073	0.060	0.053	0.046	0.033	0.020	0.013

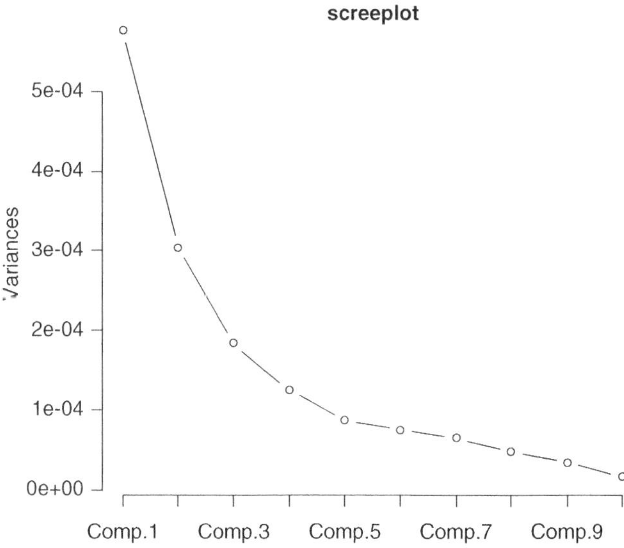

Figure 4.2 Screeplot of the PCA from the sample covariance matrix.

Figure 4.2 shows a useful plot, which is often known as a scree plot or *screeplot*. It is a plot of eigenvalues $\hat{\lambda}_i$ versus i, that is, the magnitude of an eigenvalue versus its number. To determine the suitable number, k, of components, we look for an elbow in the plot, where the component preceding the vertex of the elbow is chosen to be cutoff point k. In our example, we will choose k to be 3 or 4.

The eigenvalues of the first four components account for

$$\left(\frac{\hat{\lambda}_1+\hat{\lambda}_2+\hat{\lambda}_3+\hat{\lambda}_4}{\hat{\lambda}_1+\hat{\lambda}_2+\hat{\lambda}_3+\hat{\lambda}_4+\hat{\lambda}_5+\hat{\lambda}_6+\hat{\lambda}_7+\hat{\lambda}_8+\hat{\lambda}_9+\hat{\lambda}_{10}}\right)100\%$$
$$=\left(\frac{0.000\,58+0.0003+0.000\,18+0.000\,11}{0.001\,51}\right)100\%=77.5\%$$

of the total sample variance.

The four sample principal components are

$$\hat{Y}_1=\hat{\alpha}_1\mathbf{Z}=-0.287Z_1-0.354Z_2$$
$$-0.47Z_6-0.369Z_7-0.539Z_8-391Z_9-0.324Z_{10}$$
$$\hat{Y}_2=\hat{\alpha}_2\mathbf{Z}=-0.104Z_1-0.126Z_2-0.431Z_3-0.513Z_4-0.362Z_5$$
$$-0.329Z_6-0.167Z_7+0.289Z_8+0.175Z_9+0.379Z_{10}$$
$$\hat{Y}_3=\hat{\alpha}_3\mathbf{Z}=-0.407Z_1-0.243Z_2-0.173Z_3-0.319Z_4-0.286Z_5$$
$$+0.596Z_6+0.316Z_7-0.191Z_8-0.251Z_{10}$$
$$\hat{Y}_4=\hat{\alpha}_4\mathbf{Z}=0.576Z_1+0.552Z_2-0.354Z_3-0.231Z_4-0.197Z_5$$
$$+0.126Z_7-0.222Z_8\ -0.256Z_{10}$$

Now, let us examine the four components more carefully. The first component represents the general market other than communication technology sector. The second component represents the contrast between financial and non-financial sectors. The third component represents the contrast between health and non-health sectors. The fourth component represents the contrast between oil and non-oil industries. Thus, the PCA has provided us with four components that contain a vast amount of information for the 10 stock returns traded on the New York Stock Exchange.

4.5.1.2 The PCA based on the sample correlation matrix

Now let us try the PCA using the sample correlation matrix, which is based on the standardized variables, $\hat{\mathbf{U}} = \left[\hat{U}_1, \ \ldots, \ \hat{U}_m\right]' = \left[\frac{Z_1 - \bar{Z}_1}{\sqrt{\hat{\gamma}_{1,1}}}, \ \ldots, \ \frac{Z_m - \bar{Z}_m}{\sqrt{\hat{\gamma}_{m,m}}}\right]'$.

The eigenvalues and eigenvectors, which are also known as variances and component loadings, of the sample correlation matrix are given in Table 4.2. The screeplot is shown in Figure 4.3.

The screeplot again indicates $k = 4$. The eigenvalues of the first four components account for

$$\left(\frac{\hat{\lambda}_1 + \hat{\lambda}_2 + \hat{\lambda}_3 + \hat{\lambda}_4}{m}\right) 100\% = \left(\frac{3.393 + 2.21 + 1.196 + 0.939}{10}\right) 100\% = 77.38\%$$

which is almost the same as the one obtained using the covariance matrix. The four sample principal components are now

$$\begin{aligned}
\hat{Y}_1 &= \hat{\alpha}_1 \hat{\mathbf{U}} = -0.287U_1 - 0.354U_2 - 0.143U_4 \ -0.15U_5 \\
&\quad -0.346U_6 - 0.364U_7 - 0.433U_8 - 0.458U_9 \ -0.31U_{10} \\
\hat{Y}_2 &= \hat{\alpha}_2 \hat{\mathbf{U}} = -0.155U_1 - 0.137U_2 - 0.491U_3 - 0.506U_4 \ -0.49U_5 \\
&\quad +0.236U_8 + 0.231U_9 + 0.32U_{10} \\
\hat{Y}_3 &= \hat{\alpha}_3 \hat{\mathbf{U}} = 0.654U_1 + 0.464U_2 - 0.199U_3 - 0.418U_6 - 0.354U_7 \\
\hat{Y}_4 &= \hat{\alpha}_4 \hat{\mathbf{U}} = -0.136U_1 - 0.32U_2 + 0.233U_3 + 0.171U_4 + 0.341U_5 \\
&\quad -0.405U_6 - 0.429U_7 + 0.281U_8 + 0.206U_9 + 0.458U_{10}
\end{aligned}$$

Now, let us examine the four components. The first component now represents the general stock market. The second component represents the contrast mainly between financial and non-health related sectors. The third component represents the contrast between oil and health sectors. The fourth component now represents the contrast between financial/technology and non-financial/non-technology industry. For this data set, the PCA results from the covariance matrix and the correlation matrix are very much equivalent.

Table 4.2 Sample PCA result for the Greater New York City CPI based on the sample correlation matrix.

	Comp. 1 $\hat{\alpha}_1$	Comp. 2 $\hat{\alpha}_2$	Comp. 3 $\hat{\alpha}_3$	Comp. 4 $\hat{\alpha}_4$	Comp. 5 $\hat{\alpha}_5$	Comp. 6 $\hat{\alpha}_6$	Comp. 7 $\hat{\alpha}_7$	Comp. 8 $\hat{\alpha}_8$	Comp. 9 $\hat{\alpha}_9$	Comp. 10 $\hat{\alpha}_{10}$
CVX	−0.287	−0.155	0.654	−0.136			0.287		0.578	−0.150
XOM	−0.354	−0.137	0.464	−0.320	0.187		−0.319	0.116	−0.598	0.146
AAPL		−0.491	−0.199	0.233	0.744	−0.192	0.163	−0.204		
FB	−0.143	−0.506		0.171	−0.517	−0.416	0.355	0.308	−0.167	
MSFT	−0.150	−0.490		0.341	−0.241	0.493	−0.539	−0.131	0.102	
MRK	−0.346		−0.418	−0.405	0.164	0.318		0.602	0.207	
PFE	−0.364		−0.354	−0.429	−0.198	−0.367	−0.217	−0.538	0.143	−0.166
BAC	−0.433	0.236		0.281		0.258	0.310	−0.121	−0.367	−0.602
JPM	−0.458	0.231		0.206		0.154	0.238	−0.210		0.747
WFC	−0.310	0.320		0.458	0.144	−0.457	−0.415	0.342	0.269	
Variance $(\hat{\lambda}_i)$	3.393	2.210	1.196	0.939	0.557	0.504	0.406	0.378	0.296	0.121
Cumulative % of Total Variance	33.93%	56.03%	67.99%	77.38%	82.85%	87.99%	92.95	95.83%	98.79%	100%

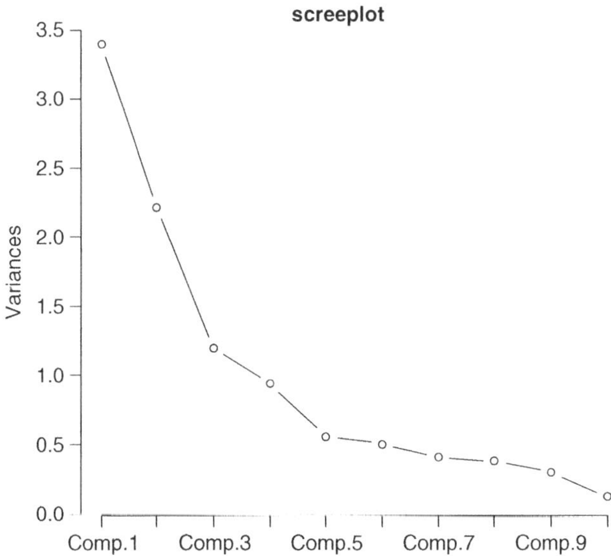

Figure 4.3 Screeplot of the PCA from the sample correlation matrix.

4.5.2 Monthly Consumer Price Index (CPI) from five sectors

Example 4.2 Let us consider the data set, WW4b, for the monthly CPI from five different sectors, energy, apparel, commodities, housing, and gas, from January 1986 to December 2014 in Greater New York City, where the January 1984 index is the reference point at 100 for all sectors. The data are plotted in Figure 4.4 and listed as WW4b in the Data Appendix. The variables are observed monthly from January 1986 through December 2014.

The plot shows that these series are seasonal and nonstationary. So, we remove seasonality using the method of Cleveland et al. (1990) and trend phenomenon with differencing. With the notations introduced before, we have $m = 5$ and $n = 348$. Let $Z_{1,t}$, $Z_{2,t}$, $Z_{3,t}$, $Z_{4,t}$, $Z_{5,t}$ be the monthly price index for energy, apparel, commodities, housing, and gas, respectively. Then, we have the sample mean vector,

$$\begin{aligned}\bar{\mathbf{Z}} &= (\bar{Z}_1, \bar{Z}_2, \ldots, \bar{Z}_5)' \\ &= (144.06, 119.59, 153.07, 195.06, 148.59)',\end{aligned}$$

the sample covariance matrix,

$$\hat{\boldsymbol{\Gamma}} = \begin{bmatrix} 2964.002 & -5.161 & 1276.581 & 2605.703 & 3919.732 \\ -5.161 & 70.674 & 38.965 & 19.095 & 18.230 \\ 1276.581 & 38.965 & 627.869 & 1250.621 & 1663.275 \\ 2605.703 & 19.095 & 1250.621 & 2567.474 & 3336.379 \\ 3919.732 & 18.230 & 1663.275 & 3336.379 & 5329.116 \end{bmatrix},$$

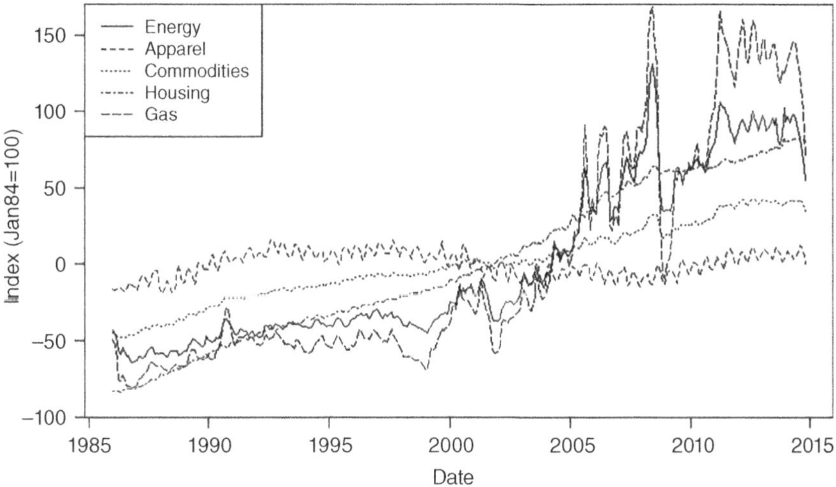

Figure 4.4 Consumer price index for five industry sectors for the Greater New York City Area between January 1986 and December 2014.

and the sample correlation matrix,

$$\hat{\boldsymbol{\rho}} = \begin{bmatrix} 1.00 & -0.011 & 0.936 & 0.945 & 0.986 \\ -0.011 & 1.00 & 0.185 & 0.040 & 0.030 \\ 0.936 & 0.185 & 1.00 & 0.985 & 0.909 \\ 0.945 & 0.045 & 0.985 & 1.00 & 0.902 \\ 0.986 & 0.030 & 0.909 & 0.902 & 1.00 \end{bmatrix}.$$

Note the large variances of energy, housing, and gas, the small variances of apparel and commodities, and the large correlation among energy, commodities, housing, and gas.

4.5.2.1 The PCA based on the sample covariance matrix

The eigenvalues and eigenvectors, which are often known as variances and component loadings of the sample variance–covariance matrix, are given in Table 4.3.

The two sample principal components are

$$\hat{Y}_1 = 0.516Z_1 + 0.003Z_2 + 0.227Z_3 + 0.461Z_4 + 0.686Z_5$$
$$\hat{Y}_2 = 0.081Z_1 - 0.057Z_2 - 0.331Z_3 - 0.755Z_4 + 0.557Z_5$$

The first component explains 95.76% of the total sample variance, and the first two explain 99.04%. Thus, sample variation is very much summarized by the first principle component or the first two principle components. Figure 4.5 shows the useful screeplot where the vertex of the elbow can be easily seen to be $k = 1$.

Table 4.3 Sample PCA results for the Greater New York City CPI based on the sample covariance matrix.

	Comp. 1	Comp. 2	Comp. 3	Comp. 4	Comp. 5
Energy	0.516	0.081	−0.301	0.782	−0.158
Apparel	0.003	−0.057	0.859	0.398	0.318
Commodities	0.227	−0.331	0.354	−0.148	−0.832
Housing	0.461	−0.755	−0.117	−0.187	0.411
Gas	0.686	0.557	0.184	−0.415	0.118
Variance $(\hat{\lambda}_i)$	11 068.83	379.65	89.84	18.11	2.70
Proportion of total variance explained by ith component	0.9576	0.0328	0.0078	0.0016	0.0002

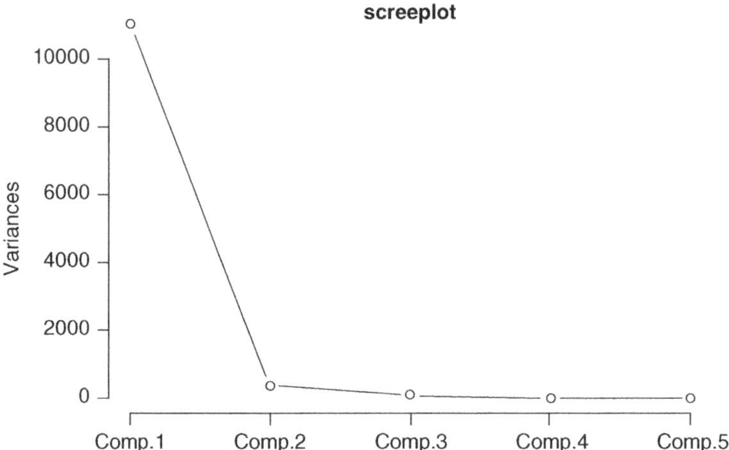

Figure 4.5 Screeplot of the PCA from the sample covariance matrix.

Now, let us examine component 1 more carefully. In this component, the loadings are all positive. The component can be regarded as the CPI growth component that grew over the time period that we observed. The five variables are combined into a composite score, which is plotted in Figure 4.6, and it follows a combination of patterns observed mainly for gasoline and energy in Figure 4.4.

Thus, the PCA has provided us with a single component that contains the vast majority of information for the five individual variables. From this, we can conclude that gasoline and energy were the true drivers of the overall economy for the Greater New York City area during the period between 1986 and 2014.

4.5.2.2 The PCA based on the sample correlation matrix

Now let us try the PCA using the sample correlation matrix. The eigenvalues and eigenvectors, which are also known as variances and component loadings, of the sample correlation matrix are given in Table 4.4.

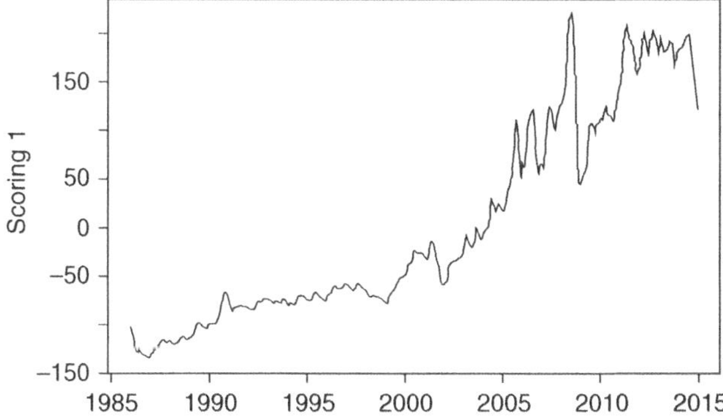

Figure 4.6 Time series plot of principle component 1.

Table 4.4 Sample PCA results for the Greater New York City CPI based on the sample correlation matrix.

	Comp. 1	Comp. 2	Comp. 3	Comp. 4	Comp. 5
Energy	0.503	0.100	0.322	0.676	−0.420
Apparel	0.044	−0.987	0.100	0.103	0.060
Commodities	0.501	−0.107	−0.417	−0.505	−0.556
Housing	0.499	0.032	−0.550	0.267	0.614
Gas	0.495	0.061	0.641	−0.455	0.366
Variance $(\hat{\lambda}_i)$	3.837	1.018	0.135	0.008	0.002
Proportion of total variance explained by *i*th component	0.7674	0.2036	0.027	0.0016	0.0004

The two sample principal components are

$$\hat{Y}_1 = 0.503Z_1 + 0.044Z_2 + 0.501Z_3 + 0.499Z_4 + 0.495Z_5$$
$$\hat{Y}_2 = 0.100Z_1 - 0.987Z_2 - 0.107Z_3 + 0.032Z_4 + 0.061Z_5$$

The first component explains 76.74% of the total sample variance, and the first two explain 97.1%. Thus, sample variation of the five industries is primarily summarized by the first two principle components. Figure 4.7 shows the screeplot, which clearly indicates $k = 2$.

From Table 4.4, we see that the loadings in component 1 are all positive, almost equal for energy, commodities, housing, and gas, and have strong positive correlations among them. It represents the CPI growth over the time period that we observed. The loadings in component 2 are relatively positive small numbers for energy, housing, and gas, and negative for apparel and commodities. It represents the market contrast between consumer goods and utility housing. Since the loading for apparel is especially dominating, component 2 can also be simply regarded as representing the apparel sector.

The five variables are combined into two composite scores, which are plotted in Figure 4.8.

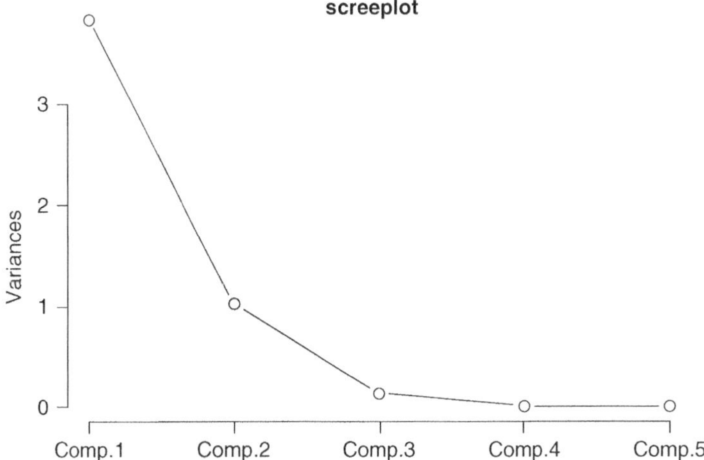

Figure 4.7 Screeplot of the PCA from the sample correlation matrix.

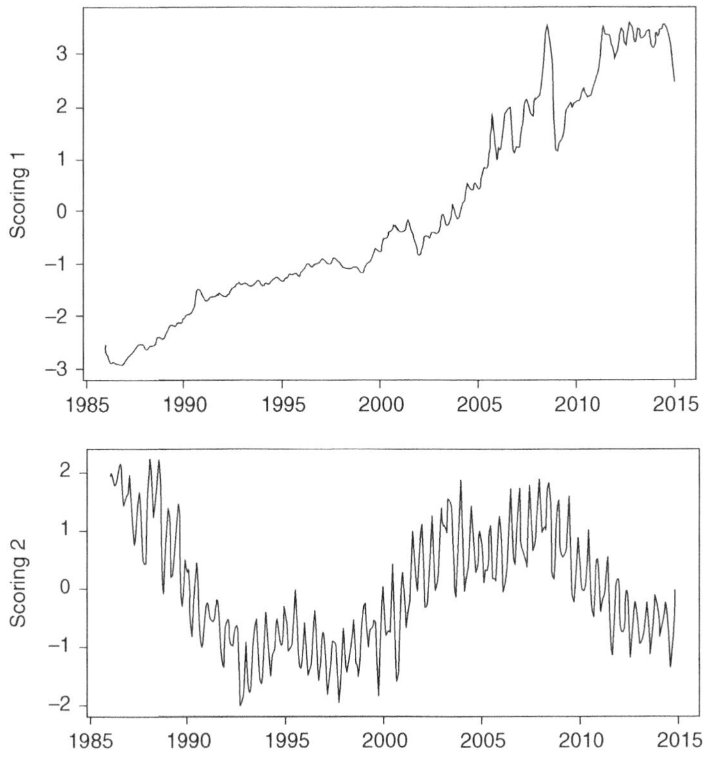

Figure 4.8 Time series plot of principle components 1 and 2.

The plots of component 1 based on sample covariance matrix and sample correlation matrix as shown in Figures 4.6 and 4.8 are almost the same. However, the proportion of total variance explained by component 1 is 96% when the original variables are used and 77% when the standardized variables are used. As shown in the example, the PCA results from the covariance matrix and the correlation matrix could be different.

For further information on PCA and applications, we refer readers to Joyeux (1992), Cubadda (1995), Ait-Sahalia and Xiu (2017), Estrada and Perron (2017), Jandarov et al. (2017), Passemier et al. (2017), Sang et al. (2017), and Zhu et al. (2017), among others.

Software code

R code for Example 4.1

```
library(ggplot2)
setwd("C:/Bookdata/")
##Import Data: daily stock returns from ten different stocks
from C:/Bookdata/WW4a.csv
d10 <- as.data.frame(read.csv("WW4a.csv")[,-1])
d10 <- t(t(d10) - colMeans(d10))
rownames(d10) <- seq(as.Date("2016/8/2"), by="day",
length=107)

cov(d10)
cor(d10)

##Plot Time Series Data
plot(seq(as.Date("2016/8/2"), by="day", length=107),d10[,1],
type='l',ylab="Stock Returns",xlab="Day")
for(i in 2:10){
  lines(seq(as.Date("2016/8/2"), by="day", length=107),
d10[,i],type='l',col=i)
}
legend("topleft",legend=colnames(d10),col=1:10,lty=1)

##Principal Component Analysis
pca <- princomp(d10)
pca <- princomp(d10,cor=F)
lds <- pca$loadings
screeplot(pca,type="lines",main="screeplot")
pca <- princomp(d10,cor=T)
lds <- pca$loadings
scs <- pca$scores
screeplot(pca,type="lines",main="screeplot")
```

```
##Plot: Sectors by Their First 2 Loadings
library(ggplot2)
C <- as.data.frame(cbind(lds[,1],lds[,2]))
ggplot(C,aes(C[,1],C[,2],label=rownames(C))) +
    geom_point(size=4,col=3) +
    geom_text(vjust=0,hjust=0,angle = 10,size=5) +
    xlab("Loading 1") +
    ylab("Loading 2")

##Plot: Time Points by Their Scores on First 2 Components
C <- as.data.frame(cbind(scs[,1],scs[,2]))
palette(rainbow(400))
ggplot(C,aes(C[,1],C[,2],label=substring(rownames(C),1,7))) +
    geom_point(size=4,col=1:nrow(C)) +
    geom_text(vjust=0,hjust=0,angle = 10,size=5) +
    xlab("Scoring 1") +
    ylab("Scoring 2")
palette("default")

##Plot: Time Points by Their Scores on First 2 Components
C <- as.data.frame(cbind(scs[,1],scs[,2]))
palette(rainbow(400))
ggplot(C,aes(C[,1],C[,2],label=substring(rownames(C),1,4))) +
    geom_path() +
    geom_point(size=4,col=1:nrow(C)) +
    geom_text(vjust=0,hjust=0,angle = 10,size=5) +
    xlab("Scoring 1") +
    ylab("Scoring 2")
palette("default")
plot(seq(as.Date("2009/4/1"), by="month", length=107),C[,1],
type="l",ylab="Scoring 1",xlab="Date")
plot(seq(as.Date("2009/4/1"), by="month", length=107),C[,2],
type="l",ylab="Scoring 2",xlab="Date")
ag <- aggregate(C[,1:2],list(substr(rownames(C),1,4)),mean)
C <- as.data.frame(ag)
ggplot(C,aes(C[,2],C[,3],label=C[,1])) +
    geom_point(size=4,col=3) +
    geom_text(vjust=0,hjust=0,angle = 10,size=5) +
    xlab("Scoring 1") +
    ylab("Scoring 2")
q()
```

R Code for Example 4.2

```
library(ggplot2)

##Import Data: monthly Consumer Price Index (CPI) from five
different sectors from C:/Bookdata/WW4b.csv
setwd("/Users/zedali/Desktop/Bookdata/")
d5 <- as.data.frame(read.csv("WW4b.csv")[,-1])
rownames(d5) <- seq(as.Date("1986/1/1"), by="month",
length=347)

##Covariance and Correlation Matrices
d5 <- t(t(d5) - colMeans(d5))
cov(d5)
cor(d5)

##Plot Time Series Data
plot(seq(as.Date("1986/1/1"), by="month", length=347),d5
[,1],type='l',ylab="Consumer Price
     Index (Jan84=100)",xlab="Date",ylim=c(-90,160))
for(i in 2:5){
        lines(seq(as.Date("1986/1/1"), by="month",
        length=347),d5[,i],type='l',lty=i)
}

legend("topleft",legend=colnames(d5),,lty=1:5)
##Principal Component Analysis
pca <- princomp(d5,cor=T)
lds <- pca$loadings
scs <- pca$scores
screeplot(pca,type="lines",main="screeplot")

##Plot: Sectors by Their First 2 Loadings
library(ggplot2)
C <- as.data.frame(cbind(lds[,1],lds[,2]))
ggplot(C,aes(C[,1],C[,2],label=rownames(C))) +
    geom_point(size=4,col=3) +
    geom_text(vjust=0,hjust=0,angle = 10,size=5) +
    xlab("Loading 1") +
    ylab("Loading 2")

##Plot: Time Points by Their Scores on First 2 Components
C <- as.data.frame(cbind(scs[,1],scs[,2]))
palette(rainbow(400))
```

```
ggplot(C,aes(C[,1],C[,2],label=substring(rownames(C),1,7))) +
    geom_point(size=4,col=1:nrow(C)) +
    geom_text(vjust=0,hjust=0,angle = 10,size=5) +
    xlab("Scoring 1") +
    ylab("Scoring 2")
palette("default")

##Plot: Time Points by Their Scores on First 2 Components
C <- as.data.frame(cbind(scs[,1],scs[,2]))
palette(rainbow(400))
ggplot(C,aes(C[,1],C[,2],label=substring(rownames(C),1,4))) +
    geom_path() +
    geom_point(size=4,col=1:nrow(C)) +
    geom_text(vjust=0,hjust=0,angle = 10,size=5) +
    xlab("Scoring 1") +
    ylab("Scoring 2")
palette("default")

plot(seq(as.Date("2009/4/1"), by="month", length=347),C[,1],
type="l",ylab="Scoring
      1",xlab="Date")
plot(seq(as.Date("2009/4/1"), by="month", length=347),C[,2],
type="l",ylab="Scoring
      2",xlab="Date")
ag <- aggregate(C[,1:2],list(substr(rownames(C),1,4)),mean)
C <- as.data.frame(ag)
ggplot(C,aes(C[,2],C[,3],label=C[,1])) +
    geom_point(size=4,col=3) +
    geom_text(vjust=0,hjust=0,angle = 10,size=5) +
    xlab("Scoring 1") +
    ylab("Scoring 2")
q()
```

Projects

1. Find a multivariate analysis book and carefully read its chapter on PCA.
2. Find an m-dimensional social science related time series data set with $m \geq 6$. Construct your principle component model based on its sample covariance matrix and evaluate your findings with a written report and analysis software code.
3. For the data set in Project 2, construct your principle component model based on its sample correlation matrix, and compare your result with that in Project 2. Write a report on your findings with associated software code.

4. Find an m-dimensional natural science related data set with $m \geq 6$. Construct your principle component model based on its sample covariance and correlation matrices separately, and evaluate your findings with a written report and analysis software code.

5. Find an m-dimensional time series data set of your interest with $m \geq 10$. Construct your principle component model based on its sample covariance and correlation matrices separately, and evaluate your findings with a written report and analysis software code.

References

Ait-Sahalia, Y. and Xiu, D. (2018). Principal component analysis of high frequency data. *Journal of American Statistical Association* 11–14. https://doi.org/10.1080/01621459.2017.1401542.

Anderson, T.W. (1963). Asymptotic theory for principal components analysis. *Annals of Mathematical Statistics* **34**: 122–148.

Cleveland, R.B., Cleveland, W.S., McRae, J.E., and Terpenning, I. (1990). A seasonal-trend decomposition procedure based on loess (with discussion). *Journal of Official Statistics* **6**: 3–73.

Cubadda, G. (1995). A note on testing for seasonal cointegration using principal components in the frequency domain. *Journal of Time Series Analysis* **16**: 499–508.

Estrada, F. and Perron, P. (2017). Extracting and analyzing the warming trend in global and hemispheric temperatures. *Journal of Time Series Analysis* **38**: 711–732.

Hotelling, H. (1933). Analysis of a complex of statistical variables into principle components. *Journal of Educational Psychology* **24**: 417–441. 498–520.

Jandarov, R.A., Sheppard, L.A., Sampson, P.D., and Szpiro, A.A. (2017). A novel principal component analysis for spatially misaligned multivariate air pollution data. *Journal of Statistical Society, Series C* **66**: 3–28.

Johnson, R.A. and Wichern, D.W. (2002). *Applied Multivariate Statistical Analysis*, 5e. Prentice Hall.

Joyeux, R. (1992). Testing for seasonal cointegration using principal components. *Journal of Time Series Analysis* **13**: 109–118.

Passemier, D., Li, Z., and Yao, J. (2017). On estimation of the noise variance in high dimensional probabilistic principal component analysis. *Journal of Royal Statistical Society, Series B* **79**: 51–67.

Pearson, K. (1901). On lines and planes of closest fit to systems of points in space. *Philosophical Magazine, 6th Series* **II**: 559–572.

Rao, C.R. (2002). *Linear Statistical Inference and Its Applications*, 2e. Wiley.

Sang, T., Wang, L., and Cao, J. (2017). Parametric functional principal component analysis. *Biometrics* **73**: 802–810.

Zhu, H., Shen, D., Peng, X., and Liu, L.Y. (2017). MWPCR: multiscale weighted principal component regression for high-dimensional prediction. *Journal of American Statistical Association* **112**: 1009–1021.

5

Factor analysis of multivariate time series

Similar to principle component analysis, factor analysis is one of the commonly used dimension reduction methods. It is a statistical technique widely used to explain a m-dimensional vector with a few underlying factors. After introducing different methods to derive factors, we will illustrate the method with empirical examples. We will also discuss its use in forecasting.

5.1 Introduction

Just like principle component analysis, the purpose of factor analysis is to approximate the covariance relationships among a set of variables. Specifically, it is used to describe the covariance relationships for many variables in terms of a relatively few underlying factors, which are unobservable random quantities. The concept was developed by the researchers in the field of psychometrics in the early twentieth century. It has become a commonly used statistical method in many areas.

5.2 The orthogonal factor model

Given a weakly stationary m-dimensional random vector at time t, $\mathbf{Z}_t = [Z_{1,t}, Z_{2,t}, \dots, Z_{m,t}]'$ with mean $\boldsymbol{\mu} = (\mu_1, \mu_2, \dots, \mu_m)'$, and covariance matrix $\boldsymbol{\Gamma}$, the factor model assumes that $\mathbf{Z}_t$ is dependent on a small number of k unobservable factors, $F_{j,t}, j = 1, 2, \dots, k$, known as common factors, and m additional noises $\varepsilon_{i,t}$, $i = 1, 2, \dots, m$, also known as specific factors, that is

$$\begin{aligned}
Z_{1,t} - \mu_1 &= \ell_{1,1}F_{1,t} + \ell_{1,2}F_{2,t} + \cdots + \ell_{1,k}F_{k,t} + \varepsilon_{1,t}, \\
Z_{2,t} - \mu_2 &= \ell_{2,1}F_{1,t} + \ell_{2,2}F_{2,t} + \cdots + \ell_{2,k}F_{k,t} + \varepsilon_{2,t}, \\
&\vdots \\
Z_{m,t} - \mu_m &= \ell_{m,1}F_{1,t} + \ell_{m,2}F_{2,t} + \cdots + \ell_{m,k}F_{k,t} + \varepsilon_{m,t}.
\end{aligned} \tag{5.1}$$

Multivariate Time Series Analysis and Applications, First Edition. William W.S. Wei.
© 2019 John Wiley & Sons Ltd. Published 2019 by John Wiley & Sons Ltd.
Companion website: www.wiley.com/go/wei/datasets

More compactly, we can write the system in following matrix form,

$$\underset{m\times 1}{\dot{\mathbf{Z}}_t} = \underset{(m\times k)(k\times 1)}{\mathbf{L}\ \mathbf{F}_t} + \underset{(m\times 1)}{\boldsymbol{\varepsilon}_t}, \tag{5.2}$$

where $\dot{\mathbf{Z}}_t = (\mathbf{Z}_t - \boldsymbol{\mu})$, $\mathbf{F}_t = (F_{1,t}, F_{2,t}, \dots, F_{k,t})'$ is a $(k\times 1)$ vector of factors at time t, $\mathbf{L} = [\ell_{i,j}]$ is a $(m\times k)$ loading matrix, with $\ell_{i,j}$ is the loading of the ith variable on the jth factor, $i = 1, 2, \dots, m$, $j = 1, 2, \dots, k$, and $\boldsymbol{\varepsilon}_t = (\varepsilon_{1,t}, \varepsilon_{2,t}, \dots, \varepsilon_{m,t})'$ is a $(m\times 1)$ vector of noises, with $E(\boldsymbol{\varepsilon}_t) = \mathbf{0}$, and $\text{Cov}(\boldsymbol{\varepsilon}_t) = diag\{\sigma_1^2, \sigma_2^2, \dots, \sigma_m^2\}$.

The factor model in Eq. (5.2) is an orthogonal factor model if it satisfies the following assumptions:

1. $E(\mathbf{F}_t) = \mathbf{0}$, and $\text{Cov}(\mathbf{F}_t) = \mathbf{I}_k$, the $(k\times k)$ identity matrix,
2. $E(\boldsymbol{\varepsilon}_t) = \mathbf{0}$, and $\text{Cov}(\varepsilon_t) = \boldsymbol{\Sigma} = diag\{\sigma_1^2, \sigma_2^2, \dots, \sigma_m^2\}$, a $(m\times m)$ diagonal matrix, and
3. $\mathbf{F}_t$ and $\boldsymbol{\varepsilon}_t$ are independent and so $\text{Cov}(\mathbf{F}_t, \boldsymbol{\varepsilon}_t) = E(\mathbf{F}_t\boldsymbol{\varepsilon}_t') = \mathbf{0}$, a $(k\times m)$ zero matrix.

It follows from Eq. (5.2) that the covariance structure of $\mathbf{Z}_t$ is

$$\begin{aligned}
\boldsymbol{\Gamma} &= \text{Cov}(\mathbf{Z}_t) = E(\mathbf{Z}_t - \boldsymbol{\mu})(\mathbf{Z}_t - \boldsymbol{\mu})' \\
&= E(\mathbf{L}\mathbf{F}_t + \boldsymbol{\varepsilon}_t)(\mathbf{L}\mathbf{F}_t + \boldsymbol{\varepsilon}_t)' \\
&= E(\mathbf{L}\mathbf{F}_t + \boldsymbol{\varepsilon}_t)(\mathbf{F}_t'\mathbf{L}' + \boldsymbol{\varepsilon}_t') \\
&= \mathbf{L}E(\mathbf{F}_t\mathbf{F}_t')\mathbf{L}' + \mathbf{L}E(\mathbf{F}_t\boldsymbol{\varepsilon}_t') + [E(\boldsymbol{\varepsilon}_t\mathbf{F}_t')]\mathbf{L}' + E(\boldsymbol{\varepsilon}_t\boldsymbol{\varepsilon}_t') \\
&= \mathbf{L}\mathbf{L}' + \boldsymbol{\Sigma}.
\end{aligned} \tag{5.3}$$

The model in Eq. (5.2) shows that the m-dimensional process $\mathbf{Z}_t$ is linear related to the k common factors. More specifically,

$$\begin{aligned}
\text{Cov}(\mathbf{Z}_t, \mathbf{F}_t) &= E(\mathbf{Z}_t - \boldsymbol{\mu})(\mathbf{F}_t - \mathbf{0})' = E(\mathbf{L}\mathbf{F}_t + \boldsymbol{\varepsilon}_t)\mathbf{F}_t' \\
&= \mathbf{L}E(\mathbf{F}_t\mathbf{F}_t') + E(\boldsymbol{\varepsilon}\mathbf{F}'t) = \mathbf{L},
\end{aligned} \tag{5.4}$$

which implies that

$$\text{Cov}(Z_{i,t}, F_{j,t}) = \ell_{i,j}. \tag{5.5}$$

Also,

$$\begin{aligned}
\text{Var}(Z_{i,t}) &= \text{Var}(\ell_{i,1}F_{1,t} + \ell_{i,2}F_{2,t} + \cdots + \ell_{i,k}F_{k,t} + \varepsilon_{i,t}) \\
&= \ell_{i,1}^2 + \ell_{i,2}^2 + \cdots + \ell_{i,k}^2 + \sigma_i^2.
\end{aligned} \tag{5.6}$$

So the variance of the ith variable $Z_{i,t}$ is the sum $c_i^2 = \left(\ell_{i,1}^2 + \ell_{i,2}^2 + \cdots + \ell_{i,k}^2\right)$ due to the k common factors, which is known as the ith communality, and σ_i^2 due to the ith specific factor, which is known as the ith specific variance.

5.3 Estimation of the factor model

5.3.1 The principal component method

Given observations $\mathbf{Z}_t = (Z_{1,t}, Z_{2,t}, \ldots, Z_{m,t})'$, for $t = 1, 2, \ldots, n$, and its $m \times m$ sample covariance matrix $\hat{\boldsymbol{\Gamma}} = \left[\hat{\gamma}_{i,j}\right]$, a natural method of estimation is simply to use the principle component analysis introduced in Chapter 4 and choose k, which is much less than m, common factors from the first k largest eigenvalue-eigenvector pairs in $\left(\hat{\lambda}_1, \hat{\boldsymbol{\alpha}}_1\right), \left(\hat{\lambda}_2, \hat{\boldsymbol{\alpha}}_2\right), \ldots, \left(\hat{\lambda}_m, \hat{\boldsymbol{\alpha}}_m\right)$, with $\hat{\lambda}_1 \geq \hat{\lambda}_2 \geq, \ldots, \geq \hat{\lambda}_m$. Let $\hat{\mathbf{L}}$ be the estimate of $\mathbf{L}$. Then,

$$\underset{m\times k}{\hat{\mathbf{L}}} = \left[\sqrt{\hat{\lambda}_1}\hat{\boldsymbol{\alpha}}_1 \ \sqrt{\hat{\lambda}_2}\hat{\boldsymbol{\alpha}}_2 \ \ldots \ \sqrt{\hat{\lambda}_k}\hat{\boldsymbol{\alpha}}_k\right], \tag{5.7}$$

and the estimated specific variances are obtained by

$$\hat{\boldsymbol{\Sigma}} = \begin{bmatrix} \hat{\sigma}_1^2 & 0 & . & \cdots & . & 0 \\ 0 & \hat{\sigma}_2^2 & 0 & \cdots & . & 0 \\ . & 0 & . & \cdots & . & . \\ \vdots & \vdots & \vdots & \ddots & \vdots & \vdots \\ 0 & . & . & \cdots & . & 0 \\ 0 & . & . & \cdots & 0 & \hat{\sigma}_m^2 \end{bmatrix}, \tag{5.8}$$

with the ith specific variance estimate being

$$\hat{\sigma}_i^2 = \hat{\gamma}_{i,i} - \left(\hat{\ell}_{i,1}^2 + \hat{\ell}_{i,2}^2 + \cdots + \hat{\ell}_{i,k}^2\right), \tag{5.9}$$

where the sum of squares is the estimate of the ith communality

$$\hat{c}_i^2 = \left(\hat{\ell}_{i,1}^2 + \hat{\ell}_{i,2}^2 + \cdots + \hat{\ell}_{i,k}^2\right). \tag{5.10}$$

The contribution to the first common factor to the total sample variance is given by

$$\left(\hat{\ell}_{1,1}^2 + \hat{\ell}_{2,1}^2 + \cdots + \hat{\ell}_{m,1}^2\right) = \left(\sqrt{\hat{\lambda}_1}\hat{\boldsymbol{\alpha}}_1\right)'\left(\sqrt{\hat{\lambda}_1}\hat{\boldsymbol{\alpha}}_1\right) = \hat{\lambda}_1, \tag{5.11}$$

where we note that the eigenvectors in the principle component analysis are standardized to the unit length. More general, if we let P be the proportion of the jth common factor to the total sample variance, we have

$$P = \frac{\hat{\lambda}_j}{\left(\hat{\gamma}_{1,1} + \hat{\gamma}_{2,2} + \cdots + \hat{\gamma}_{m,m}\right)}, \tag{5.12}$$

and it can be used to determine the desired number of common factors.

In practice, it is important to check carefully whether the units used in the component variables $Z_{i,t}$, $i = 1, 2, \ldots, m$, are comparable. If not, to avoid improperly influence of the units, we can standardize these variables, that is

$$\mathbf{U}_t = \left[U_{1,t}, \quad \ldots, \quad U_{m,t}\right]' = \left[\frac{Z_{1,t} - \bar{Z}_1}{\sqrt{\hat{\gamma}_{1,1}}}, \quad \ldots, \quad \frac{Z_{m,t} - \bar{Z}_m}{\sqrt{\hat{\gamma}_{m,m}}}\right]' = \mathbf{D}^{-1/2}(\mathbf{Z}_t - \bar{\mathbf{Z}}), \tag{5.13}$$

where $\mathbf{D}$ is the diagonal matrix in which the ith diagonal element is the sample variance of $Z_{i,t}$, that is

$$\mathbf{D} = diag\left[\hat{\gamma}_{1,1}, \hat{\gamma}_{2,2}, \ldots, \hat{\gamma}_{m,m}\right]. \tag{5.14}$$

Then, we will perform the principle component estimation method to the sample covariance matrix $\hat{\boldsymbol{\Gamma}}$ of the standardized observations $\mathbf{U}_t$, $t = 1,2, \ldots, n$, which is actually the sample correlation matrix of the original variables in $\mathbf{Z}_t$, $t = 1,2, \ldots, n$.

Again, the results from the sample covariance matrix and sample correlation matrix may not be the same, and the choice depends on applications.

5.3.2 Empirical Example 1 – Model 1 on daily stock returns from the second set of 10 stocks

Example 5.1 Principle component approach on 10 daily stock returns

Let us consider the data set, WW5, of the daily stock returns from 10 different stocks including CVX (Chevron), XOM (Exxon), AAPL (Apple), FB (Facebook), MSFT (Microsoft), JNJ (Johnson and Johnson), MRK (Merck), PFE (Pfizer), FBIOX (Fidelity Select Biotechnology), and FSPHX (Fidelity Select Health Care) that were traded in the New York Stock Exchange from August 2, 2016 to December 30, 2016. The data are plotted in Figure 5.1.

In this example, the sample $\mathbf{Z}_t = [Z_{1,t}, Z_{2,t}, Z_{3,t}, Z_{4,t}, Z_{5,t}, Z_{6,t}, Z_{7,t}, Z_{8,t}, Z_{9,t}, Z_{10,t}]$, $t = 1, 2, \ldots, n$, is the daily stock returns for Chevron ($Z_{1,t}$), Exxon ($Z_{2,t}$), Apple ($Z_{3,t}$), Facebook ($Z_{4,t}$), Microsoft ($Z_{5,t}$), Johnson and Johnson($Z_{6,t}$), Merck ($Z_{7,t}$), Pfizer($Z_{8,t}$), Fidelity Select Biotechnology ($Z_{9,t}$), and Fidelity Select Health ($Z_{10,t}$), which were traded in the New York Stock Exchange from August 2, 2016 to December 30, 2016 and so $m = 10$, and $n = 106$. We have the sample mean vector,

$$\begin{aligned}\bar{\mathbf{Z}} &= (\bar{Z}_1, \bar{Z}_2, \ldots, \bar{Z}_{10})' \\ &= (0.00168, 0.00052, 0.00089, -0.00066, 0.00094, \\ &\quad -0.00077, 0.00016, -0.00123, -0.00071, -.00102)',\end{aligned}$$

and the sample covariance matrix, $\hat{\boldsymbol{\Gamma}}$,

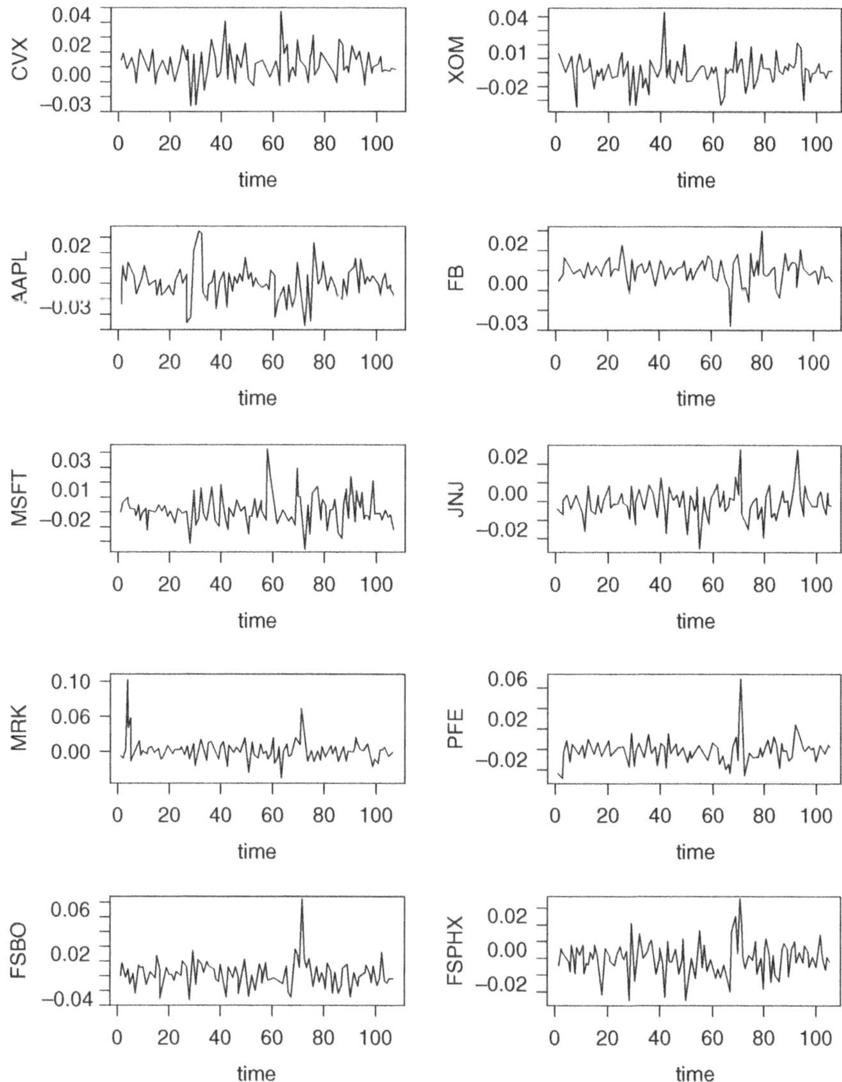

Figure 5.1 Ten daily stock returns for 10 stocks between August 2 and December 30, 2016.

	CVX	XOM	AAPL	FB	MSFT
CVX	1.087154e−04	6.983585e−05	8.576443e−06	2.841400e−05	2.349692e−05
XOM	6.983585e−05	1.063956e−04	1.159600e−05	2.475253e−05	2.188846e−05
AAPL	8.576443e−06	1.159600e−05	1.231327e−04	6.237891e−05	5.277169e−05
FB	2.841400e−05	2.475253e−05	6.237891e−05	1.487562e−04	6.697871e−05
MSFT	2.349692e−05	2.188846e−05	5.277169e−05	6.697871e−05	1.012891e−04
JNJ	1.794397e−05	2.544403e−05	1.763044e−05	1.691757e−05	1.720156e−05

MRK	2.015158e−05	5.188890e−05	3.017592e−05	3.534955e−05	2.720992e−05
PFE	1.959854e−05	4.650749e−05	1.080636e−05	3.061439e−05	1.225974e−05
FBIOX	3.440326e−05	4.943891e−05	2.463021e−05	5.788276e−05	2.019403e−05
FSPHX	2.652420e−05	3.863676e−05	2.939527e−05	4.926419e−05	2.813533e−05

	JNJ	MRK	PFE	FBIOX	FSPHX
CVX	1.794397e−05	2.015158e−05	1.959854e−05	3.440326e−05	2.652420e−05
XOM	2.544403e−05	5.188890e−05	4.650749e−05	4.943891e−05	3.863676e−05
AAPL	1.763044e−05	3.017592e−05	1.080636e−05	2.463021e−05	2.939527e−05
FB	1.691757e−05	3.534955e−05	3.061439e−05	5.788276e−05	4.926419e−05
MSFT	1.720156e−05	2.720992e−05	1.225974e−05	2.019403e−05	2.813533e−05
JNJ	6.606769e−05	6.159274e−05	5.885948e−05	6.521474e−05	4.117798e−05
MRK	6.159274e−05	2.522852e−04	1.180967e−04	1.580319e−04	8.893903e−05
PFE	5.885948e−05	1.180967e−04	1.659857e−04	1.379977e−04	8.349862e−05
FBIOX	6.521474e−05	1.580319e−04	1.379977e−04	3.597483e−04	1.729823e−04
FSPHX	4.117798e−05	8.893903e−05	8.349862e−05	1.729823e−04	1.061556e−04

For this example, we will obtain the principal component estimation for the factor model using standardized variables through the sample correlation matrix, $\hat{\boldsymbol{\rho}}$, which is given next.

	CVX	XOM	AAPL	FB	MSFT	JNJ	MRK	PFE	FBIOX	FSPHX
CVX	1.000	0.649	0.074	0.223	0.224	0.212	0.122	0.146	0.174	0.247
XOM	0.649	1.000	0.101	0.197	0.211	0.303	0.317	0.350	0.253	0.364
AAPL	0.074	0.101	1.000	0.461	0.473	0.195	0.171	0.076	0.117	0.257
FB	0.223	0.197	0.461	1.000	0.546	0.171	0.182	0.195	0.250	0.392
MSFT	0.224	0.211	0.473	0.546	1.000	0.210	0.170	0.095	0.106	0.271
JNJ	0.212	0.303	0.195	0.171	0.210	1.000	0.477	0.562	0.423	0.492
MRK	0.122	0.317	0.171	0.182	0.170	0.477	1.000	0.577	0.525	0.543
PFE	0.146	0.350	0.076	0.195	0.095	0.562	0.577	1.000	0.565	0.629
FBIOX	0.174	0.253	0.117	0.250	0.106	0.423	0.525	0.565	1.000	0.885
FSPHX	0.247	0.364	0.257	0.392	0.271	0.492	0.543	0.629	0.885	1.000

The screeplot of the PCA is given in Figure 5.2, which suggests $k = 3$ or 4.

For more details, the eigenvalues and eigenvectors, which are variances and component loadings, of the sample correlation matrix are given in Table 5.1.

The eigenvalues of the first three components are greater than 1 and they account for

$$\left(\frac{\hat{\lambda}_1+\hat{\lambda}_2+\hat{\lambda}_3}{m}\right)100\% = \left(\frac{3.987+1.7+1.326}{m}\right)100\% = 70.13\%$$

of the total standardized sample variance. So, we will simply choose them as our first, second, and third common factors F_1, F_2, F_3, and summarize the factor model estimation in Table 5.2.

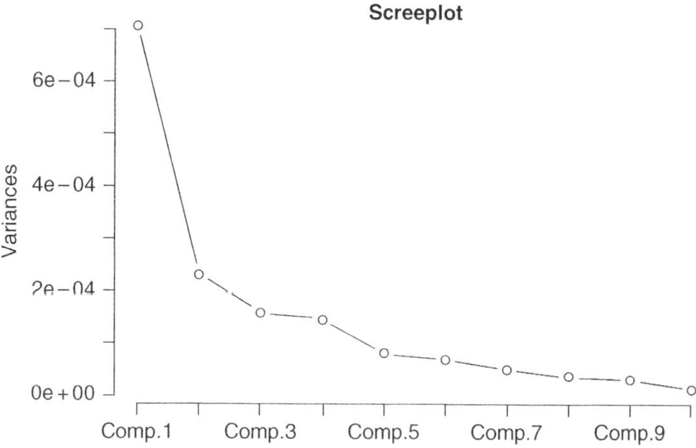

Figure 5.2 The PCA Scree plot of the 10 daily stock returns.

The first common factor represents the general stock market. The second common factor represents the contrast between the technology sector and the health-related sector. The third common factor represents the contrast between oil industry and non-oil industry.

5.3.3 The maximum likelihood method

When the common factors $\mathbf{F}_t$ and the specific factors $\boldsymbol{\varepsilon}_t$ can be assumed to be normally distributed, then $\mathbf{Z}_t$, $t = 1, 2, \dots, n$, will follow a multivariate normal distribution $\mathbf{N}(\boldsymbol{\mu}, \boldsymbol{\Gamma})$, and we can apply the maximum likelihood method to its likelihood function,

$$L(\boldsymbol{\mu},\boldsymbol{\Gamma}) = (2\pi)^{-nm/2}|\boldsymbol{\Gamma}|^{-n/2}e^{-\frac{1}{2}\sum_{t=1}^{n}(\mathbf{Z}_t-\boldsymbol{\mu})'\boldsymbol{\Gamma}^{-1}(\mathbf{Z}_t-\boldsymbol{\mu})}, \tag{5.15}$$

which is actually a function of $\mathbf{L}$ and $\boldsymbol{\Gamma}$ through the relationship $\boldsymbol{\Gamma} = \mathbf{LL}' + \boldsymbol{\Sigma}$, and $\boldsymbol{\mu}$ is estimated by the sample mean $\bar{\mathbf{Z}}$. However, because of $\boldsymbol{\Gamma} = \mathbf{LL}' + \boldsymbol{\Sigma} = \mathbf{L}\boldsymbol{\Phi}\boldsymbol{\Phi}'\mathbf{L}' + \boldsymbol{\Sigma}$, and $\dot{\mathbf{Z}}_t = \mathbf{LF}_t + \boldsymbol{\varepsilon}_t = \mathbf{L}\boldsymbol{\Phi}\boldsymbol{\Phi}'\mathbf{F}_t + \boldsymbol{\varepsilon}_t$, for any $k \times k$ orthogonal matrix $\boldsymbol{\Phi}$, to make Eq. (5.15) well defined, we also impose the following unique condition,

$$\mathbf{L}'\boldsymbol{\Sigma}^{-1}\mathbf{L} = \boldsymbol{\Delta} \text{ a diagonal matrix} \tag{5.16}$$

The maximum likelihood estimates $\hat{\mathbf{L}}$ and $\hat{\boldsymbol{\Gamma}}$ can be accomplished with numerical methods using statistical software such as EViews, MATLAB, MINITAB, R, SAS, and SPSS. However, the useful patterns of the factor loadings from the maximum likelihood solution obtained under the imposed unique condition may not be clear until the factors are rotated, which will be discussed later.

The maximum likelihood estimate of the ith communality is

$$\hat{c}_i^2 = \left(\hat{\ell}_{i,1}^2 + \hat{\ell}_{i,2}^2 + \cdots + \hat{\ell}_{i,k}^2\right), \; i = 1, 2, \dots, m \tag{5.17}$$

Table 5.1 Sample PCA result for the daily stock returns for the 10 stocks based on the sample correlation matrix.

	Comp. 1 $\hat{\alpha}_1$	Comp. 2 $\hat{\alpha}_2$	Comp. 3 $\hat{\alpha}_3$	Comp. 4 $\hat{\alpha}_4$	Comp. 5 $\hat{\alpha}_5$	Comp. 6 $\hat{\alpha}_6$	Comp. 7 $\hat{\alpha}_7$	Comp. 8 $\hat{\alpha}_8$	Comp. 9 $\hat{\alpha}_9$	Comp. 10 $\hat{\alpha}_{10}$
CVX	−0.219	−0.169	0.676	0.115		−0.168		−0.320	0.560	
XOM	−0.288		0.602		0.172	0.193	−0.117	0.340	−0.588	
AAPL	−0.195	−0.480	−0.284	−0.222	0.723	−0.177	0.129	0.112	0.125	
FB	−0.261	−0.466	−0.154	0.287	−0.346	0.104	0.563	−0.358	−0.176	
MSFT	−0.224	−0.534		−0.123	−0.419	0.156	−0.571	0.320		
JNJ	−0.340	0.145		−0.585	−0.236	−0.568		−0.285	−0.242	
MRK	−0.352	0.212	−0.121	−0.285	0.171	0.700	−0.154	−0.433		
PFE	−0.370	0.297		−0.172	−0.211		0.458	0.507	0.464	
FBIOX	−0.384	0.257	−0.166	0.486	0.123	−0.162	−0.252			−0.643
FSPHX	−0.436	0.126	−0.141	0.382		−0.152	−0.147	0.110		0.748
Variance $(\hat{\lambda}_i)$	3.987	1.700	1.326	0.753	0.536	0.518	0.443	0.368	0.285	0.084
Cumulative Percentage %	39.87	56.87	70.13	77.66	83.02	88.20	92.63	96.31	99.16	100

Table 5.2 The principal component estimation for the factor model.

Variable (j)	$\hat{\ell}_{j,1}=\sqrt{\hat{\lambda}_1}\hat{\alpha}_{1,j};\hat{F}_{1,t}$	$\hat{\ell}_{j,2}=\sqrt{\hat{\lambda}_2}\hat{\alpha}_{2,j};\hat{F}_{2,t}$	$\hat{\ell}_{j,3}=\sqrt{\hat{\lambda}_3}\hat{\alpha}_{3,j};\hat{F}_{3,t}$	$\hat{c}_i^2$	$\hat{\sigma}_i^2$
CVX	−0.437	−0.220	0.778	0.845	0.155
XOM	−0.575		0.693	0.811	0.189
AAPL	−0.389	−0.626	−0.327	0.650	0.350
FB	−0.521	−0.608	−0.177	0.672	0.328
MSFT	−0.447	−0.696		0.684	0.316
JNJ	−0.679	0.189		0.497	0.503
MRK	−0.703	0.276	−0.139	0.590	0.410
PFE	−0.739	0.387		0.696	0.304
FBIOX	−0.767	0.335	−0.191	0.737	0.263
FSPHX	−0.871	0.164	−0.162	0.812	0.188
Sum of squared loadings	3.99	1.70	1.32		
Cumulative percentage of total standdardized sample variance	39.9	56.9	70.1		

and the proportion of the contribution of the ith common factor is by

$$P=\left(\frac{\hat{\ell}_{1,i}^2+\hat{\ell}_{2,i}^2+\cdots+\hat{\ell}_{m,i}^2)}{(\hat{\gamma}_{1,1}+\hat{\gamma}_{2,2}+\cdots+\hat{\gamma}_{m,m})}\right). \tag{5.18}$$

To build a k common factor model for an m-dimensional process, under the normal assumption, we can test the adequacy of the model with the following null hypothesis,

$$\mathrm{H}_0: \underset{(m\times m)}{\mathbf{\Gamma}} = \underset{(m\times k)}{\mathbf{L}}\ \underset{(k\times m)}{\mathbf{L}'} + \underset{(m\times m)}{\mathbf{\Sigma}} \tag{5.19}$$

versus

$$\mathrm{H}_1: \underset{(m\times m)}{\mathbf{\Gamma}} = \text{any}\,(m\times m)\,\text{postive definite matrix}. \tag{5.20}$$

To test the null hypothesis, we apply the likelihood ratio test

$$\Lambda=\left(\frac{\max\limits_{H_0} L(\mathbf{\Gamma})}{\max\limits_{H_1} L(\mathbf{\Gamma})}\right)=\frac{L\left(\hat{\mathbf{\Gamma}}=\hat{\mathbf{L}}\hat{\mathbf{L}}'+\hat{\mathbf{\Sigma}}\right)}{L(\hat{\mathbf{\Gamma}})}=\left(\frac{|\hat{\mathbf{\Gamma}}|}{\left|\hat{\mathbf{L}}\hat{\mathbf{L}}'+\hat{\mathbf{\Sigma}}\right|}\right)^{n/2}. \tag{5.21}$$

It is well known that

$$-2\ln\Lambda = -n\ln\left(\frac{|\hat{\mathbf{\Gamma}}|}{|\hat{\mathbf{L}}\hat{\mathbf{L}}'+\hat{\mathbf{\Sigma}}|}\right) = n\ln\left(\frac{|\hat{\mathbf{L}}\hat{\mathbf{L}}'+\hat{\mathbf{\Sigma}}|}{|\hat{\mathbf{\Gamma}}|}\right) \tag{5.22}$$

follows approximately the chi-square distribution with degrees of freedom,

$$\upsilon-\upsilon_0 = \frac{m^2+m}{2} - \left[(mk+m)-\frac{k^2-k}{2}\right] = \frac{1}{2}\left[(m-k)^2-(m+k)\right], \tag{5.23}$$

where we note that

$$\hat{\mathbf{\Gamma}} = \frac{1}{n}\sum_{t=1}^{n}(\mathbf{Z}_t-\bar{\mathbf{Z}})(\mathbf{Z}_t-\bar{\mathbf{Z}})' = \begin{bmatrix} \hat{\gamma}_{1,1} & \hat{\gamma}_{1,2} & \cdots & \hat{\gamma}_{1,m} \\ \hat{\gamma}_{2,1} & \hat{\gamma}_{2,2} & \cdots & \hat{\gamma}_{2,m} \\ \vdots & \vdots & \vdots & \vdots \\ \hat{\gamma}_{m,1} & \hat{\gamma}_{m,2} & \cdots & \hat{\gamma}_{m,m} \end{bmatrix} \tag{5.24}$$

is the MLE of $\mathbf{\Gamma}$ without any restriction, which has $[(m^2+m)/2]$ free parameters, and

$$\hat{\mathbf{\Gamma}} = \left(\hat{\mathbf{L}}\hat{\mathbf{L}}'+\hat{\mathbf{\Sigma}}\right) \tag{5.25}$$

is the MLE of $\mathbf{\Gamma}$ under the restriction of factor model, which has $(mk+m)$ parameters but with $[(k^2-k)/2]$ constraints on the unique condition on $\hat{\mathbf{L}}'\hat{\mathbf{\Sigma}}^{-1}\hat{\mathbf{L}} = \mathbf{\Delta}$ being a $k\times k$ diagonal matrix given in Eq. (5.16). In practice, the test statistic in Eq. (5.22) will be replaced by Bartlett-Corrected Test Statistic,

$$\chi^2 = \left[n-1-\frac{1}{6}(2m+4k+5)\right]\ln\frac{|\hat{\mathbf{L}}\hat{\mathbf{L}}'+\hat{\mathbf{\Sigma}}|}{|\hat{\mathbf{\Gamma}}|}, \tag{5.26}$$

which gives a better approximation to the chi-square distribution as shown by Bartlett (1954). Hence, the null hypothesis will be rejected if

$$\chi^2 = \left[n-1-\frac{1}{6}(2m+4k+5)\right]\ln\frac{|\hat{\mathbf{L}}\hat{\mathbf{L}}'+\hat{\mathbf{\Sigma}}|}{|\hat{\mathbf{\Gamma}}|} > \chi^2_{[(m-k)^2-(m+k)]/2}(\alpha). \tag{5.27}$$

Because the degrees of freedom must be positive, we also require that

$$k < \frac{2m+1-\sqrt{8m+1}}{2}. \tag{5.28}$$

When correlation matrix $\boldsymbol{\rho}$ is used for factor analysis, the test statistic in Eq. (5.26) becomes

$$\chi^2 = \left[n-1-\frac{1}{6}(2m+4k+5)\right] \ln \frac{\left|\hat{\mathbf{L}}\hat{\mathbf{L}}' + \hat{\boldsymbol{\Sigma}}\right|}{|\hat{\boldsymbol{\rho}}|}, \tag{5.29}$$

with exactly the same degrees of freedom, $\frac{1}{2}\left[(m-k)^2-(m+k)\right]$. The reason is that the elements and computation of $\boldsymbol{\rho}$ are based on the covariance matrix $\boldsymbol{\Gamma}$. The number of its free parameters is exactly the same as that of $\boldsymbol{\Gamma}$. In terms of the m-dimensional case, it is $(m^2+m)/2$.

5.3.4 Empirical Example II – Model 2 on daily stock returns from the second set of 10 stocks

Example 5.2 Maximum likelihood method on 10 daily stock returns
Let us consider again the daily stock returns from the 10 different stocks including Chevron, Exxon, Apple, Facebook, Microsoft, Johnson and Johnson, Merck, Pfizer, Fidelity Select Biotechnology, and Fidelity Select Health that were traded in the New York Stock Exchange from August 2, 2016 to December 30, 2016, which was used in Example 5.1 as shown in Figure 5.1.

From Eq. (5.26), $k<\left(2m+1-\sqrt{8m+1}\right)/2=\left(20+1-\sqrt{80+1}\right)/2=6$. So we will consider k = 5 common factors, and obtain the following maximum likelihood estimate for the factor model based on the sample correlation matrix using the program *factanal* of software R as shown in Table 5.3.

The proportion of the contribution of the ith common factor is computed as follows,

$$P=\frac{\left(\hat{\ell}_{1,i}^2+\hat{\ell}_{2,i}^2+\cdots+\hat{\ell}_{m,i}^2\right)}{\left(\hat{\gamma}_{1,1}+\hat{\gamma}_{2,2}+\cdots+\hat{\gamma}_{m,m}\right)}=\begin{cases}0.326, i=1,\\ 0.172, i=2,\\ 0.104, i=3,\\ 0.055, i=4,\\ 0.027, i=5.\end{cases} \tag{5.30}$$

where we note that for the standardized variables, $\left(\hat{\gamma}_{1,1}+\hat{\gamma}_{2,2}+\cdots+\hat{\gamma}_{m,m}\right)=m=10,$ for the example.

With $m=10$, to test the null hypothesis $\mathrm{H}_0: \boldsymbol{\Gamma}_{(m\times m)} = \mathbf{L}_{(m\times k)}\mathbf{L}'_{(k\times m)} + \boldsymbol{\Sigma}_{(m\times m)}$ for $k=5$, we compute

$$\begin{aligned}\chi^2 &= \left[n-1-\frac{1}{6}(2m+4k+5)\right] \ln \frac{\left|\hat{\mathbf{L}}\hat{\mathbf{L}}' + \hat{\boldsymbol{\Sigma}}\right|}{|\hat{\boldsymbol{\rho}}|}\\ &= \left[106-1-\frac{1}{6}(20+20+5)\right] \ln\left(\frac{0.006493925}{0.007534897}\right)\\ &= -14.49611,\end{aligned} \tag{5.31}$$

Table 5.3 MLE of common factor loadings, communalities, and specific variances for $k = 5$.

Variable (j)	$\hat{\ell}_{j,1};\hat{F}_{1,t}$	$\hat{\ell}_{j,2};\hat{F}_{2,t}$	$\hat{\ell}_{j,3};\hat{F}_{3,t}$	$\hat{\ell}_{j,4};\hat{F}_{4,t}$	$\hat{\ell}_{j,5};\hat{F}_{5,t}$	$\hat{c}_i^2$	$\hat{\sigma}_i^2$
CVX	0.187	0.685	−0.454			0.710	0.290
XOM	0.347	0.603	−0.401	−0.106		0.656	0.343
AAPL	0.115	0.417	0.489		0.140	0.446	0.554
FB	0.259	0.540	0.479		−0.215	0.634	0.366
MSFT	0.121	0.580	0.436			0.637	0.363
JNJ	0.560	0.186		−0.153	0.293	0.457	0.543
MRK	0.625	0.103			0.314	0.500	0.500
PFE	0.880			−0.470		0.995	0.005
FBIOX	0.886			0.458		0.995	0.005
FSPHX	0.862	0.200	0.107	0.272		0.868	0.132
Sum of squared loadings	3.256	1.720	1.039	0.550	0.269		
Cumulative percentage of total standardized sample variance	32.56	49.76	60.15	65.65	68.34		

which is much less than $\chi^2_{[(10-5)^2-(10+5)]/2}(0.05) = \chi^2_5(0.05) = 11.071$. So at a 5% significance level, the null hypothesis of five common factor is not rejected.

However, after carefully examining the loadings for factors $\hat{\mathbf{F}}_4$ and $\hat{\mathbf{F}}_5$, which have sum of squared loadings much less than 1, we decide to try a model with three common factors. So, we will re-compute the maximum likelihood estimate for the factor model with $k = 3$. The result is given in Table 5.4.

The proportion of the contribution of the common factor is

$$P = \frac{\left(\hat{\ell}_{1,i}^2 + \hat{\ell}_{2,i}^2 + \cdots + \hat{\ell}_{m,i}^2\right)}{\left(\hat{\gamma}_{1,1} + \hat{\gamma}_{2,2} + \cdots + \hat{\gamma}_{m,m}\right)} = \begin{cases} 0.2425, i = 1, \\ 0.2146, i = 2, \\ 0.1272, i = 3. \end{cases} \tag{5.32}$$

From Table 5.4, it is very clear that common factor one, $F_{1,t}$, has relatively large positive loadings in the health and biotechnology related industry, common factor two, $F_{2,t}$, has large positive loadings on energy related industry, and common factor three, $F_{3,t}$, represents communication technology industry. This three common factor model explains almost the same amount of total variation as the five common factor model shown in Table 5.3. Most importantly, the simpler three factor model provides much clearer interpretations. Thus, we will choose it as our factor model to represent the stock return process of dimension $m = 10$.

Table 5.4 MLE of common factor loadings, communalities, and specific variances for $k = 3$.

Variable (j)	$\hat{\ell}_{j,1};\hat{F}_{1,t}$	$\hat{\ell}_{j,2};\hat{F}_{2,t}$	$\hat{\ell}_{j,3};\hat{F}_{3,t}$	$\hat{c}_i^2$	$\hat{\sigma}_i^2$
CVX		**0.651**	0.102	0.434	0.566
XOM		**0.996**		0.984	0.016
AAPL	0.227	0.114	**0.582**	0.403	0.597
FB	0.333	0.216	**0.612**	0.532	0.468
MSFT	0.193	0.223	**0.708**	0.585	0.415
JNJ	**0.416**	0.326		0.452	0.548
MRK	**0.479**	0.341		0.346	0.654
PFE	**0.541**	0.378	−0.118	0.449	0.551
FBIOX	**0.865**	0.296	−0.174	0.866	0.134
FSPHX	**0.883**	0.409		0.947	0.053
Sum of squared loadings	2.425	2.146	1.272		
Cumulative percentage of total standardized sample variance	24.25	45.71	58.43		

Comparing Tables 5.2 and 5.4, we note that the cumulative percentage of total standardized sample variance explained by the factors is larger for the principal component factoring than the maximum likelihood factoring. This may be expected, because the loadings obtained by the principal component method are achieved through a variance optimizing property.

5.4 Factor rotation

It should be noted that regardless of what method is used in factor analysis, from Eq. (5.2), for any $k \times k$ orthogonal matrix $\mathbf{\Phi}$, we have

$$\dot{\mathbf{Z}}_t = \mathbf{L}\mathbf{F}_t + \boldsymbol{\varepsilon}_t = \mathbf{L}^*\mathbf{F}_t^* + \boldsymbol{\varepsilon}_t, \tag{5.33}$$

where

$$\mathbf{L}^* = \mathbf{L}\mathbf{\Phi} \text{ and } \mathbf{F}_t^* = \mathbf{\Phi}'\mathbf{F}_t. \tag{5.34}$$

As a result, there are many multiple choices for $\mathbf{L}$ through orthogonal transformations, each of which is equivalent to rotating the common factors in the m-dimensional space. This leads researchers to use arithmetic to find a new set of factor loadings so that the resulting common factors have easier and nicer interpretations. The methods are commonly known as factor rotations.

5.4.1 Orthogonal rotation

Let $\mathbf{F}_t^*$ be the rotated factor and $\mathbf{L}^* = \left[\hat{\ell}_{i,j}^*\right]$ be the rotated matrix of factor loadings. One of the most widely used orthogonal rotation methods is the varimax proposed by Kaiser (1958), which finds an orthogonal transformation to maximize the sum of the variances of the squared loadings:

$$V = \frac{1}{m}\sum_{j=1}^{k}\left[\sum_{i=1}^{m}\tilde{\ell}_{i,j}^{*\,4} - \frac{1}{m}\left(\sum_{i=1}^{m}\tilde{\ell}_{i,j}^{*\,2}\right)^2\right], \tag{5.35}$$

where

$$\tilde{\ell}_{i,j}^* = \frac{\hat{\ell}_{i,j}^*}{\hat{c}_i}, \tag{5.36}$$

are the rotated coefficients scaled by the square root of communalities. The varimax will be achieved if any given variable has a high loading on a single factor but nearly zero loadings on the remaining factors or any given factor is formed by only a few variables with very high loadings and the remaining variables have nearly zero loadings on this factor. This is a widely used method for orthogonal rotation with all factors remaining uncorrelated. After the orthogonal transformation is determined, we will multiply the loadings $\tilde{\ell}_{i,j}^*$ by $\hat{c}_i$ so that the original communalities are preserved.

5.4.2 Oblique rotation

The orthogonal rotation methods like varimax assume that the factors in the analysis are independent. On the other hand, some researchers believe that the purpose of factor rotations is to achieve a simple structure with a new set of factor loadings so that the resulting common factors have simpler and nicer interpretations, and hence one should relax the independence assumption for the factors. The resulting method is often known as oblique rotation. Just like orthogonal rotations, there are many different forms of oblique rotation, see Carroll (1953, 1957), and Jennrich and Sampson (1966).

Although we introduce the rotation concept here, they are used for principal components analysis (PCA) too. There are many factor rotations available and they are implemented in statistical software like EViews, MATLAB, MINITAB, R, SAS, and SPSS. We will not spend more time on the discussion of various factor rotations. Instead, we would like to point out that the rationale of factor rotations is to simplify the factor structure with easier interpretation, and Thurstone (1947) suggested the following criteria:

1. Each variable should produce at least one zero loading on some factor.
2. Each factor should have at least as many nearly zero loadings as there are factors.
3. Each pair of factors should have variables with significant loadings on one and near zero loadings on the other.
4. Each pair of factors should have a large proportion of zero loadings on both factors.

5. Each pair of factors should have only a small number of large loadings. For more details, we refer readers to a good multivariate analysis textbook by Johnson and Wichern (2007) and some relevant statistical software manuals.

5.4.3 Empirical Example III – Model 3 on daily stock returns from the second set of 10 stocks

Example 5.3 The maximum likelihood estimation results for the factor model with three common factors given in Table 5.4 is one without rotation. Applying the orthogonal rotation method, varimax, using the program *factanal* of software R, we get the result given in Table 5.5.

The proportion of the contribution of the ith common factor is computed as follows:

$$P = \frac{\left[\left(\hat{\ell}^*_{1,i}\right)^2 + \left(\hat{\ell}^*_{2,i}\right)^2 + \cdots + \left(\hat{\ell}^*_{m,i}\right)^2\right]}{\left(\hat{\gamma}_{1,1} + \hat{\gamma}_{2,2} + \cdots + \hat{\gamma}_{m,m}\right)} = \begin{cases} 0.2813, i=1, \\ 0.1569, i=2, \\ 0.1462, i=3. \end{cases} \tag{5.37}$$

Comparing the results in Tables 5.4 and 5.5, we see that the rotation does enhance the values of factor loadings and interpretations. The m = 10-dimensional stock return process can be very well represented by the three common factors, the factor of health care related industry, the factor of communication technology related industry, and the factor of energy related industry.

Table 5.5 Varimax rotated MLE of common factor loadings, communalities, and specific variances for k = 3.

Variable (j)	$\hat{\ell}^*_{j,1};\hat{F}^*_{1,t}$	$\hat{\ell}^*_{j,2};\hat{F}^*_{2,t}$	$\hat{\ell}^*_{j,3};\hat{F}^*_{3,t}$	$\hat{c}_i^2$	$\hat{\sigma}_i^2$
CVX	0.147	0.149	**0.625**	0.434	0.566
XOM	0.253		**0.962**	0.989	0.011
AAPL		**0.627**		0.393	0.607
FB	0.217	**0.690**		0.523	0.477
MSFT		**0.750**	0.146	0.584	0.416
JNJ	**0.472**	0.160	0.180	0.281	0.719
MRK	**0.552**	0.107	0.176	0.347	0.653
PFE	**0.641**		0.192	0.448	0.552
FBIOX	**0.928**			0.742	0.258
FSPHX	**0.931**	0.258	0.114	0.946	0.054
Sum of squared loadings	2.813	1.569	1.462		
Cumulative percentage of total standardized sample variance	28.13	43.82	58.44		

5.5 Factor scores

5.5.1 Introduction

Once factor analysis is completed and parameters are estimated, we have $\hat{\mathbf{L}}$ and $\hat{\boldsymbol{\Sigma}}$. By treating these $\hat{\mathbf{L}}$ and $\hat{\boldsymbol{\Sigma}}$ as known and their elements like the common factor loadings $\hat{\ell}_{i,j}$ and the variances $\hat{\sigma}_i^2$ of specific factors as if they were the true values, we can estimate the values $\hat{\mathbf{F}}_t$ of the unobserved random factor vector $\mathbf{F}_t$, known as factor scores, from the Eq. (5.2), which repeats as follows:

$$\underset{m\times 1}{\dot{\mathbf{Z}}_t} = \underset{(m\times k)}{\mathbf{L}}\underset{(k\times 1)}{\mathbf{F}_t} + \underset{(m\times 1)}{\boldsymbol{\varepsilon}_t}, \tag{5.38}$$

where we regard the specific factors $\boldsymbol{\varepsilon}_t' = [\varepsilon_{1,t}, \varepsilon_{2,t}, \dots, \varepsilon_{m,t}]$ as errors. Because the variances σ_i^2 of $\varepsilon_{i,\,t}$ for $i = 1, 2, \dots, m$, can be unequal, Bartlett (1937) has suggested using the weighted least squares to estimate the common factor values, that is choosing the estimate $\hat{\mathbf{F}}_t$ to minimize the following weighted sum of squares of the errors,

$$\sum_{i=1}^{m} \frac{\varepsilon_i^2}{\sigma_i^2} = \boldsymbol{\varepsilon}'\boldsymbol{\Sigma}^{-1}\boldsymbol{\varepsilon} = \left(\dot{\mathbf{Z}}_t - \mathbf{L}\mathbf{F}_t\right)'\boldsymbol{\Sigma}^{-1}\left(\dot{\mathbf{Z}}_t - \mathbf{L}\mathbf{F}_t\right). \tag{5.39}$$

The solution from Chapter 3 can be easily seen to be

$$\hat{\mathbf{F}}_t = \left(\mathbf{L}'\boldsymbol{\Sigma}^{-1}\mathbf{L}\right)^{-1}\mathbf{L}'\boldsymbol{\Sigma}^{-1}\dot{\mathbf{Z}}_t = \left(\mathbf{L}'\boldsymbol{\Sigma}^{-1}\mathbf{L}\right)^{-1}\mathbf{L}'\boldsymbol{\Sigma}^{-1}(\mathbf{Z}_t - \boldsymbol{\mu}). \tag{5.40}$$

Thus, treating $\hat{\mathbf{L}}$, $\hat{\boldsymbol{\Sigma}}$, and $\hat{\boldsymbol{\mu}} = \bar{\mathbf{Z}}$ as the true values, the factor scores for time t is obtained as

$$\hat{\mathbf{F}}_t = \left(\hat{\mathbf{L}}'\hat{\boldsymbol{\Sigma}}^{-1}\hat{\mathbf{L}}\right)^{-1}\hat{\mathbf{L}}'\hat{\boldsymbol{\Sigma}}^{-1}(\mathbf{Z}_t - \bar{\mathbf{Z}}). \tag{5.41}$$

When the factor model is obtained through the standardized variables, $\mathbf{U}_t = [U_{1,t}, \dots, U_{m,t}]' = \left[Z_{1,t} - \bar{Z}_1/\sqrt{\hat{\gamma}_{1,1}}, \dots, Z_{m,t} - \bar{Z}_m/\sqrt{\hat{\gamma}_{m,m}}\right]'$, that is through the correlation matrix, it becomes

$$\hat{\mathbf{F}}_t = \left(\hat{\mathbf{L}}'_{\mathrm{U}}\hat{\boldsymbol{\Sigma}}_{\mathrm{U}}^{-1}\hat{\mathbf{L}}_{\mathrm{U}}\right)^{-1}\hat{\mathbf{L}}'_{\mathrm{U}}\hat{\boldsymbol{\Sigma}}_{\mathrm{U}}^{-1}\mathbf{U}_t. \tag{5.42}$$

When $\hat{\mathbf{L}}$ and $\hat{\boldsymbol{\Sigma}}$ are determined by the MLE, we have

$$\begin{aligned}\hat{\mathbf{F}}_t &= \left(\hat{\mathbf{L}}'\hat{\boldsymbol{\Sigma}}^{-1}\hat{\mathbf{L}}\right)^{-1}\hat{\mathbf{L}}'\hat{\boldsymbol{\Sigma}}^{-1}(\mathbf{Z}_t - \bar{\mathbf{Z}}) \\ &= \hat{\boldsymbol{\Delta}}^{-1}\hat{\mathbf{L}}'\hat{\boldsymbol{\Sigma}}^{-1}(\mathbf{Z}_t - \bar{\mathbf{Z}}),\end{aligned} \tag{5.43}$$

where $\hat{\boldsymbol{\Delta}} = \left(\hat{\mathbf{L}}'\hat{\boldsymbol{\Sigma}}^{-1}\hat{\mathbf{L}}\right)$ from Eq. (5.16). If the factor model is obtained through the standardized variables, then

$$\hat{\mathbf{F}}_t = \left(\hat{\mathbf{L}}'_{\mathrm{U}}\hat{\boldsymbol{\Sigma}}_{\mathrm{U}}^{-1}\hat{\mathbf{L}}_{\mathrm{U}}\right)^{-1}\hat{\mathbf{L}}'_{\mathrm{U}}\hat{\boldsymbol{\Sigma}}_{\mathrm{U}}^{-1}\mathbf{U}_t$$

$$= \hat{\boldsymbol{\Delta}}_{\mathrm{U}}^{-1}\hat{\mathbf{L}}'_{\mathrm{U}}\hat{\boldsymbol{\Sigma}}_{\mathrm{U}}^{-1}\mathbf{U}_t. \tag{5.44}$$

5.5.2 Empirical Example IV – Model 4 on daily stock returns from the second set of 10 stocks

Example 5.4 Let us use again the daily stock returns from the 10 different stocks (Chevron, Exxon, Apple, Facebook, Microsoft, Johnson and Johnson, Merck, Pfizer, Fidelity Biotech, and Fidelity Health) that were traded in the New York Stock Exchange from August 2, 2016 to December 30, 2016, as an example. We have obtained the maximum likelihood estimation varimax rotated three common factor model as shown in Table 5.5. Given the standardized observation at time 106,

$$\mathbf{U}'_{106} = \left[\frac{-0.00102-0.00168}{\sqrt{0.000109}}, \frac{-0.001-0.00052}{\sqrt{0.000106}}, \frac{-0.0078-0.00089}{\sqrt{0.000123}}, \frac{-0.01117-(-0.00066)}{\sqrt{0.000149}},\right.$$

$$\frac{-0.01208-0.00094}{\sqrt{0.000101}}, \frac{-0.00242-(-0.00077)}{\sqrt{0.000066}}, \frac{-0.00288-0.00016}{\sqrt{0.000252}},$$

$$\left.\frac{-0.00031-(-0.00123)}{\sqrt{0.000166}}, \frac{-0.00679-(-0.00071)}{\sqrt{0.000360}}, \frac{-0.00307-(-0.00102)}{\sqrt{0.000106}}\right]'$$

$$= [-0.25861, -0.14764, -0.78355, -0.86101, -1.29554, -0.20310, -0.19150, 0.07141,$$

$$-0.32044, -0.19911]'$$

$$\hat{\mathbf{L}}_{\mathrm{U}} = \begin{bmatrix} 0.147 & 0.149 & 0.625 \\ 0.253 & 0.0 & 0.962 \\ 0 & 0.627 & 0 \\ 0.217 & 0.690 & 0 \\ 0 & 0.750 & 0.146 \\ 0.472 & 0.160 & 0.180 \\ 0.552 & 0.107 & 0.176 \\ 0.641 & 0 & 0.192 \\ 0.928 & 0.0 & 0 \\ 0.931 & 0.258 & 0.114 \end{bmatrix},$$

and

$$\hat{\boldsymbol{\Sigma}}_{\mathrm{U}} = \mathrm{diag}[0.566, 0.016, 0.597, 0.468, 0.415, 0.548, 0.654, 0.551, 0.134, 0.053].$$

So, we can compute

$$\hat{\mathbf{F}}_t = \left(\hat{\mathbf{L}}'_{\mathrm{U}}\hat{\boldsymbol{\Sigma}}_{\mathrm{U}}^{-1}\hat{\mathbf{L}}_{\mathrm{U}}\right)^{-1}\hat{\mathbf{L}}'_{\mathrm{U}}\hat{\boldsymbol{\Sigma}}_{\mathrm{U}}^{-1}\mathbf{U}_t$$

$$= \left(\begin{bmatrix} 0.147 & 0.149 & 0.625 \\ 0.253 & 0.0 & 0.962 \\ 0 & 0.627 & 0 \\ 0.217 & 0.690 & 0 \\ 0 & 0.750 & 0.146 \\ 0.472 & 0.160 & 0.180 \\ 0.552 & 0.107 & 0.176 \\ 0.641 & 0 & 0.192 \\ 0.928 & 0.0 & 0 \\ 0.931 & 0.258 & 0.114 \end{bmatrix}' (\mathrm{diag}[0.566, 0.011, 0.607, 0.477, 0.416, 0.719, 0.653, 0.552, 0.258, 0.054])^{-1} \begin{bmatrix} 0.147 & 0.149 & 0.625 \\ 0.253 & 0.0 & 0.962 \\ 0 & 0.627 & 0 \\ 0.217 & 0.690 & 0 \\ 0 & 0.750 & 0.146 \\ 0.472 & 0.160 & 0.180 \\ 0.552 & 0.107 & 0.176 \\ 0.641 & 0 & 0.192 \\ 0.928 & 0.0 & 0 \\ 0.931 & 0.258 & 0.114 \end{bmatrix}\right)^{-1}$$

$$\begin{bmatrix} 0.147 & 0.149 & 0.625 \\ 0.253 & 0.0 & 0.962 \\ 0 & 0.627 & 0 \\ 0.217 & 0.690 & 0 \\ 0 & 0.750 & 0.146 \\ 0.472 & 0.160 & 0.180 \\ 0.552 & 0.107 & 0.176 \\ 0.641 & 0 & 0.192 \\ 0.928 & 0.0 & 0 \\ 0.931 & 0.258 & 0.114 \end{bmatrix}' (\mathrm{diag}[0.566, 0.011, 0.607, 0.477, 0.416, 0.719, 0.653, 0.552, 0.258, 0.054])^{-1} \begin{bmatrix} -0.25861 \\ -0.14764 \\ -0.78355 \\ -0.86101 \\ -1.29554 \\ -0.20310 \\ -0.19150 \\ 0.07141 \\ -0.32044 \\ -0.19911 \end{bmatrix}$$

$$= \begin{bmatrix} -0.06862 & 0.07473 & 0.01899 \\ 0.07473 & -0.31338 & -0.01786 \\ 0.01899 & -0.01786 & -0.01702 \end{bmatrix} \begin{bmatrix} -8.65224 \\ -5.48654 \\ -14.14962 \end{bmatrix} = \begin{bmatrix} -0.08512 \\ 1.32551 \\ 0.17445 \end{bmatrix}$$

With the known factor scores, $\hat{\mathbf{F}}_t$, we can estimate the values of $\mathbf{U}_t$ by

$$\hat{\mathbf{U}}_t = \hat{\mathbf{L}}_{\mathrm{U}}\hat{\mathbf{F}}_t. \tag{5.45}$$

Hence,

$$\hat{\mathbf{U}}_{106} = \hat{\mathbf{L}}_{\mathrm{U}}\hat{\mathbf{F}}_{106} = \begin{bmatrix} 0.147 & 0.149 & 0.625 \\ 0.253 & 0.0 & 0.962 \\ 0 & 0.627 & 0 \\ 0.217 & 0.690 & 0 \\ 0 & 0.750 & 0.146 \\ 0.472 & 0.160 & 0.180 \\ 0.552 & 0.107 & 0.176 \\ 0.641 & 0 & 0.192 \\ 0.928 & 0.0 & 0 \\ 0.931 & 0.258 & 0.114 \end{bmatrix} \begin{bmatrix} -0.08512 \\ 1.32551 \\ 0.17445 \end{bmatrix} = \begin{bmatrix} 0.22984 \\ 0.29402 \\ 0.14628 \\ 0.83109 \\ 0.89613 \\ 1.01960 \\ 0.20331 \\ 0.12555 \\ -0.07899 \\ 0.28263 \end{bmatrix}.$$

Similarly, we can obtain the estimate $\hat{\mathbf{F}}_t$, $\hat{\mathbf{U}}_t$, and hence $\hat{\mathbf{Z}}_t$, for $t = 1, 2, \ldots, 105$. They can be compared with the true value of $\mathbf{Z}_t$ and used to judge whether the factor model fits the data well in terms of MSE and other similar criteria. However, one should not get confused with these estimates and forecasts.

5.6 Factor models with observable factors

In many applications, we can consider a factor model, where the factors $\mathbf{F}_t$ are observable with factor scores. For example, in some economic and financial studies, a commonly used factor model is the one where some macroeconomic variables and market indices such as inflation rate, industrial production index, employment and unemployment rates, interest rate, S&P 500 Index, Dow Jones Industrial Average (DJIA), and Consumer Price Index (CPI) can be used as factors, and they are observable.

When factors are observable, the factor loadings can be estimated using both $\mathbf{Z}_t$ and $\mathbf{F}_t$. Specifically, let $\dot{\mathbf{Z}}_t = (\mathbf{Z}_t - \boldsymbol{\mu}) = (Z_{1,t}-\mu_1, Z_{2,t}-\mu_2, \ldots, Z_{m,t}-\mu_m)' = \left(\dot{Z}_{1,t}, \dot{Z}_{2,t}, \ldots, \dot{Z}_{m,t}\right)'$, we can rewrite the model in Eqs. (5.2) or (5.38) as

$$\underset{(1\times m)}{\dot{Z}_t'} = \underset{(1\times k)(k\times m)}{F_t'L'} + \underset{(1\times m)}{\varepsilon_t'} . \tag{5.46}$$

Thus, for $t = 1, 2, \ldots, n$, the whole system of the factor model can be expressed as the multivariate multiple time series regression,

$$\underset{n\times m}{\dot{Z}} = \underset{(n\times k)}{F}\ \underset{(k\times m)}{\beta} + \underset{n\times m}{\xi} , \tag{5.47}$$

where

$$\dot{\mathbf{Z}} = \begin{bmatrix} \dot{\mathbf{Z}}'_1 \\ \dot{\mathbf{Z}}'_2 \\ \vdots \\ \dot{\mathbf{Z}}'_n \end{bmatrix} = \begin{bmatrix} \dot{Z}_{1,1} & \dot{Z}_{2,1} & \cdots & \dot{Z}_{m,1} \\ \dot{Z}_{1,2} & \dot{Z}_{2,2} & \cdots & \dot{Z}_{m,2} \\ \vdots & \vdots & \vdots & \vdots \\ \dot{Z}_{1,n} & \dot{Z}_{2,n} & \cdots & \dot{Z}_{m,n} \end{bmatrix} = \left[\dot{\mathbf{Z}}_{(1)},\ \dot{\mathbf{Z}}_{(2)},\ \ldots,\ \dot{\mathbf{Z}}_{(m)}\right],$$

$$\mathbf{F} = \begin{bmatrix} \mathbf{F}'_1 \\ \mathbf{F}'_2 \\ \vdots \\ \mathbf{F}'_n \end{bmatrix} = \begin{bmatrix} F_{1,1} & F_{2,1} & \cdots & F_{k,1} \\ F_{1,2} & F_{2,2} & \cdots & F_{k,2} \\ \vdots & \vdots & \vdots & \vdots \\ F_{1,n} & F_{2,n} & \cdots & F_{k,n} \end{bmatrix},$$

$$\boldsymbol{\beta} = \mathbf{L}' = \begin{bmatrix} \ell_{1,1} & \ell_{2,1} & \cdots & \ell_{m,1} \\ \ell_{1,2} & \ell_{2,2} & \cdots & \ell_{m,2} \\ \vdots & \vdots & \vdots & \vdots \\ \ell_{1,k} & \ell_{2,k} & \cdots & \ell_{m,k} \end{bmatrix} = \left[\boldsymbol{\beta}_{(1)},\ \boldsymbol{\beta}_{(2)},\ \ldots,\ \boldsymbol{\beta}_{(m)}\right],$$

and

$$\boldsymbol{\xi}=\begin{bmatrix}\boldsymbol{\xi}_1'\\ \boldsymbol{\xi}_2'\\ \vdots\\ \boldsymbol{\xi}_n'\end{bmatrix}=\begin{bmatrix}\varepsilon_{1,1} & \varepsilon_{2,1} & \cdots & \varepsilon_{m,1}\\ \varepsilon_{1,2} & \varepsilon_{2,2} & \cdots & \varepsilon_{m,2}\\ \vdots & \vdots & \vdots & \vdots\\ \varepsilon_{1,n} & \varepsilon_{2,n} & \cdots & \varepsilon_{m,n}\end{bmatrix}=\left[\boldsymbol{\xi}_{(1)},\ \boldsymbol{\xi}_{(2)},\ \ldots,\ \boldsymbol{\xi}_{(m)}\right].$$

Each $\boldsymbol{\xi}_{(i)}$ follows a n-dimensional multivariate normal distribution $N\left(\mathbf{0},\boldsymbol{\Sigma}_{(i)}\right)=N\left(\mathbf{0},\sigma_i^2\mathbf{I}\right)$, $i=1,\ldots,$ m, and $\boldsymbol{\xi}_{(i)}$ and $\boldsymbol{\xi}_{(j)}$ are uncorrelated if $i\neq j$.

Equation (5.47) implies that

$$\dot{\mathbf{Z}}_{(i)}=\mathbf{F}\boldsymbol{\beta}_{(i)}+\boldsymbol{\xi}_{(i)}, \tag{5.48}$$

which is the standard matrix form of the multiple regression model, and the least squares estimate of $\boldsymbol{\beta}_{(i)}$ is given by

$$\hat{\boldsymbol{\beta}}_{(i)}=(\mathbf{F}'\mathbf{F})^{-1}\mathbf{F}'\dot{\mathbf{Z}}_{(i)}. \tag{5.49}$$

Hence,

$$\hat{\boldsymbol{\beta}}=\hat{\mathbf{L}}'=\begin{bmatrix}\hat{\ell}_{1,1} & \hat{\ell}_{2,1} & \cdots & \hat{\ell}_{m,1}\\ \hat{\ell}_{1,2} & \hat{\ell}_{2,2} & \cdots & \hat{\ell}_{m,2}\\ \vdots & \vdots & \vdots & \vdots\\ \hat{\ell}_{1,k} & \hat{\ell}_{2,k} & \cdots & \hat{\ell}_{m,k}\end{bmatrix}=\left[\hat{\boldsymbol{\beta}}_{(1)},\ \hat{\boldsymbol{\beta}}_{(2)},\ \ldots,\ \hat{\boldsymbol{\beta}}_{(m)}\right]. \tag{5.50}$$

The residuals of Eq. (5.47) are

$$\begin{aligned}\hat{\boldsymbol{\xi}}&=\dot{\mathbf{Z}}-\mathbf{F}\hat{\boldsymbol{\beta}}=\left[\dot{\mathbf{Z}}_{(1)}-\mathbf{F}\hat{\boldsymbol{\beta}}_{(1)},\ \dot{\mathbf{Z}}_{(2)}-\mathbf{F}\hat{\boldsymbol{\beta}}_{(2)},\ \ldots,\ \dot{\mathbf{Z}}_{(m)}-\mathbf{F}\hat{\boldsymbol{\beta}}_{(m)}\right]\\ &=\left[\hat{\boldsymbol{\xi}}_{(1)},\ \hat{\boldsymbol{\xi}}_{(2)},\ \ldots,\ \hat{\boldsymbol{\xi}}_{(m)}\right].\end{aligned} \tag{5.51}$$

Under the assumption given in Eq. (5.2), the estimate of the covariance matrix of $\boldsymbol{\varepsilon}_t$ is

$$\hat{\boldsymbol{\Sigma}}=\operatorname{diag}\left(\frac{\hat{\boldsymbol{\xi}}'\hat{\boldsymbol{\xi}}}{n-m}\right), \tag{5.52}$$

where diag(**A**) represents the diagonal matrix consisting of the diagonal elements of the matrix **A**.

In time series applications, given a m-dimensional time series, $\dot{\mathbf{Z}}_t, t=1,2,\ldots,n$, it is natural to build a factor model with lag operator and a time series model on the factors as shown in the following,

$$\dot{\mathbf{Z}}_t=\mathbf{L}_0\mathbf{F}_t+\mathbf{L}_1\mathbf{F}_{t-1}+\cdots+\mathbf{L}_p\mathbf{F}_{t-p}+\boldsymbol{\varepsilon}_t, \tag{5.53}$$

or equivalently,

$$\dot{\mathbf{Z}}_t = \mathbf{L}\mathbf{F}_t + \boldsymbol{\varepsilon}_t,$$
$$\mathbf{F}_t = \boldsymbol{\Phi}_1 \mathbf{F}_{t-1} + \boldsymbol{\Phi}_2 \mathbf{F}_{t-2} + \cdots + \boldsymbol{\Phi}_p \mathbf{F}_{t-p} + \mathbf{u}_t, \tag{5.54}$$

where all time series are assumed to be stationary, $\mathbf{u}_t$ is a k-dimensional zero mean white noise process independent of $\boldsymbol{\varepsilon}_t$. The model is known as the dynamic factor model. It was first introduced by Geweke (1977) and has been widely used in practice. Once values of factors are obtained, we can combine the lagged values of $\mathbf{Z}_t$ or other observed variables in the model and build a model for the h-step ahead forecast for $\mathbf{Z}_{t+h}$,, that is

$$\dot{\mathbf{Z}}_{t+h} = \mathbf{L}\mathbf{F}_t + \boldsymbol{\alpha}\mathbf{X}_t + \boldsymbol{\varepsilon}_{t+h}, \tag{5.55}$$

where $\mathbf{X}_t$ is a $m \times 1$ vector of lagged values of $\mathbf{Z}_t$ and/or other observed variables.

5.7 Another empirical example – Yearly U.S. sexually transmitted diseases (STD)

We will now consider a data set that contains yearly STD morbidity rates reported to National Center for HIV/AIDS, viral Hepatitis, STD, and TB Prevention (NCHHSTP), Center for HIV, and Centers for Disease Control and Prevention (CDC) from 1984 to 2013. The dataset was retrieved from CDC's website and includes 50 states plus D.C. The rates per 100000 persons are calculated as the incidence of STD reports, divided by the population, and multiple by 100000.

For the analysis, we remove data from following states, Montana, North Dakota, South Dakota, Vermont, Wyoming, Alaska, and Hawaii, due to missing data. Hence, the dimension of series $\mathbf{X}_t$ is $m = 44$ and $n = 30$. The data set is known as WW8c in the Data Appendix, and its plot is shown in Figure 5.3.

We first take difference of the series by $\mathbf{Z}_t = (1 - B)\mathbf{X}_t$, and the differenced data are plotted in Figure 5.4. The analysis will be based on differenced data. We used the first 24 data points for model fitting, and the rest of observations for evaluating the forecasting performance.

5.7.1 Principal components analysis (PCA)

As discussed in Chapter 4, PCA can be based on a covariance matrix or a correlation matrix. For a high dimensional case, the print out of a covariance or correlation matrix is tedious. Since the correlation matrix is simply the covariance matrix of standardized variables, instead of saying that PCA is based on a covariance matrix or a correlation matrix, we will simply specify whether PCA is based on unstandardized variables or standardized variables.

5.7.1.1 PCA for standardized Z_t

We first do PCA for $\mathbf{Z}_t$ where the data set is standardized. The screeplot in Figure 5.5 shows that first six principal components can explain most of the variance so that first six components will be enough.

The first six components from sample PCA for the standardized variables is given in Table 5.6.

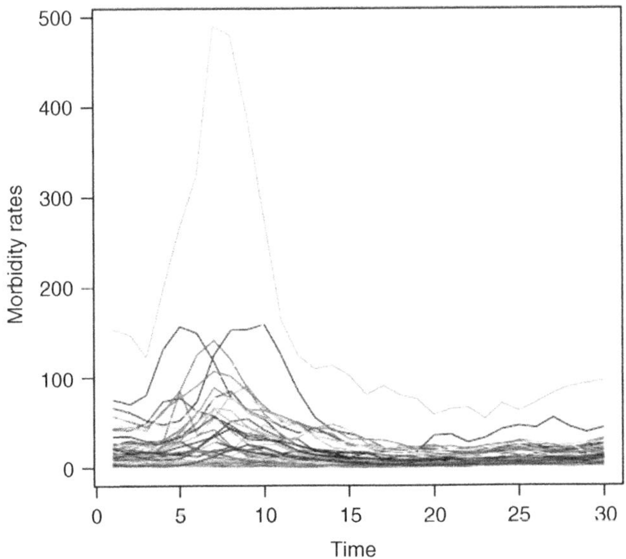

Figure 5.3 U.S. yearly STD of 43 states and D.C.

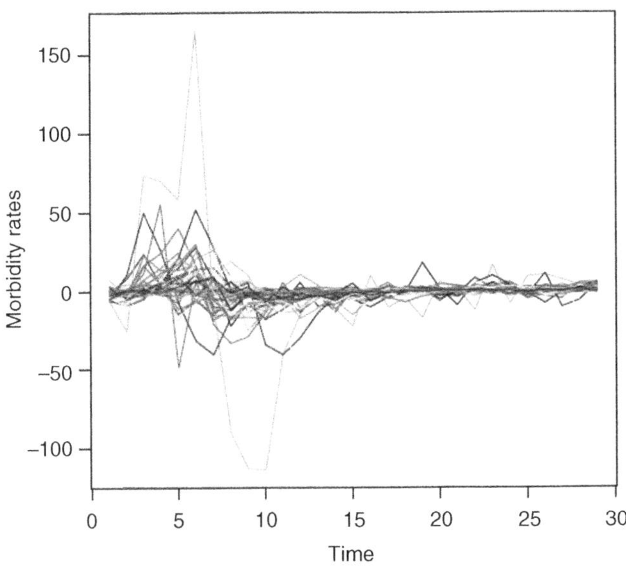

Figure 5.4 The differenced yearly STD of 43 states and D.C.

Let us look at the plot for the first and second components of these time series in Figure 5.6. It appears that states that are spatially close are also tending to be close to each other in the component plot. For examples: (i) NJ, NY, PA, DC; (ii) OH, KS, MS, MO; and (iii) CA, OR, NV, WA.

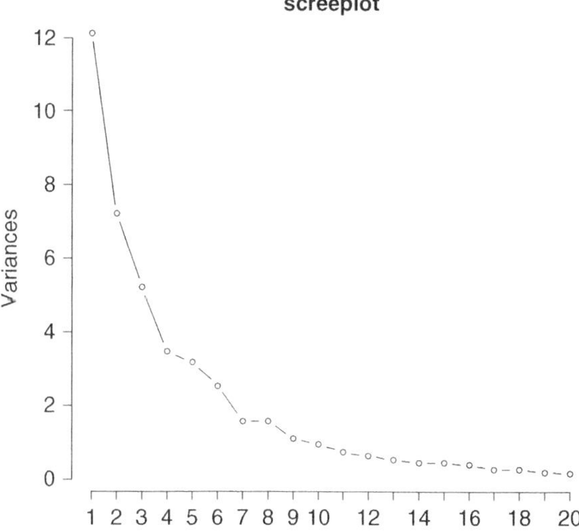

Figure 5.5 The screeplot for standardized variables.

Table 5.6 Sample PCA result for standardized variables.

	Comp. 1 $\hat{\alpha}_1$	Comp. 2 $\hat{\alpha}_2$	Comp. 3 $\hat{\alpha}_3$	Comp. 4 $\hat{\alpha}_4$	Comp. 5 $\hat{\alpha}_5$	Comp. 6 $\hat{\alpha}_6$
CT	−0.127	−0.215	0.200	−0.046	−0.193	0.191
ME	0.008	0.051	−0.048	−0.188	−0.123	0.213
MA	−0.196	−0.176	−0.071	−0.144	0.063	−0.064
NH	−0.082	0.023	−0.058	0.095	−0.329	−0.223
RI	−0.141	0.055	0.212	0.138	0.081	0.230
NJ	−0.214	−0.155	0.026	−0.216	0.020	0.089
NY	−0.233	−0.168	−0.033	0.074	−0.076	−0.015
DE	−0.137	−0.031	0.217	0.067	−0.336	−0.070
DC	−0.250	−0.131	−0.007	0.014	0.039	−0.155
MD	−0.170	−0.168	−0.011	−0.068	0.208	−0.134
PA	−0.223	−0.127	0.075	−0.138	−0.201	0.004
VA	−0.213	0.021	0.009	0.113	0.204	0.254
WV	−0.099	0.086	0.229	0.234	0.013	0.315
AL	−0.237	−0.042	0.060	−0.083	0.204	−0.025
FL	0	−0.283	−0.077	0.152	−0.251	0.134
GA	−0.208	−0.183	0.122	−0.027	−0.101	0.044
KY	−0.049	0.124	−0.264	−0.106	−0.081	0.229
MS	−0.096	0.179	−0.174	−0.128	−0.061	0.101
NC	−0.170	0.135	−0.049	0.076	−0.166	−0.196
SC	−0.203	0.113	0.043	0.228	0.162	−0.006
TN	−0.223	−0.096	−0.022	−0.255	0.012	0.004
IL	−0.212	0.178	0.007	0.105	0.003	−0.149

(continued overleaf)

Table 5.6 (*continued*)

	Comp. 1 $\hat{\alpha}_1$	Comp. 2 $\hat{\alpha}_2$	Comp. 3 $\hat{\alpha}_3$	Comp. 4 $\hat{\alpha}_4$	Comp. 5 $\hat{\alpha}_5$	Comp. 6 $\hat{\alpha}_6$
IN	−0.031	0.206	−0.171	−0.090	−0.208	0.021
MI	−0.245	0.010	0.054	0.115	0.141	−0.075
MN	−0.119	0.070	−0.110	0.008	−0.213	0.063
OH	−0.111	0.217	−0.217	−0.115	−0.058	0.143
WI	−0.199	0.160	−0.089	0.025	0.089	0.045
AR	−0.199	0.164	−0.003	0.144	0.006	−0.194
LA	−0.227	0.142	−0.121	−0.053	0.015	0.040
NM	0.029	−0.127	−0.292	0.110	0.160	−0.001
OK	−0.101	0.071	−0.178	−0.009	−0.179	−0.331
TX	−0.232	−0.031	−0.072	−0.063	0.190	0.069
IA	−0.098	0.056	−0.176	0.006	0.074	0.222
KS	−0.129	0.175	0.087	0.308	−0.054	−0.125
MO	−0.061	0.236	−0.227	−0.009	−0.198	0.064
NE	−0.053	0.070	0.154	0.242	−0.100	0.178
CO	0.017	0.034	−0.289	0.070	0.111	0.212
UT	0.007	−0.063	−0.185	0.283	0.067	0.086
AZ	−0.073	−0.227	−0.174	−0.187	0.089	−0.165
CA	−0.003	−0.245	−0.219	0.178	−0.162	0.043
NV	0.001	−0.190	−0.225	0.099	0.047	0.025
ID	0.018	−0.129	−0.17	0.276	0.153	−0.168
OR	0.045	−0.202	−0.203	0.327	−0.079	0.009
WA	−0.053	−0.256	−0.043	0.041	−0.221	0.226
Variance $\hat{\lambda}_i$	12.128	7.218	5.228	3.478	3.192	2.561
Cumulative Percentage	27.56	43.97	55.85	63.76	71.01	76.83

5.7.1.2 *PCA for unstandardized* Z_t

Now, let us look at what the result is when we apply PCA on unstandardized $\mathbf{Z}_t$, which is equivalent to the construction of the PCA model based on the covariance matrix of $\mathbf{Z}_t$. The screeplot in Figure 5.7 shows that first three principal components can explain most of the variance, so we will choose a three-component PCA model.

The first three components from a sample PCA for the unstandardized variables are given in Table 5.7.

As shown in Figure 5.8, the plot of the first and second components for the unstandardized variables is very different from the one for standardized variables. It suggests that the unstandardized data does not produce similar or good separation as the standardized data does. It also shows that PCA for standardized variables is much more robust. So, we will use standardized variables for our PCA model.

5.7.2 Factor analysis

To build a factor model, we can either use the principle components method or maximum likelihood estimation method. However, for our data set, we have a vector series with a dimension

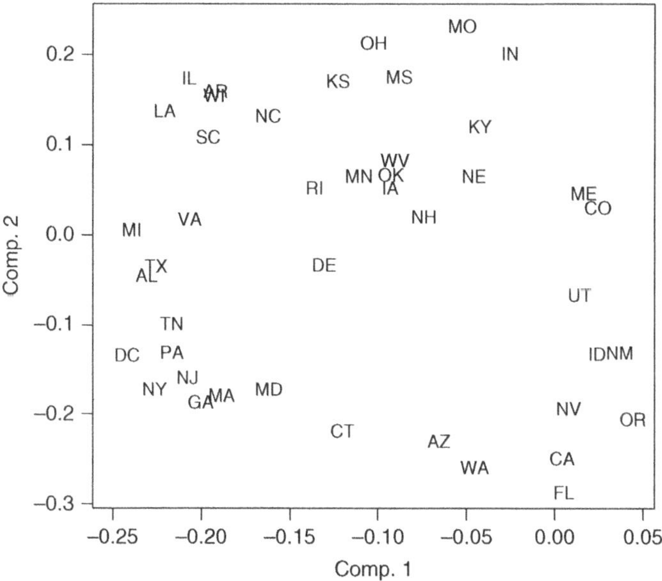

Figure 5.6 The first and second components for standardized variables.

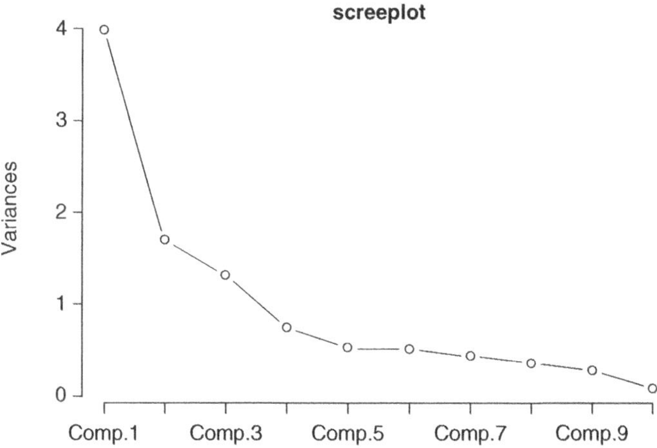

Figure 5.7 The screeplot for unstandardized variables.

$m = 44$ and $n = 30$. Its maximum likelihood equation cannot be properly solved. So, we can only use the principle component method. Thus, we will use PCA to estimate factors. In this analysis, we will follow Bai and Ng (2002), and select six factors, which is supported by the screeplot shown in Figure 5.5.

From the PCA analysis in Section 5.7.1, we will choose a factor model with six common factors,

$$F_{1,t} = Comp.1, F_{2,t} = Comp.2, F_{3,t} = Comp.3, F_{4,t} = Comp.4, F_{5,t} = Comp.5, F_{6,t} = Comp.6$$

Table 5.7 Sample PCA result for unstandardized variables.

	Comp. 1 $\hat{\alpha}_1$	Comp. 2 $\hat{\alpha}_2$	Comp. 3 $\hat{\alpha}_3$
CT	0.064	0.115	−0.016
ME	−0.002	−0.004	−0.001
MA	0.042	0.018	0.001
NH	0.005	−0.001	−0.004
RI	0.018	−0.026	0.054
NJ	0.067	0.007	−0.018
NY	0.138	0.028	−0.133
DE	0.073	0.025	0.079
DC	0.900	0.098	0.098
MD	0.097	0.058	0.059
PA	0.093	0.011	−0.024
VA	0.040	−0.043	−0.003
WV	0.007	−0.013	0.051
AL	0.117	−0.061	0.115
FL	0.054	0.396	−0.482
GA	0.188	0.115	−0.038
KY	0	−0.029	−0.045
MS	0.034	−0.597	−0.560
NC	0.057	−0.116	−0.029
SC	0.081	−0.131	0.110
TN	0.113	−0.041	−0.052
IL	0.058	−0.137	0.056
IN	−0.004	−0.047	−0.030
MI	0.059	−0.035	0.053
MN	0.006	−0.013	−0.012
OH	0.007	−0.091	−0.057
WI	0.024	−0.062	0.011
AR	0.091	−0.206	0.050
LA	0.182	−0.451	−0.094
NM	0.003	0.045	−0.135
OK	0.017	−0.038	−0.026
TX	0.079	−0.057	−0.030
IA	0.003	−0.017	−0.025
KS	0.012	−0.027	0.047
MO	−0.001	−0.146	−0.120
NE	0	−0.001	0.006
CO	−0.004	−0.010	−0.042
UT	0.001	0.010	−0.019
AZ	0.040	0.055	−0.069
CA	0.025	0.137	−0.239
NV	0.046	0.254	−0.487
ID	0.004	0.019	−0.015

Table 5.7 (*continued*)

	Comp. 1 $\hat{\alpha}_1$	Comp. 2 $\hat{\alpha}_2$	Comp. 3 $\hat{\alpha}_3$
OR	0.003	0.090	−0.124
WA	0.011	0.039	−0.052
Variance $\hat{\lambda}_i$	3472.528	717.536	332.014
Cumulative Percentage	65.55	79.10	85.37

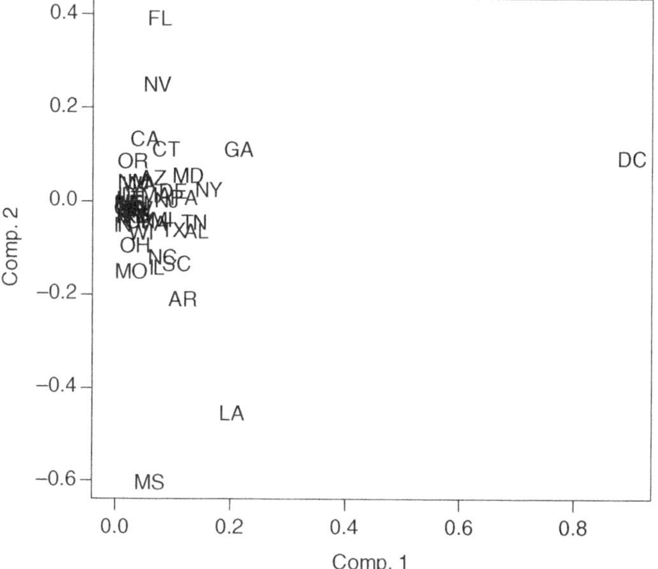

Figure 5.8 The first and second components for unstandardized variables.

based on the standardized variables. The corresponding factor loadings, communalities, and specific variables are given in Table 5.8. The six estimated factor scores are plotted in Figure 5.9.

So, we have our factor model for the STD data set,

$$\underset{44\times 1}{\mathbf{Z}_t} = \underset{(44\times 6)}{\mathbf{L}}\ \underset{(6\times 1)}{\mathbf{F}_t} + \underset{(44\times 1)}{\boldsymbol{\varepsilon}_t}, \tag{5.56}$$

where the element and specification of $\mathbf{L}$, $\mathbf{F}_t$, and $\boldsymbol{\varepsilon}_t$ are given in Table 5.8. As described in Section 5.6, once values or scores of factors are estimated, we can combine the lagged values of $\mathbf{Z}_t$ or other observed variables in the model and build a forecast equation for the h-step ahead forecast for $\mathbf{Z}_{t+h}$, that is

Table 5.8 The principal component estimation for the factor model from the STD data set.

	$\hat{\ell}_{j,1}$	$\hat{\ell}_{j,2}$	$\hat{\ell}_{j,3}$	$\hat{\ell}_{j,4}$	$\hat{\ell}_{j,5}$	$\hat{\ell}_{j,6}$	$\hat{c}_i^2$	$\hat{\sigma}_i^2$
CT	−0.442	−0.577	0.456	−0.086	−0.344	0.306	0.956	0.044
ME	0.029	0.136	−0.110	−0.350	−0.219	0.341	0.319	0.681
MA	−0.683	−0.473	−0.163	−0.269	0.113	−0.102	0.812	0.188
NH	−0.285	0.063	−0.133	0.178	−0.588	−0.357	0.607	0.393
RI	−0.492	0.149	0.485	0.257	0.145	0.368	0.722	0.278
NJ	−0.747	−0.417	0.059	−0.403	0.036	0.143	0.919	0.081
NY	−0.813	−0.451	−0.075	0.137	−0.136	−0.024	0.907	0.093
DE	−0.479	−0.082	0.495	0.124	−0.601	−0.113	0.870	0.130
DC	−0.869	−0.352	−0.015	0.026	0.070	−0.248	0.948	0.052
MD	−0.592	−0.452	−0.025	−0.128	0.372	−0.214	0.756	0.244
PA	−0.778	−0.342	0.172	−0.258	−0.359	0.007	0.947	0.053
VA	−0.743	0.055	0.021	0.211	0.365	0.407	0.900	0.100
WV	−0.345	0.231	0.525	0.436	0.022	0.503	0.892	0.108
AL	−0.825	−0.112	0.137	−0.155	0.364	−0.039	0.870	0.130
FL	−0.002	−0.761	−0.176	0.284	−0.449	0.215	0.938	0.062
GA	−0.723	−0.491	0.279	−0.050	−0.180	0.070	0.883	0.117
KY	−0.172	0.332	−0.602	−0.197	−0.144	0.366	0.697	0.303
MS	−0.336	0.482	−0.398	−0.239	−0.110	0.161	0.599	0.401
NC	−0.591	0.363	−0.111	0.142	−0.297	−0.314	0.701	0.299
SC	−0.709	0.304	0.098	0.425	0.289	−0.010	0.868	0.132
TN	−0.778	−0.258	−0.050	−0.475	0.022	0.006	0.900	0.100
IL	−0.737	0.479	0.017	0.196	0.006	−0.238	0.869	0.131
IN	−0.108	0.553	−0.391	−0.168	−0.372	0.033	0.637	0.363
MI	−0.852	0.026	0.124	0.214	0.252	−0.120	0.866	0.134
MN	−0.415	0.187	−0.253	0.016	−0.381	0.101	0.426	0.574
OH	−0.385	0.584	−0.495	−0.215	−0.104	0.230	0.844	0.156
WI	−0.694	0.429	−0.202	0.046	0.159	0.072	0.739	0.261
AR	−0.693	0.441	−0.006	0.268	0.011	−0.311	0.843	0.157
LA	−0.791	0.380	−0.276	−0.099	0.027	0.065	0.862	0.138
NM	0.100	−0.342	−0.669	0.204	0.286	−0.002	0.697	0.303
OK	−0.350	0.190	−0.408	−0.017	−0.320	−0.530	0.709	0.291
TX	−0.808	−0.083	−0.164	−0.118	0.340	0.110	0.829	0.171
IA	−0.343	0.150	−0.402	0.012	0.132	0.356	0.446	0.554
KS	−0.451	0.470	0.199	0.574	−0.096	−0.200	0.843	0.157
MO	−0.212	0.635	−0.520	−0.017	−0.354	0.102	0.855	0.145
NE	−0.186	0.187	0.352	0.451	−0.179	0.285	0.510	0.490
CO	0.058	0.090	−0.661	0.131	0.198	0.339	0.620	0.380
UT	0.026	−0.170	−0.424	0.528	0.119	0.137	0.521	0.479
AZ	−0.253	−0.609	−0.397	−0.349	0.158	−0.264	0.810	0.190
CA	−0.010	−0.658	−0.501	0.331	−0.289	0.069	0.882	0.118
NV	0.003	−0.511	−0.514	0.184	0.085	0.040	0.568	0.432
ID	0.064	−0.347	−0.389	0.514	0.274	−0.269	0.687	0.313
OR	0.157	−0.543	−0.464	0.609	−0.140	0.014	0.926	0.074
WA	−0.186	−0.687	−0.097	0.076	−0.396	0.362	0.809	0.191

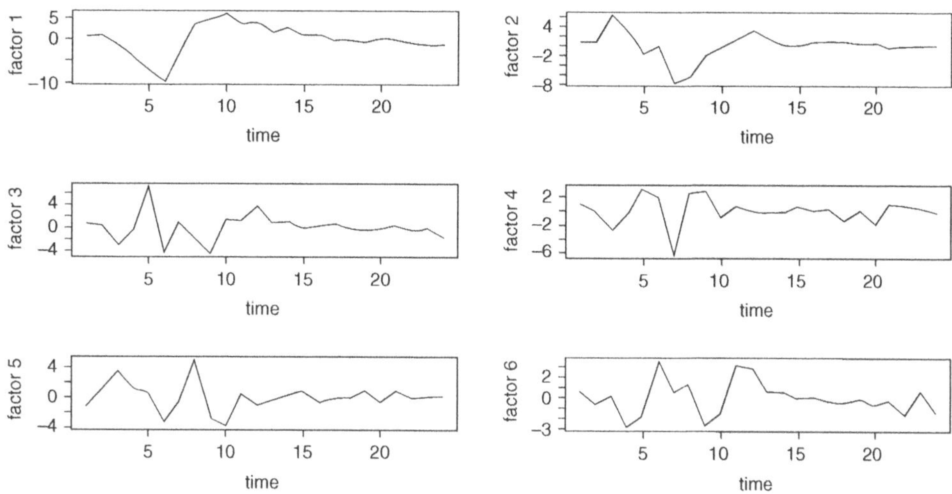

Figure 5.9 The estimated six common factor scores for the STD data set.

$$\hat{\mathbf{Z}}_t(h) = \hat{\mathbf{\Pi}}\hat{\mathbf{F}}_t + \hat{\boldsymbol{\alpha}}\mathbf{V}_t, \tag{5.57}$$

where $\mathbf{V}_t$ are lagged values of $\mathbf{Z}_t$ and/or other observed variables, and $\hat{\mathbf{\Pi}}$ and $\hat{\boldsymbol{\alpha}}$ are the parameter estimates based on $\mathbf{V}_t$ given in Eq. (5.57).

In this example, we will simply choose $\mathbf{V}_t$ to be the first lagged value of $\mathbf{Z}_t$. Specifically, our forecasting equations will be

$$\hat{Z}_{i,t}(h) = \hat{\boldsymbol{\beta}}_i'\hat{\mathbf{F}}_t + \hat{\alpha}_i\hat{Z}_{i,t}(h-1), i = 1,\ldots,44, \tag{5.58}$$

where $\hat{\boldsymbol{\beta}}_i'$ is the ith row of $\hat{\mathbf{\Pi}}$, and $\hat{\alpha}_i$ is the ith element of $\hat{\boldsymbol{\alpha}}$. For $i = 1$, we have

$$\hat{Z}_{1,t}(h) = \hat{\boldsymbol{\beta}}'_1\hat{\mathbf{F}}_t + \hat{\alpha}_1\hat{Z}_{1,t}(h-1), t = 1,\ldots,24, \tag{5.59}$$

and for $h = 1$, it becomes

$$\hat{Z}_{1,24}(1) = \hat{\boldsymbol{\beta}}'_1\hat{\mathbf{F}}_t + \hat{\alpha}_1 Z_{1,t}.$$

From the estimation results, we have

$$\hat{\boldsymbol{\beta}}'_1 = [0.054,\ \ 0.232,\ \ -0.056,\ \ -0.100,\ \ -0.208,\ \ -0.203],$$
$$\hat{\alpha}_1 = 0.234,$$

and the factor scores, $\hat{\mathbf{F}}_t$, for $t = 1, 2, \ldots, 24$, and $Z_{1,t}$ are given in Table 5.9.

Table 5.9 The estimated factor scores and standardized $Z_{1,t}$.

t	Factor 1	Factor 2	Factor 3	Factor 4	Factor 5	Factor 6	t	$Z_{1,t}$
1	0.754	0.729	0.745	0.872	−1.280	0.629	1	0.147
2	1	0.745	0.338	−0.181	1.160	−0.686	2	0.010
3	−1.165	6.198	−2.948	−2.638	3.382	0.107	3	1.067
4	−4.217	3.103	−0.599	−0.511	1.237	−2.825	4	2.033
5	−7.161	−1.663	7.076	3.145	0.621	−1.730	5	3.342
6	−9.685	−0.185	−4.299	1.901	−3.252	3.630	6	−0.734
7	−2.813	−7.658	0.829	−6.294	−0.704	0.553	7	−2.073
8	3.543	−6.566	−1.992	2.611	4.981	1.335	8	−1.996
9	4.445	−2.224	−4.613	2.797	−2.742	−2.706	9	−1.102
10	5.751	−0.298	1.360	−0.786	−3.816	−1.423	10	−0.866
11	3.709	1.050	1.099	0.630	0.403	3.209	11	−0.221
12	3.850	3.066	3.592	−0.027	−1.059	2.854	12	0.310
13	1.635	1.396	1.014	−0.166	−0.407	0.536	13	0.013
14	2.646	0.108	0.792	−0.176	0.169	0.647	14	−0.545
15	0.848	−0.276	−0.209	0.475	0.757	0.074	15	−0.153
16	0.976	0.642	0.110	−0.044	−0.578	−0.014	16	0.151
17	−0.379	0.595	0.538	0.131	−0.044	−0.359	17	0.105
18	0.043	0.731	−0.428	−1.389	−0.046	−0.573	18	0.140
19	−0.511	0.428	−0.605	−0.078	0.908	−0.134	19	0.124
20	0.371	0.566	−0.164	−1.795	−0.571	−0.642	20	−0.096
21	−0.046	−0.380	0.455	0.857	0.764	−0.215	21	0.043
22	−1.106	−0.071	−0.497	0.511	−0.117	−1.552	22	0.175
23	−1.171	0.221	0.035	0.340	0.145	0.600	23	−0.133
24	−1.317	−0.255	−1.629	−0.185	0.089	−1.316	24	0.151

Based on $Z_{1,24} = 0.151$ and $\hat{\mathbf{F}}_{24} = [-1.317, -0.255, -1.629, -0.185, 0.090, -1.316]$, the one-step-ahead forecasting of the first time series is

$$\hat{Z}_{1,24}(1) = [0.054, 0.232, -0.056, -0.100, -0.208, -0.203]\begin{bmatrix} -1.317 \\ -0.255 \\ -1.629 \\ -0.185 \\ 0.090 \\ -1.316 \end{bmatrix} + 0.234(0.151) = 0.29874,$$

which is shown as the forecast for Connecticut in Table 5.10.

The mean squared forecast error (MSFE) is 0.222. It should be noted that the series we have used in model fitting and forecasting are standardized differenced series of the original data set.

Table 5.10 The one-step forecast result for the standardized STD series.

State	Forecast	Actual	Error	State	Forecast	Actual	Error
CT	0.299	0.075	0.224	IN	0.08	−0.194	0.274
ME	−0.488	−0.922	0.434	MI	0.06	0.194	−0.134
MA	−0.313	−0.088	−0.225	MN	0.042	−0.875	0.917
NH	0.375	−0.465	0.84	OH	0.392	0.042	0.35
RI	0.8	0.297	0.503	WI	0.333	−0.143	0.476
NJ	0.198	−0.168	0.366	AR	0.38	0.14	0.24
NY	0.281	−0.456	0.737	LA	0.162	−0.077	0.239
DE	0.233	0.283	−0.05	NM	0.256	0.227	0.029
DC	0.08	0.213	−0.133	OK	−0.285	0.352	−0.637
MD	−0.254	−0.217	−0.037	TX	0.25	0.574	−0.324
PA	0.198	0.127	0.071	IA	0.689	−0.435	1.124
VA	0.587	−0.067	0.654	KS	−0.059	0.382	−0.441
WV	0.339	−0.046	0.385	MO	−0.382	−0.074	−0.308
AL	0.067	−0.066	0.133	NE	0.33	0.369	−0.039
FL	−0.03	−0.161	0.131	CO	−0.789	−1.05	0.261
GA	0.332	−0.047	0.379	UT	−0.313	0.279	−0.592
KY	0.046	0.219	−0.173	AZ	−0.381	−0.952	0.571
MS	0.995	0.073	0.922	CA	−0.525	−0.276	−0.249
NC	0.549	0.888	−0.339	NV	0.771	−0.053	0.824
SC	0.241	0.367	−0.126	ID	−0.095	0.225	−0.32
TN	0.11	0.07	0.04	OR	−0.058	0.173	−0.231
IL	0.23	0.46	−0.23	WA	0.052	−0.893	0.945

5.8 Concluding remarks

In time series applications, it is natural to consider the observable factors in terms of some time series structures and use time lag operator and time series models on the factors. As a result, many different forms of factor models have been developed. The use of time series models such as vector autoregressive models on factors has also been extended to vector time series modeling in both time domain and frequency domain approaches. Some good examples are the various dynamic factor models.

For further information on factor models and applications, we recommend to readers Sharpe (1970), Geweke (1977), Engle and Watson (1981), Geweke and Sinleton (1981), Harvey (1989), Bai and Ng (2002, 2007), Garcia-Ferrer and Poncela 2002, Bai (2003), Peña and Poncela (2004), Deistler and Zinner (2007), Hallin and Liska (2007), Doz et al. (2011), Jungbacker et al. (2011), Stock and Watson (2009, 2011, 2016), Lam and Yao (2012), Fan et al. (2016), Rockova and George (2016), Chan et al. (2017), and Gonçalves et al. (2017), among others.

Software code

R code for Examples 5.1 and 5.2

```
library(ggplot2)
##Import Data: daily stock returns from ten different stocks
from C:/Bookdata/WW5.csv
setwd("C:/Bookdata/")
d10 <- as.data.frame(read.csv("WW5.csv")[,-1])
rownames(d10) <- seq(as.Date("2016/8/2"), by="day", length=106)

##Covariance and Correlation Matrices
d10 <- t(t(d10) - colMeans(d10))
cov(d10)
cor(d10)

##Plot Time Series Data
plot(seq(as.Date("2016/8/2"), by="day", length=106),d10[,1],
type='l',ylab="Stock Returns",xlab="Day")
for(i in 2:10){
  lines(seq(as.Date("2016/8/2"), by="day", length=106),d10[,i],
 type='l',col=i)
}
legend("topleft",legend=colnames(d5),col=1:5,lty=1)

##Principal Component Analysis
pca <- princomp(d10,cor=F)
lds <- pca$loadings
scs <- pca$scores
screeplot(pca,type="lines",main="screeplot")

pca <- princomp(d10,cor=T)
lds <- pca$loadings
scs <- pca$scores
screeplot(pca,type="lines",main="screeplot")


##Plot: Sectors by Their First 2 Loadings
library(ggplot2)
C <- as.data.frame(cbind(lds[,1],lds[,2]))
ggplot(C,aes(C[,1],C[,2],label=rownames(C))) +
   geom_point(size=4,col=3) +
   geom_text(vjust=0,hjust=0,angle = 10,size=5) +
   xlab("Loading 1") +
   ylab("Loading 2")
```

```
##Plot: Time Points byTheir Scores on First 2 Components
C <- as.data.frame(cbind(scs[,1],scs[,2]))
palette(rainbow(400))
ggplot(C,aes(C[,1],C[,2],label=substring(rownames(C),1,7))) +
  geom_point(size=4,col=1:nrow(C)) +
  geom_text(vjust=0,hjust=0,angle = 10,size=5) +
  xlab("Scoring 1") +
  ylab("Scoring 2")
palette("default")

##Plot: Time Points by Their Scores on First 2 Components
C <- as.data.frame(cbind(scs[,1],scs[,2]))
palette(rainbow(400))
ggplot(C,aes(C[,1],C[,2],label=substring(rownames(C),1,4))) +
  geom_path() +
  geom_point(size=4,col=1:nrow(C)) +
  geom_text(vjust=0,hjust=0,angle = 10,size=5) +
  xlab("Scoring 1") +
  ylab("Scoring 2")
palette("default")
plot(seq(as.Date("2009/4/1"), by="month", length=106),C[,1],
type="l",ylab="Scoring 1",xlab="Date")
plot(seq(as.Date("2009/4/1"), by="month", length=106),C[,2],
type="l",ylab="Scoring 2",xlab="Date")
ag <- aggregate(C[,1:2],list(substr(rownames(C),1,4)),mean)
C <- as.data.frame(ag)
ggplot(C,aes(C[,2],C[,3],label=C[,1])) +
  geom_point(size=4,col=3) +
  geom_text(vjust=0,hjust=0,angle = 10,size=5) +
  xlab("Scoring 1") +
  ylab("Scoring 2")

##Factor Analysis
fit <- factanal(scale(d10), factors=2,rotation="none")
print(fit, digits=3, cutoff=.001, sort=TRUE)

##Plot Sectors by their First 2 Loadings
load <- fit$loadings[,1:2]
plot(load,type="n") # set up plot
text(load,labels=colnames(d5),cex=.7)
q()
```

R code for Example 5.4

```
#================================================================
#  Necessary packages and functions
#================================================================
library(forecast)
library(tseries)
library(MTS)
library(TSA)
SWfore <- function (y, x, orig, m, h)
{
    if (!is.matrix(x))
       x = as.matrix(x)
    nT = dim(x)[1]
    k = dim(x)[2]
    if (orig > nT)
        orig = nT
    if (m > k)
        m = k
    if (m < 1)
        m = 1
    #orig=orig-2
    x1 = x[1:orig,]
    me = apply(x1, 2, mean)
    se = sqrt(apply(x1, 2, var))
    x1 = x
    for (i in 1:k) {
        x1[,i] = (x1[,i] - me[i])/se[i]
    }
    V1 = cov(x1[1:orig,])
    m1 = eigen(V1)
    sdev = m1$values

    #**************************
    ff<-c()
    for (j in 2:m){
        M = m1$vectors
        M1 = M[,1:j]
        Dindex = x1 %*% M1
        tt = Dindex%*%t(M1)
        ff[j-1] <- log(mean((tt-x)^2)) + j*((orig+k)
        /(orig*k))*log((orig*k)/(orig+k))
    }
    order <- which(ff==min(ff)) + 1
    #**************************

    M = m1$vectors
    M1 = M[,1:order]
    Dindex = x1 %*% M1
```

```
    ##################################################
    y1 = y[(h+1):orig]
    DF = Dindex[1:(orig-h),]
    y1l = y[h:(orig-1)]
    dat = data.frame(cbind(y1,DF,y1l))
    #fit = arimax(y1,order=c(1,0,0),xreg=DF,include.mean=FALSE)
    fit = lm(y1~.-1,data=dat)
    resi = fit$residuals
    coef <- fit$coefficients
    #se <- fit$se
    yhat = NULL
    MSE = NULL
    #yhat=coef%*%c(Dindex[(orig - h + 1),], y[orig])
    for(j in 1:h) {
      if (j==1){
          yhat[j] = coef%*%c(Dindex[(orig - h + 1),], y[orig])
        }else{
          yhat[j] = coef%*%c(Dindex[orig - h + j,], yhat[j-1])
        }
    }
    SWfore <- list(coef = coef, yhat = yhat, loadings = M1,
            DFindex = Dindex, order=order, residuals = resi,
            DF=DF,y1=y1)
}

##Import Data: Yearly sexually transmitted disease (STD) from
C:/Bookdata/WW8c.csv
#==========================================================
#  Read the data
#==========================================================
setwd("C:/Bookdata/")
A_MMWR <- as.matrix(read.csv("WW8c-R9.csv"))
A_STD <- as.matrix(read.csv("WW8c-R4.csv"))
SexDise <- as.matrix(read.csv("WW8c.csv"))
d1 <- SexDise[,-1]

# Take difference and mean-center the data
d2 <- d1[-1,]-d1[-nrow(d1),]
d3 <- (d2 -
t(matrix(colMeans(d2),ncol(d2),nrow(d2))))/t(matrix(apply
(d2,2,sd),ncol(d2),nrow(d2)))

# Final data matrix
Z <- as.matrix(d3)
# Identify # of locations (r) and # of time points (T)
Z.fit <- Z[1:24,]
r <- ncol(Z.fit)
T <- nrow(Z.fit)
```

```
Z.MMWR <- Z%*%t(A_MMWR)
Z.MMWR.fit <- Z.MMWR[1:24,]

Z.STD <- Z%*%t(A_STD)
Z.STD.fit <- Z.STD[1:24,]

#=============================================================
# Covariance and Correlation Matrices
#=============================================================
# Covariance
print(round(cov(Z),3))
# Correlation
print(round(cor(Z),3))

#=============================================================
# Principal Component Analysis
#=============================================================
y <- Z
x <- Z

#### Standardized
m1 <- prcomp(x,scale=TRUE,center=TRUE)
#Loadings
M1=m1$rotation[,1:6]
print(round(M1,3))

# Scree plot
plot(m1, type = "l",main="screeplot",20)

# PC1 vs. PC2
plot(M1[,1:2],type="n",xlab="Comp. 1", ylab="Comp. 2",main="")
text(M1,labels=colnames(x))

# Eigenvalues and cumulative variance
round(m1$sdev^2,3)
round(cumsum(m1$sdev^2)/sum(m1$sdev^2),4)

#Table 5.6
a1=m1$sdev[1]*m1$rotation[,1]
a2=m1$sdev[2]*m1$rotation[,2]
a3=m1$sdev[3]*m1$rotation[,3]
a4=m1$sdev[4]*m1$rotation[,4]
a5=m1$sdev[5]*m1$rotation[,5]
a6=m1$sdev[6]*m1$rotation[,6]
va=a1^2+a2^2+a3^2+a4^2+a5^2+a6^2
sigma =1-va
table5.6 <- data.frame(a1=a1,a2=a2,a3=a3,a4=a4,a5=a5,
```

```
a6=a6,va=va,sigma=sigma)
write.csv(table5.6,"table5.6.csv")

#### Without standardized
m2 <- prcomp(d2,scale=FALSE,center=TRUE)
#Loadings
M2=m2$rotation[,1:3]
print(round(M2,3))
plot(m2, type = "l",main="screeplot",20)
plot(M2[,1:2],type="n",xlab="Comp. 1", ylab="Comp. 2",main="")
text(M2,labels=colnames(x))
# Eigenvalues and cumulative variance
round(m2$sdev^2,3)
round(cumsum(m2$sdev^2)/sum(m2$sdev^2),4)

#===========================================================
#  Factor Model
#===========================================================
# Factor plots
par(mfrow=c(3,2))
ts.plot(m1$x[,1],ylab="factor 1",xlab="time")
ts.plot(m1$x[,2],ylab="factor 2",xlab="time")
ts.plot(m1$x[,3],ylab="factor 3",xlab="time")
ts.plot(m1$x[,4],ylab="factor 4",xlab="time")
ts.plot(m1$x[,5],ylab="factor 5",xlab="time")
ts.plot(m1$x[,6],ylab="factor 6",xlab="time")
par(mfrow=c(1,1))

####### Coefficients for the first time series
round(SWfore(y[,1],x,24,10,1)$coef,3)
round(SWfore(y[,1],x,24,10,1)$residual,3)

####### Forecasting
# f_factor  <- matrix(0,5,r)
# for(k in 1: 44){
#     m1=SWfore(y[,k],x,24,10,5)
#     f_factor[,k] <- m1$yhat
# }
f_factor  <- matrix(0,5,r)
for(k in 1: 44){
  f_factor[1,k] <- SWfore(y[,k],x,24,10,1)$yhat[1]
  f_factor[2,k] <- SWfore(y[,k],x,24,10,2)$yhat[2]
  f_factor[3,k] <- SWfore(y[,k],x,24,10,3)$yhat[3]
  f_factor[4,k] <- SWfore(y[,k],x,24,10,4)$yhat[4]
  f_factor[5,k] <- SWfore(y[,k],x,24,10,5)$yhat[5]
}
```

```
# One-step-ahead forecast for each state
one_step <- data.frame(forecast=f_factor[1,],actual=Z[25,],
error=abs(f_factor[1,]-Z[25,]))
round(one_step,3)
write.csv(one_step,"one_step.csv")

# MMWR
f_9<-f_factor%*%t(A_MMWR)
ff <-as.vector(t(f_9))
region<-rep(1:9,5)
step <- rep(1:5,each=9)
actual <- as.vector(t(Z.MMWR[25:29,]))
error <- abs(ff-actual)
table_mmwr <- data.frame(step=step, region=region, forecast=ff,
actual=actual, error=error)
table_mmwr
write.csv(table_mmwr,"table_mmwr.csv")

# SDT
f_4<-f_factor%*%t(A_STD)
ff <-as.vector(t(f_4))
region<-rep(1:4,5)
step <- rep(1:5,each=4)
actual <- as.vector(t(Z.STD[25:29,]))
error <- abs(ff-actual)
table_std <- data.frame(step=step, region=region, forecast=ff,
actual=actual, error=error)
table_std
write.csv(table_std,"table_std.csv")

## Example
m1=SWfore(y[,1],x,24,10,1)
tdat=m1$coef%*%c(m1$DFindex[24,],y[24,1])
tdat
q()
```

Projects

1. Find a multivariate analysis book and carefully read its chapter on factor analysis.
2. Find a m-dimensional social science related time series data set with $m \geq 10$, construct your best factor models based on both principle component and likelihood ratio test methods, and compare your results with a written report and analysis software code. Email your data set and software code to your course instructor.
3. Let $\mathbf{Z}_t$ be the m-dimensional vector in Project 1. Build a forecast equation to compute three-step ahead forecasts for $\hat{\mathbf{Z}}_t(j), j = 1, 2, \text{and } 3$.

4. Find a m-dimensional natural science related time series data set with $m \geq 20$, construct your best factor models based on both principle component and likelihood ratio test methods and compare your results with a written report and analysis software code. Email your data set and software code to your course instructor.

5. Let $\mathbf{Z}_t$ be the m-dimensional vector in Project 4. Build a forecast equation to compute five-step ahead forecasts for $\hat{\mathbf{Z}}_t(j), j = 1, 2, 3, 4$, and 5.

References

Bai, J. (2003). Inferential theory for factor models of large dimensions. *Econometrica* **71**: 135–171.

Bai, J. and Ng, S. (2002). Determining the number of factors in approximate factor models. *Econometrica* **70**: 191–221.

Bai, J. and Ng, S. (2007). Determining the number of primitive shocks in factor models. American Statistical Association. *Journal of Business & Economic Statistics* **25**: 52–60.

Bartlett, M.S. (1937). The statistical concept of mental factors. *British Journal of Psychology* **28**: 97–104.

Bartlett, M.S. (1954). A note on multiplying factors for various chi-squared approximation. *Journal of the Royal Statistical Society, Series B* **16**: 296–298.

Carroll, J.B. (1953). An analytic solution for approximating simple structure in factor analysis. *Psychometrika* **18**: 23–38.

Carroll, J.B. (1957). Biquartimin criterion for rotation to oblique simple structure in factor analysis. *Science* **126**: 1114–1115.

Chan, N.H., Lu, Y., and Yau, C.Y. (2017). Factor modelling for high-dimensional time series: inference and model selection. *Journal of Time Series Analysis* **38**: 285–307.

Deistler, M. and Zinner, C. (2007). Modelling high-dimensional time series by generalized linear dynamic factor models; an introductory survey. *Communications in Information and Systems* **7**: 153–166.

Doz, C., Giannone, D., and Reichlin, L. (2011). A two-step estimator for large approximate dynamic factor models based on Kalman filtering. *Journal of Econometrics* **164**: 188–205.

Engle, R.F. and Watson, M.W. (1981). A one-factor multivariate time series model of metropolitan wage rates. *Journal of American Statistical Association* **76**: 774–781.

Fan, J., Liao, Y., and Wang, W. (2016). Projected principal component analysis in factor models. *Annals of Statistics* **44**: 219–254.

Garcia-Ferrer, A. and Poncela, P. (2002). Forecasting international GNP data through common factor models and other procedures. *Journal of Forecasting* **21**: 225–244.

Geweke, J.F. (1977). The dynamic factor analysis of economic time series. In: *Latent Variables in Socio-Economic Models* (ed. D.J. Aigner and A.S. Goldberger). North-Holland, Amsterdam: Elsevier.

Geweke, J.F. and Sinleton, K.J. (1981). Maximum likelihood confirmatory factor analysis of economic time series. *International Economic Review* **22**: 37–54.

Gonçalves, S., Perron, B., and Djogbenou, A. (2017). Bootstrap prediction intervals for factor models. *Journal of Business & Economic Statistics* **35**: 53–69.

Hallin, M. and Liska, R. (2007). The generalized dynamic factor model: determining the number of factors. *Journal of American Statistical Association* **102**: 603–617.

Harvey, A.C. (1989). *Forecasting Structural Time Series Models and the Kalman Filter*. Cambridge University Press.

Jennrich, R.I. and Sampson, P.F. (1966). Rotation for simple loadings. *Psychometrica* **31**: 313–323.

Johnson, R.A. and Wichern, D.W. (2007). *Applied Multivariate Statistical Analysis*, 6e. Englewood Cliffs, NJ: Pearson Prentice Hall.

Jungbacker, B., Koopman, S.J., and van der Wel, M. (2011). Maximum likelihood estimation for dynamic factor models with missing data. *Journal of Economic Dynamics & Control* **35**: 1358–1368.

Kaiser, H.F. (1958). The varimax criterion for analytic rotation in factor analysis. *Psychometrika* **23**: 187–200.

Lam, C. and Yao, Q. (2012). Factor modeling for high-dimensional time series: inference for the number of factors. *Annals of Statistics* **40**: 694–726.

Peña, D. and Poncela, P. (2004). Forecasting with nonstationary dynamic factor model. *Journal of Econometrics* **119**: 291–321.

Rockova, V. and George, E. (2016). Fast Bayesian factor analysis via automatic rotations to sparsity. *Journal of American Statistical Association* **111**: 1608–1622.

Sharpe, W. (1970). *Portfolio Theory and Capital Markets*. New York: McGraw-Hill.

Stock, J.H. and Watson, M.W. (2009). Forecasting in dynamic factor models subject to structural instability, Chapter 7. In: *The Methodology and Practice of Econometrics: Festschrift in Honor of D.F. Hendry* (ed. N. Shephard and J. Castle). Oxford University Press.

Stock, J.H. and Watson, M.W. (2011). Dynamic factor models. In: *The Oxford Handbook on Economic Forecasting* (ed. M.P. Clements and D.F. Hendry), 35–59. Oxford University Press.

Stock, J.H. and Watson, M.W. (2016). Dynamic factor models, factor-augmented vector autoregressions, and structural vector autoregressions in macroeconomics. *Handbook of Macroeconomics*, Elsevier 415–525.

Thurstone, L.L. (1947). *Multiple-Factor Analysis*. Chicago: University of Chicago Press.

6

Multivariate GARCH models

To study the heteroscedasticity problem, we can use the autoregressive conditional heteroscedasticity (ARCH) model and the generalized autoregressive conditional heteroscedasticity (GARCH) model, which have been introduced in many papers and books. These models can be applied to both univariate autoregressive integrated moving average (ARIMA) and vector autoregressive moving average (VARMA) models. In recent years, their multivariate extensions have become of great interest to researchers and many papers on this topic have been published. Great challenges in these extensions include the representations of the models and their estimation. In this chapter, we will introduce some useful representations of multivariate GARCH models and the estimation of these models.

6.1 Introduction

Let $Z_t = \mu_t + a_t$ be the univariate time series model, where $\mu_t = E_{t-1}(Z_t) = E(Z_t|\Psi_{t-1})$ is the conditional expectation of Z_t given the past information set, Ψ_{t-1}, up to time $(t-1)$, which corresponds to the terms related relevant covariates or a univariate ARMA structure, and a_t is the corresponding noise process. Engle (1982, 2002), Bollerslev (1986), and many other researchers introduced various conditional variance models to study the volatility phenomenon of Z_t through the following equation

$$a_t = \sigma_t e_t \tag{6.1}$$

where the e_t are i.i.d. random variables with mean 0 and variance 1, and given Ψ_{t-1}, σ_t^2 is the conditional variance of Z_t such that

$$\sigma_t^2 = c + \sum_{j=1}^{p} \phi_j \sigma_{t-j}^2 + \sum_{j=1}^{q} \theta_j a_{t-j}^2, \tag{6.2}$$

Multivariate Time Series Analysis and Applications, First Edition. William W.S. Wei.
© 2019 John Wiley & Sons Ltd. Published 2019 by John Wiley & Sons Ltd.
Companion website: www.wiley.com/go/wei/datasets

which is known as the GARCH(p,q) model. Other various univariate conditional heteroscedasticity models were summarized in Wei (2006, chapter 15).

Similarly, to study the volatility problem in a multivariate regression model or a $(m \times 1)$ vector process, $\mathbf{Z}_t$, that contains a conditional heteroscedastic phenomenon, we can express it as

$$\mathbf{Z}_t = \boldsymbol{\mu}_t + \boldsymbol{\varepsilon}_t, \tag{6.3}$$

where $\boldsymbol{\mu}_t = E(\mathbf{Z}_t|\boldsymbol{\Psi}_{t-1})$ is the conditional expectation of $\mathbf{Z}_t$ given the past information set, $\boldsymbol{\Psi}_{t-1}$, up to time $(t-1)$, which corresponds to the terms related relevant vector covariates or a vector ARMA structure, and $\boldsymbol{\varepsilon}_t = (\varepsilon_{1,t}, \ldots, \varepsilon_{m,t})'$ is the corresponding noise process. A natural extension to Eq. (6.1) is

$$\boldsymbol{\varepsilon}_t = \boldsymbol{\Sigma}_t^{1/2}\mathbf{e}_t, \tag{6.4}$$

where the $\mathbf{e}_t$ are i.i.d. m-dimensional multivariate random vectors with mean vector $\mathbf{0}$ and covariance matrix $\mathbf{I}$ so that the conditional covariance matrix of $\mathbf{Z}_t$ given information up to time $(t-1)$ is $\boldsymbol{\Sigma}_t$.

In financial applications, the variables Z_t and $\mathbf{Z}_t$ are often return series of assets. The models that are used to describe the conditional variance,σ_t, in Eq. (6.1) or the conditional covariance matrix, $\boldsymbol{\Sigma}_t$, in Eq. (6.4) are known as volatility models for these asset returns.

6.2 Representations of multivariate GARCH models

Various specifications of $\boldsymbol{\Sigma}_t$ have been proposed in literature. Some are presented in this section.

6.2.1 VEC and DVEC models

For any square matrix $\mathbf{A}$, let $vech(\mathbf{A})$ be the vector formed by stacking the elements of each column on or below the diagonal of $\mathbf{A}$. Thus,

$$vech\begin{bmatrix} \sigma_{1,1} & \sigma_{1,2} & \sigma_{1,3} \\ \sigma_{2,1} & \sigma_{2,2} & \sigma_{2,3} \\ \sigma_{3,1} & \sigma_{3,2} & \sigma_{3,3} \end{bmatrix} = \begin{bmatrix} \sigma_{1,1} \\ \sigma_{2,1} \\ \sigma_{3,1} \\ \sigma_{2,2} \\ \sigma_{3,2} \\ \sigma_{3,3} \end{bmatrix}. \tag{6.5}$$

Then the most natural extension of Eq. (6.2) to the $(m \times 1)$ vector case in Eq. (6.5) is the following vector VEC model proposed by Bollerslev, Engle, and Wooldridge (1988)

$$vech(\boldsymbol{\Sigma}_t) = \boldsymbol{\Theta}_0 + \sum_{j=1}^{p} \boldsymbol{\Phi}_j vech(\boldsymbol{\Sigma}_{t-j}) + \sum_{j=1}^{q} \boldsymbol{\Theta}_j vech\left(\boldsymbol{\varepsilon}_{t-j}\boldsymbol{\varepsilon}'_{t-j}\right). \tag{6.6}$$

where each element of $\boldsymbol{\Sigma}_t$ is a function of lagged values of $\boldsymbol{\Sigma}_t$ and the squared and cross products of lagged variables or errors, $\boldsymbol{\Theta}_0$ is a $[m(m+1)/2] \times 1$ column vector, and each of the coefficient matrices, $\mathbf{\Phi}_j$ and $\mathbf{\Theta}_j$, is a $[m(m+1)/2] \times [m(m+1)/2]$ matrix. The model in Eq. (6.6) is known as VEC(p,q) model, and shown by Engle and Kroner (1995) it is stationary if and only if all eigenvalues of $\sum_{j=1}^{p} \mathbf{\Phi}_j + \sum_{j=1}^{q} \mathbf{\Theta}_j$ are within the unit circle. For example, the VEC(1,1) model is given by

$$vech(\mathbf{\Sigma}_t) = \mathbf{\Theta}_0 + \mathbf{\Phi}_1 vech(\mathbf{\Sigma}_{t-1}) + \mathbf{\Theta}_1 vech\left(\boldsymbol{\varepsilon}_{t-1}\boldsymbol{\varepsilon}_{t-1}'\right). \tag{6.7}$$

Both VEC and VECH refer to the vectorization of a matrix. Engle and Kroner (1995) called Eq. (6.6) a VEC model and not a VECH model, and researchers simply adapted that name. More specifically, Eq. (6.7) to be stationary, the eigenvalues of ($\mathbf{\Phi}_1 + \mathbf{\Theta}_1$) must be inside the unit circle, and the unconditional covariance is given by

$$\mathbf{\Sigma} = [\mathbf{I} - \mathbf{\Phi}_1 - \mathbf{\Theta}_1]^{-1} \mathbf{\Theta}_0. \tag{6.8}$$

Explicitly, the three-dimensional VEC(1,1) model is given by

$$\begin{aligned}
\begin{bmatrix} \sigma_{1,1,t} \\ \sigma_{2,1,t} \\ \sigma_{3,1,t} \\ \sigma_{2,2,t} \\ \sigma_{3,2,t} \\ \sigma_{3,3,t} \end{bmatrix} &= \begin{bmatrix} \theta_{0,1} \\ \theta_{0,2} \\ \theta_{0,3} \\ \theta_{0,4} \\ \theta_{0,5} \\ \theta_{0,6} \end{bmatrix} + \begin{bmatrix} \phi_{1,1} & \phi_{1,2} & \phi_{1,3} & \phi_{1,4} & \phi_{1,5} & \phi_{1,6} \\ \phi_{2,1} & \phi_{2,2} & \phi_{2,3} & \phi_{2,4} & \phi_{2,5} & \phi_{2,6} \\ \phi_{3,1} & \phi_{3,2} & \phi_{3,3} & \phi_{3,4} & \phi_{3,5} & \phi_{3,6} \\ \phi_{4,1} & \phi_{4,2} & \phi_{4,3} & \phi_{4,4} & \phi_{4,5} & \phi_{4,6} \\ \phi_{5,1} & \phi_{5,2} & \phi_{5,3} & \phi_{5,4} & \phi_{5,5} & \phi_{5,6} \\ \phi_{6,1} & \phi_{6,2} & \phi_{6,3} & \phi_{6,4} & \phi_{6,5} & \phi_{6,6} \end{bmatrix} \begin{bmatrix} \sigma_{1,1,t-1} \\ \sigma_{2,1,t-1} \\ \sigma_{3,1,t-1} \\ \sigma_{2,2,t-1} \\ \sigma_{3,2,t-1} \\ \sigma_{3,3,t-1} \end{bmatrix} \\
&+ \begin{bmatrix} \theta_{1,1} & \theta_{1,2} & \theta_{1,3} & \theta_{1,4} & \theta_{1,5} & \theta_{1,6} \\ \theta_{2,1} & \theta_{2,2} & \theta_{2,3} & \theta_{2,4} & \theta_{2,5} & \theta_{2,6} \\ \theta_{3,1} & \theta_{3,2} & \theta_{3,3} & \theta_{3,4} & \theta_{3,5} & \theta_{3,6} \\ \theta_{4,1} & \theta_{4,2} & \theta_{4,3} & \theta_{4,4} & \theta_{4,5} & \theta_{4,6} \\ \theta_{5,1} & \theta_{5,2} & \theta_{5,3} & \theta_{5,4} & \theta_{5,5} & \theta_{5,6} \\ \theta_{6,1} & \theta_{6,2} & \theta_{6,3} & \theta_{6,4} & \theta_{6,5} & \theta_{6,6} \end{bmatrix} \begin{bmatrix} \varepsilon_{1,t-1}^2 \\ \varepsilon_{2,t-1}\varepsilon_{1,t-1} \\ \varepsilon_{3,t-1}\varepsilon_{1,t-1} \\ \varepsilon_{2,t-1}^2 \\ \varepsilon_{3,t-1}\varepsilon_{2,t-1} \\ \varepsilon_{3,t-1}^2 \end{bmatrix}.
\end{aligned} \tag{6.9}$$

The difficulty of this multivariate volatility model is evident because of the number of parameters involved. For a m-dimensional process, the number of parameters in the simple VEC(1,1) model in Eq. (6.7) is

$$\frac{m(m+1)}{2} + 2\left[\frac{m(m+1)}{2}\right]^2. \tag{6.10}$$

Thus, the total number of parameters for a *three*-dimensional process as shown in Eq. (6.9) is 78 and that for a *six*-dimensional process is 903.

To reduce the number of parameters in the model, Bollerslev, Engle, and Wooldridge (1988) considered a simpler representation of the model where the coefficient matrices $\boldsymbol{\Phi}_j$ and $\boldsymbol{\Theta}_j$ are assumed to be diagonal. The model is known as the Diagonal VEC or DVEC model.

6.2.2 Constant Conditional Correlation (CCC) models

Onc other way to reduce the number of parameters in the VEC model was suggested by Bollerslev (1990) who proposed a representation where the conditional correlation matrix is assumed to be constant. Under such an assumption, the conditional covariances are proportional to the product of the corresponding conditional standard deviations. The model becomes

$$\boldsymbol{\Sigma}_t = \mathbf{D}_t \mathbf{R} \mathbf{D}_t, \tag{6.11}$$

where

$$D_t = \text{diag}\left(\sigma_{1,1,t}^{1/2}, \ldots, \sigma_{m,m,t}^{1/2}\right),$$

$$R = \begin{bmatrix} 1 & \rho_{1,2} & \cdots & \cdots & \rho_{1,m} \\ \rho_{1,2} & 1 & \cdots & \cdots & \rho_{2,m} \\ \vdots & \vdots & \ddots & \vdots & \vdots \\ \vdots & \vdots & \vdots & \ddots & \vdots \\ \rho_{1,m} & \rho_{2,m} & \cdots & \cdots & 1 \end{bmatrix},$$

and $\rho_{i,j}$ is the constant conditional correlation between $\varepsilon_{i,t}$ and $\varepsilon_{j,t}$. The representation is known as the constant conditional correlation (CCC) model. Thus,

$$\sigma_{i,j,t} = \rho_{i,j}\sqrt{\sigma_{i,i,t}\sigma_{j,j,t}}, \tag{6.12}$$

and $\sigma_{i,i,t}$ can be modeled independently as univariate GARCH models like the simple GARCH (1,1) model,

$$\sigma_{i,i,t} = c_i + \alpha_i \sigma_{i,i,t-1} + \beta_i \varepsilon_{i,t-1}^2, i = 1, \ldots, m. \tag{6.13}$$

The reduction of the number of parameters under the CCC representation is clearly significant. For the m-dimensional CCC model in Eq. (6.11), from Eqs. (6.12) and (6.13), the number of parameters is

$$\frac{m(m-1)}{2} + 3m = \left[\frac{m(m+5)}{2}\right]. \tag{6.14}$$

When $m = 3$, the total number of parameters becomes 12 instead of 78, and when $m = 6$, it is 33 instead of 903, for the VEC(1,1) model. Although the CCC model reduces the number of parameters significantly, its constant correlation assumption is very restrictive and undesirable.

6.2.3 BEKK models

Other than the large number of parameters in the model, the other problem with the VEC model is that the conditional covariance matrix as formulated in Eq. (6.6) may not be positive definite. To overcome the difficulty, Engle and Kroner (1995) used a quadratic form to propose the following model, which, without including exogenous variables, is given by

$$\boldsymbol{\Sigma}_t = \mathbf{C}'\mathbf{C} + \sum_{j=1}^{p} \boldsymbol{\Phi}_j' \boldsymbol{\Sigma}_{t-1} \boldsymbol{\Phi}_j + \sum_{j=1}^{q} \boldsymbol{\Theta}_j' \boldsymbol{\varepsilon}_{t-1} \boldsymbol{\varepsilon}_{t-1}' \boldsymbol{\Theta}_j, \tag{6.15}$$

where $\mathbf{C}$ is a $m \times m$ triangular matrix, which is to ensure $\boldsymbol{\Sigma}_t$ to be definitely positive. Engle and Kroner call it BEKK model because it is related to their earlier joint work of Baba et al. (1990). For convenience, we will call the model in Eq. (6.15) as BEKK(p,q) model.

The BEKK(1,1) model is

$$\boldsymbol{\Sigma}_t = \mathbf{C}'\mathbf{C} + \boldsymbol{\Phi}_1' \boldsymbol{\Sigma}_{t-1} \boldsymbol{\Phi}_1 + \boldsymbol{\Theta}_1' \boldsymbol{\varepsilon}_{t-1} \boldsymbol{\varepsilon}_{t-1}' \boldsymbol{\Theta}_1. \tag{6.16}$$

The model will be stationary if and only if the eigenvalues of $\boldsymbol{\Phi}_1' \otimes \boldsymbol{\Phi}_1 + \boldsymbol{\Theta}_1' \otimes \boldsymbol{\Theta}_1$ are in the unit circle, where $\otimes$ is the Kronecker product of two matrices. For a two-dimensional BEKK (1,1) model, its explicit form is given by

$$\boldsymbol{\Sigma}_t = \begin{bmatrix} \sigma_{1,1,t} & \sigma_{1,2,t} \\ \sigma_{1,2,t} & \sigma_{2,2,t} \end{bmatrix} = \begin{bmatrix} c_{1,1} & c_{1,2} \\ c_{2,1} & c_{2,2} \end{bmatrix} + \begin{bmatrix} \phi_{1,1} & \phi_{1,2} \\ \phi_{2,1} & \phi_{2,2} \end{bmatrix}' \begin{bmatrix} \sigma_{1,1,t-1} & \sigma_{1,2,t-1} \\ \sigma_{1,2,t-1} & \sigma_{2,2,t-1} \end{bmatrix} \begin{bmatrix} \phi_{1,1} & \phi_{1,2} \\ \phi_{2,1} & \phi_{2,2} \end{bmatrix}$$
$$+ \begin{bmatrix} \theta_{1,1} & \theta_{1,2} \\ \theta_{2,1} & \theta_{2,2} \end{bmatrix}' \begin{bmatrix} \varepsilon_{1,t-1}^2 & \varepsilon_{1,t-1}\varepsilon_{2,t-1} \\ \varepsilon_{2,t-1}\varepsilon_{1,t-1} & \varepsilon_{2,t-1}^2 \end{bmatrix} \begin{bmatrix} \theta_{1,1} & \theta_{1,2} \\ \theta_{2,1} & \theta_{2,2} \end{bmatrix}, \tag{6.17}$$

and hence,

$$\begin{aligned}
\sigma_{1,1,t} &= c_{1,1} + \phi_{1,1}^2 \sigma_{1,1,t-1} + 2\phi_{1,1}\phi_{2,1}\sigma_{1,2,t-1} + \phi_{2,1}^2 \sigma_{2,2,t-1} \\
&\quad + \theta_{1,1}^2 \varepsilon_{1,t-1}^2 + 2\theta_{1,1}\theta_{2,1}\varepsilon_{1,t-1}\varepsilon_{2,t-1} + \theta_{2,1}^2 \varepsilon_{2,t-1}^2, \\
\sigma_{1,2,t} &= c_{1,2} + \phi_{1,1}\phi_{1,2}\sigma_{1,1,t-1} + \left(\phi_{1,1}\phi_{2,2} + \phi_{1,2}\phi_{2,1}\right)\sigma_{1,2,t-1} + \phi_{2,1}\phi_{2,2}\sigma_{2,2,t-1} \\
&\quad + \theta_{1,1}\theta_{1,2}\varepsilon_{1,t-1}^2 + (\theta_{1,1}\theta_{2,2} + \theta_{1,2}\theta_{2,1})\varepsilon_{1,t-1}\varepsilon_{2,t-1} + \theta_{2,1}\theta_{2,2}\varepsilon_{2,t-1}^2, \\
\sigma_{2,2,t} &= c_{2,2} + \phi_{1,2}^2 \sigma_{1,1,t-1} + 2\phi_{1,1}\phi_{2,2}\sigma_{1,2,t-1} + \phi_{2,2}^2 \sigma_{2,2,t-1} \\
&\quad + \theta_{1,2}^2 \varepsilon_{1,t-1}^2 + 2\theta_{1,2}\theta_{2,2}\varepsilon_{1,t-1}\varepsilon_{2,t-1} + \theta_{2,1}^2 \varepsilon_{2,t-1}^2,
\end{aligned} \tag{6.18}$$

The other advantage of the BEKK model over the VEC model is its significantly reduced number of parameters in the representation. The number of the parameters of the BEKK(1,1) model in Eq. (6.16) is

$$\frac{m(m+1)}{2} + 2m^2 = \frac{m(5m+1)}{2}, \tag{6.19}$$

which is much less than the number of parameters given in Eq. (6.7) for the VEC(1,1) model. For example, in terms of a three-dimensional process, the number of parameters for the BEKK (1,1) model is 24 instead of 78, and for a six-dimensional process it is 93 instead of 903, for the VEC(1,1) model.

Taking Vech on both side of Eq. (6.15) and using the similar result given in Equation (16.3.3) in Wei (2006) for Vech that $Vech(\mathbf{ABC}) = (\mathbf{C}' \otimes \mathbf{A})Vech(\mathbf{B})$, we can rewrite the BEKK(p,q) model in Eq. (6.15) as

$$Vech(\boldsymbol{\Sigma}_t) = (\mathbf{C}'\otimes\mathbf{C})Vech(\mathbf{I}_m) + \sum_{j=1}^{p}\left(\boldsymbol{\Phi}_j'\otimes\boldsymbol{\Phi}_j\right)Vech(\boldsymbol{\Sigma}_{t-1}) + \sum_{j=1}^{q}\left(\boldsymbol{\Theta}_j'\otimes\boldsymbol{\Theta}_j\right)Vech\left(\boldsymbol{\varepsilon}_{t-1}\boldsymbol{\varepsilon}_{t-1}'\right). \tag{6.20}$$

Thus, VEC and BEKK models can be related. Using different notations to avoid confusion, we rewrite BEKK(1,1) model in Eq. (6.16) as

$$\boldsymbol{\Sigma}_t = \mathbf{C}'\mathbf{C} + \mathbf{A}'\boldsymbol{\Sigma}_{t-1}\mathbf{A} + \mathbf{B}'\boldsymbol{\varepsilon}_{t-1}\boldsymbol{\varepsilon}_{t-1}'\mathbf{B}. \tag{6.21}$$

Then VEC(1,1) model in Eq. (6.7) and BEKK(1,1) model in Eq. (6.21) are equivalent if and only if there exist matrices **C**, **A**, and **B** such that

$$\boldsymbol{\Theta}_0 = (\mathbf{C}'\otimes\mathbf{C})Vech(\mathbf{I}_m), \boldsymbol{\Phi}_1 = (\mathbf{A}'\otimes\mathbf{A}), \text{and}\, \boldsymbol{\Theta}_1 = (\mathbf{B}'\otimes\mathbf{B}). \tag{6.22}$$

As a result, some VEC models have BEKK representation but some do not.

The interpretation of BEKK model is not easy especially for the general case of Eq. (6.15) with a large p and q. To make it simpler, the associated matrices in the model such as $\boldsymbol{\Phi}_i$ and $\boldsymbol{\Theta}_i$ are taken to be diagonal, which also leads to the equivalence of some DVEC and orthogonal BEKK models.

6.2.4 Factor models

To simplify the interpretation and representation of $\boldsymbol{\Sigma}_t$, one can also use factor models where the volatility process is assumed to be determined only by a small number of underlying common factors. For example, Engle, Ng, and Rothschild (1990) introduced the following factor model,

$$\boldsymbol{\Sigma}_t = \boldsymbol{\Pi} + \sum_{j=1}^{k}\boldsymbol{\omega}_j\boldsymbol{\omega}_j' r_{j,t}, \tag{6.23}$$

where $\boldsymbol{\Pi}$ is a $m \times m$ positive semi-definite matrix, $r_{j,t}$, $j = 1, \ldots, k$ are the factors, and $\boldsymbol{\omega}_j$, $j = 1, \ldots, k$ are the corresponding $(m \times 1)$ vectors of factor weights. The factors are often modeled with a GARCH(1,1) model,

$$r_{j,t} = c_j + \alpha_j r_{j,t-1} + \beta_j\left(\mathbf{v}_j'\boldsymbol{\varepsilon}_{t-1}\boldsymbol{\varepsilon}_{t-1}'\mathbf{v}_j\right) = c_j + \alpha_j r_{j,t-1} + \beta_j\left(\mathbf{v}_j'\boldsymbol{\varepsilon}_{t-1}\right)^2, \tag{6.24}$$

where c_j, α_j, and β_j are scalars, and $\mathbf{v}_j$ is a $(m \times 1)$ vector. If $\boldsymbol{\varepsilon}_{t-1}$ is the $(m \times 1)$ vector of our concerned variables such as some financial assets, its linear combination $\mathbf{v}_j'\boldsymbol{\varepsilon}_{t-1}$ is the underlying force for the jth factor $r_{j,t}$. The overall conditional covariance $\boldsymbol{\Sigma}_t$ is determined by weighted combination of these factors as given in Eq. (6.23). The number of factors k is normally much smaller than m so that the model is useful.

6.3 O-GARCH and GO-GARCH models

One problem with the model in Eq. (6.23) is that some of the factors may be correlated and become redundant. This leads to the introduction of the following Orthogonal GARCH (shortened to O-GARCH) model, introduced by Alexander and Chibumba (1997),

$$\boldsymbol{\varepsilon}_t = \boldsymbol{\Omega}\mathbf{r}_t, \tag{6.25}$$

where $\boldsymbol{\Omega}$ is an $m \times m$ orthogonal matrix often known as the linkage matrix, transformation matrix, or factor loading matrix, $\mathbf{r}_t = (r_{1,t}, \ldots, r_{m,t})'$, and the $r_{i,t}$'s are independent factors and each follows a univariate GARCH(p,q) such as a GARCH(1,1) model. That is,

$$\mathbf{r}_t = \boldsymbol{\Gamma}_t^{1/2}\mathbf{e}_t, \tag{6.26}$$

where the $\mathbf{e}_t$ are i.i.d. m-dimensional multivariate random vectors with mean vector $\mathbf{0}$ and covariance matrix $\mathbf{I}$,

$$\boldsymbol{\Gamma}_t = \mathrm{Var}_{t-1}(\mathbf{r}_t) = \mathrm{diag}\left(\sigma^2_{r_{1,t}}, \ldots, \sigma^2_{r_{m,t}}\right), \tag{6.27}$$

and

$$\sigma^2_{r_{i,t}} = (1 - \alpha_i - \beta_i) + \alpha_i\sigma^2_{r_{i,t-1}} + \beta_i r^2_{i,t-1}, i = 1, \ldots, m. \tag{6.28}$$

To ensure the value in Eq. (6.28) to be positive, we assume that α_i and β_i are positive, and $\alpha_i + \beta_i < 1$. Thus, the conditional variance of $\boldsymbol{\varepsilon}_t$ and hence that of $\mathbf{Z}_t$ becomes

$$\boldsymbol{\Sigma}_t = \mathrm{Var}_{t-1}(\boldsymbol{\varepsilon}_t) = \mathrm{Var}(\boldsymbol{\varepsilon}_t|\boldsymbol{\Psi}_{t-1}) = \boldsymbol{\Omega}\boldsymbol{\Gamma}_t\boldsymbol{\Omega}'. \tag{6.29}$$

The linkage matrix and the independent components in Eq. (6.25) are obtained by performing a principal component analysis (PCA) on the series through the sample covariance matrix. Alexander (2001) further illustrated the use of the O-GARCH model in her book and emphasized that the strength of the model is to choose a small number of principal components from PCA compared to the number of variables (assets).

The disadvantage of the O-GARCH model is the imposed orthogonality restriction of its associated linkage matrix. This often leads to identification problem because its estimation is based on unconditional information of the sample covariance matrix, and a true orthogonal matrix is difficult to identify when strong multicollinearity exits between the data components. To remove the restriction, van der Weide (2002) extended the O-GARCH model to the generalized O-GARCH model known as GO-GARCH model, where the linkage matrix does not have to be orthogonal and can be any possible invertible matrix. So, the main difference between the O-GARCH and GO-GARCH models lies in the linkage factor loading matrix. The linkage matrix is orthogonal in O-GARCH model and any invertible matrix in the GO-GARCH model.

6.4 Estimation of GO-GARCH models

There are many multivariate GARCH models. Because of its generality and feasibility, in this section, we will concentrate on the estimation of the GO-GARCH model, which we summarize again as follows:

$$\boldsymbol{\varepsilon}_t = \boldsymbol{\Omega}\mathbf{r}_t, \tag{6.30}$$

where $\boldsymbol{\Omega}$ is an invertible $m \times m$ linkage matrix, $\mathbf{r}_t = (r_{1,t}, \ldots, r_{m,t})'$ is a vector of independent factors, which, following van der Weide (2002), are standardized to have unit unconditional variances, that is, $E(\mathbf{r}_t\mathbf{r}_t') = \mathbf{I}$. Furthermore, the factors are typically assumed to follow a univariate GARCH(p,q) such as the following GARCH(1,1) process. So, given the information up to time $(t-1)$, we have $\mathbf{r}_t \sim N(\mathbf{0}, \boldsymbol{\Gamma}_t)$, where

$$\boldsymbol{\Gamma}_t = \text{Var}_{t-1}(\mathbf{r}_t) = \text{diag}\left(\sigma^2_{r_{1,t}}, \ldots, \sigma^2_{r_{m,t}}\right), \tag{6.31}$$

and

$$\sigma^2_{r_{i,t}} = (1-\alpha_i-\beta_i) + \alpha_i\sigma^2_{r_{i,t-1}} + \beta_i r^2_{i,t-1}, i = 1, \ldots, m. \tag{6.32}$$

Given the information up to time $(t-1)$, $\boldsymbol{\Psi}_{t-1}$, the volatility model, that is, the conditional covariance of $\boldsymbol{\varepsilon}_t$ is then given by

$$\boldsymbol{\Sigma}_t = \text{Var}_{t-1}(\boldsymbol{\varepsilon}_t) = Var(\boldsymbol{\varepsilon}_t|\boldsymbol{\Psi}_{t-1}) = \boldsymbol{\Omega}\boldsymbol{\Gamma}_t\boldsymbol{\Omega}'. \tag{6.33}$$

The unconditional variance of $\boldsymbol{\varepsilon}_t$ is:

$$\boldsymbol{\Sigma} = \text{Var}(\boldsymbol{\varepsilon}_t) = \boldsymbol{\Omega}\boldsymbol{\Omega}'. \tag{6.34}$$

6.4.1 The two-step estimation method

When the GO-GARCH model was first proposed by van der Weide (2002), he used the singular value decomposition of the linkage matrix as a parameterization, that is, $\boldsymbol{\Omega} = \mathbf{U}\boldsymbol{\Lambda}^{1/2}\mathbf{V}'$, where $\mathbf{U}$ is the orthogonal matrix containing the orthogonal eigenvectors of $\boldsymbol{\Omega}\boldsymbol{\Omega}'$, $\boldsymbol{\Lambda} = \text{diag}(\lambda_1, \ldots, \lambda_m)$ contains the corresponding eigenvalues ($\lambda_i > 0$, for all i) and $\mathbf{V}$ is the orthogonal matrix of eigenvectors of $\boldsymbol{\Omega}'\boldsymbol{\Omega}$. He proposed a two-step estimation method. In the first step, $\mathbf{U}$ and $\boldsymbol{\Lambda}$ are consistently estimated through PCA of the unconditional sample covariance:

$$\hat{\boldsymbol{\Sigma}} = T^{-1/2}\Sigma_{t=1}^{T}\boldsymbol{\varepsilon}_t\boldsymbol{\varepsilon}_t' = \hat{\mathbf{U}}\hat{\boldsymbol{\Lambda}}\hat{\mathbf{U}}', \tag{6.35}$$

where $\hat{\mathbf{U}}$ contains the orthogonal eigenvectors of $\hat{\boldsymbol{\Sigma}}$, $\hat{\boldsymbol{\Lambda}} = \left(\hat{\lambda}_1, \ldots, \hat{\lambda}_m\right)$ contains the corresponding eigenvalues, and T is the length of the series. In the second step, $\mathbf{V}$ and the univariate GARCH parameters are estimated by maximizing the following log likelihood:

$$
\begin{aligned}
L(\boldsymbol{\theta}) &= -\frac{1}{2}\sum_{t=1}^{T}\{m\log(2\pi)+\log|\hat{\boldsymbol{\Sigma}}_t|+\boldsymbol{\varepsilon}_t'\hat{\boldsymbol{\Sigma}}_t^{-1}\boldsymbol{\varepsilon}_t\} \\
&= -\frac{1}{2}\sum_{t=1}^{T}\left\{m\log(2\pi)+\log\left|\hat{\boldsymbol{\Omega}}\boldsymbol{\Gamma}_t\hat{\boldsymbol{\Omega}}'\right|+\boldsymbol{\varepsilon}_t'\left(\hat{\boldsymbol{\Omega}}\boldsymbol{\Gamma}_t\hat{\boldsymbol{\Omega}}'\right)^{-1}\boldsymbol{\varepsilon}_t\right\} \qquad (6.36)\\
&= -\frac{1}{2}\sum_{t=1}^{T}\left\{m\log(2\pi)+\log|\boldsymbol{\Gamma}_t|+\log\left|\hat{\mathbf{U}}\hat{\boldsymbol{\Lambda}}\hat{\mathbf{U}}'\right|+\boldsymbol{\varepsilon}_t'\hat{\mathbf{U}}\hat{\boldsymbol{\Lambda}}^{-1/2}\mathbf{V}'\boldsymbol{\Gamma}_t^{-1}\mathbf{V}\hat{\boldsymbol{\Lambda}}^{-1/2}\hat{\mathbf{U}}'\boldsymbol{\varepsilon}_t\right\},
\end{aligned}
$$

where $\log\left|\hat{\boldsymbol{\Omega}}\boldsymbol{\Gamma}_t\hat{\boldsymbol{\Omega}}'\right| = \log|\boldsymbol{\Gamma}_t| + \log\left|\hat{\boldsymbol{\Omega}}\hat{\boldsymbol{\Omega}}'\right|$, $\hat{\boldsymbol{\Omega}}\hat{\boldsymbol{\Omega}}' = \hat{\mathbf{U}}\hat{\boldsymbol{\Lambda}}\hat{\mathbf{U}}'$, and $\boldsymbol{\theta} = (\boldsymbol{\theta}_1', \boldsymbol{\theta}_2')$ with $\boldsymbol{\theta}_1$ being a vector of dimension $m(m-1)/2$ characterizing the $m \times m$ orthogonal matrix $\mathbf{V}$, $\boldsymbol{\theta}_2$ being the GARCH parameter vector of dimension $2m$ for $\boldsymbol{\Gamma}_t$. We note here that by using unconditional information first, van der Weide (2002) showed that the number of parameters to be estimated for $\mathbf{V}$ is $m(m-1)/2$ instead of m^2.

Fan et al. (2008) relaxed the independence assumption of factors to be uncorrelated and proposed an alternative method to estimate $\mathbf{V}$ in the second step. Let $\boldsymbol{\Psi}_{t-1}$ be the sigma field generated by the past information up to time $(t-1)$. Since being conditionally uncorrelated $E(r_{i,t}r_{j,t}|\boldsymbol{\Psi}_{t-1}) = 0$ is equivalent to

$$
\sum_{\mathbf{A}\in A_t} E\{(r_{i,t}r_{j,t}I(\mathbf{A})\} = 0, \qquad (6.37)
$$

for any $A_t \subset \boldsymbol{\Psi}_{t-1}$, such that the σ-algebra that is generated by A_t is equal to $\boldsymbol{\Psi}_{t-1}$. Thus, we have

$$
\boldsymbol{\Pi}_T(\mathbf{V}) = \sum_{1\le i<j\le m}\sum_{\mathbf{A}\in A_t} w(\mathbf{A})\sum_{k=1}^{k_0}\frac{1}{T-k}\left|\mathbf{v}_i'\left\{\sum_{t=k+1}^{T}\boldsymbol{\Lambda}^{-1/2}\mathbf{U}'\boldsymbol{\varepsilon}_t\boldsymbol{\varepsilon}_t'\mathbf{U}\boldsymbol{\Lambda}^{-1/2}I(\boldsymbol{\varepsilon}_{t-k}\in\mathbf{A})\right\}\mathbf{v}_j\right|, \qquad (6.38)
$$

where the $\mathbf{v}_j$ are column vectors of $\mathbf{V}$, $I(\cdot)$ is the indication function, $k_0 \ge 1$ is a prescribed integer, $w(\cdot)$ is a weighting function satisfying $\sum_{\mathbf{A}\in A_t} w(\mathbf{A}) < \infty$. The estimators are derived by minimizing Eq. (6.38). In practice, $\mathbf{U}$ and $\boldsymbol{\Lambda}$ in Eq. (6.38) are replaced by their estimators from the first step.

These estimation methods require numerical maximization of a criterion function over a high-dimensional parameter space. In general, they require more advanced developments in computing power and high-dimensional optimization algorithms. They are likely to encounter numerical problems when the number of dimensions is high. Boswijk and van der Weide (2011) further propose a new method of moments estimation that does not require any Newton-type optimization and involves only the identification of $\mathbf{V}$ matrix from autocorrelation coefficients of quadratic terms of errors, assuming non-zero autocorrelation. For the details, we refer readers to their paper.

6.4.2 The weighted scatter estimation method

It is well known that sample covariance is not the most efficient method to estimate the population covariance. The estimation of the linkage matrix using PCA based on sample covariance is prone to outliers, and outliers are quite common in economic and business data. In this section, we propose a new weighted scatter estimation method (WSE) to estimate the linkage matrix, which inherits many nice properties of robust estimation.

For this method, after singular value decomposition of the linkage matrix, in the first step, $\mathbf{U}$ and $\boldsymbol{\Lambda}$ are still consistently estimated through PCA of the unconditional sample covariance in Eq. (6.35). However, in the second step, $\mathbf{V}$ is estimated based on weighted multivariate scatter estimators of $\mathbf{s}_t = \mathbf{V}'\mathbf{r}_t$, and the univariate GARCH parameters in Eq. (6.32) are estimated separately in the third step. The proposed estimation of the linkage matrix does not require the complicated distribution form and any optimization of an objective function; thus, it is free of computational and convergence problems, even when the dimension is high. This property makes the new method numerically attractive and easy to apply.

Under the GO-GARCH specification Eqs. (6.30)–(6.34), we see that $\text{Var}(\mathbf{s}_t) = \text{Var}(\mathbf{r}_t) = \mathbf{I}_m$. $\mathbf{V}$ is unidentifiable through PCA on unconditional variance, as for any orthogonal matrix $\mathbf{Q}$, Var $(\mathbf{Q}\mathbf{s}_t) = \text{Var}(\mathbf{Q}\mathbf{V}'\mathbf{r}_t) = \mathbf{I}_m$. The key idea of the proposed estimation method is that $\mathbf{V}$ can be identified through PCA on weighted multivariate scatter estimators, denoted as $\hat{\mathbf{H}}_w$, that assign weights to each observation based on predefined measures with respect to the distribution hyper-contour and higher moments. There are many ways to define weighted multivariate scatter estimation. For example, we can apply the weighting scheme of M-estimation, which is used in the robust statistics criterion. The concept can be easily extended to other weighting schemes.

Define $\hat{\mathbf{H}}_w$ as the solution of the equation:

$$\frac{1}{T}\sum_{t=1}^{T} w(g_t)\mathbf{s}_t\mathbf{s}_{t-1}' = \mathbf{H}, \tag{6.39}$$

where $g_t = \mathbf{s}_t'\mathbf{H}^{-1}\mathbf{s}_t \geq 0$, the squared Mahalanobis distance, $\mathbf{s}_t = \boldsymbol{\Lambda}^{-1/2}\mathbf{U}'\boldsymbol{\varepsilon}_t = \mathbf{V}\mathbf{r}_t$, and $w(g)$, $g \geq 0$, is a weighting function with conditions given in the following Theorem 6.1. We define analogously the functional form of multivariate scatter at a distribution $F_\mathbf{s}$ in R^m, denoted as $\mathbf{H}_w(F_\mathbf{s})$, to be the solution of the equation:

$$E[w(g)\mathbf{s}\mathbf{s}'] = \mathbf{H}, \tag{6.40}$$

where $\mathbf{s}$, without the time subscript, denotes a $m \times 1$ random vector following distribution $F_\mathbf{s}$, $g = \mathbf{s}'\mathbf{H}^{-1}\mathbf{s} \geq 0$, and E is the expectation operator over $\mathbf{s}$. If $\mathbf{s}_1, \ldots, \mathbf{s}_T$ is a sample series of length T, and $F_\mathbf{s}$ is the corresponding empirical distribution with $P(\mathbf{s}_t) = T^{-1}$, $t = 1, \ldots, T$, Eq. (6.40) becomes Eq. (6.39), and $\mathbf{H}_w(F_\mathbf{s})$ becomes $\hat{\mathbf{H}}_w$. We have the following results.

Theorem 6.1 Assume that the weighting function $w(g)$, $g \geq 0$ satisfies conditions:

1. The function $w(g)$ is nonnegative and nonincreasing for $g \geq 0$.
2. $\psi(g) = gw(g)$ is a nonnegative nondecreasing bounded function with $\sup\psi(g) = B > m$. $\psi(g)$ is strictly increasing in the interval where $\psi(g) < B$.
3. For every hyperplane $\mathbf{Q}$, $P(\mathbf{Q}) < 1 - m/B$.

Then, $\hat{\mathbf{H}}_w$ and $\mathbf{H}_w(F_\mathbf{s})$, the solutions to Eqs. (6.39) and (6.40), respectively, exist uniquely, and $\hat{\mathbf{H}}_w$ is a consistent estimator of $\mathbf{H}_w(F_\mathbf{s})$.

The most commonly used ψ functions include "Huber's ψ", that is, $\psi(g) = \min(g, B)$, and $\psi(g) = (m+\nu)g/(\nu+g)$, with ν degrees of freedom, which is related to the multivariate Student t distribution with ν degrees of freedom, $t(\nu)$.

Theorem 6.2 Assume that $\mathbf{H}_w(F_\mathbf{s})$ and $\hat{\mathbf{H}}_w$ both have distinct eigenvalues and let $\hat{\mathbf{V}}_w$ be the $m \times m$ matrix, whose columns, $\hat{\mathbf{v}}_{w,1}, \ldots, \hat{\mathbf{v}}_{w,m}$, contain all eigenvectors of $\hat{\mathbf{H}}_w$. Then, $\hat{\mathbf{V}}_w$ is a consistent estimator of $\mathbf{V}$ up to multiplication of some permutation matrix of dimension m.

We refer readers to Zheng and Wei (2012) for the proof of these theorems. The assumption that $\hat{\mathbf{H}}_w$ and $\mathbf{H}_w(F_\mathbf{s})$ have distinct eigenvalues is not unreasonable. Intuitively, as long as univariate GARCH processes have different unconditional distributions $\mathbf{H}_w(F_\mathbf{s})$ will have distinct eigenvalues, and a sufficient condition for different GARCH distributions is that GARCH coefficients (α_i, β_j) are different. This is almost always true empirically for $\hat{\mathbf{H}}_w$ and it makes the method feasible.

6.5 Properties of the weighted scatter estimator

6.5.1 Asymptotic distribution and statistical inference

Properties and statistical inferences are easy to derive for distributions that are elliptically symmetric. Unfortunately, the closed-form of the distribution of $\mathbf{r}_t$ is unknown, and it is certainly not elliptically symmetric under the GO-GARCH model. We study the distribution of the estimator $\hat{\mathbf{V}}_w$ based on the influence function (*IF*), first introduced by Hampel et al. (1986). For simplicity, we assume that all functionals discussed here are Frechet differentiable. For a fixed point $\mathbf{x} \in R^m$, let $\delta_\mathbf{x}$ be the point-mass probability measure that assigns mass 1 to $\mathbf{x}$. The influence function of the functional $K(\cdot)$ at a fixed point $\mathbf{x}$ and the given distribution F is defined as:

$$IF(\mathbf{x};F,K) = \lim_{c\to 0^+} \frac{K[(1-c)F + c\delta_\mathbf{x}] - K(F)}{c}, \tag{6.41}$$

where "$\to 0^+$" stands for limit from the right of zero, $\delta_\mathbf{x}$ is the point mass at $\mathbf{x}$. Clearly, the influence function measures the relative effect on the functional $K(\cdot)$ of an infinitesimal point-mass contamination of the distribution F.

We derive the *IF* of $\mathbf{v}_{w,i}$, measuring changes of $\mathbf{v}_{w,i}$ under infinitesimal change of $F_\mathbf{s}$. Let $\mathbf{r}^* = \left(r_1^*, \ldots, r_m^*\right)' = \text{diag}\left(k_{w,1}^{-1/2}, \ldots, k_{w,m}^{-1/2}\right)\mathbf{V}_w\mathbf{s}$, where $k_{w,1}, \ldots, k_{w,m}$ are the eigenvalues of $\mathbf{H}_w$, ordered corresponding to the associated columns of $\mathbf{V}_w$, and $\mathbf{s}$ is a $(m \times 1)$ random vector. Define $\mathbf{W}$ to be a $m \times m$ matrix with diagonal elements $W_{i,i} = 2 + E\left[w\left(\|\mathbf{r}^*\|^2 (r_i^*)^4\right)/\|\mathbf{r}^*\|\right]$ and off-diagonal elements $W_{i,j} = E\left[w\left(\|\mathbf{r}^*\|^2 (r_i^*)^2 (r_j^*)^2\right)/\|\mathbf{r}^*\|\right], i,j = 1,\ldots,m, i \neq j$, where $\|\cdot\|$ denotes the Euclidean norm. Assuming that $E\left(r_i^4\right) < \infty$ and $\mathbf{W}$ is nonsingular, we have the following theorem:

Theorem 6.3 The influence function of $\mathbf{v}_{w,\ i}$ at $F_{\mathbf{s}}$ and a fixed point $\mathbf{s}_0 \in R^m$ is given by:

$$IF(\mathbf{s}_0;F_{\mathbf{s}},\mathbf{v}_{w,i}) = \sum_{j\neq 1, j\neq i}^{m} \frac{\sqrt{k_{w,i}k_{w,j}}w\left(\|\mathbf{r}_0^*\|^2\right) r_{i,0}^* r_{j,0}^*}{\left(k_{w,i}-k_{w,j}\right)\left[1+E\left\{\frac{w(\|\mathbf{r}^*\|^2)}{\|\mathbf{r}^*\|}(r_i^*)^2(r_j^*)^2\right\}\right]}\mathbf{v}_{w,j}, \tag{6.42}$$

where $\mathbf{r}_0^* = \left(r_{1,0}^*,\ldots,r_{m,0}^*\right)' = \text{diag}\left(k_{w,1}^{-1/2},\ldots,k_{w,m}^{-1/2}\right)\mathbf{V}_w\mathbf{s}_0$. For the proof, please see Zheng and Wei (2012).

We derive the asymptotic distribution of our weighted scatter estimators based on the influence function. Under regularity conditions (Hampel et al. (1986), and Huber (1981), we have:

$$T^{1/2}(\hat{\mathbf{v}}_{w,i}-\mathbf{v}_i)\xrightarrow{d}N\{\mathbf{0},ASV(\mathbf{v}_{w,i},F_{\mathrm{s}})\}, \tag{6.43}$$

where

$$\begin{aligned} ASV(\mathbf{v}_{w,i},F_{\mathrm{s}}) &= E\{[IF(\mathbf{s}_0;F_{\mathrm{s}},\mathbf{v}_{w,i})][IF(\mathbf{s}_0;F_{\mathbf{s}},\mathbf{v}_{w,i})]'\} \\ &= \sum_{j\neq 1, j\neq i}^{m} \frac{k_{w,i}k_{w,j}E\left\{\left[w\left(\|\mathbf{r}_0^*\|^2\right)\right]^2(r_{i,0}^*)^2(r_{j,0}^*)^2\right\}}{\left(k_{w,i}-k_{w,j}\right)^2\left[1+E\left\{\frac{w'\left(\|\mathbf{r}^*\|^2\right)}{\|\mathbf{r}^*\|}(r_i^*)^2(r_j^*)^2\right\}\right]^2}\mathbf{v}_{w,j}\mathbf{v}_{w,j}'. \end{aligned}$$

We can use the asymptotic distribution to obtain an asymptotic confidence region for $\mathbf{v}_i$ and perform hypothesis tests. Let $ave\{.\}$ to denote the average taken over the sample. It is easy to show that:

$$T\mathbf{v}_i'\sum_{j\neq 1, j\neq i}^{m} \frac{\left(k_{w,i}-k_{w,j}\right)^2\left[1+ave\left\{\frac{w(\|\mathbf{r}^*\|^2)}{\|\mathbf{r}^*\|}(r_i^*)^2(r_j^*)^2\right\}\right]^2}{k_{w,i}k_{w,j}ave\left\{\left[w\left(\|\mathbf{r}_0^*\|^2\right)\right]^2(r_i^*)^2(r_j^*)^2\right\}}\hat{\mathbf{v}}_{w,j}\hat{\mathbf{v}}'_{w,j}\mathbf{v}_i\xrightarrow{d}\chi_{m-1}^2, \tag{6.44}$$

where $\hat{k}_{w,i}$ and $\hat{\mathbf{v}}_{w,j}$ are eigenvalues and eigenvectors of $\hat{\mathbf{H}}_w$. A confidence region for $\mathbf{v}_i$ with confidence $(1-\alpha)$ consists of the intersection of $\mathbf{v}_i'\mathbf{v}_i = 1$ and the set of $\mathbf{v}_i$ such that the left-hand side of Eq. (6.44) is less than $(1-\alpha)\%$-quantile of the χ^2 distribution with $(m-1)$ degrees of freedom.

We can also perform a hypothesis test with the null hypothesis being $\mathbf{v}_i = \mathbf{v}_{i,0}$ such that $\mathbf{v}_{i,0}'\mathbf{v}_{i,0} = 1$. The hypothesis is rejected if the left-hand side of Eq. (6.44) with $\mathbf{v}_i$ replaced by $\mathbf{v}_{i,0}$ is larger than $(1-\alpha)-$*quantile* of the χ^2 distribution with $(m-1)$ degrees of freedom.

6.5.2 Combining information from different weighting functions

While the determination of the weighting function is arbitrary, we may obtain more efficient estimation results by combining information from different weighting functions. Denote

$\hat{\mathbf{V}}_{w,1},\ldots,\hat{\mathbf{V}}_{w,k}$ to be k estimators of $\mathbf{V}$ matrix based on k different weighting functions. We can pool them together using the Cayley transformation (see Liebeck and Osborne [1991]):

$$\hat{\mathbf{V}}_w=\left\{\mathbf{I}_m-\sum_{i=1}^{k}\left[p_i\left(\mathbf{I}_m-\hat{\mathbf{V}}_{w,i}\right)\right]\left(\mathbf{I}_m-\hat{\mathbf{V}}_{w,i}\right)^{-1}\right\}\left\{\mathbf{I}_m+\sum_{i=1}^{k}\left[p_i\left(\mathbf{I}_m-\hat{\mathbf{V}}_{w,i}\right)\right]\left(\mathbf{I}_m-\hat{\mathbf{V}}_{w,i}\right)^{-1}\right\}^{-1}, \tag{6.45}$$

where p_i's are some weighting functions satisfying $\sum_{i=1}^{k}p_i=1$. The weights p_i's are chosen to be dependent on the closeness of eigenvalues of the k weighted scatter estimators, $\hat{\mathbf{H}}_{w,1},\ldots,$ and $\hat{\mathbf{H}}_{w,k}$. For example, let the eigenvalues of $\hat{\mathbf{H}}_{w,i}$ to be $\hat{\lambda}_{1,i},\ldots,\hat{\lambda}_{m,i}$, $i=1,\ldots,k$, we choose:

$$p_i=\frac{\min_{1\le \ell<j\le m}\left(\hat{\lambda}_{\ell,i}-\hat{\lambda}_{j,i}\right)^2}{\sum_{i=1}^{k}\min_{1\le \ell<j\le m}\left(\hat{\lambda}_{\ell,i}-\hat{\lambda}_{j,i}\right)^2}. \tag{6.46}$$

Thus, if any two eigenvalues of a weighted multivariate scatter estimator are close, the weight of the associated estimator $\hat{\mathbf{V}}_{w,i}$ receives a small weight. This provides protection against possible identification problems. Since the estimation of $\mathbf{V}$ is up to multiplying a permutation matrix with entries of 0, +1 or −1, care must be taken. Orthogonal matrices $\hat{\mathbf{V}}_{w,\ell}$, $\ell=2,\ldots,k$, are matched as closely as possible to $\hat{\mathbf{V}}_{w,1}$. The ith row of the matched matrix $\hat{\mathbf{V}}_{w,\ell}$, denoted $\hat{\mathbf{v}}'_{w,\ell,i}$, satisfies:

$$\left|\hat{\mathbf{v}}'_{w,\ell,i}\hat{\mathbf{v}}_{w,\ell,i}\right|=\max_{i\le j\le m}\left|\hat{\mathbf{v}}'_{w,\ell,j}\hat{\mathbf{v}}_{w,\ell,i}\right|. \tag{6.47}$$

The sign of the first element in each row of $\hat{\mathbf{V}}_{w,\ell}$'s, $\ell=2,\ldots,k$, is set to be consistent with the sign of the first element in each row of $\hat{\mathbf{V}}_{w,1}$. If the resulting matrix has the determinant of −1, the row of $\hat{\mathbf{V}}_{w,\ell}$ that has the smallest absolute inner product with the corresponding row of $\hat{\mathbf{V}}_{w,1}$ is multiplied by −1. Using this procedure, we first match $\hat{\mathbf{V}}_{w,1}$ to the identity matrix $\mathbf{I}_m$ to guarantee the existence of inverse in Eq. (6.47).

6.6 Empirical examples

6.6.1 U.S. weekly interest over time on six exercise items

Example 6.1 In this example, I will use the project result from one of the best students in our program, Mr. Qingyang (Kevin) Liu, using the data obtained from Google Trends, which is a public web facility of Google Inc., based on Google Search, which shows how often a particular search-term is entered relative to the total search-volume across various regions in the world. For this project, he collected the data within the United States on people's feeling/motivation about six exercise related topics. Specifically, he obtained people's "interest over time" once every week between April 15, 2012 and April 9, 2017 on fitness, home workout, gym,

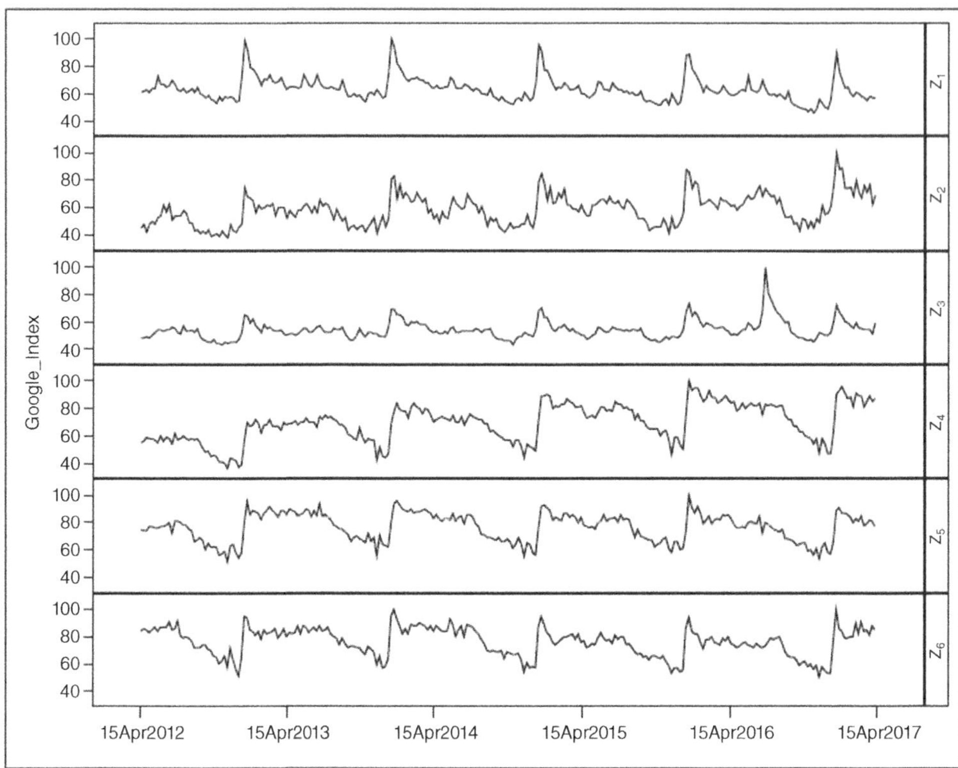

Figure 6.1 U.S. weekly interest over time on six exercise items from April 15, 2012 to April 9, 2017.

bodybuilding supplement, cardio, and weight loss. So, we have data, $\mathbf{Z}_t = [Z_{1,t}, Z_{2,t}, \ldots, Z_{6,t}]'$, $t = 1, 2, \ldots, 262$, as listed on WW6a in the Data Appendix.

Based on the definition from https://trends.google.com/trends. The "interest over time" are numbers represent search interest relative to the highest point on the chart for the given region (United States) and time (one week). A value of 100 is the peak popularity for the term. A value of 50 means that the term is half as popular. Likewise, a score of 0 means the term was less than 1% as popular as the peak. The plot of the six series is given in Figure 6.1.

From Figure 6.1, we note that all six topics become very popular around the week of December 30 and/or January 1 each year. It could be because this is the big holiday and most people are free at home.

6.6.1.1 Choose a best VAR/VARMA model

Based on various information criteria given in Table 6.1, a VARMA(1,1) is tentatively chosen.

The estimation result from VARMA (1,1) is shown in Equation (6.48).

Table 6.1 Information criteria on various model fitting.

VAR(1)		VAR(2)		VAR(3)		VMA(1)		VMA(2)		VARMA(1,1)	
AICC	5 695.40	**AICC**	5 612.43	**AICC**	5 748.33	**AICC**	6 722.21	**AICC**	Unavailable	**AICC**	5 563.32
HQC	5 744.44	**HQC**	5 629.48	**HQC**	5 640.22	**HQC**	6 771.25	**HQC**	Unavailable	**HQC**	5 581.28
AIC	5 654.26	**AIC**	5 487.90	**AIC**	5 447.35	**AIC**	6 681.07	**AIC**	Unavailable	**AIC**	5 439.57
SBC	5 878.58	**SBC**	5 840.03	**SBC**	5 927.00	**SBC**	6 905.39	**SBC**	Unavailable	**SBC**	5 792.08
FPEC	5 871 933	**FPEC**	3 359 056	**FPEC**	3 117 615	**FPEC**	3.86E + 08	**FPEC**	Unavailable	**FPEC**	2 536 797

AICC: Corrected Akaike information criterion.
HQC: Hanna–Quinn criterion.
AIC: Akaike information criterion
SBC: Schwarz Bayesian criterion. Also known as BIC.
FPEC: Final Prediction Error criterion.

$$\mathbf{Z}_t = \begin{bmatrix} 19.89848 \\ (5.10010) \\ 19.48988 \\ (4.91947) \\ 17.85950 \\ (3.70957) \\ 6.61368 \\ (4.56801) \\ 6.76711 \\ (3.91188) \\ 11.06914 \\ (5.41315) \end{bmatrix} + \begin{bmatrix} 0.63303 & 0.19688 & 0.07904 & -0.39947 & 0.65876 & -0.46134 \\ (0.13478) & (0.16448) & (0.14681) & (0.10303) & (0.18235) & (0.12938) \\ -0.20053 & 1.24826 & -0.08465 & -0.34413 & 0.51166 & -0.41896 \\ (0.13972) & (0.15745) & (0.14415) & (0.10515) & (0.18284) & (0.12499) \\ -0.17356 & 0.34388 & 0.71257 & -0.26346 & 0.35765 & -0.26891 \\ (0.10209) & (0.11698) & (0.11435) & (0.07747) & (0.14657) & (0.10001) \\ -0.10562 & 0.36286 & 0.00732 & 0.63517 & 0.40680 & -0.35727 \\ (0.12596) & (0.15207) & (0.14384) & (0.09370) & (0.18505) & (0.12426) \\ -0.00858 & 0.17014 & 0.04149 & -0.24434 & 1.32236 & -0.34299 \\ (0.11337) & (0.12935) & (0.12761) & (0.08462) & (0.16345) & (0.10479) \\ -0.26311 & 0.23987 & 0.13605 & -0.39308 & 0.63188 & 0.51625 \\ (0.14456) & (0.18327) & (0.16408) & (0.10778) & (0.20422) & (0.13906) \end{bmatrix} \mathbf{Z}_{t-1}$$

$$+ \boldsymbol{\epsilon}_t - \begin{bmatrix} -0.18192 & 0.07857 & 0.12208 & -0.48582 & 0.98713 & -0.50093 \\ (0.14485) & (0.16551) & (0.14827) & (0.12989) & (0.19787) & (0.15320) \\ -0.86433 & 0.92333 & -0.24713 & -0.25386 & 0.59241 & -0.44581 \\ (0.17747) & (0.16351) & (0.18084) & (0.16375) & (0.22379) & (0.16511) \\ -0.35239 & 0.27536 & -0.01838 & -0.18695 & 0.68211 & -0.37447 \\ (0.11652) & (0.12618) & (0.13405) & (0.10636) & (0.16908) & (0.11891) \\ -0.83784 & 0.28153 & 0.05873 & 0.25101 & 0.71752 & -0.54789 \\ (0.15336) & (0.16809) & (0.17224) & (0.13461) & (0.22201) & (0.16148) \\ -0.62581 & -0.00394 & 0.18855 & -0.21271 & 1.35509 & -0.56432 \\ (0.14374) & (0.15724) & (0.16979) & (0.12924) & (0.21009) & (0.12867) \\ -1.00884 & 0.18904 & 0.24802 & -0.33788 & 0.86778 & 0.07108 \\ (0.17502) & (0.19822) & (0.17736) & (0.14376) & (0.24316) & (0.17991) \end{bmatrix} \boldsymbol{\epsilon}_{t-1}. \tag{6.48}$$

The portmanteau test for cross correlations of residuals is shown in Table 6.2. Since all p-values are smaller than 0.01, we believe that the residuals are correlated, which implies that a VARMA-GARCH model may be more appropriate.

Carefully examining Table 6.3, we see that p-values on the normality test in the third column are smaller than 0.01. Together with the results from the ARCH test shown in the last column, we will consider a VARMA-GARCH model.

6.6.1.2 Finding a VARMA-ARCH/GARCH model

We have tried the VARMA(1,1)-ARCH(1) and VARMA(1,1)-GARCH(1,1) with BEKK representation or CCC representation, and the result from SAS software indicates that estimations are N/A. So, we try the VAR (1)-ARCH (1) model with BEKK, CCC, and DCC representations. The results are given in Table 6.4. Clearly, the BEKK representation works best for this data set.

Table 6.2 Portmanteau test for cross correlations of residuals.

Portmanteau Test for Cross Correlations of Residuals			
Up to lag	DF	Chi-Square	Pr > ChiSq
3	36	103.02	<0.0001
4	72	138.11	<0.0001
5	108	181.74	<0.0001
6	144	250.78	<0.0001
7	180	292.41	<0.0001
8	216	328.80	<0.0001
9	252	358.59	<0.0001
10	288	381.50	0.0002
11	324	419.74	0.0003
12	360	450.26	0.0008

Table 6.3 Univariate model white noise diagnostics.

Univariate Model White Noise Diagnostics					
		Normality		ARCH	
Variable	Durbin Watson	Chi-Square	Pr > ChiSq	F Value	Pr > F
fitness	1.89268	660.80	<0.0001	1.99	0.1593
home_workout	2.02870	82.01	<0.0001	0.01	0.9176
gym	1.98574	5265.50	<0.0001	10.32	0.0015
Bodybuilding_supplement	1.93495	38.21	<0.0001	2.36	0.1261
cardio	1.91137	14.99	0.0006	0.23	0.6344
weight_loss	1.91890	178.10	<0.0001	2.33	0.1281

Table 6.4 Estimation result from VAR(1)-ARCH(1) model.

Information Criteria – BEKK		Information Criteria – CCC		Information Criteria – DCC	
AICC	5 333.919	AICC	5 426.821	AICC	5 514.731
HQC	5 351.882	HQC	5 474.748	HQC	5 561.98
AIC	5 210.169	AIC	5 375.979	AIC	5 460.348
SBC	5 562.677	SBC	5 621.666	SBC	5 713.156
FPEC	9 614 860	FPEC	6 756 409	FPEC	6 430 731

The estimation results are given in the following.

$$\begin{bmatrix} Z_{1,t} \\ Z_{2,t} \\ Z_{3,t} \\ Z_{4,t} \\ Z_{5,t} \\ Z_{6,t} \end{bmatrix} = \begin{bmatrix} 5.82964 \\ -3.60400 \\ 9.20221 \\ -1.29564 \\ 2.83663 \\ 1.05316 \end{bmatrix}$$

$$+ \begin{bmatrix} 0.58771 & -0.04919 & 0.08093 & -0.05466 & 0.22532 & 0.05692 \\ 0.09790 & 0.53528 & 0.15322 & -0.18215 & 0.06743 & -0.03017 \\ -0.0320 & 0.06813 & 0.73881 & -0.01472 & 0.02443 & 0.02039 \\ 0.17349 & 0.20726 & -0.03494 & 0.87769 & -0.11659 & -0.03274 \\ 0.30207 & 0.146576 & -0.1033 & 0.03132 & 0.57924 & 0.06700 \\ -0.06191 & -0.00277 & 0.11212 & -0.06947 & 0.23525 & 0.77694 \end{bmatrix} \begin{bmatrix} Z_{1,t-1} \\ Z_{2,t-1} \\ Z_{3,t-1} \\ Z_{4,t-1} \\ Z_{5,t-1} \\ Z_{6,t-1} \end{bmatrix} + \boldsymbol{\epsilon}_t \tag{6.49}$$

where $\boldsymbol{\epsilon}_t \sim N(\mathbf{0}, \boldsymbol{\Sigma}_t)$ and

$$\boldsymbol{\Sigma}_t = \begin{bmatrix} 9.80234 & 3.35024 & 4.80129 & -0.18923 & 1.68804 & 1.92194 \\ 3.35024 & 22.20685 & 3.07953 & 5.09859 & 5.86953 & 4.76385 \\ 4.80129 & 3.07953 & 3.65150 & 1.12157 & 1.37264 & 1.19117 \\ -0.18923 & 5.09859 & 1.12157 & 13.55704 & 8.94010 & 6.79630 \\ 1.68804 & 5.86953 & 1.37264 & 8.94010 & 13.43418 & 8.34803 \\ 1.92194 & 4.76385 & 1.19117 & 6.79630 & 8.34803 & 13.87160 \end{bmatrix}$$

$$+ \begin{bmatrix} -0.93493 & -0.99433 & -0.60933 & -0.59776 & -0.39312 & -1.07987 \\ -0.30103 & -0.31432 & -0.27980 & -0.12160 & -0.04586 & -0.16552 \\ 0.83822 & 0.96991 & 0.92086 & 0.58621 & 0.42032 & 0.95090 \\ 0.12057 & -0.25423 & -0.19626 & -0.44719 & -0.32223 & -0.31706 \\ 0.28636 & 0.44698 & -0.00464 & 0.46160 & 0.40561 & 0.50146 \\ -0.05777 & 0.10827 & 0.16004 & 0.02246 & -0.13215 & 0.06454 \end{bmatrix}' \boldsymbol{\epsilon}_{t-1}\boldsymbol{\epsilon}'_{t-1}$$

$$\begin{bmatrix} -0.93493 & -0.99433 & -0.60933 & -0.59776 & -0.39312 & -1.07987 \\ -0.30103 & -0.31432 & -0.27980 & -0.12160 & -0.04586 & -0.16552 \\ 0.83822 & 0.96991 & 0.92086 & 0.58621 & 0.42032 & 0.95090 \\ 0.12057 & -0.25423 & -0.19626 & -0.44719 & -0.32223 & -0.31706 \\ 0.28636 & 0.44698 & -0.00464 & 0.46160 & 0.40561 & 0.50146 \\ -0.05777 & 0.10827 & 0.16004 & 0.02246 & -0.13215 & 0.06454 \end{bmatrix}.$$

6.6.1.3 The fitted values from VAR(1)-ARCH(1) model

The plot of the fitted values is given in Figure 6.2 with the real data. The model successfully predicts the surcharge/peak of interest over time at the end of each year.

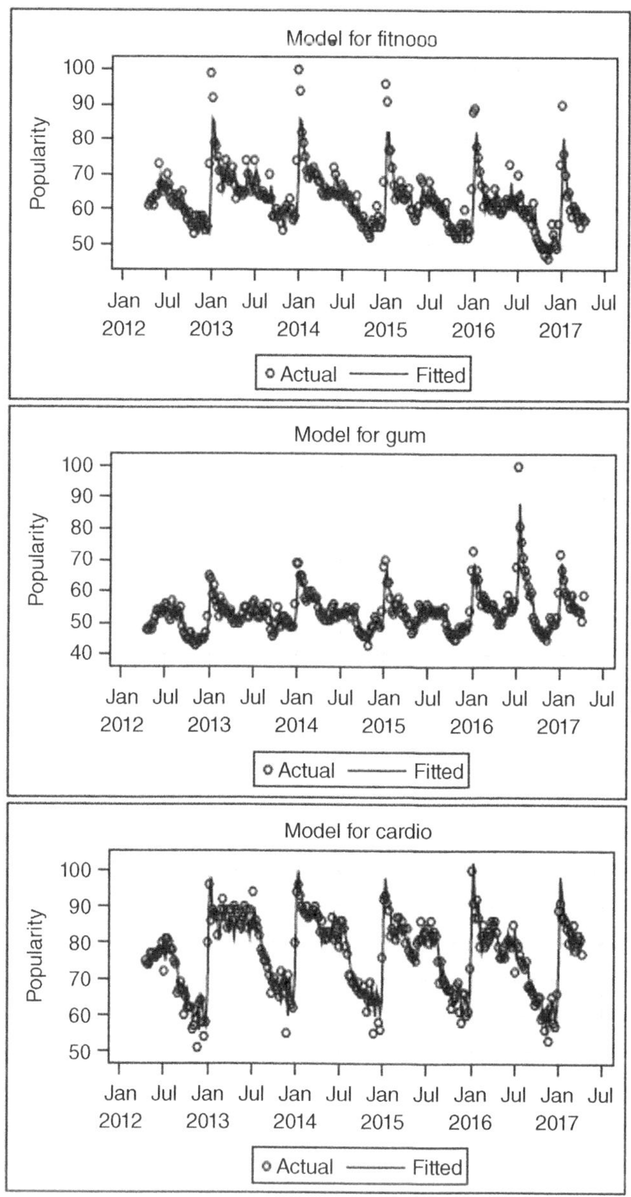

Figure 6.2 The fitted values from VAR(1)–ARCH (1) model.

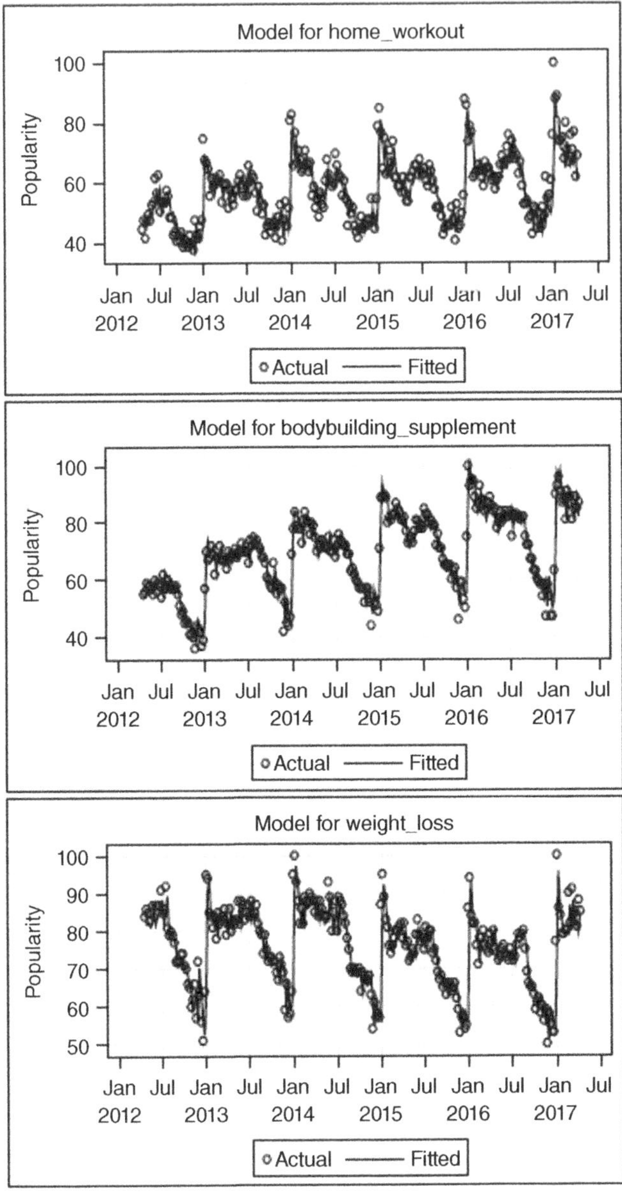

Figure 6.2 (Continued)

6.6.2 Daily log-returns of the SP 500 index and three financial stocks

Example 6.2 We have successfully applied the proposed method to many high-dimensional economic and financial data. The first example we illustrate in this section involves analyzing daily log-returns of the Standard & Poor 500 Index (SP) and three financial stocks: Bank of America (BAC), JPMorgan Chase & Co. (JPM), and Citigroup Inc. (C). The four-variate sample covers the period between January 4, 1994, and December 30, 2005, yielding $T = 3021$ daily returns. The data set is listed as WW6b in the Data Appendix. We study the conditional variances and correlations of these financial variables, SP, BAC, JPM, and C before the housing market started to peak and the sub-prime crisis began to emerge.

Descriptive statistics of the data are displayed in Tables 6.5 and 6.6. Table 6.5 shows mean, standard deviation (sd), skewness, and kurtosis for daily log-returns of SP, BAC, JPM, and C. All the means are essentially zero. The standard deviation shows little variation across the three financial stocks, while the market (SP) displays relatively small volatility. The empirical densities for SP, BAC, JPM, and C appear almost symmetric and exhibit heavier tails. Table 6.6 shows sample correlations between SP, BAC, JPM, and C. We observe that the unconditional correlations are fairly strong with values ranging from 0.601 to 0.692.

Figure 6.3 displays daily log-returns of SP, BAC, JPM, and C between January 4, 1994, and December 30, 2005. The figure clearly shows volatility clustering. We construct the GO-GARCH model and use the method of WSE, combining information from weighting functions $t(\nu)$, $\nu = 5, 6, \ldots, 10$ to estimate the coefficients.

The estimated linkage matrix and GARCH coefficients are summarized in Table 6.7.

With these estimates in Table 6.7, we can obtain the vector of four factors, $\mathbf{r}_t = (r_{1,t}, \ldots, r_{m,t})'$, from Eq. (6.31), $\mathbf{r}_t = \boldsymbol{\Omega}^{-1}\boldsymbol{\varepsilon}_t$, where $\boldsymbol{\varepsilon}_t$ is now the returns of the four equities at time t. Next, we compute $\boldsymbol{\Gamma}_t$ from Eqs. (6.31) and (6.32), and the conditional covariance $\boldsymbol{\Sigma}_t = \mathrm{Var}_{t-1}(\boldsymbol{\varepsilon}_t)$ from Eq. (6.33). For an illustration, Figure 6.4 displays the plot of the estimated volatilities (conditional variances) of the S&P 500 daily log-return with the GO-GARCH(1,1) model between January 4, 1994, and December 30, 2005.

Figure 6.5 illustrates the time plots of the estimated conditional correlation between the returns of S&P 500 and Bank of America. Recall that the unconditional correlation between

Table 6.5 Mean, standard deviation (sd), skewness, and kurtosis for daily log-returns of SP, BAC, JPM, and C.

	SP	BAC	JPM	C
mean	0.000	0.000	0.000	0.000
sd	0.011	0.019	0.022	0.022
skewness	−0.158	−0.157	0.116	0.045
kurtosis	3.625	3.079	5.390	4.789

Table 6.6 Correlations among daily log-returns of SP, BAC, JPM, and C.

	SP	BAC	JPM	C
SP	1.000	0.601	0.675	0.692
BAC		1.000	0.650	0.630
JPM			1.000	0.677
C				1.000

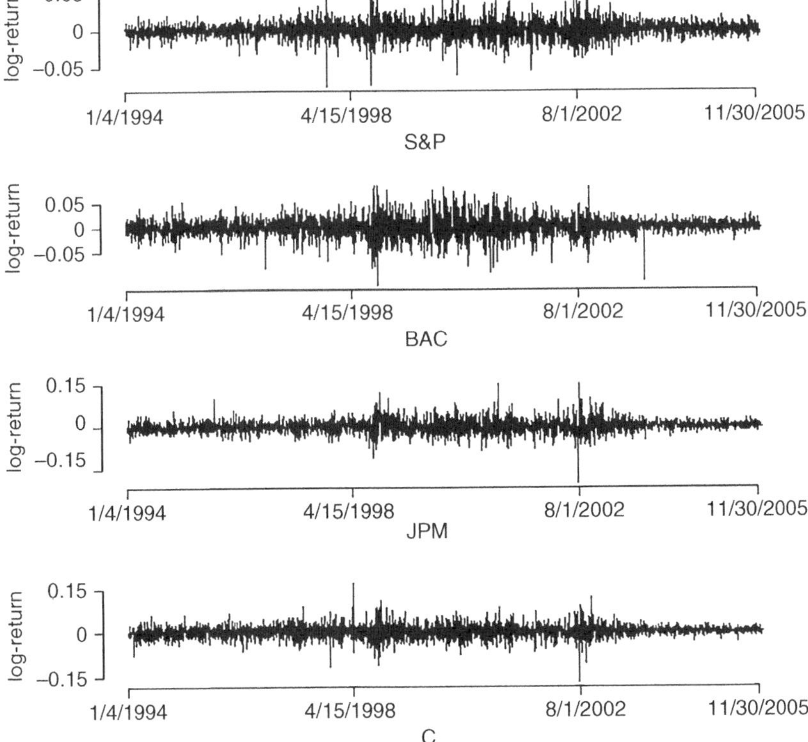

Figure 6.3 Daily log-returns of SP, BAC, JPM, and C between January 4, 1994, and December 30, 2005.

Table 6.7 Estimated linkage matrix and GARCH coefficients.

$\hat{\Omega}$	$\hat{\alpha}$	$\hat{\beta}$
$\begin{bmatrix} -0.0055 & 0.0089 & -0.0040 & 0.0016 \\ -0.0147 & 0.0089 & 0.0075 & -0.0014 \\ -0.0200 & 0.0038 & -0.0081 & -0.0037 \\ -0.0098 & 0.0139 & -0.0036 & -0.0129 \end{bmatrix}$	$\begin{bmatrix} 0.0405 \\ 0.0325 \\ 0.0556 \\ 0.0609 \end{bmatrix}$	$\begin{bmatrix} 0.9593 \\ 0.9659 \\ 0.9448 \\ 0.9391 \end{bmatrix}$

8e+04
6e+04
4e+04
2e+04
0e+00
1/4/1994 4/15/1998 8/1/2002 11/30/2005

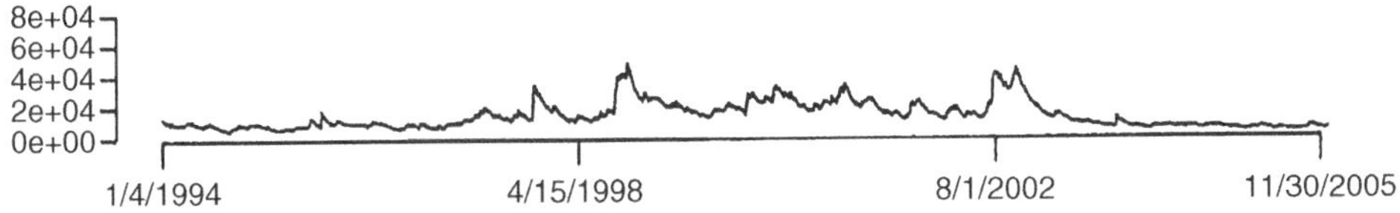

Figure 6.4 The estimated volatilities of the S&P 500 daily log-return between January 4, 1994 and December 30, 2005.

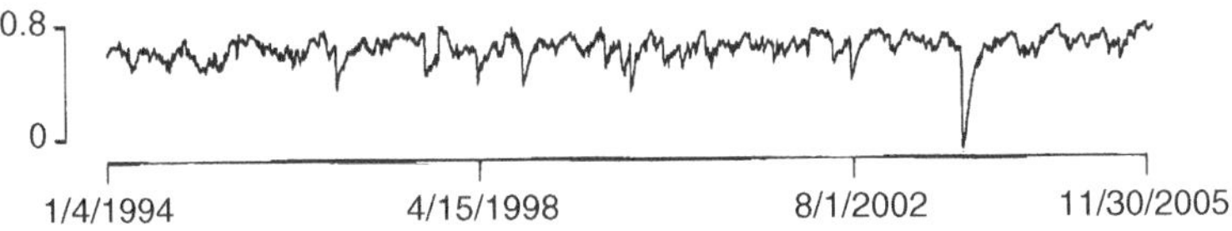

Figure 6.5 The estimated conditional correlation between the returns of S&P 500 and BAC from January 4, 1994 to December 30, 2005.

SP and BAC is 0.601. We notice that there is a significant drop in estimated conditional correlation close to the end of 2003. This can be well explained by the original log-return series of SP and BAC in Figure 6.3. The drop is caused by the extreme negative return of Bank of America during that time, while the same pattern is not found in the S&P 500. A similar phenomenon can also be found in the estimated conditional correlation in other pairs of SP, BAC, JPM, and C. To save space, their figures are not reported here.

Volatility forecasts can be obtained by using the forecasting methods discussed in earlier chapters for VARMA models. Once the GO-GARCH model is found, we can use it to forecast the conditional covariance matrices for the next 10 periods, that is, $\hat{\boldsymbol{\Sigma}}_{3021}(l), l = 1, \ldots, 10$. In an effort to save space, we only show here the one-step-ahead forecast of the conditional covariance, $\hat{\boldsymbol{\Sigma}}_{3021}(1)$.

$$\hat{\boldsymbol{\Sigma}}_{3021}(1) = \begin{bmatrix} & \text{S\&P} & \text{BAC} & \text{JPM} & \text{C} \\ \text{S\&P} & 0.0000613 & 0.0000609 & 0.0000684 & 0.0000830 \\ \text{BAC} & & 0.0001461 & 0.0001071 & 0.0001196 \\ \text{JPM} & & & 0.0001871 & 0.0001347 \\ \text{C} & & & & 0.0002246 \end{bmatrix}.$$

6.6.3 The analysis of the Dow Jones Industrial Average of 30 stocks

Example 6.3 The advantage of the proposed WSE method is its robust property and fast algorithm. To further demonstrate how powerful and easy it is to use the WSE method in high-dimensional data, we use the method to estimate the GO-GARCH model for the Dow Jones Industrial Average 30 stocks. The data contains 1006 daily log-return observations from July 1, 2006 to June 30, 2010 for each of the 30 Dow Jones stocks. The data was downloaded from the finance link, http://yahoo.com, and is listed as WW6c in the Data Appendix. The new method is very easy and extremely fast. Recall that under the GO-GARCH model, we have $\boldsymbol{\varepsilon}_t = \boldsymbol{\Omega}\mathbf{r}_t$, where $\boldsymbol{\Omega}$ is a 30×30 invertible matrix, $\mathbf{r}_t = (r_{1,t}, \ldots, r_{m,t})'$ being a 30×1 random vector, whose components are independent and each of them follows a univariate GARCH(1,1) model with GARCH coefficientsα_i and β_i, $i = 1, \ldots, 30$. The estimated (30×30) linkage matrix $\hat{\boldsymbol{\Omega}}$ multiplied by 100 is shown in Table 6.8 and the estimated GARCH coefficients are shown in Table 6.9.

From these results, we can easily compute the dynamic conditional covariance structure of the Dow Jones 30 stocks. Figure 6.6 shows the estimated volatilities (conditional variances) of the daily log-returns for each of these stocks. The patterns clearly indicate unusual market fluctuations during late 2008 and early 2009 following the economic crisis that began in late 2007. The market volatility gradually reduced only after worldwide governments and their central banks implemented their stimulus and rescue policies.

These examples clearly show that the WSE method provides an easy to apply and reliable estimation method in fitting the dynamics of conditional covariance processes for the GO-GARCH model.

In concluding the chapter, it should be pointed out that other than these presentations and cited references, there are many other useful references available on the representations and

Table 6.8 The estimated (30×30) linkage matrix $\hat{\boldsymbol{\Omega}}$.

0.10	−0.46	0.14	−0.84	0.43	−0.53	0.49	−0.23	−0.24	−0.81	0.22	0.76	−0.16	0.10	0.14	0.59	0.96	−0.33	0.35	−1.11	−0.03	0.66	0.51	−0.68	−0.74	1.55	0.08	−0.98	−0.49	−1.73
0.16	−0.42	0.22	0.03	−0.41	0.20	0.06	0.41	−0.36	−0.37	0.27	0.19	−0.28	−0.41	0.16	0.26	−0.06	0.43	0.11	−0.60	−0.45	−0.01	1.79	−1.65	−0.82	0.15	0.06	−0.53	0.29	−1.82
0.20	−0.37	−0.86	−0.22	−0.50	−0.28	−0.15	0.48	−0.69	−0.75	0.00	0.00	0.12	0.08	0.00	0.21	0.10	0.70	0.07	−0.57	0.08	0.18	0.37	−0.28	−0.17	0.50	0.07	−0.64	−0.21	−0.80
0.09	−0.40	0.06	−0.04	−0.23	0.10	−0.11	0.13	−0.40	−0.58	0.22	0.05	−0.13	−0.42	−0.03	−0.01	−0.02	0.12	0.31	−0.33	−0.48	−0.05	0.50	−0.95	−0.33	0.41	0.31	−1.21	2.47	−3.94
0.31	−0.08	−0.03	0.17	−0.04	−0.27	0.16	0.29	−0.83	−0.58	0.21	−0.04	−0.15	−0.66	−0.38	0.11	0.63	−0.13	1.11	−0.97	−0.39	0.28	0.44	−0.53	−0.61	0.36	0.30	−0.32	−0.20	−0.99
0.13	−0.40	0.14	0.34	−0.53	−0.24	0.06	0.19	−0.48	−0.52	−0.06	0.18	−0.48	−0.67	0.24	0.50	−0.41	−0.32	−0.05	−0.31	−0.32	0.51	0.59	−0.03	−0.37	0.57	0.35	−0.74	−0.10	−0.82
0.12	0.49	−0.03	−0.13	0.12	−0.13	0.23	−0.03	0.11	−0.61	−0.17	−0.19	−0.44	−0.33	−0.20	0.31	0.08	0.36	0.20	−0.55	−0.19	0.03	0.54	−0.15	−0.34	0.49	0.11	−1.09	−0.50	−0.98
0.36	−0.40	−0.29	−0.16	−0.11	0.21	0.01	0.44	0.08	−0.55	−0.37	0.55	−0.15	−0.40	−0.11	0.29	0.01	−0.27	0.75	−0.48	−0.16	0.01	0.64	−0.32	−0.63	0.45	0.38	−0.83	0.00	−0.85
−0.12	−0.50	−0.09	0.02	−0.24	0.37	0.18	0.25	−0.48	0.08	−0.11	−0.32	−0.14	−0.34	−0.52	0.42	0.10	−0.07	0.09	−0.54	0.10	−0.10	0.55	−0.39	−0.38	0.73	0.20	−0.92	−0.23	−0.99
0.20	−0.23	−0.04	0.13	−0.21	0.01	−0.03	0.21	−0.22	−0.48	0.01	−0.15	0.00	−0.24	0.00	0.36	−0.08	0.00	0.50	−0.73	−0.35	0.07	0.51	−0.62	−0.77	0.90	−1.02	−0.91	0.99	−0.76
−0.10	−0.68	0.48	−0.12	−0.58	−0.08	−0.47	0.41	0.15	−0.59	0.16	−0.02	−0.10	−0.49	−0.57	−0.42	−0.07	0.10	0.21	−0.48	−0.28	0.18	0.44	−0.69	−0.11	0.58	0.21	−0.87	0.06	−0.52
0.36	−0.23	−0.01	0.06	−0.09	−0.01	0.03	0.40	−0.21	−0.64	0.64	−0.47	−0.45	−0.25	0.20	0.20	−0.25	−0.36	0.23	−0.13	0.52	−0.05	0.36	−0.36	−0.61	0.62	0.32	−0.54	−0.22	−0.73
0.25	0.05	0.30	0.28	−0.42	0.34	0.09	0.42	−0.13	−0.47	0.21	0.14	−0.15	−0.04	0.04	0.11	0.11	−0.17	−0.06	−0.73	0.25	0.19	0.27	−0.10	−0.27	0.33	0.26	−0.71	0.08	−0.44
0.52	−0.21	0.42	−0.25	−0.86	0.44	0.22	−0.29	−0.61	−0.78	0.28	0.13	−0.43	−0.02	−0.21	0.33	−0.40	−0.08	0.55	−0.20	−0.23	0.33	0.32	−0.38	−0.55	0.50	0.14	−0.69	−0.13	−0.75
0.10	−0.04	−0.04	0.10	−0.34	−0.01	0.06	0.05	−0.01	0.03	0.10	−0.05	0.12	−0.02	0.24	−0.14	−0.01	−0.08	0.23	−0.24	−0.29	−0.23	0.17	−0.07	−0.04	0.32	−0.05	−0.65	−0.37	−0.49
0.11	−0.34	0.04	−0.03	−0.23	0.09	0.02	0.13	−0.47	−0.52	0.41	−0.06	−0.06	−0.42	0.17	0.25	0.05	0.19	0.25	−0.30	−0.51	−0.12	0.91	−1.47	−0.43	0.28	1.06	−1.86	1.87	−1.32
0.79	0.10	0.16	−0.46	−0.05	0.01	−0.32	0.36	−0.31	0.12	0.01	−0.01	0.05	−0.37	0.00	0.20	0.01	−0.07	−0.17	−0.22	−0.28	−0.21	0.19	−0.21	−0.09	0.43	0.01	−0.62	−0.11	−0.42
0.15	0.07	0.05	0.33	−0.10	0.25	−0.08	0.30	−0.23	−0.25	−0.18	0.09	−0.06	−0.02	0.11	0.07	0.13	0.03	0.04	0.17	−0.39	−0.03	−0.08	−0.54	−0.02	0.76	0.11	−0.49	−0.42	−0.41
0.44	−0.41	0.05	0.19	−0.26	−0.07	0.36	0.45	0.05	−0.07	0.21	0.24	−0.41	−0.20	−0.24	0.28	0.12	0.49	0.24	0.01	−0.14	−0.11	0.13	−0.16	0.09	0.30	−0.05	−0.62	−0.04	−0.37
0.21	−0.21	−0.06	−0.02	−0.05	0.09	0.16	0.25	−0.24	−0.51	0.01	−0.10	−0.08	−0.16	−0.04	0.26	0.02	−0.38	0.39	−0.39	−0.09	0.20	0.63	−0.46	0.52	0.32	−0.25	−0.65	−0.07	−0.65
0.20	−0.25	0.01	0.07	−0.16	0.36	0.22	0.23	−0.17	−0.52	0.34	−0.21	0.40	0.00	−0.13	−0.07	0.40	−0.08	−0.12	0.35	−0.81	−0.22	0.94	0.43	−0.46	0.52	−0.07	−0.86	−0.32	−0.59
0.20	−0.07	0.06	0.20	−0.23	0.20	0.02	0.29	0.01	−0.22	0.54	−0.02	0.39	−0.23	−0.07	0.71	−0.47	0.18	0.40	−0.21	−0.14	0.80	0.29	−0.38	−0.29	0.50	0.35	−0.74	−0.42	−0.87
0.09	−0.26	−0.08	−0.06	−0.24	0.19	0.28	0.25	0.03	−0.64	0.25	−0.11	0.17	−0.31	−0.05	0.37	−0.02	−0.18	−0.17	−0.41	−0.78	−0.51	−0.02	−0.31	−0.22	0.12	−0.02	−0.57	−0.24	−0.71
0.19	−0.08	−0.01	0.35	−0.07	−0.09	−0.28	0.02	−0.18	−0.17	0.18	0.26	0.00	0.04	−0.23	0.28	−0.13	−0.06	0.12	−0.18	−0.10	−0.51	0.38	−0.14	0.02	0.41	0.03	−0.64	−0.22	−0.48
0.47	−0.48	0.21	0.25	0.30	0.11	0.05	0.09	−0.21	−0.33	−0.09	−0.20	−0.12	0.22	0.00	−0.09	−0.43	0.24	0.06	−0.67	−0.50	0.03	0.35	−0.12	−0.20	0.47	0.39	−0.76	−0.18	−0.67
0.11	−0.32	0.11	0.16	−0.01	0.15	−0.16	0.28	−0.38	−0.29	0.22	0.18	−0.22	−0.12	−0.06	0.03	0.06	−0.09	−0.06	−0.14	−0.29	0.25	0.24	−0.80	−0.61	−0.58	−0.36	−1.78	−0.23	−1.09
0.18	−0.12	−0.20	0.03	−0.07	−0.06	0.44	0.35	−0.57	−0.49	0.19	0.32	0.15	−0.44	−0.26	−0.13	−0.38	0.07	0.23	−0.36	0.05	0.02	0.37	−0.46	−0.20	0.58	−0.06	−0.82	−0.10	−0.65
0.10	−0.12	−0.10	0.04	0.11	0.25	−0.07	0.38	−0.20	−0.11	0.42	0.06	−0.53	0.21	−0.15	−0.03	−0.39	−0.07	0.11	−0.57	−0.78	0.14	0.39	−0.11	−0.09	0.56	0.17	−0.66	−0.06	−0.47
0.03	−0.41	0.10	0.09	0.01	0.52	−0.27	−0.01	−0.20	−0.26	0.38	0.19	−0.17	−0.64	0.24	−0.07	0.00	0.31	0.15	−0.34	−0.16	−0.19	0.12	0.00	0.14	0.30	−0.09	−0.63	−0.22	−0.28
−0.12	0.33	0.31	−0.25	0.15	−0.12	0.17	0.35	−0.20	−0.58	−0.16	−0.10	−0.29	−0.11	0.04	0.50	−0.10	0.36	0.44	−0.41	−0.20	−0.19	0.47	0.02	−0.23	0.42	0.08	−1.00	−0.52	−0.82

Table 6.9 The estimated GARCH coefficients.

	1	2	3	4	5	6	7	8	9	10	11	12	13	14	15
α_i	0.007	0.020	0.032	0.033	0.053	0.028	0.019	0.071	0.025	0.026	0.029	0.043	0.033	0.065	0.028
β_i	0.985	0.973	0.960	0.933	0.897	0.959	0.973	0.884	0.965	0.965	0.965	0.941	0.961	0.913	0.965

	16	17	18	19	20	21	22	23	24	25	26	27	28	29	30
α_i	0.027	0.031	0.197	0.034	0.038	0.056	0.114	0.150	0.099	0.047	0.171	0.057	0.108	0.205	0.113
β_i	0.966	0.965	0.646	0.957	0.958	0.932	0.863	0.809	0.880	0.924	0.805	0.940	0.880	0.795	0.877

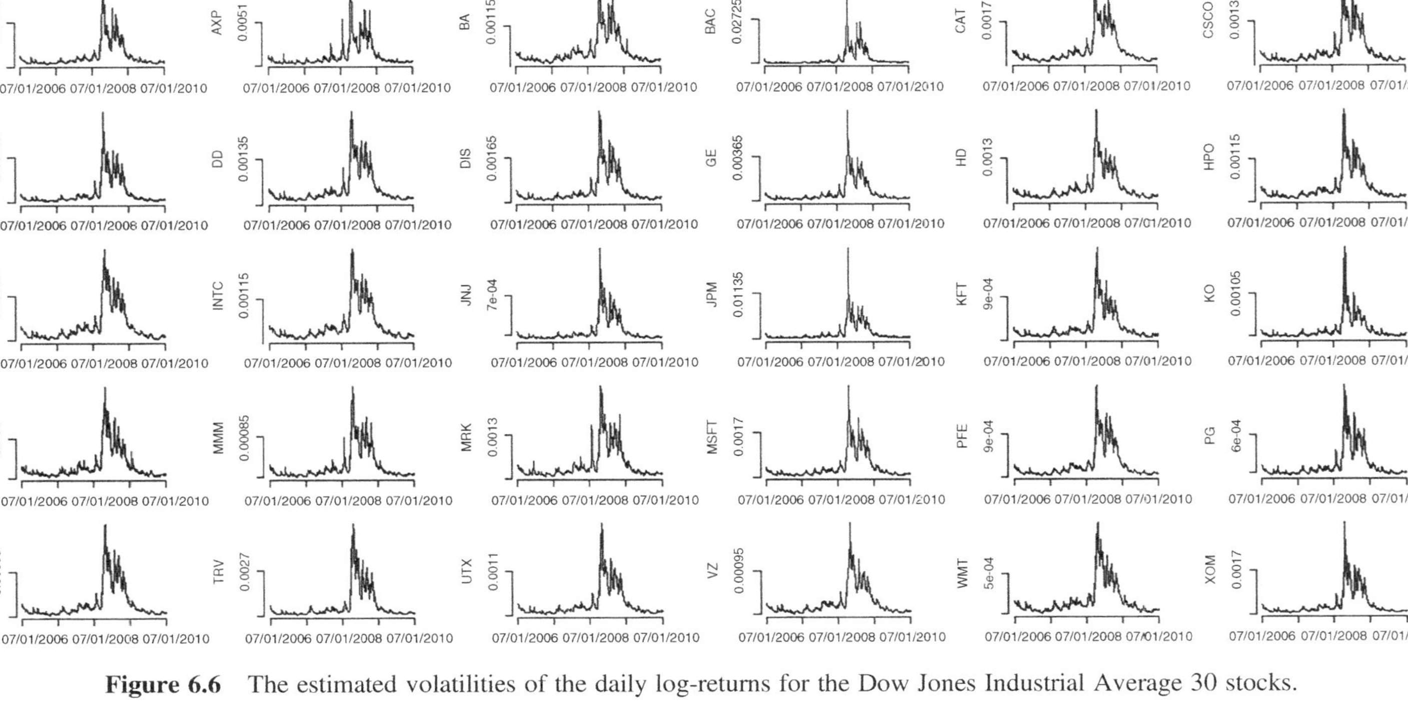

Figure 6.6 The estimated volatilities of the daily log-returns for the Dow Jones Industrial Average 30 stocks.

applications of multivariate GARCH models. They include Lin (1992), Sentana (1998), Vrontos, Dellaportas, and Politis (2003), Bauwens, Laurent, and Rombouts (2006), Boswijk and van der Weide (2006, 2011), Lanne and Saikkonen (2007), Fan, Wang, and Yao (2008), Silvennoinen and Terasvita (2009), Bondon and Bahamonde (2012), Pedersen and Rahbek (2014), Zheng and Wei (2014), Lin, Lee, and Guo (2016), Ahmed, Haider, and Zaman (2017), Albert et al (2017), Aue, Horvath, and Pellatt (2017), Grublyte, Surgailis, and Skarnulis (2017), Hafner and Preminger (2017), Harvey and Lange (2017), Truquet (2017), Yu, Li, and Ng (2017), Zhu, Li, and Yu (2017), and Bibi and Ghezal (2018), among others.

Software code

SAS code for Example 6.1

```
/*Data Import*/
proc import file = "C:\Bookdata\WW6a.csv" out = df1;
run;

/*The following codes are tested in SAS/ETS 14.1*/

/*Table 6.1*/
ods select InfoCriteria; /*show information criteria only*/
proc varmax data = df1;
id week interval = week;
model fitness home_workout gym bodybuilding_supplement
cardio weight_loss/ p = 1;
run;

ods select InfoCriteria;
proc varmax data = df1;
id week interval = week;
model fitness home_workout gym bodybuilding_supplement
cardio weight_loss/ p = 2;
run;

ods select InfoCriteria;
proc varmax data = df1;
id week interval = week;
model fitness home_workout gym bodybuilding_supplement
cardio weight_loss/ p = 3;
run;

ods select InfoCriteria;
proc varmax data = df1;
id week interval = week;
model fitness home_workout gym bodybuilding_supplement
```

```
cardio weight_loss/ q = 1;
run;

ods select InfoCriteria;
proc varmax data = df1;
id week interval = week;
model fitness home_workout gym bodybuilding_supplement
cardio weight_loss/ q = 2;
run;

ods select InfoCriteria;
proc varmax data = df1;
nloptions tech=qn maxiter = 20000 GCONV=1E-8 noprint;
id week interval = week;
model fitness home_workout gym bodybuilding_supplement
cardio weight_loss/ p = 1 q = 1;
run;
/*End of Table 6.1*/

/*Equation 6.26 + Table 6.2 + Table6.3*/
proc varmax data = df1;
nloptions tech=qn maxiter = 20000 GCONV=1E-8 noprint;
id week interval = week;
model fitness home_workout gym bodybuilding_supplement
cardio weight_loss/ p = 1 q = 1 print = (corry pcorr
DIAGNOSE) printform = both ;
run;
/*End of Equation 6.26 + Table 6.2 + Table6.3*/

/*Table 6.4*/
ods select InfoCriteria;
proc varmax data = df1;
nloptions tech=qn maxiter = 20000 GCONV=1E-8 noprint;
id week interval = week;
model fitness home_workout gym bodybuilding_supplement
cardio weight_loss/ p = 1 print =
(DIAGNOSE) printform = both;
garch q = 1 form = BEKK;
run;

ods select InfoCriteria;
proc varmax data = df1;
nloptions tech=qn maxiter = 20000 GCONV=1E-8 noprint;
id week interval = week;
model fitness home_workout gym bodybuilding_supplement
cardio weight_loss/ p = 1 print =
(DIAGNOSE) printform = both;
```

```
garch q = 1 form = CCC;
run;

ods select InfoCriteria;
proc varmax data = df1;
nloptions tech=qn maxiter = 20000 GCONV=1E-8 noprint;
id week interval = week;
model fitness home_workout gym bodybuilding_supplement
cardio weight_loss/ p = 1 print =
(DIAGNOSE) printform = both;
garch q = 1 form = DCC;
run;
/*End of Table 6.4*/

/*Equation 6.27 + Forecast*/
proc varmax data = df1 plots = (model residual forecasts);
nloptions tech=qn maxiter = 20000 GCONV=1E-8 noprint;
id week interval = week;
model fitness home_workout gym bodybuilding_supplement
cardio weight_loss/ p = 1 print =
(DIAGNOSE) printform = both;
garch q = 1 form = BEKK;
output out = for lead = 2;
run;
/*End of Equation 6.27 + Forecast*/
/*Figure 6.1*/
proc import file = "C:\Bookdata\WW6a.csv" out = df1 dbms =
csv replace;
run;

data df2;
set df1;
array vars{*} fitness home_workout gym
Bodybuilding_supplement cardio weight_loss;
do i = 1 to dim(vars);
Google_Index = vars{i};
Category = vname(vars{i});
output;
end;
drop i fitness home_workout gym Bodybuilding_supplement
cardio weight_loss;
run;

data df2;
set df2;
if Category = "fitness" then Category = "y1";
else if Category = "home_workout" then Category = "y2";
```

```
else if Category = "gym" then Category = "y3";
else if Category = "Bodybuilding_supplement" then Category =
"y4";
else if Category = "cardio" then Category = "y5";
else if Category = "weight_loss" then Category = "y6";
run;

ODS GRAPHICS / RESET HEIGHT = 5IN WIDTH = 6.5IN;

ods rtf file = "C:\Bookdata\figure6.1.rtf" style = JOURNAL
startpage = no;
proc sgpanel data = df2;
panelby Category/onepanel layout=rowlattice UNISCALE = ALL
novarname;
format Week Date9.;
series x =Week y = Google_Index;
colaxis values = ('15Apr2012'd TO '15Apr2017'd BY YEAR)
display = (nolabel);
run;
ods rtf close;
/*End of Figure 6.1*/



/*Figure 6.2*/
proc import file = "C:\Bookdata\WW6a.csv" out = df1;
run;

data df1;
set df1;
date = week;
run;

ODS GRAPHICS / RESET HEIGHT = 2IN WIDTH = 3.25IN;

ods trace on;
ods graphics on;

proc template;
   link Ets.Varmax.Graphics.ModelPlot to Statgraph.Forecast.
ModelPlot;
   define statgraph Statgraph.Forecast.ModelPlot;
      dynamic Title Time Variable VariableLabel ID IDLabel
IDFormat IDType
         Interval IntegerTime ConfidenceLabel _byline_
_bytitle_ _byfootnote_;
      BeginGraph;
```

```
        EntryTitle TITLE;
        Layout Overlay / YAxisOpts=(shortlabel=VARIABLE
label='Popularity'
          Display=all GridDisplay=Auto_On) XAxisOpts=
(shortlabel=ID label=
          IDLABEL type=IDTYPE timeopts=(tickvalueformat
=IDFORMAT) Display=
          (LINE ticks tickvalues) GridDisplay=Auto_On
linearopts=(integer=INTEGERTIME));
          ScatterPlot x=TIME y=ACTUAL / markerattrs
=GRAPHDATADEFAULT name=
            "Actual" legendlabel="Actual" tip=(Time Actual
Predict Lower Upper);
          SeriesPlot x=TIME y=PREDICT / lineattrs
=GRAPHPREDICTION (pattern=
            solid thickness=1) name="Fitted" legendlabel
="Fitted" tip
            =(Time Actual Predict Lower Upper);
          DiscreteLegend "Actual" "Fitted"/;
        EndLayout;
        if (_BYTITLE_)
          entrytitle _BYLINE_ / textattrs=GRAPHVALUETEXT;
        else
          if (_BYFOOTNOTE_)
            entryfootnote halign=left _BYLINE_;
          endif;
        endif;
      EndGraph;
    end;
run;

ods rtf file = "C:\Bookdata\figure6.2.rtf" style = JOURNAL
startpage = no;
proc varmax data = df1 plots = (model residual);
nloptions tech=qn maxiter = 20000 GCONV=1E-8 noprint;
id date interval = week;
model fitness home_workout gym bodybuilding_supplement
cardio weight_loss/ p = 1 print =
(DIAGNOSE) printform = both;
garch q = 1 form = BEKK;
run;
ods rtf close;

ods trace off;
/*End of Figure 6.2*/
```

Projects

1. Find and carefully read the papers written by (i) Bollerslev, Engle, and Wooldridge (1988), and (ii) Engle and Kroner (1995).
2. Find a m-dimensional social science related time series data set with $m \geq 6$. Try to construct VEC and DVEC models with a written report and analysis software code. Email your data set and software code to your course instructor.
3. For the data set in Project 2, construct a BEKK(p, q) model, and compare and comment the two results with a written report and associated software code. Email your report to your course instructor.
4. Find a m-dimensional natural science related time series data set with $m \geq 6$. Construct a factor model for its variance–covariance matrix, $\mathbf{\Sigma}_t$, with a written report and analysis software code. Email your data set and software code to your course instructor.
5. Find a m-dimensional time series data set of your interest with $m \geq 10$. Construct a GO-GARCH model with a written report and analysis software code. Email your data set and software code to your course instructor.

References

Ahmed, M., Haider, G., and Zaman, A. (2017). Detecting structural change with heteroskedasticity. *Communication of Statistics, Theory and Methods* **46**: 10446–10455.

Albert, S., Messer, M., Schiemann, J., Roeper, J., and Schneidar, G. (2017). Multi-scale detection of variance changes in renewal processes in the presence of rate change points. *Journal of Time Series Analysis* **38**: 1028–1052.

Alexander, C.O. (2001). *Market Models*. Wiley.

Alexander, C.O. and Chibumba, A. (1997). *Multivariate Orthogonal Factor GARCH*. Mimeo. University of Sussex.

Aue, A., Horvath, L., and Pellatt, D.F. (2017). Functional generalized autoregressive conditional heteroskedasticity. *Journal of Time Series Analysis* **38**: 3–21.

Baba, Y., Engle, R.F., Kraft, D., and Kroner, K.F. (1990). *Multivariate Simultaneous Generalized ARCH*. San Diego: Mimeo. Department of Economics, University of California.

Bauwens, L., Laurent, S., and Rombouts, J.V.K. (2006). Multivariate GARCH models: a survey. *Journal of Applied Econometrics* **21**: 79–109.

Bibi, A. and Ghezal, A. (2018). Markov-switching bilinear-GARCH models: structure and estimation. *Communication of Statistics, Theory and Methods* **47**: 307–323.

Bollerslev, T. (1986). Generalized autoregressive conditional heteroskedasticity. *The Review of Economics and Statistics* **31**: 307–327.

Bollerslev, T. (1990). Modeling the coherence in short-run nominal exchange rates: a multivariate generalized ARCH model. *The Review of Economics and Statistics* **72**: 498–505.

Bollerslev, T., Engle, R.F., and Wooldridge, F.M. (1988). A capital asset pricing model with time-varying covariances. *The Journal of Political Economy* **96** (1): 116–131.

Bondon, P. and Bahamonde, N. (2012). Least squares estimation of ARCH models with missing observations. *Journal of Time Series Analysis* **33**: 880–891.

Boswijk, H.P., and van der Weide, R. (2006). Wake me up before you GO-GARCH. *UvA-Econometrics Discussion Paper*, http://www.ase.uva.nl/pp/bin/381fulltext.pdf.

Boswijk, H.P. and van der Weide, R. (2011). Method of moments estimation of GO-GARCH models. *Journal of Econometrics* **163**: 118–126.

van der Weide, R. (2002). A multivariate generalized orthogonal GARCH model. *Journal of Applied Econometrics* **17**: 549–564.

Engle, R.F. (1982). Autoregressive conditional heteroscedasticity with estimates of the variance of United Kingdom inflation. *Econometrica* **50**: 987–1007.

Engle, R.F. (2002). Dynamic conditional correlation: a simple class of multivariate GARCH models. *Journal of Business and Economic Statistics* **20**: 339–350.

Engle, R.F. and Kroner, K.F. (1995). Multivariate simultaneous generalized ARCH. *Econometric Theory* **11**: 122–150.

Engle, R.F., Ng, V.K., and Rothschild, M. (1990). Asset pricing with a factor-ARCH covariance structure: empirical estimates for treasury bills. *Journal of Econometrics* **45**: 213–238.

Fan, J., Wang, M., and Yao, Q. (2008). Modeling multivariate volatilities via conditionally uncorrelated components. *Journal of the Royal Statistical Society, Series B* **70**: 679–702.

Grublyte, I., Surgailis, D., and Skarnulis, A. (2017). QMLE for quadratic ARCH model with long memory. *Journal of Time Series Analysis* **38**: 535–551.

Hafner, C.M. and Preminger, A. (2017). On asymptotic theory for ARCH (∞) models. *Journal of Time Series Analysis* **38**: 865–879.

Hampel, F.R., Ronchetti, E.M., Rousseeuw, P.J., and Stahel, W.A. (1986). *Robust Statistics, the Approach Based on Influence Functions*. Wiley.

Harvey, A. and Lange, R. (2017). Volatility modeling with a generalized t distribution. *Journal of Time Series Analysis* **38**: 175–190.

Huber, P.J. (1981). *Robust Statistics*. Wiley.

Lanne, M. and Saikkonen, P. (2007). A multivariate generalized orthogonal factor GARCH model. *Journal of Business and Economic Statistics* **25**: 61–75.

Liebeck, H. and Osborne, A. (1991). The generation of all rational orthogonal matrices. *The American Mathematical Monthly* **98**: 131–133.

Lin, W. (1992). Alternative estimators for factor GARCH models, a Monte Carlo comparison. *Journal of Applied Econometrics* **3**: 259–279.

Lin, L.C., Lee, S., and Guo, M. (2016). Goodness-of-fit test for stochastic volatility models based on noisy observations. *Statistica Sinica* **26**: 1305–1329.

Pedersen, R.S. and Rahbek, A. (2014). Multivariate variance targeting in the BEKK–GARCH model. *The Econometrics Journal* **17**: 24–55.

Sentana, E. (1998). The relation between conditionally heteroskedastic factor models and factor GARCH models. *The Econometrics Journal* **1**: 1–9.

Silvennoinen, A. and Teräsvirta, T. (2009). Multivariate GARCH models. In: *Handbook of Financial Time Series* (ed. T.G. Andersen, R.A. Davis, J.P. Kreiss and T. Mikosch), 201–229. Springer.

Truquet, L. (2017). Parametric stability and semiparametric inference in time varying auto-regressive conditional heteroscedasticity models. *Journal of the Royal Statistical Society, Series B* **79**: 1391–1414.

Vrontos, I.D., Dellaportas, P., and Politis, D.N. (2003). A full-factor multivariate GARCH model. *The Econometrics Journal* **6**: 311–333.

Wei, W.W.S. (2006). *Time Series analysis: Univariate and Multivariate Methods*, 2e. Pearson-Addison Wesley.

Yu, P.L.H., Li, W.K., and Ng, F.C. (2017). The generalized conditional autoregressive Wishart model for multivariate realized volatility. *Journal of Business & Economic Statistics* **35**: 513–527.

Zheng, L. and Wei, W.W.S. (2012). Weighted scatter estimation method of the GO-GARCH models. *Journal of Time Series Analysis* **33**: 81–95.

Zheng, L. and Wei, W.W.S. (2014). Robust estimation of the linkage matrix in O-GARCH model. *Journal of Applied Statistical Science* **21**: 43–62.

Zhu, K., Li, W.K., and Yu, P.L.H. (2017). Buffered autoregressive models with conditional heteroscedasticity: an application to exchange rates. *Journal of Business & Economic Statistics* **35**: 528–542.

7

Repeated measurements

Repeated measurements occur in many areas of study. The measurements repeat in time, space, or both. However, they are normally relatively short, and therefore the standard time series methods cannot be used. In this chapter, we will introduce some methods and various models that are useful for analyzing repeated measures data. Empirical examples will be used for illustration.

7.1 Introduction

Many fields of study, such as medical and biological science, social science, and education, involve sets of relatively short time series where the application of standard time series methods introduced earlier is difficult, if not impossible. For instance, an experiment may involve measurements taken at some selected times (or locations) from subjects associated with several treatments. The term "subject" is often used because the phenomenon of repeated measurements commonly occurs in the areas of medical, social, and educational studies, where human subjects are involved. However, the term may refer to an animal, a company, or even a tool. For example, the following is a study involving the growth curve data on the body weights of 27 rats from Box (1950) given in Table 7.1. The subject is a rat that is assigned to one of three treatment groups (Control, Thiouracil, Thyroxin) and its weight is measured weekly for 5 weeks. The objective of the study is to test whether there are differences in growth rates between groups.

The type of time series data described previously differs from other time series data that we have studied earlier. First, although a series of measurements on a subject over time does constitute a univariate time series, it is often relatively short. More importantly, the main interest of the study is normally not about the stochastic nature of the series of any subject. Additionally, when several series of measurements were observed from several subjects, they do look like vectors of multiple time series, especially when the measurements of these subjects were observed at the same time points. However, the components of a vector constructed at each time point are measurements of the same phenomenon. For example, the components of the vector obtained from Table 7.1 will all be body weights. Certainly, the interest of the study

Multivariate Time Series Analysis and Applications, First Edition. William W.S. Wei.
© 2019 John Wiley & Sons Ltd. Published 2019 by John Wiley & Sons Ltd.
Companion website: www.wiley.com/go/wei/datasets

Table 7.1 Body weight of rats under three different treatments (1 = Control, 2 = Thyroxin, 3 = Thiouracil) with weight at five different time points (Week 0, Week 1, Week 2, Week 3, and Week 4).

	W0	W1	W2	W3	W4
1	57	86	114	139	172
1	60	93	123	146	177
1	52	77	111	144	185
1	49	67	100	129	164
1	56	81	104	121	151
1	46	70	102	131	153
1	51	71	94	110	141
1	63	91	112	130	154
1	49	67	90	112	140
1	57	82	110	139	169
2	59	85	121	146	181
2	54	71	90	110	138
2	56	75	108	151	189
2	59	85	116	148	177
2	57	72	97	120	144
2	52	73	97	116	140
2	52	70	105	138	171
3	61	86	109	120	129
3	59	80	101	111	122
3	53	79	100	106	133
3	59	88	100	111	122
3	51	75	101	123	140
3	51	75	92	100	119
3	56	78	95	103	108
3	58	69	93	116	140
3	46	61	78	90	107
3	53	72	89	104	122

is not the cross-correlational structure of different component series from these subjects. Rather, subjects in each group constitute a random sample. The main interest is to study whether there are any differences between groups based on sample information from these groups. Since these series are measurements of the same phenomenon from several subjects that are often assumed to be independent, we refer to them as "repeated measurement data." Other names such as "longitudinal data" or "panel data" are also used.

There are two factors in the experiment of repeated measurements, treatments, and time. Treatment is the first factor, often known as the between-subjects factor because its levels change only between subjects. Time is the second factor, often known as within-subjects factor because its values are changing over time on the same subject. In this factorial experiment, we are interested in finding whether (i) treatment means are different, (ii) treatment means are changing over time, and (iii) there are interactions between treatment and time.

In this chapter, we give a simple introduction to some normal distribution-based methods. For more general treatments, we refer readers to Crowder and Hand (1990), Lindsey (1993), and Diggle, Liang, and Zeger (2013), among others.

7.2 Multivariate analysis of variance

7.2.1 Test treatment effects

To study the phenomenon described in Table 7.1, we can, in general, use $\mathbf{Z}_{i,j}(i = 1, 2, \ldots, m;$ $j = 1, 2, \ldots, n_i)$ to denote the response of the jth subject for the ith treatment. Clearly, each $\mathbf{Z}_{ij}$ is a p-dimensional random vector, where p is the total time period of the response. The subjects are elements of random samples collected from each of m populations (treatments) and can be arranged as follows:

$$\begin{aligned}&\text{Population 1}: \mathbf{Z}_{1,1}, \mathbf{Z}_{1,2}, \ldots, \mathbf{Z}_{1,n_1}\\&\text{Population 2}: \mathbf{Z}_{2,1}, \mathbf{Z}_{2,2}, \ldots, \mathbf{Z}_{2,n_2}\\&\vdots\\&\text{Population m}: \mathbf{Z}_{m,1}, \mathbf{Z}_{m,2}, \ldots, \mathbf{Z}_{m,n_m}.\end{aligned}$$

Assume that (i) $\mathbf{Z}_{i,1}, \mathbf{Z}_{i,2}, \ldots, \mathbf{Z}_{i,n_i}$ is a random sample of size n_i from a multivariate normal population with mean vector $\boldsymbol{\mu}_i$, $i = 1, 2, \ldots, m$; (ii) the random samples from different populations are independent; and (iii) all populations have a common variance–covariance matrix $\boldsymbol{\Sigma}$. We can use multivariate analysis of variance (MANOVA) to investigate whether the population mean vectors are the same. Thus,

$$\mathbf{Z}_{i,j} = \boldsymbol{\mu}_i + \mathbf{e}_{i,j} = \boldsymbol{\mu} + \boldsymbol{\theta}_i + \mathbf{e}_{i,j}, \tag{7.1}$$

for $i = 1, 2, \ldots, m$ and $j = 1, 2, \ldots, n_i$ where the errors $\mathbf{e}_{ij}$ are independent $N_p(\mathbf{0}, \boldsymbol{\Sigma})$ variables, $\boldsymbol{\mu}$ is an overall mean, and $\boldsymbol{\theta}_i$ represents the ith treatment effect with $\sum_{i=1}^{m} n_i\boldsymbol{\theta}_i = \mathbf{0}$.

Note that a vector of observations can be written as

$$\mathbf{Z}_{i,j} = \bar{\mathbf{Z}} + (\bar{\mathbf{Z}}_i - \bar{\mathbf{Z}}) + \left(\mathbf{Z}_{i,j} - \bar{\mathbf{Z}}_i\right), \tag{7.2}$$

and it leads to the following decomposition

$$\begin{aligned}&\sum_{i=1}^{m}\sum_{j=1}^{n_i}\left(\mathbf{Z}_{i,j} - \bar{\mathbf{Z}}\right)\left(\mathbf{Z}_{i,j} - \bar{\mathbf{Z}}\right)' = \sum_{i=1}^{m} n_i(\bar{\mathbf{Z}}_i - \bar{\mathbf{Z}})(\bar{\mathbf{Z}}_i - \bar{\mathbf{Z}})' + \sum_{i=1}^{m}\sum_{j=1}^{n_i}\left(\mathbf{Z}_{i,j} - \bar{\mathbf{Z}}_i\right)\left(\mathbf{Z}_{i,j} - \bar{\mathbf{Z}}_i\right)',\\&\begin{pmatrix}\text{total (corrected)}\\ \text{sum of squares}\\ \text{and cross products}\end{pmatrix} = \begin{pmatrix}\text{treatment (between)}\\ \text{sum of squares}\\ \text{and cross products}\end{pmatrix} + \begin{pmatrix}\text{residual (within)}\\ \text{sum of squares}\\ \text{and cross products}\end{pmatrix}.\end{aligned} \tag{7.3}$$

This is clearly the extension of the univariate analysis variance where we have

Total (corrected) sum of squares = Treatment sum of squares + Residual sum of squares.

For computation, the within-sum of squares and cross product matrix can also be written as

$$\sum_{i=1}^{m}\sum_{j=1}^{n_i}(\mathbf{Z}_{i,j}-\bar{\mathbf{Z}}_i)(\mathbf{Z}_{i,j}-\bar{\mathbf{Z}}_i)'=(n_1-1)\mathbf{S}_1+(n_2-1)\mathbf{S}_2+\cdots+(n_m-1)\mathbf{S}_m \tag{7.4}$$

where $\mathbf{S}_i$ is the sample covariance matrix for the ith sample.

Thus, the hypothesis of no treatment effects

$$\mathrm{H}_0:\boldsymbol{\theta}_1=\boldsymbol{\theta}_2=\ldots=\boldsymbol{\theta}_m=\mathbf{0} \tag{7.5}$$

can be tested by comparing the relative sizes of the treatment (between) and residual (within) sums of squares and cross products summarized in the MANOVA table (Table 7.2):

In the univariate analysis of variance, we reject the null hypothesis when the treatment sum of squares is significantly larger than the residual sum of squares, or equivalently, when the residual sum of squares is a much smaller portion of the total sum of squares. Following the same logic, in multivariate analysis of variance, we reject H_0 if the following ratio of generalized variances

$$\Lambda=\frac{|\mathbf{R}|}{|\mathbf{T}+\mathbf{R}|}=\frac{\left|\sum_{i=1}^{m}\sum_{j=1}^{n_i}(\mathbf{Z}_{i,j}-\bar{\mathbf{Z}}_i)(\mathbf{Z}_{i,j}-\bar{\mathbf{Z}}_i)'\right|}{\left|\sum_{i=1}^{m}\sum_{j=1}^{n_i}(\mathbf{Z}_{i,j}-\bar{\mathbf{Z}})(\mathbf{Z}_{i,j}-\bar{\mathbf{Z}})'\right|} \tag{7.6}$$

is too small. The statistic Λ, originally proposed by Wilks (1932), is often known as Wilk's lambda. The sampling distribution of Wilk's lambda depends on the values of m and p. For example, with $m = 3$ and $p \geq 1$, we have the following sampling distribution (see Johnson and Wichern, 2007, p. 303)

$$\left(\frac{\sum n_i-p-2}{p}\right)\left(\frac{1-\sqrt{\Lambda}}{\sqrt{\Lambda}}\right)\sim F\left(2p,2\left(\sum n_i-p-2\right)\right). \tag{7.7}$$

Table 7.2 MANOVA table for comparing population mean vectors.

Source of variation	Matrix of sum of squares and cross products (SSCP)	Degrees of freedom (d.f.)
Treatment (between)	$\mathbf{T}=\sum_{i=1}^{m}n_i(\bar{\mathbf{Z}}_i-\bar{\mathbf{Z}})(\bar{\mathbf{Z}}_i-\bar{\mathbf{Z}})'$	$m-1$
Residual (within)	$\mathbf{R}=\sum_{i=1}^{m}\sum_{j=1}^{n_i}(\mathbf{Z}_{i,j}-\bar{\mathbf{Z}}_i)(\mathbf{Z}_{i,j}-\bar{\mathbf{Z}}_i)'$	$\sum_{i=1}^{m}n_i-m$
Total (corrected for mean)	$\mathbf{T}+\mathbf{R}=\sum_{i=1}^{m}\sum_{j=1}^{n_i}(\mathbf{Z}_{i,j}-\bar{\mathbf{Z}})(\mathbf{Z}_{i,j}-\bar{\mathbf{Z}})'$	$\sum_{i=1}^{m}n_i-1$

Alternatively, under H_0, when $\sum n_i = n$ is large, we can also use the following Bartlett's approximation (Bartlett, 1954):

$$-\left(n-1-\frac{(p+m)}{2}\right)\ln\Lambda = -\left(n-1-\frac{(p+m)}{2}\right)\ln\frac{|\mathbf{R}|}{|\mathbf{T}+\mathbf{R}|} \tag{7.8}$$

which has a chi-square distribution with $p(m-1)$ degrees of freedom. Thus, we reject H_0 if

$$-\left(n-1-\frac{(p+m)}{2}\right)\ln\frac{|\mathbf{R}|}{|\mathbf{T}+\mathbf{R}|} > \chi^2_\alpha(p(m-1)). \tag{7.9}$$

7.2.2 Empirical Example 1 – First analysis on body weight of rats under three different treatments

Example 7.1 From the experiment, the interest of the study is to find out whether the growth of rats in terms of body weights in the three treatment groups (control, thyroxin, and thiouracil) is different in the following model,

$$\mathbf{Z}_{i,j} = \boldsymbol{\mu}_i + \mathbf{e}_{i,j} = \boldsymbol{\mu} + \boldsymbol{\theta}_i + \mathbf{e}_{i,j}, \tag{7.10}$$

where we have $m = 3$, $i = 1, 2, 3$, $j = 1, 2, \ldots, n_i$, $n_1 = 10$, $n_2 = 7$, $n_3 = 10$, and $p = 5$, and $\mathbf{e}_{i,j}$ are independent $N_5(\mathbf{0}, \boldsymbol{\Sigma})$ variables.

The observed mean vectors for the three treatment groups are given in Eq. (7.11) with their corresponding standard deviations listed in the parentheses. They are illustrated in Figure 7.1.

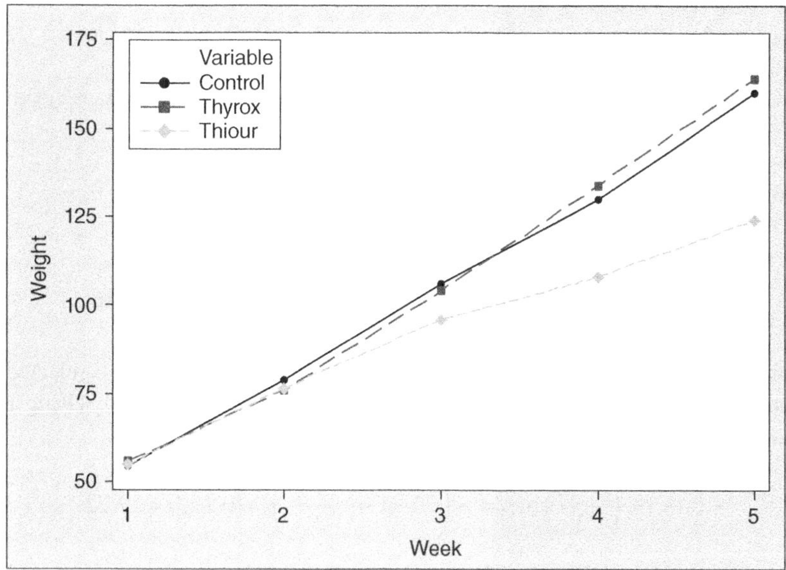

Figure 7.1 Average body weights of three treatment groups.

$$\hat{\boldsymbol{\mu}}_1 = \left[\underset{(5.44)}{54}, \underset{(9.64)}{78.5}, \underset{(9.92)}{106}, \underset{(12.56)}{130}, \underset{(15.2)}{160.6}\right]$$

$$\hat{\boldsymbol{\mu}}_2 = \left[\underset{(2.99)}{55.57}, \underset{(6.44)}{75.86}, \underset{(11.1)}{104.86}, \underset{(16.98)}{132.71}, \underset{(21.51)}{162.85}\right] \tag{7.11}$$

$$\hat{\boldsymbol{\mu}}_3 = \left[\underset{(4.69)}{54.7}, \underset{(7.92)}{76.3}, \underset{(8.5)}{95.8}, \underset{(9.9)}{108.4}, \underset{(11.55)}{124.2}\right]$$

where 1 = control group, 2 = thyroxin group, and 3 = thiouracil.

The figure seems to indicate that the average body weights for all three groups are increasing over time. The initial averages for the three treatment groups are close to each other; the average body weights for the thiouracil group remained lowest after week 1, and the average weight for the thyroxin starts to overtake the other two groups between the second and third weeks. More rigorously, we now test the following hypothesis:

$$H_0 : \boldsymbol{\mu}_1 = \boldsymbol{\mu}_2 = \boldsymbol{\mu}_3 \tag{7.12}$$

or equivalently,

$$H_0 : \boldsymbol{\theta}_1 = \boldsymbol{\theta}_2 = \boldsymbol{\theta}_3 = \mathbf{0}. \tag{7.13}$$

Using available multivariate analysis variance software, for example, the SAS (2015) general linear model (GLM),

```
data Rat;
      infile 'c:/Bookdata/WW7a.csv';
      input Treatment Week0 Week1 Week2 Week3 Week4;
/* GLM procedure with MANOVA */
proc glm data=Rat;
      class Treatment;
      model Week0 Week1 Week2 week3 week4 = Treatment /NOUNI;
      means Treatment;
      manova h = Treatment/printh printe htype=1 etype=1;
run;
```

we get the following multivariate MANOVA table (Table 7.3).

Thus, the Wilk's lambda from Table 7.3 is $\Lambda = |\mathbf{R}|/|\mathbf{T}+\mathbf{R}|=0.259$, and

$$\left(\frac{\sum n_i - p - 2}{p}\right)\left(\frac{1-\sqrt{\Lambda}}{\sqrt{\Lambda}}\right) = \left(\frac{27-5-2}{5}\right)\left(\frac{1-\sqrt{0.259}}{\sqrt{0.259}}\right) = 3.86,$$

which is much greater than the $F_{0.05}(10, 40)=2.08$. Thus, the null hypothesis is rejected, and we conclude that the mean vectors of the three treatment groups are different. Using Bartlett's approximation, we have

$$-\left(n-1-\left(\frac{p+m}{2}\right)\ln\Lambda = -\left(27-1-\frac{(5+3)}{2}\right)\ln(0.259) = 29.72,\right.$$

which is greater than $\chi^2_{0.05}(10) = 18.307$, and hence the null hypothesis is rejected.

Table 7.3 MANOVA table of comparing treatment mean vectors for body weights of rats.

Source	Matrix of sum of squares and cross products (SSCP)	DF
Treatment	$\mathbf{T}=\begin{bmatrix} 10.186 & -17.471 & -10.638 & 9.234 & 2.162 \\ -17.471 & 36.543 & 80.543 & 130.986 & 236.743 \\ -10.638 & 80.543 & 601.395 & 1382.967 & 2276.150 \\ 9.234 & 130.986 & 1382.967 & 3294.467 & 5377.612 \\ 2.162 & 236.743 & 2276.150 & 5377.612 & 8794.772 \end{bmatrix}$	2
Residual	$\mathbf{R}=\begin{bmatrix} 517.814 & 792.471 & 757.971 & 677.343 & 564.171 \\ 792.471 & 1649.457 & 1657.457 & 1443.014 & 1255.257 \\ 757.971 & 1657.457 & 2274.457 & 2625.514 & 2782.257 \\ 677.343 & 1443.014 & 2625.514 & 4032.729 & 4616.314 \\ 564.171 & 1255.257 & 2782.257 & 4616.314 & 6052.857 \end{bmatrix}$	24
Total	$\mathbf{T}+\mathbf{R}=\begin{bmatrix} 528 & 775 & 747.333 & 686.577 & 566.333 \\ 775 & 1686 & 1738 & 1574 & 1492 \\ 747.333 & 1738 & 2875.852 & 4008.481 & 5058.407 \\ 686.577 & 1574 & 4008.481 & 7327.96 & 9993.926 \\ 566.333 & 1492 & 5058.407 & 9993.926 & 14847.629 \end{bmatrix}$	26

We now try to find out exactly when these mean differences occur in the experiment. The box plots for the three treatment groups over the four week-long periods is given in Figure 7.2. It shows that the difference of the mean body weights for the three groups occurred in week 3 and week 4, which was also supported by the test results from the SAS GLM procedure.

If ODS Graphics are enabled, GLM also displays by default an interaction plot for this analysis. The following statements, which are the same as in the previous analysis but with ODS Graphics enabled, additionally produce Figure 7.3. The plot shows that there is no interaction between the groups.

7.3 Models utilizing time series structure

7.3.1 Fixed effects model

In the previous analysis, we assume that $e_{i,j}$ in model (7.1) are independent Gaussian vector white noise $N_p(\mathbf{0}, \mathbf{\Sigma})$, where $\mathbf{\Sigma}$ is a general $p \times p$ variance–covariance matrix with $p(p+1)/2$ elements. We do not utilize the fact that the elements in each p-dimensional vector, $\mathbf{e}_{i,j} = [e_{i,j,1}, e_{i,j,2}, \ldots, e_{i,j,p}]'$, are time series, and that they are autocorrelated. In a repeated measurement study, to be more rigorous, we should first estimate the variance–covariance structure $\mathbf{\Sigma}$ and then incorporate this structure in the analysis.

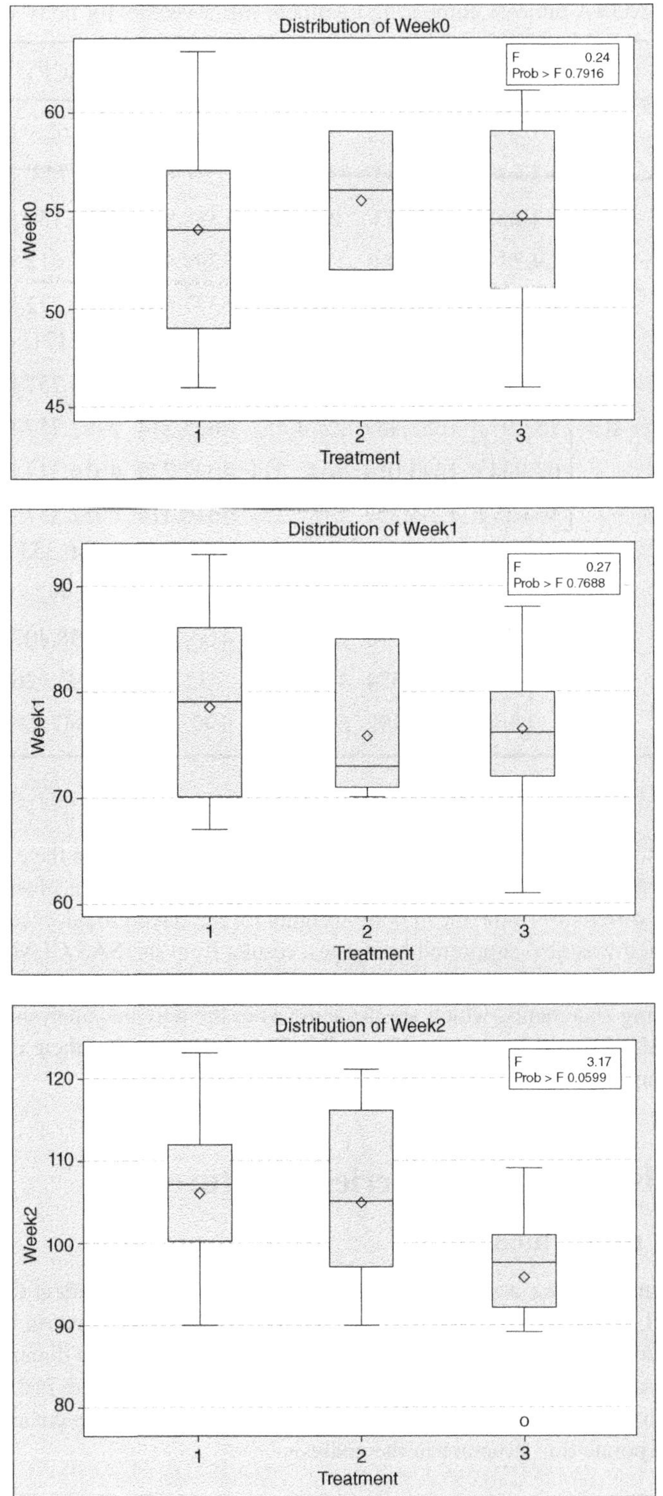

Figure 7.2 The box plots for the three treatment groups over time.

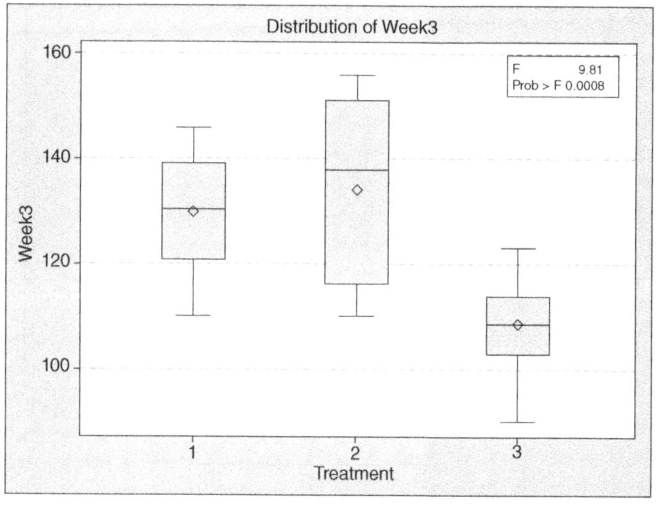

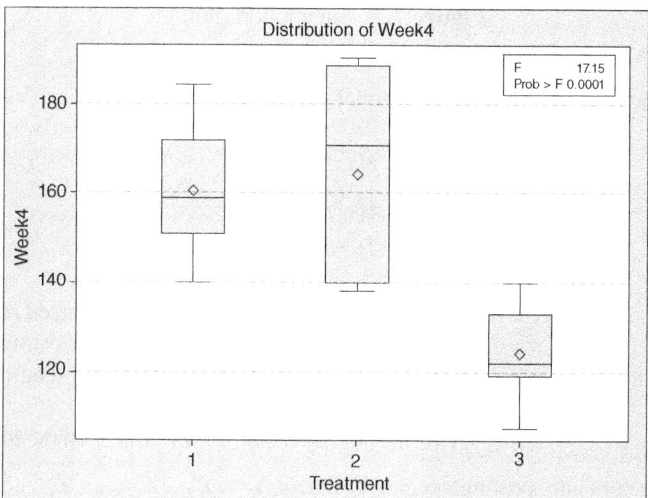

Figure 7.2 (Continued)

More generally, we will consider a two-factor fixed effects model that not only includes treatment effect but also time effect and treatment/time interaction effect as follows:

$$Z_{i,j,t} = \mu + \alpha_i + \beta_t + \gamma_{i,t} + e_{i,j,t}, \tag{7.14}$$

where α_i, $i = 1, \ldots, m$, represents the treatment effect, β_t, $t = 1, \ldots, p$, represents the time effect, $\gamma_{i,t}$ represents time/treatment interaction effect, $j = 1, \ldots, n_i$, represents subject, and $\sum_{i=1}^{m} n_i = n$. For the error term $e_{i,j,t}$, we assume that $\mathrm{Cov}\left(e_{i,j,t}, e_{i',j',t'}\right) = 0$, if either $i \neq i'$ or $j \neq j'$. However, even though it is logical to assume the same variance–covariance structure for all subjects, we cannot assume $\mathrm{Cov}\left(e_{i,j,t}, e_{i,j,t'}\right) = 0$. In fact, $\mathrm{Cov}\left(e_{i,j,t}, e_{i,j,t'}\right) \neq 0$, when $t \neq t'$, and its structure for the same subject clearly depends on time.

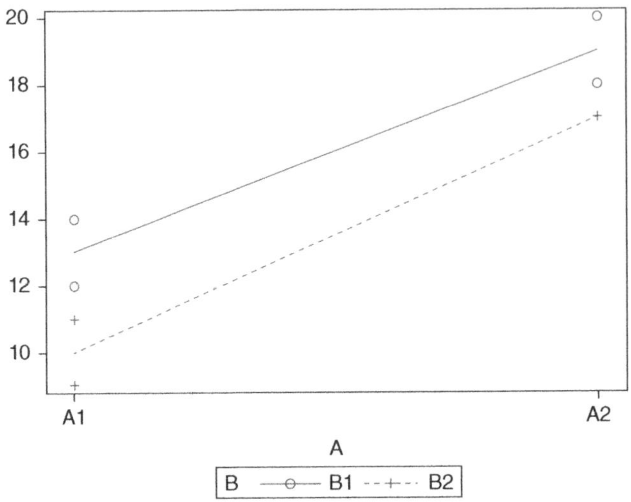

Figure 7.3 Interaction plot.

The null hypotheses to be tested in Model (7.14) are:

$$\begin{aligned} &1.\ H_0 : \alpha_i = 0, \\ &2.\ H_0 : \beta_t = 0, \\ &3.\ H_0 : \lambda_{i,t} = 0. \end{aligned} \tag{7.15}$$

The analysis of variance table used to test these hypotheses is summarized in Table 7.4. We normally test the interaction term first. When the interaction term is significantly different from 0, it implies that the factor effects are not additive, which makes the interpretation of the results complicated.

By letting $\mathbf{Z}_{i,j} = [Z_{i,j,1}, \ldots, Z_{i,j,p}]'$ be the $(1 \times p)$ vector for jth subject on the ith treatment and stacking these vectors into a column vector $\mathbf{Y}$, that is, $\mathbf{Y} = \left[\mathbf{Z}'_{1,1}, \mathbf{Z}'_{1,2}, \ldots, \mathbf{Z}'_{m,n_m}\right]'$, we can write Model in Eq. (7.14) in the following matrix form,

$$\mathbf{Y} = \mathbf{X}\boldsymbol{\beta} + \boldsymbol{\varepsilon}, \tag{7.16}$$

Table 7.4 Repeated measurement analysis of variance table.

Source	SS	DF	Mean square
Treatment	SST	$(m-1)$	SST/$(m-1)$
Time (location)	SSL	$(p-1)$	SSL/$(p-1)$
Time∗treatment	SSLT	$(m-1)(p-1)$	$SSLT/[(m-1)(p-1)]$
Error (time)	SSE	$(p-1)\left(\sum_{i=1}^{m} n_i - m\right)$	$SSE/\left[(p-1)\left(\sum_{i=1}^{m} n_i - m\right)\right]$

where $\mathbf{Y}$ is the $(np \times 1)$ vector of observations for $Z_{i,j,t}$, $\mathbf{X}$ is the $[np \times (1+m)(1+p)]$matrix of corresponding known constants, $\boldsymbol{\beta}$ is the $[(1+m)(1+p) \times 1]$ vector of parameters, and $\boldsymbol{\varepsilon}$ is a $(np \times 1)$ vector of random errors with mean vector $\mathbf{0}$ and variance–covariance matrix $\boldsymbol{\Omega} = \mathbf{I}_n \otimes \boldsymbol{\Sigma}$, $\mathbf{I}_n$ is the $n \times n$ identity matrix,$n = (n_1 + \cdots + n_m)$ is the number of subjects, and $\boldsymbol{\Sigma}$ is the $(p \times p)$ common variance–covariance structure for all subjects. The generalized least square (GLS) estimator of $\boldsymbol{\beta}$ is

$$\hat{\boldsymbol{\beta}} = \left(\mathbf{X}'\boldsymbol{\Omega}^{-1}\mathbf{X}\right)\mathbf{X}\boldsymbol{\Omega}^{-1}\mathbf{Y}, \tag{7.17}$$

and variance–covariance matrix of $\hat{\boldsymbol{\beta}}$ is

$$Var\left(\hat{\boldsymbol{\beta}}\right) = \left(\mathbf{X}'\boldsymbol{\Omega}^{-1}\mathbf{X}\right)^{-1}. \tag{7.18}$$

Under normal assumptions, it is well known that $\hat{\boldsymbol{\beta}}$ follows a multivariate vector normal distribution with mean $\boldsymbol{\beta}$ and variance–covariance matrix $(\mathbf{X}'\boldsymbol{\Omega}^{-1}\mathbf{X})^{-1}$, that is, $N(\boldsymbol{\beta}, (\mathbf{X}'\boldsymbol{\Omega}^{-1}\mathbf{X})^{-1})$.

7.3.2 Some common variance–covariance structures

Since a large number of parameters in a variance–covariance matrix will unfavorably affect the estimation efficiency, we should use the correlation pattern of the time series to simplify its form. The following are some commonly used variance–covariance matrices used in repeated measurement studies. Except for the first unstructured matrix, we introduce some simple and useful structures that contain only a small number of parameters.

1. The unstructured matrix:

$$\boldsymbol{\Sigma} = \begin{bmatrix} \sigma_1^2 & \sigma_{1,2} & \cdots & \cdots & \sigma_{1,p} \\ & \sigma_2^2 & \sigma_{2,3} & \cdots & \sigma_{2,p} \\ & & \ddots & \vdots & \vdots \\ & & & \sigma_{p-1}^2 & \sigma_{(p-1),p} \\ & & & & \sigma_p^2 \end{bmatrix}. \tag{7.19}$$

The form implies that variances and covariances at different times are not necessarily equal. There are *p(p + 1)/2* parameters in the matrix.

2. The identical and independent structure:

$$\boldsymbol{\Sigma} = \begin{bmatrix} \sigma^2 & 0 & 0 & \cdots & 0 \\ & \sigma^2 & 0 & \cdots & 0 \\ & & \ddots & \vdots & \vdots \\ & & & \sigma^2 & 0 \\ & & & & \sigma^2 \end{bmatrix} = \sigma^2\mathbf{I}. \tag{7.20}$$

The form in Eq. (7.20) is the simplest one and contains only one parameter. It may be applicable in some applications especially when the repeated measurements are taken far apart such that the correlation between different times is effectively zero relative to the other variation.

3. The independent but non-identical structure:

$$\boldsymbol{\Sigma} = \begin{bmatrix} \sigma_1^2 & 0 & 0 & \cdots & 0 \\ & \sigma_2^2 & 0 & \cdots & 0 \\ & & \ddots & \vdots & \vdots \\ & & & \sigma_{p-1}^2 & 0 \\ & & & 0 & \sigma_p^2 \end{bmatrix}. \tag{7.21}$$

This is a generalized form of Eq. (7.20), where the variances at different times are not necessarily equal. It contains p parameters.

4. The structure of common symmetry:

$$\boldsymbol{\Sigma} = \begin{bmatrix} \sigma^2 & \sigma^2\rho & \cdots & \cdots & \sigma^2\rho \\ & \sigma^2 & \sigma^2\rho & \cdots & \sigma^2\rho \\ & & \ddots & \vdots & \vdots \\ & & & \sigma^2 & \sigma^2\rho \\ & & & & \sigma^2 \end{bmatrix}. \tag{7.22}$$

The form in Eq. (7.22) assumes that $E(e_{i,j,k}e_{i,j,\ell}) = \sigma^2$ if $k = \ell$, and $E(e_{i,j,k}e_{i,j,\ell}) = \sigma^2\rho$ if $k \neq \ell$. There are only two parameters. However, it implies that (i) variances are equal at all times, and (ii) covariances and hence correlations are equal at all pairs of times. This strong assumption may not hold in many situations.

5. The structure of heterogeneous common symmetry:

$$\boldsymbol{\Sigma} = \begin{bmatrix} \sigma_1^2 & \sigma_1\sigma_2\rho & \cdots & \cdots & \sigma_1\sigma_p\rho \\ & \sigma_2^2 & \sigma_2\sigma_3\rho & \cdots & \sigma_2\sigma_p\rho \\ & & \ddots & \vdots & \vdots \\ & & & \ddots & \sigma_{(p-1)}\sigma_p\rho \\ & & & & \sigma_p^2 \end{bmatrix}. \tag{7.23}$$

The form in Eq. (7.23) assumes that the variances at different times may be distinct, but the correlations are equal at all pairs of times. In this case, $\boldsymbol{\Sigma}$ contains $(p + 1)$ parameters.

6. The Toeplitz structure:

$$\boldsymbol{\Sigma}=\sigma^2\begin{bmatrix} 1 & \rho_1 & \rho_2 & \cdots & \rho_{(p-1)} \\ & 1 & \rho_1 & \cdots & \rho_{(p-2)} \\ & & \ddots & \vdots & \vdots \\ & & & \ddots & \rho_1 \\ & & & & 1 \end{bmatrix}=\begin{bmatrix} \sigma^2 & \sigma_{1,2} & \sigma_{1,3} & \cdots & \sigma_{1,p} \\ & \sigma^2 & \sigma_{1,2} & \cdots & \sigma_{1,(p-1)} \\ & & \ddots & \vdots & \vdots \\ & & & \sigma^2 & \sigma_{1,2} \\ & & & & \sigma^2 \end{bmatrix}. \tag{7.24}$$

The form in Eq. (7.24) assumes that the correlations for any pairs separated by the same time lag are the same. It contains p parameters.

7. The heterogeneous Toeplitz structure:

$$\boldsymbol{\Sigma}=\begin{bmatrix} \sigma_1^2 & \sigma_1\sigma_2\rho_1 & \sigma_1\sigma_3\rho_2 & \cdots & \sigma_1\sigma_p\rho_{(p-1)} \\ & \sigma_2^2 & \sigma_2\sigma_3\rho_1 & \cdots & \sigma_2\sigma_p\rho_{(p-2)} \\ & & \ddots & \vdots & \vdots \\ & & & \sigma_{(p-1)}^2 & \sigma_{(p-1)}\sigma_p\rho_1 \\ & & & & \sigma_p^2 \end{bmatrix}. \tag{7.25}$$

The form in Eq. (7.25) allows unequal variances and it contains $(2p - 1)$ parameters.

8. The AR(1) structure:

$$\boldsymbol{\Sigma}=\begin{bmatrix} \sigma^2 & \sigma^2\rho & \sigma^2\rho^2 & \cdots & \sigma^2\rho^{p-1} \\ & \sigma^2 & \sigma^2\rho & \cdots & \sigma^2\rho^{p-2} \\ & & \ddots & \vdots & \vdots \\ & & & \sigma^2 & \sigma^2\rho \\ & & & & \sigma^2 \end{bmatrix}. \tag{7.26}$$

The form assumes the first-order autoregressive structure, that is, the correlation with k time lags apart are ρ^k. It greatly simplifies the form and contains only two parameters.

9. The heterogeneous AR(1) structure:

$$\boldsymbol{\Sigma}=\begin{bmatrix} \sigma_1^2 & \sigma_1\sigma_2\rho & \sigma_1\sigma_3\rho^2 & \cdots & \sigma_1\sigma_p\rho^{p-1} \\ & \sigma_2^2 & \sigma_2\sigma_3\rho & \cdots & \sigma_2\sigma_p\rho^{p-2} \\ & & \ddots & \vdots & \vdots \\ & & & \sigma_{(p-1)}^2 & \sigma_{(p-1)}\sigma_p\rho \\ & & & & \sigma_p^2 \end{bmatrix}. \tag{7.27}$$

The form in Eq. (7.27) assumes the first-order autoregressive structure but allows unequal variances. It contains ($p + 1$) parameters.

10. The ARMA($1,1$) structure:

$$\boldsymbol{\Sigma}=\begin{bmatrix} \sigma^2 & \sigma^2\gamma & \sigma^2\gamma\rho & \cdots & \sigma^2\gamma\rho^{p-2} \\ & \sigma^2 & \sigma^2\gamma & \cdots & \sigma^2\gamma\rho^{p-3} \\ & & \ddots & \vdots & \vdots \\ & & & \sigma^2 & \sigma^2\gamma \\ & & & & \sigma^2 \end{bmatrix}. \tag{7.28}$$

The form in Eq. (7.28) assumes a general autoregressive moving average of order ($1,1$) structure, and it contains only three parameters.

More generally, AR(2), and ARMA(p, q) models can be used to represent the variance–covariance structure, which contain a much smaller number of parameters than the general form of $\boldsymbol{\Sigma}$.

Since each mean vector of $\boldsymbol{\mu}_i$, $i = 1, 2, \ldots, m$, is p-dimensional, more explicitly we can write $\boldsymbol{\mu}_i = \left[\mu_{i,1},\ \mu_{i,2},\ \ldots,\ \mu_{i,p}\right]'$, and after the hypotheses of no treatment, time, and treatment/time interaction effects are rejected, a very natural next step is to test

$$\mathrm{H}_0 : \mu_{1,t} = \mu_{2,t} = \cdots = \mu_{m,t},$$

for each time t with $t = 1, \ldots, p$. Through these careful further tests, one can find out exactly when these mean differences occur in the experiment. We will illustrate these in the following example.

7.3.3 Empirical Example II – Further analysis on body weight of rats under three different treatments

Example 7.2 We now further study the growth data of rats under three treatment groups (control, thyroxin, and thiouracil) by incorporating a stationary variance–covariance structure in the analysis. To determine a proper matrix structure, we can compute the sample cross-correlation matrix from the data $Z_{i,j,t}$. We can also obtain the estimate using the residuals from Eq. (7.10), which equals

$$\hat{\boldsymbol{\Sigma}} = \begin{bmatrix} 1 & 0.88 & 0.78 & 0.69 & 0.61 \\ 0.88 & 1 & 0.88 & 0.78 & 0.69 \\ 0.78 & 0.88 & 1 & 0.88 & 0.78 \\ 0.69 & 0.78 & 0.88 & 1 & 0.88 \\ 0.61 & 0.69 & 0.78 & 0.88 & 1 \end{bmatrix}. \tag{7.29}$$

The pattern strongly suggests an AR(*1*) structure. Thus, we fit Model (7.14) with the following SAS code for mixed models, where an AR(*1*) covariance matrix is specified.

```
data rat;
        infile 'c:/Bookdata/WW7a.csv';
        input Treatment$ Week0 Week1 Week2 Week3 Week4;
        rat=_n_;
        weight=Week0; week=0; output;
        weight=Week1; week=1; output;
        weight=Week2; week=2; output;
        weight=Week3; week=3; output;
        weight=Week4; week=4; output;
        drop Week0-Week4;
/*AR(1)covariance matrix*/
proc mixed data=rat;
        class Treatment week rat;
        model weight = Treatment week Treatment*week;
        repeated/type=ar(1) sub=rat r rcorr;
run;
```

The associated analysis of variance table is given in Table 7.5.

Table 7.5 Repeated measurement analysis of variance of body weights of rats with AR (*1*) covariance matrix.

	The *mixed* procedure				
Source	SS	DF	Mean Square	F Value	Pr > F
Treatment	1 606.3504	2	803.1752	5.84	0.0086
Week	300 596.5704	4	75 149.1426	546.42	<0.0001
Treatment*week	15 370.3528	8	1 921.2941	13.97	<0.0001
Error	13 202.88	96	137.53		

Table 7.6 Weekly treatment effect of body weights of rats.

Week	Source	SS	DF	MS	F Value	Pr > F
0	Treatment	10.1857	2	5.0929	0.24	0.7916
1	Treatment	36.5429	2	8.2715	0.27	0.7688
2	Treatment	601.3947	2	300.6974	3.17	0.0599
3	Treatment	3563.6058	2	781.8029	9.81	0.0008
4	Treatment	9204.1714	2	4602.0857	17.15	<0.0001

Table 7.7 Repeated measurement analysis of variance for body weights of rats with an unspecified general covariance matrix.

	The *mixed* procedure				
Source	SS	DF	Mean Square	F Value	Pr > F
Treatment	6 635.0486	2	3 317.5243	7.73	0.0026
Week	573 549.7372	4	143 387.4343	334.10	<0.0001
Treatment∗week	22 935.1224	8	2 866.8903	6.68	0.0001
Error	10 300.2048	24	429.1752		

The results clearly reject the null hypotheses of Eq. (7.15). In addition to the treatment effect, there are strong time effect and treatment/time interaction effects, which can also be seen from Figure 7.1. More explicitly, we can summarize the test results for each week from the previous SAS repeated measurement analysis of variance in Table 7.6. They show that the treatments began to produce different results after week 2.

For comparison, if we fit the model with general covariance matrix $\boldsymbol{\Sigma}$, the associated analysis of variance table is given in Table 7.7. The degrees of freedom for the error term are reduced significantly.

7.3.4 Random effects and mixed effects models

The model in Eq. (7.14) is a special case of the fixed effects model with two factors. When the levels of these factors are randomly selected, and we want to generalize the result from analysis to a much larger population, then the model becomes a random effects model. The model becomes a mixed effects model when some factors are random and some are fixed. For example, in Model (7.14), if treatments are randomly assigned, we have

$$Z_{i,j,t} = \mu + \alpha_i + \beta_t + \gamma_{i,t} + e_{i,j,t}, \tag{7.30}$$

where α_i is random, i.i.d. $N(0, \sigma_\alpha^2)$, independent of $e_{i,j,t}$. The analysis of variance table for the fixed effects model, the random effects model, and the mixed effects model are the same as Table 7.7, but when a model contains a random factor like Eq. (7.30), it is important to note the following:

1. The variance of $Z_{i,j,t}$ is no longer equal to the variance of $e_{i,j,t}$. Instead, if we also assume that $\text{Var}(\alpha_i) = \sigma_\alpha^2$, we have

$$\text{Var}\left(Z_{i,j,t}\right) = \text{Var}(\alpha_i) + \text{Var}\left(e_{i,j,t}\right) = \sigma_\alpha^2 + \sigma^2. \tag{7.31}$$

2. The model in (7.31) can also be written as

$$Z_{i,j,t} = \mu + \beta_t + \gamma_{i,t} + \varepsilon_{i,j,t}, \tag{7.32}$$

where the $\varepsilon_{i,j,t}$ are now i.i.d. $N\left(0, \sigma_\varepsilon^2\right), \sigma_\varepsilon^2 = \sigma_\alpha^2 + \sigma^2$.

3. The Expected Mean Squares (EMS) for treatment is $\sigma^2 + n\sigma_\alpha^2$. Hence, we can estimate the variance of the random treatment term using

$$\hat{\sigma}_\alpha^2 = \frac{\text{Mean Squares for Treatment} - s^2}{n}, \tag{7.33}$$

where s^2 is the residual mean square error.

4. The null and alternative hypotheses for the random treatment in Model (7.30) are now

$$H_0: \sigma_\alpha^2 = 0,$$
$$H_a: \sigma_\alpha^2 > 0.$$

7.4 Nested random effects model

In some applications, subjects are randomly selected from a population. For example, in agricultural studies, where researchers want to compare the effects of three different fertilizers in terms of the yield of a certain product such as tomatoes. In this case, "subjects" refers to plots of land. By realizing the effects of land, and more importantly, being interested in the effects of fertilizers on a wide variety of plots, the researchers may randomly select a certain number of plots of land from a population of plots when assigning fertilizers within their experiments. In such a case, we will consider the following nested random effects model:

$$Z_{i,j,t} = \mu + \alpha_i + \theta_{j(i)} + \beta_t + \gamma_{i,t} + e_{i,j,t}, \tag{7.34}$$

where α_i, β_t, and $\gamma_{i,t}$ are fixed effects defined in Eq. (7.14), but $\theta_{j(i)}$ is a random effect for subject j associated with treatment i. We assume that the $\theta_{j(i)}$ are i.i.d. $N\left(0, \sigma_\theta^2\right)$, which are independent of $e_{i,j,t}$. The variance of $Z_{i,j,t}$ is no longer equal to the variance of $e_{i,j,t}$. It becomes the sum of the variances of $\theta_{j(i)}$ and $e_{i,j,t}$. Hence, the variance–covariance matrix of $\mathbf{Z}_{i,j} = [Z_{i,j,1}, \ldots, Z_{i,j,p}]'$ becomes

$$\sigma_\theta^2 \mathbf{H} + \boldsymbol{\Sigma}, \tag{7.35}$$

where $\mathbf{H}$ is a matrix of ones.

Table 7.8 Repeated measurement analysis of variance table with nested random effects.

Source	SS	df	Mean square
Treatment	SST	$(m-1)$	SST/$(m-1)$
Subjects (treatment)	SSB(T)	$(n-m)$	SSB(T)/(n-m)
Time (location)	SSL	$(p-1)$	SSL/$(p-1)$
Treatment∗time	SSLT	$(m-1)(p-1)$	$SSLT/[(m-1)(p-1)]$
Error (time)	SSE	$(p-1)(n-m)$	$SSE/[(p-1)(n-m)]$

where $n=\sum_{i=1}^{m} n_i$.

Equivalently, we can rewrite the model in Eq. (7.34) as

$$Z_{i,j,t}=\mu+\alpha_i+\beta_t+\gamma_{i,t}+\varepsilon_{i,j,t}, \tag{7.36}$$

where $\varepsilon_{i,j,t}=\theta_{j(i)}+e_{i,j,t}$. If the $e_{i,j,t}$ are i.i.d. $N(0, \sigma^2)$, then it can be shown that the variance and covariance of $\varepsilon_{i,j,t}$ and hence $Z_{i,j,t}$ will follow the structure of common symmetry given in Eq. (7.22) of Section 7.3.2.

The analysis of variance table for this nested random effects model is now modified as given in Table 7.8.

7.5 Further generalization and remarks

More generally, we can include some covariates in all the models introduced in Sections 7.3 and 7.4. For example, when subjects in the model are people, we may want to include related information, such as age, gender, education level, and others, in the model. Thus, we have

$$Z_{i,j,t}=\mu+c_1X_{1,j}+\cdots+c_kX_{k,j}+\alpha_i+\beta_t+\gamma_{i,t}+\varepsilon_{i,j,t}, \tag{7.37}$$

where the X's are covariates with associated coefficients c's.

With proper modifications, the model in (7.37) can be written in matrix form,

$$\mathbf{Y}=\mathbf{X}\boldsymbol{\beta}+\boldsymbol{\varepsilon}, \tag{7.38}$$

where the matrix $\mathbf{X}$ will now contain the values of covariates in addition to the values of 0 and 1 for factors, $\boldsymbol{\beta}$ is the vector of the associated parameters, and $\boldsymbol{\varepsilon}$ is a vector of normal random errors with mean vector $\mathbf{0}$ and variance–covariance matrix $\boldsymbol{\Omega}=\mathbf{I}_n\otimes\boldsymbol{\Sigma}$, $\mathbf{I}_n$ is the $n\times n$ identity matrix, n is the number of subjects, and $\boldsymbol{\Sigma}$ is the $(p\times p)$ common variance–covariance structure for all subjects. The GLS estimator of $\boldsymbol{\beta}$, $\hat{\boldsymbol{\beta}}=\left(\mathbf{X}'\boldsymbol{\Omega}^{-1}\mathbf{X}\right)\mathbf{X}\boldsymbol{\Omega}^{-1}\mathbf{Y}$, is the best linear unbiased estimator that follows a multivariate vector normal distribution $N(\boldsymbol{\beta}, (\mathbf{X}'\boldsymbol{\Omega}^{-1}\mathbf{X})^{-1})$.

It should be noted that although the examples illustrated in this chapter are equally spaced repeated measurements, the methods introduced can also be applied to cases when repeated measurements are unequally spaced. This is true even when a covariance structure such as an unstructured general covariance, a covariance of independent case, a covariance of common symmetry, or a covariance of heterogeneous common symmetry is used in the analysis. It should be noted, however, that most software such as SAS Proc Mixed may assume equally spaced times when a time series covariance structure like AR(*1*), ARMA(*1*, *1*), or the Toeplitz form is specified in the analysis. In such a case, using the software may require some adjustment.

7.6 Another empirical example – the oral condition of neck cancer patients

Example 7.3 In this example, we will consider the data set listed in Table 7.9 from the Mid-Michigan Medical Center for its study of the oral condition of neck cancer patients in 1999, which is also listed as WW7b in the Data Appendix. The study randomly divided patients into two groups; Group 0 received a placebo and Group 1 received a treatment of aloe juice. The oral conditions of the patients were measured and recorded at the initial stage, and thereafter at the

Table 7.9 Background and oral condition of neck cancer patients under two treatments (0 = placebo, 1 = aloe juice) with total condition at initial stage (Week 0), Week 2, Week 4, and Week 6.

ID	TRT	AGE	WEIGHT	STAGE	W0	W2	W4	W6
1	0	52	124	2	6	6	6	7
5	0	77	160	1	9	6	10	9
6	0	60	136.5	4	7	9	17	19
9	0	61	179.6	1	6	7	9	3
11	0	59	175.8	2	6	7	16	13
15	0	69	167.6	1	6	6	6	11
21	0	67	186	1	6	11	11	10
26	0	56	158	3	6	11	15	15
31	0	61	212.8	1	6	9	6	8
35	0	51	189	1	6	4	8	7
39	0	46	149	4	7	8	11	11
41	0	65	157	1	6	6	9	6
45	0	67	186	1	8	8	9	10
2	0	46	163.8	2	7	16	9	10
12	1	56	227.2	4	6	10	11	9
14	1	42	162.6	1	4	6	8	7
16	1	44	261.4	2	6	11	11	14
22	1	27	225.4	1	6	7	6	6
24	1	68	226	4	12	11	12	9
34	1	77	164	2	5	7	13	12
37	1	86	140	1	6	7	7	7
42	1	73	181.5	0	8	11	16	
44	1	67	187	1	5	7	7	7
50	1	60	164	2	6	8	16	
58	1	54	172.8	4	7	8	10	8

end of week 2, week 4, and week 6, respectively. The measurement values varied between 3 and 19 with higher values indicating a better condition. In addition, the age, the initial weight, and the initial cancer stage of patients were also collected. There were a total 25 patients with missing values for patient 42 and 50. Specifically, we will consider the following model suggested in Eq. (7.37),

$$Z_{i,j,t} = \mu + c_1 X_{1,j} + c_2 X_{2,j} + c_3 X_{3,j} + \alpha_i + \beta_t + \gamma_{i,t} + \varepsilon_{i,j,t}, \tag{7.39}$$

where $Z_{i,j,t}$ is the oral condition, $X_{1,j}$ is the patient's age, $X_{2,j}$ is the patient's weight at the initial stage, $X_{3,j}$ is the patient's initial cancer stage, α_i represents the random treatment effect that follows i.i.d. $N(0,\sigma_\alpha^2)$, β_t represents the time effect, $\gamma_{i,t}$ represents time/treatment interaction effect, $\varepsilon_{i,j,t}$ is the error term, and α_i and $\varepsilon_{i,j,t}$ are independent, $i = 1, 2$, $n_1 = 14$, $n_2 = 9$, $j = 1, \ldots, n_i$, and $t = 1, 2, 3, 4$.

We fit model (7.39) with the following SAS code for mixed models, where an unspecified covariance matrix is specified.

```
data oral;
       infile 'c:/Bookdata/WW7b.csv';
       input id trt age weight stage w0 w2 w4 w6;
       tc=w0; week=0; output;
       tc=w2; week=2; output;
       tc=w4; week=4; output;
       tc=w6; week=6; output;
       drop w0 w2 w4 w6;
proc mixed data=oral;
  class trt age weight stage week;
  model tc = age weight stage trt week trt*week;
  random trt;
run;
```

The associated analysis of variance table is given in Table 7.10. The results show that age, weight, and interaction term are all not significant. The time period, a week, is the only significant variable.

In concluding this chapter, it should be pointed out that repeated measurements, clustered data, and longitudinal data are related, and we can use a mixed model to analyze all of them. However, the issues involved in these data sets are different, and some of the specifications used are also different. For further information and applications, we refer readers to Hand and Taylor (1987), Icaza and Jones (1999), Singer and Willett (2003), Fress (2004), Chen (2006), Dehlendorff (2007), Menard (2008), Coke and Tsao (2010), Davis and Ensor (2007), Fokianos (2010), Stram (2014), West, Welch, and Galecki et al. (2014), Bravo (2016), Heyse and Chan (2016), Li, Qian, and Su (2016), Ando and Bai (2017), Arellano, Blundell, and Bonhomme (2017), Broemeling (2017), Chalikias and Kounias (2017), Gile and Handcock (2017), Giordano, Rocca, and Parrella (2017), Islam and Chowdhury (2017), Nakashima (2017), Suarez et al. (2017), and Vogt and Linton (2017), among others.

Table 7.10 Repeated measurement analysis of variance for the oral conditions of neck cancer patients with covariates and an unspecified general covariance matrix.

	Type 3 Tests of Fixed Effects			
Effect	Num df	Den df	F Value	Pr > F
Age	1	68	0.28	0.5966
Weight	2	68	0.32	0.7280
Stage	0			
Treatment	0			
Week	3	68	13.13	<0.0001
Treatment∗Week	3	68	0.27	0.8480

Software code

SAS code for Example 7.1

```
data Rat;
      infile 'c:/Bookdata/WW7a.csv';
      input Treatment Week0 Week1 Week2 Week3 Week4;
/* GLM procedure with MANOVA */
proc glm data=Rat;
      class Treatment;
      model Week0 Week1 Week2 week3 week4 = Treatment
      /NOUNI;
      means Treatment;
      manova h = Treatment/printh printe htype=1 etype=1;
run;
```

SAS Code for Example 7.2

```
data rat;
      infile 'c:/Bookdata/WW7a.csv';
      input Treatment$ Week0 Week1 Week2 Week3 Week4;
      rat=_n_;
      weight=Week0; week=0; output;
      weight=Week1; week=1; output;
      weight=Week2; week=2; output;
      weight=Week3; week=3; output;
      weight=Week4; week=4; output;
      drop Week0-Week4;
/*AR(1)covariance matrix*/
proc mixed data=rat;
      class Treatment week rat;
      model weight = Treatment week Treatment*week;
      repeated/type=ar(1) sub=rat r rcorr;
run;
```

SAS Code for Example 7.3

```
data oral;
        infile 'c:/Bookdata/WW7b.csv';
        input id trt age weight stage w0 w2 w4 w6;
        tc=w0; week=0; output;
        tc=w2; week=2; output;
        tc=w4; week=4; output;
        tc=w6; week=6; output;
        drop w0 w2 w4 w6;
proc mixed data=oral;
        class trt age weight stage week;
        model tc = age weight stage trt week trt*week;
        random trt;
run;
```

Projects

1. Find a repeated measurement data set of your interest, carry out its multivariate analysis variance, and complete your report with an appendix that contains your data set and analysis software code.
2. Find a repeated measurement data set of your interest, carry out its analysis by incorporating a proper variance–covariance structure in the analysis, and complete your report with an appendix that contains your data set and analysis software code.
3. Find a repeated measurement data set in a social science field, carry out its analysis with both fixed effect and random effect models, make comparisons, and complete your report with an appendix that contains your data set and analysis software code.
4. Find a repeated measurement data set in a natural science related field, carry out its multivariate analysis variance, and complete your report with an appendix that contains your data set and analysis software code.
5. For the data set in Project 4, carry out its analysis with both fixed effect and random effect models, make comparisons, and complete your report with an appendix that contains your data set and analysis software code.

References

Ando, T. and Bai, J. (2017). Clustering huge number of financial time series: a panel data approach with high-dimensional predictors and factor structures. *Journal of the American Statistical Association* **112**: 1182–1198.

Arellano, M., Blundell, R., and Bonhomme, S. (2017). Earnings and consumption dynamics: a nonlinear panel data framework. *Econometrica* **85**: 693–734.

Bartlett, M.S. (1954). A note on the multiplying factors for various χ^2 approximation. *Journal of the Royal Statistical Society Series B* **16**: 296–298.

Box, G.E.P. (1950). Problems in the analysis of growth and wear curves. *Biometrics* **6**: 262–289.

Bravo, F. (2016). Local information theoretic methods for smooth coefficients dynamic panel data models. *Journal of Time Series Analysis* **37**: 690–708.

Broemeling, L.D. (2017). *Bayesian Methods for Repeated Measures*. Chapman and Hall/CRC.

Chalikias, M. and Kounias, S. (2017). Optimal two treatment repeated measurement designs for three periods. *Communications in Statistics: Theory and Methods* **46**: 200–209.

Chen, W. (2006). An approximate likelihood function for panel data with a mixed ARMA(p,q) remainder disturbance model. *Journal of Time Series Analysis* **27**: 911–921.

Coke, G. and Tsao, M. (2010). Random effects mixture models for clustering electrical load series. *Journal of Time Series Analysis* **31**: 451–464.

Crowder, M.J. and Hand, D.J. (1990). *Analysis of Repeated Measures*. Chapman and Hall.

Davis, G.M. and Ensor, K.B. (2007). Multivariate time-series analysis with categorical and continuous variables in an LSTR model. *Journal of Time Series Analysis* **28**: 867–885.

Dehlendorff, C. (2007). *Longitudinal Data Analysis of Asthma and Wheezing in Children*. Technical University of Denmark.

Diggle, P.J., Liang, K.Y., and Zeger, S.L. (2013). *Analysis of Longitudinal Data*, Oxford Statistical Science Series, 2e. Oxford University Press.

Fokianos, K. (2010). Antedependence models for longitudinal data. *Journal of Time Series Analysis* **31**: 494.

Fress, E.W. (2004). *Longitudinal and Panel Data – Analysis and Applications in the Social Sciences*. Cambridge University Press.

Gile, K. and Handcock, M.S. (2017). Analysis of networks with missing data with application to the national longitudinal study of adolescent health. *Journal of the Royal Statistical Society Series C* **66**: 501–519.

Giordano, F., Rocca, M.L., and Parrella, M.L. (2017). Clustering complex time-series databases by using periodic components. *Statistical Analysis and Data Mining* **10**: 89–106.

Hand, D.J. and Taylor, C.C. (1987). *Multivariate Analysis of Variance and Repeated Measures: A Practical Approach for Behavioural Scientists*, 1e. Chapman & Hall /CRC.

Heyse, J. and Chan, I. (2016). Review of statistical innovations in trials supporting vaccine clinical development. *Statistics in Biopharmaceutical Research* **8**: 128–142.

Icaza, G. and Jones, R. (1999). A state-space EM algorithm for longitudinal data. *Journal of Time Series Analysis* **20**: 537–550.

Islam, M.A. and Chowdhury, R.I. (2017). *Analysis of Repeated Measures Data*. Springer.

Johnson, R.A. and Wichern, D.W. (2007). *Applied Multivariate Statistical Analysis*, 6e. Prentice Hall.

Li, D., Qian, J., and Su, L. (2016). Panel data models with interactive fixed effects and multiple structural breaks. *Journal of the American Statistical Association* **111**: 1804–1819.

Lindsey, J.K. (1993). *Models for Repeated Measurements*. New York: Oxford University Press.

Menard, S. (2008). *Handbook of Longitudinal Research, Design, Measurement, and Analysis*. Academic Press, Elsevier.

Nakashima, E. (2017). Modification of GEE1 and linear mixed-effects models for heteroscedastic longitudinal Gaussian data. *Communications in Statistics: Theory and Methods* **46**: 11110–11122.

SAS Institute, Inc. (2015). *SAS for Windows, 9.4*, Cary, North Carolina.

Singer, J.D. and Willett, J.B. (2003). *Applied Longitudinal Data Analysis – Modeling Change and Event Occurrence*. Oxford University Press.

Stram, D.O. (2014). *Design, Analysis, and Interpretation of Genome-Wide Association Scans*. Springer.

Suarez, C.C., Klein, N., Kneib, T., Molenberghs, G., and Rizopoulos, D. (2017). Editorial "joint modeling of longitudinal and time-to-event data and beyond". *Biomedical Journal* **59**: 1101–1103.

Vogt, M. and Linton, O. (2017). Classification of non-parametric regression functions in longitudinal data models. *Journal of the Royal Statistical Society Series B* **79**: 5–27.

West, B.T., Welch, K.B., and Galecki, A.T. (2014). *Linear Mixed Models: A Practical Guide Using Statistical Software*, 2e. Chapman and Hall/CRC.

Wilks, S.S. (1932). On the sampling distribution of the multiple correlation coefficient. *Annals of Mathematical Statistics* **3**: 196–203.

8

Space–time series models

As indicated in Chapter 1, although the ordering of a time series is usually through time, it may also be ordered through space. If distances of spaces between observations are equal, the time series models and methods discussed earlier are equally applicable. However, in practice, we often see the phenomenon where regular time series are observed in different regions, and the point of interest is the study of the spatial patterns of these series. For example, in studying a traffic pattern of a certain location, traffic flow data is collected at constant intervals of time not only at that location but also at neighboring locations. In studying the prices of a certain commodity in a city, we not only want to study the price series of that commodity in that city but also the prices of the same commodity in nearby cities. In this chapter, we will present methods and models that can be used to describe spatial time series that are related to both time and space.

8.1 Introduction

One way to model the space–time series is to treat them as a vector time series where the ith component corresponds to the ith location. Thus,

$$\begin{bmatrix} Z_{1,t} \\ Z_{2,t} \\ \vdots \\ Z_{m,t} \end{bmatrix} = \begin{bmatrix} \text{Time series for space 1} \\ \text{Time series for space 2} \\ \vdots \\ \text{Time series for space } m \end{bmatrix}. \tag{8.1}$$

One can use the methods that were introduced in Chapter 2 to construct a vector autoregressive moving average VARMA(p,q) model or a vector autoregressive VAR(p) model for the series and use it to forecast and make inferences.

Alternatively, by classifying the spaces (locations) into different groups (regions), these spatial series are often measurements of the same phenomenon from several locations. For example, in studying the prices of a certain commodity in a city, we may divide the city into

Multivariate Time Series Analysis and Applications, First Edition. William W.S. Wei.
© 2019 John Wiley & Sons Ltd. Published 2019 by John Wiley & Sons Ltd.
Companion website: www.wiley.com/go/wei/datasets

different regions where each region includes certain city blocks and the measurements are prices of the commodity under study. Hence, we can treat the spatial series at different locations in these groups (regions) at time t as repeated measurements at time t.

More specifically, we can present the spatial series for time from 1 to n as $(n \times 1)$ vector time series $\mathbf{Z}_{i,j}$, where i refers to the ith group (region), and j refers to the jth location. If the purpose of the study is to find out whether there are any differences between groups (regions), we can use the methods discussed in the repeated measurements in Chapter 7 to analyze the data set, where locations in each group (region) constitute a random sample.

In this chapter, we will introduce another modeling method specifically designed for space–time series and illustrate the method with detailed examples.

8.2 Space–time autoregressive integrated moving average (STARIMA) models

In this section, we will discuss an approach where we build a model that deals not only the serial dependence but also spatial dependence between observations at each time point. Specifically, we will introduce a class of time series models where the structure of spatial dependence is incorporated into the commonly used ARIMA(p,d,q) models, which are found to be useful in many applications.

8.2.1 Spatial weighting matrix

To study space and time dependence, first we need to specify the order of spatial neighbors that reflects their distances to a particular location. Conventionally, the first order neighbors are closer than the second order ones, and the second order neighbors are closer than the third order neighbors, and so on. A standard definition of spatial order is sometimes available, but other times a model builder may need to define these orders so that they are suitable to the particular application.

Suppose that there are a total of m locations and we let $\mathbf{Z}_t = [Z_{1,t}, \ Z_{2,t}, \ \dots \ , \ Z_{m,t}]'$ be the vector of times series of these m locations. Based on the spatial orders, with respect to the time series at location i, we will assign weights related to this location, $w_{i,j}^{(\ell)}$, such that its value is nonzero only when the location j is the ℓth order neighbors of i, and the sum of these weights is equal to 1. In other words, with respect to location i, we have $\sum_{j=1}^{m} w_{i,j}^{(\ell)} = 1$, where

$$w_{i,j}^{(\ell)} = \begin{cases} (0,1], \text{if location } j \text{ is the } \ell^{\text{th}} \text{order neighbor of } i, \\ \qquad 0, \ \text{otherwise.} \end{cases}$$

Combining these weights, $w_{i,j}^{(\ell)}$, for all m locations, we have the spatial weight matrix for the neighborhood, $\mathbf{W}^{(\ell)} = \left[w_{(i,j)}^{(\ell)} \right]$, which is a $m \times m$ matrix with $w_{(i,j)}^{(\ell)}$ being nonzero if and only if locations i and j are in the same ℓ^{th} order neighbor and each row sums to 1. The weight can be chosen to reflect physical properties such as border length or distance of neighboring locations. One can also assign equal weights to all the locations of the same spatial order. Clearly, $\mathbf{W}^{(0)} = \mathbf{I}$, an identity matrix, because each location is its own zeroth order neighbor.

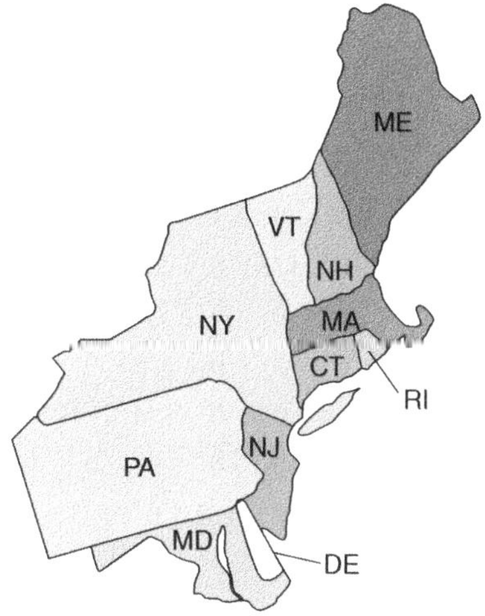

Figure 8.1 Northeastern states of the USA.

Example 8.1 To illustrate the use of a weighting matrix, let us consider an example of weather forecasting for the northeastern area of the United States, where we may want to consider the temperatures series in all these northeast states shown in Figure 8.1.

If we define the first order neighborhoods of a region to be those sharing a border with it, the second order neighborhoods of a region to be those sharing a border with the first order neighborhoods, and the third order neighborhoods of a region to be those sharing a border with the second order neighborhoods, then for the northeast states of USA given in Figure 8.1, we will have Table 8.1 of the three-order neighborhoods.

If we assign equal weight to all the locations of the same spatial order, the weighting matrix of $\mathbf{W}^{(1)}$ will be

Table 8.1 Spatial neighborhoods of 11 states in the Northeast United States.

Region (m)	First Order Neighborhood	Second Order Neighborhood	Third Order Neighborhood
CT(1)	MA, NY, RI	NH, NJ, PA, VT	DE, MD, ME
DE(2)	MD, NJ, PA	NY	CT, MA, VT
MA(3)	CT, NH, NY, RI, VT	ME, NJ, PA	DE, MD
MD(4)	DE, PA	NJ, NY	CT, MA, VT
ME(5)	NH	MA, VT	CT, NY, RI
NH(6)	MA, ME, VT	CT, NY, RI	NJ, PA
NJ(7)	DE, NY, PA	CT, MA, MD, VT	NH, RI
NY(8)	CT, MA, NJ, PA, VT	DE, MD, NH, RI	ME
PA(9)	DE, MD, NJ, NY	CT, MA, VT	NH, RI
RI(10)	CT, MA	NH, NY, VT	ME, NJ, PA
VT(11)	MA, NH, NY	CT, ME, NJ, PA, RI,	DE, MD

$$
\mathbf{W}^{(1)} = \left[\begin{array}{ccccccccccc}
0 & 0 & 1/3 & 0 & 0 & 0 & 0 & 1/3 & 0 & 1/3 & 0 \\
0 & 0 & 0 & 1/3 & 0 & 0 & 1/3 & 0 & 1/3 & 0 & 0 \\
1/5 & 0 & 0 & 0 & 1/5 & 0 & 1/5 & 0 & 0 & 1/5 & 1/5 \\
0 & 1/2 & 0 & 0 & 0 & 0 & 0 & 0 & 1/2 & 0 & 0 \\
0 & 0 & 0 & 0 & 0 & 1 & 0 & 0 & 0 & 0 & 0 \\
0 & 0 & 1/3 & 0 & 1/3 & 0 & 0 & 0 & 0 & 0 & 1/3 \\
0 & 1/3 & 0 & 0 & 0 & 0 & 0 & 1/3 & 1/3 & 0 & 0 \\
1/5 & 0 & 1/5 & 0 & 0 & 0 & 1/5 & 0 & 1/5 & 0 & 1/5 \\
0 & 1/4 & 0 & 1/4 & 0 & 0 & 1/4 & 1/4 & 0 & 0 & 0 \\
1/2 & 0 & 1/2 & 0 & 0 & 0 & 0 & 0 & 0 & 0 & 0 \\
0 & 0 & 1/3 & 0 & 0 & 1/3 & 0 & 1/3 & 0 & 0 & 0
\end{array}\right]. \tag{8.2a}
$$

Similarly, we have

$$
\mathbf{W}^{(2)} = \left[\begin{array}{ccccccccccc}
0 & 0 & 0 & 0 & 0 & 1/4 & 1/4 & 0 & 1/4 & 0 & 1/4 \\
0 & 0 & 0 & 0 & 0 & 0 & 0 & 1 & 0 & 0 & 0 \\
0 & 0 & 0 & 0 & 1/3 & 0 & 1/3 & 0 & 1/3 & 0 & 0 \\
0 & 0 & 0 & 0 & 0 & 0 & 1/2 & 1/2 & 0 & 0 & 0 \\
0 & 0 & 1/2 & 0 & 0 & 0 & 0 & 0 & 0 & 0 & 1/2 \\
1/3 & 0 & 0 & 0 & 0 & 0 & 0 & 1/3 & 0 & 1/3 & 0 \\
1/4 & 0 & 1/4 & 1/4 & 0 & 0 & 0 & 0 & 0 & 0 & 1/4 \\
0 & 1/4 & 0 & 1/4 & 0 & 1/4 & 0 & 0 & 0 & 1/4 & 0 \\
1/3 & 0 & 1/3 & 0 & 0 & 0 & 0 & 0 & 0 & 0 & 1/3 \\
0 & 0 & 0 & 0 & 0 & 1/3 & 0 & 1/3 & 0 & 0 & 1/3 \\
1/5 & 0 & 0 & 0 & 1/5 & 0 & 1/5 & 0 & 1/5 & 1/5 & 0
\end{array}\right], \tag{8.2b}
$$

and

$$
\mathbf{W}^{(3)} = \left[\begin{array}{ccccccccccc}
0 & 1/3 & 0 & 1/3 & 1/3 & 0 & 0 & 0 & 0 & 0 & 0 \\
1/3 & 0 & 0 & 1/3 & 0 & 0 & 0 & 0 & 0 & 0 & 1/3 \\
0 & 1/2 & 0 & 1/2 & 0 & 0 & 0 & 0 & 0 & 0 & 0 \\
1/3 & 0 & 1/3 & 0 & 0 & 0 & 0 & 0 & 0 & 0 & 1/3 \\
1/3 & 0 & 0 & 0 & 0 & 0 & 0 & 1/3 & 0 & 1/3 & 0 \\
0 & 0 & 0 & 0 & 0 & 0 & 1/2 & 0 & 1/2 & 0 & 0 \\
0 & 0 & 0 & 0 & 0 & 1/2 & 0 & 0 & 0 & 1/2 & 0 \\
0 & 0 & 0 & 0 & 1 & 0 & 0 & 0 & 0 & 0 & 0 \\
0 & 0 & 0 & 0 & 0 & 1/2 & 0 & 0 & 0 & 1/2 & 0 \\
0 & 0 & 0 & 0 & 1/3 & 0 & 1/3 & 0 & 1/3 & 0 & 0 \\
0 & 1/2 & 0 & 1/2 & 0 & 0 & 0 & 0 & 0 & 0 & 0
\end{array}\right]. \tag{8.2c}
$$

$\triangleleft$

8.2.2 STARIMA models

Thus, incorporating the structure of spatial dependence into a vector VARIMA model, we have the following space–time autoregressive integrated moving average (STARIMA) model,

$$\phi_p(B)\mathbf{D}(B)\mathbf{Z}_t = \theta_0 + \theta_q(B)\mathbf{a}_t, \tag{8.3}$$

where

$$\phi_p(B) = \mathbf{I} - \sum_{k=1}^{p}\sum_{\ell=0}^{a_k}\phi_{k,\ell}\mathbf{W}^{(\ell)}B^k,$$

$$\theta_q(B) = \mathbf{I} - \sum_{k=1}^{q}\sum_{\ell=0}^{m_k}\theta_{k,\ell}\mathbf{W}^{(\ell)}B^k,$$

$\mathbf{a}_t$ is a Gaussian vector white noise process with zero-mean vector $\mathbf{0}$, and covariance matrix structure

$$E\left[\mathbf{a}_t\mathbf{a}'_{t+k}\right] = \begin{cases} \mathbf{\Sigma}, \text{if } k = 0, \\ \mathbf{0}, \text{if } k \neq 0, \end{cases} \tag{8.4}$$

and $\mathbf{\Sigma}$ is a $m \times m$ symmetric positive definite matrix. $\mathbf{D}(B)$ is a proper differencing operator defined similarly as in Eq. (2.59) of Chapter 2. In this application, since component series are observations of the same nature in different locations, the differencing operator could simply be $\mathbf{D}(B) = (1 - B)^d\mathbf{I}$. $\phi_{k,\ell}$ and $\theta_{k,\ell}$ are autoregressive and moving average parameters at time lag k and space lag ℓ. respectively. p is the autoregressive order, q is the moving average order, a_k is the spatial order for the kth autoregressive term, and m_k is the spatial order for the kth moving average term.

It should be noted that the variance of each time series $Z_{i,t}$ is assumed to be constant. If not, a suitable variance stabilization transformation is assumed to be used to obtain $Z_{i,t}$. For simplicity, we will simply denote the model as a STARIMA$\left(p_{a_1,\ldots,a_p}, d, q_{m_1,\ldots,m_q}\right)$ model.

The STARIMA model in Eq. (8.3) can be extended to a seasonal STARIMA model that contains both seasonal and non-seasonal AR and MA polynomials as follows,

$$\alpha_P(B^s)\phi_p(B)\mathbf{D}(B^s)\mathbf{D}(B)\mathbf{Z}_t = \theta_0 + \beta_Q(B^s)\theta_q(B)\mathbf{a}_t, \tag{8.5}$$

where

$$\alpha_P(B^s) = \mathbf{I} - \sum_{k=1}^{P}\sum_{\ell=0}^{\lambda_k}\alpha_{k,\ell}\mathbf{W}^{(\ell)}B^{ks},$$

$$\beta_Q(B^s) = \mathbf{I} - \sum_{k=1}^{Q}\sum_{\ell=0}^{\tau_k}\beta_{k,\ell}\mathbf{W}^{(\ell)}B^{ks},$$

$\mathbf{D}(\mathrm{B}^s) = (1 - \mathrm{B}^s)^D\mathbf{I}$, and s is a seasonal period. $\alpha_{k,\ell}$ and $\beta_{k,\ell}$ are seasonal autoregressive and seasonal moving average parameters at time lag k and space lag ℓ, respectively. P is the seasonal autoregressive order, Q is the seasonal moving average order, λ_k is the spatial order for the kth seasonal autoregressive term, and τ_k is the spatial order for the kth seasonal moving average term. For simplicity, we will denote the seasonal STARIMA model with a seasonal period s in Eq. (8.5) as STARIMA$\left(p_{a_1,\ldots,a_p}, d, q_{m_1,\ldots,m_q}\right) \times \left(P_{\lambda_1,\ldots,\lambda_P}, D, Q_{\tau_1,\ldots,\tau_Q}\right)_s$.

8.2.3 STARMA models

For a zero-mean stationary spatial time series, the model in Eq. (8.3) reduces to the following space–time autoregressive moving average STARMA$\left(p_{a_1,\ldots,a_p}, q_{m_1,\ldots,m_q}\right)$ model,

$$\mathbf{Z}_t = \sum_{k=1}^{p}\sum_{\ell=0}^{a_k}\phi_{k,\ell}\mathbf{W}^{(\ell)}\mathbf{Z}_{t-k} + \mathbf{a}_t - \sum_{k=1}^{q}\sum_{\ell=0}^{m_k}\theta_{k,\ell}\mathbf{W}^{(\ell)}\mathbf{a}_{t-k}, \tag{8.6}$$

where the zeros of $\det\left(\mathbf{I} - \sum_{k=1}^{p}\sum_{\ell=0}^{a_k}\phi_{k,\ell}\mathbf{W}^{(\ell)}B^k\right) = 0$ lie outside the unit circle.

The STARMA$\left(p_{a_1,\ldots,a_p}, q_{m_1,\ldots,m_q}\right)$ model becomes a space–time autoregressive (STAR$\left(p_{a_1,\ldots,a_p}\right)$) model when $q = 0$. It becomes a space–time moving average (STMA$\left(q_{m_1,\ldots,m_q}\right)$) model when $p = 0$. The STAR models were first introduced by Cliff and Ord (1975a, 1975b), and Martin and Oeppen (1975), and further extended to STARMA models by Pfeifer and Deutsch (1980a, b, c).

Since a stationary model can be approximated by an autoregressive model, because of its easier interpretation, the most widely used STARMA models in practice are STAR$\left(p_{a_1,\ldots,a_p}\right)$ models,

$$\mathbf{Z}_t = \sum_{k=1}^{p}\sum_{\ell=0}^{a_k}\phi_{k,\ell}\mathbf{W}^{(\ell)}\mathbf{Z}_{t-k} + \mathbf{a}_t, \tag{8.7}$$

where $\mathbf{Z}_t$ is a zero-mean stationary spatial time series or a proper transformed and differenced series of a nonstationary spatial time series.

Example 8.2 Let us consider a simple but very popular STAR(1_1) model

$$\mathbf{Z}_t = \phi_{1,0}\mathbf{Z}_{t-1} + \phi_{1,1}\mathbf{W}^{(1)}\mathbf{Z}_{t-1} + \mathbf{a}_t. \tag{8.8}$$

For a four-location series in a lot design shown in Figure 8.2, if we assign equal weights to locations sharing a border and 0 otherwise, then the weighting matrix will be given in Eq. (8.9).

1	2
3	4

Figure 8.2 A four-location lot.

$$\mathbf{W}^{(1)} = \begin{bmatrix} 0 & \frac{1}{2} & \frac{1}{2} & 0 \\ \frac{1}{2} & 0 & 0 & \frac{1}{2} \\ \frac{1}{2} & 0 & 0 & \frac{1}{2} \\ 0 & \frac{1}{2} & \frac{1}{2} & 0 \end{bmatrix}. \tag{8.9}$$

Note that the model in Eq. (8.8) with the weighting matrix in Eq. (8.9) can be rewritten as

$$\begin{bmatrix} Z_{1,t} \\ Z_{2,t} \\ Z_{3,t} \\ Z_{4,t} \end{bmatrix} = \begin{bmatrix} \phi_{1,0} & \frac{1}{2}\phi_{1,1} & \frac{1}{2}\phi_{1,1} & 0 \\ \frac{1}{2}\phi_{1,1} & \phi_{1,0} & 0 & \frac{1}{2}\phi_{1,1} \\ \frac{1}{2}\phi_{1,1} & 0 & \phi_{1,0} & \frac{1}{2}\phi_{1,1} \\ 0 & \frac{1}{2}\phi_{1,1} & \frac{1}{2}\phi_{1,1} & \phi_{1,0} \end{bmatrix} \begin{bmatrix} Z_{1,t-1} \\ Z_{2,t-1} \\ Z_{3,t-1} \\ Z_{4,t-1} \end{bmatrix} + \begin{bmatrix} a_{1,t} \\ a_{2,t} \\ a_{3,t} \\ a_{4,t} \end{bmatrix}, \tag{8.10}$$

which can be seen to be a VAR(1) model. In fact, it can be easily shown that the STARMA model in Eq. (8.6) is actually a special case of the VARMA model. Similarly, the STARIMA model in Eq. (8.3) is also a special case of the VARIMA model. ◁

8.2.4 ST-ACF and ST-PACF

Similar to the vector time series models, the important characteristics of STARMA models are shown through their space–time autocorrelation function (ST-ACF) and space–time partial autocorrelation function (ST-PACF).

Without loss of generality, we will assume a zero-mean spatial time series. To properly incorporate the space weighting matrices in the analysis, the ST-ACF for ℓ^{th} neighbors at time lag k is given by

$$\rho_\ell(k) = \frac{\gamma_{\ell,0}(k)}{\left[\gamma_{\ell,\ell}(0)\gamma_{0,0}(0)\right]^{1/2}}, \tag{8.11}$$

where the space–time autocovariance function between i^{th} and j^{th} neighbors at time lag k for the m-dimensional space–time process is defined by:

$$\gamma_{i,j}(k) = E\left[\frac{\left(\mathbf{W}^{(i)}\mathbf{Z}_{t-k}\right)'\left(\mathbf{W}^{(j)}\mathbf{Z}_t\right)}{m}\right]. \tag{8.12}$$

The sample ST-ACF, $\hat{\rho}_\ell(k)$, is thus given by

$$\hat{\rho}_\ell(k) = \frac{\sum_{t=k+1}^{n} (\mathbf{W}^{(\ell)}\mathbf{Z}_{t-k})'(\mathbf{Z}_t)}{\left[\sum_{t=1}^{n} (\mathbf{W}^{(\ell)}\mathbf{Z}_t)'(\mathbf{W}^{(\ell)}\mathbf{Z}_t) \sum_{t=1}^{n} (\mathbf{Z}_t)'(\mathbf{Z}_t)\right]^{1/2}}. \tag{8.13}$$

Similar to the ARMA models, we can use space–time analog of the Yule–Walker equations to derive the ST-PACF. Specifically, in Eq. (8.7), we replace $a_1, \ldots,$ and a_p by $\lambda = \max(a_1, \ldots, a_p)$. In other words, we consider the following STAR($p_{\lambda,\ldots,\lambda}$) model

$$\mathbf{Z}_t = \sum_{k=1}^{p} \sum_{\ell=0}^{\lambda} \phi_{k,\ell} \mathbf{W}^{(\ell)} \mathbf{Z}_{t-k} + \mathbf{a}_t. \tag{8.14}$$

Pre-multiplying both sides of Eq. (8.14) by $[\mathbf{W}^{(h)}\mathbf{Z}_{t-s}]'$ and taking expectation, we have the space–time analog of the Yule–Walker equations,

$$\gamma_{h,0}(s) = \sum_{k=1}^{p} \sum_{\ell=0}^{\lambda} \phi_{k,\ell} \gamma_{h,\ell}(s-k). \tag{8.15}$$

The ST-PACF is obtained as the last coefficient, $\phi_{p,\ell}$, for $\ell = 0, 1, \ldots, \lambda$, from the solution of the successively solving the system of equations for each $p = 1, 2, \ldots, s = p$, each set of $\ell = \{0\}$, $\{0, 1\}, \ldots, \{0, 1, \ldots \lambda\}$, and $h = \{0\}, \{0, 1\}, \ldots, \{0, 1, \ldots \lambda\}$.

As an illustration, let us first consider the computation of some specific STAR models.

Example 8.3 Let us consider the SP-PACF of the *STAR*(1_3) model. In this case, $p = 1$, and $\lambda = 3$. Hence, we will successively consider the following systems of equations:

1. $p = 1$, $s = 1$, $\ell = 0$, and $h = 0$, we have

$$\gamma_{0,0}(1) = \phi_{1,0}, \gamma_{0,0}(0),$$

which will lead to $\phi_{1,0}$;

2. $p = 1$, $s = 1$, $\ell = 0, 1$, $h = 0, 1$, and hence

$$\gamma_{0,0}(1) = \phi_{1,0}\gamma_{0,0}(0) + \phi_{1,1}\gamma_{0,1}(0),$$
$$\gamma_{1,0}(1) = \phi_{1,0}\gamma_{1,0}(0) + \phi_{1,1}\gamma_{1,1}(0),$$

which will lead to $\phi_{1,1}$;

3. $p = 1$, $s = 1$, $\ell = 0, 1, 2$, $h = 0, 1, 2$, and hence

$$\gamma_{0,0}(1) = \phi_{1,0}\gamma_{0,0}(0) + \phi_{1,1}\gamma_{0,1}(0) + \phi_{1,2}\gamma_{0,2}(0),$$
$$\gamma_{1,0}(1) = \phi_{1,0}\gamma_{1,0}(0) + \phi_{1,1}\gamma_{1,1}(0) + \phi_{1,2}\gamma_{1,2}(0),$$
$$\gamma_{2,0}(1) = \phi_{1,0}\gamma_{2,0}(0) + \phi_{1,1}\gamma_{2,1}(0) + \phi_{1,2}\gamma_{2,2}(0),$$

which will lead to $\phi_{1,2}$;

4. $p = 1$, $s = 1$, $\ell = 0, 1, 2, 3$, $h = 0, 1, 2, 3$, and hence

$$\gamma_{0,0}(1) = \phi_{1,0}\gamma_{0,0}(0) + \phi_{1,1}\gamma_{0,1}(0) + \phi_{1,2}\gamma_{0,2}(0) + \phi_{1,3}\gamma_{0,3}(0),$$
$$\gamma_{1,0}(1) = \phi_{1,0}\gamma_{1,0}(0) + \phi_{1,1}\gamma_{1,1}(0) + \phi_{1,2}\gamma_{1,2}(0) + \phi_{1,3}\gamma_{1,3}(0),$$
$$\gamma_{2,0}(1) = \phi_{1,0}\gamma_{2,0}(0) + \phi_{1,1}\gamma_{2,1}(0) + \phi_{1,2}\gamma_{2,2}(0) + \phi_{1,3}\gamma_{2,3}(0),$$
$$\gamma_{3,0}(1) = \phi_{1,0}\gamma_{3,0}(0) + \phi_{1,1}\gamma_{3,1}(0) + \phi_{1,2}\gamma_{3,2}(0) + \phi_{1,3}\gamma_{3,3}(0),$$

which will lead to $\phi_{1,3}$.

Hence, ST-PACF at time lag 1 and space lag 1, 2, and 3 will be: $\phi_{1,\,0}, \phi_{1,\,1}, \phi_{1,\,2}, \phi_{1,\,3}$, and for the $STAR(1_3)$ model, $\phi_{1,j} = 0$ with $j > 3$.

Example 8.4 Now let us consider the ST-PACF of the $STAR(2_{3,\,2})$ model. In this case, $p = 2$, and $\lambda = 3$, and $\ell = 0, 1, 2, 3$. The solution of the successive systems of equations becomes:

1. $p = 1$, $s = 1$, $\ell = 0$, $h = 0$, and hence

$$\gamma_{0,0}(1) = \phi_{1,0}\gamma_{0,0}(0)$$

   which will lead to $\phi_{1,0}$;

2. $p = 1$, $s = 1$, $\ell = 0, 1$, $h = 0, 1$, and hence

$$\gamma_{0,0}(1) = \phi_{1,0}\gamma_{0,0}(0) + \phi_{1,1}\gamma_{0,1}(0)$$
$$\gamma_{1,0}(1) = \phi_{1,0}\gamma_{1,0}(0) + \phi_{1,1}\gamma_{1,1}(0)$$

   which will lead to $\phi_{1,1}$;

3. $p = 1$, $s = 1$, $\ell = 0, 1, 2$, $h = 0, 1, 2$, and hence

$$\gamma_{0,0}(1) = \phi_{1,0}\gamma_{0,0}(0) + \phi_{1,1}\gamma_{0,1}(0) + \phi_{1,2}\gamma_{0,2}(0),$$
$$\gamma_{1,0}(1) = \phi_{1,0}\gamma_{1,0}(0) + \phi_{1,1}\gamma_{1,1}(0) + \phi_{1,2}\gamma_{1,2}(0),$$
$$\gamma_{2,0}(1) = \phi_{1,0}\gamma_{2,0}(0) + \phi_{1,1}\gamma_{2,1}(0) + \phi_{1,2}\gamma_{2,2}(0),$$

   which will lead to $\phi_{1,2}$;

4. $p = 1$, $s = 1$, $\ell = 0, 1, 2, 3$, $h = 0, 1, 2, 3$, and hence

$$\gamma_{0,0}(1) = \phi_{1,0}\gamma_{0,0}(0) + \phi_{1,1}\gamma_{0,1}(0) + \phi_{1,2}\gamma_{0,2}(0) + \phi_{1,3}\gamma_{0,3}(0),$$
$$\gamma_{1,0}(1) = \phi_{1,0}\gamma_{1,0}(0) + \phi_{1,1}\gamma_{1,1}(0) + \phi_{1,2}\gamma_{1,2}(0) + \phi_{1,3}\gamma_{1,3}(0),$$
$$\gamma_{2,0}(1) = \phi_{1,0}\gamma_{2,0}(0) + \phi_{1,1}\gamma_{2,1}(0) + \phi_{1,2}\gamma_{2,2}(0) + \phi_{1,3}\gamma_{2,3}(0),$$
$$\gamma_{3,0}(1) = \phi_{1,0}\gamma_{3,0}(0) + \phi_{1,1}\gamma_{3,1}(0) + \phi_{1,2}\gamma_{3,2}(0) + \phi_{1,3}\gamma_{3,3}(0),$$

   which will lead to $\phi_{1,3}$.

5. $p = 2$, $s = 2$, $\ell = 0$, $h = 0$, and hence

$$\gamma_{0,0}(2) = \phi_{1,0}\gamma_{0,0}(1) + \phi_{2,0}\gamma_{0,0}(0).$$

Since $\phi_{1,0}$ is known from (1), this leads to $\phi_{2,0}$;

6. $p = 2$, $s = 2$, $\ell = 0, 1$, $h = 0, 1$, and hence

$$\gamma_{0,0}(2) = \phi_{1,0}\gamma_{0,0}(1) + \phi_{1,1}\gamma_{0,1}(1) + \phi_{2,0}\gamma_{0,0}(0) + \phi_{2,1}\gamma_{0,1}(0),$$
$$\gamma_{1,0}(2) = \phi_{1,0}\gamma_{1,0}(1) + \phi_{1,1}\gamma_{1,1}(1) + \phi_{2,0}\gamma_{1,0}(0) + \phi_{2,1}\gamma_{1,1}(0),$$

Since $\phi_{1,0}$ and $\phi_{1,1}$ are known from (1) and (2), this leads to $\phi_{2,1}$;

7. $p = 2$, $s = 2$, $\ell = 0, 1, 2$, $h = 0, 1, 2$, and hence

$$\gamma_{0,0}(2) = \phi_{1,0}\gamma_{0,0}(1) + \phi_{1,1}\gamma_{0,1}(1) + \phi_{1,2}\gamma_{0,2}(1) + \phi_{2,0}\gamma_{0,0}(0) + \phi_{2,1}\gamma_{0,1}(0) + \phi_{2,2}\gamma_{0,2}(0),$$
$$\gamma_{1,0}(2) = \phi_{1,0}\gamma_{1,0}(1) + \phi_{1,1}\gamma_{1,1}(1) + \phi_{1,2}\gamma_{1,2}(1) + \phi_{2,0}\gamma_{1,0}(0) + \phi_{2,1}\gamma_{1,1}(0) + \phi_{2,2}\gamma_{1,2}(0),$$
$$\gamma_{2,0}(2) = \phi_{1,0}\gamma_{2,0}(1) + \phi_{1,1}\gamma_{2,1}(1) + \phi_{1,2}\gamma_{2,2}(1) + \phi_{2,0}\gamma_{2,0}(0) + \phi_{2,1}\gamma_{2,1}(0) + \phi_{2,2}\gamma_{2,2}(0),$$

Since $\phi_{1,\,0}$, $\phi_{1,\,1}$ and $\phi_{1,\,2}$ are known from (1), (2), and (3), this leads to $\phi_{2,2}$. For the *STAR* $(2_{3,2})$ model, we have $\phi_{1,0}$, $\phi_{1,1}$, $\phi_{1,2}$, $\phi_{1,3}$, with $\phi_{1,j} = 0$, for $j > 3$, and $\phi_{2,0}$, $\phi_{2,1}$, $\phi_{2,2}$, with $\phi_{2,j} = 0$, for $j > 2$.

The sample ST-PACF is the solution when the sample space–time autocovariances, $\hat{\gamma}_{h,\ell}$, are used in Eq. (8.15). Following Wei (2006, p. 403), this sample ST-PACF can also be obtained from the last estimate $\hat{\phi}_{p,\ell}$ by successively fitting the $\text{STAR}\left(p_{a_1,a_2,\ldots,a_p}\right)$ model in Eq. (8.7) for $p = 1, 2, \ldots$, when ℓ varies from 0, 1, … to a, where a is the maximum spatial order to be considered.

In practice, λ and a are normally unknown, and their values are chosen based on a model builder's decision. However, in most cases, the maximum space order is normally no more than 4. For further references, we refer readers to Pfeifer and Deutsch (1980a).

Since these functions involve both time and space, to make the identification easier, we will consider the following two-way tables and graphs for the sample SP-ACF and SP-PACF.

Sample ST-ACF Table

Time lag/Space Lag	0	1	2	3	…
1	$\hat{\rho}_0(1)$	$\hat{\rho}_1(1)$	$\hat{\rho}_2(1)$	$\hat{\rho}_3(1)$	…
2	$\hat{\rho}_0(2)$	$\hat{\rho}_1(2)$	$\hat{\rho}_2(2)$	$\hat{\rho}_3(2)$	…
3	$\hat{\rho}_0(3)$	$\hat{\rho}_1(3)$	$\hat{\rho}_2(3)$	$\hat{\rho}_3(3)$	…
4	$\hat{\rho}_0(4)$	$\hat{\rho}_1(4)$	$\hat{\rho}_2(4)$	$\hat{\rho}_3(4)$	…
.	.	.	.	.	.
.	.	.	.	.	.

Sample ST-ACF Graph

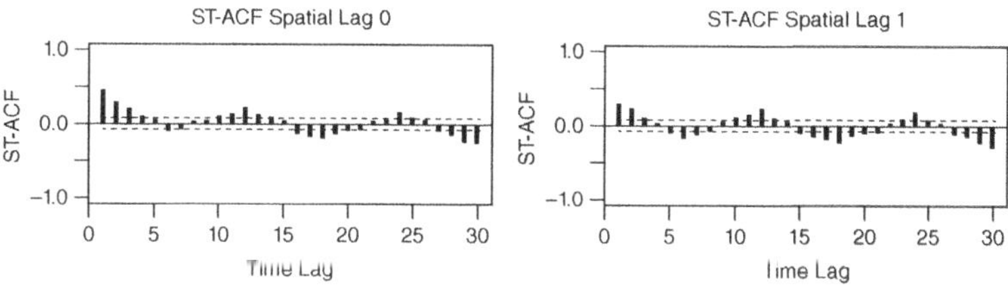

Sample ST-PACF Table

Time lag/Space Lag	0	1	2	3	...
1	$\hat{\phi}_{1,0}$	$\hat{\phi}_{1,1}$	$\hat{\phi}_{1,2}$	$\hat{\phi}_{1,3}$	...
2	$\hat{\phi}_{2,0}$	$\hat{\phi}_{2,1}$	$\hat{\phi}_{2,2}$	$\hat{\phi}_{2,3}$	...
3	$\hat{\phi}_{3,0}$	$\hat{\phi}_{3,1}$	$\hat{\phi}_{3,2}$	$\hat{\phi}_{3,3}$	...
4	$\hat{\phi}_{4,0}$	$\hat{\phi}_{4,1}$	$\hat{\phi}_{4,2}$	$\hat{\phi}_{4,3}$	...
.	.	.	.	.	.
.	.	.	.	.	.

Sample ST-PACF Graph

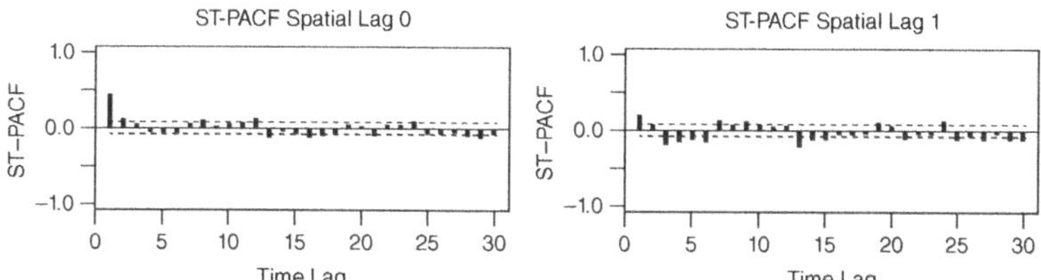

Under the assumption that the $\mathbf{a}_t$ are i.i.d. random vectors with $E(\mathbf{a}_t) = \mathbf{0}$ and variance–covariance matrix $\sigma^2\mathbf{I}$, the variance of the sample ST-ACF and ST-PACF can be approximated by

$$\operatorname{Var}(\hat{\rho}_{\ell}(s)) \approx \frac{1}{m(n-s)}, \tag{8.16}$$

and

$$\operatorname{Var}\left(\hat{\phi}_{p,a}(s)\right) \approx \frac{1}{m(n-s)}, \tag{8.17}$$

where m is the number of locations and n is the length of the time series in each location.

8.3 Generalized space–time autoregressive integrated moving average (GSTARIMA) models

It should be pointed out that the parameters $\phi_{k,\ell}$ and $\theta_{k,\ell}$ in the STARMA or STARIMA models discussed in Section 8.2 are all scalar, which is the same for all locations. To remove this undesirable restriction, the generalized STAR models, denoted by GSTAR models, were first introduced by Borovkova, Lopuhaa, and Ruchjana (2002), and later extended to the following generalized STARMA, denoted by GSTARMA models, by Di Giacinto (2006),

$$\mathbf{Z}_t = \sum_{k=1}^{p}\sum_{\ell=0}^{a_k} \mathbf{\Phi}_{k,\ell}\mathbf{W}^{(\ell)}\mathbf{Z}_{t-k} + \mathbf{a}_t - \sum_{k=1}^{q}\sum_{\ell=0}^{m_k} \mathbf{\Theta}_{k,\ell}\mathbf{W}^{(\ell)}\mathbf{a}_{t-k}, \tag{8.18}$$

where the assumption of $\mathbf{a}_t$ is the same as Eq. (8.3). However, the time lag k and space lag ℓ parameters, $\mathbf{\Phi}_{k,\ell}$ and $\mathbf{\Theta}_{k,\ell}$, are diagonal matrices instead of scalars. These parameters can vary depending on locations and hence offer more flexibility.

More generally, the GSTARMA$\left(p_{a_1,\ldots,a_p}, q_{m_1,\ldots,m_q}\right)$ model in Eq. (8.18) can be extended to generalized GSTARIMA$\left(p_{a_1,\ldots,a_p}, d, q_{m_1,\ldots,m_q}\right)$ models,

$$\mathbf{\Phi}_p(B)\mathbf{D}(B)\mathbf{Z}_t = \mathbf{\theta}_0 + \mathbf{\Theta}_q(B)\mathbf{a}_t, \tag{8.19}$$

where

$$\mathbf{\Phi}_p(B) = \mathbf{I} - \sum_{k=1}^{p}\sum_{\ell=0}^{a_k} \mathbf{\Phi}_{k,\ell}\mathbf{W}^{(\ell)}B^k,$$

and

$$\mathbf{\Theta}_q(B) = \mathbf{I} - \sum_{k=1}^{q}\sum_{\ell=0}^{m_k} \mathbf{\Theta}_{k,\ell}\mathbf{W}^{(\ell)}B^k.$$

To include a seasonal phenomenon, we can also extend the seasonal STARIMA $\left(p_{a_1,\ldots,a_p}, d, q_{m_1,\ldots,m_q}\right) \times \left(P_{\lambda_1,\ldots,\lambda_P}, D, Q_{\tau_1,\ldots,\tau_Q}\right)_s$ model in Eq. (8.5) to the following generalized seasonal GSTARIMA$\left(p_{a_1,\ldots,a_p}, d, q_{m_1,\ldots,m_q}\right) \times \left(P_{\lambda_1,\ldots,\lambda_P}, D, Q_{\tau_1,\ldots,\tau_Q}\right)_s$ model,

$$\mathbf{\alpha}_P(B^s)\mathbf{\Phi}_p(B)\mathbf{D}(B^s)\mathbf{D}(B)\mathbf{Z}_t = \mathbf{\theta}_0 + \mathbf{\beta}_Q(B^s)\mathbf{\Theta}_q(B)\mathbf{a}_t, \tag{8.20}$$

where

$$\mathbf{\alpha}_P(B^s) = \mathbf{I} - \sum_{k=1}^{P}\sum_{\ell=0}^{\lambda_k} \mathbf{\alpha}_{k,\ell}\mathbf{W}^{(\ell)}B^{ks},$$

and

$$\mathbf{\beta}_Q(B^s) = \mathbf{I} - \sum_{k=1}^{Q}\sum_{\ell=0}^{\tau_k} \mathbf{\beta}_{k,\ell}\mathbf{W}^{(\ell)}B^{ks}.$$

The coefficients, $\mathbf{\alpha}_{k,\ell}$ and $\mathbf{\beta}_{k,\ell}$, are $(m \times m)$ diagonal matrices of parameters at time lag k and space lag ℓ.

8.4 Iterative model building of STARMA and GSTARMA models

As shown in Example 8.2, a STAR(1_1) model in Eq. (8.8) is equivalent to a first order vector autoregressive VAR(1) model in Eq. (8.10), that is

$$\mathbf{Z}_t = \mathbf{\Phi}_1 \mathbf{Z}_{t-1} + \mathbf{a}_t.$$

Thus, one can treat STARMA models as special cases of the vector VARMA models, where the diagonal elements of the $m \times m$ autoregressive and moving average parameter matrices are assumed to be equal because the m series represent a single process operating at different locations and the off-diagonal elements are assumed to be a linear combination of the $\mathbf{W}^{(\ell)}$ weighting matrices. Obviously, GSTARMA models are also special cases of vector VARMA models.

Similar to the vector VARMA models, STARMA and GSTARMA models are characterized by the ST-ACF and ST-PACF. For a STAR or a GSTAR model, its ST-PACF cuts off after certain temporal and space lags. Similarly, for a STMA or a GSTMA model, its ST-ACF cuts off after certain temporal and space lags. If neither of these ST-ACF and ST-PACF cuts off over time and space, it will be a STARMA or a GSTARMA model.

Once a STARMA model is identified, the parameters are estimated by minimizing

$$\mathbf{S}(\mathbf{\Phi}, \mathbf{\Theta}) = \sum_{t=p+1}^{n} \mathbf{a}_t \mathbf{a}_t', \tag{8.21}$$

using various nonlinear optimization techniques, which is equivalent to maximize the likelihood function of

$$\mathbf{a}_t = \mathbf{Z}_t - \sum_{k=1}^{p} \sum_{\ell=0}^{a_k} \phi_{k,\ell} \mathbf{W}^{(\ell)} \mathbf{Z}_{t-k} + \sum_{k=1}^{q} \sum_{\ell=0}^{m_k} \phi_{k,\ell} \mathbf{W}^{(\ell)} \mathbf{a}_{t-k}, \tag{8.22}$$

discussed in Chapter 2. The insignificant parameters should be removed and a simpler model be re-estimated until all parameters are statistically significant.

Before a model can be used for forecasting and making inferences, a diagnostic check must be carefully done to make sure that the residuals are approximately white noise. Specifically, we can calculate the residual sample ST-ACF and ST-PACF and compare with their estimated standard deviation. If they are not white noise, based on the results of the residual analysis, we need to re-identify a new model, re-estimate the parameters, and perform a diagnostic check, which constitutes an iterative process.

8.5 Empirical examples

8.5.1 Vehicular theft data

Example 8.5 From January 2006 through December 2014, over 750,000 crimes were reported among the 21 police districts in Philadelphia, Pennsylvania, shown on the map in Figure 8.3. The most common crimes reported were thefts and burglaries. The total crime followed a seasonal pattern with more crimes in the summer than in the winter.

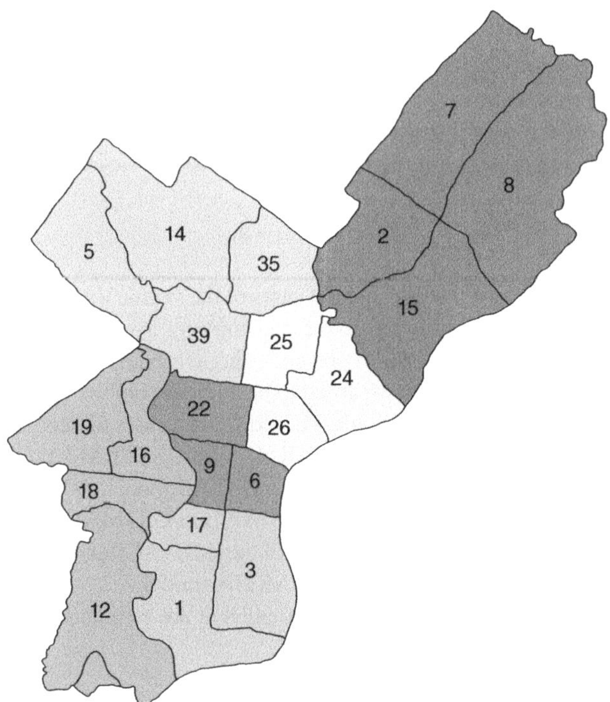

Figure 8.3 The map of 21 police districts of Philadelphia, PA, USA.

For this analysis, we consider the monthly total of vehicle thefts recorded at the 21 police districts in Philadelphia from January 2006 to December 2014. The data set consists of $m = 21$ dimensions corresponding to the 21 police districts and is listed as WW8a in the Data Appendix. The monthly total of vehicle thefts from January 2006 to April 2014 was used for model fitting with a total of $n = 100$ observations. The monthly total of the thefts from May 2014 to December 2014 was held out for the forecast comparison. Figure 8.4 displays these raw counts for the in-sample time points.

Visually, each time series shows a constant mean but varies among districts. For the variance, we compare several values for the Box-Cox transformation, and $\lambda = 0.5$ is chosen for most locations to stabilize the variance. Thus, the square root of each series was performed. Neither regular nor seasonal differencing was required for any of the 21 series. After being square-root-transformed and mean centered, the resulting 21 series for $t = 1, 2, \ldots, 100$, is given in Figure 8.5.

Bordering districts are defined as first neighbors, and the weighting matrix for $\mathbf{W}^{(1)}$ is given in Table 8.2. Models involving additional spatial lags were considered but the parameter estimates were not significant different from 0, and hence the weighting matrices $\mathbf{W}^{(\ell)}$ for $\ell > 1$ are not given.

The patterns in Figure 8.6 show that the sample ST-ACFs exponentially decay at both regular time lags, seasonal time lags of multiple of 12, and that ST-PACFs are significant only at time lags 2, 12, and 24, and spatial lag at 1. For a monthly series of seasonal period of 12, we

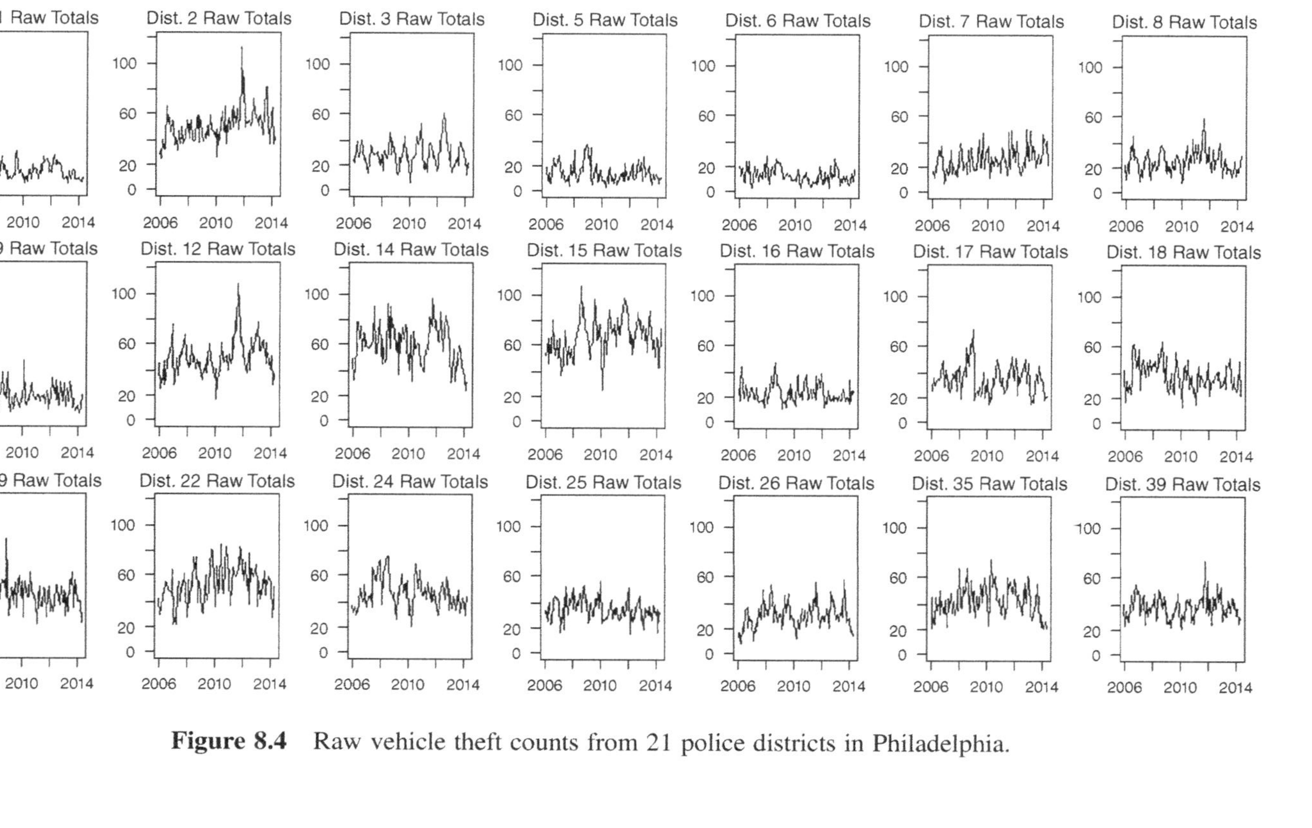

Figure 8.4 Raw vehicle theft counts from 21 police districts in Philadelphia.

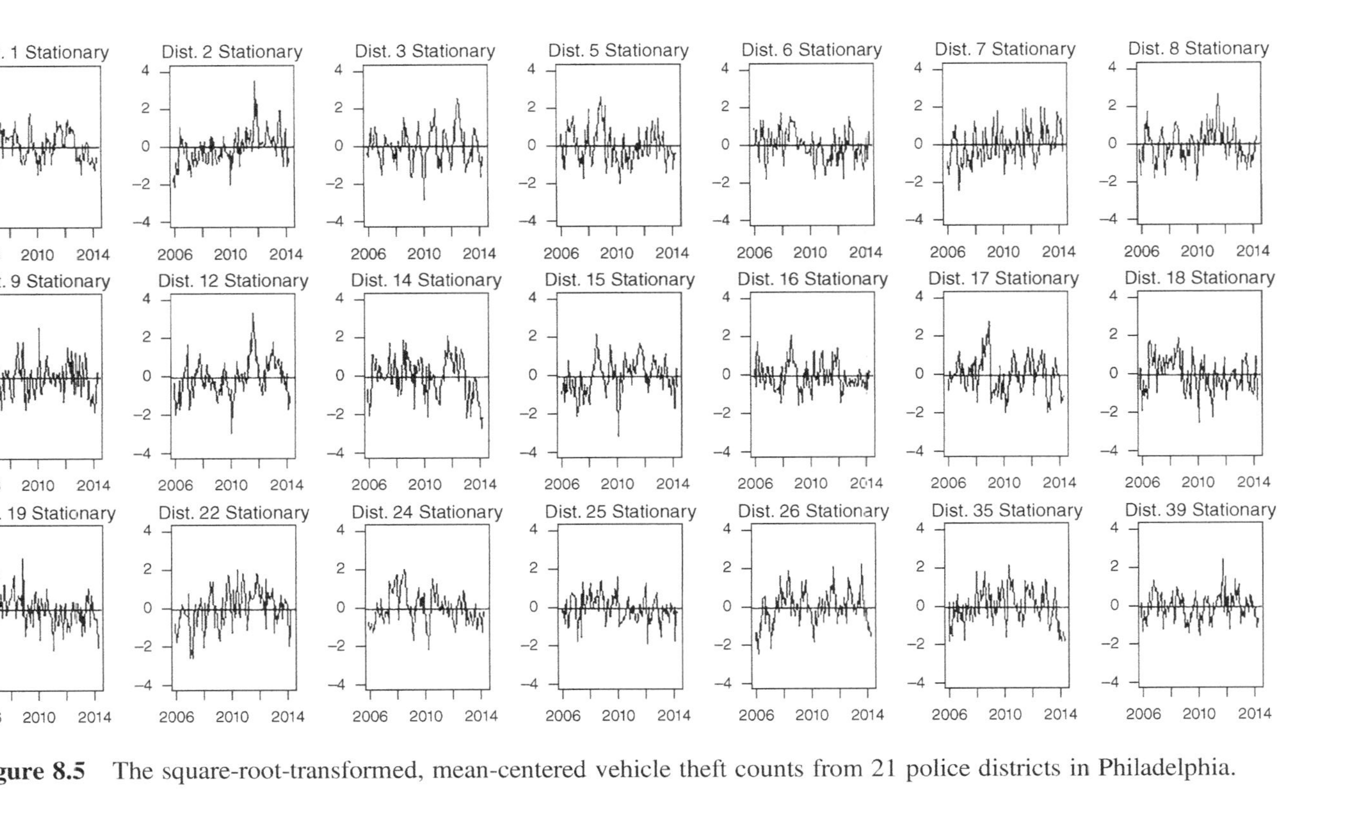

Figure 8.5 The square-root-transformed, mean-centered vehicle theft counts from 21 police districts in Philadelphia.

Table 8.2 The weighting matrix **W**(1) for the 21 police districts in Philadelphia, PA, USA.

$W_{z,1}$	D1	D2	D3	D5	D6	D7	D8	D9	D12	D14	D15	D16	D17	D18	D19	D22	D24	D25	D26	D35	D39
D1	0	0	1/3	0	0	0	0	0	1/3	0	0	0	1/3	0	0	0	0	0	0	0	0
D2	0	0	0	0	0	1/3	0	0	0	0	1/3	0	0	0	0	0	0	0	0	1/3	0
D3	1/3	0	0	0	1/3	0	0	0	0	0	0	0	1/3	0	0	0	0	0	0	0	0
D5	0	0	0	0	0	0	0	0	0	1/2	0	0	0	0	0	0	0	0	0	0	1/2
D6	0	0	1/3	0	0	0	0	1/3	0	0	0	0	0	0	0	0	0	0	1/3	0	0
D7	0	1/2	0	0	0	0	1/2	0	0	0	0	0	0	0	0	0	0	0	0	0	0
D8	0	0	0	0	0	1/2	0	0	0	0	1/2	0	0	0	0	0	0	0	0	0	0
D9	0	0	0	0	1/4	0	0	0	0	0	0	1/4	1/4	0	0	1/4	0	0	0	0	0
D12	1/2	0	0	0	0	0	0	0	0	0	0	0	0	1/2	0	0	0	0	0	0	0
D14	0	0	0	1/3	0	0	0	0	0	0	0	0	0	0	0	0	0	0	0	1/3	1/3
D15	0	1/3	0	0	0	0	1/3	0		0	0	0	0	0	0	0	1/3	0	0	0	0
D16	0	0	0	0	0	0	0	1/5	0	0	0	0	0	1/5	1/5	1/5	0	0	0	0	1/5
D17	1/4	0	1/4	0	0	0	0	1/4	0	0	0	0	0	1/4	0	0	0	0	0	0	0
D18	0	0	0	0	0	0	0	0	1/4	0	0	1/4	1/4	0	1/4	0	0	0	0	0	0
D19	0	0	0	0	0	0	0	0	0	0	0	1/2	0	1/2	0	0	0	0	0	0	0
D22	0	0	0	0	0	0	0	1/4	0	0	0	1/4	0	0	0	0	0	0	1/4	1/4	1/4
D24	0	0	0	0	0	0	0	0	0	0	1/3	0	0	0	0	0	0	1/3	1/3	0	0
D25	0	0	0	0	0	0	0	0	0	0	0	0	0	0	0	0	1/4	0	1/4	1/4	1/4
D26	0	0	0	0	14	0	0	0	0	0	0	0	0	0	0	1/4	1/4	1/4	0	0	0
D35	0	1/4	0	0	0	0	0	0	0	1/4	0	0	0	0	0	0	0	1/4	0	0	1/4
D39	0	0	0	1/6	0	0	0	0	0	1/6	0	1/6	0	0	0	1/6	0	1/6	0	1/6	0

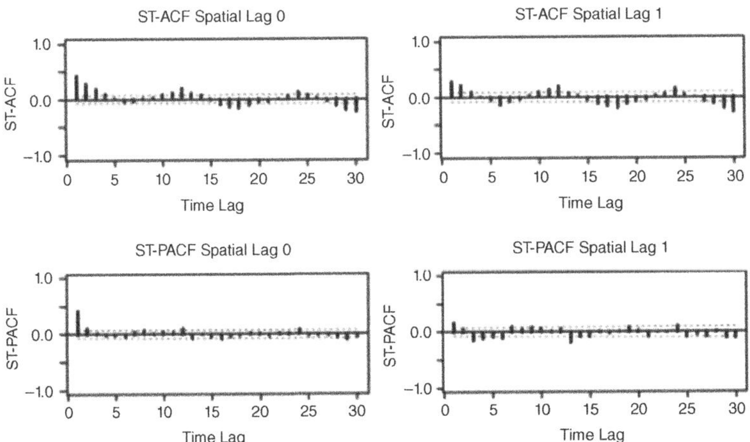

Figure 8.6 The sample ST-ACFs and ST-PACFs of the data $\mathbf{Z}_t$

tentatively choose a STAR$(2_{1,1}) \times (2_{1,1})_{12}$ as its possible underlying model. The estimation result is given as follows:

$$\left(\mathbf{I}_{21} - \underset{(.031)}{0.130}\mathbf{W}^{(0)}B^{12} - \underset{(.064)}{0.146}\mathbf{W}^{(1)}B^{12} - \underset{(.033)}{0.060}\mathbf{W}^{(0)}B^{24} - \underset{(.068)}{0.152}\mathbf{W}^{(1)}B^{24}\right)$$
$$\left(\mathbf{I}_{21} - \underset{(.0291)}{0.337}\mathbf{W}^{(0)}B - \underset{(.060)}{0.133}\mathbf{W}^{(1)}B - \underset{(.029)}{0.106}\mathbf{W}^{(0)}B^2 - \underset{(.061)}{0.089}\mathbf{W}^{(1)}B^2\right)\mathbf{Z}_t = \mathbf{a}_t. \tag{8.23}$$

Since the estimate of $\phi_{2,1}$ is only 1.46 standard errors away from 0, we re-fit a STAR $(2_{1,0}) \times (2_{1,1})_{12}$ model with the following result,

$$\left(\mathbf{I}_{21} - \underset{(.031)}{0.128}\mathbf{W}^{(0)}B^{12} - \underset{(.063)}{0.145}\mathbf{W}^{(1)}B^{12} - \underset{(.033)}{0.058}\mathbf{W}^{(0)}B^{24} - \underset{(.067)}{0.149}\mathbf{W}^{(1)}B^{24}\right)$$
$$\left(\mathbf{I}_{21} - \underset{(.029)}{0.336}\mathbf{W}^{(0)}B - \underset{(.056)}{0.171}\mathbf{W}^{(1)}B - \underset{(.034)}{0.121}\mathbf{W}^{(0)}B^2\right)\mathbf{Z}_t = \mathbf{a}_t. \tag{8.24}$$

Now, the estimate of $\alpha_{2,0}$ is insignificant, so we reduce it to the following STAR$(2_{1,0}) \times (2_{0,1})_{12}$ model,

$$\left(\mathbf{I}_{21} - \underset{(.030)}{0.134}\mathbf{W}^{(0)}B^{12} - \underset{(.064)}{0.148}\mathbf{W}^{(1)}B^{12} - \underset{(.068)}{0.177}\mathbf{W}^{(1)}B^{24}\right)$$
$$\left(\mathbf{I}_{21} - \underset{(.029)}{0.337}\mathbf{W}^{(0)}B - \underset{(.056)}{0.170}\mathbf{W}^{(1)}B - \underset{(.034)}{0.120}\mathbf{W}^{(0)}B^2\right)\mathbf{Z}_t = \mathbf{a}_t. \tag{8.25}$$

Now all estimates are significant. In addition, the white noise pattern of the ST-ACFs and ST-PACFs of residuals from the fitted model shown in Figure 8.7 indicates that the model is adequate.

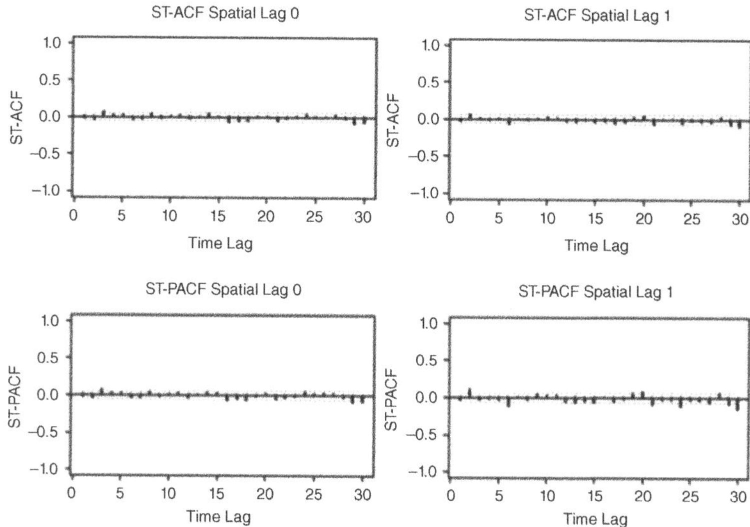

Figure 8.7 The sample ST-ACFs and ST-PACFs of the residuals from the fitted STAR $(2_{1,0}) \times (2_{0,1})_{12}$ model.

8.5.2 The annual U.S. labor force count

Example 8.6 The United States Department of Labor provides dozens of variables through the Bureau of Labor Statistics (BLS), many of which could be fitted with various STARMA models. We select non-seasonally adjusted labor force count for the 48 contiguous states and Washington D.C., for a total of $m = 49$ locations. The data are annual from 1976 through 2014. This constitutes our raw data, which is listed as WW8b in the Data Appendix.

To arrive at the variable $\mathbf{Z}_t$, we already tested and determined that most of the individual time series variables were variance stationary, but appear to have a single unit root. Thus, we took the difference of the counts, and then mean centered the variable for each location. This gives us our zero-mean variable $\mathbf{Z}_t$, with dimension $m = 49$ and time points $n = 39$. As a side note, one reason for a STARMA-type model to be beneficial for this data is that a VAR(1) model would require $49 \times 49 = 2401$ parameters, which are more than the number of observations in the data set, and makes the regular VAR fitting impossible. However, the same first order STAR(1_1) model involves only two parameters.

First, we construct the weighting matrix $\mathbf{W}^{(1)}$ according to the number of bordering states (first neighbors) each state has. Though too large to include for each state in this example, the number of border states ranges from 1 (Maine) to 8 (Missouri and Tennessee). And although not shown, we also construct $\mathbf{W}^{(2)}$ and $\mathbf{W}^{(3)}$ regarding second and third neighbors. Second neighbors are states that do not border each other but have at least one first neighbor in common. These terms are not needed for the chosen STAR(1_1) model in this example but are included to see if the spatial order is indeed only 1. Actually, these matrices allow us to construct the sample ST-ACF and ST-PACF, as shown in Tables 8.3 and 8.4.

The sample ST-ACF decays either exponentially or in a sinusoidal fashion. In addition, the sample ST-PACFs cut-off after time lag 1 and spatial lag 1, indicating a STAR(1_1) model.

Table 8.3 Sample ST-ACF of $\mathbf{Z}_t$.

k	$\hat{\rho}_{z,0}(k)$		$\hat{\rho}_{z,1}(k)$		$\hat{\rho}_{z,2}(k)$		$\hat{\rho}_{z,3}(k)$	
1	0.480	(0.081)	0.220	(0.048)	0.192	(0.046)	−0.051	(0.041)
2	0.210	(0.082)	0.144	(0.049)	0.117	(0.047)	0.024	(0.042)
3	0.051	(0.084)	0.036	(0.050)	0.037	(0.047)	0.052	(0.042)
4	−0.052	(0.085)	0.006	(0.051)	0.001	(0.048)	0.066	(0.043)
5	−0.162	(0.086)	−0.012	(0.051)	0.006	(0.049)	0.125	(0.043)
6	−0.149	(0.087)	−0.006	(0.052)	−0.001	(0.050)	0.110	(0.044)
7	−0.018	(0.089)	0.009	(0.053)	0.047	(0.050)	0.092	(0.045)
8	0.089	(0.090)	0.045	(0.054)	0.072	(0.051)	0.043	(0.046)
9	0.055	(0.092)	0.061	(0.055)	0.061	(0.052)	0.060	(0.046)
10	0.206	(0.093)	0.072	(0.056)	0.071	(0.053)	−0.016	(0.047)

Table 8.4 Sample ST-PACF of $\mathbf{Z}_t$.

k	$\hat{\phi}_{z,0}(k)$		$\hat{\phi}_{z,1}(k)$		$\hat{\phi}_{z,2}(k)$		$\hat{\phi}_{z,3}(k)$	
1	0.480	(0.081)	0.111	(0.048)	0.087	(0.046)	−0.024	(0.041)
2	−0.034	(0.082)	0.033	(0.049)	−0.028	(0.047)	0.077	(0.042)
3	−0.058	(0.084)	−0.100	(0.050)	−0.032	(0.047)	0.053	(0.042)
4	−0.065	(0.085)	0.063	(0.051)	−0.008	(0.048)	0.035	(0.043)
5	−0.131	(0.086)	0.053	(0.051)	0.118	(0.049)	0.143	(0.043)
6	−0.026	(0.087)	0.042	(0.052)	−0.046	(0.050)	0.000	(0.044)
7	0.099	(0.089)	−0.002	(0.053)	0.114	(0.050)	0.073	(0.045)
8	0.078	(0.090)	0.016	(0.054)	0.001	(0.051)	−0.004	(0.046)
9	−0.067	(0.092)	0.025	(0.055)	−0.034	(0.052)	0.095	(0.046)
10	0.207	(0.093)	−0.059	(0.056)	0.012	(0.053)	−0.069	(0.047)

Table 8.5 Sample ST-ACF of residuals.

k	$\hat{\rho}_{z,0}(k)$		$\hat{\rho}_{z,1}(k)$		$\hat{\rho}_{z,2}(k)$		$\hat{\rho}_{z,3}(k)$	
1	0.011	(0.074)	−0.007	(0.039)	0.035	(0.035)	−0.014	(0.036)
2	0.008	(0.075)	0.050	(0.040)	0.034	(0.035)	0.029	(0.036)
3	−0.030	(0.076)	−0.047	(0.040)	−0.027	(0.036)	0.022	(0.037)
4	−0.017	(0.077)	−0.001	(0.041)	−0.027	(0.036)	−0.009	(0.037)
5	−0.134	(0.078)	−0.014	(0.041)	0.014	(0.037)	0.097	(0.038)
6	−0.107	(0.079)	0.011	(0.042)	−0.022	(0.037)	0.043	(0.039)
7	0.007	(0.081)	−0.010	(0.043)	0.030	(0.038)	0.051	(0.039)
8	0.117	(0.082)	0.039	(0.043)	0.051	(0.039)	−0.024	(0.040)
9	−0.097	(0.083)	−0.002	(0.044)	−0.002	(0.039)	0.071	(0.041)
10	0.164	(0.085)	0.014	(0.045)	0.009	(0.040)	−0.051	(0.041)

Table 8.6 Sample ST-PACF of residuals.

k	$\hat{\phi}_{z,0}(k)$		$\hat{\phi}_{z,1}(k)$		$\hat{\phi}_{z,2}(k)$		$\hat{\phi}_{z,3}(k)$	
1	0.011	(0.074)	−0.019	(0.039)	0.099	(0.035)	−0.055	(0.036)
2	0.006	(0.075)	0.093	(0.040)	0.033	(0.035)	0.027	(0.036)
3	−0.031	(0.076)	−0.080	(0.040)	−0.009	(0.036)	0.050	(0.037)
4	−0.020	(0.077)	0.007	(0.041)	−0.066	(0.036)	−0.015	(0.037)
5	0.128	(0.078)	0.012	(0.041)	0.103	(0.037)	0.135	(0.038)
6	−0.109	(0.079)	0.065	(0.042)	−0.022	(0.037)	0.035	(0.039)
7	0.001	(0.081)	−0.022	(0.043)	0.100	(0.038)	0.073	(0.039)
8	0.107	(0.082)	0.020	(0.043)	0.061	(0.039)	0.002	(0.040)
9	−0.112	(0.083)	0.039	(0.044)	−0.005	(0.039)	0.110	(0.041)
10	0.144	(0.085)	−0.086	(0.045)	−0.009	(0.040)	0.000	(0.041)

Estimation through minimizing the trace of the error covariance matrix (the determinant was not easily worked with for such a large matrix) leads to the following STAR(1_1) model:

$$\underset{(49\times 1)}{\mathbf{Z}_t} = \underset{(0.083)}{0.457}\,\mathbf{Z}_{t-1} + \underset{(0.073)}{0.120}\,\mathbf{W}^{(1)}\mathbf{Z}_{t-1} + \underset{(49\times 1)}{\mathbf{a}_t}\,. \tag{8.26}$$

Both parameters are significant, and almost all the sample ST-ACF and ST-PACF of the model residual are small or at least within two standard errors of 0 as shown in Tables 8.5 and 8.6, indicating an adequate model.

For more related discussions of Examples 8.5 and 8.6, we refer readers to a Ph.D. dissertation by Gehman (2016), and papers by Gehman and Wei (2014, 2015a, b).

8.5.3 U.S. yearly sexually transmitted disease data

Example 8.7 In this example, we will consider a data set that contains yearly STD morbidity rates reported to National Center for HIV/AIDS, viral Hepatitis, STD, and TB Prevention (NCHHSTP), Center for HIV, and Centers for Disease Control and Prevention (CDC) from 1984 to 2013. The dataset was retrieved from CDC's website and includes 50 states plus D.C. The rates per 100 000 persons are calculated as the incidence of STD reports, divided by the population, and multiplied by 100,000.

For the analysis, we standardized each time series and removed all data from the following states, Montana, North Dakota, South Dakota, Vermont, Wyoming, Alaska, and Hawaii, due to missing data. Hence, the dimension of series $\mathbf{Z}_t$ is $m = 44$ and $n = 29$, and the data set is known as WW8c in the Data Appendix, and its plot is shown in Figure 8.8. We used the first 24 observations for model fitting, and the rest of the observations for evaluating the forecasting performance. In this example, we will also demonstrate the use of spatial aggregation, which is very natural in space–time series data.

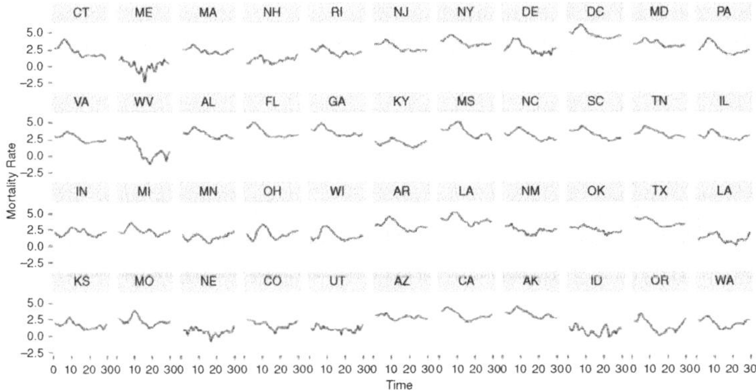

Figure 8.8 U.S. yearly STD of 43 states and D.C.

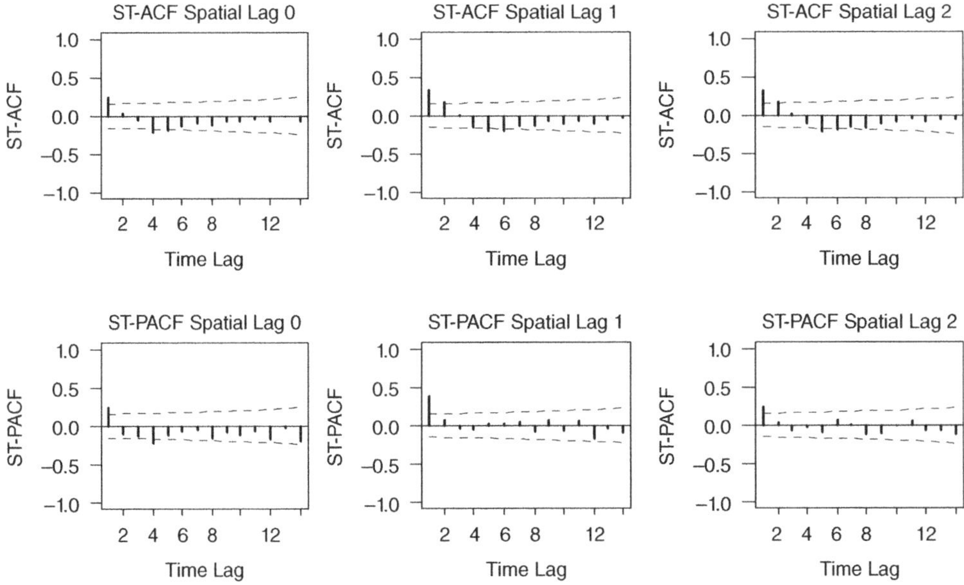

Figure 8.9 The sample ST-ACFs and ST-PACFs for the series $\mathbf{Z}_t$.

The patterns in Figure 8.9 show that the sample ST-ACFs follows a sine/cosine pattern and ST-PACFs are significant only at time lag 1. Although possibly overfitting, we will start by choosing a STAR($2_{2,\ 1}$) model as a possible underlying model.

For the weighting functions, in this example, the first order neighborhoods of a state include the states sharing a border with it, and the second order neighborhoods of a state include the states sharing a border with the first order neighborhoods. The associated weighting matrices $\mathbf{W}^{(1)}$ and $\mathbf{W}^{(2)}$ are (44×44) each, and they are given next.

$$W^{(1)} = $$ [illegible]

$$W^{(2)} = $$ [illegible]

For the STAR($2_{2,\ 1}$) model,

$$\mathbf{Z}_t = \sum_{k=1}^{2}\sum_{\ell=0}^{a_k} \phi_{k,\ell}\mathbf{W}^{(\ell)}\mathbf{Z}_{t-k} + \mathbf{a}_t, \tag{8.27}$$

where $a_1 = 2$ and $a_2 = 1$, the estimation for $\phi_{k,\ell}$ is given in Table 8.7.

Table 8.7 Parameter estimation.

Parameters	$\hat{\phi}_{1,0}$	$\hat{\phi}_{1,1}$	$\hat{\phi}_{1,2}$	$\hat{\phi}_{2,0}$	$\hat{\phi}_{2,1}$
Estimates	0.074	0.286	0.239	–0.126	0.065

Table 8.8 One-step forecast from Eq. (8.29).

State	Forecast	Actual	Error	State	Forecast	Actual	Error
CT	0.147	0.075	−0.072	IN	0.003	−0.194	−0.197
ME	0.164	−0.922	−1.086	MI	0.006	0.194	0.188
MA	0.164	−0.088	−0.252	MN	0.047	−0.875	−0.922
NH	0.154	−0.465	−0.619	OH	0.083	0.042	−0.041
RI	0.141	0.297	0.156	WI	0.019	−0.143	−0.162
NJ	0.075	−0.168	−0.243	AR	0.116	0.140	0.024
NY	0.058	−0.456	−0.514	LA	0.155	−0.077	−0.232
DE	0.053	0.283	0.230	NM	−0.001	0.227	0.228
DC	0.149	0.213	0.064	OK	−0.074	0.352	0.426
MD	0.099	−0.217	−0.316	TX	0.076	0.574	0.498
PA	0.057	0.127	0.070	IA	−0.172	−0.435	−0.263
VA	0.085	−0.067	−0.152	KS	−0.028	0.382	0.410
WV	0.016	−0.046	−0.062	MO	0.019	−0.074	−0.093
AL	0.081	−0.066	−0.147	NE	−0.152	0.369	0.521
FL	0.198	−0.161	−0.359	CO	−0.081	−1.050	−0.969
GA	0.192	−0.047	−0.239	UT	−0.110	0.279	0.389
KY	−0.068	0.219	0.287	AZ	−0.042	−0.952	−0.910
MS	0.187	0.073	−0.114	CA	−0.054	−0.276	−0.222
NC	0.085	0.888	0.803	NV	−0.117	−0.053	0.064
SC	0.216	0.367	0.151	ID	0.218	0.225	0.007
TN	0.124	0.070	−0.054	OR	−0.002	0.173	0.175
IL	−0.002	0.460	0.462	WA	−0.219	−0.893	−0.674

The sample ST-ACF and ST-PACF of the residuals of the model are all small and follow no identifiable pattern, and thus are white noise. As a result, the STAR($2_{2,1}$) model is deemed adequate, so we can use the fitted model to compute the forecast. To be exact, the one-step forecast is given by

$$\hat{\mathbf{Z}}_t(\ell) = \left(0.074\mathbf{I} + 0.286\mathbf{W}^{(1)} + 0.239\mathbf{W}^{(2)}\right)\hat{\mathbf{Z}}_t(\ell-1) + \left(-0.126\mathbf{I} + 0.065\mathbf{W}^{(1)}\right)\hat{\mathbf{Z}}_t(\ell-2). \tag{8.28}$$

The one-step ahead forecast is

$$\hat{\mathbf{Z}}_t(1) = \left(0.074\mathbf{I} + 0.286\mathbf{W}^{(1)} + 0.239\mathbf{W}^{(2)}\right)\mathbf{Z}_t + \left(-0.126\mathbf{I} + 0.065\mathbf{W}^{(1)}\right)\mathbf{Z}_{t-1}, \tag{8.29}$$

and is shown in Table 8.8.

The information in Table 8.8 is only for a one-step ahead forecast. A table for a five-step forecast may not be very practical. In many applications, cancer researchers are interested in the

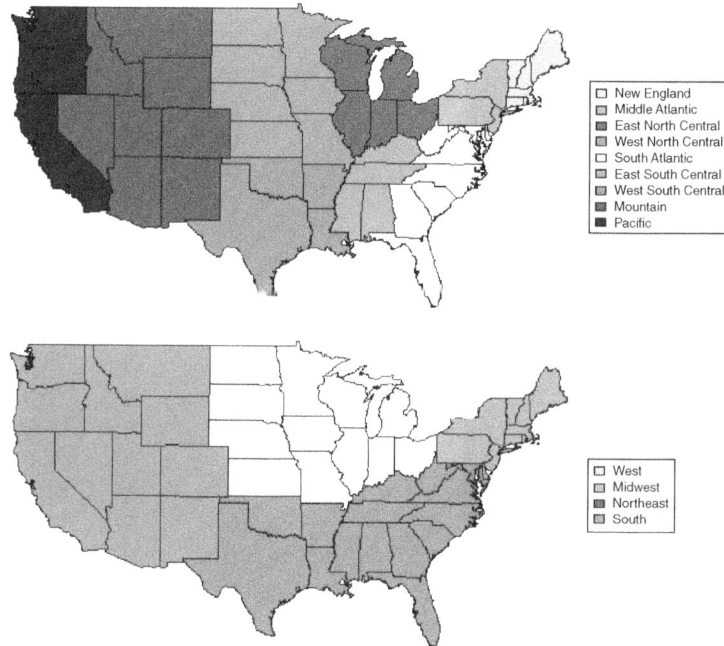

Figure 8.10 Top: U.S. states grouped into nine MMWR regions; bottom: U.S. states grouped into four STD regions.

forecasts for the nine Morbidity and Mortality Weekly Report (MMWR) regions or four STD regions shown in Figure 8.10. They can easily be obtained from the forecasts from Eq. (8.28) through aggregation, and the five-step forecasts are shown in Table 8.9 for the nine MMWR regions and Table 8.10 for the four STD regions, respectively.

The forecasts for the nine MMWR regions and four STD regions shown in Tables 8.9 and 8.10 are examples of spatial aggregation. With the forecast $\hat{\mathbf{Z}}_t(\ell)$ from Eq. (8.28), the forecast for the nine regions is obtained by

$$\mathbf{A}\hat{\mathbf{Z}}_t(\ell), \tag{8.30}$$

where $\mathbf{A}\hat{\mathbf{Z}}_t(\ell)$ is a (44×1) vector, $\mathbf{A}$ is the (9×44) spatial aggregation matrix shown next by its transpose.

State	Row 1	Row 2	Row 3	Row 4	Row 5	Row 6	Row 7	Row 8	Row 9
CT	1	0	0	0	0	0	0	0	0
ME	1	0	0	0	0	0	0	0	0
MA	1	0	0	0	0	0	0	0	0
NH	1	0	0	0	0	0	0	0	0
RI	1	0	0	0	0	0	0	0	0
NJ	0	1	0	0	0	0	0	0	0

NY	0	1	0	0	0	0	0	0	0
DE	0	0	0	0	1	0	0	0	0
DC	0	0	0	0	1	0	0	0	0
MD	0	0	0	0	1	0	0	0	0
PA	0	1	0	0	0	0	0	0	0
VA	0	0	0	0	1	0	0	0	0
WV	0	0	0	0	1	0	0	0	0
AL	0	0	0	0	0	1	0	0	0
FL	0	0	0	0	1	0	0	0	0
GA	0	0	0	0	1	0	0	0	0
KY	0	0	0	0	0	1	0	0	0
MS	0	0	0	0	0	1	0	0	0
NC	0	0	0	0	1	0	0	0	0
SC	0	0	0	0	1	0	0	0	0
TN	0	0	0	0	0	1	0	0	0
IL	0	0	1	0	0	0	0	0	0
IN	0	0	1	0	0	0	0	0	0
MI	0	0	1	0	0	0	0	0	0
MN	0	0	0	1	0	0	0	0	0
OH	0	0	1	0	0	0	0	0	0
WI	0	0	1	0	0	0	0	0	0
AR	0	0	0	0	0	0	1	0	0
LA	0	0	0	0	0	0	1	0	0
NM	0	0	0	0	0	0	0	1	0
OK	0	0	0	0	0	0	1	0	0
TX	0	0	0	0	0	0	1	0	0
IA	0	0	0	1	0	0	0	0	0
KS	0	0	0	1	0	0	0	0	0
MO	0	0	0	1	0	0	0	0	0
NE	0	0	0	1	0	0	0	0	0
CO	0	0	0	0	0	0	0	1	0
UT	0	0	0	0	0	0	0	1	0
AZ	0	0	0	0	0	0	0	1	0
CA	0	0	0	0	0	0	0	0	1
NV	0	0	0	0	0	0	0	1	0
ID	0	0	0	0	0	0	0	1	0
OR	0	0	0	0	0	0	0	0	1
WA	0	0	0	0	0	0	0	0	1

Similarly, the forecast for the four regions is obtained by $\mathbf{A}\hat{\mathbf{Z}}_t(\ell)$, with the (4×44) spatial aggregation matrix $\mathbf{A}$ shown next by its transpose.

State	Row 1	Row 2	Row 3	Row 4
CT	0	0	1	0
ME	0	0	1	0
MA	0	0	1	0

NH	0	0	1	0
RI	0	0	1	0
NJ	0	0	1	0
NY	0	0	1	0
DE	0	0	0	1
DC	0	0	0	1
MD	0	0	0	1
PA	0	0	1	0
VA	0	0	0	1
WV	0	0	0	1
AL	0	0	0	1
FL	0	0	0	1
GA	0	0	0	1
KY	0	0	0	1
MS	0	0	0	1
NC	0	0	0	1
SC	0	0	0	1
TN	0	0	0	1
IL	0	1	0	0
IN	0	1	0	0
MI	0	1	0	0
MN	0	1	0	0
OH	0	1	0	0
WI	0	1	0	0
AR	0	0	0	1
LA	0	0	0	1
NM	1	0	0	0
OK	0	0	0	1
TX	0	0	0	1
IA	0	1	0	0
KS	0	1	0	0
MO	0	1	0	0
NE	0	1	0	0
CO	1	0	0	0
UT	1	0	0	0
AZ	1	0	0	0
CA	1	0	0	0
NV	1	0	0	0
ID	1	0	0	0
OR	1	0	0	0
WA	1	0	0	0

Clearly, the nine-region and four-region forecasts can also be achieved by first getting the nine-region and four-region time series through spatial data aggregation,

$$\mathbf{Y}_t = \mathbf{A}\mathbf{Z}_t, \tag{8.31}$$

Table 8.9 Five-step forecasts for nine MMWR regions.

Forecast Step	Region	Forecast	Actual	Error	Forecast Step	Region	Forecast	Actual	Error
1	1	0.769	−1.102	−1.871	3	6	0.107	−0.072	−0.179
1	2	0.190	−0.497	−0.687	3	7	−0.007	−0.761	−0.754
1	3	0.109	0.359	0.250	3	8	−0.386	0.209	0.595
1	4	−0.286	−0.633	−0.347	3	9	−0.138	1.938	2.076
1	5	1.093	1.213	0.120	4	1	−0.085	2.272	2.357
1	6	0.323	0.295	−0.028	4	2	0.150	0.468	0.318
1	7	0.273	0.988	0.715	4	3	0.641	1.322	0.681
1	8	−0.132	−1.324	−1.192	4	4	0.828	1.887	1.059
1	9	−0.274	−0.996	−0.722	4	5	0.661	1.286	0.625
2	1	−0.202	3.936	4.138	4	6	0.420	0.275	−0.145
2	2	0.243	0.374	0.131	4	7	0.565	0.388	−0.177
2	3	1.186	1.967	0.781	4	8	0.962	2.822	1.860
2	4	1.399	0.402	−0.997	4	9	0.441	1.444	1.003
2	5	1.046	−0.126	−1.172	5	1	0.256	1.861	1.605
2	6	0.737	0.038	−0.699	5	2	0.055	0.962	0.907
2	7	0.938	0.224	−0.714	5	3	−0.092	0.806	0.898
2	8	1.820	1.521	−0.299	5	4	−0.170	9.339	9.509
2	9	0.446	1.835	1.389	5	5	0.182	3.097	2.915
3	1	0.431	−1.537	−1.968	5	6	0.024	−0.243	−0.267
3	2	0.100	0.270	0.170	5	7	−0.066	1.825	1.891
3	3	−0.074	0.488	0.562	5	8	−0.320	1.776	2.096
3	4	−0.168	−0.692	−0.524	5	9	−0.072	1.426	1.498
3	5	0.521	1.540	1.019					

Table 8.10 Five-step forecasts for four STD regions.

Forecast Step	Region	Forecast	Actual	Error	Forecast Step	Region	Forecast	Actual	Error
1	1	−0.406	−2.320	−1.914	3	3	0.531	−1.267	−1.798
1	2	−0.178	−0.274	−0.096	3	4	0.621	0.708	0.087
1	3	0.959	−1.599	−2.558	4	1	1.404	4.266	2.862
1	4	1.690	2.496	0.806	4	2	1.469	3.209	1.740
2	1	2.266	3.357	1.091	4	3	0.065	2.740	2.675
2	2	2.585	2.370	−0.215	4	4	1.646	1.950	0.304
2	3	0.040	4.309	4.269	5	1	−0.392	3.202	3.594
2	4	2.721	0.135	−2.586	5	2	−0.262	10.146	10.408
3	1	−0.524	2.147	2.671	5	3	0.312	2.824	2.512
3	2	−0.243	−0.204	0.039	5	4	0.139	4.679	4.540

and then computing the forecast using the aggregate models constructed from the aggregate series obtained from Eq. (8.31). We will leave this as a data analysis project for readers. The related theory and concept of aggregation will be discussed further in Chapter 10.

For further information on space–time models and applications, we refer readers to Stoffer (1986), Smirnov and Anselin (2001), Lichstein et al. (2002), Zhou and Buongiorno (2006), Craigmile and Guttorp (2011), Eshel (2012), Chang et al. (2014), Clements et al. (2014), Blangiardo and Cameletti (2015), Casals et al. (2016), Cseke et al. (2016), Jalbert et al. (2017), Lin et al. (2017), Rao and Terdik (2017), Shand and Li (2017), Quick et al. (2018), and Singh and Lalitha (2018), among others.

Software code

R code for Example 8.7 (U.S. yearly sexually transmitted disease data)

```
 if("ggplot2" %in% rownames(installed.packages()) == FALSE)
{install.packages("ggplot2")}
 if("expm" %in% rownames(installed.packages()) == FALSE)
{install.packages("expm")}
 if("gtools" %in% rownames(installed.packages()) == FALSE)
{install.packages("gtools")}
 if("reshape2" %in% rownames(installed.packages()) == FALSE)
{install.packages("reshape2")}
 if("psych" %in% rownames(installed.packages()) == FALSE)
{install.packages("psych")}
 if("devtools" %in% rownames(installed.packages()) == FALSE)
{install.packages("devtools")}

 library(ggplot2)
 library(expm)
 library(gtools)
 library(reshape2)
 library(psych)
 library(devtools)

#=====================================================
Read the data, U.S. yearly STD morbidity rates, 9-region
and 4-region aggregation matrices, and a neighboring
information from U.S. Health and Human Services for
preparing spatial weighting matrices
#=====================================================
 SexDise <- read.csv("C:/Bookdata/WW8C.csv", header=TRUE)
 A_MMWR <- as.matrix(read.csv("C:/Bookdata/WW8c-R9.csv"))
 A_STD <- as.matrix(read.csv("C:/Bookdata/WW8c-R4.csv"))
 wmat <- as.matrix(read.csv("C:/Bookdata/WW8c-HHS.csv"))
```

```
 d1 <- as.matrix(SexDise[,-1])
 rownames(d1) <- SexDise[,1]

# Take difference and mean-center the data
 d2 <- diff(d1)
 d3 <- (d2 - t(matrix(colMeans(d2),ncol(d2),nrow(d2))))/t
(matrix(apply(d2,2,sd),ncol(d2),nrow(d2)))

# Final data matrix
 Z <- as.matrix(d3)

# Identify # of locations (r) and # of time points (T)
 Z.fit <- Z[1:24,]
 r <- ncol(Z.fit)
 T <- nrow(Z.fit)
 Z.MMWR <- Z%*%t(A_MMWR)
 Z.MMWR.fit <- Z.MMWR[1:24,]

 Z.STD <- Z%*%t(A_STD)
 Z.STD.fit <- Z.STD[1:24,]

#================================================
#  Necessary functions for modelling
#==================================================

# starest - Estimating STAR model parameters
# Author - Andrew Gehman
# Function that calculates parameter estimates for specified
#    STAR Model by minimizing determinant of covariance matrix
#    of model residuals
# Function called through R function constrOptim which takes
#    - sq: vector of possible parameter values for R function
#        constrOptim (not user specified)
#    - Z: numerical matrix; mean centered data matrix with rows
#       corresponding to time points and columns corresponding
#       to spatial locations
#    - W: list of matrices; each element of list is spatial
#       weighting matrix (non-zero rows sum to 0); the 0th order
#       weighting matrix is first, followed by the 1st order,
#       and so on; ST-ACFs and ST-PACFs are provided for each
#       spatial order of weighting matrix provided
#    - M: the STAR model order as described below

 starest <- function(sq,Z,W,M){
   T <- nrow(Z)
   r <- ncol(Z)
```

```
    ms <- sum(M[-1]+1)
    pp <- c(1,sq)
    cc <- 1
    X <- matrix(NA,T*r,ms+1)
    PS <- c()
    for(k in 1:length(M)){
       for(l in 0:M[k]){
          ks <- (k-1)
          if(ks==0){
             X[,cc] <- cbind(c(t(Z)))
             PS[cc] <- pp[cc]
          } else {
            X[,cc] <-cbind(c(rep(0,r*ks),t(Z[-c((T-(ks-1)):T),]
            %*%t(W[[l+1]]))))
            PS[cc] <- pp[cc]
          }
          cc <- cc + 1
       }
    }
    ZY <- cbind(c(t(Z)))
    XS <- X[,-1]
    PS <- PS[-1]
    E1 <- ZY-XS%*%t(t(PS))
    E <- t(matrix(c(E1),r,T))
    GE <- cov(E)
    de <- tr(GE)
    return(de)
}

# starfit - Calculating STAR model residuals
# Author - Andrew Gehman
# Function that calculates model residuals for given parameter
  values
# Function takes
#   - sq: vector of estimated parameter values from "starest"
        function
#   - Z: numerical matrix; mean centered data matrix with rows
#        corresponding to time points and columns corresponding
#        to spatial locations
#   - W: list of matrices; each element of list is spatial
#        weighting matrix (non-zero rows sum to 0); the 0th order
#        weighting matrix is first, followed by the 1st order,
#        and so on; ST-ACFs and ST-PACFs are provided for each
#        spatial order of weighting matrix provided
#   - M: the STAR model order as described below
```

```
starfit <- function(sq,Z,W,M){
  T <- nrow(Z)
  r <- ncol(Z)
  ms <- sum(M[-1]+1)
  pp <- c(1,sq)
  cc <- 1
  X <- matrix(NA,T*r,ms+1)
  PS <- c()
  for(k in 1:length(M)){
    for(l in 0:M[k]){
       ks <- (k-1)
       if(ks==0){
         X[,cc] <- cbind(c(t(Z)))
         PS[cc] <- pp[cc]
       } else {
           X[,cc] <- cbind(c(rep(0,r*ks),t(Z[-c((T-(ks-1)):T),]
           %*%t(W[[l+1]]))))
           PS[cc] <- pp[cc]
       }
       cc <- cc + 1
    }
  }
  ZY <- cbind(c(t(Z)))
  XS <- X[,-1]
  PS <- PS[-1]
  E1 <- ZY-XS%*%t(t(PS))
  E <- t(matrix(c(E1),r,T))
  return(E)
}

# acfs - Space-Time Autocorrelation and Partial Autocorrelation
  Functions
# Author - Andrew Gehman
# Function that returns the ST-ACFs, ST-PACFs, their variances, and
#    which are significantly different from 0 at alpha = 0.05.
#    Provided as a list of matrices. Also plots the ST-ACFs and
     ST-PACFs.
# Function takes
#   - Z: numerical matrix; mean centered data matrix with rows
#        corresponding to time points and columns corresponding
#        to spatial locations
#   - W: list of matrices; each element of list is spatial
#        weighting matrix (non-zero rows sum to 0); the 0th order
#        weighting matrix is first, followed by the 1st order,
#        and so on; ST-ACFs and ST-PACFs are provided for each
#        spatial order of weighting matrix provided
#   - t: the maximal temporal order for the ST-ACFs and ST-PACFs
```

```
acfs <- function(Z,W,t){
  G <- cov(Z)
  nobs <- nrow(Z)
  nloc <- ncol(Z)
  ntim <- t
  nnei <- length(W)

  Y <- list()
  for(s in 0:(ntim-1)){
     y <- matrix(0,nnei,nnei)
     for(k in 1:nnei){
        for(l in 1:nnei){
           yy <- 0
           for(t in 1:(nobs-s)){
              yy<-yy+t(W[[l]]%*%t(t(Z[t,])))%*%W[[k]]%*%t
              (t(Z[t+s,]))
             }
             y[l,k] <- yy/(nloc*(nobs-s))
           }
        }
        Y[[s+1]] <- y
     }
     P <- list()
    PP <- matrix(0,ntim,nnei)
    for(s in 0:(ntim-1)){
        p <- matrix(0,nnei,nnei)
        for(l in 1:nnei){
            for(k in 1:nnei){
                p[l,k]<-Y[[s+1]][l,k]/sqrt(Y[[1]][l,l]%*
                %Y[[1]][k,k])
            }
        }
        P[[s+1]] <- p
        for(l in 1:nnei){
           PP[s+1,l] <- P[[s+1]][l,l]
        }
    }

    QQ <- matrix(NA,nnei*(ntim-1),nnei*(ntim-1))
    for(k in 1:(ntim-1)){
      for(l in 1:(ntim-1)){
         if(k > l){
             QQ[(k*nnei-nnei+1):(k*nnei),(l*nnei-nnei+1):
             (l*nnei)]<-Y[[k-l+1]]
         } else {
             QQ[(k*nnei-nnei+1):(k*nnei),(l*nnei-nnei+1):
             (l*nnei)] <- t(Y[[l-k+1]])
```

```
          }
        }
      }
      QQZ <- rbind(t(t(QQ[(nnei+1):(nnei*(ntim-1)),1])),
  t(t(Y[[ntim]][,1])))
      sol <- c()
      for(k in 1:(nnei*(ntim-1))){
        sol[k] <- (solve(QQ[1:k,1:k])%*%t(t(QQZ[1:k])))[k]
      }
      Q <- t(matrix(sol,nnei,ntim-1))

      V <- matrix(0,ntim,nnei)
      for(s in 0:(ntim-1)){
        for(l in 1:nnei){
          V[s+1,l] <- tr(t(W[[l]])%*%G%*%W[[l]]%*%G)/(tr(G)
  *tr(t(W[[l]])%*%W[[l]]%*%G))/(nobs-s)
          }
      }

      par(mfrow=c(2,nnei))
      title<-c("ST-ACF Spatial Lag 0","ST-ACF Spatial Lag 1",
  "ST-ACF Spatial Lag 2","ST-ACF Spatial
        Lag 3")
      for(i in 1:nnei){
        plot(PP[-1,i],type='h',lwd=3,ylim=c(-1,1),ylab="ST-ACF",
  xlab="Time Lag",main=title[i])
        lines(1.96*sqrt(V[-1,i]),lty=2,type='l')
        lines(-1.96*sqrt(V[-1,i]),lty=2,type='l')
        abline(0,0)
      }
    title<-c("ST-PACF Spatial Lag 0","ST-PACF Spatial Lag 1",
  "ST-PACF Spatial Lag 2","ST-PACF
        Spatial Lag 3")
      for(i in 1:nnei){
        plot(Q[,i],type='h',lwd=3,ylim=c(-1,1),ylab="ST-PACF",
  xlab="Time Lag",main=title[i])
        lines(1.96*sqrt(V[-1,i]),lty=2,type='l')
        lines(-1.96*sqrt(V[-1,i]),lty=2,type='l')
        abline(0,0)
      }
      ret<-list(Z,W,PP,Q,V,1*(abs(PP)>(1.96*sqrt(V))),
  1*(abs(Q)>(1.96*sqrt(V[-1,]))))
      names(ret) <- c("data","wmat","ACF","PACF","VAR",
  "SigA","SigP")
      return(ret)
  }
```

```
# starfore - Forecasting from estimated STAR model
# Author - Andrew Gehman
# Function that produces model predictions and one-step ahead
  forecasts from given estimated model
# Function takes
#    - sq: vector of estimated parameter values from "starest"
       function
#    - Z: numerical matrix; mean centered data matrix with rows
#        corresponding to time points and columns corresponding
#        to spatial locations
#    - W: list of matrices; each element of list is spatial
#        weighting matrix (non-zero rows sum to 0); the 0th order
#        weighting matrix is first, followed by the 1st order,
#        and so on; ST-ACFs and ST-PACFs are provided for each
#        spatial order of weighting matrix provided
#    - M: the STAR model order as described below
#    - t: the number of one-step-ahead forecasts to calculate

starfore <- function(sq,Z,W,M,t){
   T <- nrow(Z)
   r <- ncol(Z)
   ms <- sum(M[-1]+1)
   pp <- c(1,sq)
   cc <- 1
   X <- matrix(NA,T*r,ms+1)
   PS <- c()
   for(k in 1:length(M)){
      for(l in 0:M[k]){
         ks <- (k-1)
         if(ks==0){
           X[,cc] <- cbind(c(t(Z)))
           PS[cc] <- pp[cc]
         } else {
          X[,cc] <-cbind(c(rep(0,r*ks),t(Z[-c((T-(ks-1)):T),]
%*%t(W[[l+1]]))))
          PS[cc] <- pp[cc]
         }
         cc <- cc + 1
      }
    }
    XS <- X[,-1]
    PS <- PS[-1]
    Z1 <- XS%*%t(t(PS))
    Z2 <- t(matrix(c(Z1),r,T))
    Z3 <- rbind(Z,matrix(0,t,r))
    for(h in 1:t){
```

```
      cc <- 2
      for(k in 2:length(M)){
         for(l in 0:M[k]){
            Z3[T+h,] <- Z3[T+h,] + pp[cc]*c(t(Z3[T+h-k,]
%*%t(W[[l+1]])))
            cc <- cc + 1
         }
      }
   }
   return(rbind(Z2,Z3[(T+1):(T+t),]))
}
#==============================================================
#  Start the modelling process
#==============================================================

# User Input: Upload binary spatial location adjacency matrix
 wmat <- as.matrix(read.csv("C:/Bookdata/WW8c-HHS.csv"))
 wfun <- function(W1,n){
  r <- nrow(W1)
  WW <- list(diag(r))
  for(i in 1:n){
     Wa <- 1*(W1%^%i>0)
     Wb <- matrix(0,r,r)
     for(j in 0:(i-1)){
        Wb <- Wb + 1*(W1%^%j>0)
     }
     WWs <- 1*(Wa-Wb>0)
     rs <- rowSums(WWs)
     rs[which(rs==0)] <- 1
     WW[[i+1]] <- WWs/rs
  }
    return(WW)
}

# Create spatial weighting matrices for this data up to spatial
order 3
W <- wfun(wmat,2)

# Weighting matrix, W[[1]] is W^(0), W[[2]] is W^(1), and W[[3]] is W^(2).
 W

# attached function "acfs" to calculate ST-ACFs and ST-PACFs
up to time 20
 one <- acfs(Z.fit,W,15)
 round(cbind(one$ACF[-1,],sqrt(one$VAR[-1,]),one$PACF,sqrt
(one$VAR[-1,])),3)

# User Input: STAR model order as a vector of integers
# First element is always 0
```

```
# Number of integers after first is maximum temporal order
# Each integer after first is maximum spatial order at that
temporal order
# 'M<-c(0,2,1)' means that a STAR(2_{2,1}) model will be fitted

 M <- c(0,2,1)
# Function to specify number of parameters and restricted
values estimated parameters can take
 pm <- list()
 for(i in 1:10){
   pm[[i]] <- permutations(2,i,c(-1,1),repeats.allowed=T)
}
 ms <- sum(M[-1]+1)
 bm <- M[-1]+1
 bmat <- list()
 for(i in 1:length(bm)){
   bmat[[i]] <- pm[[bm[i]]]
}
 mat <- as.matrix(bdiag(bmat))

# Calls attached function "starest" to estimate specified
STAR model
 pp <- constrOptim(rep(0,ms),f=starest,NULL,ui=mat,ci=rep
(-1,nrow(mat)),control=list(reltol = 1e-10),Z=Z.fit,W=W,M=M)

# The estimated parameters in order of temporal order then
spatial order
 bb <- pp$par
 bb
 spatial.matrix <- cbind(bb[2]*W[[1]],bb[2]*W[[2]],bb[3]
*W[[3]],bb[4]*W[[3]])
 spatial.matrix

# forecast and errors
#======================
 Ft <- starfore(bb,Z.fit,W,M,5)

# calculate error
# one-step ahead forecast
 one_step = cbind(Ft[25,], Z[25,], Z[25,]-Ft[25,])
 colnames(one_step) = c("Forecast", "Actual", "Error")
 one_step

# 9 MMWR regions
 forecast_MMWR = cbind(rep(1:5,each=9),rep(1:9,5),as.vector
(t(Ft[25:29,]%*% t(A_MMWR))),
   as.vector(t(Z.MMWR[25:29,])), as.vector(t(Z.MMWR[25:29,]
- Ft[25:29,]%*% t(A_MMWR))))
 colnames(forecast_MMWR) = c("Forecast Step", "Region",
```

```
"Forecast", "Actual", "Error")
 forecast_MMWR
# 4 STD regions
 forecast_STD = cbind(rep(1:5,each=4),rep(1:4,5),
as.vector(t(Ft[25:29,]%*% t(A_STD))),
   as.vector(t(Z.STD[25:29,])), as.vector(t(Z.STD[25:29,]
- Ft[25:29,]%*% t(A_STD))))
 colnames(forecast_STD) = c("Forecast Step", "Region",
"Forecast", "Actual", "Error")
 forecast_STD
```

Projects

1. For the yearly U.S. STD data discussed in Example 8.7, use the nine-MMWR-region aggregation matrix, which is saved as Bookdata set, WW8c-R9, to create the nine-dimensional aggregate series. Build a nine-dimensional model for the aggregate series and use it to compute the next three-step forecasts.
2. For the yearly U.S. STD data discussed in Example 8.7, use the four-STD-region aggregation matrix, which is saved as Bookdata set, WW8c-R4, to create the four-dimensional aggregate series. Build a four-dimensional model for the aggregate series and use it to compute the next three-step forecasts.
3. Build your best aggregate models with $(n-5)$ observations for the aggregate series described in Projects 1 and 2. Use your models to compute the next five-step forecasts for the total STD cases for the combined 43 states and D.C. Compare your forecast results between the two models.
4. Find a social science related space–time data set of your interest, build a space–time series model, and use it to forecast. Complete your analysis with a detailed written report and attach your data set and software code.
5. Find a natural science related space–time data set of your interest, construct a space–time series model, and use it to forecast. Complete your analysis with a detailed written report and attach your data set and software code.

References

Blangiardo, M. and Cameletti, M. (2015). *Spatial and Spatio-Temporal Bayesian Models with R – INLA*, 1e. Wiley.

Borovkova, S.A., Lopuhaa, H.P., and Ruchjana, B.N. (2002). Generalized STAR models with experimental weights. *Proceedings of the 17th International Workshop on Statistical Modelling*. Chania, Greece, 139–147.

Casals, J., Garcia-Hiernaux, A., Jerez, M., Sotoca, S., and Trindade, A.A. (2016). *State–Space Methods for Time Series Analysis: Theory, Applications and Software*. CRC Press.

Chang, C.H., Huang, H.C., and Ing, C.K. (2014). Asymptotic theory of generalized information criterion for geostatistical regression model selection. *Annals of Mathematical Statistics* **42**: 2441–2468.

Clements, N., Sarkar, S., and Wei, W.W.S. (2014). Multiplicative spatio-temporal models for remotely sensed normalized difference vegetation index data. *Journal of International Energy Policy* **3**: 1–14.

Cliff, A.D. and Ord, J. (1975a). Model building and the analysis of spatial pattern in human geography. *Journal of the Royal Statistical Society Series B* **37**: 297–348.

Cliff, A.D. and Ord, J. (1975b). Space-time modelling with an application to regional forecasting. *Transactions of the Institute of British Geographers* **66**: 119–128.

Craigmile, P.F. and Guttorp, P. (2011). Space-time modelling of trends in temperature series. *Journal of Time Series Analysis* **32**: 378–395.

Cseke, B., Zammit-Mangion, A., Heskes, T., and Sanguinetti, G. (2016). Sparse approximation inference for spatio-temporal point process models. *Journal of the American Statistical Association* **111**: 1746–1763.

Di Giacinto, V. (2006). A generalized space-time ARMA model with an application to regional unemployment analysis in Italy. *International Regional Science Review* **29**: 159–198.

Eshel, G. (2012). *Spatiotemporal Data Analysis*. Princeton University Press

Gehman, A. (2016). *The Effects of Spatial Aggregation on Spatial Time Series Modeling and Forecasting*. PhD dissertation, Temple University.

Gehman, A., and Wei, W.W.S. (2014). The effects of spatial aggregation on model parameters and forecasts of space-time autoregressive moving average models. A manuscript.

Gehman, A., and Wei, W.W.S. (2015a). Testing for poolability of the space-time autoregressive moving-average model. A manuscript.

Gehman, A., and Wei, W.W.S. (2015b). Determining the model order of the spatial aggregate of a STAR (1_1) model. A manuscript.

Jalbert, J., Favre, A., Belisle, C., and Angers, J. (2017). A spatiotemporal model for extreme precipitation simulated by a climate model, with an application to assessing changes in return levels over North America. *Journal of the Royal Statistical Society Series C* **66**: 941–962.

Lichstein, J.W., Simons, T.R., Shriner, S.A., and Franzreb, K.E. (2002). Spatial autocorrelation and autoregressive models in ecology. *Ecology Monographs* **72**: 445–463.

Lin, Z., Wang, T., Yang, C., and Zhao, H. (2017). On joint estimation of Gaussian graphical models for spatial and temporal data. *Biometrics* **73**: 769–779.

Martin, R.L. and Oeppen, J.E. (1975). The identification of regional forecasting models using space-time correlation functions. *Transactions of the Institute of British Geographers* **66**: 95–118.

Pfeifer, P.E. and Deutsch, S.J. (1980a). A three-stage iterative procedure for space-time modeling. *Technometrics* **22**: 35–47.

Pfeifer, P.E. and Deutsch, S.J. (1980b). Identification and interpretation of the first order space-time ARMA models. *Technometrics* **22**: 397–408.

Pfeifer, P.E. and Deutsch, S.J. (1980c). Stationary and invertibility regions for low order STARMA models. *Communications in Statistics: Part B Simulation and Computation* **9**: 551–562.

Quick, H., Waller, L.A., and Casper, M. (2018). A multivariate space-time model for analysing county level heart disease death rates by race and sex. *Journal of the Royal Statistical Society Series C* **67**: 291–304.

Rao, T.S. and Terdik, G. (2017). A new covariance function and spatio-temporal prediction (Kriging) for a stationary spatio-temporal random process. *Journal of Time Series Analysis* **38**: 936–959.

Shand, L. and Li, B. (2017). Modeling nonstationarity in space and time. *Biometrics* **73**: 759–768.

Singh, A.K. and Lalitha, S. (2018). A novel spatial outlier detection technique. *Communications in Statistics - Theory and Methods* **47**: 247–257.

Smirnov, O. and Anselin, L. (2001). Fast maximum likelihood estimation of very large spatial autoregressive models: a characteristic polynomial approach. *Computational Statistics & Data Analysis* **35**: 301–319.

Stoffer, D.S. (1986). Estimation and identification of space-time ARMAX models in the presence of missing data. *Journal of the American Statistical Association* **81**: 762–772.

Wei, W.W.S. (2006). *Time Series Analysis – Univariate and Multivariate Methods*, 2e. Pearson Addison-Wesley.

Zhou, M. and Buongiorno, J. (2006). Space-time modeling of timber prices. *Journal of Agricultural and Resource Economics* **31**: 40–56.

9

Multivariate spectral analysis of time series

Similar to the univariate time series analysis, where one can study a univariate time series through its autocovariance/autocorrelation functions and lag relationships or through its spectrum properties, we can study a multivariate time series through a time domain approach or a frequency domain approach. In the time domain approach, we use the covariance/correlation matrices, and in frequency domain approach we will use the spectrum matrices. In this chapter, after a brief review of the univariate frequency domain method, we will introduce the spectral analysis for both stationary and nonstationary vector time series. With no loss of generality, we will assume a zero-mean time series in the following discussion.

9.1 Introduction

Recall that for a univariate stationary time series process, Z_t, its spectral representation is given by

$$Z_t = \int_{-\pi}^{\pi} e^{i\omega t} dU(\omega), -\pi \le \omega \le \pi, \tag{9.1}$$

where $dU(\omega)$ is a complex-valued orthogonal stochastic process for each ω such that

$$E[dU(\omega)] = 0, \tag{9.2}$$

$$E[dU(\omega)dU^*(\lambda)] = 0, \text{ for all } \omega \neq \lambda, \tag{9.3}$$

and

$$E\left[|dU(\omega)|^2\right] = E[dU(\omega)dU^*(\omega)] = dF(\omega), \tag{9.4}$$

Multivariate Time Series Analysis and Applications, First Edition. William W.S. Wei.
© 2019 John Wiley & Sons Ltd. Published 2019 by John Wiley & Sons Ltd.
Companion website: www.wiley.com/go/wei/datasets

where $U^*(\omega)$ is the complex conjugate of $U(\omega)$. Let γ_k be the autocovariance function of Z_t. Its spectral representation is given by

$$\gamma_k = \int_{-\pi}^{\pi} e^{i\omega k} dF(\omega), -\pi \le \omega \le \pi, \tag{9.5}$$

where $F(\omega)$ is the spectral distribution function.

When the autocovariance function is absolutely summable, the spectrum or the spectral density exists and is equal to

$$f(\omega)d\omega = dF(\omega). \tag{9.6}$$

In this case, we have

$$f(\omega) = \frac{1}{2\pi} \sum_{k=-\infty}^{\infty} \gamma_k e^{-i\omega k}, -\pi \le \omega \le \pi, \tag{9.7}$$

with

$$\gamma_k = \int_{-\pi}^{\pi} f(\omega) e^{-i\omega k} d\omega. \tag{9.8}$$

Given a time series Z_1, Z_2, …, Z_n, its Fourier transform at the Fourier frequencies $\omega_j = 2\pi j/n, -[(n-1)/2] \le j \le [n/2]$, is

$$y(\omega_j) = \frac{1}{\sqrt{2\pi n}} \sum_{t=1}^{n} Z_t e^{-i\omega_j t}. \tag{9.9}$$

It can be shown that the sample spectrum is given by

$$\begin{aligned} \tilde{f}(\omega_j) &= y(\omega_j) y^*(\omega_j) \\ &= \frac{1}{2\pi n} \left[\sum_{t=1}^{n} Z_t \exp(-i\omega_j t) \right] \left[\sum_{r=1}^{n} Z_r \exp(i\omega_j r) \right] \\ &= \frac{1}{2\pi n} \sum_{t=1}^{n} \sum_{r=1}^{n} Z_t Z_r e^{-i\omega_j (t-r)}. \end{aligned} \tag{9.10}$$

Looking at the following expression of the previous double sums for $Z_t Z_r$

$$\begin{aligned} &Z_1Z_1 + Z_1Z_2 + Z_1Z_3 + \cdots + Z_1Z_n \\ &+ Z_2Z_1 + Z_2Z_2 + Z_2Z_3 + \cdots + Z_2Z_n \\ &+ Z_3Z_1 + Z_3Z_2 + Z_3Z_3 + \cdots + Z_3Z_n \\ &\vdots \\ &+ Z_nZ_1 + Z_nZ_2 + Z_nZ_3 + \cdots + Z_nZ_n \end{aligned}$$

and letting $k = t - r$, we see that

$$\begin{aligned}\widetilde{f}(\omega_j) &= \frac{1}{2\pi}\sum_{k=-(n-1)}^{n-1}\hat{\gamma}_k e^{-i\omega_j k}\\ &= \frac{1}{2\pi}\left[\hat{\gamma}_0 + 2\sum_{k=1}^{n-1}\hat{\gamma}_k e^{-i\omega_j k}\right],\end{aligned} \tag{9.11}$$

where

$$\hat{\gamma}_k = \frac{1}{n}\sum_{t=1}^{n-k} Z_t Z_{t+k}$$

is the sample autocovariance function. If Z_t is a Gaussian white noise process with mean 0 and variance σ^2, then $\widetilde{f}(\omega_j)$ for $j = 1, 2, \ldots, (n-1)/2$, are distributed independently and identically as the following Chi-square distribution, that is

$$\widetilde{f}(\omega_j) \sim \frac{\sigma^2}{2\pi}\frac{\chi^2(2)}{2}, \tag{9.12}$$

where $(\sigma^2/2\pi)$ is actually the spectrum of Z_t.

Although $\widetilde{f}(\omega_j)$ is an unbiased estimate, since the sample spectrum is only defined at the Fourier frequencies and its variance is independent of the sample size n, it is a rather poor estimate of the spectrum. To solve this problem, we introduce a suitable kernel or spectral window to smooth the sample spectrum, that is,

$$\hat{f}(\omega_p) = \sum_{j=-M_n}^{M_n} W_n(\omega_j)\widetilde{f}(\omega_p - \omega_j), \tag{9.13}$$

where $\omega_j = 2\pi j/n$, $j = 0, \pm 1, \ldots, \pm [n/2]$, are Fourier frequencies; M_n is a function of n such that $M_n \to \infty$ but $M_n/n \to 0$ as $n \to \infty$; and $W_n(\omega_j)$ is the kernel or spectral window that is a weighting function with the following properties

$$\begin{aligned}&\sum_{j=-M_n}^{M_n} W_n(\omega_j) = 1,\\ &W_n(\omega_j) = W_n(\omega_{-j}),\\ &\lim_{n\to\infty}\sum_{j=-M_n}^{M_n} W_n^2(\omega_j) = 0.\end{aligned} \tag{9.14}$$

From Eq. (9.7), we see that the spectrum is the Fourier transform of the autocovariance function. So, we can also apply a weighting function to the sample autocovariances, that is,

$$\hat{f}(\omega) = \frac{1}{2\pi}\sum_{k=-M}^{M} W(k)\hat{\gamma}_k e^{-i\omega k}, \tag{9.15}$$

where M is the truncation point that is a function of n such that $M/n \to 0$ as $n \to \infty$; $W(k)$ is the lag window chosen to be an absolutely summable sequence

$$W(k) = W\left(\frac{k}{M}\right), \tag{9.16}$$

which is derived from a bounded even continuous function $W(x)$ satisfying

$$\begin{aligned} &|W(x)| \le 1, \\ &W(0) = 1, \\ &W(x) = W(-x), \\ &W(x) = 0, |x| > 1. \end{aligned} \tag{9.17}$$

Some commonly used spectral and lag windows include the Daniell (1946) Rectangular, Bartlett (1950), Blackman–Tukey (1959), and Parzen (1961, 1963) windows. For more details, we refer readers to Priestley (1981), and Wei (2006, 2008).

When Z_t is a Gaussian process with the spectrum $f(\omega)$, under a properly chosen lag window, we have

$$\hat{f}(\omega) \sim f(\omega)\frac{\chi^2(\nu)}{\nu}, \tag{9.18}$$

with

$$\nu = \frac{2n}{M\int_{-1}^{1} W^2(x)dx,}$$

and $W(x)$ is the continuous weighting function used in the associated lag window. This smoothed spectrum $\hat{f}(\omega)$ has a distribution related to the Chi-square distribution. $\hat{f}(\omega)$ is an asymptotically unbiased estimate of $f(\omega)$ so that

$$E\left[\hat{f}(\omega)\right] \cong f(\omega),$$

and

$$\mathrm{Var}\left[\hat{f}(\omega)\right] \to 0, \text{ as } n \to \infty.$$

The estimation procedure for series observed at high sampling frequency that does not require equally spaced observations, we refer readers to a recent study by Chang, Hall, and Tang (2017).

9.2 Spectral representations of multivariate time series processes

These univariate time series results can be readily generalized to the m-dimensional vector process. Let $\mathbf{Z}_t = [Z_{1,t}, Z_{2,\,t}, \ldots, Z_{m,t}]'$ be a zero-mean jointly stationary m-dimensional vector process with the covariance matrix function, $\mathbf{\Gamma}(k) = [\gamma_{i,j}(k)]$, the spectral representation of $\mathbf{Z}_t$ is given by

$$\mathbf{Z}_t = \int_{-\pi}^{\pi} e^{i\omega t} d\mathbf{U}(\omega), \tag{9.19}$$

where $d\mathbf{U}(\omega) = [dU_1(\omega), dU_2(\omega), \ldots, dU_m(\omega)]'$ is a m-dimensional complex-valued process with $dU_i(\omega)$, for $i = 1, 2, \ldots, m$, being both orthogonal as well as cross-orthogonal such that

$$E[d\mathbf{U}(\omega)] = \mathbf{0}, -\pi \leq \omega \leq \pi, \tag{9.20}$$

and

$$E\{d\mathbf{U}(\omega)[d\mathbf{U}^*(\lambda)]'\} = \mathbf{0}, \text{ for all } \omega \neq \lambda. \tag{9.21}$$

The spectral representation of the covariance matrix function is given by

$$\mathbf{\Gamma}(k) = \int_{-\pi}^{\pi} e^{i\omega k} d\mathbf{F}(\omega), \tag{9.22}$$

where

$$\begin{aligned} d\mathbf{F}(\omega) &= E\{d\mathbf{U}(\omega)[d\mathbf{U}^*(\omega)]'\} \\ &= \left[E\left\{dU_i(\omega) dU_j^*(\omega)\right\}\right] \\ &= \left[dF_{i,j}(\omega)\right], \end{aligned} \tag{9.23}$$

and $\mathbf{F}(\omega)$ is the spectral distribution matrix function of $\mathbf{Z}_t$. The diagonal elements $F_{i,i}(\omega)$ are the spectral distribution functions of the $Z_{i,t}$ and the off-diagonal elements $F_{i,j}(\omega)$ are the cross-spectral distribution functions between the $Z_{i,t}$ and the $Z_{j,t}$.

If the covariance matrix function is absolutely summable in the sense that each of the $m \times m$ sequence $\gamma_{i,j}(k)$ is absolutely summable, then the spectrum matrix or the spectral density matrix function exists and is given by

$$\begin{aligned} \mathbf{f}(\omega) d\omega &= d\mathbf{F}(\omega) \\ &= \left[dF_{i,j}(\omega)\right] \\ &= \left[f_{i,j}(\omega) d\omega\right]. \end{aligned} \tag{9.24}$$

Thus, we can write

$$\mathbf{\Gamma}(k) = \int_{-\pi}^{\pi} e^{i\omega k} \mathbf{f}(\omega) d\omega, \tag{9.25}$$

and

$$\mathbf{f}(\omega) = \frac{1}{2\pi} \sum_{k=-\infty}^{\infty} \mathbf{\Gamma}(k) e^{-i\omega k} = \left[f_{i,j}(\omega)\right], \tag{9.26}$$

where

$$f_{i,j}(\omega) = \frac{1}{2\pi} \sum_{k=-\infty}^{\infty} \gamma_{i,j}(k) e^{-i\omega k}. \tag{9.27}$$

When $k = 0$ in Eq. (9.22), we have

$$\mathbf{\Gamma}(0) = \int_{-\pi}^{\pi} \mathbf{f}(\omega)d\omega. \tag{9.28}$$

So, the area under the multivariate spectrum is the variance–covariance matrix of $\mathbf{Z}_t$. The element $f_{i,i}(\omega)$ in Eq. (9.26) is the spectrum or the spectral density of $Z_{i,t}$, and the element $f_{i,j}(\omega)$ in Eq. (9.26) is the cross-spectrum or the cross-spectral density of $Z_{i,t}$ and $Z_{j,t}$. It is easily seen that the spectral density matrix function $\mathbf{f}(\omega)$ is positive semidefinite, that is, $\mathbf{c}'\mathbf{f}(\omega)\mathbf{c} \geq 0$ for any nonzero m-dimensional complex vector $\mathbf{c}$. Also, the matrix $\mathbf{f}(\omega)$ is Hermitian, that is,

$$\mathbf{f}^*(\omega) = \mathbf{f}(\omega). \tag{9.29}$$

Hence, $f_{i,j}(\omega) = f^*_{j,i}(\omega)$ for all i and j.

Since $f_{i,j}(\omega)$ is in general complex, we can write it as

$$f_{i,j}(\omega) = c_{i,j}(\omega) - iq_{i,j}(\omega), \tag{9.30}$$

where $c_{i,j}(\omega)$ and $-iq_{i,j}(\omega)$ are the real and imaginary parts of $f_{i,j}(\omega)$, that is,

$$c_{i,j}(\omega) = \frac{1}{2\pi}\sum_{k=-\infty}^{\infty} \gamma_{i,j}(k)\cos(\omega k), \tag{9.31}$$

and

$$q_{i,j}(\omega) = \frac{1}{2\pi}\sum_{k=-\infty}^{\infty} \gamma_{i,j}(k)\sin(\omega k). \tag{9.32}$$

The function $c_{i,j}(\omega)$ is known as the co-spectrum, and $q_{i,j}(\omega)$ is known as the quadrature spectrum, between $Z_{i,t}$ and $Z_{j,t}$. We can also write $f_{i,j}(\omega)$ in the polar form,

$$f_{i,j}(\omega) = \alpha_{i,j}(\omega)e^{i\phi_{i,j}(\omega)}, \tag{9.33}$$

where

$$\alpha_{i,j}(\omega) = |f_{i,j}(\omega)| = \left[c^2_{i,j}(\omega) + q^2_{i,j}(\omega)\right]^{1/2} \tag{9.34}$$

and

$$\phi_{i,j}(\omega) = \tan^{-1}\left[\frac{-q_{i,j}(\omega)}{c_{i,j}(\omega)}\right]. \tag{9.35}$$

The function $\alpha_{i,j}(\omega)$ is called the cross-amplitude spectrum, and the function $\phi_{i,j}(\omega)$ is called the phase spectrum.

To properly understand these spectra, we note that for any ω, the function $dU_i(\omega)$ is a complex-valued random variable, and we can write it as

$$dU_i(\omega) = \alpha_i(\omega)e^{i\phi_i(\omega)}, \tag{9.36}$$

where $\alpha_i(\omega)$ and $\phi_i(\omega)$ are the amplitude spectrum and the phase spectrum of the $Z_{i,t}$ series. From Eqs. (9.23), (9.24), and (9.33), since

$$\begin{aligned}\alpha_{i,j}(\omega)e^{i\phi_{i,j}(\omega)}d\omega = f_{i,j}(\omega)d\omega &= E\left[dU_i(\omega)dU_j^*(\omega)\right] \\ &= E\left[\alpha_i(\omega)\alpha_j(\omega)\right]E\left[e^{i\left[\phi_i(\omega)-\phi_j(\omega)\right]}\right],\end{aligned} \tag{9.37}$$

where for simplicity we assume that the amplitude and phase spectra are independent. Thus, $\alpha_{i,j}(\omega)$ can be thought as the average value of the product of amplitudes of the ω – frequency components of $Z_{i,t}$ and $Z_{j,t}$, and the phase spectrum $\phi_{i,j}(\omega)$ represents the average phase shift, $\phi_i(\omega) - \phi_j(\omega)$, between the ω – frequency components of $Z_{i,t}$ and $Z_{j,t}$. In terms of a causal relationship, $Z_{j,t} = \alpha Z_{i,t-\tau} + e_t$, or $Z_{i,t} = \alpha Z_{j,t-\tau} + e_t$, where there is no feedback relationship between them, the phase spectrum is a measure of the extent to which each frequency component of one series leads the other. The ω – frequency component of $Z_{i,t}$ leads the ω – frequencycomponent of $Z_{j,t}$ if the phase $\phi_{i,j}(\omega)$ is negative. The ω – frequency component of $Z_{i,t}$ lags the ω – frequency component of $Z_{j,t}$ if the phase $\phi_{i,j}(\omega)$ is positive. For a given $\phi_{i,j}(\omega)$, the shift in time units is $\phi_{i,j}(\omega)/\omega$, and the actual time delay of the ω – frequency component of $Z_{j,t}$ is equal to

$$\tau = -\frac{\phi_{i,j}(\omega)}{\omega}. \tag{9.38}$$

Other useful functions in the multivariate spectral analysis are the gain function defined as

$$G_{i,j}(\omega) = \frac{|f_{i,j}(\omega)|}{f_i(\omega)} = \frac{\alpha_{i,j}(\omega)}{f_i(\omega)}, \tag{9.39}$$

and the squared coherency function defined as

$$K_{i,j}^2(\omega) = \frac{|f_{i,j}(\omega)|^2}{f_i(\omega)f_j(\omega)}. \tag{9.40}$$

From Eqs. (9.23) and (9.24), it is easy to see that

$$K_{i,j}^2(\omega) = \frac{\{\text{Cov}\left[dU_i(\omega), dU_j(\omega)\right]\}^2}{\text{Var}[dU_i(\omega)]\text{Var}\left[dU_j(\omega)\right]}, \tag{9.41}$$

which is actually square of the correlation coefficient between the ω – frequency components of $Z_{i,t}$ and $Z_{j,t}$. A value of $K_{i,j}^2(\omega)$ close to 1 implies that the ω – frequency components of the two series are strongly linearly related, and a value of $K^2_{i,j}(\omega)$ close to 0 implies that they are very weakly linearly related. It should be noted that just like the correlation coefficient between two random variables, the square coherency is invariant under linear transformations.

Example 9.1 Given a simple linear model related to $Z_{i,t}$ and $Z_{j,t}$ as follows

$$Z_{j,t} = \alpha Z_{i,t-\ell} + e_t, \tag{9.42}$$

where $\ell > 0$, and $Z_{i,t}$ and e_t are jointly independent zero-mean stationary processes. We have

$$\begin{aligned}\gamma_{i,j}(k) &= E\left(Z_{i,t}Z_{j,t+k}\right)\\ &= E[Z_{i,t}(\alpha Z_{i,t+k-\ell} + e_{t+k})]\\ &= \alpha\gamma_i(k-\ell).\end{aligned}$$

So,

$$\begin{aligned}f_{i,j}(\omega) &= \frac{1}{2\pi}\sum_{k=-\infty}^{\infty}\gamma_{i,j}(k)e^{-i\omega k}\\ &= \frac{\alpha}{2\pi}\sum_{k=-\infty}^{\infty}\gamma_i(k-\ell)e^{-i\omega k}\\ &= \frac{\alpha}{2\pi}\sum_{s=-\infty}^{\infty}\gamma_i(s)e^{-i\omega(s+\ell)}\\ &= \alpha e^{-i\omega\ell}f_i(\omega).\end{aligned} \tag{9.43}$$

It implies that

$$\begin{aligned}c_{i,j}(\omega) &= \alpha\cos(\omega\ell)f_i(\omega),\\ q_{i,j}(\omega) &= \alpha\sin(\omega\ell)f_i(\omega),\\ \alpha_{i,j}(\omega) &= |\alpha|\,f_i(\omega),\\ \phi_{i,j}(\omega) &= \tan^{-1}\left(\frac{-q_{i,j}(\omega)}{c_{i,j}(\omega)}\right) = -\omega\ell,\end{aligned}$$

and from Eq. (9.38), we have

$$\tau = \ell.$$

Also,

$$\begin{aligned}G_{i,j}(\omega) &= \frac{\alpha_{i,j}(\omega)}{f_i(\omega)} = |\alpha|,\\ f_j(\omega) &= \alpha^2 f_i(\omega) + f_e(\omega),\end{aligned}$$

and

$$K_{i,j}^2(\omega) = \frac{|f_{i,j}(\omega)|^2}{f_i(\omega)f_j(\omega)}$$

$$= \frac{[\alpha e^{-i\omega\ell} f_i(\omega)][\alpha e^{i\omega\ell} f_i(\omega)]}{f_i(\omega)[\alpha^2 f_i(\omega) + f_e(\omega)]}$$

$$= \left[1 + \frac{f_e(\omega)}{\alpha^2 f_i(\omega)}\right]^{-1}.$$

If e_t is identically 0, $K_{i,j}^2(\omega) = 1$, for all ω. This is also expected since in this case, $Z_{j,t} = \alpha Z_{i,t-\ell}$, and $Z_{i,t}$ and $Z_{j,t}$ satisfy the exact linear relationship.

Example 9.2 Let us consider a m-dimensional Gaussian vector white noise, $\mathbf{Z}_t = \mathbf{a}_t$, with $\mathbf{a}_t$ being i.i.d. $N(\mathbf{0}, \mathbf{\Sigma})$ and

$$\mathbf{\Sigma} = \begin{bmatrix} \sigma_1^2 & 0 & . & . & . & 0 & 0 \\ 0 & \sigma_2^2 & 0 & . & . & . & 0 \\ 0 & 0 & . & . & . & . & . \\ . & . & . & . & . & . & . \\ . & . & . & . & . & 0 & 0 \\ 0 & . & . & . & 0 & \sigma_{m-1}^2 & 0 \\ 0 & 0 & . & . & . & 0 & \sigma_m^2 \end{bmatrix}.$$

For $i \neq j$, since $\gamma_{i,j}(k) = 0$, for all k, we have $f_{i,j}(\omega) = 0$, for all ω. Hence, $c_{i,j}(\omega) = 0$, $q_{i,j}(\omega) = 0$, $\alpha_{i,j}(\omega) = 0$, $\phi_{i,j}(\omega) = 0$, and for all ω, as expected.

9.3 The estimation of the spectral density matrix

9.3.1 The smoothed spectrum matrix

Given a zero-mean m-dimensional time series, $\mathbf{Z}_1, \mathbf{Z}_2, \ldots$, and $\mathbf{Z}_n$, its Fourier transform at the Fourier frequencies $\omega_p = 2\pi p/n$, $-[(n-1)/2] \leq p \leq [n/2]$, is

$$\mathbf{Y}(\omega_p) = \frac{1}{\sqrt{2\pi n}} \sum_{t=1}^{n} \mathbf{Z}_t \exp(-i\omega_p t). \tag{9.44}$$

Then, the $m \times m$ sample spectrum matrix, which is also known as periodogram matrix, is simply the extension of Eqs. (9.9)–(9.11). Thus,

$$\begin{aligned}
\widetilde{\mathbf{f}}(\omega_p) &= \mathbf{Y}(\omega_p)\mathbf{Y}^*(\omega_p) = |\mathbf{Y}(\omega_p)|^2 = \frac{1}{2\pi n}\left|\sum_{t=1}^{n}\mathbf{Z}_t \exp(-i\omega_p t)\right|^2 \\
&= \frac{1}{2\pi n}\left[\sum_{t=1}^{n}\mathbf{Z}_t \exp(-i\omega_p t)\right]\left[\sum_{r=1}^{n}\mathbf{Z}_r' \exp(i\omega_p r)\right] \\
&= \frac{1}{2\pi n}\sum_{t=1}^{n}\sum_{r=1}^{n}\mathbf{Z}_t\mathbf{Z}_r' e^{-i\omega_p(t-r)} \\
&= \frac{1}{2\pi}\sum_{k=-(n-1)}^{(n-1)} \hat{\mathbf{\Gamma}}(k)e^{-i\omega_p k} \\
&= \frac{1}{2\pi}\left[\hat{\mathbf{\Gamma}}(0) + 2\sum_{k=1}^{(n-1)}\hat{\mathbf{\Gamma}}(k)e^{-i\omega_p k}\right] = \left[\widetilde{f}_{i,j}(\omega_p)\right],
\end{aligned} \tag{9.45}$$

where

$$\hat{\mathbf{\Gamma}}(k) = [\hat{\gamma}_{i,j}(k)],$$

$$\widetilde{f}_{i,j}(\omega_p) = \frac{1}{2\pi}\sum_{k=-(n-1)}^{(n-1)} \hat{\gamma}_{i,j}(k)e^{-i\omega_p k} = y_i(\omega_p)y_j^*(\omega_p),$$

and

$$y_i(\omega_p) = \frac{1}{\sqrt{2\pi n}}\sum_{t=1}^{n} Z_{i,t}e^{-i\omega_p t}.$$

When $\mathbf{Z}_t$ is a multivariate Gaussian process with mean vector $\mathbf{0}$ and variance–covariance matrix $\mathbf{\Sigma}$, $\widetilde{\mathbf{f}}(\omega_p)$ has a distribution related to the sample variance–covariance matrix that is known as Wishart distribution with n degrees of freedom, which is the multivariate analog of the Chi-square distribution. We refer readers to Goodman (1963), Hannan (1970), and Brillinger (2002) for further discussion of the properties of the periodogram and Wishart distribution.

Similar to the univariate extension of the sample spectral density discussed in Section 9.1, the sample spectrum matrix or periodogram matrix is also a poor estimate. So, we replace it by the following smoothed spectrum matrix

$$\hat{\mathbf{f}}(\omega) = [\hat{f}_{i,j}(\omega)], \tag{9.46}$$

where

$$\hat{f}_{i,i}(\omega_p) = \hat{f}_i(\omega_p)\sum_{k=-M_i}^{M_i} W_i(\omega_k)\widetilde{f}_{i,i}(\omega_p - \omega_k), \tag{9.47}$$

$W_i(\omega)$ is a smoothing function, also known as kernel or spectral window, and M_i is the bandwidth of the spectral window, and

$$\hat{f}_{i,j}(\omega_p) = \sum_{k=-M_{i,j}}^{M_{i,j}} W_{i,j}(\omega_k)\tilde{f}_{i,j}(\omega_p - \omega_k), \tag{9.48}$$

where $W_{i,j}(\omega)$ is a spectral window, and $M_{i,j}$ is the corresponding bandwidth. Similar to the extension of the univariate smoothed spectrum, the smoothed spectrum matrix can also be approximated by the Wishart distribution.

Once $f_{i,i}(\omega)$ and $f_{i,j}(\omega)$ are estimated, we can estimate the co-spectrum, $c_{i,j}(\omega)$, the quadrature spectrum, $q_{i,j}(\omega)$, the cross-amplitude spectrum, $\alpha_{i,j}(\omega)$, phase spectrum, $\phi_{i,j}(\omega)$, the gain function, $G_{i,j}(\omega)$, and the squared coherency function, $K^2_{i,j}(\omega)$.

Note that the spectrum matrix is the Fourier transform of the covariance function, $\mathbf{\Gamma}(k) = [\gamma_{i,j}(k)]$, and the sample spectrum matrix is

$$\tilde{\mathbf{f}}(\omega_p) = \frac{1}{2\pi}\sum_{k=-(n-1)}^{(n-1)} \hat{\mathbf{\Gamma}}(k)e^{-i\omega_p k} = \left[\tilde{f}_{i,j}(\omega_p)\right]. \tag{9.49}$$

Instead of spectrum smoothing, we can also apply the smoothing function to the sample covariance matrices, that is,

$$\hat{\mathbf{f}}(\omega) = \left[\hat{f}_{i,j}(\omega)\right], \tag{9.50}$$

where

$$\hat{f}_{i,i}(\omega) = \hat{f}_i(\omega) = \frac{1}{2\pi}\sum_{k=-M_i}^{M_i} W_i(k)\hat{\gamma}_i(k)e^{-i\omega k}, \tag{9.51}$$

$\hat{\gamma}_i(k)$ is the sample autocovariance for the $Z_{i,t}$ series, $W_i(k)$ is a suitable lag window, and M_i is the truncation point, and

$$\hat{f}_{i,j}(\omega) = \frac{1}{2\pi}\sum_{k=-M_{i,j}}^{M_{i,j}} W_{i,j}(k)\hat{\gamma}_{i,j}(k)e^{-i\omega k}, \tag{9.52}$$

$\hat{\gamma}_{i,j}(k)$ is the sample cross-covariance function between $Z_{i,t}$ and $Z_{j,t}$, $W_{i,j}(k)$ is a suitable lag window, and $M_{i,j}$ is the corresponding truncation point. In this case, the estimations of the co-spectrum, the quadrature spectrum, the cross-amplitude spectrum, the phase spectrum, the gain function, and the squared coherency function are given by

$$\begin{aligned}\hat{c}_{i,j}(\omega) &= \frac{1}{2\pi}\sum_{k=-M_{i,j}}^{M_{i,j}} W_{i,j}(k)\hat{\gamma}_{i,j}(k)\cos(\omega k) \\ &= \frac{1}{2\pi}\left\{W_{i,j}(0)\hat{\gamma}_{i,j}(0) + \sum_{k=1}^{M_{i,j}} W_{i,j}(k)\left[\hat{\gamma}_{i,j}(k) + \hat{\gamma}_{i,j}(-k)\right]\cos(\omega k)\right\},\end{aligned} \tag{9.53}$$

$$\hat{q}_{i,j}(\omega) = \frac{1}{2\pi} \sum_{k=-M_{i,j}}^{M_{i,j}} W_{i,j}(k)\hat{\gamma}_{i,j}(k)\sin(\omega k)$$
$$= \frac{1}{2\pi}\left\{\sum_{k=1}^{M_{i,j}} W_{i,j}(k)\left[\hat{\gamma}_{i,j}(k) + \hat{\gamma}_{i,j}(-k)\right]\sin(\omega k)\right\}, \tag{9.54}$$

$$\hat{\alpha}_{i,j}(\omega) = \left[\hat{c}_{i,j}^2(\omega) + \hat{q}_{i,j}^2(\omega)\right]^{1/2}, \tag{9.55}$$

$$\hat{\phi}_{i,j}(\omega) = \tan^{-1}\left[\frac{-\hat{q}_{i,j}(\omega)}{\hat{c}_{i,j}(\omega)}\right], \tag{9.56}$$

$$\hat{G}_{i,j}(\omega) = \frac{\hat{\alpha}_{i,j}(\omega)}{\hat{f}_i(\omega)}, \tag{9.57}$$

and

$$\hat{K}_{i,j}^2(\omega) = \frac{\left|\hat{f}_{i,j}(\omega)\right|^2}{\hat{f}_i(\omega)\hat{f}_j(\omega)}. \tag{9.58}$$

For commonly used spectral or lag windows and their properties, we refer readers to Priestley (1981), Brillinger (2002), and Wei (2006).

Some important remarks are in order here.

1. The smoothed lag windows and the associated truncation points used in estimating $f_{i,i}(\omega)$ and $f_{i,j}(\omega)$ are not necessarily the same for all i and j. Even the same lag window is used, the associated truncation points can be different. However, the use of the same lag window and truncation point for all estimates will in general make the study of sampling properties of the estimates simpler.
2. In estimating the cross-spectrum, the lag window places heavier weight on the sample cross-covariances around the zero lag. If one series leads another and the cross-covariance does not peak at zero lag, then the estimated cross-spectrum will be biased. This bias is especially severe for the estimated square coherency. Because the square coherency is invariant under the linear transformation, to reduce the bias, one can properly align two series by shifting one of the series by a time lag τ so that the peak in the cross-spectrum function of the aligned series occurs at the zero lag. In practice, the time lag τ can be estimated by the lag corresponding to the maximum cross-correlation.
3. The estimates of the cross-amplitude, phase, and gain functions are not reliable when the square coherency is small.
4. It should be noted that when we write the spectral density as $f(\omega) = \frac{1}{2\pi}\sum_{k=-\infty}^{\infty}\gamma_k e^{-i\omega k}$, it implies that the range of the frequencies is from $-\pi$ to π. Given a time series of length n, due to symmetry, the Fourier frequencies used in the sample spectrum will be $\omega_j = 2\pi j/n$, with $j = 0, 1, \ldots, [n/2]$. However, if one writes the spectral density as

$f(\omega)=\sum_{k=-\infty}^{\infty}\gamma_k e^{-2\pi i\omega k}$, it implies that the range of the frequencies will be −1/2 to 1/2. In this case, the frequencies used in the sample spectrum will be $\omega_j = j/n$, with $j = 0, 1, \ldots, [n/2]$. In the following discussions, we may use either one depending on what is used in software and the referenced papers.

9.3.2 Multitaper smoothing

Developed by Thompson (1982) for univariate processes and extended by Walden (2000) for multivariate processes, multitaper smoothing is another useful way to estimate power spectrum density that balances the bias and variance of nonparametric spectral estimation. The multitaper reduces estimation bias by averaging modified periodograms obtained using a family of mutually orthogonal tapers from the same sample data. Let $h_j(t)$ for $t = 1, \ldots, n$ and $j = 1, \ldots, n$, be n orthonormal tapers such that

$$\sum_{t=1}^{n} h_j^2(t) = 1, \text{ and}$$
$$\sum_{t=1}^{n} h_i(t)h_j(t) = 0, (i \neq j). \tag{9.59}$$

From Eq. (9.45), we note that

$$\begin{aligned}\tilde{\mathbf{f}}(\omega) &= \frac{1}{2\pi}\sum_{k=-(n-1)}^{(n-1)} \hat{\boldsymbol{\Gamma}}(k)e^{-i\omega k} \\ &= \frac{1}{\sqrt{2\pi n}}\sum_{t=1}^{n}\mathbf{Z}_t e^{-i\omega t}\frac{1}{\sqrt{2\pi n}}\sum_{t=1}^{n}\mathbf{Z}_t' e^{i\omega t} \\ &= \tilde{\mathbf{Y}}(\omega)\tilde{\mathbf{Y}}^*(\omega),\end{aligned} \tag{9.60}$$

where $\tilde{\mathbf{Y}}(\omega) = \frac{1}{\sqrt{2\pi n}}\sum_{t=1}^{n}\mathbf{Z}_t e^{-i\omega t}$ is the discrete Fourier transform of $\mathbf{Z}_t$. The multitaper power spectral estimator at frequency ω is

$$\hat{\mathbf{f}}_M(\omega) = \frac{1}{K}\sum_{j=1}^{K}\hat{\mathbf{Y}}_j(\omega)\hat{\mathbf{Y}}_j^*(\omega), \tag{9.61}$$

where K is chosen through the method shown below and $\hat{\mathbf{Y}}_j(\omega)$ is the tapered Fourier transform such that

$$\hat{\mathbf{Y}}_j(\omega) = \frac{1}{\sqrt{2\pi n}}\sum_{t=1}^{n} h_j(t)\mathbf{Z}_t \exp(-i\omega t). \tag{9.62}$$

The multitaper spectral estimator is asymptotically unbiased and is consistent when the number of tapers K increase with the number of observations n.

One of the most commonly used multitapers is the Discrete Prolate Spheroidal Sequences (DPSS) or tapers suggested by Thompson (1982), also known as Slepian sequences or tapers because the term DPSS was given in Slepian (1978). The Slepian tapers are sequences of functions, which are the Fourier transform of the DPSS, that were designed to maximize the spectral concentration defined as

$$\lambda_j(n, W) = \frac{\int_{-W}^{W} |\mathrm{U}_j(\omega)|^2 d\omega}{\int_{-\pi/2}^{\pi/2} |\mathrm{U}_j(\omega)|^2 d\omega} \tag{9.63}$$

for a frequency W, $|W| < \pi/2$ is the bandwidth defining local and normally in the order of 1/n, where $\mathrm{U}_j(\omega) = \frac{1}{\sqrt{2\pi n}} \sum_{t=1}^{n} h_j(t) \exp(-i\omega t)$. Consequently, the first of the sequences, $\{h_1(t), t = 1, \ldots, n\}$, is chosen such that its corresponding spectral window or taper maximizing the concentration ratio in Eq. (9.63) over the interval $(-W, W)$, which is equal to $\lambda_1(n,W)$. This is done by maximizing the power

$$\int_{-W}^{W} |\mathrm{U}_j(\omega)|^2 d\omega$$

subject to the constraint that the total power is normalized such that

$$\int_{-\pi/2}^{\pi/2} |\mathrm{U}_j(\omega)|^2 d\omega = 1,$$

which leads to the following eigenvalue equation,

$$\sum_{t=1}^{n} \frac{\sin(2\pi W(t-t'))}{\pi(t-t')} h_j(t) = \lambda_j(n, W) h_j(t) \tag{9.64}$$

for $t' = 1, \ldots, n$. So, $\lambda_1(n,W)$ is the largest eigenvalue and $\{h_1(t), t = 1, \ldots, n\}$ is the first taper. The second taper, $\{h_2(t), t = 1, \ldots, n\}$ is selected to maximize the corresponding concentration ratio but subject to being orthogonal to the first. Going further, the third taper maximizes the concentration ratio, but subject to being orthogonal to the first two tapers. In general, the jth taper corresponds to the jth largest eigenvalue, $\lambda_j(n,W)$.

By maximizing concentration ratio, we are essentially minimizing the sidelobe energy outside a frequency band $(-W,W)$. The eigenvalues should be close to unity for well-behaved tapers. Let $\mathbf{M}$ be the matrix formed by those orthogonal tapers. Then $\mathbf{M}$ is an $n \times n$ positive–definite matrix with K dominant eigenvalues, $\lambda_1(n,W) > \lambda_2(n,W) > \cdots > \lambda_K(n,W) > \cdots > \lambda_n(n,W)$, which are close to 1.

Another widely used set of tapers is called the Sine tapers (Riedel and Sidorenko, 1995), which are in the form of

$$h_j(t) = \left(\frac{2}{n+1}\right)^{1/2} \sin\left(\pi t \frac{j+1}{n+1}\right), \tag{9.65}$$

for $t = 1, \ldots, n$. They were used by Dai and Guo (2004) to obtain a preliminary estimate of the spectral matrix in the smoothing spline framework discussed in the next section.

9.3.3 Smoothing spline

The main disadvantage of the kernel smoothing method and the multitaper smoothing method is that they cannot guarantee that the final estimate is positive semidefinite while allowing flexible smoothing for each element of the spectral matrix. Thus, the same bandwidth is often applied to smoothing all the spectral components. However, in many applications, different components of the spectral matrix may need different smoothnesses, and require different smoothing parameters to get optimal estimates. To overcome this difficulty, Dai and Guo (2004) proposed a Cholesky decomposition based smoothing spline method for the spectrum estimation. The method models each Cholesky component separately by using different smoothing parameters. The method first obtains positive–definite and asymptotically unbiased initial spectral estimator $\widetilde{\mathbf{f}}_M(\omega)$ through sine multitapers as shown in Eq. (9.65). Then, it further smooths the Cholesky components of the spectral matrix via the smoothing spline and penalized sum of squares, which allows different degrees of smoothness for different Cholesky elements.

Suppose the spectral matrix $\widetilde{\mathbf{f}}(\omega)$ has Cholesky decomposition such that $\widetilde{\mathbf{f}}(\omega) = \mathbf{\Gamma\Gamma}^*$, where $\mathbf{\Gamma}$ is $m \times m$ lower triangular matrix. To obtain unique decomposition, the diagonal elements of $\mathbf{\Gamma}$ are constrained to be positive. The diagonal elements $\gamma_{j,j}$, $j = 1, \ldots, m$, the real part of $\gamma_{j,k}$, $\Re(\gamma_{j,k})$, and imaginary part of $\gamma_{j,k}$, $\Im(\gamma_{j,k})$, $j > k$ are smoothed by spline with different smoothing parameters. Suppose $\gamma \in \{\gamma_{j,j}, \Re(\gamma_{j,k}), \Im(\gamma_{j,k}), j > k, \text{ for } j,k = 1, \ldots, m\}$, we have

$$\gamma(\omega_\ell) = a(\omega_\ell) + e(\omega_\ell), \tag{9.66}$$

where $a(\omega_\ell) = \mathrm{E}\{\gamma(\omega_\ell)\}$, and the $e(\omega_\ell)$, $\ell = 1, \ldots, n$,are independent errors with zero means and the variances depending on the frequency point ω_ℓ. $a(\cdot)$ is periodic and is fitted by periodic smoothing spline (Wahba, 1990) of the form,

$$a(\omega) = c_0 + \sum_{\nu=1}^{n/2-1} c_\nu \sqrt{2}\cos(2\pi\nu\omega) + \sum_{\nu=1}^{n/2-1} d_\nu \sqrt{2}\sin(2\pi\nu\omega) + c_{n/2}\cos(\pi n\omega). \tag{9.67}$$

The estimator of power spectrum component is obtained by minimizing the following

$$\frac{1}{\tau}\sum_{\ell=1}^{n} \{\gamma(\omega_\ell) - \hat{\gamma}(\omega_\ell)\}^2 + \lambda \int_0^1 [\hat{\gamma}''(\omega)]^2 d\omega, \tag{9.68}$$

where because of issues related asymptotic distributions, $\omega_\ell = 0, 0.5,$ and 1 are not used in the estimation, $\tau = n - 1$ if n is odd, and $\tau = n - 2$ if n is even; $\hat{\gamma}(\omega_\ell)$ is the estimator of $\gamma(\omega_\ell)$; and λ is a smoothing parameter. The final form of the estimator is given by

$$\hat{\gamma}(\omega) = \frac{1}{\tau}\sum_{\ell=1}^{n} \sum_{t=-[(n-1)/2]}^{[n/2]} \frac{1}{1 + \lambda(2\pi t)^4} \exp\{2\pi i t(\omega - \omega_\ell)\}\gamma(\omega_\ell), \tag{9.69}$$

where $[\cdot]$ is the greatest integer function. The smoothing parameter λ can be chosen by the generalized cross-validation or the generalized maximum likelihood. For more details, we refer readers to Wahba (1990) and Dai and Guo (2004). Qin and Wang (2008) suggested, through simulations, that the generalized maximum likelihood is stable and performs better than the generalized cross-validation.

9.3.4 Bayesian method

Recall that the discrete Fourier transform $\widetilde{\mathbf{Y}}_\ell = \widetilde{\mathbf{Y}}(\omega_\ell)$, $\ell = 0, \ldots, (n-1)/2$, are approximately independent complex multivariate normal random variables. The large-sample distribution of $\widetilde{\mathbf{Y}}_\ell$ leads to the Whittle likelihood (Whittle, 1953, 1954),

$$L(\mathbf{f}) = \prod_{\ell=1}^{L} |\mathbf{f}(\omega_\ell)|^{-1} \exp\left\{-\widetilde{\mathbf{Y}}_\ell^* [\mathbf{f}(\omega_\ell)]^{-1} \widetilde{\mathbf{Y}}_\ell\right\}, \tag{9.70}$$

where $\omega_k = \ell/n$ and $L = [(n-1)/2]$. Based on the Whittle likelihood, Rosen and Stoffer (2007) proposed a Bayesian method to the estimate spectrum of the second-order stationary multivariate time series. They model the spectrum $\mathbf{f}(\omega)$ by the modified complex Cholesky factorization, such that,

$$\mathbf{f}^{-1}(\omega_\ell) = \boldsymbol{\Gamma}_\ell^* \mathbf{D}_\ell^{-1} \boldsymbol{\Gamma}_\ell, \tag{9.71}$$

where $\boldsymbol{\Gamma}_\ell$ is a complex-valued lower triangular matrix with one on its diagonal,

$$\boldsymbol{\Gamma}_\ell = \begin{bmatrix} 1 & 0 & \cdots & \cdots & \cdots & \cdots & 0 \\ -\theta_{2,1}^{(\ell)} & 1 & 0 & \cdots & \cdots & \cdots & 0 \\ -\theta_{3,1}^{(\ell)} & -\theta_{3,2}^{(\ell)} & 1 & 0 & \cdots & \cdots & \vdots \\ \vdots & \cdots & \vdots & \ddots & \ddots & \vdots & \vdots \\ \vdots & \vdots & \vdots & \cdots & \ddots & 0 & \vdots \\ \vdots & \cdots & \cdots & \cdots & \cdots & 1 & 0 \\ -\theta_{m,1}^{(\ell)} & -\theta_{m,2}^{(\ell)} & \cdots & \cdots & \cdots & -\theta_{m,m-1}^{(\ell)} & 1 \end{bmatrix},$$

and $\mathbf{D}_\ell$ is a diagonal matrix with positive real values, that is, $\mathbf{D}_\ell = \text{diag}\left(d_{1,\ell}^2, \ldots, d_{m,\ell}^2\right)$. Let $\boldsymbol{\theta}_\ell = \left(\theta_{2,1}^{(\ell)}, \theta_{3,1}^{(\ell)}, \theta_{3,2}^{(\ell)}, \ldots, \theta_{m,m-1}^{(\ell)}\right)$, $\boldsymbol{\Theta} = (\boldsymbol{\theta}_1, \ldots, \boldsymbol{\theta}_L)$, and $\mathbf{D} = \{\mathbf{D}_1, \ldots, \mathbf{D}_L\}$,the modified Cholesky representation facilitates development of Bayesian sampler by noticing that the Whittle likelihood can be rewritten as

$$L(\mathbf{Y}|\mathbf{D}, \boldsymbol{\Theta}) \approx \prod_{\ell=1}^{L}\prod_{j=1}^{m} d_{j,\ell}^{-2} \exp\left\{-\left(\widetilde{\mathbf{Y}}(\omega_\ell) - \mathbf{R}_\ell \boldsymbol{\theta}_\ell\right)^* \mathbf{D}_\ell^{-1} \left(\widetilde{\mathbf{Y}}(\omega_\ell) - \mathbf{R}_\ell \boldsymbol{\theta}_\ell\right)\right\}, \tag{9.72}$$

where $\boldsymbol{\theta}_\ell$ is an $m(m-1)/2$–dimensional vector and $\mathbf{R}_\ell$ is a $m \times m(m-1)/2$ design matrix such that

$$\mathbf{R}_\ell = \begin{bmatrix} 0 & 0 & 0 & 0 & 0 & 0 & 0 & \dots & 0 & 0 & 0 & \dots & 0 \\ Y_{1,\ell} & 0 & 0 & 0 & 0 & 0 & 0 & \dots & 0 & 0 & 0 & \dots & 0 \\ 0 & Y_{1,\ell} & Y_{2,\ell} & 0 & 0 & 0 & 0 & \dots & 0 & 0 & 0 & \dots & 0 \\ 0 & 0 & 0 & Y_{1,\ell} & Y_{2,\ell} & Y_{3,\ell} & 0 & \dots & 0 & 0 & 0 & \dots & 0 \\ 0 & 0 & 0 & 0 & 0 & 0 & \dots & \dots & \dots & \dots & \dots & \dots & \vdots \\ \vdots & \vdots & \vdots & \vdots & \vdots & \vdots & \vdots & \dots & \dots & \dots & \dots & \dots & 0 \\ 0 & 0 & 0 & 0 & 0 & 0 & 0 & \dots & 0 & Y_{1,\ell} & Y_{2,\ell} & \dots & Y_{m-1,\ell} \end{bmatrix}$$

and $Y_{j,\ell}$ is the jth entry of $\widetilde{\mathbf{Y}}(\omega_\ell)$. Let $\boldsymbol{\Delta}_j = \left(d_{j,1}^2, \dots, d_{j,L}^2\right)'$ and $\boldsymbol{\theta}_{j,k} = \left(\theta_{j,k}^{(1)}, \dots, \theta_{j,k}^{(L)}\right)'$. Each component of the Cholesky decomposition is fitted by the Demmler–Reinsch basis functions for linear smoothing splines of Eubank and Hsing (2008) as follows

$$c_0 + c_1\omega_\ell + \sum_{k=1}^{L} \sqrt{2}\cos\{(k-1)\pi\omega_\ell\}d_k \tag{9.73}$$

for each frequency ω_ℓ. Let $\mathbf{X}$ be the design matrix of the basis functions and $\boldsymbol{\beta}_j = \left(\mathbf{c}_j', \mathbf{d}_j'\right)'$ be the associated parameters. We then have

$$\log\left(\boldsymbol{\Delta}_j\right) = \mathbf{X}\boldsymbol{\beta}_j,\ \ \Re\left(\boldsymbol{\theta}_{j,k}\right) = -\mathbf{X}\boldsymbol{\beta}_{j,k,(re)},\ \ \Im\left(\boldsymbol{\theta}_{j,k}\right) = -\mathbf{X}\boldsymbol{\beta}_{j,k,(im)}, \tag{9.74}$$

for $j = 1, \dots, m$, $k = 1, \dots, j-1$. The priors on $\mathbf{c}_j$, $\mathbf{c}_{j,k,(re)}$, and $\mathbf{c}_{j,k,(im)}$ are chosen to be bivariate normal distributions $N\left(\mathbf{0}, \sigma_{\mathbf{c}_j}^2\mathbf{I}_2\right), N\left(\mathbf{0}, \sigma_{\mathbf{c}_{j,k,(re)}}^2\mathbf{I}_2\right), N\left(\mathbf{0}, \sigma_{\mathbf{c}_{j,k,(im)}}^2\mathbf{I}_2\right)$, and the priors on $\mathbf{d}_j$, $\mathbf{d}_{j,k,(re)}$, and $\mathbf{d}_{j,k,(im)}$ are chosen to be L–dimensional normal distributions $N\left(\mathbf{0}, \lambda_j^2\mathbf{I}_L\right)$, $N\left(\mathbf{0}, \lambda_{j,k,(re)}^2\mathbf{I}_L\right)$, and $N\left(\mathbf{0}, \lambda_{j,k,(im)}^2\mathbf{I}_L\right)$, respectively. The hyperparameters λ_j^2, $\lambda_{j,k,(re)}^2$, and $\lambda_{j,k,(im)}^2$ are smoothing parameters, which control the amount of smoothness. As the smoothing parameters tend to zero, the spline becomes a linear fit; as the smoothing parameters tend to infinity, the spline will be an interpolating spline. Gibbs sampling with the Metropolis–Hastings algorithm is used to draw parameters from posterior distribution.

9.3.5 Penalized Whittle likelihood

Penalized methods have been one of the most popular research topics for the past decade. In the area of spectrum estimation, Pawitan and O'Sullivan (1994) developed penalized likelihood method for the nonparametric estimation of the power spectrum of a univariate time series. Pawitan (1996) developed a penalized Whittle likelihood estimator for a bivariate time series. However, their approach cannot be easily generalized to higher dimension. More recently, Krafty and Collinge (2013) proposed a penalized Whittle likelihood method to estimate the power spectrum of a vector-valued time series. The method allows for varying levels of smoothness among spectral components while accounting for the positive definiteness of spectral matrices and the Hermitian and periodic structures of power spectra as functions of frequency. In this section, we briefly discuss the method by Krafty and Collinge (2013).

Recall that $\widetilde{\mathbf{Y}}_\ell = \widetilde{\mathbf{Y}}(\omega_\ell)$ are discrete Fourier transform of the time series, and define the negative log Whittle likelihood as

$$L(\mathbf{f}) = \sum_{\ell=1}^{L} \left\{ \log|\mathbf{f}(\omega_\ell)| + \widetilde{\mathbf{Y}}_\ell^* [\mathbf{f}(\omega_\ell)]^{-1} \widetilde{\mathbf{Y}}_\ell \right\}, \tag{9.75}$$

where $\omega_\ell = \ell/n$ and $L = [(n-1)/2]$. Based on L, the method penalizes the roughness of the estimated spectrum. The spectral matrix $\mathbf{f}(\omega)$ is modeled by the Cholesky decomposition such that $\mathbf{f}(\omega) = [\boldsymbol{\Gamma}\boldsymbol{\Gamma}^*]^{-1}$, where $\boldsymbol{\Gamma}$ is a $m \times m$ lower triangular matrix with real-valued diagonal elements. Further, denote $\boldsymbol{\Gamma}_{i,j,R}(\omega;\mathbf{f})$ and $\boldsymbol{\Gamma}_{i,j,I}(\omega;\mathbf{f})$ as the real and imaginary parts of the (i,j) element of $\boldsymbol{\Gamma}$. The method proposes a measure of roughness of a power spectrum through the integrated squared kth derivatives of the $m(m+1)/2$ real and $m(m-1)/2$ imaginary components of the Cholesky decomposition. Suppose the smoothing parameters, $\boldsymbol{\lambda} = \{\rho_{i,j}, \theta_{i,j} : i \le j = 1, \dots, m\}$,of $\boldsymbol{\Gamma}_{i,j,R}$ and $\boldsymbol{\Gamma}_{i,j,I}$ that control the roughness penalty are $\rho_{i,j} > 0$ and $\theta_{i,j} > 0$, the roughness measure for a spectrum $\widetilde{\mathbf{f}}(\omega)$ is

$$J_\lambda(\mathbf{f}) = \sum_{i \le j=1}^{m} \rho_{i,j} \int_0^{1/2} \left\{ \boldsymbol{\Gamma}_{i,j,R}^{(k)}(\omega;\mathbf{f}) \right\}^2 d\omega + \sum_{i<j=1}^{m} \theta_{i,j} \int_0^{1/2} \left\{ \boldsymbol{\Gamma}_{i,j,I}^{(k)}(\omega;\mathbf{f}) \right\}^2 d\omega. \tag{9.76}$$

The interpretation is that the penalty function J shrinks the estimates of power spectra toward real-valued matrix functions that are constant across frequency. Consequently, we consider minimizing the penalized Whittle negative loglikelihood

$$Q_\lambda(\mathbf{f}) = L(\mathbf{f}) + J_\lambda(\mathbf{f}).$$

The estimation procedure is based on an iterative algorithm adopted from Wood (2011) and the confidence intervals are obtained via bootstrap.

9.3.6 VARMA spectral estimation

As shown in Chapter 2, the underlying vector time series process is often described by a vector autoregressive moving average (VARMA) model. Specifically, the m-dimensional VARMA process of order p and q, VARMA(p, q) is given by

$$\mathbf{Z}_t = \boldsymbol{\theta}_0 + \boldsymbol{\Phi}_1 \mathbf{Z}_{t-1} + \ldots + \boldsymbol{\Phi}_p \mathbf{Z}_{t-p} + \mathbf{a}_t - \boldsymbol{\Theta}_1 \mathbf{a}_{t-1} - \ldots - \boldsymbol{\Theta}_q \mathbf{a}_{t-q}, \tag{9.77}$$

or

$$\boldsymbol{\Phi}_p(B)\dot{\mathbf{Z}}_t = \boldsymbol{\Theta}_q(B)\mathbf{a}_t, \tag{9.78}$$

where $\dot{\mathbf{Z}}_t = \mathbf{Z}_t - \boldsymbol{\mu}$, $\mathbf{a}_t$ is a sequence of m-dimensional vector white noise process, **VWN** $(\mathbf{0}, \boldsymbol{\Sigma})$, and

$$\begin{aligned}\mathbf{\Phi}_p(B) &= \mathbf{I}-\mathbf{\Phi}_1 B - \ldots - \mathbf{\Phi}_p B^p,\\ \mathbf{\Theta}_q(B) &= \mathbf{I}-\mathbf{\Theta}_1 B - \ldots - \mathbf{\Theta}_q B^q.\end{aligned} \tag{9.79}$$

We assume that all zeros of $|\mathbf{\Phi}_p(B)|$ and $|\mathbf{\Theta}_q(B)|$ lie outside of the unit circle, so that we can also represent it as

$$\dot{\mathbf{Z}}_t = \mathbf{\Psi}(B)\mathbf{a}_t, \tag{9.80}$$

where $\mathbf{\Psi}(B) = [\mathbf{\Phi}_p(B)]^{-1}\mathbf{\Theta}_q(B) = \sum_{j=0}^{\infty}\mathbf{\Psi}_j B^j$ and the sequence $\mathbf{\Psi}_j$ is square summable. The spectral density matrix of VARMA(p, q) model is given by

$$\mathbf{f}(\omega) = \mathbf{\Psi}\left(e^{-i\omega}\right)\mathbf{\Sigma}\mathbf{\Psi}^*\left(e^{-i\omega}\right), \tag{9.81}$$

where $\mathbf{\Psi}(e^{-i\omega}) = \sum_{j=0}^{\infty}\mathbf{\Psi}_j e^{-i\omega j}$, and $\mathbf{\Psi}^*(e^{-i\omega})$ is its conjugate transpose.

Given a vector time series of n observations, $\mathbf{Z} = (\mathbf{Z}_1, \mathbf{Z}_2, \ldots, \mathbf{Z}_n)$, we will first build its **VARMA**(p,q) model including the estimation of its parameter matrices. Let $\hat{\boldsymbol{\mu}} = \bar{\mathbf{Z}}$, $\hat{\mathbf{\Phi}}_1, \ldots, \hat{\mathbf{\Phi}}_p, \hat{\mathbf{\Theta}}_1, \ldots, \hat{\mathbf{\Theta}}_q$ and $\hat{\mathbf{\Sigma}}$ be the corresponding estimates of the parameter matrices. We have

$$\widehat{\dot{\mathbf{Z}}}_t = \hat{\mathbf{\Psi}}(B)\mathbf{a}_t, \tag{9.82}$$

where $\widehat{\dot{\mathbf{Z}}}_t = \mathbf{Z}_t - \bar{\mathbf{Z}}$, $\hat{\mathbf{\Psi}}(B) = \sum_{s=0}^{\infty}\hat{\mathbf{\Psi}}_s B^s = \left[\mathbf{I}-\hat{\mathbf{\Phi}}_1 B - \ldots - \hat{\mathbf{\Phi}}_p B^p\right]^{-1}\left[\mathbf{I}-\hat{\mathbf{\Theta}}_1 B - \ldots - \hat{\mathbf{\Theta}}_q B^q\right]$, and $\mathbf{a}_t$ is a sequence of m-dimensional vector white noise, $\mathbf{VWN}(\mathbf{0}, \hat{\mathbf{\Sigma}})$. Then, the spectral density matrix estimation of the underlying process is given by

$$\hat{\mathbf{f}}(\omega) = \hat{\mathbf{\Psi}}\left(e^{-i\omega}\right)\hat{\mathbf{\Sigma}}\hat{\mathbf{\Psi}}^*\left(e^{-i\omega}\right), \tag{9.83}$$

where

$$\begin{aligned}\hat{\mathbf{\Psi}}\left(e^{-i\omega}\right) &= \sum_{j=0}^{\infty}\hat{\mathbf{\Psi}}_j e^{-i\omega j}\\ &= \left[\hat{\mathbf{\Phi}}_p\left(e^{-i\omega}\right)\right]^{-1}\hat{\mathbf{\Theta}}_q\left(e^{-i\omega}\right)\\ &= \left[\left(\mathbf{I}-\hat{\mathbf{\Phi}}_1 e^{-i\omega} - \ldots - \hat{\mathbf{\Phi}}_p e^{-i\omega p}\right)\right]^{-1}\left[\left(\mathbf{I}-\hat{\mathbf{\Theta}}_1 e^{-i\omega} - \ldots - \hat{\mathbf{\Theta}}_q e^{-i\omega q}\right)\right].\end{aligned} \tag{9.84}$$

In practice, given a vector time series $\mathbf{Z}_t$, $t = 1, 2, \ldots, n$, one can often approximate the underlying process with a VAR(p) model,

$$\hat{\mathbf{\Phi}}_P(B)\dot{\mathbf{Z}}_t = \mathbf{a}_t. \tag{9.85}$$

So, the commonly used parametric estimation of spectral matrix is

$$\hat{\mathbf{f}}(\omega) = \left[\hat{\mathbf{\Phi}}_p\left(e^{-i\omega}\right)\right]^{-1}\hat{\mathbf{\Sigma}}\left\{\left[\hat{\mathbf{\Phi}}_p\left(e^{-i\omega}\right)\right]^{-1}\right\}^*, \tag{9.86}$$

where the order p can be determined through some kind of model selection criteria such as Akaike's information criterion (AIC) or the Bayesian information criterion (BIC).

9.4 Empirical examples of stationary vector time series

Example 9.3 In this example, we consider the monthly U.S. sales from June 2, 2009 to November 11, 2016 in five categories: AUT (automobile), BUM (building materials), GEM (general merchandise), GRO (grocery), and COM (consumer materials) introduced as WW2b in Chapter 2, where COM is the sum of CLO (clothing), BWL (beer wine and liquor), and FUR (furniture). The only difference is that we take the difference of these five series and work on the monthly changes of these sales. There are 90 points for each original sales, so in the following analysis, we will have 89 data points for each series, as shown in Figure 9.1. It is clear that the time series contain some periodic patterns.

9.4.1 Sample spectrum

Example 9.3(a) Based on the given *five*-dimensional vector time series $\mathbf{Z}_t = [Z_{1,t}, Z_{2,t}, \ldots, Z_{5,t}]'$ for $t = 1, 2, \ldots, 89$, we can compute its sample covariance matrix function, $\hat{\boldsymbol{\Gamma}}(k) = [\hat{\gamma}_{i,j}(k)]$. Then, the sample spectral matrix is given by

$$\tilde{\mathbf{f}}(\omega_j) = \frac{1}{2\pi}\sum_{k=-88}^{88} \hat{\boldsymbol{\Gamma}}(k)e^{-i\omega_j k} = \left[\tilde{f}_{i,j}(\omega_j)\right], \tag{9.87}$$

where the $\omega_j = 2\pi j/89$ are Fourier frequencies.

Let us first examine the sample spectrum of each series, which as shown in Wei (2006) and Priestley (1981) can be used to test hidden periodic components. They are displayed in Figure 9.2 and they are noisy.

So, we will compute a smoothed kernel sample spectrum matrix

$$\hat{\mathbf{f}}(\omega) = [\hat{f}_{i,j}(\omega)], \tag{9.88}$$

where

$$\hat{f}_{i,j}(\omega_p) = \sum_{k=-M}^{M} W(\omega_k)\tilde{f}_{i,j}(\omega_p - \omega_k), \tag{9.89}$$

where $W(\omega)$ is a spectral window, and M is the corresponding bandwidth. Let us simply consider Daniell's window, that is,

$$\hat{\mathbf{f}}(\omega_p) = \frac{1}{2M+1}\sum_{k=-M}^{M} \tilde{\mathbf{f}}(\omega_p - \omega_k) = [\hat{f}_{i,j}(\omega_p)], \tag{9.90}$$

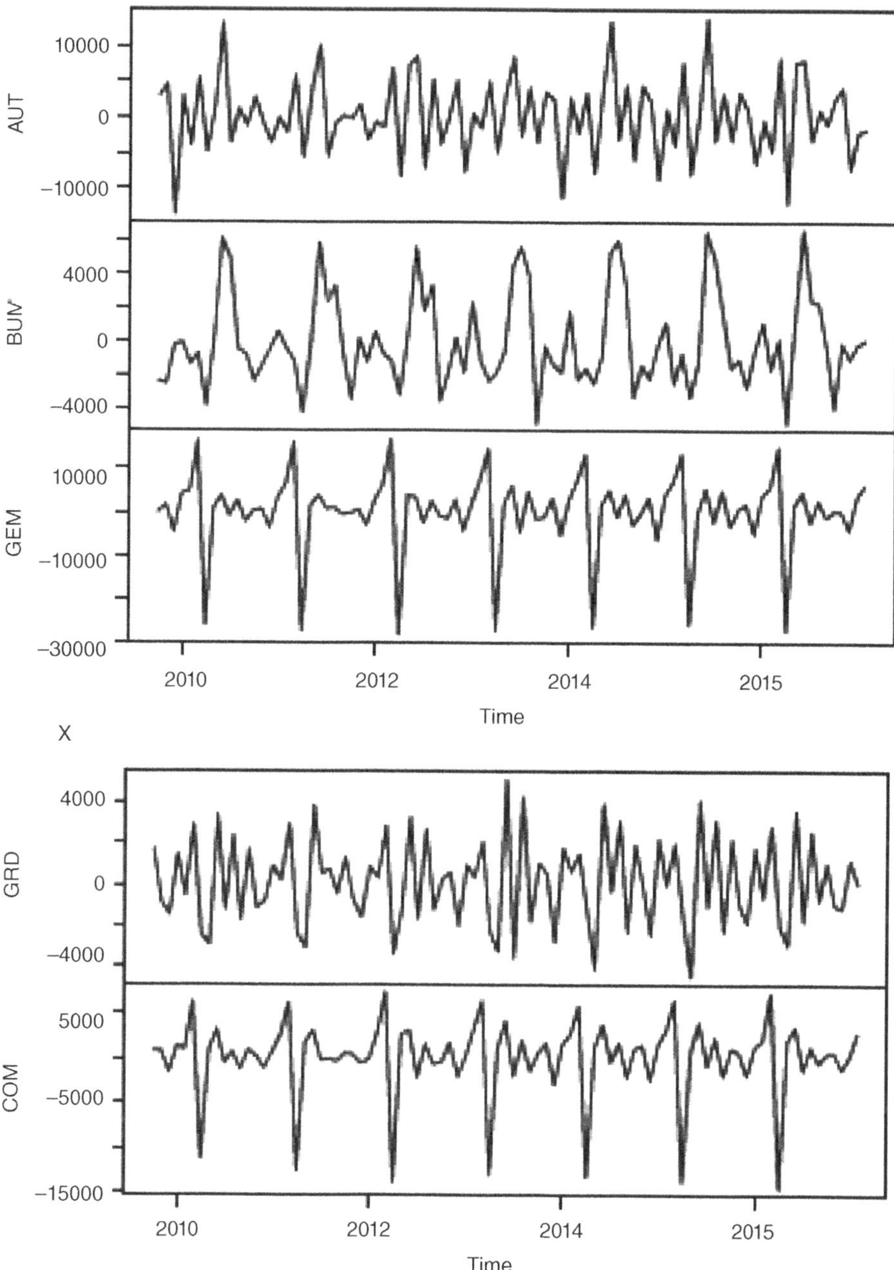

Figure 9.1 Five U.S. monthly changes of sales from June 2, 2009 to November 11, 2016 in AUT (automobile), BUM (building materials), GEM (general merchandise), GRO (grocery), and COM (consumer materials).

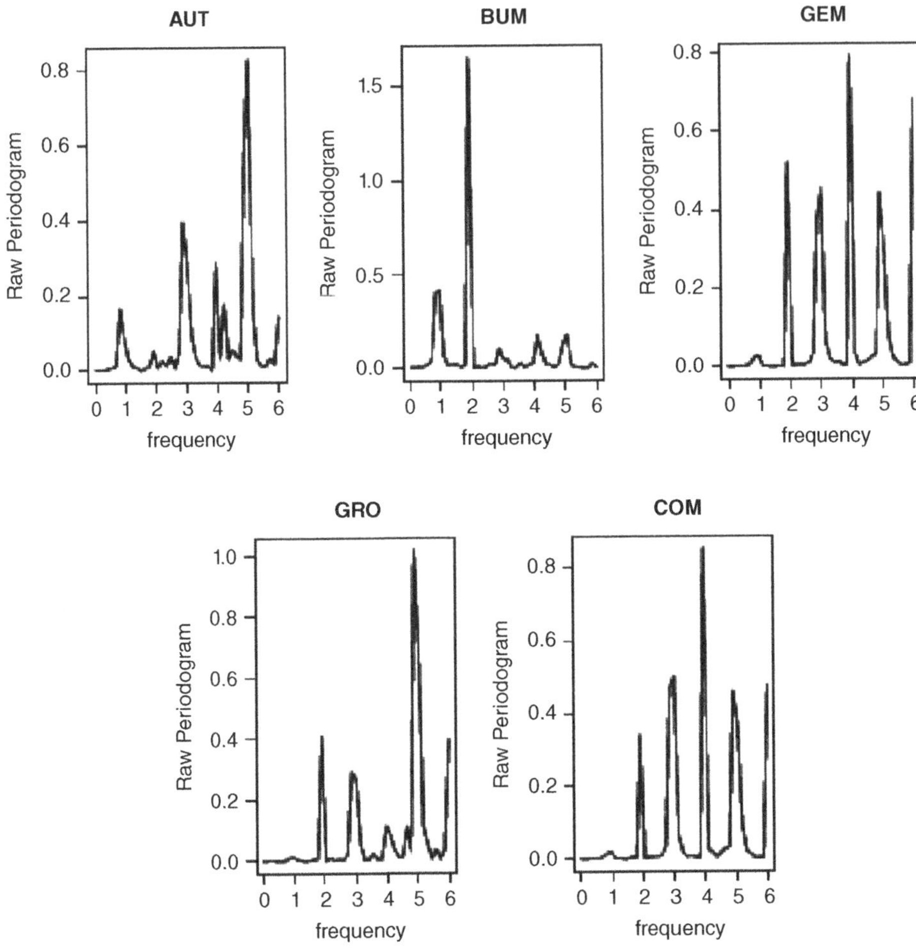

Figure 9.2 The sample spectrum of the five U.S. monthly sales changes.

where

$$\hat{f}_{i,j}(\omega_p) = \frac{1}{2M+1} \sum_{k=-M}^{M} \widetilde{f}_{i,j}(\omega_p - \omega_k). \tag{9.91}$$

The smoothed five individual sample spectra with $M = 5$ are shown on the first row in Figure 9.3. We then increase the truncation point M to 7. The corresponding five smoothed sample spectra are shown on the second row in Figure 9.3, and they are much smoother.

It is clear that, except AUT, almost all other estimated spectra contain a large peak at two cycles per year, which corresponds to a half-year periodic pattern. In addition, the estimated spectra of AUT and BUM have a peak at one cycle per year that is related to a yearly pattern. Finally, the spectra of GEM and COM also suggest a peak at four cycles per year, which is related to seasonal effects. In general, the power of AUT and GRO concentrates at higher frequencies; the power of BUM concentrates at lower frequencies; and the power of GEM and COM seems evenly spread out from medium to high frequencies.

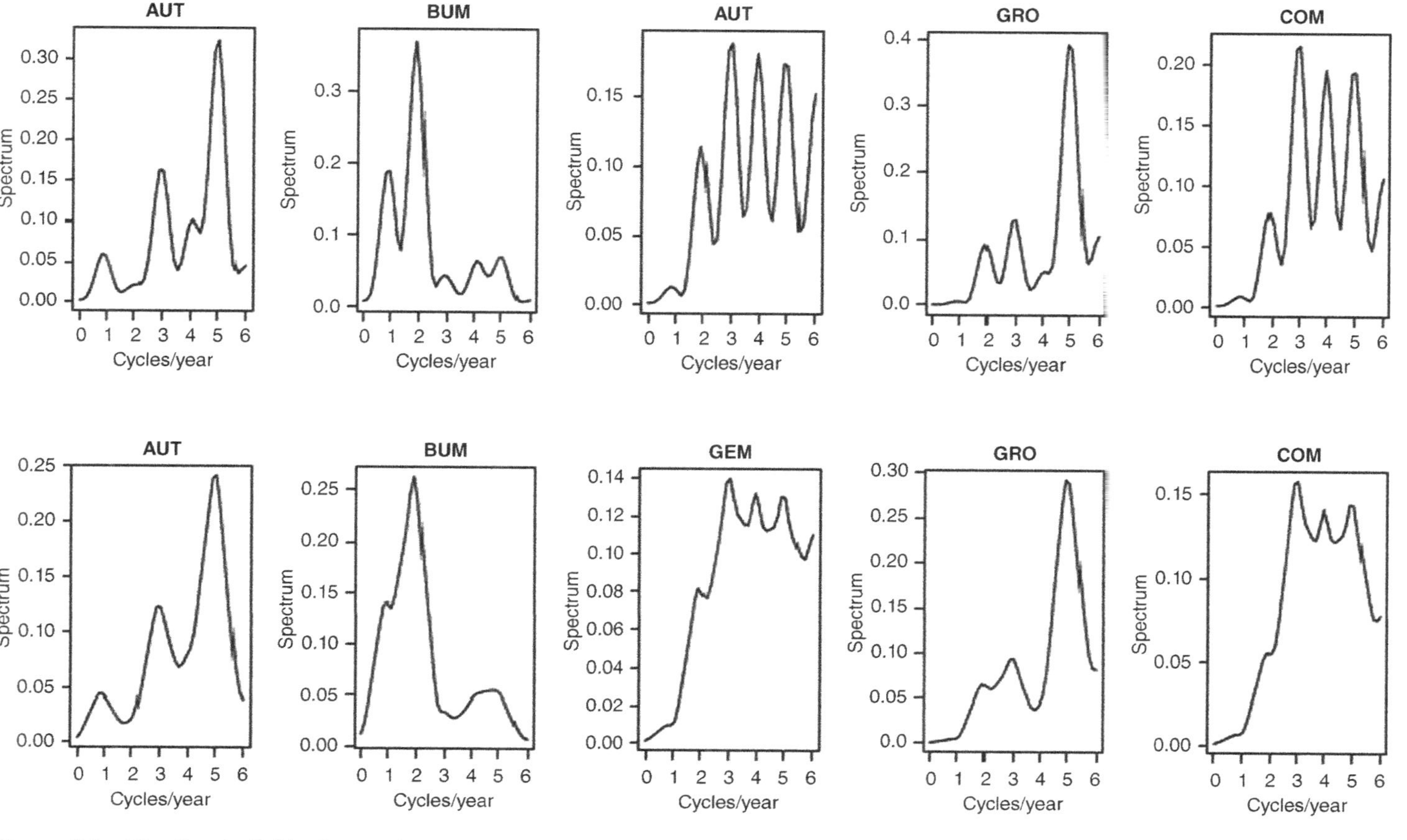

Figure 9.3 The five individual smoothed sample spectrum using Daniell window with $M = 5$: first row; Daniell window with $M = 7$: second row.

From Eqs. (9.31) and (9.32), we have

$$c_{i,j}(\omega) = \frac{1}{2}\left\{f_{i,j}(\omega) + f_{i,j}^{*}(\omega)\right\}, \tag{9.92}$$

and

$$q_{i,j}(\omega) = \frac{1}{2}\left\{f_{i,j}^{*}(\omega) - f_{i,j}(\omega)\right\}. \tag{9.93}$$

Once the smoothed $\hat{f}_{i,j}(\omega)$ are obtained, we can compute the estimated co-spectra, $\hat{c}_{i,j}(\omega)$, and the estimated quadrature spectra, $\hat{q}_{i,j}(\omega)$, as follows.

$$\hat{c}_{i,j}(\omega) = \frac{1}{2}\left\{\hat{f}_{i,j}(\omega) + \hat{f}_{i,j}^{*}(\omega)\right\}, \tag{9.94}$$

and

$$\hat{q}_{i,j}(\omega) = \frac{1}{2}\left\{\hat{f}_{i,j}^{*}(\omega) - \hat{f}_{i,j}(\omega)\right\}, \tag{9.95}$$

for $i = 1, \ldots, 5$ and $j = 1, \ldots, 5$. For an illustration, we use Daniell window with $M = 7$, and Figure 9.4 shows the estimated co-spectra, $\hat{c}_{5,j}(\omega)$, and quadrature spectra, $\hat{q}_{5,j}(\omega)$, for $j = 1, \ldots, 4$.

The estimated squared coherences can be obtained by

$$\hat{K}_{i,j}^{2}(\omega) = \frac{\left|\hat{f}_{i,j}(\omega)\right|^{2}}{\hat{f}_{i}(\omega)\hat{f}_{j}(\omega)} = \frac{\hat{c}_{i,j}^{2}(\omega) + \hat{q}_{i,j}^{2}(\omega)}{\hat{f}_{i}(\omega)\hat{f}_{j}(\omega)}. \tag{9.96}$$

They are shown in Figure 9.5. An unexpected observation is that the squared coherence between COM and GEM is large and is constant across all frequencies. This indicates a strong relationship between sales changes of consumer materials and general merchandise.

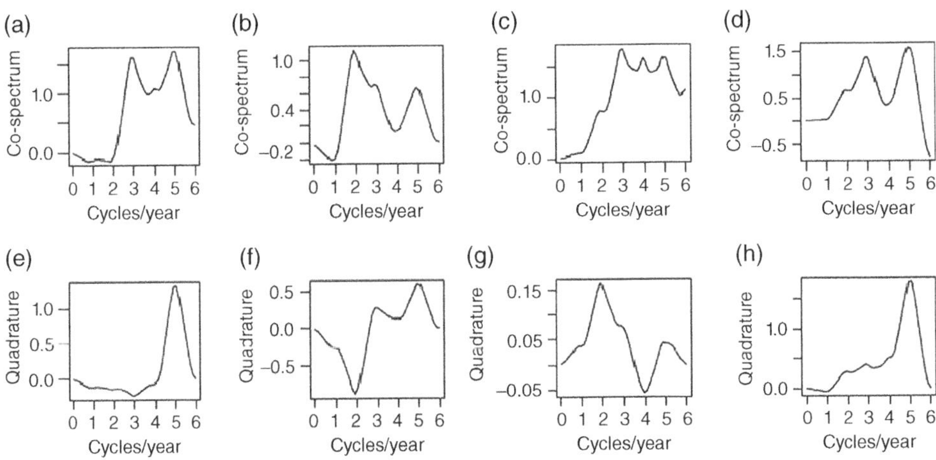

Figure 9.4 Estimated co-spectra (a–d): $\hat{c}_{5,1}(\omega)$, $\hat{c}_{5,2}(\omega)$, $\hat{c}_{5,3}(\omega)$, and $\hat{c}_{5,4}(\omega)$; estimated quadrature spectra (e–h): $\hat{q}_{5,1}(\omega)$, $\hat{q}_{5,2}(\omega)$, $\hat{q}_{5,3}(\omega)$, and $\hat{q}_{5,4}(\omega)$.

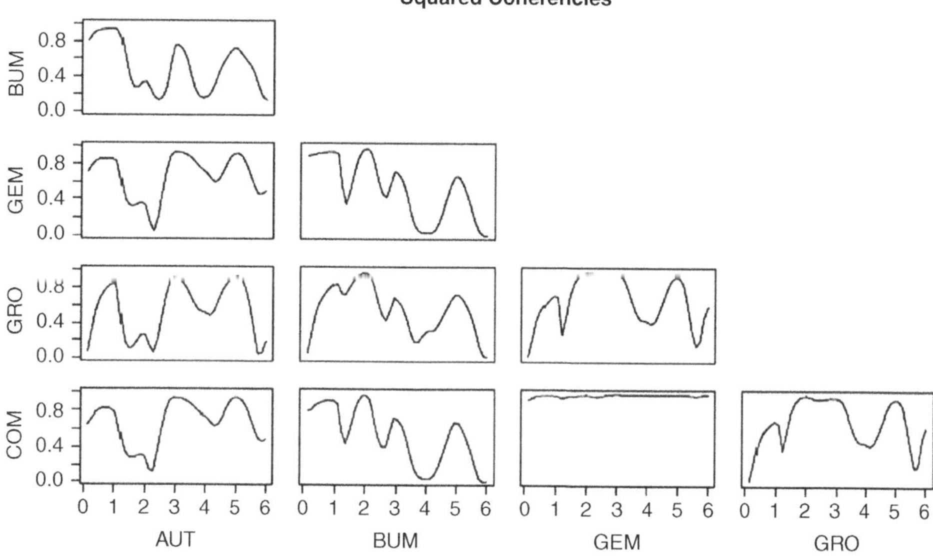

Figure 9.5 The estimated squared coherences with the Daniell window and bandwidth of 7.

9.4.2 Bayesian method

Example 9.3(b) Next, we apply the Bayesian spectral estimation method of Rosen and Stoffer (2007) to this dataset. Since the computer program is able to analyze at the maximum a three-dimensional time series, we select AUT, BUM, and GEM to construct a trivariate time series in this analysis. The estimated spectra are shown in Figure 9.6. The results are largely consistent with the results of the kernel smoothing method. However, it appears that the spectra obtained from the Bayesian method are smoother compared to that of the kernel smoothing method. In addition, it seems that the GEM series only have one peak around three to four cycles per year, and its other harmonic peaks at higher frequencies disappear according to the Bayesian spectral analysis.

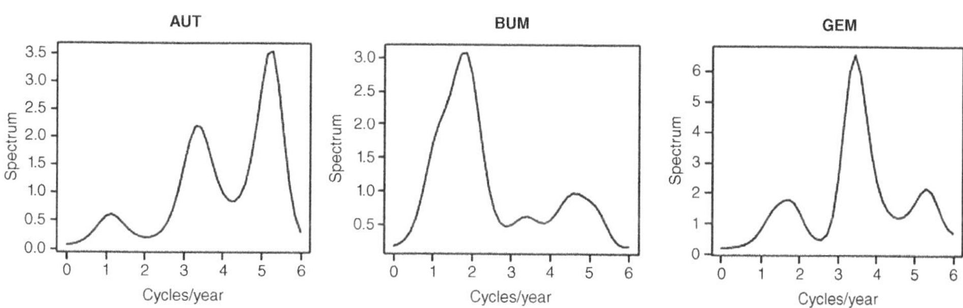

Figure 9.6 The three estimated spectra using the Bayesian method by Rosen and Stoffer (2007).

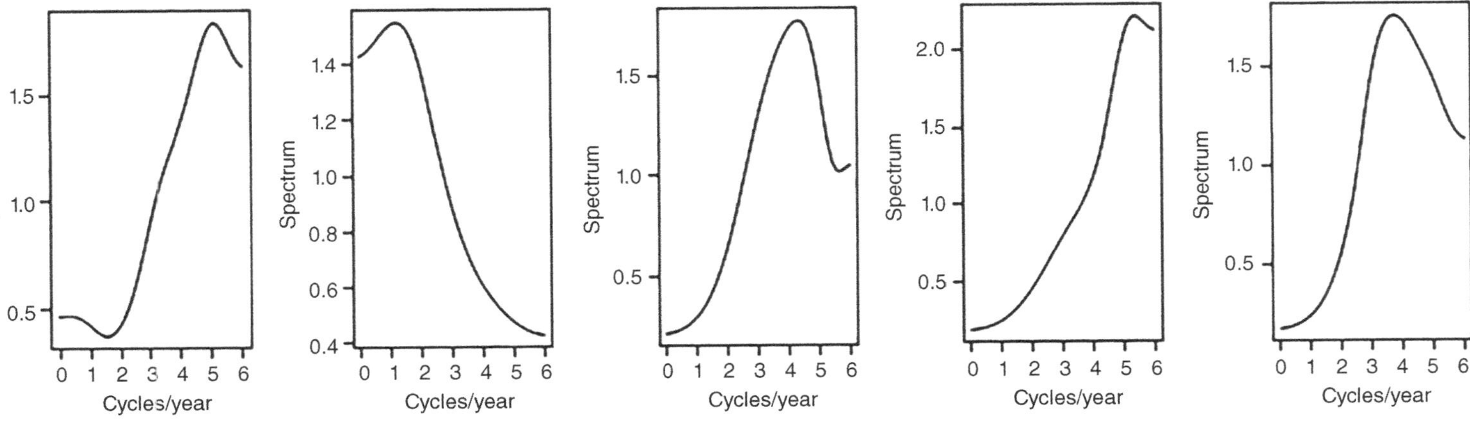

Figure 9.7 The estimated spectra by the penalized Whittle likelihood method of Krafty and Collinge (2013).

9.4.3 Penalized Whittle likelihood method

Example 9.3(c) In this section, we apply the penalized Whittle likelihood by Krafty and Collinge (2013) to the five U.S. monthly sale changes. Figure 9.7 shows the estimated spectra. Comparing to the kernel smoothing and Bayesian methods, the penalized Whittle likelihood method produces the smoothest estimates.

9.4.4 Example of VAR spectrum estimation

Example 9.3(d) Now, we will consider the same five monthly changes of U.S. sales and find the best vector autoregressive (VAR) model for the data set. Based on AIC, a VAR(4) model is chosen,

$$\mathbf{\Phi}_4(B)\mathbf{Z}_t = \mathbf{a}_t, \tag{9.97}$$

where $\mathbf{\Phi}_4(B) = (\mathbf{I} - \mathbf{\Phi}_1 B - \mathbf{\Phi}_2 B^2 - \mathbf{\Phi}_3 B^3 - \mathbf{\Phi}_4 B^4)$, and $\mathbf{a}_t$ is a Gaussian vector white noise, $N(\mathbf{0}, \mathbf{\Sigma})$. The estimated coefficient matrices and noise variance–covariance matrix are both given next:

$$\widehat{\mathbf{\Phi}}_1 = \begin{pmatrix} -0.403 & 0.106 & 1.220 & -0.466 & -1.750 \\ 0.309 & -0.237 & 0.320 & -0.204 & -1.029 \\ -0.613 & 0.075 & -0.868 & 0.061 & 0.139 \\ -0.353 & 0.303 & 0.479 & -1.186 & -0.021 \\ -0.561 & 0.053 & -0.265 & 0.067 & -0.543 \end{pmatrix},$$

$$\widehat{\mathbf{\Phi}}_2 = \begin{pmatrix} -0.147 & 0.265 & 1.809 & -0.608 & -2.675 \\ 0.372 & -0.158 & 0.449 & -0.148 & -1.680 \\ -0.658 & 0.161 & -1.067 & -0.533 & 0.323 \\ -0.183 & 0.278 & 0.886 & -1.007 & -1.073 \\ -0.543 & 0.116 & -0.723 & -0.456 & -0.216 \end{pmatrix},$$

$$\widehat{\mathbf{\Phi}}_3 = \begin{pmatrix} -0.267 & 0.427 & 1.567 & -0.868 & -1.890 \\ 0.425 & 0.058 & 0.733 & -0.056 & -2.008 \\ -0.703 & 0.120 & -1.927 & -1.143 & 2.046 \\ -0.392 & 0.326 & -0.350 & -0.675 & 0.365 \\ -0.577 & 0.167 & -1.533 & -1.086 & 1.477 \end{pmatrix},$$

$$\widehat{\mathbf{\Phi}}_4 = \begin{pmatrix} -0.063 & 0.346 & 1.509 & -0.784 & -1.529 \\ 0.144 & -0.350 & -0.006 & -0.241 & -0.406 \\ 0.177 & 0.426 & 0.957 & -0.917 & -0.935 \\ -0.010 & 0.265 & 0.564 & -0.322 & -0.843 \\ 0.206 & 0.439 & 1.196 & -0.849 & -1.243 \end{pmatrix},$$

and

$$\widehat{\mathbf{\Sigma}} = \begin{pmatrix} 0.093 & 0.037 & 0.018 & 0.033 & 0.023 \\ 0.037 & 0.108 & -0.029 & 0.007 & -0.025 \\ 0.018 & -0.029 & 0.097 & 0.035 & 0.093 \\ 0.033 & 0.007 & 0.035 & 0.067 & 0.031 \\ 0.023 & -0.025 & 0.093 & 0.031 & 0.095 \end{pmatrix}.$$

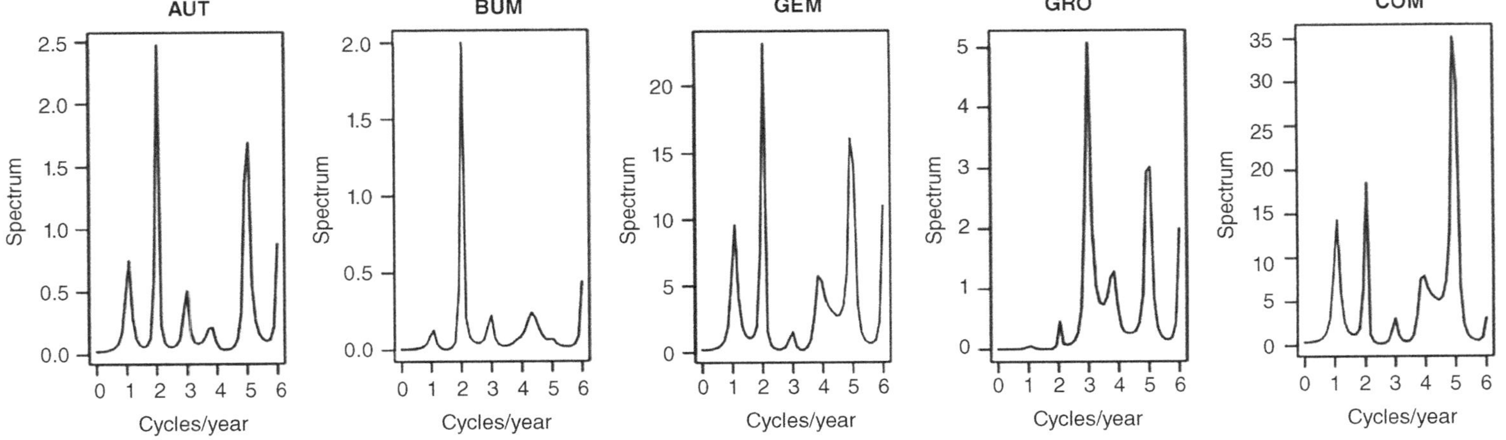

Figure 9.8 The estimated power spectra by using the VAR(4) model.

According to Eq. (9.86), the estimated spectral matrix is

$$\hat{\mathbf{f}}(\omega) = \left[\hat{\mathbf{\Phi}}_4\left(e^{-i\omega}\right)\right]^{-1}\hat{\mathbf{\Sigma}}\left\{\left[\hat{\mathbf{\Phi}}_4\left(e^{-i\omega}\right)\right]^{-1}\right\}^{*}, \tag{9.98}$$

where $\widehat{\mathbf{\Phi}}_4\left(e^{-i\omega}\right) = \left(\mathbf{I} - \sum_{j=1}^{4}\widehat{\mathbf{\Phi}}_j e^{-i\omega j}\right)$.

Figure 9.8 shows the estimated spectrum from the VAR(4) model. Comparing with the kernel smoothed estimations given in Figure 9.3, we see that there are some similarities but also some differences. Similar to the nonparametric estimation, only the spectra of GEM and COM have peak at four cycles per year, which is related to seasonal effects. BUM concentrates at lower frequencies, GRO concentrates at higher frequencies, and the powers of GEM and COM are more evenly spread out in frequencies. From Figure 9.3, we see that the estimated spectra contain a large peak at two cycles per year except AUT. However, in terms of the VAR estimation shown in Figure 9.8, the estimated spectra contain a large peak at two cycles per year including AUT. Also, all estimated spectra have peak at one cycle per year, which is related to a yearly pattern, and unlike the kernel estimation, only AUT and BUM have peak at one cycle per year. Overall, the estimates of kernel window smoothing are much smoother than that of the VAR model. One possible reason is that the sample size of this particular example is small, which leads to unstable VAR parameter estimation.

Figure 9.9 shows the estimated co-spectra, $\hat{c}_{5,j}(\omega)$, and quadrature spectra, $\hat{q}_{5,j}(\omega)$, for $j = 1, \ldots, 4$. Comparing with the kernel smoothed estimations given in Figure 9.4, again we see that there are some similarities and differences. However, the VAR estimation of the squared coherences as shown in Figure 9.10 is quite similar to the kernel estimation from Figure 9.5. Both estimation methods show that the estimated squared coherence between COM and GEM is large and constant across all frequencies, indicating a strong relationship between sales changes of consumer materials and general merchandise.

9.5 Spectrum analysis of a nonstationary vector time series

9.5.1 Introduction

For a nonstationary time series, we normally use some transformation like variance stabilization and/or differencing to reduce it to stationary before performing its spectral matrix estimation. However, there are many kinds of nonstationary time series that cannot be reduced to stationary by these transformations. Let us first consider a univariate case. There are many univariate nonstationary processes Z_t, which cannot be represented by $Z_t = \int_{-\pi}^{\pi} e^{i\omega t}dU(\omega)$ given in Eq. (9.1), because the function $\phi(\omega) = e^{i\omega t}$ as a sine and cosine waves is stationary. Priestley (1965, 1966, and 1967) has pointed out that in this case, instead of using $\phi(\omega) = e^{i\omega t}$, we need to consider an oscillatory function, which is a generalized Fourier transform,

$$\phi(t,\omega) = A(t,\omega)e^{i\omega t}, \tag{9.99}$$

so that

$$Z_t = \int_{-\pi}^{\pi} A(t,\omega)e^{i\omega t}dU(\omega) \tag{9.100}$$

where $A(t,\omega)$ is a time-varying modulating or transfer function with absolute maximum at zero frequency. In other words, Z_t is an oscillatory process with an evolutionary spectrum, which has the same type of physical interpretation as the spectrum of a stationary process. The main

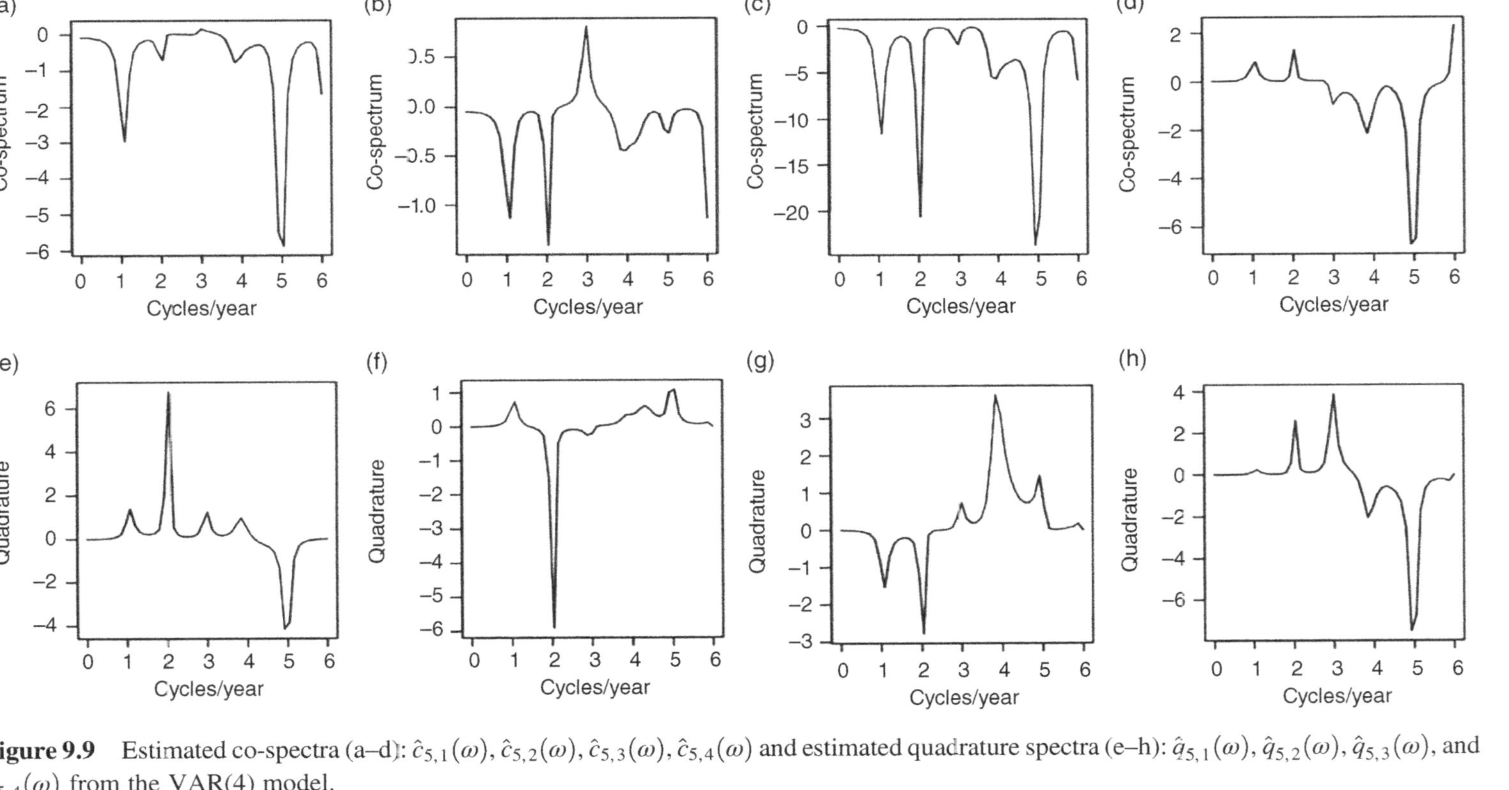

Figure 9.9 Estimated co-spectra (a–d): $\hat{c}_{5,1}(\omega), \hat{c}_{5,2}(\omega), \hat{c}_{5,3}(\omega), \hat{c}_{5,4}(\omega)$ and estimated quadrature spectra (e–h): $\hat{q}_{5,1}(\omega), \hat{q}_{5,2}(\omega), \hat{q}_{5,3}(\omega)$, and $\hat{q}_{5,4}(\omega)$ from the VAR(4) model.

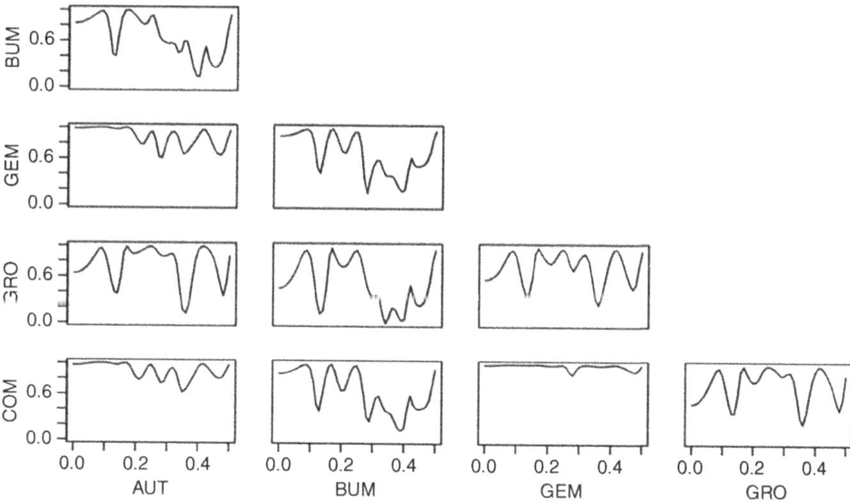

Figure 9.10 The estimated squared coherences by the VAR(4) model.

difference is that while the spectrum of a stationary process describes the power distribution across frequencies over all time, the evolutionary spectrum describes power distribution over frequency at instantaneous time. However, within this framework, letting the length of series $n \to \infty$ does not increase our knowledge about local behavior of spectrum since the pattern of the forthcoming series is different. As a result, the formulation does not allow the development of rigorous asymptotic theory of statistical inference. To overcome this problem, Dahlhaus (1996, 2000) introduced locally stationary time series that allows theoretical asymptotic analysis of the evolutionary spectrum for a univariate case, and further extended it to the multivariate case.

9.5.2 Spectrum representations of a nonstationary multivariate process

Let $\{\mathbf{Z}_t : t = 1, \ldots, n\}$ be a m-dimensional time series. The idea of locally stationary process is to rescale the transfer function to unit time scale, such that

$$\mathbf{Z}_{t,n} = \int_{-\pi}^{\pi} \mathbf{A}(t/n, \omega) e^{(i\omega t)} d\mathbf{U}(\omega). \tag{9.101}$$

More rigorously, we have following definition.

Definition 9.1 The m-dimensional zero-mean time series $\{\mathbf{Z}_t : t = 1, \ldots, n\}$ is called locally stationary with an $(m \times m)$ matrix-valued transfer function $\mathbf{A}^0 = \left[A^0_{i,j}(t/n, \omega)\right]$ and mean function vector $\boldsymbol{\mu}$ if there exists a representation

$$\mathbf{Z}_t = \boldsymbol{\mu}(t/n) + \int_{-\pi}^{\pi} \mathbf{A}^0(t/n, \omega) \exp(i\omega t) d\mathbf{U}(\omega), \tag{9.102}$$

with the following properties:

1. $\mathbf{U}(\omega)$ is a complex-valued zero-mean vector process with $E\{d\mathbf{U}(\omega) d\mathbf{U}^*(\zeta)\}$ being the identity matrix if $\omega = \zeta$ and zero otherwise.

2. There exists a constant K and a $(m \times m)$ matrix-valued function $\mathbf{A}(u,\omega) = [A_{i,j}(u,\omega)]$, with $\mathbf{A}(u, \omega) = \mathbf{A}(u, -\omega)^*$ and

$$\sup_{t,\omega} \left| A_{j,k}^{0}(t/n,\omega) - A_{j,k}(t/n,\omega) \right| \le Kn^{-1}$$

for all $j,k = 1, \ldots, m$. Also, $\mathbf{A}(u,\omega)$ are assumed to be continuous in u.

Based on Definition 9.1, the time-varying power spectrum of the process is given by

$$\mathbf{f}(u,\omega) - \mathbf{A}(u,\omega)\mathbf{A}(u,\omega)^*. \tag{9.103}$$

Developed from locally stationary time series, methods for estimating the time-varying spectrum of a multivariate time series can be roughly grouped into three categories. The first category consists of estimators in which second-order frequency domain structures evolve continuously over time. This includes the slowly evolving multivariate locally stationary process, which can be analyzed parametrically by fitting time series models with time-varying parameters (Dahlhaus, 2000), and nonparametrically via the bivariate smoothing of spectral components as functions of frequency and time (Guo and Dai, 2006). The second category of methods consist of estimators that are piecewise stationary. Approaches within this second category typically divide a time series into approximately stationary segments, then obtain the estimates of local spectrum within segments. These methods include both parametric approaches, such as fitting piecewise vector autoregressive models (Davis et al., 2006), and nonparametric approaches, such as using the multivariate smooth localized complex exponential (SLEX) library (Ombao von Sachs and Guo, 2005). The final category are methods that can automatically approximate both abrupt and slowly varying changes by averaging over piecewise stationary models, and they include the smoothing stochastic approximation Monte Carlo (SSAMC) based method of Zhang (2016) and the reversible jump Monte Carlo and Hamiltonian Monte Carlo (HMC) based method by Li and Krafty (2018).

9.5.2.1 *Time-varying autoregressive model*

One of the methods used in Dahlhaus (2000) is the time-varying VARMA(p,q) model. For a vector autoregressive model VAR(p), it is defined by

$$\mathbf{Z}_t = \boldsymbol{\mu}(t/n) + \sum_{j=1}^{p} \boldsymbol{\Phi}_j(t/n) \left[\mathbf{Z}_{t-j} - \boldsymbol{\mu}[(t-j)]\right] + \boldsymbol{\Sigma}(t/n)\boldsymbol{\varepsilon}_t, \; t = 1, \ldots, n, \tag{9.104}$$

where $\boldsymbol{\varepsilon}_t$ are m-dimensional independent random variable with mean zero and unit variance $\mathbf{I}$. In addition, we assume some smoothness conditions on $\boldsymbol{\Sigma}(\cdot)$ and $\boldsymbol{\Phi}_j(\cdot)$. In some neighborhood of a fixed time point $u_0 = t_0/n$, the process $\mathbf{Z}_t$ can be approximated by the stationary process $\mathbf{Z}_t(u_0)$ given by

$$\mathbf{Z}_t(u_0) = \boldsymbol{\mu}(u_0) + \sum_{j=1}^{p} \boldsymbol{\Phi}_j(u_0) \left[\mathbf{Z}_{t-j}(u_0) - \boldsymbol{\mu}(u_0) + \boldsymbol{\Sigma}(u_0)\boldsymbol{\varepsilon}_t, \; t = 1, \ldots, n. \tag{9.105}$$

$\mathbf{Z}_t$ has a unique time-varying power spectrum, which is locally the same as the power spectrum of $\mathbf{Z}_t(u)$, that is,

$$\mathbf{f}(u, \omega) = \left[\boldsymbol{\Phi}\left(u, e^{-i\omega}\right)\right]^{-1} \boldsymbol{\Sigma}(u) \left\{ \left[\boldsymbol{\Phi}\left(u, e^{-i\omega}\right)\right]^{-1} \right\}^*, \tag{9.106}$$

where $u = t/n$, and

$$\boldsymbol{\Phi}(u,B) = \left(\mathbf{I} - \boldsymbol{\Phi}_1(u,B) - \cdots - \boldsymbol{\Phi}_p(u,B^p)\right).$$

Similarly, the locally covariance matrix is

$$\boldsymbol{\Gamma}(u,j) = \int_{-\pi}^{\pi} \mathbf{f}(u,\omega)\exp(ji\omega)d\omega. \tag{9.107}$$

Based on this statement, the time-varying spectrum can be obtained by estimating the time-varying parameters of the VAR model. For more properties of the estimation based on time-varying VARMA(p,q), VAR(p), and VMA(q) models, we refer readers to Dahlhaus (2000).

9.5.2.2 Smoothing spline ANOVA model

Based on the locally stationary process, the time-varying spectrum can also be estimated nonparametrically via the smoothing spline Analysis of Variance (ANOVA) model by Guo and Dai (2006). However, their definition of locally stationary process is slightly different from Dahlhaus (2000). In Section 9.5.2, we mentioned that Dahlhaus (2000) assumes a series of transfer functions $\mathbf{A}^0(t/n,\omega)$ that converge to a large-sample transfer function $\mathbf{A}(u,\omega)$ in order to allow for the fitting of parametric models. Since Guo and Dai (2006) considered a nonparametric estimation, they used $\mathbf{A}(u,\omega)$ directly.

Definition 9.2 Without loss of generality, the m-dimensional zero-mean time series of length n, $\{\mathbf{Z}_t: t = 1, \ldots, n\}$, is called locally stationary if

$$\mathbf{Z}_t = \int_{-\pi}^{\pi} \mathbf{A}(t/n,\omega)\exp(i\omega t)d\mathbf{U}(\omega),$$

where we assume that the cumulants of $d\,\mathbf{U}(\omega)$ exists and are bounded for all orders. For the details, please see Brillinger (2002), and Guo and Dai (2006).

Based on this definition, the smoothing ANOVA model takes a two-stage estimation procedure. At the first stage, the locally stationary process is approximated by piecewise stationary time series with small blocks to obtain initial spectrum estimates and the Cholesky decomposition. The initial spectrum estimates are obtained by the multitaper method to reduce variance. At the second stage, each element of the Cholesky decomposition is treated as a bivariate smooth function of time and frequency and is modeled by the smoothing spline ANOVA model by Gu and Wahba (1993). The final estimated time-varying spectrum is reconstructed from the smoothed elements of the Cholesky decomposition. Thus, the method provides a way to ensure the final estimate of the multivariate spectrum is positive–definite while allowing enough flexibility in the smoothness of its elements.

We shall briefly discuss the smoothing spline ANOVA step. Suppose the spectral matrix $\widetilde{\mathbf{f}}(u,\omega)$ has the Cholesky decomposition such that $\widetilde{\mathbf{f}}(u,\omega) = \mathbf{L}(u,\omega)\mathbf{L}(u,\omega)^*$, where $\mathbf{L}(u,\omega)$ is a $m \times m$ lower triangular matrix. The method smooths the diagonal elements $\gamma_{j,j}(u,\omega), j = 1, \ldots, m$, the real part of $\gamma_{j,k}(u,\omega)$, $\Re\{\gamma_{j,k}(u,\omega)\}$, and the imaginary part of $\gamma_{j,k}(u,\omega)$, $\Im\{\gamma_{j,k}(u,\omega)\}$, for $j > k$ separately with their own smoothing parameters. Let $\gamma(u,\omega) \in \{\gamma_{j,j}(u,\omega), \Re(\gamma_{j,k})(u,\omega), \Im(\gamma_{j,k})(u,\omega), j > k$, for $j, k = 1, \ldots, m\}$. We have

$$\gamma(u,\omega) = a(u,\omega) + \varepsilon(u,\omega),$$

where $a(u,\omega)$ is the corresponding Cholesky decomposition element of the spectrum, such that $a(u,\omega)=E\{\gamma(u,\omega)\}$, the $\varepsilon(u,\omega)$ are independent errors with zero-mean and the variance depending on the time-frequency point (u,ω).

The smoothing spline ANOVA model is defined by the corresponding reproducing kernels (RKs) for the time and frequency domains. In the frequency domain, the reproducing kernel Hilbert space (RKHS) W_1 for ω can be decomposed as $W_1 = 1 \oplus H_1$. The reproducing kernel for H_1 is $R_1(\omega_1, \omega_2) = -K_4(|\omega_1 - \omega_2|)/24$, where $K_k(\cdot)$ is the kth order Bernoulli polynomial. In the time domain, the RKHS W_2 can be decomposed as $W_2 = 1 \oplus (u-0.5) \oplus H_2$. The reproducing kernel for H_2 is $R_2(u_1, u_2) = K_2(u_1)K_2(u_2)/4 - K_4(|u_1 - u_2|)/24$. The full tensor product RKHS for (u,ω) is then

$$\begin{aligned} W &= W_1 \otimes W_2 && (9.108)\\ &= \{1 \oplus H_1\} \otimes \{1 \oplus (u-0.5) \oplus H_2\} && (9.109)\\ &= 1 \oplus (u-0.5) \oplus H_1 \oplus H_2 \oplus \{H_1 \otimes (u-0.5)\} \oplus \{H_1 \otimes H_2\}, && (9.110)\end{aligned}$$

where the RK for $H_3 = H_1 \otimes (u-0.5)$ is $R_3\{(u_1, \omega_1), (u_2, \omega_2)\} = R_1(\omega_1, \omega_2)(u_1 - 0.5)(u_2 - 0.5)$, the RK for $H_4 = H_1 \otimes H_2$ is $R_4\{(u_1, \omega_1), (u_2, \omega_2)\} = R_1(\omega_1, \omega_2)R_2(u_1, u_2)$. Thus, $a(u,\omega)$ has the ANOVA-like decomposition

$$a(u,\omega) = d_1 + d_2(u-0.5) + a_1(\omega) + a_2(u) + a_3(u,\omega) + a_4(u,\omega), \tag{9.111}$$

where $d_1 + d_2(u-0.5)$ is the linear trend, $a_1(\omega)$ is the smooth main effect for frequency, $a_2(u)$ is the smooth main effect across time, $a_3(u,\omega)$ is the smooth in frequency and linear in time, and $a_4(u,\omega)$ is the interaction that is smooth at both time and frequency. The least square estimation can be used to obtain the estimates $\hat{a}(u,\omega)$. For the details, see Guo and Dai (2006).

9.5.2.3 *Piecewise vector autoregressive model*

Davis, Lee, and Rodriguez-Yam (2006) considered modeling nonstationary time series using piecewise VAR processes. The number and locations of the piecewise VAR segments, and the orders of the corresponding VAR process, are assumed to be unknown. Based on the minimum description length (MDL) principle, the method penalizes complexity of the model, and thus provides the criteria to define the best fitting model. We refer readers to Rissanen (1989), Hansen and Yu (2001) for more details about the MDL principle.

Let $\mathbf{Z}_t$ be a m-dimensional time series with k segments, and assume that there are partitions or changepoints $\boldsymbol{\delta} = (\delta_0, \ldots, \delta_k)'$ with $\delta_0 = 0$ and $\delta_k = n$. The time series within the jth segment is modeled by VAR(p_j) process, such that

$$\mathbf{Z}_t^{(j)} = \boldsymbol{\Phi}_{j,1}\mathbf{Z}_{t-1}^{(j)} + \cdots + \boldsymbol{\Phi}_{j,p_j}\mathbf{Z}_{t-p_j}^{(j)} + \boldsymbol{\varepsilon}_t^{(j)}, \tag{9.112}$$

where $(\boldsymbol{\Phi}_{j,1}, \ldots, \boldsymbol{\Phi}_{j,p_j})'$ are $m \times m$ dimensional coefficient matrices of the VAR process. We define the entire class of piecewise VAR models as $\mathbf{M}$ and a model from this class as $\mathbf{F} \in \mathbf{M}$. The principle of MDL is to find the best fitting model from $\mathbf{M}$ as the one that produces the shortest code length. The code length of an object is the amount of memory space required to store the data $\mathbf{Z}_t$. The MDL has two components, a fitted model $\hat{\mathbf{F}}$ and the portion that is unexplained by $\hat{\mathbf{F}}$. The later component can be defined as the residuals $\hat{\boldsymbol{\varepsilon}}_t = \mathbf{Z}_t - \hat{\mathbf{Z}}_t$.

Let $L\left(\hat{\mathbf{F}} \mid \mathbf{Z}_t\right)$ be the code length of fitted model $\hat{\mathbf{F}}$ and $L\left(\hat{\boldsymbol{\varepsilon}}_j \mid \hat{\mathbf{F}}\right)$ be the code length of the residuals of the jth segment. Then, the total code length of the data can be decomposed to

$$L(\mathbf{Z}_t) = L\left(\hat{\mathbf{F}} \mid \mathbf{Z}_t\right) + L\left(\hat{\boldsymbol{\varepsilon}}_j \mid \hat{\mathbf{F}}\right). \tag{9.113}$$

The MDL principle suggests that a best fitting model by minimizing $L(\mathbf{Z}_t)$.

Let n_j be the sample size in the jth segment and $\boldsymbol{\delta}$ be the information set about the partition of segments. Since the piecewise VAR model can be characterized by following parameters: k, $\boldsymbol{\delta}$, $\mathbf{p} = (p_1, \ldots, p_k)$, and $\hat{\boldsymbol{\Psi}}_j = \left(\hat{\boldsymbol{\Phi}}_{j,1}, \ldots, \hat{\boldsymbol{\Phi}}_{j,p_j}, \hat{\boldsymbol{\Sigma}}_j\right)$, we can further decompose $L\left(\hat{\mathbf{F}} \mid \mathbf{Z}_t\right)$ by

$$L\left(\hat{\mathbf{F}} \mid \mathbf{Z}_t\right) = L(k) + L(\boldsymbol{\delta}) + L(\mathbf{p}) + L\left(\hat{\boldsymbol{\Psi}}_1\right) + \cdots + L\left(\hat{\boldsymbol{\Psi}}_k\right) \tag{9.114}$$

$$= L(k) + L(n_1, \ldots, n_k) + L(\mathbf{p}) + L\left(\hat{\boldsymbol{\Psi}}_1\right) + \cdots + L\left(\hat{\boldsymbol{\Psi}}_k\right). \tag{9.115}$$

The equality is due to the fact that the complete knowledge of $\boldsymbol{\delta}$ implies complete knowledge of $n_1, \ldots, n_k$. In order to store an integer I whose value is not bounded, approximately $\log_2 I$ bits are needed. Thus, we have $L(k) = \log_2 k$, $L(p_j) = \log_2 p_j$, and $L(\mathbf{p}) = \sum_{j=1}^{k} \log_2 p_j$. On the other hand, if integer I is upper-bounded say I_U, then $\log_2 I_U$ bits are needed. Since n_j is bounded by n, we have $L(n_j) = \log_2 n$ and $L(n_1, \ldots, n_k) = k\log_2 n$ (Hansen and Yu, 2001). It is known that a maximum likelihood estimate of a real parameter computed from n observations can be effectively encoded with $(1/2)\log_2 n$ bits. Because the total number of parameters in $\hat{\boldsymbol{\Psi}}_j$ is $(p_j + 1)m^2$, we have

$$L\left(\hat{\boldsymbol{\Psi}}_j\right) = \frac{(p_j + 1)m^2}{2} \log_2 n_j. \tag{9.116}$$

Combining the results, we have

$$L\left(\hat{\mathbf{F}} \mid \mathbf{Z}_t\right) = \log_2 k + \sum_{j=1}^{k} \log_2 p_j + k\log_2 n + \sum_{j=1}^{k} \frac{(p_j + 1)m^2}{2} \log_2 n_j. \tag{9.117}$$

Rissanen (1989) showed that the code length of $\hat{\boldsymbol{\varepsilon}}_j$ is approximated by the negative of the loglikelihood of the fitted model $\hat{\mathbf{F}}$. Thus, we have

$$L\left(\hat{\boldsymbol{\varepsilon}}_j \mid \hat{\mathbf{F}}\right) = -\sum_{j=1}^{k} \log L\left(\widehat{\boldsymbol{\Phi}}_{j,1}, \ldots, \widehat{\boldsymbol{\Phi}}_{j,p_j}, \widehat{\boldsymbol{\Sigma}}_j\right), \tag{9.118}$$

where L denotes the Whittle likelihood. Consequently, from Eqs. (9.114) and (9.115), the MDL objective function is defined as

$$\begin{aligned} L(\mathbf{Z}_t) = {} & \log_2(k) + (k)\log_2(n) + \sum_{j=1}^{k} \left\{ \log_2(p_j) + \frac{(p_j + 1)m^2}{2} \log_2 n_j \right\} \\ & - \sum_{j=1}^{k} \log L\left(\hat{\boldsymbol{\Phi}}_{j,1}, \hat{\boldsymbol{\Phi}}_{j,2}, \ldots, \hat{\boldsymbol{\Phi}}_{j,p_j}, \hat{\boldsymbol{\Sigma}}_j\right). \end{aligned} \tag{9.119}$$

The best fitting model is the one that minimizes the MDL. To overcome difficulties raised in optimization, the method proposes using a genetic algorithm.

9.5.2.4 *Bayesian methods*

There are two recently proposed Bayesian methods to estimate multivariate time-varying spectrum, including Zhang (2016) and Li and Krafty (2018). In this section, we focus on the method of Li and Krafty (2018). The method is also based on locally stationary time series to estimate the time-varying spectrum

$$\mathbf{f}(u,\omega)=\mathbf{A}(u,\omega)\mathbf{A}(u,\omega)^{*}. \tag{9.120}$$

The method assumes that for every $u \in (0, 1)$, each component of $\mathbf{f}(u, \cdot)$ possesses a square-integrable first derivative as a function of frequency; for every ω, each component of $\mathbf{f}(\cdot, \omega)$ is continuous as a function of scaled time at all but a possible finite number of points. The assumption on transfer function and time-varying spectrum are slightly different from Definitions 9.1 and 9.2 in two ways. First, Definition 9.1 assumes a series of transfer functions $\mathbf{A}^0(u,\omega)$ that converge to a large-sample transfer function $\mathbf{A}(u,\omega)$ in order to allow for the fitting of parametric models. Since the method considers nonparametric estimation, in a manner similar to the smoothing ANOVA method, $\mathbf{A}(u,\omega)$ is used. Second, Definitions 9.1 and 9.2 require the time-varying spectrum to be continuous in both time and frequency. The method is more flexible and allows for components of spectrum to evolve not only continuously, but also abruptly in time.

The analysis of a locally stationary time series $\{\mathbf{Z}_t : t = 1, \dots, n\}$ begins by using the piecewise stationary approximation. Consider a partition of the time series into k segments defined by partition points $\boldsymbol{\delta} = (\delta_0, \dots, \delta_k)$ with $\delta_0 = 0$ and $\delta_k = n$ such that $\mathbf{Z}_t$ is approximately stationary within the segments $\{t : \delta_{q-1} < t \le \delta_q\}$ for $q = 1, \dots, k$. Then

$$\mathbf{Z}_t \approx \sum_{q=1}^{k}\int_{-\pi}^{\pi}\mathbf{A}_q(\omega)\exp(i\omega t)d\mathbf{U}(\omega), \tag{9.121}$$

where $\mathbf{A}_q(\omega) = \mathbf{A}(u_q, \omega)I(\delta_{q-1} < t \le \delta_q)$, $I(\cdot)$ is the indicator function, and $u_q = (\delta_q + \delta_{q-1})/2n$ is the scaled midpoint of the qth segment. Within the qth segment, the time series is approximately second-order stationary with local power spectrum $\mathbf{f}(u_q, \omega) = \mathbf{A}_q(\omega)\mathbf{A}_q(\omega)^*$.

Conditional on an approximately stationary partition $\boldsymbol{\delta}$, known approaches and properties for stationary time series can be applied. In particular, the large-sample distribution of the discrete Fourier transform of a stationary time series that provides the Whittle likelihood allows for the formulation of a product of Whittle likelihoods. Let us define the local discrete Fourier transform at frequency ℓ within segment q as

$$\mathbf{y}_{q,\ell}=n_q^{-1/2}\sum_{t=\delta_{q-1}+1}^{\delta_q}\mathbf{Z}_t\exp\left(-i\omega_{q,\ell}t\right),\quad \ell=1,\dots,L_q, q=1,\dots,k, \tag{9.122}$$

where $n_q = \delta_q - \delta_{q-1}$ is the number of observations in segment q, $\omega_{q,\ell} = \ell/n_q$ are the Fourier frequencies, and $L_q = [(n_q - 1)/2]$.Given a partition of k segments, $\boldsymbol{\delta}$, the $\mathbf{y}_{q,\ell}$ are approximately independent zero-mean complex multivariate Gaussian random variables with covariance matrices, which equals spectral matrices $\mathbf{f}(u_q, \omega_{q,\ell})$. This leads to a loglikelihood that can be approximated by a sum of log Whittle likelihoods

$$L(\mathbf{Y} \mid \mathbf{f}, \boldsymbol{\delta}, k) \approx -\sum_{q=1}^{k} \sum_{\ell=1}^{L_q} \left\{ \log \left| \mathbf{f}(u_q, \omega_{q,\ell}) \right| + \mathbf{y}_{q,\ell}^{*} \mathbf{f}^{-1}(u_q, \omega_{q,\ell}) \mathbf{y}_{q,\ell} \right\}, \tag{9.123}$$

where $\mathbf{Y}$ is collection of Fourier transformations.

The spectral matrix is positive–definite and, to nonparametrically allow for flexible smoothing among the different components while preserving positive definiteness, the procedure modified the Cholesky components of local spectra via linear penalized splines. The modified Cholesky decomposition represents a time-varying spectral matrix as

$$\mathbf{f}^{-1}(u, \omega) = \boldsymbol{\Theta}(u, \omega) \boldsymbol{\Psi}(u, \omega)^{-1} \boldsymbol{\Theta}(u, \omega)^{*} \tag{9.124}$$

for a complex-valued $m \times m$ lower triangular matrix $\boldsymbol{\Theta}(u,\omega)$ with ones on the diagonal and a positive diagonal matrix $\boldsymbol{\Psi}(u,\omega)$. For a piecewise stationary approximation with partition $\boldsymbol{\delta}$ into k segments, we define the local modified Cholesky decomposition as $\mathbf{f}^{-1}(u_q, \omega) = \boldsymbol{\Theta}(u_q, \omega)\boldsymbol{\Psi}(u_q, \omega)^{-1}\boldsymbol{\Theta}(u_q, \omega)^{*}$, for $q = 1, \ldots, k$, and let $\theta_{j,k,q}(\omega)$ and $\psi_{j,j,q}(\omega)$ be the (j,k) and (j,j) elements of $\boldsymbol{\Theta}(u_q, \omega)$ and $\boldsymbol{\Psi}(u_q, \omega)$, respectively. Then, for each segment, there are m^2 components to estimate: $\Re\{\theta_{j,k,q}(\omega)\}$ for $(j > k) = 1, \ldots, m, \Im\{\theta_{j,k,q}(\omega)\}$ for $(j > k) = 1, \ldots, m$, and $\psi_{j,j,q}(\omega)$ for $j = 1, \ldots, m$. The Cholskey components are modeled by periodic even and odd linear splines by considering

$$\Re\{\theta_{j,k,q}(\omega)\} = c_{j,k,q_0} + \sum_{s=1}^{L_q - 1} c_{j,k,q_s} \cos(2\pi s\omega), \tag{9.125}$$

$$\Im\{\theta_{j,k,q}(\omega)\} = \sum_{s=1}^{L_q} b_{j,k,q_s} \sin(2\pi s\omega), \tag{9.126}$$

and

$$\log\{\psi_{j,j,q}(\omega)\} = d_{j,j,q_0} + \sum_{s=1}^{L_q - 1} d_{j,j,q_s} \cos(2\pi s\omega). \tag{9.127}$$

The Fourier frequencies for each segment form an equally spaced grid, so that Demmler–Reinsch bases for periodic even and odd smoothing splines for local periodograms are given by $\{\cos(2\pi s\omega) : s = 0, 1, \ldots, (L_q - 1)\}$ and $\{\sin(2\pi s\omega) : s = 1, \ldots, L_q\}$, respectively. For the details, we refer readers to Schwarz and Krivobokova (2016).

The method relies on reversible jump Markov chain and HMC methods to sample from posterior distributions. The estimates of time-varying spectrum components are obtained by averaging over the distribution of partitions so that both abrupt and slowly varying changes can be recovered.

9.6 Empirical spectrum example of nonstationary vector time series

Example 9.4 In this section, we consider a trivariate time series: weekly log returns of the Dow Jones Industrial Average (DJIA), National Association of Securities Dealers Automated Quotations (NASDAQ), and S&P 500. The time period we consider is from April 1990 to

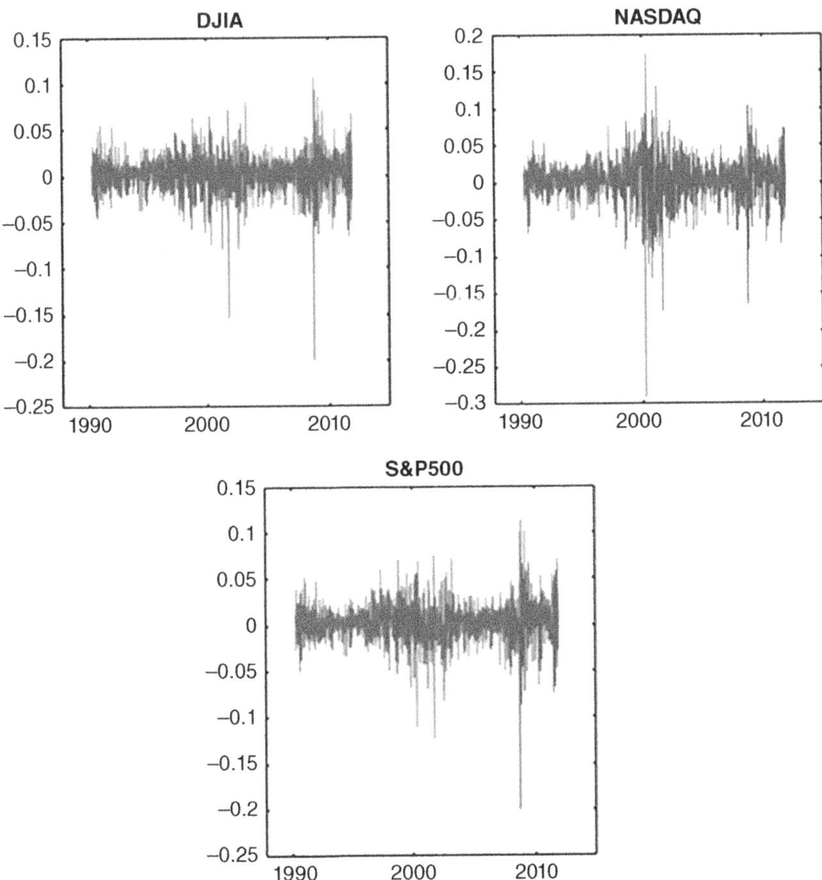

Figure 9.11 Weekly log returns for the Dow Jones, National Association of Securities Dealers Automated Quotations (NASDAQ), and S&P 500 from April 1990 to December 2011.

December 2011, thus the total length of the time series is $n = 1134$. The data set is the WW9 series listed in the Data Appendix and is shown in Figure 9.11.

We first applied the method of Li and Krafty (2018) to the time series. Figures 9.12 and 9.13 displays estimated spectra and pairwise squared coherence respectively. There are several obvious changes in the spectra and coherences, including (i) the period 1997–1998, which corresponds to the Asia financial crisis in 1997, which also affected the U.S. economy; (ii) the year 2003, which is the period of the U.S. invasion of Iraq; and (iii) the period 2007–2009, which corresponds to the global financial crisis that began in 2007. All the coherences had a big jump during this period instead of drop as in 1997–1998, indicating that the financial crisis in 2007–2009 had broad impacts on both the DJIA and the NASDAQ.

When a finite number of stationary segments are identified with a high probability, the results can be displayed as estimated local spectra at certain time point. For example, we obtain the local spectra of DJIA and NASDAQ and their coherence at the week of October 18, 2007 in the Figure 9.14. It appears that the spectra and coherence are all flat, and they clearly indicate a well-known phenomenon that the log returns of stock time series are white noise.

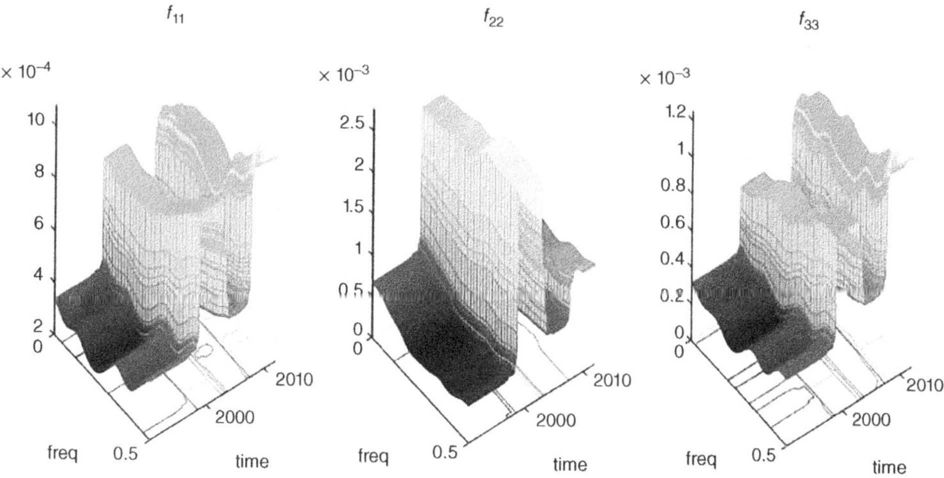

Figure 9.12 Estimated spectrum of Dow Jones Industrial Average (DJIA) (left), NASDAQ (middle), and S&P 500 (right).

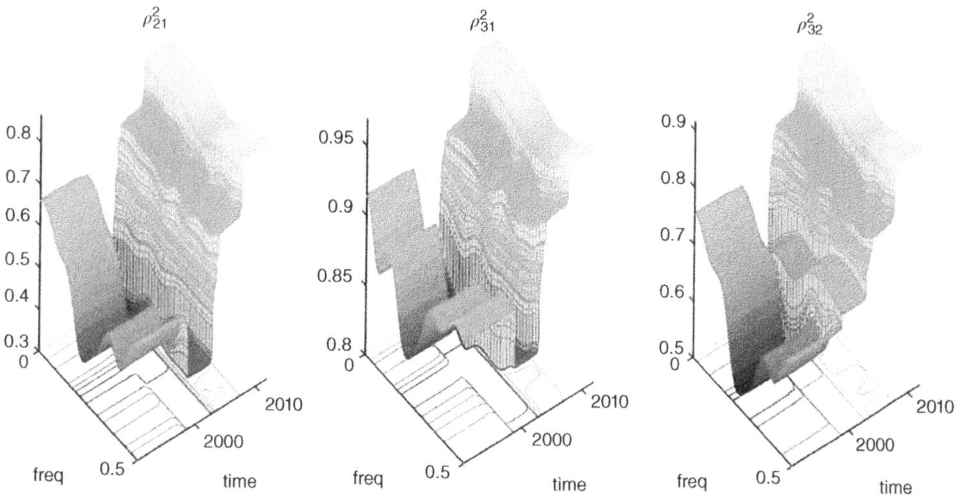

Figure 9.13 Estimated squared pairwise coherence between DJIA and NASDAQ (left), DJIA and S&P 500 (middle), and NASDAQ and S&P 500 (right).

We then applied the piecewise vector autoregressive method of Davis, Lee, and Rodriguez-Yam (2006) to the same data set. The procedure produces changepoints at the following dates: July 13, 1998, July 15, 2002, September 8, 2003, and June 25, 2007. Figure 9.15 presents the DJIA time series along with the changepoints found by the approach of Davis, et al. (2006). The changepoints found by the piecewise autoregressive model are somewhat similar to that found by the Bayesian method. However, one of the main differences between the assumptions of the Bayesian method and the piecewise vector autoregressive method is that the number and location of partitions (changepoints) is random for the Bayesian method, while they are fixed for the piecewise autoregressive method.

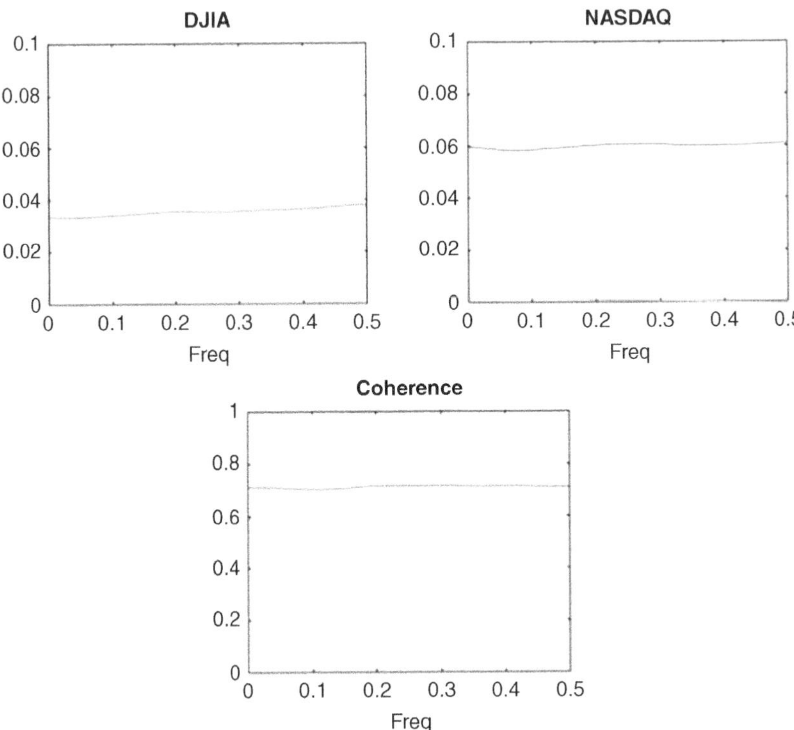

Figure 9.14 Local spectrum of DJIA (top left), NASDAQ (top right), and their coherences (bottom) at October 18, 2007.

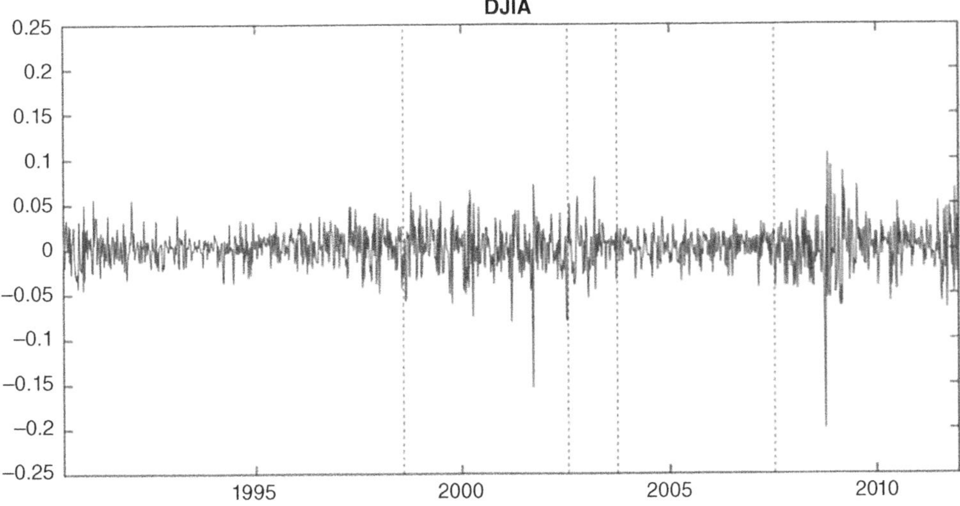

Figure 9.15 Weekly log returns for the Dow Jones from April 1990 to December 2011 along with the changepoints found by the piecewise vector autoregressive model of Davis, Lee, and Rodriguez-Yam (2006).

Before concluding this chapter on multivariate spectral analysis and its applications, I would like to mention some more recent references including Grant and Quinn (2017), Ray et al. (2017), and Wilson (2017), among others, and point out that multivariate spectral analysis can also be used to analyze space–time data sets. However, it should be stated that, so far, the developed procedures can only apply to a stationary or properly transformed stationary space–time series. For references on spectral analysis with space–time series, we refer readers to Bandyopadhyay, Jentsch, and Rao (2017), Rao and Terdick (2017), among others.

Software code

The software program for this chapter is very long, because there are no standard software programs available in the market for many spectral procedures introduced in this book. I sincerely appreciate Mr. Zeda Li who helped me use R and/or MATLAB to create numerous programs for the implementation of the procedures introduced in the chapter.

Nonparametric multivariate kernel spectrum estimation – Example 9.3(a) using R

```
library(astsa)

# data preparation

ms = as.data.frame(read.csv("C:/Bookdata/WW2b.csv")[,-1])
rownames(ms) = seq(as.Date("2009/6/1"), by="month", length=90)
z.orig = ts(ms,start=c(2009,6),frequency=12)
z = diff(z.orig,1)
plot(z)
z = ts(apply(z,2,scale), freq=1)   # scaling strips ts attributes
n = nrow(z)

# get raw periodogram using fft

h = mvfft(z)/sqrt(n) # Fourier Transformation
nf = floor(n/2)+1
h = h[1:nf,]    # Take half of the frequencies
I = matrix(h*Conj(h),c(nf,5))/(2*pi)
nn = dim(I)[1]-1
par(mfrow=c(1,5))
plot((0:nn)/88*12,Re(I[,1]),type='l',lty=1,ylab="Raw Periodogram",
xlab="frequency",main=colnames(ms)[1])
plot((0:nn)/88*12,Re(I[,2]),type='l',lty=1,ylab="Raw Periodogram",
xlab="frequency",main=colnames(ms)[2])
plot((0:nn)/88*12,Re(I[,3]),type='l',lty=1,ylab="Raw Periodogram",
xlab="frequency",main=colnames(ms)[3])
plot((0:nn)/88*12,Re(I[,4]),type='l',lty=1,ylab="Raw Periodogram",
xlab="frequency",main=colnames(ms)[4])
plot((0:nn)/88*12,Re(I[,5]),type='l',lty=1,ylab="Raw Periodogram",
xlab="frequency",main=colnames(ms)[5])

# window span 5

L = c(5,5)   # degree of smoothing
sales.spec = mvspec(z, spans=L, kernel="daniel", detrend=FALSE, taper=0,
plot=FALSE)
f = sales.spec$fxx/(2*pi)   # estimated spectral matrix
freq = sales.spec$freq
par(mfrow=c(1,5))
```

```
plot(sales.spec$freq*12,Re(f[1,1,]),type='l',lty=1,ylab="Spectrum",
xlab="Cycles/year",main=colnames(ms)[1])
plot(sales.spec$freq*12,Re(f[2,2,]),type='l',lty=1,ylab="Spectrum",
xlab="Cycles/year",main=colnames(ms)[2])
plot(sales.spec$freq*12,Re(f[3,3,]),type='l',lty=1,ylab="Spectrum",
xlab="Cycles/year",main=colnames(ms)[3])
plot(sales.spec$freq*12,Re(f[4,4,]),type='l',lty=1,ylab="Spectrum",
xlab="Cycles/year",main=colnames(ms)[4])
plot(sales.spec$freq*12,Re(f[5,5,]),type='l',lty=1,ylab="Spectrum",
xlab="Cycles/year",main=colnames(ms)[5])

# quadrature spectrum
c51 = Re(f[5,1,])
c52 = Re(f[5,2,])
c53 = Re(f[5,3,])
c54 = Re(f[5,4,])
c41 = Re(f[4,1,])
c42 = Re(f[4,2,])
c43 = Re(f[4,3,])
c31 = Re(f[3,1,])
c32 = Re(f[3,2,])
c21 = Re(f[2,1,])

# plot part of quadrature spectrum
par(mfrow=c(1,4))
plot(freq*12, c51, type='l',lty=1,ylab="Co-spectrum",xlab="Cycles/year",
main="(a)")
plot(freq*12, c52, type='l',lty=1,ylab="Co-spectrum",xlab="Cycles/year",
main="(b)")
plot(freq*12, c53, type='l',lty=1,ylab="Co-spectrum",xlab="Cycles/year",
main="(c)")
plot(freq*12, c54, type='l',lty=1,ylab="Co-spectrum",xlab="Cycles/year",
main="(d)")

# co spectrum
q51 = Im(f[5,1,])
q52 = Im(f[5,2,])
q53 = Im(f[5,3,])
q54 = Im(f[5,4,])
q41 = Im(f[4,1,])
q42 = Im(f[4,2,])
q43 = Im(f[4,3,])
q31 = Im(f[3,1,])
q32 = Im(f[3,2,])
q21 = Im(f[2,1,])

# plot part of co spectrum
par(mfrow=c(1,4))
plot(freq*12,q51,type='l',lty=1,ylab="Quadrature",xlab="Cycles/year",main="(a)")
plot(freq*12,q52,type='l',lty=1,ylab="Quadrature",xlab="Cycles/year",main="(b)")
plot(freq*12,q53,type='l',lty=1,ylab="Quadrature",xlab="Cycles/year",main="(c)")
plot(freq*12,q54,type='l',lty=1,ylab="Quadrature",xlab="Cycles/year",main="(d)")

# phase spectrum
phi = sales.spec$phase  # estimated phase spectrum

phi[,1]  # between 2-1
phi[,2]  # between 3-1
```

```
phi[,3]  # between 3-2
phi[,4]  # between 4-1
phi[,5]  # between 4-2
phi[,6]  # between 4-3
phi[,7]  # between 5-1
phi[,8]  # between 5-2
phi[,9]  # between 5-3
phi[,10] # between 5-4

# plot part of phase spectrum
par(mfrow=c(1,4))
plot(freq*12,phi[,7],type='l',lty=1,ylab="Quadrature",xlab="Cycles/year",
main="5,1")
plot(freq*12,phi[,8],type='l',lty=1,ylab="Quadrature",xlab="Cycles/year",
main="5,2")
plot(freq*12,phi[,9],type='l',lty=1,ylab="Quadrature",xlab="Cycles/year",
main="5,3")
plot(freq*12,phi[,10],type='l',lty=1,ylab="Quadrature",xlab="Cycles/year",
main="5,4")

# coherency
K = sales.spec$coh
plot.spec.coherency(sales.spec, ci=NA, main="Squared Coherencies")

# window span 7

L = c(7,7) # degree of smoothing
sales.spec = mvspec(z, spans=L, kernel="daniel", detrend=FALSE, taper=0,
plot=FALSE)
f = sales.spec$fxx/(2*pi)  # estimated spectral matrix
freq = sales.spec$freq
par(mfrow=c(1,5))
plot(sales.spec$freq*12,Re(f[1,1,]),type='l',lty=1,ylab="Spectrum",
xlab="Cycles/year",main=colnames(ms)[1])
plot(sales.spec$freq*12,Re(f[2,2,]),type='l',lty=1,ylab="Spectrum",
xlab="Cycles/year",main=colnames(ms)[2])
plot(sales.spec$freq*12,Re(f[3,3,]),type='l',lty=1,ylab="Spectrum",
xlab="Cycles/year",main=colnames(ms)[3])
plot(sales.spec$freq*12,Re(f[4,4,]),type='l',lty=1,ylab="Spectrum",
xlab="Cycles/year",main=colnames(ms)[4])
plot(sales.spec$freq*12,Re(f[5,5,]),type='l',lty=1,ylab="Spectrum",
xlab="Cycles/year",main=colnames(ms)[5])

# quadrature spectrum
c51 = Re(f[5,1,])
c52 = Re(f[5,2,])
c53 = Re(f[5,3,])
c54 = Re(f[5,4,])
c41 = Re(f[4,1,])
c42 = Re(f[4,2,])
c43 = Re(f[4,3,])
c31 = Re(f[3,1,])
c32 = Re(f[3,2,])
c21 = Re(f[2,1,])

# plot part of quadrature spectrum
par(mfrow=c(1,4))
plot(freq*12, c51, type='l',lty=1,ylab="Co-spectrum",xlab="Cycles/year",main="(a)")
plot(freq*12, c52, type='l',lty=1,ylab="Co-spectrum",xlab="Cycles/year",main="(b)")
```

```
plot(freq*12, c53, type='l',lty=1,ylab="Co-spectrum",xlab="Cycles/year",main="(c)")
plot(freq*12, c54, type='l',lty=1,ylab="Co-spectrum",xlab="Cycles/year",main="(d)")

# co spectrum
q51 = Im(f[5,1,])
q52 = Im(f[5,2,])
q53 = Im(f[5,3,])
q54 = Im(f[5,4,])
q41 = Im(f[4,1,])
q42 = Im(f[4,2,])
q43 = Im(f[4,3,])
q31 = Im(f[3,1,])
q32 = Im(f[3,2,])
q21 = Im(f[2,1,])

# plot part of co spectrum
par(mfrow=c(1,4))
plot(freq*12,q51,type='l',lty=1,ylab="Quadrature",xlab="Cycles/year",main="(a)")
plot(freq*12,q52,type='l',lty=1,ylab="Quadrature",xlab="Cycles/year",main="(b)")
plot(freq*12,q53,type='l',lty=1,ylab="Quadrature",xlab="Cycles/year",main="(c)")
plot(freq*12,q54,type='l',lty=1,ylab="Quadrature",xlab="Cycles/year",main="(d)")

# phase spectrum
phi = sales.spec$phase  # estimated phase spectrum

phi[,1]  # between 2-1
phi[,2]  # between 3-1
phi[,3]  # between 3-2
phi[,4]  # between 4-1
phi[,5]  # between 4-2
phi[,6]  # between 4-3
phi[,7]  # between 5-1
phi[,8]  # between 5-2
phi[,9]  # between 5-3
phi[,10] # between 5-4

# plot part of phase spectrum
par(mfrow=c(1,4))
plot(freq*12,phi[,7],type='l',lty=1,ylab="Quadrature",xlab="Cycles/year",
main="5,1")
plot(freq*12,phi[,8],type='l',lty=1,ylab="Quadrature",xlab="Cycles/year",
main="5,2")
plot(freq*12,phi[,9],type='l',lty=1,ylab="Quadrature",xlab="Cycles/year",
main="5,3")
plot(freq*12,phi[,10],type='l',lty=1,ylab="Quadrature",xlab="Cycles/year",
main="5,4")

# coherency
K = sales.spec$coh
plot.spec.coherency(sales.spec, ci=NA,  main="Squared Coherencies")
```

Bayesian spectral estimation – Example 9.3(b) using MATLAB

For the Bayesian spectral estimation, the first part of the code contains required functions, and the second part of the code is to read in a data set and implement the procedure. I would like to thank Professor Ori Rosen from University of Texas, El Paso for sharing the Bayesian spectral estimation code.

```
function [f,g,h]=gamma_deriv(param,xx,ngamma,nf,nbasis,tau,sigalpha,coef)
  f=-sum(xx*param+exp(-xx*param).*coef)-0.5*(param(1:2)'*param(1:2))/...
    sigalpha-0.5*(param(3:ngamma)'*param(3:ngamma))/tau;
  g=zeros(ngamma,1);
  g(1:2)=-param(1:2)/sigalpha;
  g(3:ngamma)=-param(3:ngamma)/tau;
  h=zeros(ngamma,ngamma);
  h(1:2,1:2)=-1/sigalpha*eye(2);
  h(3:ngamma,3:ngamma)=-1/tau*eye(nbasis);
  for k=1:nf
    g=g-xx(k,:)'+coef(k)*exp(-xx(k,:)*param)*xx(k,:)';
    h=h-coef(k)*exp(-xx(k,:)*param)*xx(k,:)'*xx(k,:);
  end
  f=-f;
  g=-g;
  h=-h;

%%%%%%%%%%%%%%%%%%%%%%%%%%%%%%%%%%%%%%%%%%%%%%%%%%%%%%%%%%%%%%%%%%%%%%%%%%
%  read in data and implement the analysis
%%%%%%%%%%%%%%%%%%%%%%%%%%%%%%%%%%%%%%%%%%%%%%%%%%%%%%%%%%%%%%%%%%%%%%%%%%

sales = csvread('C:/Bookdata/WW2b.csv',1,1);
sales = diff(sales);
ts = sales(:,1:3);
dim=size(ts);
nobs=dim(1);
n=nobs;
ts=ts(1:n,:); %use the first n observations only
yy=fft(ts)/sqrt(n); %find the dft
y=yy(2:round(n/2)-1,:);%drop first and last Fourier frequencies
nf=length(y); %number of Fourier freqs
z=ones(nf,2);
z(:,2)=(1:round(n/2)-2)/n;
tt=(1:round(n/2)-2)/n;
%%for spectral envelope calculation
 var_ts=cov(ts);
[evec,evl]=eig(var_ts);
invroot=evec*diag(diag(evl).^-0.5)*evec';
%%
nbasis=10;
ngamma=nbasis+2;
sigma=10; %variance of prior on theta
sigalpha=100;
taua=nbasis/2;
iter=2000;
nwarmup=1000;
epsilon=zeros(3,iter);
delta_sq=zeros(3,nf,iter);
gamma=zeros(ngamma,9,iter+1);
tau=100*ones(9,iter+1); %corresponding
to the gammas
```

```
theta=zeros(nf,3,iter+1);
cubic=0;
xx=ones(nf,ngamma);
xx(:,2)=tt';
if(cubic==0)
   for j=3:ngamma
     xx(:,j)=sqrt(2)*cos((j-1)*pi*tt);
   end
else
omega=ones(nf,nf);
for i=1:nf
   for j=1:nf
     if(tt(i) <= tt(j))
        omega(i,j)=tt(i)^2*(tt(j)-tt(i)/3)/2;
        omega(j,i)=omega(i,j);
     end
   end
end
[Q,D]=eig(omega);
xx=Q*D^.5;
xx=[ones(nf,1) reshape(tt,nf,1) xx(:,nf-nbasis+1:nf)];
end
spectmat=zeros(3,3,nf,iter-nwarmup);
%spectenv=zeros(nf,iter-nwarmup);
for p=1:iter
   p
   for j=1:3
     delta_sq(j,:,p)=exp(xx*gamma(:,j,p))';
   end
   %Drawing the gammas to smooth the thetas
   %Drawing first gamma4 and gamma7 corresponding to real(theta_1)
   %and imag(theta_1)
   A4=zeros(ngamma,ngamma);
   A4(1:2,1:2)=0.5/sigalpha*eye(2);
   A4(3:ngamma,3:ngamma)=0.5/tau(4,p)*eye(nbasis);
   A7=zeros(ngamma,ngamma);
   A7(1:2,1:2)=0.5/sigalpha*eye(2);
   A7(3:ngamma,3:ngamma)=0.5/tau(7,p)*eye(nbasis);
   B4=zeros(ngamma,ngamma);
   v4=zeros(1,ngamma);
   v7=zeros(1,ngamma);
   for k=1:nf
     ck=conj(y(k,2))*y(k,1)+conj(y(k,1))*y(k,2);
     dk=abs(y(k,1))^2;
     gk=sqrt(-1)*(conj(y(k,2))*y(k,1)-conj(y(k,1))*y(k,2));
     B4=B4+1/delta_sq(2,k,p)*dk*xx(k,:)'*xx(k,:);
     v4=v4+1/delta_sq(2,k,p)*ck*xx(k,:);
     v7=v7+1/delta_sq(2,k,p)*gk*xx(k,:);
   end
   var_gamma4=0.5*inv(B4+A4);
   mu_gamma4=var_gamma4*v4';
   B7=B4;
   var_gamma7=0.5*inv(B7+A7);
   mu_gamma7=var_gamma7*v7';
   var_gamma4=0.5*(var_gamma4+var_gamma4');
   var_gamma7=0.5*(var_gamma7+var_gamma7');
   gamma(:,4,p+1)=mvnrnd(mu_gamma4,var_gamma4)';
   gamma(:,7,p+1)=mvnrnd(mu_gamma7,var_gamma7)';
```

```
theta(:,1,p+1)=xx*gamma(:,4,p+1)+sqrt(-1)*xx*gamma(:,7,p+1);
%Drawing next gamma5 and gamma8 corresponding to real(theta_2)
%and imag(theta_2)
A5=zeros(ngamma,ngamma);
A5(1:2,1:2)=0.5/sigalpha*eye(2);
A5(3:ngamma,3:ngamma)=0.5/tau(5,p)*eye(nbasis);
A8=zeros(ngamma,ngamma);
A8(1:2,1:2)=0.5/sigalpha*eye(2);
A8(3:ngamma,3:ngamma)=0.5/tau(8,p)*eye(nbasis);
v5=zeros(1,ngamma);
v8=zeros(1,ngamma);
B5=zeros(ngamma,ngamma);
B8=zeros(ngamma,ngamma);
for k=1:nf
  ck=real(theta(k,3,p))*(conj(y(k,1))*y(k,2)+conj(y(k,2))*y(k,1))-...
    (conj(y(k,3))*y(k,1)+y(k,3)*conj(y(k,1)));
  dk=abs(y(k,1))^2;
  k=imag(theta(k,3,p))*(conj(y(k,1))*y(k,2)+conj(y(k,2))*y(k,1))-...
    sqrt(-1)*(conj(y(k,3))*y(k,1)-y(k,3)*conj(y(k,1)));
  B5=B5+1/delta_sq(3,k,p)*dk*xx(k,:)'*xx(k,:);
  v5=v5-1/delta_sq(3,k,p)*ck*xx(k,:);
  v8=v8-1/delta_sq(3,k,p)*gk*xx(k,:);
end
B8=B5;
var_gamma5=0.5*inv(B5+A5);
mu_gamma5=var_gamma5*v5';
var_gamma8=0.5*inv(B8+A8);
mu_gamma8=var_gamma8*v8';
var_gamma5=0.5*(var_gamma5+var_gamma5');
var_gamma8=0.5*(var_gamma8+var_gamma8');
gamma(:,5,p+1)=mvnrnd(mu_gamma5,var_gamma5)';
gamma(:,8,p+1)=mvnrnd(mu_gamma8,var_gamma8)';
theta(:,2,p+1)=xx*gamma(:,5,p+1)+sqrt(-1)*xx*gamma(:,8,p+1);
%Drawing gamma6 and gamma9 corresponding to real(theta_3)
%and imag(theta_3)
A6=zeros(ngamma,ngamma);
A6(1:2,1:2)=0.5/sigalpha*eye(2);
A6(3:ngamma,3:ngamma)=0.5/tau(6,p)*eye(nbasis);
A9=zeros(ngamma,ngamma);
A9(1:2,1:2)=0.5/sigalpha*eye(2);
A9(3:ngamma,3:ngamma)=0.5/tau(9,p)*eye(nbasis);
v6=zeros(1,ngamma);
v9=zeros(1,ngamma);
B6=zeros(ngamma,ngamma);
B9=zeros(ngamma,ngamma);
for k=1:nf
  ck=real(theta(k,2,p))*(conj(y(k,1))*y(k,2)+conj(y(k,2))*y(k,1))-...
    (conj(y(k,3))*y(k,2)+y(k,3)*conj(y(k,2)));
  dk=abs(y(k,2))^2;
  gk=imag(theta(k,2,p))*(conj(y(k,1))*y(k,2)+conj(y(k,2))*y(k,1))-...
    sqrt(-1)*(conj(y(k,3))*y(k,2)-y(k,3)*conj(y(k,2)));
  B6=B6+1/delta_sq(3,k,p)*dk*xx(k,:)'*xx(k,:);
  v6=v6-1/delta_sq(3,k,p)*ck*xx(k,:);
  v9=v9-1/delta_sq(3,k,p)*gk*xx(k,:);
end
B9=B6;
var_gamma6=0.5*inv(B6+A6);
```

```
    mu_gamma6=var_gamma6*v6';
    var_gamma9=0.5*inv(B9+A9);
    mu_gamma9=var_gamma9*v9';
    var_gamma6=0.5*(var_gamma6+var_gamma6');
    var_gamma9=0.5*(var_gamma9+var_gamma9');
    gamma(:,6,p+1)=mvnrnd(mu_gamma6,var_gamma6)';
    gamma(:,9,p+1)=mvnrnd(mu_gamma9,var_gamma9)';
    theta(:,3,p+1)=xx*gamma(:,6,p+1)+sqrt(-1)*xx*gamma(:,9,p+1);
    %Drawing the gammas corresponding to the deltas
    for j=1:3
      if(j==1)
        coef=abs(y(:,1)).^2;
      elseif(j==2)
        coef=abs(y(:,2)-theta(:,1,p+1).*y(:,1)).^2;
      else
        coef=abs(y(:,3)-theta(:,2,p+1).*y(:,1)-theta(:,3,p+1).*y(:,2)).^2;
    end
      param=gamma(:,j,p);
      if verLessThan('matlab','9.2')
        options = optimset('Display','off','GradObj','on','Hessian','on',
'MaxIter',10000,...
         'MaxFunEvals',10000,'TolFun',0.001,'TolX',0.00001);
      else
        options = optimoptions(@fminunc,'Display','off','GradObj','on',
'Hessian','on','MaxIter',10000,...
         'MaxFunEvals',10000,'TolFun',0.001,'TolX',0.00001,'Algorithm',
'trust-region');
      end
      [gamma_mean,fval,exitflag,output,grad,hessian]=...
      fminunc(@gamma_deriv,param,options,xx,ngamma,nf,nbasis,tau(j,p),...
        sigalpha,coef );
      if(p<nwarmup)
        gamma_var=4*inv(hessian);
      else
        gamma_var=1.6*inv(hessian);
      end
      gamma_var=0.5*(gamma_var+gamma_var');
      gamma(:,j,p+1)=mvnrnd(gamma_mean,gamma_var)';
      log_prop_gamma_new=-0.5*(gamma(:,j,p+1)-gamma_mean)'*inv(gamma_var)*...
            (gamma(:,j,p+1)-gamma_mean);
      log_prop_gamma_old=-0.5*(gamma(:,j,p)-gamma_mean)'*inv(gamma_var)*...
            (gamma(:,j,p)-gamma_mean);
      log_target_new=-sum(xx*gamma(:,j,p+1)+exp(-xx*gamma(:,j,p+1)).*coef )-...
            0.5*(gamma(1:2,j,p+1)'*gamma(1:2,j,p+1))/sigalpha-...
            0.5*(gamma(3:ngamma,j,p+1)'*gamma(3:ngamma,j,p+1))/tau(j,p);
      log_target_old=-sum(xx*gamma(:,j,p)+exp(-xx*gamma(:,j,p)).*coef )-...
            0.5*(gamma(1:2,j,p)'*gamma(1:2,j,p))/sigalpha-...
            0.5*(gamma(3:ngamma,j,p)'*gamma(3:ngamma,j,p))/tau(j,p);
      met_rat=log_target_new-log_target_old+log_prop_gamma_old-...
          log_prop_gamma_new;
      u=rand;
      epsilon(j,p)=min(1,exp(met_rat));
      if(u>epsilon(j,p)&p>1)
        gamma(:,j,p+1)=gamma(:,j,p);
      end
    end
```

```
    %Drawing the taus
    for j=1:9
      taub=2/(gamma(3:ngamma,j,p+1)'*gamma(3:ngamma,j,p+1));
      tau(j,p+1)=1/gamrnd(taua,taub);
    end
    if(p>nwarmup)
      for k=1:nf
        T=eye(3);
        T(2,1)=-theta(k,1,p);
        T(3,1)=-theta(k,2,p);
        T(3,2)=-theta(k,3,p);
        spectmat(:,:,k,p-nwarmup)=inv(T)*diag(delta_sq(:,k,p))*inv(T');
        %spectenv(k,p-nwarmup)=max(eig(2*invroot*...
        %          real(spectmat(:,:,k,p-nwarmup))*invroot./n));
      end
    end
end
mean_spectmat=mean(spectmat,4);
s11=reshape(mean_spectmat(1,1,:),nf,1);
s22=reshape(mean_spectmat(2,2,:),nf,1);
s33=reshape(mean_spectmat(3,3,:),nf,1);
s21=reshape(mean_spectmat(2,1,:),nf,1);
s31=reshape(mean_spectmat(3,1,:),nf,1);
s32=reshape(mean_spectmat(3,2,:),nf,1);
coh21=abs(s21).^2./(s11.*s22);
coh31=abs(s31).^2./(s11.*s33);
coh32=abs(s32).^2./(s22.*s33);
%%%%%%%%%%%%%%%%%%%%%%%%%%%%%%%%%%%%%%%%%%%%%%%%%%%%%%%%%%%%%%%%%%%%%%%%%%%%%%%%%%%
freq=(0:(nf-1))./(2*nf);
figure
subplot(1,3,1)
plot(freq*12,real(s11));xlabel('cycles/year');title('AUT')
subplot(1,3,2)
plot(freq*12,real(s22));xlabel('cycles/year');title('BUM')
subplot(1,3,3)
plot(freq*12,real(s33));xlabel('cycles/year');title('GEM')
```

Penalized Whittle likelihood method – Example 9.3(c) using MATLAB

The first part of the code is for various required functions, which can be placed in a different order. The second part of the code is to read in a data set and implement the procedure. I would like to thank Dr. Robert Krafty from University of Pittsburgh for sharing the penalized Whittle likelihood estimation codes.

```
function indexer=CholIndexer(P)
%%%%%%%%%%%%%%%%%%%%%%%%%%%%%%%%%%%%%%%%%%%%%%%%%%%%%%%%%%%%%%%%%%%%%%%%%%%%%%%%%%%%%
% Used for the general program to index the conversion of a NxN Cholesky
% matrix  into a vector of size P^2.

Nreal = (P*(P+1))/2;
Nimag = (P*(P-1))/2;

indexer = zeros(P^2,5);
indexer(:,1) = 1:(P^2);
indexer(1:Nreal,2) =1;
```

```
bigguy = [kron(ones(P,1), (1:P)'), kron((1:P)',ones(P,1))];
indexer(1:Nreal, 3:4) = bigguy(bigguy(:,1)>=bigguy(:,2), :);
indexer((Nreal+1):end, 3:4) = bigguy(bigguy(:,1)>bigguy(:,2), :);

indexer(indexer(:,3)==indexer(:,4), 5)=1;
function  spec =inChol2spec(inChol,indexer);
%%%%%%%%%%%%%%%%%%%%%%%%%%%%%%%%%%%%%%%%%%%%%%%%%%%%%%%%%%%%%%%%%%%%%%%%%%%%%%%%
% Takes the inverse Cholesky components and returns the components of the
% spectral matrix

[P2,K] = size(inChol);
P=sqrt(P2);
Nreal = P*(P+1)/2;
spec = zeros(P2,K);

for k = 1:K
  % inv cholesky at freq k
  cholMat = zeros(P,P);
  for p = 1:Nreal
    cholMat(indexer(p,3), indexer(p,4)) = inChol(p,k);
  end;
  for p=(Nreal+1):P2
     cholMat(indexer(p,3), indexer(p,4)) = cholMat(indexer(p,3), indexer(p,4)) +
1i*inChol(p,k);
  end;
  % spec matrix freq k
  F_k = inv(cholMat*cholMat');
  for p=1:(Nreal)
      spec(p,k) = real( F_k( indexer(p,3), indexer(p,4)));
  end;
  for p=(Nreal+1):P2
      spec(p,k) =  imag( F_k( indexer(p,3), indexer(p,4)));
  end;
end;

function [hatF, gcv]=NWspec(omegas,perds,logh)
%%%%%%%%%%%%%%%%%%%%%%%%%%%%%%%%%%%%%%%%%%%%%%%%%%%%%%%%%%%%%%%%%%%%%%%%%%%%%%%%
% Computes the Gaussian Nadaraya-Watson kernel regression estimate of the
% components spectral matrix % by smoothing the components of the raw
% periodograms.

[N,K] = size(perds);
h=exp(logh);

hatF=zeros(N,K);
gcv = 0;
I_A = 0;

for k=1:K
  z=kerf((omegas(k)-omegas)/h);
  hatF(:,k) = perds*z'/sum(z);
  I_A = I_A + 1 - z(k)/sum(z);
end

gcv = log(sum(sum( (perds - hatF).^2))) - log(I_A*I_A);
```

```
function kerval=kerf(z)
kerval=exp(-z.*z/2)/sqrt(2*pi);

function  inChol =spec2inChol(spec,indexer);
%%%%%%%%%%%%%%%%%%%%%%%%%%%%%%%%%%%%%%%%%%%%%%%%%%%%%%%%%%%%%%%%%%%%%%%%%%%%%%
% Takes spectral components and gives the inverse Cholesky components

[P2,K] = size(spec);
P=sqrt(P2);
Nreal = P*(P+1)/2;
inChol = zeros(P2,K);

for k = 1:K
   % spectral matrix at freq k
   F_k = zeros(P,P);
   for p = 1:Nreal
      F_k(indexer(p,3), indexer(p,4)) = spec(p,k);
      F_k(indexer(p,4), indexer(p,3)) = spec(p,k);
   end;
   for p=(Nreal+1):P2
      F_k(indexer(p,3), indexer(p,4)) = F_k(indexer(p,3), indexer(p,4)) + 1i*spec
(p,k);
      F_k(indexer(p,4), indexer(p,3)) = F_k(indexer(p,4), indexer(p,3)) - 1i*spec
(p,k);
   end;
   % inv cholesky at freq k
   cholMat = chol(inv(F_k))';
   for p = 1:Nreal
      inChol(p,k) = real(cholMat(indexer(p,3), indexer(p,4)));
   end;
   for p = (Nreal+1):P2
      inChol(p,k) = imag(cholMat(indexer(p,3), indexer(p,4)));
   end;
end;

function  [tildeX, perds, omegas] =time2freq(X);
%%%%%%%%%%%%%%%%%%%%%%%%%%%%%%%%%%%%%%%%%%%%%%%%%%%%%%%%%%%%%%%%%%%%%%%%%%%%%%
% Takes a time series X and returns its DFT and the periodogram

[P,T] = size(X);
P2 = P^2;
K = floor((T-1)/2);
omegas = (1:K)/T;

indexer=CholIndexer(size(X,1));

tildeX = zeros(P,T);
for p=1:P
     tildeX(p,:) = fft(X(p,:))/sqrt(T);
end;
tildeX = tildeX(:,2:(K+1));

perds = zeros(P2,K);
Nreal = P*(P+1)/2;
for k = 1:K
  P_mat = tildeX(:,k)*tildeX(:,k)';
  for p=1:Nreal
     perds(p,k) = real(P_mat(indexer(p,3), indexer(p,4)));
  end;
```

```
  for p = (Nreal+1):P2
    perds(p,k) = imag(P_mat(indexer(p,3), indexer(p,4)));
  end;
end;

function  [workvals, eW] = WorkingFish(startvals,tildeX,indexer);
%%%%%%%%%%%%%%%%%%%%%%%%%%%%%%%%%%%%%%%%%%%%%%%%%%%%%%%%%%%%%%%%%%%%%%%%%%%%%
% Gets the working variables and weight function for an iteration using
% the inverse Cholesky parameterization.

[P,K] = size(tildeX);
P2 = P*P;
Nreal = P*(P+1)/2;
spec = inChol2spec(startvals,indexer);

W = zeros(P2*K, P2*K);
eW = zeros(P2*K, P2*K);
workvals = zeros(P2,K);
for k = 1:K
  %
  % get the G and F matrices
  G = zeros(P,P);
  F = zeros(P,P);
  for p=1:Nreal;
     G(indexer(p,3), indexer(p,4))=startvals(p,k);
     F(indexer(p,3), indexer(p,4))=spec(p,k);
     F(indexer(p,4), indexer(p,3))=spec(p,k);
  end;
  for p=(Nreal+1):P2;
     G(indexer(p,3), indexer(p,4))=G(indexer(p,3), indexer(p,4)) + 1i*startvals(p,k);
     F(indexer(p,3), indexer(p,4))= F(indexer(p,3), indexer(p,4)) + 1i*spec(p,k);
     F(indexer(p,4), indexer(p,3))= F(indexer(p,4), indexer(p,3)) - 1i*spec(p,k);
  end;
  %
  % get the score
  U = zeros(P2,1);
  for p=1:P2
     dG = zeros(P,P);
     dG(indexer(p,3), indexer(p,4)) = indexer(p,2) + (1-indexer(p,2))*1i;
     %
     U(p) = tildeX(:,k)'*(dG*G' + G*dG')*tildeX(:,k) ...
         + indexer(p,5)*(-2/startvals(p,k));
  end;
  %
  % get the Hessian
  Hsmall = zeros(P2,P2);
  Fishmat = zeros(P2,P2);
  for p=1:P2
     dGp = zeros(P,P);
     dGp(indexer(p,3), indexer(p,4)) = indexer(p,2) + (1-indexer(p,2))*1i;
     for q=1:P2
        dGq = zeros(P,P);
        dGq(indexer(q,3), indexer(q,4)) = indexer(q,2) + (1-indexer(q,2))*1i;
        %
        Hsmall(p,q) = tildeX(:,k)'*(dGp*dGq' + dGq*dGp')*tildeX(:,k) ...
         +  indexer(p,5)*(p==q)*(2/startvals(p,k)^2);
        Fishmat(p,q) = trace((dGp*dGq' + dGq*dGp')*F) ...
         +  indexer(p,5)*(p==q)*(2/startvals(p,k)^2);
     end;
  end;
```

```
    W( (1+P2*(k-1)):(P2*k), (1+P2*(k-1)):(P2*k) ) = real(Hsmall);
    eW( (1+P2*(k-1)):(P2*k), (1+P2*(k-1)):(P2*k) ) = real(Fishmat);
    %
    % get the working variables
    workvals(:,k) = real(startvals(:,k) - real(Fishmat)\U);
end;

function  [hatF,  GML] = penWhit(tildeX,omegavals,logLambda0);
%%%%%%%%%%%%%%%%%%%%%%%%%%%%%%%%%%%%%%%%%%%%%%%%%%%%%%%%%%%%%%%%%%%%%%%%%%%%%%%%%%
%  Fits the penalized multivariate Whittle likelihood

tolval = .0001,
MaxIt = 2;
[N,K] = size(tildeX);
P=N*N;
indexer=CholIndexer(N);
Preal=N*(N+1)/2;
Pimag=N*(N-1)/2;

perds = zeros(P,K);
for k = 1:K
   P_mat = tildeX(:,k)*tildeX(:,k)';
   for p=1:Preal
      perds(p,k) = real(P_mat(indexer(p,3), indexer(p,4)));
   end;
   for p = (Preal+1):P
      perds(p,k) = imag(P_mat(indexer(p,3), indexer(p,4)));
   end;
end;
%
%   smooth and get initial values
[hatF_NW, G] = NWspec(omegavals', perds,2*log(20/K));
current = spec2inChol(hatF_NW,indexer);

nb = min( floor(K/4),25 );
Q = zeros(P*K, P*nb+Preal);
S = zeros(P*nb+Preal, P*nb+Preal);
Qc = [ones(K,1), sqrt(2)*cos(2*pi*omegavals*(1:nb))];
for p=1:Preal
  ep=zeros(P,1);
  ep(p)=1;
  locp = ((p-1)*(nb+1)+1):p*(nb+1);
  Q(:, locp ) = kron(Qc,ep);
  S(locp,locp) = diag(exp(logLambda0(p))*(2*pi*(0:nb)).^4);
end;
Qs = sqrt(2)*sin(2*pi*omegavals*(1:nb));
for p=(Preal+1):P
  ep=zeros(P,1);
  ep(p)=1;
  pc=p-Preal;
  locp = ((pc-1)*nb + Preal*(nb+1)+1):(pc*nb+Preal*(nb+1));
  Q(:, locp ) = kron(Qs,ep);
  S(locp,locp) = diag(exp(logLambda0(p))*(2*pi*(1:nb)).^4);
end;
R=blkdiag( kron(eye(Preal,Preal),Qc), kron(eye(Pimag,Pimag),Qs));

old = current;
tol=1;
ct=0;
```

```
while (tol>tolval) && (ct<=MaxIt-1)
  [workvars, H] = WorkingFish(current,tildeX,indexer);
  tildeBeta = reshape(workvars, K*P ,1);
  bigest = (Q'*H*Q + S)\Q'*H*tildeBeta;
  current = reshape(R*bigest, K,P)';
  %
  tol = sum(sum((current-old).^2))/sum(sum(old.^2));
  old=current;
  ct = ct+1;
end;

hatF =inChol2spec(current,indexer);

Ell = 0;
for k = 1:K;
  % get the inverse cholesky matrix
  Cholmat = zeros(N,N);
  for p = 1:Preal
    Cholmat(indexer(p,3), indexer(p,4)) = current(p,k);
  end;
  for p=(Preal+1):P
    Cholmat(indexer(p,3), indexer(p,4)) = Cholmat(indexer(p,3), indexer(p,4)) +
1i*current(p,k);
  end;
  %
  % add to the -2loglike
  Ell = Ell - log(det(Cholmat*Cholmat')) + tildeX(:,k)'*Cholmat*Cholmat'*tildeX
(:,k);
end;

GML =  real(Ell) ...
    -nb*sum(logLambda0)...
    + bigest'*S*bigest ...
    + sum(log(eig(Q'*H*Q + S)));

function [logLamHat,G]= optGML(tildeX, perds, omegas, logLamStart, options);
%%%%%%%%%%%%%%%%%%%%%%%%%%%%%%%%%%%%%%%%%%%%%%%%%%%%%%%%%%%%%%%%%%%%%%%%%%%%%%
%  Used for selecting smoothing parameters by optimizing the GML

indexer=CholIndexer(size(tildeX,1));

[Fstart, Gval] = NWspec(omegas, perds,2*log(20/length(omegas)));
startvals = spec2inChol(Fstart,indexer);

[logLamHat,G] = fminunc(@(llam)
GLMfish(llam,tildeX,omegas',startvals,indexer),logLamStart, options);

function  GML = GLMfish(logLambda0,tildeX,omegavals, startvals,indexer);

tolval = .0001;
MaxIt = 2;

[N,K] = size(tildeX);
P=N*N;
Preal=N*(N+1)/2;
Pimag=N*(N-1)/2;
current = startvals;
```

```
nb = min( floor(K/4),25 );
Q = zeros(P*K, P*nb+Preal);
S = zeros(P*nb+Preal, P*nb+Preal);
Qc = [ones(K,1), sqrt(2)*cos(2*pi*omegavals*(1:nb))];
for p=1:Preal
  ep=zeros(P,1);
  ep(p)=1;
  locp = ((p-1)*(nb+1)+1):p*(nb+1);
  Q(:, locp ) = kron(Qc,ep);
  S(locp,locp) = diag(exp(logLambda0(p))*(2*pi*(0:nb)).^4);
end;
Qs = sqrt(2)*sin(2*pi*omegavals*(1:nb));
for p=(Preal+1):P
  ep=zeros(P,1);
  ep(p)=1;
  pc=p-Preal;
  locp = ((pc-1)*nb + Preal*(nb+1)+1):(pc*nb+Preal*(nb+1));
  Q(:, locp ) = kron(Qs,ep);
  S(locp,locp) = diag(exp(logLambda0(p))*(2*pi*(1:nb)).^4);
end;
R=blkdiag( kron(eye(Preal,Preal),Qc), kron(eye(Pimag,Pimag),Qs));

old = current;
tol=1;
ct=0;
while (tol>tolval)  && (ct<=MaxIt-1)
  [workvars, H] = WorkingFish(current,tildeX,indexer);
  tildeBeta = reshape(workvars, K*P ,1);
  bigest = (Q'*H*Q + S)\Q'*H*tildeBeta;
  current = reshape(R*bigest, K,P)';
  %
  tol = sum(sum((current-old).^2))/sum(sum(old.^2));
  old=current;
  ct = ct+1;
end;
hatBeta = current;

hatF =inChol2spec(current,indexer);

Ell = 0;
for k = 1:K;
  % get the inverse cholesky matrix
  Cholmat = zeros(N,N);
  for p = 1:Preal
    Cholmat(indexer(p,3), indexer(p,4)) = current(p,k);
  end;
  for p=(Preal+1):P
    Cholmat(indexer(p,3), indexer(p,4)) = Cholmat(indexer(p,3), indexer(p,4)) +
1i*current(p,k);
  end;
  %
  % add to the -2loglike
  Ell = Ell - log(det(Cholmat*Cholmat')) + tildeX(:,k)'*Cholmat*Cholmat'*tildeX
(:,k);
end;

GML =  real(Ell) ...
     -nb*sum(logLambda0) ...
     + bigest'*S*bigest ...
     + sum(log(eig(Q'*H*Q + S)));
```

```
function  bootcell = penWhitBoot(hatF,omegavals, logLambda0, nBoot, display);
%%%%%%%%%%%%%%%%%%%%%%%%%%%%%%%%%%%%%%%%%%%%%%%%%%%%%%%%%%%%%%%%%%%%%%%%%%%%%%%%%%
%  Get bootstrap estimates from the penalized multivariate Whittle
%  likelihood

tolval = .0001;
MaxIt = 2;
%
[P,K]=size(hatF);
N=sqrt(P);
Preal = N*(N+1)/2;
Pimag = P - Preal;
bootcell=cell(P,1);
for j = 1:P
  bootcell{j} =zeros(nBoot,K);
end;

indexer=CholIndexer(N);

Fsqrt = cell(K,1);
for k = 1:K
  F =  zeros(N,N);
  for p = 1:Preal
     F(indexer(p,3), indexer(p,4)) = hatF(p,k);
     F(indexer(p,4), indexer(p,3)) = hatF(p,k);
  end;
  for p = (Preal+1):P
     F(indexer(p,3), indexer(p,4)) = F(indexer(p,3), indexer(p,4)) + 1i*hatF(p,k);
     F(indexer(p,4), indexer(p,3)) = F(indexer(p,4), indexer(p,3)) - 1i*hatF(p,k);
  end;
  Fsqrt{k} = chol(F)';
end;

nb = min( floor(K/4),25 );
Q = zeros(P*K, P*nb+Preal);
S = zeros(P*nb+Preal, P*nb+Preal);
Qc = [ones(K,1), sqrt(2)*cos(2*pi*omegavals*(1:nb))];
for p=1:Preal
  ep=zeros(P,1);
  ep(p)=1;
  locp = ((p-1)*(nb+1)+1):p*(nb+1);
  Q(:, locp ) = kron(Qc,ep);
  S(locp,locp) = diag(exp(logLambda0(p))*(2*pi*(0:nb)).^4);
end;
Qs = sqrt(2)*sin(2*pi*omegavals*(1:nb));
for p=(Preal+1):P
  ep=zeros(P,1);
  ep(p)=1;
  pc=p-Preal;
  locp = ((pc-1)*nb + Preal*(nb+1)+1):(pc*nb+Preal*(nb+1));
  Q(:, locp ) = kron(Qs,ep);
  S(locp,locp) = diag(exp(logLambda0(p))*(2*pi*(1:nb)).^4);
end;
R=blkdiag( kron(eye(Preal,Preal),Qc), kron(eye(Pimag,Pimag),Qs));
```

```
tic
for b=1:nBoot
  %%%%%% display options
  if display==1
     b
  end;
  if display==2
     [b, toc]
     tic
  end;

  %%%%%% pull the DFT residuals
  tildeXBoot=zeros(N,K);
  resids = normrnd(0,sqrt(1/2),N,K)+1i*normrnd(0,sqrt(1/2),N,K);

  % get the fourier transform and periodograms
  perds = zeros(P,K);
  for k = 1:K
     tildeXBoot(:,k) = Fsqrt{k}*resids(:,k);
     P_mat = tildeXBoot(:,k)*tildeXBoot(:,k)';
     for p=1:Preal
        perds(p,k) = real(P_mat(indexer(p,3), indexer(p,4)));
     end;
     for p = (Preal+1):P
        perds(p,k) = imag(P_mat(indexer(p,3), indexer(p,4)));
     end;
  end;

  % run iterative least squares
  % use NWKern to get the starting values
  [hatF_NW, Gval] = NWspec(omegavals', perds,2*log(20/K));
     current = spec2inChol(hatF_NW,indexer);

     old = current;
     tol=1;
     ct=0;
     while (tol>tolval) && (ct<=MaxIt-1)
        [workvars, H] = WorkingFish(current,tildeXBoot,indexer);
        tildeBeta = reshape(workvars, K*P ,1);
        bigest = (Q'*H*Q + S)\Q'*H*tildeBeta;
        current = reshape(R*bigest, K,P)';
        %
        tol = sum(sum((current-old).^2))/sum(sum(old.^2));
        old=current;
        ct = ct+1;
        end;
     %%%%%%%%%%%%%%%%%%%%%%%%%%%%%%%%%%%%%%%%%%%%%%%%%%%%%%%%%%%%%%%%%%
     % get hatF and save
     hatFboot =inChol2spec(current,indexer);
     for p = 1:P
        bootcell{p}(b,:) = hatFboot(p,:);
     end;
     %%%%%%%%%%%%%%%%%%%%%%%%%%%
end;
```

```
%%%%%%%%%%%%%%%%%%%%%%%%%%%%%%%%%%%%%%%%%%%%%%%%%%%%%%%%%%%%%%%%%%%%%%%%%
% The code reproduces the analysis of the simulated bivariate AR(2) from
% the supplement to "Penalized Multivariate Whittle Likelihood for Power
% Spectrum Estimation" by Krafty and Collinge
%%%%%%%%%%%%%%%%%%%%%%%%%%%%%%%%%%%%%%%%%%%%%%%%%%%%%%%%%%%%%%%%%%%%%%%%%

sales = csvread('C:/Bookdata/WW2b.csv',1,1);
sales = diff(sales);
X = sales';

[tildeX, perds, omegas] =time2freq(X);

logLamStart = -2*ones(25,1); %starting values
options = optimset('LargeScale','off','Display','iter', 'TolX', .01); %options
for the optimization
[logLamHat,G]= optGML(tildeX, perds, omegas, logLamStart, options);

[hatF, GLM] = penWhit(tildeX,omegas',logLamHat);

figure
subplot(1,5,1); plot((0:43)/88*12,hatF(1,:))
subplot(1,5,2); plot((0:43)/88*12,hatF(6,:))
subplot(1,5,3); plot((0:43)/88*12,hatF(10,:))
subplot(1,5,4); plot((0:43)/88*12,hatF(13,:))
subplot(1,5,5); plot((0:43)/88*12,hatF(15,:))
```

VAR estimation of multivariate spectrum – Example 9.3(d) using R

```
library(MTS)
library(astsa)
#=================
#VAR analysis
#=================

# Two necessary functions

# Obtain Spectrum from autoregressive model
ARspect <- function(phi,sigma,freq){
  dimen <- dim(phi)[2]
  len <- dim(phi)[1]

  bigphi <- array(0,dim=c(dimen,dimen,length(freq)))
  spect <- array(0,dim=c(dimen,dimen,length(freq)))
  for(k in 1:length(freq)){
     bigphi[,,k] <- diag(dimen)
     for(j in 1:(len/dimen)){
       if(j==1){
          bigmat <- phi[1:dimen,]*exp(-2*pi*(1i)*freq[k])
       }else{
           bigmat <- phi[((dimen*j-(dimen-1)):(dimen*j)),]*exp(-2*j*pi*(1i)*freq
[k])
       }
       bigphi[,,k] = bigphi[,,k] - bigmat
     }
     spect[,,k] = solve(bigphi[,,k])%*%sigma%*%solve(Conj(t(bigphi[,,k])))
  }
  return(spect/(2*pi))
}
```

```
# Plot the coherence
plot.coherency<-function (x, ci = 0.95, xlab = "frequency", ylab = "squared
coherency",
                ylim = c(0, 1), type = "l", main = NULL, ci.col = "blue",
                ci.lty = 3, ...)
{
  nser <- NCOL(x$spec)
  gg <- 2/x$df
  se <- sqrt(gg/2)
  z <- -qnorm((1 - ci)/2)
  if (is.null(main))
      main <- paste(paste("Series:", x$series), "Squared Coherency",
                sep = " - ")
  if (nser == 2) {
      plot(x$freq, x$coh, type = type, xlab = xlab, ylab = ylab,
        ylim = ylim, ...)
      coh <- pmin(0.99999, sqrt(x$coh))
      lines(x$freq, (tanh(atanh(coh) + z * se))^2, lty = ci.lty,
        col = ci.col)
      lines(x$freq, (pmax(0, tanh(atanh(coh) - z * se)))^2,
        lty = ci.lty, col = ci.col)
      title(main)
  }
  else {
      dev.hold()
      on.exit(dev.flush())
      opar <- par(mfrow = c(nser - 1, nser - 1), mar = c(1.5,
                                      1.5, 0.5, 0.5), oma = c(4, 4, 6, 4))
      on.exit(par(opar), add = TRUE)
      plot.new()
      for (j in 2:nser) for (i in 1L:(j - 1)) {
        par(mfg = c(j - 1, i, nser - 1, nser - 1))
        ind <- i + (j - 1) * (j - 2)/2
        plot(x$freq, x$coh[, ind], type = type, ylim = ylim,
          axes = FALSE, xlab = "", ylab = "", ...)
        box()
        if (i == 1) {
            axis(2, xpd = NA)
            title(ylab = x$snames[j], xpd = NA)
        }
        if (j == nser) {
            axis(1, xpd = NA)
            title(xlab = x$snames[i], xpd = NA)
        }
        mtext(main, 3, 3, TRUE, 0.5, cex = par("cex.main"),
            font = par("font.main"))
      }
  }
  invisible()
}

# data preparation

ms <- as.data.frame(read.csv("C:/Bookdata/WW2b.csv")[,-1])
rownames(ms) <- seq(as.Date("2009/6/1"), by="month", length=90)
z.orig <- ts(ms,start=c(2009,6),frequency=12)
z <- diff(z.orig,1)
plot(z)
```

```
z = ts(apply(z,2,scale), freq=1)   # scaling strips its attributes
n = nrow(z)

# VAR(4) model fitting and spectrum estimation

model <- VARMA(z,p=4,include.mean=FALSE)
nfreq = floor(n/2); freq = (0:nfreq)/(2*nfreq);
f <- ARspect(model$coef,model$Sigma,freq)

> par(mfrow=c(1,5))
plot(freq*12,Re(f[1,1,]),type='l',lty=1,ylab="Spectrum",xlab="Cycles/year",
main=colnames(ms)[1])
plot(freq*12,Re(f[2,2,]),type='l',lty=1,ylab="Spectrum",xlab="Cycles/year",
main=colnames(ms)[2])
plot(freq*12,Re(f[3,3,]),type='l',lty=1,ylab="Spectrum",xlab="Cycles/year",
main=colnames(ms)[3])
plot(freq*12,Re(f[4,4,]),type='l',lty=1,ylab="Spectrum",xlab="Cycles/year",
main=colnames(ms)[4])
plot(freq*12,Re(f[5,5,]),type='l',lty=1,ylab="Spectrum",xlab="Cycles/year",
main=colnames(ms)[5])

# quadrature spectrum
c51 = Re(f[5,1,])
c52 = Re(f[5,2,])
c53 = Re(f[5,3,])
c54 = Re(f[5,4,])
c41 = Re(f[4,1,])
c42 = Re(f[4,2,])
c43 = Re(f[4,3,])
c31 = Re(f[3,1,])
c32 = Re(f[3,2,])
c21 = Re(f[2,1,])

# plot part of quadrature spectrum
par(mfrow=c(1,4))
plot(freq*12, c51, type='l',lty=1,ylab="Co-spectrum",xlab="Cycles/year",
main="(a)")
plot(freq*12, c52, type='l',lty=1,ylab="Co-spectrum",xlab="Cycles/year",
main="(b)")
plot(freq*12, c53, type='l',lty=1,ylab="Co-spectrum",xlab="Cycles/year",
main="(c)")
plot(freq*12, c54, type='l',lty=1,ylab="Co-spectrum",xlab="Cycles/year",
main="(d)")

# co spectrum
q51 = Im(f[5,1,])
q52 = Im(f[5,2,])
q53 = Im(f[5,3,])
q54 = Im(f[5,4,])
q41 = Im(f[4,1,])
q42 = Im(f[4,2,])
q43 = Im(f[4,3,])
q31 = Im(f[3,1,])
q32 = Im(f[3,2,])
q21 = Im(f[2,1,])
# plot part of co spectrum
par(mfrow=c(1,4))
```

```
plot(freq*12,q51,type='l',lty=1,ylab="Quadrature",xlab="Cycles/year",main="(a)")
plot(freq*12,q52,type='l',lty=1,ylab="Quadrature",xlab="Cycles/year",main="(b)")
plot(freq*12,q53,type='l',lty=1,ylab="Quadrature",xlab="Cycles/year",main="(c)")
plot(freq*12,q54,type='l',lty=1,ylab="Quadrature",xlab="Cycles/year",main="(d)")

# phase spectrum
phi = matrix(0,length(freq),10)
phi[,1] = atan(-Im(f[2,1,])/Re(f[2,1,]))  # between 2-1
phi[,2] = atan(-Im(f[3,1,])/Re(f[3,1,]))  # between 3-1
phi[,3] = atan(-Im(f[3,2,])/Re(f[3,2,]))  # between 3-2
phi[,4] = atan(-Im(f[4,1,])/Re(f[4,1,]))  # between 4-1
phi[,5] = atan(-Im(f[4,2,])/Re(f[4,2,]))  # between 4-2
phi[,6] = atan(-Im(f[4,3,])/Re(f[4,3,]))  # between 4-3
phi[,7] = atan(-Im(f[5,1,])/Re(f[5,1,]))  # between 5-1
phi[,8] = atan(-Im(f[5,2,])/Re(f[5,2,]))  # between 5-2
phi[,9] = atan(-Im(f[5,3,])/Re(f[5,3,]))  # between 5-3
phi[,10] = atan(-Im(f[5,4,])/Re(f[5,4,])) # between 5-4

# plot part of phase spectrum
par(mfrow=c(1,4))
plot(freq*12, phi[,7] ,type='l',lty=1,ylab="Quadrature",xlab="Cycles/year",
main="5,1")
plot(freq*12, phi[,8] ,type='l',lty=1,ylab="Quadrature",xlab="Cycles/year",
main="5,2")
plot(freq*12, phi[,9] ,type='l',lty=1,ylab="Quadrature",xlab="Cycles/year",
main="5,3")
plot(freq*12, phi[,10] ,type='l',lty=1,ylab="Quadrature",xlab="Cycles/year",
main="5,4")

# coherency
coh <- matrix(0,length(freq),10)
coh[,1] <- Re(Mod(f[2,1,])^2/(f[1,1,] * f[2,2,]))
coh[,2] <- Re(Mod(f[3,1,])^2/(f[1,1,] * f[3,3,]))
coh[,3] <- Re(Mod(f[3,2,])^2/(f[2,2,]  * f[3,3,]))
coh[,4] <- Re(Mod(f[4,1,])^2/(f[1,1,]  * f[4,4,]))
coh[,5] <- Re(Mod(f[4,2,])^2/(f[2,2,]  * f[4,4,]))
coh[,6] <- Re(Mod(f[4,3,])^2/(f[3,3,]  * f[4,4,]))
coh[,7] <- Re(Mod(f[5,1,])^2/(f[1,1,]  * f[5,5,]))
coh[,8] <- Re(Mod(f[5,2,])^2/(f[2,2,]  * f[5,5,]))
coh[,9] <- Re(Mod(f[5,3,])^2/(f[3,3,]  * f[5,5,]))
coh[,10] <- Re(Mod(f[5,4,])^2/(f[4,4,] * f[5,5,]))
spec <- list()
spec$spec <- f
spec$freq <- freq
spec$coh <- coh
spec$snames <- colnames(z)
plot.coherency(spec, ci=NA,  main="Squared Coherencies")
plot.coherency(spec, ci=NA,  main="Squared Coherencies")
```

Code for multivariate nonstationary spectral procedures using MATLAB

Again, the first part of the following code is for various required functions, which can be placed in a different order. The last part is to read in a data set and implement the procedures.

Empirical Example 9.4 of nonstationary vector time series – analysis of financial data

```
function [f,gr,h] = Beta_derivel(x, yobs_tmp, chol_index, Phi_temp, tau_temp,...
                         Beta_temp, sigmasqalpha, nbasis)
global dimen
%%%%%%%%%%%%%%%%%%%%%%%%%%%%%%%%%%%%%%%%%%%%%%%%%%%%%%%%%%%%%%%%%%%%%%%%%%%%%%%%%%
% Function used for optimization process for coefficients selected to
% be changed
%
% Input:
%    1) x - initial values for coefficient of basis functions need to
%    be optimized
%    2) yobs_tmp - time series data within the segment
%    3) chol_index - index matrix
%    4) Phi_temp - which component changed
%    5) tau_temp - smoothing parameters
%    6) Beta_temp - current coefficients
%    7) sigmasqalpha - smoothing parameters for the constant in real
%    components
%    8) nbasis - number of basis functions used
% Main Outputs:
%   1) f - log posterior probability based on input parameters
%   2) gr - gradients for optimization process
%   3) h - Hessian matrix for optimization process
%
% Required programs: lin_basis_func
%%%%%%%%%%%%%%%%%%%%%%%%%%%%%%%%%%%%%%%%%%%%%%%%%%%%%%%%%%%%%%%%%%%%%%%%%%%%%%%%%%
%initilize Beta_1 and Beta_2: Beta_2 is for imaginary components
nBeta = nbasis + 1;
Beta_1 = zeros(nBeta,(dimen + dimen*(dimen-1)/2));
Beta_2 = zeros(nBeta,dimen*(dimen-1)/2);

select = chol_index(Phi_temp,:).*(1:dimen^2);
select = select(select~=0);

x = reshape(x,nBeta,length(select));
Beta_temp(:,select) = x;
Beta_1(:,:) = Beta_temp(:,1:(dimen + dimen*(dimen-1)/2));
Beta_2(:,:) = Beta_temp(1:nBeta,(dimen + dimen*(dimen-1)/2 + 1): end);

dim = size(yobs_tmp); n = dim(1);
nfreq = floor(n/2); tt = (0:nfreq)/(2*nfreq);
yy = fft(yobs_tmp)/sqrt(n); y = yy(1:(nfreq+1),:); nf = length(y);
[xx_r, xx_i]=lin_basis_func(tt);

%theta's
theta = zeros(dimen*(dimen-1)/2,nf );
for i=1:dimen*(dimen-1)/2
  theta_real = xx_r * Beta_1(:,i+dimen);
```

```
  theta_imag = xx_i * Beta_2(:,i);
  theta(i,:) = theta_real + sqrt(-1)*theta_imag;
end
%delta's
delta_sq = zeros(dimen,nf);
for i=1:dimen
  delta_sq(i,:) = exp(xx_r * Beta_1(:,i));
end

if dimen==2  %Bivariate Time Series
  if (mod(n,2)==1)%odd n
     f=-sum(log(delta_sq(1,2:end))'+log(delta_sq(2,2:end))'+abs(y(2:end,1)).
^2.*exp(-xx_r(2:end,:)*Beta_1(:,1)) + ...
        abs(y(2:end,2) - theta(2:end).'.*y(2:end,1)).^2.*exp(-xx_r(2:end,:)
*Beta_1(:,2))) - ...
       0.5*(log(delta_sq(1,1))'+log(delta_sq(2,1))'+abs(y(1,1)).^2.*exp(-xx_r(1,:)
*Beta_1(:,1)) + ...
        abs(y(1,2) - theta(1).'.*y(1,1))^2.*exp(-xx_r(1,:)*Beta_1(:,2)));
  else
     f=-sum(log(delta_sq(1,2:nfreq))'+log(delta_sq(2,2:nfreq))'+abs(y(2:nfreq,
1)).^2.*exp(-xx_r(2:nfreq,:)*Beta_1(:,1)) + ...
     abs(y(2:nfreq,2) - theta(2:nfreq).'.*y(2:nfreq,1)).^2.*exp(-xx_r(2:nfreq,:)
*Beta_1(:,2))) - ...
     0.5*(log(delta_sq(1,1))+log(delta_sq(2,1))+abs(y(1,1)).^2.*exp(-xx_r(1,:)
*Beta_1(:,1))+...
     abs(y(1,2) - theta(1).'.*y(1,1)).^2.*exp(-xx_r(1,:)*Beta_1(:,2))) - ...
     0.5*(log(delta_sq(1,end))+log(delta_sq(2,end))+abs(y(end,1)).^2.*exp
(-xx_r(end,:)*Beta_1(:,1)) + ...
     abs(y(end,2) - theta(end).'.*y(end,1))^2.*exp(-xx_r(end,:)*Beta_1(:,2)));
  end
  f = f - (0.5.*(Beta_1(1,1)* Beta_1(1,1)')/sigmasqalpha +
0.5.*(Beta_1(2:nBeta,1)'*Beta_1(2:nBeta,1))/tau_temp(1))*chol_index(Phi_temp,1) -...
     (0.5.*(Beta_1(1,2)* Beta_1(1,2)')/sigmasqalpha +
0.5.*(Beta_1(2:nBeta,2)'*Beta_1(2:nBeta,2))/tau_temp(2))*chol_index(Phi_temp,2) -...
     (0.5.*(Beta_1(1,3)* Beta_1(1,3)')/sigmasqalpha +
0.5.*(Beta_1(2:nBeta,3)'*Beta_1(2:nBeta,3))/tau_temp(3))*chol_index(Phi_temp,3) -...
     (0.5.*(Beta_2(1:nBeta,1)'*Beta_2(1:nBeta,1))/tau_temp(4))*chol_index
(Phi_temp,4);

  gr1=zeros(nBeta,1); gr2=zeros(nBeta,1); gr3=zeros(nBeta,1); gr4=zeros(nBeta,1);
  gr1(1) =Beta_1(1,1)/sigmasqalpha; gr1(2:nBeta,1) =Beta_1(2:nBeta,1)/tau_temp(1);
  gr2(1) =Beta_1(1,2)/sigmasqalpha; gr2(2:nBeta,1) =Beta_1(2:nBeta,2)/tau_temp(2);
  gr3(1) =Beta_1(1,3)/sigmasqalpha; gr3(2:nBeta,1) =Beta_1(2:nBeta,3)/tau_temp(3);
  gr4(1:nBeta,1) = Beta_2(1:nBeta,1)/tau_temp(4);
  h11(1,1)=1/sigmasqalpha; h11(2:nBeta,2:nBeta)=1/tau_temp(1)*eye(nbasis);
  h22(1,1)=1/sigmasqalpha; h22(2:nBeta,2:nBeta)= 1/tau_temp(2)*eye(nbasis);
  h33(1,1)=1/sigmasqalpha; h33(2:nBeta,2:nBeta)= 1/tau_temp(3)*eye(nbasis);
  h44(1:nBeta,1:nBeta)=1/tau_temp(4)*eye(nBeta);
  h42 = zeros(nBeta,nBeta); h32 = zeros(nBeta,nBeta);

  if (mod(n,2)==1)
     %%%%%%%%%%%%%%%%%%%%%%%%%%%%%%%%%%%%%%%
     %gradient
     %%%%%%%%%%%%%%%%%%%%%%%%%%%%%%%%%%%%%%%
     rk = -y(2:end,1).*conj(y(2:end,2)) - y(2:end,2).*conj(y(2:end,1));
     ik=sqrt(-1)*(-y(2:end,1).*conj(y(2:end,2))+y(2:end,2).*conj(y(2:end,1)));
     ck = 2*abs(y(2:end,1)).^2;

     gr1=gr1+xx_r(2:end,:)'*(1-abs(y(2:end,1)).^2.*exp(-xx_r(2:end,:)*Beta_1
(:,1))) + ...
```

```
            0.5*(xx_r(1,:)'*(1-abs(y(1,1)).^2.*exp(-xx_r(1,:)*Beta_1(:,1))));
      gr2 = gr2 + xx_r(2:end,:)'*(1 - abs(y(2:end,2)-theta(2:end).'.*y(2:end,1)).^2.
*exp(-xx_r(2:end,:)*Beta_1(:,2))) + ...
            0.5*(xx_r(1,:)'*(1 - abs(y(1,2)-theta(1).'.*y(1,1)).^2.*exp(-xx_r(1,:)
*Beta_1(:,2))));
      temp_mat_31 = bsxfun(@times, xx_r(2:end,:),rk);
      temp_mat_32 = bsxfun(@times, ck, bsxfun(@times,
xx_r(2:end,:),xx_r(2:end,:)*Beta_1(:,3)));
      gr3 = gr3 + sum( bsxfun(@times, (temp_mat_31+temp_mat_32), exp(-xx_r(2:
end,:)*Beta_1(:,2)) ))' +...
            0.5*(exp(-xx_r(1,:)*Beta_1(:,2))*(-y(1,1).*conj(y(1,2)) - y(1,2).*conj
(y(1,1)))*xx_r(1,:)' +...
            exp(-xx_r(1,:)*Beta_1(:,2))*2*abs(y(1,1)).^2*(xx_r(1,:)*Beta_1(:,3))
*xx_r(1,:)');
      temp_mat_41 = bsxfun(@times, ik, xx_i(2:end,:));
      temp_mat_42 = bsxfun(@times, ck,
bsxfun(@times,xx_i(2:end,:),xx_i(2:end,:)*Beta_2(:,1)));
      gr4 = gr4 + sum( bsxfun(@times, (temp_mat_41 + temp_mat_42), exp(-xx_r(2:
end,:)*Beta_1(:,2))))' + ...
              0.5*(exp(-xx_r(1,:)*Beta_1(:,2))*(sqrt(-1)*(-y(1,1).*conj(y(1,2)) +
y(1,2).*conj(y(1,1))))*xx_i(1,:)' +...
            exp(-xx_r(1,:)*Beta_1(:,2))*2*abs(y(1,1)).^2*(xx_i(1,:)*Beta_2(:,1))
*xx_i(1,:)');
      %%%%%%%%%%%%%%%%%%%%%%%%%%%%%%%%%%%%%%%%
      %Hessian
      %%%%%%%%%%%%%%%%%%%%%%%%%%%%%%%%%%%%%%%%
      bigmat_h11 = kron(bsxfun(@times, abs(y(2:end,1)).^2.*exp(-xx_r(2:end,:)
*Beta_1(:,1)), xx_r(2:end,:)),ones(nBeta,1)');
      coefmat_h11 = repmat(xx_r(2:end,:), 1,nBeta);
      h11 = h11 + reshape(sum(bsxfun(@times, bigmat_h11, coefmat_h11),1),nBeta,
nBeta) +...
           0.5*(abs(y(1,1)).^2.*exp(-xx_r(1,:)*Beta_1(:,1))*xx_r(1,:)'*xx_r(1,:));

      bigmat_h22 = kron(bsxfun(@times, abs(y(2:end,2)-theta(2:end).'.*y(2:end,1)).
^2.*exp(-xx_r(2:end,:)*Beta_1(:,2)),...
                                   xx_r(2:end,:)), ones(nBeta,1)');
      coefmat_h22 = repmat(xx_r(2:end,:), 1,nBeta);
      h22 = h22 + reshape(sum(bsxfun(@times, bigmat_h22, coefmat_h22),1),nBeta,
nBeta) +...
           0.5*(abs(y(1,2)-theta(1).'.*y(1,1)).^2.*exp(-xx_r(1,:)*Beta_1(:,2))
*xx_r(1,:)'*xx_r(1,:));

      bigmat_h33 = kron(bsxfun(@times, exp(-xx_r(2:end,:)*Beta_1(:,2)).*ck, xx_r
(2:end,:)),ones(nBeta,1)');
      coefmat_h33 = repmat(xx_r(2:end,:), 1,nBeta);
      h33 = h33 + reshape(sum(bsxfun(@times, bigmat_h33, coefmat_h33),1),nBeta,
nBeta) +...
           0.5*(exp(-xx_r(1,:)*Beta_1(:,2))*2*abs(y(1,1)).^2*xx_r(1,:)'*xx_r(1,:));

      bigmat_h44 = kron(bsxfun(@times, exp(-xx_r(2:end,:)*Beta_1(:,2)).*ck, xx_i
(2:end,:)),ones(nBeta,1)');
      coefmat_h44 = repmat(xx_i(2:end,:), 1,nBeta);
      h44 = h44 + reshape(sum(bsxfun(@times, bigmat_h44, coefmat_h44),1),nBeta,
nBeta) +...
           0.5*(exp(-xx_r(1,:)*Beta_1(:,2))*2*abs(y(1,1)).^2*xx_i(1,:)'*xx_i(1,:));

      bigmat_h42_1 = kron(bsxfun(@times, exp(-xx_r(2:end,:)*Beta_1(:,2)).*ck.*
(xx_i(2:end,:)*Beta_2(:,1)),...
           xx_i(2:end,:)), ones(nBeta,1)');
```

```
     coefmat_h42_1 = repmat(xx_r(2:end,:), 1,nBeta);
     bigmat_h42_2 = kron(bsxfun(@times,exp(-xx_r(2:end,:)*Beta_1(:,2)).*ik, xx_i
(2:end,:)), ones(nBeta,1)');
     coefmat_h42_2 = repmat(xx_r(2:end,:), 1,nBeta);
     h42 = h42 + reshape(sum(bsxfun(@times, bigmat_h42_1, coefmat_h42_1) +...
     bsxfun(@times, bigmat_h42_2, coefmat_h42_2),1),nBeta,nBeta)' +...
         0.5*(exp(-xx_r(1,:)*Beta_1(:,2))*(2*abs(y(1,1)).^2*(xx_i(1,:)*Beta_2
(:,1))*xx_i(1,:)+...
     sqrt(-1)*(-y(1,1).*conj(y(1,2)) + y(1,2).*conj(y(1,1)))*xx_i(1,:))'*xx_r
(1,:));
     bigmat_h32_1 = kron(bsxfun(@times, exp(-xx_r(2:end,:)*Beta_1(:,2)).*ck.*
(xx_r(2:end,:)*Beta_1(:,3)),...
         xx_r(2:end,:)), ones(nBeta,1)');
     coefmat_h32_1 = repmat(xx_r(2:end,:), 1,nBeta);
     bigmat_h32_2 = kron(bsxfun(@times,exp(-xx_r(2:end,:)*Beta_1(:,2)).*rk, xx_r
(2:end,:)), ones(nBeta,1)');
     coefmat_h32_2 = repmat(xx_r(2:end,:), 1,nBeta);
     h32 = h32 + reshape(sum(bsxfun(@times, bigmat_h32_1, coefmat_h32_1) +...
     bsxfun(@times, bigmat_h32_2, coefmat_h32_2),1),nBeta,nBeta)' +...
         0.5*(exp(-xx_r(1,:)*Beta_1(:,2))*(2*abs(y(1,1)).^2*(xx_r(1,:)*Beta_1
(:,3))*xx_r(1,:)+...
         (-y(1,1).*conj(y(1,2)) - y(1,2).*conj(y(1,1)))*xx_r(1,:))'*xx_r(1,:));
     h24=h42'; h23=h32';
  else
     %%%%%%%%%%%%%%%%%%%%%%%%%%%%%%
     %gradient
     %%%%%%%%%%%%%%%%%%%%%%%%%%%%%%
     rk = -y(2:nfreq,1).*conj(y(2:nfreq,2)) - y(2:nfreq,2).*conj(y(2:nfreq,1));
     ik = sqrt(-1)*(-y(2:nfreq,1).*conj(y(2:nfreq,2))+y(2:nfreq,2).*conj(y(2:
nfreq,1)));
     ck = 2*abs(y(2:nfreq,1)).^2;

     gr1=gr1+xx_r(2:nfreq,:)'*(1-abs(y(2:nfreq,1)).^2.*exp(-xx_r(2:nfreq,:)
*Beta_1(:,1))) + ...
         0.5*(xx_r(1,:)'*(1-abs(y(1,1)).^2.*exp(-xx_r(1,:)*Beta_1(:,1)))) +...
         0.5*(xx_r(end,:)'*(1-abs(y(end,1)).^2.*exp(-xx_r(end,:)*Beta_1(:,1))));
     gr2 = gr2 + xx_r(2:nfreq,:)'*(1 - abs(y(2:nfreq,2)-theta(2:nfreq).'.*y(2:
nfreq,1)).^2.*exp(-xx_r(2:nfreq,:)*Beta_1(:,2))) + ...
         0.5*(xx_r(1,:)'*(1 - abs(y(1,2)-theta(1).'.*y(1,1)).^2.*exp(-xx_r(1,:)
*Beta_1(:,2)))) + ...
         0.5*(xx_r(end,:)'*(1-abs(y(end,2)-theta(end).'.*y(end,1)).^2.*exp(-xx_r
(end,:)*Beta_1(:,2))));
     temp_mat_31 = bsxfun(@times,rk, xx_r(2:nfreq,:));
     temp_mat_32 = bsxfun(@times,ck,bsxfun(@times,
xx_r(2:nfreq,:),xx_r(2:nfreq,:)*Beta_1(:,3)));
     gr3 = gr3 + sum(bsxfun(@times, (temp_mat_31 + temp_mat_32), exp(-xx_r(2:
nfreq,:)*Beta_1(:,2)) ))' +...
         0.5*(exp(-xx_r(1,:)*Beta_1(:,2))*(-y(1,1).*conj(y(1,2)) - y(1,2).*conj
(y(1,1)))*xx_r(1,:)' +...
         exp(-xx_r(1,:)*Beta_1(:,2))*2*abs(y(1,1)).^2*(xx_r(1,:)*Beta_1(:,3))
*xx_r(1,:)') +...
         0.5*(exp(-xx_r(end,:)*Beta_1(:,2))*(-y(end,1).*conj(y(end,2)) - y
(end,2).*conj(y(end,1)))*xx_r(end,:)' +...
         exp(-xx_r(end,:)*Beta_1(:,2))*2*abs(y(end,1)).^2*(xx_r(end,:)*Beta_1
(:,3))*xx_r(end,:)');
     temp_mat_41 = bsxfun(@times, ik, xx_i(2:nfreq,:));
     temp_mat_42 = bsxfun(@times, ck,
bsxfun(@times,xx_i(2:nfreq,:),xx_i(2:nfreq,:)*Beta_2(:,1)));
```

```
    gr4 = gr4 + sum( bsxfun(@times, (temp_mat_41 + temp_mat_42), exp(-xx_r(2:
nfreq,:)*Beta_1(:,2)))) ' + ...
            0.5*(exp(-xx_r(1,:)*Beta_1(:,2))*(sqrt(-1)*(-y(1,1).*conj(y(1,2)) + y
(1,2).*conj(y(1,1))))*xx_i(1,:)' +...
            exp(-xx_r(1,:)*Beta_1(:,2))*2*abs(y(1,1)).^2*(xx_i(1,:)*Beta_2(:,1))
*xx_i(1,:)') +...
            0.5*(exp(-xx_r(end,:)*Beta_1(:,2))*(sqrt(-1)*(-y(end,1).*conj(y(end,2))
+ y(end,2).*conj(y(end,1))))*xx_i(end,:)' +...
            exp(-xx_r(end,:)*Beta_1(:,2))*2*abs(y(end,1)).^2*(xx_i(end,:)*Beta_2
(:,1))*xx_i(end,:)');
    %%%%%%%%%%%%%%%%%%%%%%%%%%%%%%%%%%
    %Hessian
    %%%%%%%%%%%%%%%%%%%%%%%%%%%%%%%%%%
    bigmat_h11 = kron(bsxfun(@times, abs(y(2:nfreq,1)).^2.*exp(-xx_r(2:nfreq,:)
*Beta_1(:,1)), xx_r(2:nfreq,:)),ones(nBeta,1)');
    coefmat_h11 = repmat(xx_r(2:nfreq,:), 1,nBeta);
    h11 = h11 + reshape(sum(bsxfun(@times, bigmat_h11, coefmat_h11),1),nBeta,
nBeta) +...
            0.5*(abs(y(1,1)).^2.*exp(-xx_r(1,:)*Beta_1(:,1))*xx_r(1,:)'
*xx_r(1,:))+...
            0.5*(abs(y(end,1)).^2.*exp(-xx_r(end,:)*Beta_1(:,1))*xx_r(end,:)'
*xx_r(end,:));

    bigmat_h22 = kron(bsxfun(@times,  abs(y(2:nfreq,2)-theta(2:nfreq).'.*y(2:
nfreq,1)).^2.*exp(-xx_r(2:nfreq,:)*Beta_1(:,2)),...
                    xx_r(2:nfreq,:)), ones(nBeta,1)');
    coefmat_h22 = repmat(xx_r(2:nfreq,:), 1,nBeta);
    h22 = h22 + reshape(sum(bsxfun(@times, bigmat_h22, coefmat_h22),1),nBeta,
nBeta) +...
            0.5*(abs(y(1,2)-theta(1).'.*y(1,1)).^2.*exp(-xx_r(1,:)*Beta_1(:,2))
*xx_r(1,:)'*xx_r(1,:))+...
            0.5*(abs(y(end,2)-theta(end).'.*y(end,1)).^2.*exp(-xx_r(end,:)
*Beta_1(:,2))*xx_r(end,:)'*xx_r(end,:));

    bigmat_h33 = kron(bsxfun(@times, exp(-xx_r(2:nfreq,:)*Beta_1(:,2)).*ck,
xx_r(2:nfreq,:)),ones(nBeta,1)');
    coefmat_h33 = repmat(xx_r(2:nfreq,:), 1,nBeta);
    h33 = h33 + reshape(sum(bsxfun(@times, bigmat_h33, coefmat_h33),1),nBeta,
nBeta) +...
            0.5*(exp(-xx_r(1,:)*Beta_1(:,2))*2*abs(y(1,1)).^2*xx_r(1,:)'*xx_r
(1,:))+...
            0.5*(exp(-xx_r(end,:)*Beta_1(:,2))*2*abs(y(end,1)).^2*xx_r(end,:)'*xx_r
(end,:));

    bigmat_h44 = kron(bsxfun(@times, exp(-xx_r(2:nfreq,:)*Beta_1(:,2)).*ck,
xx_i(2:nfreq,:)),ones(nBeta,1)');
    coefmat_h44 = repmat(xx_i(2:nfreq,:), 1,nBeta);
    h44 = h44 + reshape(sum(bsxfun(@times, bigmat_h44, coefmat_h44),1),nBeta,
nBeta) +...
            0.5*(exp(-xx_r(1,:)*Beta_1(:,2))*2*abs(y(1,1)).^2*xx_i(1,:)'*xx_i
(1,:))+...
            0.5*(exp(-xx_r(end,:)*Beta_1(:,2))*2*abs(y(end,1)).^2*xx_i
(end,:)'*xx_i(end,:));

    bigmat_h42_1 = kron(bsxfun(@times, exp(-xx_r(2:nfreq,:)*Beta_1(:,2)).*ck.*
(xx_i(2:nfreq,:)*Beta_2(:,1)),...
            xx_i(2:nfreq,:)), ones(nBeta,1)');
    coefmat_h42_1 = repmat(xx_r(2:nfreq,:), 1,nBeta);
    bigmat_h42_2 = kron(bsxfun(@times,exp(-xx_r(2:nfreq,:)*Beta_1(:,2)).*ik,
xx_i(2:nfreq,:)), ones(nBeta,1)');
```

```
        coefmat_h42_2 = repmat(xx_r(2:nfreq,:), 1,nBeta);
        h42 = h42 + reshape(sum(bsxfun(@times, bigmat_h42_1, coefmat_h42_1) +...
                bsxfun(@times, bigmat_h42_2, coefmat_h42_2),1),nBeta,nBeta)' +...
                0.5*(exp(-xx_r(1,:)*Beta_1(:,2))*(2*abs(y(1,1)).^2*(xx_i(1,:)*Beta_2
(:,1))*xx_i(1,:)+...
                sqrt(-1)*(-y(1,1).*conj(y(1,2))+y(1,2).*conj(y(1,1)))*xx_i(1,:))'*xx_r
(1,:))+...
                0.5*(exp(-xx_r(end,:)*Beta_1(:,2))*(2*abs(y(end,1)).^2*(xx_i(end,:)
*Beta_2(:,1))*xx_i(end,:)+...
                sqrt(-1)*(-y(end,1).*conj(y(end,2)) + y(end,2).*conj(y(end,1)))*xx_i
(end,:))'*xx_r(end,:));

        bigmat_h32_1 = kron(bsxfun(@times, exp(-xx_r(2:nfreq,:)*Beta_1(:,2)).*ck.*
(xx_r(2:nfreq,:)*Beta_1(:,3)),...
                xx_r(2:nfreq,:)), ones(nBeta,1)');
        coefmat_h32_1 = repmat(xx_r(2:nfreq,:), 1,nBeta);
        bigmat_h32_2 = kron(bsxfun(@times,exp(-xx_r(2:nfreq,:)*Beta_1(:,2)).*rk,
xx_r(2:nfreq,:)), ones(nBeta,1)');
        coefmat_h32_2 = repmat(xx_r(2:nfreq,:), 1,nBeta);
        h32 = h32 + reshape(sum(bsxfun(@times, bigmat_h32_1, coefmat_h32_1) +...
                bsxfun(@times, bigmat_h32_2, coefmat_h32_2),1),nBeta,nBeta)' +...
                0.5*(exp(-xx_r(1,:)*Beta_1(:,2))*(2*abs(y(1,1)).^2*(xx_r(1,:)*Beta_1
(:,3))*xx_r(1,:)+...
        (-y(1,1).*conj(y(1,2)) - y(1,2).*conj(y(1,1)))*xx_r(1,:))'*xx_r(1,:)) +...
                0.5*(exp(-xx_r(end,:)*Beta_1(:,2))*(2*abs(y(end,1)).^2*(xx_r(end,:)
*Beta_1(:,3))*xx_r(end,:)+...
        (-y(end,1).*conj(y(end,2)) - y(end,2).*conj(y(end,1)))*xx_r(end,:))'*xx_r
(end,:));
        h24=h42'; h23=h32';
    end
    h1 = [h11,zeros(nBeta,2*nBeta+nBeta)];
    h2 = [zeros(nBeta,nBeta),h22,-h23,-h24];
    h3 = [zeros(nBeta,nBeta),-h32,h33,zeros(nBeta,nBeta)];
    h4 = [zeros(nBeta,nBeta),-h42,zeros(nBeta,nBeta),h44];
    gr = [gr1;gr2;gr3;gr4]; h = [h1;h2;h3;h4]; f = -f;
    gr_index = (1:(4*nBeta)).*[kron(chol_index(Phi_temp,1:3),ones(nBeta,1)'),
kron(chol_index(Phi_temp,4),ones(nBeta,1)')];
    gr_index = gr_index(find(gr_index~=0));
    gr = gr(gr_index); h = h(gr_index,gr_index);

elseif dimen==3

    if (mod(n,2)==1) %odd n
        f = -sum(log(delta_sq(1,2:end))' + log(delta_sq(2,2:end))' + log(delta_sq
(3,2:end))' + ...
                conj(y(2:end,1)).*y(2:end,1).*exp(-xx_r(2:end,:)*Beta_1(:,1)) + ...
                conj(y(2:end,2) - theta(1,2:end).'.*y(2:end,1)).*(y(2:end,2) - theta
(1,2:end).'.*y(2:end,1)).*exp(-xx_r(2:end,:)*Beta_1(:,2))+...
                conj(y(2:end,3) -(theta(2,2:end).'.*y(2:end,1)+ theta(3,2:end).'.
*y(2:end,2))).*(y(2:end,3) -(theta(2,2:end).'.*y(2:end,1)+ theta(3,2:end).'.*y
(2:end,2))).*...
                exp(-xx_r(2:end,:)*Beta_1(:,3))) - ...
                0.5*(log(delta_sq(1,1))'+log(delta_sq(2,1))'+log(delta_sq(3,1))'+...
                conj(y(1,1)).*y(1,1).*exp(-xx_r(1,:)*Beta_1(:,1)) + ...
                conj(y(1,2) - theta(1,1).'.*y(1,1)).*(y(1,2) - theta(1,1).'.
*y(1,1)).*exp(-xx_r(1,:)*Beta_1(:,2))+...
                conj(y(1,3) -(theta(2,1).'.*y(1,1)+ theta(3,1).'.*y(1,2))).*(y(1,3)
-(theta(2,1).'.*y(1,1)+ theta(3,1).'.*y(1,2))).*...
                exp(-xx_r(1,:)*Beta_1(:,3)));
```

```
  else
    f=-sum(log(delta_sq(1,2:nfreq))'+log(delta_sq(2,2:nfreq))'+log(delta_sq(3,2:
nfreq))' + ...
        conj(y(2:nfreq,1)).*y(2:nfreq,1).*exp(-xx_r(2:nfreq,:)*Beta_1(:,1)) + ...
        conj(y(2:nfreq,2) - theta(1,2:nfreq).'.*y(2:nfreq,1)).*(y(2:nfreq,2) -
theta(1,2:nfreq).'.*y(2:nfreq,1)).*exp(-xx_r(2:nfreq,:)*Beta_1(:,2))+...
        conj(y(2:nfreq,3) -(theta(2,2:nfreq).'.*y(2:nfreq,1)+ theta(3,2:
nfreq).'.*y(2:nfreq,2))).*(y(2:nfreq,3) -(theta(2,2:nfreq).'.*y(2:nfreq,1)+
theta(3,2:nfreq).'.*y(2:nfreq,2))).*...
        exp(-xx_r(2:nfreq,:)*Beta_1(:,3))) - ...
        0.5*(log(delta_sq(1,1))' + log(delta_sq(2,1))' + log(delta_sq(3,1))' + ..
        conj(y(1,1)).*y(1,1).*exp(-xx_r(1,:)*Beta_1(:,1)) + ...
        conj(y(1,2) - theta(1,1).'.*y(1,1)).*(y(1,2) - theta(1,1).'.*y(1,1)).
*exp(-xx_r(1,:)*Beta_1(:,2))+...
        conj(y(1,3) -(theta(2,1).'.*y(1,1)+ theta(3,1).'.*y(1,2))).*(y(1,3)
-(theta(2,1).'.*y(1,1)+ theta(3,1).'.*y(1,2))).*...
        exp(-xx_r(1,:)*Beta_1(:,3)))-...
        0.5*(log(delta_sq(1,1))' + log(delta_sq(2,1))' + log(delta_sq(3,1))' + ...
        conj(y(end,1)).*y(end,1).*exp(-xx_r(end,:)*Beta_1(:,1)) + ...
        conj(y(end,2) - theta(1,end).'.*y(end,1)).*(y(end,2) - theta(1,end).
'.*y(end,1)).*exp(-xx_r(end,:)*Beta_1(:,2))+...
        conj(y(end,3) -(theta(2,end).'.*y(end,1)+ theta(3,end).'.*y(end,2))).*
(y(end,3) -(theta(2,end).'.*y(end,1)+ theta(3,end).'.*y(end,2))).*...
        exp(-xx_r(end,:)*Beta_1(:,3)));
  end
  f = f - (0.5.*(Beta_1(1,1)* Beta_1(1,1)')/sigmasqalpha + 0.5.*(Beta_1(2:
nBeta,1)'*Beta_1(2:nBeta,1))/tau_temp(1))*chol_index(Phi_temp,1) -...
        (0.5.*(Beta_1(1,2)* Beta_1(1,2)')/sigmasqalpha + 0.5.*(Beta_1(2:
nBeta,2)'*Beta_1(2:nBeta,2))/tau_temp(2))*chol_index(Phi_temp,2) -...
        (0.5.*(Beta_1(1,3)* Beta_1(1,3)')/sigmasqalpha + 0.5.*(Beta_1(2:
nBeta,3)'*Beta_1(2:nBeta,3))/tau_temp(3))*chol_index(Phi_temp,3) -...
        (0.5.*(Beta_1(1,4)* Beta_1(1,4)')/sigmasqalpha + 0.5.*(Beta_1(2:
nBeta,4)'*Beta_1(2:nBeta,4))/tau_temp(4))*chol_index(Phi_temp,4) -...
        (0.5.*(Beta_1(1,5)* Beta_1(1,5)')/sigmasqalpha + 0.5.*(Beta_1(2:
nBeta,5)'*Beta_1(2:nBeta,5))/tau_temp(5))*chol_index(Phi_temp,5) -...
        (0.5.*(Beta_1(1,6)* Beta_1(1,6)')/sigmasqalpha + 0.5.*(Beta_1(2:
nBeta,6)'*Beta_1(2:nBeta,6))/tau_temp(6))*chol_index(Phi_temp,6) -...
        (0.5.*(Beta_2(1:nBeta,1)'*Beta_2(1:nBeta,1))/tau_temp(7))*chol_index
(Phi_temp,7)-...
        (0.5.*(Beta_2(1:nBeta,2)'*Beta_2(1:nBeta,2))/tau_temp(8))*chol_index
(Phi_temp,8)-...
        (0.5.*(Beta_2(1:nBeta,3)'*Beta_2(1:nBeta,3))/tau_temp(9))*chol_index
(Phi_temp,9);

  gr1 = zeros(nBeta,1); gr2 = zeros(nBeta,1); gr3 = zeros(nBeta,1); gr4 = zeros
(nBeta,1);
  gr5 = zeros(nBeta,1); gr6 = zeros(nBeta,1); gr7 = zeros(nBeta,1); gr8 = zeros
(nBeta,1);
  gr9 = zeros(nBeta,1);

  gr1(1) = Beta_1(1,1)/sigmasqalpha; gr1(2:nBeta) = Beta_1(2:nBeta,1)/tau_temp(1);
  gr2(1) = Beta_1(1,2)/sigmasqalpha; gr2(2:nBeta) = Beta_1(2:nBeta,2)/tau_temp(2);
  gr3(1) = Beta_1(1,3)/sigmasqalpha; gr3(2:nBeta) = Beta_1(2:nBeta,3)/tau_temp(3);
  gr4(1) = Beta_1(1,4)/sigmasqalpha; gr4(2:nBeta) = Beta_1(2:nBeta,4)/tau_temp(4);
  gr5(1) = Beta_1(1,5)/sigmasqalpha; gr5(2:nBeta) = Beta_1(2:nBeta,5)/tau_temp(5);
  gr6(1) = Beta_1(1,6)/sigmasqalpha; gr6(2:nBeta) = Beta_1(2:nBeta,6)/tau_temp(6);
  gr7(1:nBeta) = Beta_2(1:nBeta,1)/tau_temp(7);
```

```
    gr8(1:nBeta) = Beta_2(1:nBeta,2)/tau_temp(8);
    gr9(1:nBeta) = Beta_2(1:nBeta,3)/tau_temp(9);

    h11(1,1)=1/sigmasqalpha; h11(2:nBeta,2:nBeta)=1/tau_temp(1)*eye(nbasis);
    h22(1,1)=1/sigmasqalpha; h22(2:nBeta,2:nBeta)=1/tau_temp(2)*eye(nbasis);
    h33(1,1)=1/sigmasqalpha; h33(2:nBeta,2:nBeta)=1/tau_temp(3)*eye(nbasis);
    h44(1,1)=1/sigmasqalpha; h44(2:nBeta,2:nBeta)=1/tau_temp(4)*eye(nbasis);
    h55(1,1)=1/sigmasqalpha; h55(2:nBeta,2:nBeta)=1/tau_temp(5)*eye(nbasis);
    h66(1,1)=1/sigmasqalpha; h66(2:nBeta,2:nBeta)=1/tau_temp(6)*eye(nbasis);
    h77(1:nBeta,1:nBeta)=1/tau_temp(7)*eye(nBeta);
    h88(1:nBeta,1:nBeta)=1/tau_temp(8)*eye(nBeta);
    h99(1:nBeta,1:nBeta)=1/tau_temp(9)*eye(nBeta);

    h42 = zeros(nBeta,nBeta);
    h53 = zeros(nBeta,nBeta);
    h56 = zeros(nBeta,nBeta);
    h59 = zeros(nBeta,nBeta);
    h63 = zeros(nBeta,nBeta);
    h68 = zeros(nBeta,nBeta);
    h72 = zeros(nBeta,nBeta);
    h83 = zeros(nBeta,nBeta);
    h93 = zeros(nBeta,nBeta);
    h98 = zeros(nBeta,nBeta);
    if (mod(n,2)==1)
        %%%%%%%%%%%%%%%%%%%%%%%%
        %gradient
        %%%%%%%%%%%%%%%%%%%%%%%%
        rk4 = -y(2:end,1).*conj(y(2:end,2)) - y(2:end,2).*conj(y(2:end,1));
        ck4 = 2*abs(y(2:end,1)).^2;
        rk5 = -y(2:end,1).*conj(y(2:end,3)) - y(2:end,3).*conj(y(2:end,1));
        ck5 = 2*abs(y(2:end,1)).^2;
        b = theta(3,:);
        dk5 =  y(2:end,2).*conj(y(2:end,1)).*(b(2:end).') + conj(y(2:end,2)).*y(2:
end,1).*conj(b(2:end).');
        rk6 = -y(2:end,2).*conj(y(2:end,3)) - y(2:end,3).*conj(y(2:end,2));
        ck6 = 2*abs(y(2:end,2)).^2;
        a = theta(2,:);
        dk6 =  y(2:end,1).*conj(y(2:end,2)).*(a(2:end).') + conj(y(2:end,1)).*y(2:
end,2).*conj(a(2:end).');
        ik7 = sqrt(-1)*(-y(2:end,1).*conj(y(2:end,2)) + y(2:end,2).*conj(y(2:
end,1)));
        ck7 = 2*abs(y(2:end,1)).^2;
        ik8 = sqrt(-1)*(-y(2:end,1).*conj(y(2:end,3)) + y(2:end,3).*conj(y(2:
end,1)));
        ck8 = 2*abs(y(2:end,1)).^2;
        dk8 = sqrt(-1)*(-y(2:end,2).*conj(y(2:end,1)).*(b(2:end).') + conj(y(2:
end,2).*conj(y(2:end,1)).*(b(2:end).'))) ;
        ik9 = sqrt(-1)*(-y(2:end,2).*conj(y(2:end,3)) + y(2:end,3).*conj(y(2:
end,2)));
        ck9 = 2*abs(y(2:end,2)).^2;
        dk9 = sqrt(-1)*(-y(2:end,1).*conj(y(2:end,2)).*(a(2:end).') + conj(y(2:
end,1).*conj(y(2:end,2)).*(a(2:end).'))) ;

        gr1 = gr1 + xx_r(2:end,:)'*(1-abs(y(2:end,1)).^2.*exp(-xx_r(2:end,:)*Beta_1
(:,1))) + ...
              0.5*(xx_r(1,:)'*(1-abs(y(1,1)).^2.*exp(-xx_r(1,:)*Beta_1(:,1))));
        gr2 = gr2 + xx_r(2:end,:)'*(1 - abs(y(2:end,2)-theta(1,2:end).'.*y(2:
end,1)).^2.*exp(-xx_r(2:end,:)*Beta_1(:,2))) + ...
```

```
            0.5*(xx_r(1,:)'*(1 - abs(y(1,2)-theta(1,1).*y(1,1)).^2.*exp(-xx_r(1,:)
*Beta_1(:,2))));
      gr3 = gr3 + xx_r(2:end,:)'*(1 - abs(y(2:end,3) - theta(2,2:end).'.*y(2:
end,1) - theta(3,2:end).'.*y(2:end,2)).^2.*exp(-xx_r(2:end,:)*Beta_1(:,3)))+...
            0.5*(xx_r(1,:)'*(1 - abs(y(1,3) - theta(2,1).*y(1,1) - theta(3,1).
*y(1,2))^2.*exp(-xx_r(1,:)*Beta_1(:,3))));
      temp_mat_41 = bsxfun(@times, xx_r(2:end,:),rk4);
      temp_mat_42 = bsxfun(@times, ck4, bsxfun(@times,
xx_r(2:end,:), xx_r(2:end,:)*Beta_1(:,4)));
      gr4 = gr4 + sum( bsxfun(@times, (temp_mat_41+temp_mat_42), exp(-xx_r(2:
end,:)*Beta_1(:,2))))' +...
            0.5*(exp(-xx_r(1,:)*Beta_1(:,2))*(-y(1,1).*conj(y(1,2)) - y(1,2).*conj
(y(1,1)))*xx_r(1,:)' +...
            exp(-xx_r(1,:)*Beta_1(:,2))*2*abs(y(1,1)).^2*(xx_r(1,:)*Beta_1(:,4))
*xx_r(1,:)');
      temp_mat_51 = bsxfun(@times, xx_r(2:end,:),rk5);
      temp_mat_52 = bsxfun(@times, ck5, bsxfun(@times,
xx_r(2:end,:), xx_r(2:end,:)*Beta_1(:,5)));
      temp_mat_53 = bsxfun(@times, xx_r(2:end,:),dk5);
      gr5 = gr5 + sum( bsxfun(@times, (temp_mat_51 + temp_mat_52 + temp_mat_53),
exp(-xx_r(2:end,:)*Beta_1(:,3)) ))'+...
            0.5* (exp(-xx_r(1,:)*Beta_1(:,3))*(-y(1,1).*conj(y(1,3)) - y(1,3).
*conj(y(1,1)))*xx_r(1,:)' +...
            exp(-xx_r(1,:)*Beta_1(:,3))*2*abs(y(1,1)).^2*(xx_r(1,:)*Beta_1(:,5))
*xx_r(1,:)'+...
            exp(-xx_r(1,:)*Beta_1(:,3))*(y(1,2).*conj(y(1,1)).*(b(1).') + conj(y
(1,2).*conj(y(1,1)).*(b(1).')))*xx_r(1,:)');
      temp_mat_61 = bsxfun(@times, xx_r(2:end,:),rk6);
      temp_mat_62 = bsxfun(@times, ck6, bsxfun(@times,
xx_r(2:end,:), xx_r(2:end,:)*Beta_1(:,6)));
      temp_mat_63 = bsxfun(@times, xx_r(2:end,:),dk6);
      gr6 = gr6 + sum(bsxfun(@times, (temp_mat_61 + temp_mat_62 + temp_mat_63), exp
(-xx_r(2:end,:)*Beta_1(:,3))))'+...
            0.5* (exp(-xx_r(1,:)*Beta_1(:,3))*(-y(1,2).*conj(y(1,3)) - y(1,3).
*conj(y(1,2)))*xx_r(1,:)' +...
            exp(-xx_r(1,:)*Beta_1(:,3))*2*abs(y(1,2)).^2*(xx_r(1,:)*Beta_1(:,6))
*xx_r(1,:)'+...
            exp(-xx_r(1,:)*Beta_1(:,3))*(y(1,1).*conj(y(1,2)).*(a(1).') + conj(y
(1,1).*conj(y(1,2)).*(a(1).')))*xx_r(1,:)');
      temp_mat_71 = bsxfun(@times, ik7, xx_i(2:end,:));
      temp_mat_72 = bsxfun(@times, ck7,
bsxfun(@times,xx_i(2:end,:),xx_i(2:end,:)*Beta_2(:,1)));
      gr7 = gr7 + sum( bsxfun(@times, (temp_mat_71 + temp_mat_72), exp(-xx_r(2:
end,:)*Beta_1(:,2))))' + ...
            0.5*(exp(-xx_r(1,:)*Beta_1(:,2))*(imag(-y(1,1).*conj(y(1,2))
+ y(1,2).*conj(y(1,1))))*xx_i(1,:)' +...
            exp(-xx_r(1,:)*Beta_1(:,2))*2*abs(y(1,1)).^2*(xx_i(1,:)*Beta_2(:,1))
*xx_i(1,:)');
      temp_mat_81 = bsxfun(@times, ik8, xx_i(2:end,:));
      temp_mat_82 = bsxfun(@times, ck8, bsxfun(@times,xx_i(2:end,:),xx_i(2:end,:)
*Beta_2(:,2)));
      temp_mat_83 = bsxfun(@times, xx_i(2:end,:),dk8);
      gr8 = gr8 + sum( bsxfun(@times, (temp_mat_81 + temp_mat_82 + temp_mat_83),
exp(-xx_r(2:end,:)*Beta_1(:,3))))' + ...
            0.5*(exp(-xx_r(1,:)*Beta_1(:,3))*(imag(-y(1,1).*conj(y(1,3))
+ y(1,3).*conj(y(1,1))))*xx_i(1,:)' +...
```

```
    exp(-xx_r(1,:)*Beta_1(:,3))*2*abs(y(1,1)).^2*(xx_i(1,:)*Beta_2(:,2))*xx_i
(1,:)'+...
    sqrt(-1)*exp(-xx_r(1,:)*Beta_1(:,3))*(-y(1,1).*conj(y(1,2)).*(b(1).') +
conj(y(1,1).*conj(y(1,2)).*(b(1).')))*xx_i(1,:)');
    temp_mat_91 = bsxfun(@times, ik9, xx_i(2:end,:));
    temp_mat_92 = bsxfun(@times, ck9, bsxfun(@times,xx_i(2:end,:),xx_i(2:end,:)
*Beta_2(:,3)));
    temp_mat_93 = bsxfun(@times, xx_i(2:end,:),dk9);
    gr9 = gr9 + sum( bsxfun(@times, (temp_mat_91 + temp_mat_92 + temp_mat_93),
exp(-xx_r(2:end,:)*Beta_1(:,3))))' + ...
            0.5 * (exp(-xx_r(1,:)*Beta_1(:,3))*(imag(-y(1,2).*conj(y(1,3)) + y
(1,3).*conj(y(1,2))))*xx_i(1,:)' +...
            exp(-xx_r(1,:)*Beta_1(:,3))*2*abs(y(1,2)).^2*(xx_i(1,:)*Beta_2(:,3))
*xx_i(1,:)'+...
            sqrt(-1)*exp(-xx_r(1,:)*Beta_1(:,3))*(-y(1,1).*conj(y(1,2)).*(a(1).')
+ conj(y(1,1).*conj(y(1,2)).*(a(1).')))*xx_i(1,:)');
    %%%%%%%%%%%%%%%%%
    %Hessian
    %%%%%%%%%%%%%%%%%
    bigmat_h11 = kron(bsxfun(@times, abs(y(2:end,1)).^2.*exp(-xx_r(2:end,:)
*Beta_1(:,1)), xx_r(2:end,:)),ones(nBeta,1)');
    coefmat_h11 = repmat(xx_r(2:end,:), 1,nBeta);
    h11 = h11 + reshape(sum(bsxfun(@times, bigmat_h11, coefmat_h11),1),nBeta,
nBeta) +...
            0.5*(abs(y(1,1)).^2.*exp(-xx_r(1,:)*Beta_1(:,1))*xx_r(1,:)'
*xx_r(1,:));

    bigmat_h22 = kron(bsxfun(@times,  abs(y(2:end,2)-theta(1,2:end).'.*y(2:
end,1)).^2.*exp(-xx_r(2:end,:)*Beta_1(:,2)),xx_r(2:end,:)), ones(nBeta,1)');
    coefmat_h22 = repmat(xx_r(2:end,:), 1,nBeta);
    h22 = h22 + reshape(sum(bsxfun(@times, bigmat_h22, coefmat_h22),1),nBeta,
nBeta) +...
            0.5*(abs(y(1,2)-theta(1,1).*y(1,1)).^2.*exp(-xx_r(1,:)*Beta_1(:,2))
*xx_r(1,:)'*xx_r(1,:));

    bigmat_h33 = kron(bsxfun(@times,  abs(y(2:end,3) - theta(2,2:end).'.*y(2:
end,1) - theta(3,2:end).'.*y(2:end,2)).^2.*...
    exp(-xx_r(2:end,:)*Beta_1(:,3)),xx_r(2:end,:)), ones(nBeta,1)');
    coefmat_h33 = repmat(xx_r(2:end,:), 1,nBeta);
    h33 = h33 + reshape(sum(bsxfun(@times, bigmat_h33, coefmat_h33),1),nBeta,
nBeta) +...
            0.5*(abs(y(1,3) - theta(2,1).'.*y(1,1) - theta(3,1).'.*y(1,2)).^2.*...
            exp(-xx_r(1,:)*Beta_1(:,3))*xx_r(1,:)'*xx_r(1,:));

    bigmat_h44 = kron(bsxfun(@times, exp(-xx_r(2:end,:)*Beta_1(:,2)).*ck4, xx_r
(2:end,:)),ones(nBeta,1)');
    coefmat_h44 = repmat(xx_r(2:end,:), 1,nBeta);
    h44 = h44 + reshape(sum(bsxfun(@times, bigmat_h44, coefmat_h44),1),nBeta,
nBeta) +...
            0.5*(exp(-xx_r(1,:)*Beta_1(:,2))*2*abs(y(1,1)).^2*xx_r(1,:)'
*xx_r(1,:));

    bigmat_h55 = kron(bsxfun(@times, exp(-xx_r(2:end,:)*Beta_1(:,3)).*ck5, xx_r
(2:end,:)),ones(nBeta,1)');
    coefmat_h55 = repmat(xx_r(2:end,:), 1,nBeta);
    h55 = h55 + reshape(sum(bsxfun(@times, bigmat_h55, coefmat_h55),1),nBeta,
nBeta) +...
```

```
        0.5*(exp(-xx_r(1,:)*Beta_1(:,3))*2*abs(y(1,1)).^2*xx_r(1,:)'*xx_r
(1,:));

    bigmat_h66 = kron(bsxfun(@times, exp(-xx_r(2:end,:)*Beta_1(:,3)).*ck6, xx_r
(2:end,:)),ones(nBeta,1)');
    coefmat_h66 = repmat(xx_r(2:end,:), 1,nBeta);
    h66 = h66 + reshape(sum(bsxfun(@times, bigmat_h66, coefmat_h66),1),nBeta,
nBeta) +...
        0.5*(exp(-xx_r(1,:)*Beta_1(:,3))*2*abs(y(1,2)).^2*xx_r(1,:)'*xx_r
(1,:));

    bigmat_h77 = kron(bsxfun(@times, exp(-xx_r(2:end,:)*Beta_1(:,2)).*ck7, xx_i
(2:end,:)),ones(nBeta,1)');
    coefmat_h77 = repmat(xx_i(2:end,:), 1,nBeta);
    h77 = h77 + reshape(sum(bsxfun(@times, bigmat_h77, coefmat_h77),1),nBeta,
nBeta) +...
        0.5*(exp(-xx_r(1,:)*Beta_1(:,2))*2*abs(y(1,1)).^2*xx_i(1,:)'*xx_i
(1,:));

    bigmat_h88 = kron(bsxfun(@times, exp(-xx_r(2:end,:)*Beta_1(:,3)).*ck8, xx_i
(2:end,:)),ones(nBeta,1)');
    coefmat_h88 = repmat(xx_i(2:end,:), 1,nBeta);
    h88 = h88 + reshape(sum(bsxfun(@times, bigmat_h88, coefmat_h88),1),nBeta,
nBeta) +...
        0.5*(exp(-xx_r(1,:)*Beta_1(:,3))*2*abs(y(1,1)).^2*xx_i(1,:)'*xx_i
(1,:));

    bigmat_h99 = kron(bsxfun(@times, exp(-xx_r(2:end,:)*Beta_1(:,3)).*ck9, xx_i
(2:end,:)),ones(nBeta,1)');
    coefmat_h99 = repmat(xx_i(2:end,:), 1,nBeta);
    h99 = h99 + reshape(sum(bsxfun(@times, bigmat_h99, coefmat_h99),1),nBeta,
nBeta) +...
        0.5*(exp(-xx_r(1,:)*Beta_1(:,3))*2*abs(y(1,2)).^2*xx_i(1,:)'*xx_i
(1,:));

    bigmat_h42_1 = kron(bsxfun(@times, exp(-xx_r(2:end,:)*Beta_1(:,2)).*ck4.*
(xx_r(2:end,:)*Beta_1(:,4)),...
            xx_r(2:end,:)), ones(nBeta,1)');
    coefmat_h42_1 = repmat(xx_r(2:end,:), 1,nBeta);
    bigmat_h42_2 = kron(bsxfun(@times,exp(-xx_r(2:end,:)*Beta_1(:,2)).*rk4,
xx_r(2:end,:)), ones(nBeta,1)');
    coefmat_h42_2 = repmat(xx_r(2:end,:), 1,nBeta);
    h42 = h42 + reshape(sum(bsxfun(@times, bigmat_h42_1, coefmat_h42_1) +...
    bsxfun(@times, bigmat_h42_2, coefmat_h42_2),1),nBeta,nBeta)' +...
        0.5*(exp(-xx_r(1,:)*Beta_1(:,2))*(2*abs(y(1,1)).^2*(xx_r(1,:)*Beta_1
(:,4))*xx_r(1,:)+...
    (-y(1,1).*conj(y(1,2)) - y(1,2).*conj(y(1,1)))*xx_r(1,:))'*xx_r(1,:));

    bigmat_h53_1 = kron(bsxfun(@times, exp(-xx_r(2:end,:)*Beta_1(:,3)).*ck5.*
(xx_r(2:end,:)*Beta_1(:,5)),...
            xx_r(2:end,:)), ones(nBeta,1)');
    coefmat_h53_1 = repmat(xx_r(2:end,:), 1,nBeta);
    bigmat_h53_2 = kron(bsxfun(@times,exp(-xx_r(2:end,:)*Beta_1(:,3)).*rk5,
xx_r(2:end,:)), ones(nBeta,1)');
    coefmat_h53_2 = repmat(xx_r(2:end,:), 1,nBeta);
    bigmat_h53_3 = kron(bsxfun(@times,exp(-xx_r(2:end,:)*Beta_1(:,3)).*dk5,
xx_r(2:end,:)), ones(nBeta,1)');
    coefmat_h53_3 = repmat(xx_r(2:end,:), 1,nBeta);
    h53 = h53 + reshape(sum(bsxfun(@times, bigmat_h53_1, coefmat_h53_1) +...
```

```
        bsxfun(@times, bigmat_h53_2, coefmat_h53_2)+...
        bsxfun(@times, bigmat_h53_3, coefmat_h53_3),1),nBeta,nBeta)' +...
        0.5*(exp(-xx_r(1,:)*Beta_1(:,3))*(2*abs(y(1,1)).^2*(xx_r(1,:)*Beta_1
(:,5))*xx_r(1,:)+...
       (-y(1,1).*conj(y(1,3)) - y(1,3).*conj(y(1,1)))*xx_r(1,:)+...
       (y(1,2).*conj(y(1,1)).*(b(1).') + conj(y(1,2).*conj(y(1,1)).
*(b(1).')))*xx_r(1,:))'*xx_r(1,:));

    bigmat_h56_1 = kron(bsxfun(@times, exp(-xx_r(2:end,:)*Beta_1(:,3)).*conj(y
(2:end,1)).*y(2:end,2),...
        xx_r(2:end,:)), ones(nBeta,1)');
    coefmat_h56_1 = repmat(xx_r(2:end,:), 1,nBeta);
    bigmat_h56_2 = kron(bsxfun(@times, exp(-xx_r(2:end,:)*Beta_1(:,3)).*conj(y
(2:end,2)).*y(2:end,1),...
        xx_r(2:end,:)), ones(nBeta,1)');
    coefmat_h56_2 = repmat(xx_r(2:end,:), 1,nBeta);
    h56 = real(h56 + reshape(sum(bsxfun(@times, bigmat_h56_1, coefmat_h56_1) +...
        bsxfun(@times, bigmat_h56_2, coefmat_h56_2),1),nBeta,nBeta)' +...
        0.5*(exp(-xx_r(1,:)*Beta_1(:,3))*((conj(y(1,1)).*y(1,2))*xx_r
(1,:)'*xx_r(1,:)+...
        (conj(y(1,2)).*y(1,1))*xx_r(1,:)'*xx_r(1,:))));

    bigmat_h59_1 = kron(bsxfun(@times, exp(-xx_r(2:end,:)*Beta_1(:,3)).*conj(y
(2:end,1)).*y(2:end,2),...
        xx_r(2:end,:)), ones(nBeta,1)');
    coefmat_h59_1 = repmat(xx_i(2:end,:), 1,nBeta);
    bigmat_h59_2 = kron(bsxfun(@times, exp(-xx_r(2:end,:)*Beta_1(:,3)).*conj(y
(2:end,2)).*y(2:end,1),...
        xx_r(2:end,:)), ones(nBeta,1)');
    coefmat_h59_2 = repmat(xx_i(2:end,:), 1,nBeta);
    h59 = imag(h59 + reshape(sum(bsxfun(@times, bigmat_h59_1, coefmat_h59_1) -...
        bsxfun(@times, bigmat_h59_2, coefmat_h59_2),1),nBeta,nBeta)' +...
        0.5*(exp(-xx_r(1,:)*Beta_1(:,3))*((conj(y(1,1)).*y(1,2))*xx_r
(1,:)'*xx_i(1,:) -...
        (conj(y(1,2)).*y(1,1))*xx_r(1,:)'*xx_i(1,:))));

    bigmat_h63_1 = kron(bsxfun(@times, exp(-xx_r(2:end,:)*Beta_1(:,3)).*ck6.*
(xx_r(2:end,:)*Beta_1(:,6)),...
        xx_r(2:end,:)), ones(nBeta,1)');
    coefmat_h63_1 = repmat(xx_r(2:end,:), 1,nBeta);
    bigmat_h63_2 = kron(bsxfun(@times,exp(-xx_r(2:end,:)*Beta_1(:,3)).*rk6,
xx_r(2:end,:)), ones(nBeta,1)');
    coefmat_h63_2 = repmat(xx_r(2:end,:), 1,nBeta);
    bigmat_h63_3 = kron(bsxfun(@times,exp(-xx_r(2:end,:)*Beta_1(:,3)).*dk6 ,
xx_r(2:end,:)), ones(nBeta,1)');
    coefmat_h63_3 = repmat(xx_r(2:end,:), 1,nBeta);
    h63 = h63 + reshape(sum(bsxfun(@times, bigmat_h63_1, coefmat_h63_1) +...
        bsxfun(@times, bigmat_h63_2, coefmat_h63_2)+...
        bsxfun(@times, bigmat_h63_3, coefmat_h63_3),1),nBeta,nBeta)' +...
        0.5*(exp(-xx_r(1,:)*Beta_1(:,3))*(2*abs(y(1,2)).^2*(xx_r(1,:)*Beta_1
(:,6))*xx_r(1,:)+...
       (-y(1,2).*conj(y(1,3)) - y(1,3).*conj(y(1,2)))*xx_r(1,:)+...
       (y(1,2).*conj(y(1,1)).*(a(1).') + conj(y(1,2).*conj(y(1,1)).*(a
(1).')))*xx_r(1,:))'*xx_r(1,:));

    bigmat_h68_1 = kron(bsxfun(@times, exp(-xx_r(2:end,:)*Beta_1(:,3)).*conj(y
(2:end,2)).*y(2:end,1),...
```

```
            xx_r(2:end,:)), ones(nBeta,1)');
      coefmat_h68_1 = repmat(xx_i(2:end,:), 1,nBeta);
      bigmat_h68_2 = kron(bsxfun(@times, exp(-xx_r(2:end,:)*Beta_1(:,3)).*conj(y
(2:end,1)).*y(2:end,2),...
            xx_r(2:end,:)), ones(nBeta,1)');
      coefmat_h68_2 = repmat(xx_i(2:end,:), 1,nBeta);
      h68 = imag(h68 + reshape(sum(bsxfun(@times, bigmat_h68_1, coefmat_h68_1) -...
      bsxfun(@times, bigmat_h68_2, coefmat_h68_2),1),nBeta,nBeta)' +...
            0.5*(exp(-xx_r(1,:)*Beta_1(:,3))*((conj(y(1,2)).*y(1,1))*xx_r
(1,:)'*xx_i(1,:)-...
            (conj(y(1,1)).*y(1,2))*xx_r(1,:)'*xx_i(1,:))));

      bigmat_h72_1 = kron(bsxfun(@times, exp(-xx_r(2:end,:)*Beta_1(:,2)).*ck7.*
(xx_i(2:end,:)*Beta_2(:,1)),...
            xx_i(2:end,:)), ones(nBeta,1)');
      coefmat_h72_1 = repmat(xx_r(2:end,:), 1,nBeta);
      bigmat_h72_2 = kron(bsxfun(@times,exp(-xx_r(2:end,:)*Beta_1(:,2)).*ik7,
xx_i(2:end,:)), ones(nBeta,1)');
      coefmat_h72_2 = repmat(xx_r(2:end,:), 1,nBeta);
      h72 = h72 + reshape(sum(bsxfun(@times, bigmat_h72_1, coefmat_h72_1) +...
            bsxfun(@times, bigmat_h72_2, coefmat_h72_2),1),nBeta,nBeta)' +...
            0.5*(exp(-xx_r(1,:)*Beta_1(:,2))*(2*abs(y(1,1)).^2*(xx_i(1,:)*Beta_2
(:,1))*xx_i(1,:)+...
            imag(-y(1,1).*conj(y(1,2)) + y(1,2).*conj(y(1,1))))*xx_i(1,:))'*xx_r
(1,:));

      bigmat_h83_1 = kron(bsxfun(@times, exp(-xx_r(2:end,:)*Beta_1(:,3)).*ck8.
*(xx_i(2:end,:)*Beta_2(:,2)),...
            xx_i(2:end,:)), ones(nBeta,1)');
      coefmat_h83_1 = repmat(xx_r(2:end,:), 1,nBeta);
      bigmat_h83_2 = kron(bsxfun(@times,exp(-xx_r(2:end,:)*Beta_1(:,3)).*ik8,
xx_i(2:end,:)), ones(nBeta,1)');
      coefmat_h83_2 = repmat(xx_r(2:end,:), 1,nBeta);
      bigmat_h83_3 = kron(bsxfun(@times,exp(-xx_r(2:end,:)*Beta_1(:,3)).*(dk8),
xx_i(2:end,:)), ones(nBeta,1)');
      coefmat_h83_3 = repmat(xx_r(2:end,:), 1,nBeta);
      h83 = h83 + reshape(sum(bsxfun(@times, bigmat_h83_1, coefmat_h83_1) +...
            bsxfun(@times, bigmat_h83_2, coefmat_h83_2)+...
            bsxfun(@times, bigmat_h83_3, coefmat_h83_3),1),nBeta,nBeta)' +...
            0.5*(exp(-xx_r(1,:)*Beta_1(:,3))*(2*abs(y(1,1)).^2*(xx_i(1,:)
*Beta_2(:,2))*xx_i(1,:)+...
            (-y(1,1).*conj(y(1,3)) + y(1,3).*conj(y(1,1))))*xx_i(1,:)+...
            imag(-y(1,2).*conj(y(1,1)).*(b(1).') + conj(y(1,2).*conj(y(1,1)).
*(b(1).')))*xx_i(1,:))'*xx_r(1,:));

      bigmat_h93_1 = kron(bsxfun(@times, exp(-xx_r(2:end,:)*Beta_1(:,3)).*ck9.
*(xx_i(2:end,:)*Beta_2(:,3)),...
            xx_i(2:end,:)), ones(nBeta,1)');
      coefmat_h93_1 = repmat(xx_r(2:end,:), 1,nBeta);
      bigmat_h93_2 = kron(bsxfun(@times,exp(-xx_r(2:end,:)*Beta_1(:,3)).*ik9,
xx_i(2:end,:)), ones(nBeta,1)');
      coefmat_h93_2 = repmat(xx_r(2:end,:), 1,nBeta);
      bigmat_h93_3 = kron(bsxfun(@times,exp(-xx_r(2:end,:)*Beta_1(:,3)).*(dk9),
xx_i(2:end,:)), ones(nBeta,1)');
      coefmat_h93_3 = repmat(xx_r(2:end,:), 1,nBeta);
      h93 = h93 + reshape(sum(bsxfun(@times, bigmat_h93_1, coefmat_h93_1) +...
            bsxfun(@times, bigmat_h93_2, coefmat_h93_2)+...
            bsxfun(@times, bigmat_h93_3, coefmat_h93_3),1),nBeta,nBeta)' +...
            0.5*(exp(-xx_r(1,:)*Beta_1(:,3))*(2*abs(y(1,2)).^2*(xx_i(1,:)*Beta_2
(:,3))*xx_i(1,:)+...
```

```
           (-y(1,2).*conj(y(1,3)) + y(1,3).*conj(y(1,2)))*xx_i(1,:)+...
           imag(-y(1,1).*conj(y(1,2)).*(a(1).') + conj(y(1,1).*conj(y(1,2)).*(a
(1).')))*xx_i(1,:))'*xx_r(1,:));

      bigmat_h98_1 = kron(bsxfun(@times, exp(-xx_r(2:end,:)*Beta_1(:,3)).*conj(y
(2:end,1)).*y(2:end,2),...
           xx_i(2:end,:)), ones(nBeta,1)');
      coefmat_h98_1 = repmat(xx_i(2:end,:), 1,nBeta);
      bigmat_h98_2 = kron(bsxfun(@times, exp(-xx_r(2:end,:)*Beta_1(:,3)).*conj(y
(2:end,2)).*y(2:end,1),...
           xx_i(2:end,:)), ones(nBeta,1)');
      coefmat_h98_2 = repmat(xx_i(2:end,:), 1,nBeta);
      h98 = real(h98 + reshape(sum(bsxfun(@times, bigmat_h98_1, coefmat_h98_1) +...
           bsxfun(@times, bigmat_h98_2, coefmat_h98_2),1),nBeta,nBeta)' +...
           0.5*(exp(-xx_r(1,:)*Beta_1(:,3))*((conj(y(1,1)).*y(1,2))*xx_i
(1,:)'*xx_i(1,:)+...
           (conj(y(1,2)).*y(1,1))*xx_i(1,:)'*xx_i(1,:))));

      h24=h42'; h35=h53'; h65=h56'; h95=h59'; h36=h63';
      h86=h68'; h27=h72'; h38=h83'; h39=h93'; h89=h98';
   else
      %%%%%%%%%%%%%%%%%%%%%%%%%%%%%%%%%%%%%%%%%%%%%%%%%%%%%%%%%%%%%%%%%%%%%%
      %gradient
      %%%%%%%%%%%%%%%%%%%%%%%%%%%%%%%%%%%%%%%%%%%%%%%%%%%%%%%%%%%%%%%%%%%%%%
      rk4 = -y(2:nfreq,1).*conj(y(2:nfreq,2)) - y(2:nfreq,2).*conj(y(2:nfreq,1));
      ck4 = 2*abs(y(2:nfreq,1)).^2;
      rk5 = -y(2:nfreq,1).*conj(y(2:nfreq,3)) - y(2:nfreq,3).*conj(y(2:nfreq,1));
      ck5 = 2*abs(y(2:nfreq,1)).^2;
      b = theta(3,:);
      dk5 =  y(2:nfreq,2).*conj(y(2:nfreq,1)).*(b(2:nfreq).') + conj(y(2:
nfreq,2)).*y(2:nfreq,1).*conj(b(2:nfreq).');
      rk6 = -y(2:nfreq,2).*conj(y(2:nfreq,3)) - y(2:nfreq,3).*conj(y(2:nfreq,2));
      ck6 = 2*abs(y(2:nfreq,2)).^2;
      a = theta(2,:);
      dk6 =  y(2:nfreq,1).*conj(y(2:nfreq,2)).*(a(2:nfreq).') + conj(y(2:
nfreq,1).*conj(y(2:nfreq,2)).*(a(2:nfreq).'));
      ik7 = sqrt(-1)*(-y(2:nfreq,1).*conj(y(2:nfreq,2)) + y(2:nfreq,2).*conj(y(2:
nfreq,1)));
      ck7 = 2*abs(y(2:nfreq,1)).^2;
      ik8 = sqrt(-1)*(-y(2:nfreq,1).*conj(y(2:nfreq,3)) + y(2:nfreq,3).*conj(y(2:
nfreq,1)));
      ck8 = 2*abs(y(2:nfreq,1)).^2;
      dk8 = sqrt(-1)*( -y(2:nfreq,2).*conj(y(2:nfreq,1)).*(b(2:nfreq).') + conj(y
(2:nfreq,2).*conj(y(2:nfreq,1)).*(b(2:nfreq).')));
      ik9 = sqrt(-1)*(-y(2:nfreq,2).*conj(y(2:nfreq,3)) + y(2:nfreq,3).*conj(y(2:
nfreq,2)));
      ck9 = 2*abs(y(2:nfreq,2)).^2;
      dk9 = sqrt(-1)*( -y(2:nfreq,1).*conj(y(2:nfreq,2)).*(a(2:nfreq).') + conj(y
(2:nfreq,1).*conj(y(2:nfreq,2)).*(a(2:nfreq).')));

     gr1 = gr1 + xx_r(2:nfreq,:)'*(1-abs(y(2:nfreq,1)).^2.*exp(-xx_r(2:nfreq,:)
*Beta_1(:,1))) + ...
           0.5*(xx_r(1,:)'*(1-abs(y(1,1)).^2.*exp(-xx_r(1,:)*Beta_1(:,1))))+...
           0.5*(xx_r(end,:)'*(1-abs(y(end,1)).^2.*exp(-xx_r(end,:)*Beta_1
(:,1))));
     gr2 = gr2 + xx_r(2:nfreq,:)'*(1 - abs(y(2:nfreq,2)-theta(1,2:nfreq).'.*y(2:
nfreq,1)).^2.*exp(-xx_r(2:nfreq,:)*Beta_1(:,2))) + ...
           0.5*(xx_r(1,:)'*(1 - abs(y(1,2)-theta(1,1).*y(1,1)).^2.*exp(-xx_r(1,:)
*Beta_1(:,2)))) +...
```

```
            0.5*(xx_r(end,:)'*(1 - abs(y(end,2)-theta(1,end).*y(end,1)).^2.*exp
(-xx_r(end,:)*Beta_1(:,2))));
        gr3 = gr3 + xx_r(2:nfreq,:)'*(1 - abs(y(2:nfreq,3) - theta(2,2:nfreq).'.*y
(2:nfreq,1) - theta(3,2:nfreq).'.*y(2:nfreq,2)).^2.*exp(-xx_r(2:nfreq,:)*Beta_1
(:,3)))+...
            0.5*(xx_r(1,:)'*(1 - abs(y(1,3) - theta(2,1).*y(1,1) - theta(3,1).*y
(1,2))^2.*exp(-xx_r(1,:)*Beta_1(:,3))))+...
            0.5*(xx_r(end,:)'*(1 - abs(y(end,3) - theta(2,end).*y(end,1) - theta
(3,end).*y(end,2))^2.*exp(-xx_r(end,:)*Beta_1(:,3))));
        temp_mat_41 = bsxfun(@times, xx_r(2:nfreq,:),rk4);
        temp_mat_42 = bsxfun(@times, ck4, bsxfun(@times, xx_r(2:nfreq,:),xx_r(2:
nfreq,:)*Beta_1(:,4)));
        gr4 = gr4 + sum( bsxfun(@times, (temp_mat_41+temp_mat_42), exp(-xx_r(2:nfreq,:)
*Beta_1(:,2)) ))'+...
            0.5*(exp(-xx_r(1,:)*Beta_1(:,2))*(-y(1,1).*conj(y(1,2)) - y(1,2).*conj
(y(1,1)))*xx_r(1,:)' +...
        exp(-xx_r(1,:)*Beta_1(:,2))*2*abs(y(1,1)).^2*(xx_r(1,:)*Beta_1(:,4))*xx_r
(1,:)')+...
            0.5*(exp(-xx_r(end,:)*Beta_1(:,2))*(-y(end,1).*conj(y(end,2)) - y
(end,2).*conj(y(end,1)))*xx_r(end,:)' +...
            exp(-xx_r(end,:)*Beta_1(:,2))*2*abs(y(end,1)).^2*(xx_r(end,:)*Beta_1
(:,4))*xx_r(end,:)');
        temp_mat_51 = bsxfun(@times, xx_r(2:nfreq,:),rk5);
        temp_mat_52 = bsxfun(@times, ck5, bsxfun(@times, xx_r(2:nfreq,:),xx_r(2:
nfreq,:)*Beta_1(:,5)));
        temp_mat_53 = bsxfun(@times, xx_r(2:nfreq,:),dk5);
        gr5 = gr5 + sum( bsxfun(@times, (temp_mat_51 + temp_mat_52 + temp_mat_53),
exp(-xx_r(2:nfreq,:)*Beta_1(:,3)) ))'+...
            0.5*(exp(-xx_r(1,:)*Beta_1(:,3))*(-y(1,1).*conj(y(1,3)) - y(1,3).*conj
(y(1,1)))*xx_r(1,:)' +...
            exp(-xx_r(1,:)*Beta_1(:,3))*2*abs(y(1,1)).^2*(xx_r(1,:)*Beta_1(:,5))
*xx_r(1,:)'+...
            exp(-xx_r(1,:)*Beta_1(:,3))*(y(1,2).*conj(y(1,1)).*(b(1).')+conj(y
(1,2).*conj(y(1,1)).*(b(1).')))*xx_r(1,:)')+...
            0.5*(exp(-xx_r(end,:)*Beta_1(:,3))*(-y(end,1).*conj(y(end,3)) - y
(end,3).*conj(y(end,1)))*xx_r(end,:)' +...
            exp(-xx_r(end,:)*Beta_1(:,3))*2*abs(y(end,1)).^2*(xx_r(end,:)*Beta_1
(:,5))*xx_r(end,:)'+...
            exp(-xx_r(end,:)*Beta_1(:,3))*(y(end,2).*conj(y(end,1)).*(b(end).') +
conj(y(end,2).*conj(y(end,1)).*(b(end).')))*xx_r(end,:)');
        temp_mat_61 = bsxfun(@times, xx_r(2:nfreq,:),rk6);
        temp_mat_62 = bsxfun(@times, ck6, bsxfun(@times, xx_r(2:nfreq,:),xx_r(2:
nfreq,:)*Beta_1(:,6)));
        temp_mat_63 = bsxfun(@times, xx_r(2:nfreq,:),dk6);
        gr6 = gr6 + sum( bsxfun(@times, (temp_mat_61 + temp_mat_62 + temp_mat_63),
exp(-xx_r(2:nfreq,:)*Beta_1(:,3)) ))'+...
            0.5*(exp(-xx_r(1,:)*Beta_1(:,3))*(-y(1,2).*conj(y(1,3)) - y(1,3).*conj
(y(1,2)))*xx_r(1,:)' +...
            exp(-xx_r(1,:)*Beta_1(:,3))*2*abs(y(1,2)).^2*(xx_r(1,:)*Beta_1(:,6))
*xx_r(1,:)'+...
            exp(-xx_r(1,:)*Beta_1(:,3))*(y(1,1).*conj(y(1,2)).*(a(1).') + conj(y
(1,1).*conj(y(1,2)).*(a(1).')))*xx_r(1,:)')+...
            0.5*(exp(-xx_r(end,:)*Beta_1(:,3))*(-y(end,2).*conj(y(end,3)) - y
(end,3).*conj(y(end,2)))*xx_r(end,:)' +...
            exp(-xx_r(end,:)*Beta_1(:,3))*2*abs(y(end,2)).^2*(xx_r(end,:)*Beta_1
(:,6))*xx_r(end,:)'+...
```

```
            exp(-xx_r(end,:)*Beta_1(:,3))*(y(end,1).*conj(y(end,2)).*(a(end).') +
conj(y(end,1).*conj(y(end,2)).*(a(end).')))*xx_r(end,:)');
        temp_mat_71 = bsxfun(@times, ik7, xx_i(2:nfreq,:));
        temp_mat_72 = bsxfun(@times, ck7, bsxfun(@times,xx_i(2:nfreq,:),xx_i(2:
nfreq,:)*Beta_2(:,1)));
        gr7 = gr7 + sum( bsxfun(@times, (temp_mat_71 + temp_mat_72), exp(-xx_r(2:
nfreq,:)*Beta_1(:,2))))' + ...
            0.5*(exp(-xx_r(1,:)*Beta_1(:,2))*(imag(-y(1,1).*conj(y(1,2)) + y(1,2).
*conj(y(1,1))))*xx_i(1,:)' +...
            exp(-xx_r(1,:)*Beta_1(:,2))*2*abs(y(1,1)).^2*(xx_i(1,:)*Beta_2(:,1))
*xx_i(1,:)')+...
            0.5*(exp(-xx_r(end,:)*Beta_1(:,2))*(imag(-y(end,1).*conj(y(end,2)) + y
(end,2).*conj(y(end,1))))*xx_i(end,:)' +...
            exp(-xx_r(end,:)*Beta_1(:,2))*2*abs(y(end,1)).^2*(xx_i(end,:)*Beta_2
(:,1))*xx_i(end,:)');
        temp_mat_81 = bsxfun(@times, ik8, xx_i(2:nfreq,:));
        temp_mat_82 = bsxfun(@times, ck8, bsxfun(@times,xx_i(2:nfreq,:),xx_i(2:
nfreq,:)*Beta_2(:,2)));
        temp_mat_83 = bsxfun(@times, xx_i(2:nfreq,:),dk8);
        gr8 = gr8 + sum( bsxfun(@times, (temp_mat_81 + temp_mat_82 + temp_mat_83),
exp(-xx_r(2:nfreq,:)*Beta_1(:,3))))' + ...
            0.5 * (exp(-xx_r(1,:)*Beta_1(:,3))*(imag(-y(1,1).*conj(y(1,3)) + y
(1,3).*conj(y(1,1))))*xx_i(1,:)' +...
            exp(-xx_r(1,:)*Beta_1(:,3))*2*abs(y(1,1)).^2*(xx_i(1,:)*Beta_2(:,2))
*xx_i(1,:)'+...
            sqrt(-1)*exp(-xx_r(1,:)*Beta_1(:,3))*(-y(1,1).*conj(y(1,2)).*(b(1).')
+ conj(y(1,1).*conj(y(1,2)).*(b(1).')))*xx_i(1,:)')+...
            0.5 * (exp(-xx_r(end,:)*Beta_1(:,3))*(imag(-y(end,1).*conj(y(end,3)) +
y(end,3).*conj(y(end,1))))*xx_i(end,:)' +...
            exp(-xx_r(end,:)*Beta_1(:,3))*2*abs(y(end,1)).^2*(xx_i(end,:)*Beta_2
(:,2))*xx_i(end,:)'+...
            sqrt(-1)*exp(-xx_r(end,:)*Beta_1(:,3))*(-y(end,1).*conj(y(end,2)).*(b
(end).') + conj(y(end,1).*conj(y(end,2)).*(b(end).')))*xx_i(end,:)');
        temp_mat_91 = bsxfun(@times, ik9, xx_i(2:nfreq,:));
        temp_mat_92 = bsxfun(@times, ck9, bsxfun(@times,xx_i(2:nfreq,:),xx_i(2:
nfreq,:)*Beta_2(:,3)));
        temp_mat_93 = bsxfun(@times, xx_i(2:nfreq,:),dk9);
        gr9 = gr9 + sum( bsxfun(@times, (temp_mat_91 + temp_mat_92 + temp_mat_93),
exp(-xx_r(2:nfreq,:)*Beta_1(:,3))))' + ...
            0.5 * (exp(-xx_r(1,:)*Beta_1(:,3))*(imag(-y(1,2).*conj(y(1,3)) + y
(1,3).*conj(y(1,2))))*xx_i(1,:)' +...
            exp(-xx_r(1,:)*Beta_1(:,3))*2*abs(y(1,2)).^2*(xx_i(1,:)*Beta_2(:,3))
*xx_i(1,:)'+...
            sqrt(-1)*exp(-xx_r(1,:)*Beta_1(:,3))*(-y(1,1).*conj(y(1,2)).*(a(1).')
+ conj(y(1,1).*conj(y(1,2)).*(a(1).')))*xx_i(1,:)')+...
            0.5 * (exp(-xx_r(end,:)*Beta_1(:,3))*(imag(-y(end,2).*conj(y(end,3)) +
y(end,3).*conj(y(end,2))))*xx_i(end,:)' +...
            exp(-xx_r(end,:)*Beta_1(:,3))*2*abs(y(end,2)).^2*(xx_i(end,:)*Beta_2
(:,3))*xx_i(end,:)'+...
            sqrt(-1)*exp(-xx_r(end,:)*Beta_1(:,3))*(-y(end,1).*conj(y(end,2)).*(a
(end).') + conj(y(end,1).*conj(y(end,2)).*(a(end).')))*xx_i(end,:)');
        %%%%%%%%%%%%%%%%%%%%%%%%%%%%%%%%%%%%%%%%%%%%%%%%%%%%%%%%
        %Hessian
        %%%%%%%%%%%%%%%%%%%%%%%%%%%%%%%%%%%%%%%%%%%%%%%%%%%%%%%%
        bigmat_h11 = kron(bsxfun(@times, abs(y(2:nfreq,1)).^2.*exp(-xx_r(2:nfreq,:)
*Beta_1(:,1)), xx_r(2:nfreq,:)),ones(nBeta,1)');
```

```
      coefmat_h11 = repmat(xx_r(2:nfreq,:), 1,nBeta);
      h11 = h11 + reshape(sum(bsxfun(@times, bigmat_h11, coefmat_h11),1),nBeta,
nBeta) +...
            0.5*(abs(y(1,1)).^2.*exp(-xx_r(1,:)*Beta_1(:,1))*xx_r(1,:)'
*xx_r(1,:))+...
            0.5*(abs(y(end,1)).^2.*exp(-xx_r(end,:)*Beta_1(:,1))*xx_r(end,:)'*
xx_r(end,:));

      bigmat_h22 = kron(bsxfun(@times,  abs(y(2:nfreq,2)-theta(1,2:nfreq).'.*y(2:
nfreq,1)).^2.*exp(-xx_r(2:nfreq,:)*Beta_1(:,2)),xx_r(2:nfreq,:)), ones
(nBeta,1)');
      coefmat_h22 = repmat(xx_r(2:nfreq,:), 1,nBeta);
      h22 = h22 + reshape(sum(bsxfun(@times, bigmat_h22, coefmat_h22),1),nBeta,
nBeta) +...
            0.5*(abs(y(1,2)-theta(1,1).*y(1,1)).^2.*exp(-xx_r(1,:)*Beta_1(:,2))
*xx_r(1,:)'*xx_r(1,:))+...
            0.5*(abs(y(end,2)-theta(1,end).*y(end,1)).^2.*exp(-xx_r(end,:)*Beta_1
(:,2))*xx_r(end,:)'*xx_r(end,:));

      bigmat_h33 = kron(bsxfun(@times,  abs(y(2:nfreq,3) - theta(2,2:nfreq).'.*y
(2:nfreq,1) - theta(3,2:nfreq).'.*y(2:nfreq,2)).^2.*...
            exp(-xx_r(2:nfreq,:)*Beta_1(:,3)),xx_r(2:nfreq,:)), ones(nBeta,1)');
      coefmat_h33 = repmat(xx_r(2:nfreq,:), 1,nBeta);
      h33 = h33 + reshape(sum(bsxfun(@times, bigmat_h33, coefmat_h33),1),nBeta,
nBeta) +...
            0.5*(abs(y(1,3) - theta(2,1).'.*y(1,1) - theta(3,1).'.*y(1,2)).^2.*...
            exp(-xx_r(1,:)*Beta_1(:,3))*xx_r(1,:)'*xx_r(1,:))+...
            0.5*(abs(y(end,3) - theta(2,end).'.*y(end,1) - theta(3,end).'.
*y(end,2)).^2.*...
            exp(-xx_r(end,:)*Beta_1(:,3))*xx_r(end,:)'*xx_r(end,:));

      bigmat_h44 = kron(bsxfun(@times, exp(-xx_r(2:nfreq,:)*Beta_1(:,2)).*ck4,
xx_r(2:nfreq,:)),ones(nBeta,1)');
      coefmat_h44 = repmat(xx_r(2:nfreq,:), 1,nBeta);
      h44 = h44 + reshape(sum(bsxfun(@times, bigmat_h44, coefmat_h44),1),nBeta,
nBeta) +...
            0.5*(exp(-xx_r(1,:)*Beta_1(:,2))*2*abs(y(1,1)).^2*xx_r(1,:)'*xx_r
(1,:))+...
            0.5*(exp(-xx_r(end,:)*Beta_1(:,2))*2*abs(y(end,1)).^2*xx_r
(end,:)'*xx_r(end,:));

      bigmat_h55 = kron(bsxfun(@times, exp(-xx_r(2:nfreq,:)*Beta_1(:,3)).*ck5,
xx_r(2:nfreq,:)),ones(nBeta,1)');
      coefmat_h55 = repmat(xx_r(2:nfreq,:), 1,nBeta);
      h55 = h55 + reshape(sum(bsxfun(@times, bigmat_h55, coefmat_h55),1),nBeta,
nBeta) +...
            0.5*(exp(-xx_r(1,:)*Beta_1(:,3))*2*abs(y(1,1)).^2*xx_r(1,:)'*xx_r
(1,:))+...
            0.5*(exp(-xx_r(end,:)*Beta_1(:,3))*2*abs(y(end,1)).^2*xx_r
(end,:)'*xx_r(end,:));

      bigmat_h66 = kron(bsxfun(@times, exp(-xx_r(2:nfreq,:)*Beta_1(:,3)).*ck6,
xx_r(2:nfreq,:)),ones(nBeta,1)');
      coefmat_h66 = repmat(xx_r(2:nfreq,:), 1,nBeta);
      h66 = h66 + reshape(sum(bsxfun(@times, bigmat_h66, coefmat_h66),1),nBeta,
nBeta) +...
```

```
            0.5*(exp(-xx_r(1,:)*Beta_1(:,3))*2*abs(y(1,2)).^2*xx_r(1,:)'*xx_r
(1,:))+...
            0.5*(exp(-xx_r(end,:)*Beta_1(:,3))*2*abs(y(end,2)).^2*xx_r
(end,:)'*xx_r(end,:));

        bigmat_h77 = kron(bsxfun(@times, exp(-xx_r(2:nfreq,:)*Beta_1(:,2)).*ck7,
xx_i(2:nfreq,:)),ones(nBeta,1)');
        coefmat_h77 = repmat(xx_i(2:nfreq,:), 1,nBeta);
        h77 = h77 + reshape(sum(bsxfun(@times, bigmat_h77, coefmat_h77),1),nBeta,
nBeta) +...
            0.5*(exp(-xx_r(1,:)*Beta_1(:,2))*2*abs(y(1,1)).^2*xx_i(1,:)'*xx_i
(1,:))+...
            0.5*(exp(-xx_r(end,:)*Beta_1(:,2))*2*abs(y(end,1)).^2*xx_i
(end,:)'*xx_i(end,:));

        bigmat_h88 = kron(bsxfun(@times, exp(-xx_r(2:nfreq,:)*Beta_1(:,3)).*ck8,
xx_i(2:nfreq,:)),ones(nBeta,1)');
        coefmat_h88 = repmat(xx_i(2:nfreq,:), 1,nBeta);
        h88 = h88 + reshape(sum(bsxfun(@times, bigmat_h88, coefmat_h88),1),nBeta,
nBeta) +...
            0.5*(exp(-xx_r(1,:)*Beta_1(:,3))*2*abs(y(1,1)).^2*xx_i(1,:)'*xx_i
(1,:))+...
            0.5*(exp(-xx_r(end,:)*Beta_1(:,3))*2*abs(y(end,1)).^2*xx_i
(end,:)'*xx_i(end,:));

        bigmat_h99 = kron(bsxfun(@times, exp(-xx_r(2:nfreq,:)*Beta_1(:,3)).*ck9,
xx_i(2:nfreq,:)),ones(nBeta,1)');
        coefmat_h99 = repmat(xx_i(2:nfreq,:), 1,nBeta);
        h99 = h99 + reshape(sum(bsxfun(@times, bigmat_h99, coefmat_h99),1),nBeta,
nBeta) +...
            0.5*(exp(-xx_r(1,:)*Beta_1(:,3))*2*abs(y(1,2)).^2*xx_i(1,:)'*xx_i
(1,:))+...
            0.5*(exp(-xx_r(end,:)*Beta_1(:,3))*2*abs(y(end,2)).^2*xx_i
(end,:)'*xx_i(end,:));

        bigmat_h42_1 = kron(bsxfun(@times, exp(-xx_r(2:nfreq,:)*Beta_1(:,2)).*ck4.
*(xx_r(2:nfreq,:)*Beta_1(:,4)),...
            xx_r(2:nfreq,:)), ones(nBeta,1)');
        coefmat_h42_1 = repmat(xx_r(2:nfreq,:), 1,nBeta);
        bigmat_h42_2 = kron(bsxfun(@times,exp(-xx_r(2:nfreq,:)*Beta_1(:,2)).*rk4,
xx_r(2:nfreq,:)), ones(nBeta,1)');
        coefmat_h42_2 = repmat(xx_r(2:nfreq,:), 1,nBeta);
        h42 = h42 + reshape(sum(bsxfun(@times, bigmat_h42_1, coefmat_h42_1) +...
            bsxfun(@times, bigmat_h42_2, coefmat_h42_2),1),nBeta,nBeta)' +...
            0.5*(exp(-xx_r(1,:)*Beta_1(:,2))*(2*abs(y(1,1)).^2*(xx_r(1,:)*Beta_1
(:,4))*xx_r(1,:)+...
            (-y(1,1).*conj(y(1,2)) - y(1,2).*conj(y(1,1)))*xx_r(1,:))'
*xx_r(1,:))+...
            0.5*(exp(-xx_r(end,:)*Beta_1(:,2))*(2*abs(y(end,1)).^2*(xx_r(end,:)
*Beta_1(:,4))*xx_r(end,:)+...
            (-y(end,1).*conj(y(end,2)) - y(end,2).*conj(y(end,1)))*xx_r
(end,:))'*xx_r(end,:));

        bigmat_h53_1 = kron(bsxfun(@times, exp(-xx_r(2:nfreq,:)*Beta_1(:,3)).*ck5.
*(xx_r(2:nfreq,:)*Beta_1(:,5)),...
            xx_r(2:nfreq,:)), ones(nBeta,1)');
        coefmat_h53_1 = repmat(xx_r(2:nfreq,:), 1,nBeta);
        bigmat_h53_2 = kron(bsxfun(@times,exp(-xx_r(2:nfreq,:)*Beta_1(:,3)).*rk5,
xx_r(2:nfreq,:)), ones(nBeta,1)');
```

```
    coefmat_h53_2 = repmat(xx_r(2:nfreq,:), 1,nBeta);
    bigmat_h53_3 = kron(bsxfun(@times,exp(-xx_r(2:nfreq,:)*Beta_1(:,3)).*dk5,
xx_r(2:nfreq,:)), ones(nBeta,1)');
    coefmat_h53_3 = repmat(xx_r(2:nfreq,:), 1,nBeta);
    h53 = h53 + reshape(sum(bsxfun(@times, bigmat_h53_1, coefmat_h53_1) +...
        bsxfun(@times, bigmat_h53_2, coefmat_h53_2)+...
        bsxfun(@times, bigmat_h53_3, coefmat_h53_3),1),nBeta,nBeta)' +...
        0.5*(exp(-xx_r(1,:)*Beta_1(:,3))*(2*abs(y(1,1)).^2*(xx_r(1,:)*Beta_1
(:,5))*xx_r(1,:)+...
        (-y(1,1).*conj(y(1,3)) - y(1,3).*conj(y(1,1)))*xx_r(1,:)+...
        (y(1,2).*conj(y(1,1)).*(b(1).') + conj(y(1,2).*conj(y(1,1)).
*(b(1).')))*xx_r(1,:))'*xx_r(1,:))+...
        0.5*(exp(-xx_r(end,:)*Beta_1(:,3))*(2*abs(y(end,1)).^2*(xx_r(end,:)
*Beta_1(:,5))*xx_r(end,:)+...
        (-y(end,1).*conj(y(end,3)) - y(end,3).*conj(y(end,1)))*xx_r(end,:)+...
        (y(end,2).*conj(y(end,1)).*(b(end).') + conj(y(end,2).*conj(y(end,1)).
*(b(end).')))*xx_r(end,:))'*xx_r(end,:));

    bigmat_h56_1 = kron(bsxfun(@times, exp(-xx_r(2:nfreq,:)*Beta_1(:,3)).*conj
(y(2:nfreq,1)).*y(2:nfreq,2),...
        xx_r(2:nfreq,:)), ones(nBeta,1)');
    coefmat_h56_1 = repmat(xx_r(2:nfreq,:), 1,nBeta);
    bigmat_h56_2 = kron(bsxfun(@times, exp(-xx_r(2:nfreq,:)*Beta_1(:,3)).*conj
(y(2:nfreq,2)).*y(2:nfreq,1),...
        xx_r(2:nfreq,:)), ones(nBeta,1)');
    coefmat_h56_2 = repmat(xx_r(2:nfreq,:), 1,nBeta);
    h56 = real(h56 + reshape(sum(bsxfun(@times, bigmat_h56_1, coefmat_h56_1)
+...
        bsxfun(@times, bigmat_h56_2, coefmat_h56_2),1),nBeta,nBeta)' +...
        0.5*(exp(-xx_r(1,:)*Beta_1(:,3))*((conj(y(1,1)).*y(1,2))*xx_r
(1,:)'*xx_r(1,:)+...
        (conj(y(1,2)).*y(1,1))*xx_r(1,:)'*xx_r(1,:)))+...
        0.5*(exp(-xx_r(end,:)*Beta_1(:,3))*((conj(y(end,1)).*y(end,2))*xx_r
(end,:)'*xx_r(end,:)+...
        (conj(y(end,2)).*y(end,1))*xx_r(end,:)'*xx_r(end,:))));

    bigmat_h59_1 = kron(bsxfun(@times, exp(-xx_r(2:nfreq,:)*Beta_1(:,3)).*conj
(y(2:nfreq,1)).*y(2:nfreq,2),...
        xx_r(2:nfreq,:)), ones(nBeta,1)');
    coefmat_h59_1 = repmat(xx_i(2:nfreq,:), 1,nBeta);
    bigmat_h59_2 = kron(bsxfun(@times, exp(-xx_r(2:nfreq,:)*Beta_1(:,3)).*conj(y
(2:nfreq,2)).*y(2:nfreq,1),
        xx_r(2:nfreq,:)), ones(nBeta,1)');
    coefmat_h59_2 = repmat(xx_i(2:nfreq,:), 1,nBeta);
    h59 = imag(h59 + reshape(sum(bsxfun(@times, bigmat_h59_1, coefmat_h59_1) -...
        bsxfun(@times, bigmat_h59_2, coefmat_h59_2),1),nBeta,nBeta)' +...
        0.5*(exp(-xx_r(1,:)*Beta_1(:,3))*((conj(y(1,1)).*y(1,2))*xx_r
(1,:)'*xx_i(1,:)-...
        (conj(y(1,2)).*y(1,1))*xx_r(1,:)'*xx_i(1,:)))+...
        0.5*(exp(-xx_r(end,:)*Beta_1(:,3))*((conj(y(end,1)).*y(end,2))*xx_r
(end,:)'*xx_i(end,:)-...
        (conj(y(end,2)).*y(end,1))*xx_r(end,:)'*xx_i(end,:))));

    bigmat_h63_1 = kron(bsxfun(@times, exp(-xx_r(2:nfreq,:)*Beta_1(:,3)).*ck6.
*(xx_r(2:nfreq,:)*Beta_1(:,6)),...
        xx_r(2:nfreq,:)), ones(nBeta,1)');
    coefmat_h63_1 = repmat(xx_r(2:nfreq,:), 1,nBeta);
```

```
        bigmat_h63_2 = kron(bsxfun(@times,exp(-xx_r(2:nfreq,:)*Beta_1(:,3)).*rk6,
xx_r(2:nfreq,:)), ones(nBeta,1)');
        coefmat_h63_2 = repmat(xx_r(2:nfreq,:), 1,nBeta);
        bigmat_h63_3 = kron(bsxfun(@times,exp(-xx_r(2:nfreq,:)*Beta_1(:,3)).*dk6,
xx_r(2:nfreq,:)), ones(nBeta,1)');
        coefmat_h63_3 = repmat(xx_r(2:nfreq,:), 1,nBeta);
        h63 = h63 + reshape(sum(bsxfun(@times, bigmat_h63_1, coefmat_h63_1) +...
              bsxfun(@times, bigmat_h63_2, coefmat_h63_2)+...
              bsxfun(@times, bigmat_h63_3, coefmat_h63_3),1),nBeta,nBeta)' +...
              0.5*(exp(-xx_r(1,:)*Beta_1(:,3))*(2*abs(y(1,2)).^2*(xx_r(1,:)*Beta_1
(:,6))*xx_r(1,:)+...
              (-y(1,2).*conj(y(1,3)) - y(1,3).*conj(y(1,2)))*xx_r(1,:)+...
              (y(1,2).*conj(y(1,1)).*(a(1).') + conj(y(1,2).*conj(y(1,1)).*(a
(1).')))*xx_r(1,:))'*xx_r(1,:))+...
              0.5*(exp(-xx_r(end,:)*Beta_1(:,3))*(2*abs(y(end,2)).^2*(xx_r(end,:)
*Beta_1(:,6))*xx_r(end,:)+...
              (-y(end,2).*conj(y(end,3)) - y(end,3).*conj(y(end,2)))*xx_r(end,:)+...
              (y(end,2).*conj(y(end,1)).*(a(end).') + conj(y(end,2).*conj(y(end,1)).*
(a(end).')))*xx_r(end,:))'*xx_r(end,:));

        bigmat_h68_1 = kron(bsxfun(@times, exp(-xx_r(2:nfreq,:)*Beta_1(:,3)).*conj(y
(2:nfreq,2)).*y(2:nfreq,1),...
              xx_r(2:nfreq,:)), ones(nBeta,1)');
        coefmat_h68_1 = repmat(xx_i(2:nfreq,:), 1,nBeta);
        bigmat_h68_2 = kron(bsxfun(@times, exp(-xx_r(2:nfreq,:)*Beta_1(:,3)).*conj(y
(2:nfreq,1)).*y(2:nfreq,2),...
              xx_r(2:nfreq,:)), ones(nBeta,1)');
        coefmat_h68_2 = repmat(xx_i(2:nfreq,:), 1,nBeta);
        h68 = imag(h68 + reshape(sum(bsxfun(@times, bigmat_h68_1, coefmat_h68_1) -...
        bsxfun(@times, bigmat_h68_2, coefmat_h68_2),1),nBeta,nBeta)' +...
              0.5*(exp(-xx_r(1,:)*Beta_1(:,3))*((conj(y(1,2)).*y(1,1))*xx_r
(1,:)'*xx_i(1,:)-...
              (conj(y(1,1)).*y(1,2))*xx_r(1,:)'*xx_i(1,:)))+...
              0.5*(exp(-xx_r(end,:)*Beta_1(:,3))*((conj(y(end,2)).*y(end,1))*xx_r
(end,:)'*xx_i(end,:)-...
              (conj(y(end,1)).*y(end,2))*xx_r(end,:)'*xx_i(end,:))));

       bigmat_h72_1 = kron(bsxfun(@times, exp(-xx_r(2:nfreq,:)*Beta_1(:,2)).*ck7.*
(xx_i(2:nfreq,:)*Beta_2(:,1)),...
              xx_i(2:nfreq,:)), ones(nBeta,1)');
        coefmat_h72_1 = repmat(xx_r(2:nfreq,:), 1,nBeta);
        bigmat_h72_2 = kron(bsxfun(@times,exp(-xx_r(2:nfreq,:)*Beta_1(:,2)).*ik7,
xx_i(2:nfreq,:)), ones(nBeta,1)');
        coefmat_h72_2 = repmat(xx_r(2:nfreq,:), 1,nBeta);
        h72 = h72 + reshape(sum(bsxfun(@times, bigmat_h72_1, coefmat_h72_1) +...
              bsxfun(@times, bigmat_h72_2, coefmat_h72_2),1),nBeta,nBeta)' +...
              0.5*(exp(-xx_r(1,:)*Beta_1(:,2))*(2*abs(y(1,1)).^2*(xx_i(1,:)*Beta_2
(:,1))*xx_i(1,:)+...
              imag(-y(1,1).*conj(y(1,2)) + y(1,2).*conj(y(1,1)))*xx_i(1,:))'*xx_r
(1,:))+...
              0.5*(exp(-xx_r(end,:)*Beta_1(:,2))*(2*abs(y(end,1)).^2*(xx_i(end,:)
*Beta_2(:,1))*xx_i(end,:)+...
              imag(-y(end,1).*conj(y(end,2)) + y(end,2).*conj(y(end,1)))*xx_i
(end,:))'*xx_r(end,:));

        bigmat_h83_1 = kron(bsxfun(@times, exp(-xx_r(2:nfreq,:)*Beta_1(:,3)).*ck8.*
(xx_i(2:nfreq,:)*Beta_2(:,2)),...
              xx_i(2:nfreq,:)), ones(nBeta,1)');
        coefmat_h83_1 = repmat(xx_r(2:nfreq,:), 1,nBeta);
```

```
      bigmat_h83_2 = kron(bsxfun(@times,exp(-xx_r(2:nfreq,:)*Beta_1(:,3)).*ik8,
xx_i(2:nfreq,:)), ones(nBeta,1)');
      coefmat_h83_2 = repmat(xx_r(2:nfreq,:), 1,nBeta);
      bigmat_h83_3 = kron(bsxfun(@times,exp(-xx_r(2:nfreq,:)*Beta_1(:,3)).*(dk8),
xx_i(2:nfreq,:)), ones(nBeta,1)');
      coefmat_h83_3 = repmat(xx_r(2:nfreq,:), 1,nBeta);
      h83 = h83 + reshape(sum(bsxfun(@times, bigmat_h83_1, coefmat_h83_1) +...
          bsxfun(@times, bigmat_h83_2, coefmat_h83_2)+...
          bsxfun(@times, bigmat_h83_3, coefmat_h83_3),1),nBeta,nBeta)' +...
          0.5*(exp(-xx_r(1,:)*Beta_1(:,3))*(2*abs(y(1,1)).^2*(xx_i(1,:)*Beta_2
(:,2))*xx_i(1,:)+...
          (-y(1,1).*conj(y(1,3)) + y(1,3).*conj(y(1,1)))*xx_i(1,:)+...
          imag(-y(1,2).*conj(y(1,1)).*(b(1).') + conj(y(1,2).*conj(y(1,1)).*(b
(1).')))*xx_i(1,:))'*xx_r(1,:))+...
          0.5*(exp(-xx_r(end,:)*Beta_1(:,3))*(2*abs(y(end,1)).^2*(xx_i(end,:)
*Beta_2(:,2))*xx_i(end,:)+...
          (-y(end,1).*conj(y(end,3)) + y(end,3).*conj(y(end,1)))*xx_i(end,:)+...
          imag(-y(end,2).*conj(y(end,1)).*(b(end).') + conj(y(end,2).*conj(y
(end,1)).*(b(end).')))*xx_i(end,:))'*xx_r(end,:));

      bigmat_h93_1 = kron(bsxfun(@times, exp(-xx_r(2:nfreq,:)*Beta_1(:,3)).*ck9.*
(xx_i(2:nfreq,:)*Beta_2(:,3)),...
          xx_i(2:nfreq,:)), ones(nBeta,1)');
      coefmat_h93_1 = repmat(xx_r(2:nfreq,:), 1,nBeta);
      bigmat_h93_2 = kron(bsxfun(@times,exp(-xx_r(2:nfreq,:)*Beta_1(:,3)).*ik9,
xx_i(2:nfreq,:)), ones(nBeta,1)');
      coefmat_h93_2 = repmat(xx_r(2:nfreq,:), 1,nBeta);
      bigmat_h93_3 = kron(bsxfun(@times,exp(-xx_r(2:nfreq,:)*Beta_1(:,3)).*(dk9),
xx_i(2:nfreq,:)), ones(nBeta,1)');
      coefmat_h93_3 = repmat(xx_r(2:nfreq,:), 1,nBeta);
      h93 = h93 + reshape(sum(bsxfun(@times, bigmat_h93_1, coefmat_h93_1) +...
          bsxfun(@times, bigmat_h93_2, coefmat_h93_2)+...
          bsxfun(@times, bigmat_h93_3, coefmat_h93_3),1),nBeta,nBeta)' +...
          0.5*(exp(-xx_r(1,:)*Beta_1(:,3))*(2*abs(y(1,2)).^2*(xx_i(1,:)*Beta_2
(:,3))*xx_i(1,:)+...
          (-y(1,2).*conj(y(1,3)) + y(1,3).*conj(y(1,2)))*xx_i(1,:)+...
          imag(-y(1,1).*conj(y(1,2)).*(a(1).') + conj(y(1,1).*conj(y(1,2)).*(a
(1).')))*xx_i(1,:))'*xx_r(1,:))+...
          0.5*(exp(-xx_r(end,:)*Beta_1(:,3))*(2*abs(y(end,2)).^2*(xx_i(end,:)
*Beta_2(:,3))*xx_i(end,:)+...
          (-y(end,2).*conj(y(end,3)) + y(end,3).*conj(y(end,2)))*xx_i(end,:)+...
          imag(-y(end,1).*conj(y(end,2)).*(a(end).') + conj(y(end,1).*conj(y
(end,2)).*(a(end).')))*xx_i(end,:))'*xx_r(end,:));

      bigmat_h98_1 = kron(bsxfun(@times, exp(-xx_r(2:nfreq,:)*Beta_1(:,3)).*conj(y
(2:nfreq,1)).*y(2:nfreq,2),...
          xx_i(2:nfreq,:)), ones(nBeta,1)');
      coefmat_h98_1 = repmat(xx_i(2:nfreq,:), 1,nBeta);
      bigmat_h98_2 = kron(bsxfun(@times, exp(-xx_r(2:nfreq,:)*Beta_1(:,3)).*conj(y
(2:nfreq,2)).*y(2:nfreq,1),...
          xx_i(2:nfreq,:)), ones(nBeta,1)');
      coefmat_h98_2 = repmat(xx_i(2:nfreq,:), 1,nBeta);
      h98 = real(h98 + reshape(sum(bsxfun(@times, bigmat_h98_1, coefmat_h98_1) +...
          bsxfun(@times, bigmat_h98_2, coefmat_h98_2),1),nBeta,nBeta)' +...
          0.5*(exp(-xx_r(1,:)*Beta_1(:,3))*((conj(y(1,1)).*y(1,2))*xx_i
(1,:)'*xx_i(1,:)+...
          (conj(y(1,2)).*y(1,1))*xx_i(1,:)'*xx_i(1,:)))+...
```

```
            0.5*(exp(-xx_r(end,:)*Beta_1(:,3))*((conj(y(end,1)).*y(end,2))*xx_i
(end,:)'*xx_i(end,:)+...
            (conj(y(end,2)).*y(end,1))*xx_i(end,:)'*xx_i(end,:))));
        h24=h42'; h35=h53'; h65=h56'; h95=h59'; h36=h63';
        h86=h68'; h27=h72'; h38=h83'; h39=h93'; h89=h98';
    end
    ze = zeros(nBeta,nBeta);
    h1 = [h11,repmat(ze,1,8)];
    h2 = [ze,h22,ze,-h24,ze,ze,-h27,ze,ze];
    h3 = [ze,ze,h33,ze,-h35,-h36,ze,-h38,-h39];
    h4 = [ze,-h42,ze,h44,repmat(ze,1,5)];
    h5 = [ze,ze,-h53,ze,h55,h56,ze,ze,h59];
    h6 = [ze,ze,-h63,ze,h65,h66,ze,h68,ze];
    h7 = [ze,-h72,ze,ze,ze,ze,h77,ze,ze];
    h8 = [ze,ze,-h83,ze,ze,h86,ze,h88,h89];
    h9 = [ze,ze,-h93,ze,h95,ze,ze,h98,h99];

    gr = [gr1;gr2;gr3;gr4;gr5;gr6;gr7;gr8;gr9]; h = [h1;h2;h3;h4;h5;h6;h7;h8;h9];
f = -f;
    gr_index= (1:(dimen^2*nBeta)).*kron(chol_index(Phi_temp,:),ones(nBeta,1)');
    gr_index = gr_index(find(gr_index~=0));
    gr = gr(gr_index); h = h(gr_index,gr_index);
end

function [f,gr,h] = Beta_derive2(x, yobs_tmp, chol_index, Phi_temp,
tau_temp_1,...
            tau_temp_2, Beta_temp_1, Beta_temp_2, sigmasqalpha, nbasis, nseg)

%%%%%%%%%%%%%%%%%%%%%%%%%%%%%%%%%%%%%%%%%%%%%%%%%%%%%%%%%%%%%%%%%%%%%%%%%%%%
% Function used for optimization process for coefficients are the same
% across segments
%
%   Input:
%       1) x - initial values for coefficient of basis functions need to
%       be optimized
%       2) yobs_tmp - time series data within the segment
%       3) chol_index - index matrix
%       4) Phi_temp - which component changed
%       5) tau_temp_1 - smoothing parameters for the first segment
%       6) tau_temp_2 - smoothing parameters for the second segment
%       7) Beta_temp_1 - current coefficients for the first segment
%       8) Beta_temp_2 - current coefficients for the second segment
%       9) sigmasqalpha - smoothing parameters for the constant in real
%       components
%       10) nbasis - number of basis function used
%       11) nseg - number of observation in the first segment
%   Main Outputs:
%       1) f - log posterior probability based on input parameters
%       2) gr - gradients for optimization process
%       3) h - Hessian matrix for optimization process
%
%   Required programs: lin_basis_func, Beta_derive1
yobs_tmp_1 = yobs_tmp(1:nseg,:);
yobs_tmp_2 = yobs_tmp(nseg+1:end,:);
[f1,grad1,hes1] = Beta_derive1(x, yobs_tmp_1, chol_index, Phi_temp, tau_temp_1,
Beta_temp_1, sigmasqalpha, nbasis);
```

```
[f2,grad2,hes2] = Beta_derivel(x, yobs_tmp_2, chol_index, Phi_temp, tau_temp_2,
Beta_temp_2, sigmasqalpha, nbasis);

f = f1 + f2;
gr = grad1 + grad2;
h = hes1 + hes2;

function[PI,nseg_prop,xi_prop,tau_prop,Beta_prop,Phi_prop]=...
       birth(chol_index,ts,nexp_curr,nexp_prop,...
           tau_curr_temp,xi_curr_temp,nseg_curr_temp,Beta_curr_temp,
Phi_curr_temp,....
           log_move_curr,log_move_prop)

%%%%%%%%%%%%%%%%%%%%%%%%%%%%%%%%%%%%%%%%%%%%%%%%%%%%%%%%%%%%%%%%%%%%%%%%%%%%
% Does the birth step in the paper
%
%   Input:
%       1) chol_index - index matrix
%       2) ts - TxN matrix of time series data
%       3) nexp_curr - current number of segment
%       4) nexp_prop - proposed number of segment
%       5) tau_curr_temp - current smoothing parameters
%       6) xi_curr_temp - current partitions
%       7) nseg_curr_temp - current number of observations in each segment
%       8) Beta_curr_temp - current coefficients
%       9) Phi_curr_temp - which component changed
%       10) log_move_curr - probability: proposed to current
%       11) log_move_prop - probability: current to proposed
%   Main Outputs:
%       1) A - acceptance probability
%       2) nseg_prop - proposed number of observations in each segment
%       3) xi_prop - proposed partitions
%       4) tau_prop - proposed smoothing parameters
%       5) Beta_prop - proposed coefficients
%       6) Phi_prop - proposed indicator variable
%
%   Required programs: postBeta1, Beta_derive1, whittle_like
%%%%%%%%%%%%%%%%%%%%%%%%%%%%%%%%%%%%%%%%%%%%%%%%%%%%%%%%%%%%%%%%%%%%%%%%%%%%

global nobs dimen nbasis nBeta sigmasqalpha tmin tau_up_limit

Beta_prop = zeros(nBeta,dimen^2,nexp_prop);
tau_prop = ones(dimen^2,nexp_prop,1);
nseg_prop = zeros(nexp_prop,1);
xi_prop = zeros(nexp_prop,1);
Phi_prop = zeros(nexp_prop,1);

%****************************************
%Drawing segment to split
%****************************************
kk = find(nseg_curr_temp>2*tmin); %Number of segments available for splitting
nposs_seg = length(kk);
seg_cut = kk(unidrnd(nposs_seg)); %Drawing segment to split
nposs_cut = nseg_curr_temp(seg_cut)-2*tmin+1; %Drawing new birthed partition
```

```
%*********************************************
% Proposing new parameters, Beta, Phi, tau, xi
%*********************************************
for jj=1:nexp_curr
  if jj<seg_cut %nothing updated or proposed here
      xi_prop(jj) = xi_curr_temp(jj);
      tau_prop(:,jj) = tau_curr_temp(:,jj);
      nseg_prop(jj) = nseg_curr_temp(jj);
      Beta_prop(:,:,jj) = Beta_curr_temp(:,:,jj);
      Phi_prop(jj) = Phi_curr_temp(jj);
  elseif jj==seg_cut  %updating parameters in the selected paritions
      index = unidrnd(nposs_cut);
      if (seg_cut==1)
         xi_prop(seg_cut)=index+tmin-1;
      else
         xi_prop(seg_cut)=xi_curr_temp(jj-1)-1+tmin+index;
      end
      xi_prop(seg_cut+1) = xi_curr_temp(jj);

      %Determine which Cholesky components should change here
      Phi_prop(seg_cut) = randsample(2:2^(dimen^2),1);
      Phi_prop(seg_cut+1) = Phi_curr_temp(jj);

      %Drawing new tausq
      select = find(chol_index(Phi_prop(seg_cut),:)~=0);
      select_inv = find(chol_index(Phi_prop(seg_cut),:)==0);
      zz = rand(dimen^2,1)'.*chol_index(Phi_prop(seg_cut),:);
      uu = zz./(1-zz);
      uu(find(uu==0))= 1;
      tau_prop(:,seg_cut)= tau_curr_temp(:,seg_cut).*uu';
      tau_prop(:,seg_cut+1)=tau_curr_temp(:,seg_cut).*(1./uu)';

      %Drawing new values for coefficient of basis function for new birthed segments.
      nseg_prop(seg_cut) = index+tmin-1;
      nseg_prop(seg_cut+1) = nseg_curr_temp(jj)-nseg_prop(seg_cut);
      Phi_need = Phi_prop(seg_cut);
      for k=jj:(jj+1)
         if k==jj
            [Beta_mean_1, Beta_var_1,yobs_tmp_1]=postBeta1(chol_index, Phi_need, ...
            k, ts, tau_prop(:,k), Beta_curr_temp(:,:,jj), xi_prop);
            Beta_prop(:,select,k) = reshape(mvnrnd(Beta_mean_1,...
                   0.5*(Beta_var_1+Beta_var_1')),nBeta,length(select));
            Beta_prop(:,select_inv,k) = Beta_curr_temp(:,select_inv,jj);

       else
            [Beta_mean_2, Beta_var_2,yobs_tmp_2]=postBeta1(chol_index, Phi_need, ...
            k, ts, tau_prop(:,k), Beta_curr_temp(:,:,jj), xi_prop);
            Beta_prop(:,select,k)=reshape(mvnrnd(Beta_mean_2,...
                   0.5*(Beta_var_2+Beta_var_2')),nBeta,length(select));
           Beta_prop(:,select_inv,k) = Beta_curr_temp(:,select_inv,jj);
       end
    end
  else %nothing updated or proposed here
     xi_prop(jj+1) = xi_curr_temp(jj);
            tau_prop(:,jj+1) = tau_curr_temp(:,jj);
            nseg_prop(jj+1) = nseg_curr_temp(jj);
            Beta_prop(:,:,jj+1) = Beta_curr_temp(:,:,jj);
     Phi_prop(jj+1) = Phi_curr_temp(jj);
  end
end
```

```
%Calculating Jacobian
ja = tau_curr_temp(:,seg_cut)./(zz.*(1-zz))'; ja = ja(ja~=Inf);
log_jacobian = sum(log(2*ja));

%*************************************************************
%Calculations related to proposed values
%*************************************************************

%=====================================================================
%Evaluating the Likelihood, Proposal and Prior Densities at the Proposed values
%=====================================================================

log_Beta_prop = 0;
log_tau_prior_prop = 0;
log_Beta_prior_prop = 0;
loglike_prop = 0;

for jj=seg_cut:seg_cut+1
  if jj==seg_cut
      Beta_mean = Beta_mean_1;
      Beta_var = Beta_var_1;
      yobs_tmp = yobs_tmp_1;
  else
      Beta_mean = Beta_mean_2;
      Beta_var = Beta_var_2;
      yobs_tmp = yobs_tmp_2;
  end
  pb = reshape(Beta_prop(:,select,jj), numel(Beta_prop(:,select,jj)),1);
  %Proposed density for coefficient of basis functions
  log_Beta_prop = log_Beta_prop - 0.5*(pb-Beta_mean)'*matpower(0.5*(Beta_var
+Beta_var'),-1)*(pb-Beta_mean);

  %Prior density for coefficient of basis functions
  prior_tau = reshape([[repmat(sigmasqalpha,1,dimen^2-dimen*(dimen-1)/2) tau_prop
((dimen + dimen*(dimen-1)/2 + 1):end,jj)'];...
            reshape(kron(tau_prop(:,jj),ones(nbasis,1)), nbasis, dimen^2)],
nBeta,(dimen^2));
  prior_tau = reshape(prior_tau(:,select),length(select)*nBeta,1);

  log_Beta_prior_prop=log_Beta_prior_prop-0.5*(pb)'*matpower(diag(prior_tau),-1)*(pb);

  log_tau_prior_prop = log_tau_prior_prop-length(select)*log(tau_up_limit);
% Prior Density of tausq
  [log_prop_spec_dens] = whittle_like(yobs_tmp,Beta_prop(:,:,jj));
  loglike_prop = loglike_prop + log_prop_spec_dens; %Loglikelihood at proposed values
end

log_seg_prop = -log(nposs_seg);%Proposal density for segment choice
log_cut_prop = -log(nposs_cut);%Proposal density for partition choice
log_Phi_prop = -log(2^(dimen^2)-1); %proposal density for component choice
log_prior_Phi_prop = -log(2^(dimen^2)); %prior density for component choice

%Evaluating prior density for cut points at proposed values
log_prior_cut_prop = 0;
for k=1:nexp_prop-1
      if k==1
            log_prior_cut_prop=-log(nobs-(nexp_prop-k+1)*tmin+1);
      else
            log_prior_cut_prop=log_prior_cut_prop-log(nobs-xi_prop(k-1)-
(nexp_prop-k+1)*tmin+1);
      end
end
```

```
%Calculating Log Proposal density at Proposed values
log_proposal_prop = log_Beta_prop + log_seg_prop + log_move_prop + log_cut_prop +
log_Phi_prop;
%Calculating Log Prior density at Proposed values
log_prior_prop = log_Beta_prior_prop + log_tau_prior_prop + log_prior_cut_prop +
log_prior_Phi_prop;
%Calculating Target density at Proposed values
log_target_prop = loglike_prop + log_prior_prop;

%************************************************************
%Calculations related to current values
%************************************************************

%==========================================================================
%Evaluating the Likelihood, Proposal and Prior Densities at the Current values
%==========================================================================

[Beta_mean, Beta_var, yobs_tmp] = postBeta1(chol_index, Phi_need, seg_cut, ts,
tau_curr_temp(:,seg_cut),...
                            Beta_curr_temp(:,:,seg_cut), xi_curr_temp);
pb = reshape(Beta_curr_temp(:,select,seg_cut),numel(Beta_curr_temp(:,select,
seg_cut)),1);

%Current density for coefficient of basis functions
log_Beta_curr = -0.5*(pb-Beta_mean)'*matpower(0.5*(Beta_var+Beta_var'),-1)*(pb-
Beta_mean);

%Prior density for coefficient of basis functions at current values
prior_tau = reshape([[repmat(sigmasqalpha,1,dimen^2-dimen*(dimen-1)/2)
tau_curr_temp((dimen + dimen*(dimen-1)/2 + 1):end,seg_cut)'];...
           reshape(kron(tau_curr_temp(:,seg_cut),ones(nbasis,1)), nbasis,
dimen^2)], nBeta,(dimen^2));
prior_tau = reshape(prior_tau(:,select),length(select)*nBeta,1);

log_Beta_prior_curr = -0.5*(pb)'*matpower(diag(prior_tau),-1)*(pb);

log_tau_prior_curr = -length(select)*log(tau_up_limit); %prior density for
smoothing parameters
[log_curr_spec_dens] = whittle_like(yobs_tmp,Beta_curr_temp(:,:,seg_cut));
loglike_curr = log_curr_spec_dens; %Loglikelihood at current values

log_Phi_curr = -log(1); %proposal for component choice
%Calculating Log Proposal density at current values
log_proposal_curr = log_Beta_curr + log_move_curr + log_Phi_curr;

%Evaluating prior density for partition current values
log_prior_cut_curr = 0;
for k=1:nexp_curr-1
       if k==1
              log_prior_cut_curr = -log(nobs-(nexp_curr-k+1)*tmin+1);
       else
              log_prior_cut_curr = log_prior_cut_curr-log(nobs-xi_curr_temp(k-1)-
(nexp_curr-k+1)*tmin+1);
       end
end

log_prior_Phi_curr = -log(2^(dimen^2)); %prior density for component choice
%Calculating Priors at Current Values
log_prior_curr = log_Beta_prior_curr + log_tau_prior_curr + log_prior_cut_curr +
```

```
log_prior_Phi_curr;
%Evalulating Target densities at current values
log_target_curr = loglike_curr + log_prior_curr;

%***********************************************************
%Calculations acceptance probability
%***********************************************************
 PI = min(1,exp(log_target_prop - log_target_curr +...
       log_proposal_curr - log_proposal_prop + log_jacobian));
end

function[chol_index]=chol_ind(dimen)

chol_index = zeros(2^(dimen^2), dimen^2 );
k=0;
for i=0:dimen^2
  C = nchoosek(1:dimen^2,i);
  dim = size(C);
  if i==0
     k=k+1;
     chol_index(1,:) = 0;
  else
     for j=1:dim(1)
       k=k+1;
       chol_index(k,C(j,:)) = 1;
     end
  end
end

end

function [result] = conv_diag(nobs, nBeta, nseg,nexp_curr, Beta_curr,
tausq_curr, dimen)

%   conv_diag Calculate convergence diagnostics for beta and tau sq params
%   Get max eigenvalue of beta and tausq matrices across time and cov

idx = cumsum(nseg);
tausq_diag = zeros(nobs,dimen^2);
Beta_diag = zeros(nobs,nBeta,dimen^2);

for i=1:nexp_curr
  if i==1
   for k=1:dimen^2
     tausq_diag(1:idx(i),k) = repmat(tausq_curr(k,i),nseg(i),1);

     for j=1:nBeta
       Beta_diag(1:idx(i),j,k)=repmat(Beta_curr(j,k),nseg(i),1);
     end
   end
  else
   for k=1:dimen^2
     tausq_diag((idx(i-1)+1):idx(i),k) = repmat(tausq_curr(k,i),nseg(i),1);

     for j=1:nBeta
       Beta_diag((idx(i-1)+1):idx(i),j,k)=repmat(Beta_curr(j,k),nseg(i),1);
     end
   end
  end

end
```

```
    %calculate eigenval of symmetric transform
    result = zeros(1,dimen^2*(nBeta+1));
    for k=1:dimen^2
      result(k) = real(max(eig(tausq_diag(:,k)*tausq_diag(:,k)')));
    end
    k=1;
    for i=1:dimen^2
      for j=1:nBeta
        result(k+dimen^2) = real(max(eig(Beta_diag(:,j,i)*Beta_diag(:,j,i)')));
        k=k+1;
      end
   end
end

function[PI,nseg_prop,xi_prop,tau_prop,Beta_prop,Phi_prop]=...
death(chol_index,ts,nexp_curr,nexp_prop,...
      tau_curr_temp,xi_curr_temp,nseg_curr_temp,Beta_curr_temp,Phi_curr_temp,
log_move_curr,log_move_prop)

%%%%%%%%%%%%%%%%%%%%%%%%%%%%%%%%%%%%%%%%%%%%
% Does the death step in the paper
%
%  Input:
%     1) chol_index - index matrix
%     2) ts - TxN matrix of time series data
%     3) nexp_curr - current number of segment
%     4) nexp_prop - proposed number of segment
%     5) tau_curr_temp - current smoothing parameters
%     6) xi_curr_temp - current partitions
%     7) nseg_curr_temp - current number of observations in each segment
%     8) Beta_curr_temp - current coefficients
%     9) Phi_curr_temp - which component changed
%     10) log_move_curr - probability: proposed to current
%     11) log_move_prop - probability: current to proposed
%   Main Outputs:
%     1) A - acceptance probability
%     2) nseg_prop - proposed number of observations in each segment
%     3) xi_prop - proposed partitions
%     4) tau_prop - proposed smoothing parameters
%     5) Beta_prop - proposed coefficients
%     6) Phi_prop - proposed indicator variable
%
%   Required programs: postBeta1, Beta_derive_1, whittle_like
%%%%%%%%%%%%%%%%%%%%%%%%%%%%%%%%%%%%%%%%%%%%%%%%%%%%%%%%%%%%%%%%%%

global nobs dimen nBeta nbasis sigmasqalpha tmin tau_up_limit

Beta_prop = zeros(nBeta,dimen^2,nexp_prop);
tau_prop = ones(dimen^2,nexp_prop,1);
nseg_prop = zeros(nexp_prop,1);
xi_prop = zeros(nexp_prop,1);
Phi_prop = zeros(nexp_prop,1);

%****************************************
%Draw a partition to delete
%****************************************
cut_del=unidrnd(nexp_curr-1);

j=0;
for k = 1:nexp_prop
  j = j+1;
  if k==cut_del
```

```
    %*********************************************************
         %Calculations related to proposed values
    %*********************************************************
    xi_prop(k) = xi_curr_temp(j+1);
    tau_prop(:,k) = sqrt(tau_curr_temp(:,j).*tau_curr_temp(:,j+1)); %Combine
two taus into one
    nseg_prop(k) = nseg_curr_temp(j) + nseg_curr_temp(j+1); %Combine two
segments into one
    Phi_prop(k) = Phi_curr_temp(j+1);

%==============================================================
    %Evaluate the Likelihood at proposed values
%==============================================================
    need = sum(Beta_curr_temp(:,:,k)-Beta_curr_temp(:,:,k+1));
    need = need./need;
    need(isnan(need))=0;
    aa = zeros(2^(dimen^2),1);
    for i=1:2^(dimen^2)
      aa(i)=sum(chol_index(i,:)==need);
    end
    Phi_need = find(aa==dimen^2);
    select = find(chol_index(Phi_need,:)~=0);
    select_inv = find(chol_index(Phi_need,:)==0);

    %Compute mean and variances for coefficents of basis functions
    [Beta_mean, Beta_var, yobs_tmp]= postBeta1(chol_index, Phi_need, k, ts,
tau_prop(:,k),...
                              Beta_curr_temp(:,:,k), xi_prop);
    Beta_prop(:,select,k) = reshape(mvnrnd(Beta_mean,0.5*(Beta_var+Beta_var')),
nBeta,length(select));
    Beta_prop(:,select_inv,k) = Beta_curr_temp(:,select_inv,k);

    %Loglikelihood at proposed values
    [loglike_prop]=whittle_like(yobs_tmp,Beta_prop(:,:,k));
  %=====================================================================
  %Evaluate the Proposal Densities at the Proposed values for tau, Phi, and Beta
  %=====================================================================
    pb = reshape(Beta_prop(:,select,k), numel(Beta_prop(:,select,k)),1);
    %Proposed density for coefficient of basis functions
    log_Beta_prop = - 0.5*(pb-Beta_mean)'*matpower(0.5*(Beta_var+Beta_var'),-1)
*(pb-Beta_mean);

    log_seg_prop = -log(nexp_curr-1);  %Proposal for segment choice
    log_Phi_prop = -log(1); %proposal for component choice
    %Calcualte Jacobian
            log_jacobian = -sum(log(2*(sqrt(tau_curr_temp(select,j)) + sqrt
(tau_curr_temp(select,j+1))).^2));

    %log proposal probabililty
    log_proposal_prop = log_Beta_prop + log_seg_prop + log_move_prop +
log_Phi_prop;

%===============================================================
% Evaluate the PRIOR Densities at the Proposed values for tau, Phi, and Beta
%===============================================================
    prior_tau = reshape([[repmat(sigmasqalpha,1,dimen^2-dimen*(dimen-1)/2)
tau_prop((dimen + dimen*(dimen-1)/2 + 1):end,k)'];...
               reshape(kron(tau_prop(:,k),ones(nbasis,1)), nbasis,...
               dimen^2)], nBeta,(dimen^2));
    prior_tau = reshape(prior_tau(:,select),length(select)*nBeta,1);
```

```
    %Prior density for coefficient of basis functions
    log_Beta_prior_prop = -0.5*(pb)'*matpower(diag(prior_tau),-1)*(pb);

            %Prior Density of tausq
            log_tau_prior_prop = -length(select)*log(tau_up_limit);
    %Prior Density of Phi
    log_Phi_prior_prop = -log(2^(dimen^2));
    log_prior_prop = log_tau_prior_prop + log_Beta_prior_prop + log_Phi_prior_prop;
%*********************************************************
%Calculations related to current values
%*********************************************************

%=======================================================================
%Evaluate the Likelihood, Proposal and Prior Densities at the Current values
%=======================================================================
            log_Beta_curr = 0;
            log_tau_prior_curr = 0;
            log_Beta_prior_curr = 0;
            loglike_curr=0;
    for jj=j:j+1
      [Beta_mean, Beta_var, yobs_tmp]= postBeta1(chol_index, Phi_need, jj, ts,
tau_curr_temp(:,jj),...
                          Beta_curr_temp(:,:,jj), xi_curr_temp);

        pb = reshape(Beta_curr_temp(:,select,jj), numel(Beta_curr_temp
(:,select,jj)),1);
        %Current density for coefficient of basis functions
        log_Beta_curr = log_Beta_curr -0.5*(pb-Beta_mean)'*matpower(0.5*(Beta_var
+Beta_var'),-1)*...
                                                  (pb-Beta_mean);

        prior_tau = reshape([[repmat(sigmasqalpha,1,dimen^2-dimen*(dimen-1)/2)
tau_curr_temp((dimen + dimen*(dimen-1)/2 + 1):end,jj)'];...
            reshape(kron(tau_curr_temp(:,jj),ones(nbasis,1)), nbasis,...
            dimen^2)], nBeta,(dimen^2));
        prior_tau = reshape(prior_tau(:,select),length(select)*nBeta,1);

        %Prior density for coefficient of basis functions at current values
        log_Beta_prior_curr = log_Beta_prior_curr - 0.5*(pb)'*matpower(diag
(prior_tau),-1)*(pb);
        [log_curr_spec_dens] = whittle_like(yobs_tmp,Beta_curr_temp(:,:,jj));
        %Loglikelihood at proposed values
        loglike_curr = loglike_curr + log_curr_spec_dens;

        %prior density for smoothing parameters
        log_tau_prior_curr = log_tau_prior_curr - length(select)*log
(tau_up_limit);
    end

    log_Phi_curr = -log(2^(dimen^2)-1); %proposal for component choice
    log_Phi_prior_curr = -log(2^(dimen^2)); %prior for component choice

    %Calculate Log proposal density at current values
    log_proposal_curr = log_move_curr + log_Beta_curr + log_Phi_curr;

    %Calculate Priors at Current Vlaues
    log_prior_curr = log_Beta_prior_curr + log_tau_prior_curr + log_Phi_prior_curr;
    j=j+1;
  else
```

```
      xi_prop(k) = xi_curr_temp(j);
              tau_prop(:,k) = tau_curr_temp(:,j);
              nseg_prop(k) = nseg_curr_temp(j);
      Beta_prop(:,:,k) = Beta_curr_temp(:,:,j);
      Phi_prop(k) = Phi_curr_temp(j);
  end
end

%==================================================
%Evaluate Target density at proposed values
%==================================================
log_prior_cut_prop=0;
for k=1:nexp_prop-1
       if k==1
              log_prior_cut_prop=-log(nobs-(nexp_prop-k+1)*tmin+1);
       else
              log_prior_cut_prop=log_prior_cut_prop-log(nobs-xi_prop(k-1)-
(nexp_prop-k+1)*tmin+1);
       end
end
log_target_prop = loglike_prop + log_prior_prop + log_prior_cut_prop;

%==================================================
%Evaluate Target density at current values
%==================================================
log_prior_cut_curr=0;
for k=1:nexp_curr-1
       if k==1
              log_prior_cut_curr=-log(nobs-(nexp_curr-k+1)*tmin+1);
       else
              log_prior_cut_curr=log_prior_cut_curr-log(nobs-xi_curr_temp(k-1)-
(nexp_curr-k+1)*tmin+1);
       end
end
log_target_curr = loglike_curr + log_prior_curr + log_prior_cut_curr;

%**********************************************************
%Calculations acceptance probability
%**********************************************************
PI = min(1,exp(log_target_prop - log_target_curr + ...
       log_proposal_curr - log_proposal_prop + log_jacobian));

function [gr] = Gradient1(yobs_tmp, chol_index, Phi_temp, tau_temp,...
                       Beta_temp, sigmasqalpha, nbasis)
global dimen
%%%%%%%%%%%%%%%%%%%%%%%%%%%%%%%%%%%%%%%%%%%%
% Calculate log gradients for coefficients selected to be different
%
%  Input:
%      1) x - initial values for coefficient of basis functions need to
%      be optimized
%      2) yobs_tmp - time series data within the segment
%      3) chol_index - index matrix
%      4) Phi_temp - which component changed
%      5) tau_temp - smoothing parameters
%      6) Beta_temp - current coefficients
%      7) sigmasqalpha - smoothing parameters for the constant in real
%      components
%      8) nbasis - number of basis functions used
```

```
%   Main Outputs:
%      2) gr - gradients for optimization process
%
%   Required programs: lin_basis_func
%%%%%%%%%%%%%%%%%%%%%%%%%%%%%%%%%%%%%%%%%%%%

%initilize Beta_1 and Beta_2: Beta_2 is for imaginary components
nBeta = nbasis + 1;

Beta_1(:,:) = Beta_temp(:,1:(dimen + dimen*(dimen-1)/2));
Beta_2(:,:) = Beta_temp(1:nBeta,(dimen + dimen*(dimen-1)/2 + 1):end);

dim = size(yobs_tmp); n = dim(1);
nfreq = floor(n/2); tt = (0:nfreq)/(2*nfreq);
yy = fft(yobs_tmp)/sqrt(n); y = yy(1:(nfreq+1),:); nf = length(y);
[xx_r, xx_i]=lin_basis_func(tt);

%theta's
theta = zeros(dimen*(dimen-1)/2,nf);
for i=1:dimen*(dimen-1)/2
  theta_real = xx_r * Beta_1(:,i+dimen);
  theta_imag = xx_i * Beta_2(:,i);
  theta(i,:) = theta_real + sqrt(-1)*theta_imag;
end
%delta's
delta_sq = zeros(dimen,nf);
for i=1:dimen
   delta_sq(i,:) = exp(xx_r * Beta_1(:,i));
end

if dimen==2  %Bivariate Time Series

   gr1 = zeros(nBeta,1); gr2 = zeros(nBeta,1); gr3 = zeros(nBeta,1); gr4 = zeros(nBeta,1);
   gr1(1) = Beta_1(1,1)/sigmasqalpha; gr1(2:nBeta,1) = Beta_1(2:nBeta,1)/tau_temp(1);
   gr2(1) = Beta_1(1,2)/sigmasqalpha; gr2(2:nBeta,1) = Beta_1(2:nBeta,2)/tau_temp(2);
   gr3(1) = Beta_1(1,3)/sigmasqalpha; gr3(2:nBeta,1) = Beta_1(2:nBeta,3)/tau_temp(3);
   gr4(1:nBeta,1) = Beta_2(1:nBeta,1)/tau_temp(4);

  if (mod(n,2)==1)
     %%%%%%%%%%%%%%%%%%%%%%%%
     %gradient
     %%%%%%%%%%%%%%%%%%%%%%%%
     rk = -y(2:end,1).*conj(y(2:end,2)) - y(2:end,2).*conj(y(2:end,1));
     ik = sqrt(-1)*(-y(2:end,1).*conj(y(2:end,2)) + y(2:end,2).*conj(y(2:end,1)));
     ck = 2*abs(y(2:end,1)).^2;

     gr1 = gr1 + xx_r(2:end,:)'*(1-abs(y(2:end,1)).^2.*exp(-xx_r(2:end,:)*Beta_1
(:,1))) + ...
             0.5*(xx_r(1,:)'*(1-abs(y(1,1)).^2.*exp(-xx_r(1,:)*Beta_1(:,1))));
     gr2 = gr2 + xx_r(2:end,:)'*(1 - abs(y(2:end,2)-theta(2:end).'.*y(2:end,1)).
^2.*exp(-xx_r(2:end,:)*Beta_1(:,2))) + ...
             0.5*(xx_r(1,:)'*(1 - abs(y(1,2)-theta(1).'.*y(1,1)).^2.*exp(-xx_r
(1,:)*Beta_1(:,2))));
     temp_mat_31 = bsxfun(@times, xx_r(2:end,:),rk);
     temp_mat_32 = bsxfun(@times, ck, bsxfun(@times,
xx_r(2:end,:),xx_r(2:end,:)*Beta_1(:,3)));
     gr3 = gr3 + sum( bsxfun(@times, (temp_mat_31+temp_mat_32), exp(-xx_r(2:
end,:)*Beta_1(:,2)) ))' +...
             0.5*(exp(-xx_r(1,:)*Beta_1(:,2))*(-y(1,1).*conj(y(1,2)) - y(1,2).
*conj(y(1,1)))*xx_r(1,:)' +...
```

```
            exp(-xx_r(1,:)*Beta_1(:,2))*2*abs(y(1,1)).^2*(xx_r(1,:)*Beta_1(:,3))
*xx_r(1,:)');
      temp_mat_41 = bsxfun(@times, ik, xx_i(2:end,:));
      temp_mat_42 = bsxfun(@times, ck, bsxfun(@times,xx_i(2:end,:),xx_i(2:end,:)
*Beta_2(:,1)));
      gr4 = gr4 + sum( bsxfun(@times, (temp_mat_41 + temp_mat_42), exp(-xx_r(2:
end,:)*Beta_1(:,2))))' + ...
            0.5*(exp(-xx_r(1,:)*Beta_1(:,2))*(sqrt(-1)*(-y(1,1).*conj(y(1,2)) +
y(1,2).*conj(y(1,1))))*xx_i(1,:)' +...
            exp(-xx_r(1,:)*Beta_1(:,2))*2*abs(y(1,1)).^2*(xx_i(1,:)
*Beta_2(:,1))*xx_i(1,:)');
   else
       %%%%%%%%%%%%%%%%%%%%%%%%%
       %gradient
       %%%%%%%%%%%%%%%%%%%%%%%%%
      rk = -y(2:nfreq,1).*conj(y(2:nfreq,2)) - y(2:nfreq,2).*conj(y(2:nfreq,1));
      ik = sqrt(-1)*(-y(2:nfreq,1).*conj(y(2:nfreq,2)) + y(2:nfreq,2).*conj(y(2:
nfreq,1)));
      ck = 2*abs(y(2:nfreq,1)).^2;

      gr1 = gr1 + xx_r(2:nfreq,:)'*(1-abs(y(2:nfreq,1)).^2.*exp(-xx_r(2:nfreq,:)
*Beta_1(:,1))) + ...
            0.5*(xx_r(1,:)'*(1-abs(y(1,1)).^2.*exp(-xx_r(1,:)*Beta_1(:,1))))+...
            0.5*(xx_r(end,:)'*(1-abs(y(end,1)).^2.*exp(-xx_r(end,:)*Beta_1(:,1))));
      gr2 = gr2 + xx_r(2:nfreq,:)'*(1 - abs(y(2:nfreq,2)-theta(2:nfreq).'.*y(2:
nfreq,1)).^2.*exp(-xx_r(2:nfreq,:)*Beta_1(:,2))) + ...
            0.5*(xx_r(1,:)'*(1 - abs(y(1,2)-theta(1).'.*y(1,1)).^2.*exp(-xx_r(1,:)
*Beta_1(:,2)))) + ...
            0.5*(xx_r(end,:)'*(1 - abs(y(end,2)-theta(end).'.*y(end,1)).^2.*exp
(-xx_r(end,:)*Beta_1(:,2))));
      temp_mat_31 = bsxfun(@times,rk, xx_r(2:nfreq,:));
      temp_mat_32 = bsxfun(@times,ck,bsxfun(@times, xx_r(2:nfreq,:),xx_r(2:
nfreq,:)*Beta_1(:,3)));
       gr3 = gr3 + sum( bsxfun(@times, (temp_mat_31 + temp_mat_32), exp(-xx_r(2:
nfreq,:)*Beta_1(:,2)) ))' +...
            0.5*(exp(-xx_r(1,:)*Beta_1(:,2))*(-y(1,1).*conj(y(1,2)) - y(1,2).
*conj(y(1,1)))*xx_r(1,:)' +...
            exp(-xx_r(1,:)*Beta_1(:,2))*2*abs(y(1,1)).^2*(xx_r(1,:)*Beta_1(:,3))
*xx_r(1,:)') +...
            0.5*(exp(-xx_r(end,:)*Beta_1(:,2))*(-y(end,1).*conj(y(end,2)) - y
(end,2).*conj(y(end,1)))*xx_r(end,:)' +...
            exp(-xx_r(end,:)*Beta_1(:,2))*2*abs(y(end,1)).^2*(xx_r(end,:)*Beta_1
(:,3))*xx_r(end,:)');
      temp_mat_41 = bsxfun(@times, ik, xx_i(2:nfreq,:));
      temp_mat_42 = bsxfun(@times, ck, bsxfun(@times,xx_i(2:nfreq,:),xx_i(2:
nfreq,:)*Beta_2(:,1)));
      gr4 = gr4 + sum( bsxfun(@times, (temp_mat_41 + temp_mat_42), exp(-xx_r(2:nfreq,:)
*Beta_1(:,2))))' + ...
            0.5*(exp(-xx_r(1,:)*Beta_1(:,2))*(sqrt(-1)*(-y(1,1).*conj(y(1,2))
+ y(1,2).*conj(y(1,1))))*xx_i(1,:)' +...
            exp(-xx_r(1,:)*Beta_1(:,2))*2*abs(y(1,1)).^2*(xx_i(1,:)*Beta_2(:,1))
*xx_i(1,:)') +...
            0.5*(exp(-xx_r(end,:)*Beta_1(:,2))*(sqrt(-1)*(-y(end,1).
*conj(y(end,2)) + y(end,2).*conj(y(end,1))))*xx_i(end,:)' +...
            exp(-xx_r(end,:)*Beta_1(:,2))*2*abs(y(end,1)).^2*(xx_i(end,:)
*Beta_2(:,1))*xx_i(end,:)');
   end
   gr = [gr1;gr2;gr3;gr4];
```

```
gr_index = (1:(4*nBeta)).*[kron(chol_index(Phi_temp,1:3),ones(nBeta,1)'),kron
(chol_index(Phi_temp,4),ones(nBeta,1)')];
  gr_index = gr_index(find(gr_index~=0));
  gr = gr(gr_index);

elseif dimen==3  %trivariate time series

  gr1 = zeros(nBeta,1); gr2 = zeros(nBeta,1); gr3 = zeros(nBeta,1); gr4 = zeros(nBeta,1);
  gr5 = zeros(nBeta,1); gr6 = zeros(nBeta,1); gr7 = zeros(nBeta,1); gr8 = zeros(nBeta,1);
  gr9 = zeros(nBeta,1);

  gr1(1) = Beta_1(1,1)/sigmasqalpha; gr1(2:nBeta) = Beta_1(2:nBeta,1)/tau_temp(1);
  gr2(1) = Beta_1(1,2)/sigmasqalpha; gr2(2:nBeta) = Beta_1(2:nBeta,2)/tau_temp(2);
  gr3(1) = Beta_1(1,3)/sigmasqalpha; gr3(2:nBeta) = Beta_1(2:nBeta,3)/tau_temp(3);
  gr4(1) = Beta_1(1,4)/sigmasqalpha; gr4(2:nBeta) = Beta_1(2:nBeta,4)/tau_temp(4);
  gr5(1) = Beta_1(1,5)/sigmasqalpha; gr5(2:nBeta) = Beta_1(2:nBeta,5)/tau_temp(5);
  gr6(1) = Beta_1(1,6)/sigmasqalpha; gr6(2:nBeta) = Beta_1(2:nBeta,6)/tau_temp(6);
  gr7(1:nBeta) = Beta_2(1:nBeta,1)/tau_temp(7);
  gr8(1:nBeta) = Beta_2(1:nBeta,2)/tau_temp(8);
  gr9(1:nBeta) = Beta_2(1:nBeta,3)/tau_temp(9);

  if (mod(n,2)==1)
      %%%%%%%%%%%%%%%%%%%%%%%%%
      %gradient
      %%%%%%%%%%%%%%%%%%%%%%%%%
      rk4 = -y(2:end,1).*conj(y(2:end,2)) - y(2:end,2).*conj(y(2:end,1));
      ck4 = 2*abs(y(2:end,1)).^2;
      rk5 = -y(2:end,1).*conj(y(2:end,3)) - y(2:end,3).*conj(y(2:end,1));
      ck5 = 2*abs(y(2:end,1)).^2;
      b = theta(3,:);
      dk5 =  y(2:end,2).*conj(y(2:end,1)).*(b(2:end).') + conj(y(2:end,2)).*y(2:
end,1).*conj(b(2:end).');
      rk6 = -y(2:end,2).*conj(y(2:end,3)) - y(2:end,3).*conj(y(2:end,2));
      ck6 = 2*abs(y(2:end,2)).^2;
      a = theta(2,:);
      dk6 =  y(2:end,1).*conj(y(2:end,2)).*(a(2:end).') + conj(y(2:end,1)).*y(2:
end,2).*conj(a(2:end).');
      ik7 = sqrt(-1)*(-y(2:end,1).*conj(y(2:end,2)) + y(2:end,2).*conj(y(2:end,1)));
      ck7 = 2*abs(y(2:end,1)).^2;
      ik8 = sqrt(-1)*(-y(2:end,1).*conj(y(2:end,3)) + y(2:end,3).*conj(y(2:end,1)));
      ck8 = 2*abs(y(2:end,1)).^2;
      dk8 = sqrt(-1)*(-y(2:end,2).*conj(y(2:end,1)).*(b(2:end).') + conj(y(2:
end,2).*conj(y(2:end,1)).*(b(2:end).'))) ;
      ik9 = sqrt(-1)*(-y(2:end,2).*conj(y(2:end,3)) + y(2:end,3).*conj(y(2:end,2)));
      ck9 = 2*abs(y(2:end,2)).^2;
      dk9 = sqrt(-1)*(-y(2:end,1).*conj(y(2:end,2)).*(a(2:end).') + conj(y(2:end,1).
*conj(y(2:end,2)).*(a(2:end).'))) ;

      gr1 = gr1 + xx_r(2:end,:)'*(1-abs(y(2:end,1)).^2.*exp(-xx_r(2:end,:)*Beta_1
(:,1))) + ...
              0.5*(xx_r(1,:)'*(1-abs(y(1,1)).^2.*exp(-xx_r(1,:)*Beta_1(:,1))));
      gr2 = gr2 + xx_r(2:end,:)'*(1 - abs(y(2:end,2)-theta(1,2:end).'.*y(2:
end,1)).^2.*exp(-xx_r(2:end,:)*Beta_1(:,2))) + ...
              0.5*(xx_r(1,:)'*(1 - abs(y(1,2)-theta(1,1).*y(1,1)).^2.*exp(-xx_r(1,:)
*Beta_1(:,2))));
      gr3 = gr3 + xx_r(2:end,:)'*(1 - abs(y(2:end,3) - theta(2,2:end).'.*y(2:
end,1) - theta(3,2:end).'.*y(2:end,2)).^2.*exp(-xx_r(2:end,:)*Beta_1(:,3)))+...
              0.5*(xx_r(1,:)'*(1 - abs(y(1,3) - theta(2,1).*y(1,1) - theta(3,1).*y
(1,2))^2.*exp(-xx_r(1,:)*Beta_1(:,3))));
      temp_mat_41 = bsxfun(@times, xx_r(2:end,:),rk4);
```

```
      temp_mat_42 = bsxfun(@times, ck4, bsxfun(@times, xx_r(2:end,:),xx_r(2:
end,:)*Beta_1(:,4)));
      gr4 = gr4 + sum(bsxfun(@times, (temp_mat_41+temp_mat_42), exp(-xx_r(2:end,:)
*Beta_1(:,2)) ))'+...
                0.5*(exp(-xx_r(1,:)*Beta_1(:,2))*(-y(1,1).*conj(y(1,2)) - y(1,2).*conj
(y(1,1)))*xx_r(1,:)' +...
                exp(-xx_r(1,:)*Beta_1(:,2))*2*abs(y(1,1)).^2*(xx_r(1,:)*Beta_1(:,4))
*xx_r(1,:)');
      temp_mat_51 = bsxfun(@times, xx_r(2:end,:),rk5);
      temp_mat_52 = bsxfun(@times, ck5, bsxfun(@times, xx_r(2:end,:),xx_r(2:
end,:)*Beta_1(:,5)));
      temp_mat_53 = bsxfun(@times, xx_r(2:end,:),dk5);
      gr5 = gr5 + sum(bsxfun(@times, (temp_mat_51 + temp_mat_52 + temp_mat_53), exp(-xx_r
(2:end,:)*Beta_1(:,3)) ))'+...
                0.5* ( exp(-xx_r(1,:)*Beta_1(:,3))*(-y(1,1).*conj(y(1,3)) - y(1,3).
*conj(y(1,1)))*xx_r(1,:)' +...
                exp(-xx_r(1,:)*Beta_1(:,3))*2*abs(y(1,1)).^2*(xx_r(1,:)*Beta_1(:,5))
*xx_r(1,:)'+...
                exp(-xx_r(1,:)*Beta_1(:,3))*(y(1,2).*conj(y(1,1)).*(b(1).') + conj(y
(1,2).*conj(y(1,1)).*(b(1).')))*xx_r(1,:)');
      temp_mat_61 = bsxfun(@times, xx_r(2:end,:),rk6);
      temp_mat_62 = bsxfun(@times, ck6, bsxfun(@times, xx_r(2:end,:),xx_r
(2:end,:)*Beta_1(:,6)));
      temp_mat_63 = bsxfun(@times, xx_r(2:end,:),dk6);
      gr6 = gr6 + sum(bsxfun(@times, (temp_mat_61 + temp_mat_62 + temp_mat_63),
exp(-xx_r(2:end,:)*Beta_1(:,3)) ))'+...
                0.5* (exp(-xx_r(1,:)*Beta_1(:,3))*(-y(1,2).*conj(y(1,3)) - y(1,3).
*conj(y(1,2)))*xx_r(1,:)' +...
                exp(-xx_r(1,:)*Beta_1(:,3))*2*abs(y(1,2)).^2*(xx_r(1,:)*Beta_1(:,6))
*xx_r(1,:)'+...
                exp(-xx_r(1,:)*Beta_1(:,3))*(y(1,1).*conj(y(1,2)).*(a(1).') + conj
(y(1,1).*conj(y(1,2)).*(a(1).')))*xx_r(1,:)');
                temp_mat_71 = bsxfun(@times, ik7, xx_i(2:end,:));
                temp_mat_72 = bsxfun(@times, ck7, bsxfun(@times,xx_i(2:end,:),xx_i
(2:end,:)*Beta_2(:,1)));
                gr7 = gr7 + sum(bsxfun(@times, (temp_mat_71 + temp_mat_72), exp(-xx_r
(2:end,:)*Beta_1(:,2))))' + ...
                0.5*(exp(-xx_r(1,:)*Beta_1(:,2))*(imag(-y(1,1).*conj(y(1,2)) + y(1,2).
*conj(y(1,1))))*xx_i(1,:)' +...
                exp(-xx_r(1,:)*Beta_1(:,2))*2*abs(y(1,1)).^2*(xx_i(1,:)*Beta_2(:,1))
*xx_i(1,:)');
      temp_mat_81 = bsxfun(@times, ik8, xx_i(2:end,:));
      temp_mat_82 = bsxfun(@times, ck8, bsxfun(@times,xx_i(2:end,:),
xx_i(2:end,:)*Beta_2(:,2)));
      temp_mat_83 = bsxfun(@times, xx_i(2:end,:),dk8);
      gr8 = gr8 + sum(bsxfun(@times, (temp_mat_81 + temp_mat_82 + temp_mat_83),
exp(-xx_r(2:end,:)*Beta_1(:,3))))' + ...
                0.5*(exp(-xx_r(1,:)*Beta_1(:,3))*(imag(-y(1,1).*conj(y(1,3)) + y
(1,3).*conj(y(1,1))))*xx_i(1,:)' +...
                exp(-xx_r(1,:)*Beta_1(:,3))*2*abs(y(1,1)).^2*(xx_i(1,:)*Beta_2(:,2))
*xx_i(1,:)'+...
                sqrt(-1)*exp(-xx_r(1,:)*Beta_1(:,3))*(-y(1,1).*conj(y(1,2)).
*(b(1).') + conj(y(1,1).*conj(y(1,2)).*(b(1).')))*xx_i(1,:)');
      temp_mat_91 = bsxfun(@times, ik9, xx_i(2:end,:));
      temp_mat_92 = bsxfun(@times, ck9, bsxfun(@times,xx_i(2:end,:),xx_i(2:end,:)
*Beta_2(:,3)));
      temp_mat_93 = bsxfun(@times, xx_i(2:end,:),dk9);
      gr9 = gr9 + sum(bsxfun(@times, (temp_mat_91 + temp_mat_92 + temp_mat_93),
exp(-xx_r(2:end,:)*Beta_1(:,3))))' + ...
                0.5 * (exp(-xx_r(1,:)*Beta_1(:,3))*(imag(-y(1,2).*conj(y(1,3)) + y
(1,3).*conj(y(1,2))))*xx_i(1,:)' +...
```

```
                exp(-xx_r(1,:)*Beta_1(:,3))*2*abs(y(1,2)).^2*(xx_i(1,:)*Beta_2(:,3))
*xx_i(1,:)'+...
                sqrt(-1)*exp(-xx_r(1,:)*Beta_1(:,3))*(-y(1,1).*conj(y(1,2)).*(a
(1).') + conj(y(1,1).*conj(y(1,2)).*(a(1).')))*xx_i(1,:)');
   else
    %%%%%%%%%%%%%%%%%%%%%%%%
    %gradient
    %%%%%%%%%%%%%%%%%%%%%%%%
        rk4 = -y(2:nfreq,1).*conj(y(2:nfreq,2)) - y(2:nfreq,2).*conj(y(2:nfreq,1));
        ck4 = 2*abs(y(2:nfreq,1)).^2;
        rk5 = -y(2:nfreq,1).*conj(y(2:nfreq,3)) - y(2:nfreq,3).*conj(y(2:nfreq,1));
        ck5 = 2*abs(y(2:nfreq,1)).^2;
        b = theta(3,:);
        dk5 =  y(2:nfreq,2).*conj(y(2:nfreq,1)).*(b(2:nfreq).') +
conj(y(2:nfreq,2)).*y(2:nfreq,1).*conj(b(2:nfreq).');
        rk6 = -y(2:nfreq,2).*conj(y(2:nfreq,3)) - y(2:nfreq,3).*conj(y(2:nfreq,2));
        ck6 = 2*abs(y(2:nfreq,2)).^2;
        a = theta(2,:);
        dk6 = y(2:nfreq,1).*conj(y(2:nfreq,2)).*(a(2:nfreq).') + conj(y(2:nfreq,1).
*conj(y(2:nfreq,2)).*(a(2:nfreq).'));
        ik7 = sqrt(-1)*(-y(2:nfreq,1).*conj(y(2:nfreq,2)) + y(2:nfreq,2).
*conj(y(2:nfreq,1)));
        ck7 = 2*abs(y(2:nfreq,1)).^2;
        ik8 = sqrt(-1)*(-y(2:nfreq,1).*conj(y(2:nfreq,3)) + y(2:nfreq,3).
*conj(y(2:nfreq,1)));
        ck8 = 2*abs(y(2:nfreq,1)).^2;
         dk8 = sqrt(-1)*(-y(2:nfreq,2).*conj(y(2:nfreq,1)).*(b(2:nfreq).') +
conj(y(2:nfreq,2).*conj(y(2:nfreq,1)).*(b(2:nfreq).')));
        ik9 = sqrt(-1)*(-y(2:nfreq,2).*conj(y(2:nfreq,3)) + y(2:nfreq,3).*conj(y(2:
nfreq,2)));
        ck9 = 2*abs(y(2:nfreq,2)).^2;
        dk9 = sqrt(-1)*(-y(2:nfreq,1).*conj(y(2:nfreq,2)).*(a(2:nfreq).') + conj(y
(2:nfreq,1).*conj(y(2:nfreq,2)).*(a(2:nfreq).')));

        gr1 = gr1 + xx_r(2:nfreq,:)'*(1-abs(y(2:nfreq,1)).^2.*exp(-xx_r(2:nfreq,:)
*Beta_1(:,1))) + ...
                0.5*(xx_r(1,:)'*(1-abs(y(1,1)).^2.*exp(-xx_r(1,:)*Beta_1(:,1))))+...
                0.5*(xx_r(end,:)'*(1-abs(y(end,1)).^2.*exp(-xx_r(end,:)*Beta_1(:,1))));
        gr2 = gr2 + xx_r(2:nfreq,:)'*(1 - abs(y(2:nfreq,2)-theta(1,2:nfreq).'.
*y(2:nfreq,1)).^2.*exp(-xx_r(2:nfreq,:)*Beta_1(:,2))) + ...
                0.5*(xx_r(1,:)'*(1 - abs(y(1,2)-theta(1,1).*y(1,1)).^2.*exp(-xx_r
(1,:)*Beta_1(:,2)))) +...
                0.5*(xx_r(end,:)'*(1 - abs(y(end,2)-theta(1,end).*y(end,1)).^2.
*exp(-xx_r(end,:)*Beta_1(:,2))));
        gr3 = gr3 + xx_r(2:nfreq,:)'*(1 - abs(y(2:nfreq,3) - theta(2,2:nfreq).'.*y
(2:nfreq,1) - theta(3,2:nfreq).'.*y(2:nfreq,2)).^2.*exp(-xx_r(2:nfreq,:)*Beta_1
(:,3)))+...
                0.5*(xx_r(1,:)'*(1 - abs(y(1,3) - theta(2,1).*y(1,1) - theta(3,1).
*y(1,2))^2.*exp(-xx_r(1,:)*Beta_1(:,3))))+...
                0.5*(xx_r(end,:)'*(1 - abs(y(end,3) - theta(2,end).*y(end,1) - theta
(3,end).*y(end,2))^2.*exp(-xx_r(end,:)*Beta_1(:,3))));
        temp_mat_41 = bsxfun(@times, xx_r(2:nfreq,:),rk4);
        temp_mat_42 = bsxfun(@times, ck4, bsxfun(@times, xx_r(2:nfreq,:),xx_r(2:
nfreq,:)*Beta_1(:,4)));
        gr4 = gr4 + sum(bsxfun(@times, (temp_mat_41+temp_mat_42), exp(-xx_r(2:nfreq,:)
*Beta_1(:,2)) ))' +...
                0.5*(exp(-xx_r(1,:)*Beta_1(:,2))*(-y(1,1).*conj(y(1,2)) - y(1,2).
*conj(y(1,1)))*xx_r(1,:)' +...
                exp(-xx_r(1,:)*Beta_1(:,2))*2*abs(y(1,1)).^2*(xx_r(1,:)
*Beta_1(:,4))*xx_r(1,:)')+...
```

```
                0.5*(exp(-xx_r(end,:)*Beta_1(:,2))*(-y(end,1).*conj(y(end,2)) -
y(end,2).*conj(y(end,1)))*xx_r(end,:)' +...
                exp(-xx_r(end,:)*Beta_1(:,2))*2*abs(y(end,1)).^2*(xx_r(end,:)
*Beta_1(:,4))*xx_r(end,:)');
        temp_mat_51 = bsxfun(@times, xx_r(2:nfreq,:),rk5);
        temp_mat_52 = bsxfun(@times, ck5, bsxfun(@times, xx_r(2:nfreq,:),xx_r(2:
nfreq,:)*Beta_1(:,5)));
        temp_mat_53 = bsxfun(@times, xx_r(2:nfreq,:),dk5);
        gr5 = gr5 + sum( bsxfun(@times, (temp_mat_51 + temp_mat_52 + temp_mat_53),
exp(-xx_r(2:nfreq,:)*Beta_1(:,3)) ))'+...
                0.5*(exp(-xx_r(1,:)*Beta_1(:,3))*(-y(1,1).*conj(y(1,3)) - y(1,3).
*conj(y(1,1)))*xx_r(1,:)'+...
                exp(-xx_r(1,:)*Beta_1(:,3))*2*abs(y(1,1)).^2*(xx_r(1,:)*Beta_1(:,5))
*xx_r(1,:)'+...
                exp(-xx_r(1,:)*Beta_1(:,3))*(y(1,2).*conj(y(1,1)).*(b(1).') +
conj(y(1,2).*conj(y(1,1)).*(b(1).')))*xx_r(1,:)')+...
                0.5*(exp(-xx_r(end,:)*Beta_1(:,3))*(-y(end,1).*conj(y(end,3)) - y
(end,3).*conj(y(end,1)))*xx_r(end,:)' +...
                exp(-xx_r(end,:)*Beta_1(:,3))*2*abs(y(end,1)).^2*(xx_r(end,:)
*Beta_1(:,5))*xx_r(end,:)'+...
                exp(-xx_r(end,:)*Beta_1(:,3))*(y(end,2).*conj(y(end,1)).
*(b(end).') + conj(y(end,2).*conj(y(end,1)).*(b(end).')))*xx_r(end,:)');
        temp_mat_61 = bsxfun(@times, xx_r(2:nfreq,:),rk6);
        temp_mat_62 = bsxfun(@times, ck6, bsxfun(@times, xx_r(2:nfreq,:),xx_r(2:
nfreq,:)*Beta_1(:,6)));
        temp_mat_63 = bsxfun(@times, xx_r(2:nfreq,:),dk6);
        gr6 = gr6 + sum(bsxfun(@times, (temp_mat_61 + temp_mat_62 + temp_mat_63),
exp(-xx_r(2:nfreq,:)*Beta_1(:,3)) ))'+...
                0.5*(exp(-xx_r(1,:)*Beta_1(:,3))*(-y(1,2).*conj(y(1,3)) - y(1,3).
*conj(y(1,2)))*xx_r(1,:)'+...
                exp(-xx_r(1,:)*Beta_1(:,3))*2*abs(y(1,2)).^2*(xx_r(1,:)
*Beta_1(:,6))*xx_r(1,:)'+...
                exp(-xx_r(1,:)*Beta_1(:,3))*(y(1,1).*conj(y(1,2)).
*(a(1).') + conj(y(1,1).*conj(y(1,2)).*(a(1).')))*xx_r(1,:)')+...
                0.5*(exp(-xx_r(end,:)*Beta_1(:,3))*(-y(end,2).*conj(y(end,3))
- y(end,3).*conj(y(end,2)))*xx_r(end,:)' +...
                exp(-xx_r(end,:)*Beta_1(:,3))*2*abs(y(end,2)).^2*(xx_r(end,:)
*Beta_1(:,6))*xx_r(end,:)'+...
                exp(-xx_r(end,:)*Beta_1(:,3))*(y(end,1).*conj(y(end,2)).*(a(end).')
+ conj(y(end,1).*conj(y(end,2)).*(a(end).')))*xx_r(end,:)');
        temp_mat_71 = bsxfun(@times, ik7, xx_i(2:nfreq,:));
        temp_mat_72 = bsxfun(@times, ck7, bsxfun(@times,xx_i(2:nfreq,:),xx_i(2:
nfreq,:)*Beta_2(:,1)));
        gr7 = gr7 + sum(bsxfun(@times, (temp_mat_71 + temp_mat_72), exp(-xx_r(2:nfreq,:)
*Beta_1(:,2))))' +...
                0.5*(exp(-xx_r(1,:)*Beta_1(:,2))*(imag(-y(1,1).*conj(y(1,2)) + y
(1,2).*conj(y(1,1))))*xx_i(1,:)' +...
                exp(-xx_r(1,:)*Beta_1(:,2))*2*abs(y(1,1)).^2*(xx_i(1,:)*Beta_2(:,1))
*xx_i(1,:)')+...
                0.5*(exp(-xx_r(end,:)*Beta_1(:,2))*(imag(-y(end,1).*conj(y(end,2))
+ y(end,2).*conj(y(end,1))))*xx_i(end,:)' +...
                exp(-xx_r(end,:)*Beta_1(:,2))*2*abs(y(end,1)).^2*(xx_i(end,:)
*Beta_2(:,1))*xx_i(end,:)');
        temp_mat_81 = bsxfun(@times, ik8, xx_i(2:nfreq,:));
        temp_mat_82 = bsxfun(@times, ck8, bsxfun(@times,xx_i(2:nfreq,:),xx_i
(2:nfreq,:)*Beta_2(:,2)));
        temp_mat_83 = bsxfun(@times, xx_i(2:nfreq,:),dk8);
```

```
        gr8 = gr8 + sum(bsxfun(@times, (temp_mat_81 + temp_mat_82 + temp_mat_83),
exp(-xx_r(2:nfreq,:)*Beta_1(:,3))))' + ...
              0.5 * (exp(-xx_r(1,:)*Beta_1(:,3))*(imag(-y(1,1).*conj(y(1,3)) + y
(1,3).*conj(y(1,1))))*xx_i(1,:)' +...
              exp(-xx_r(1,:)*Beta_1(:,3))*2*abs(y(1,1)).^2*(xx_i(1,:)*Beta_2(:,2))
*xx_i(1,:)'+...
              sqrt(-1)*exp(-xx_r(1,:)*Beta_1(:,3))*(-y(1,1).*conj(y(1,2)).
*(b(1).') + conj(y(1,1).*conj(y(1,2)).*(b(1).')))*xx_i(1,:)')+...
              0.5 * (exp(-xx_r(end,:)*Beta_1(:,3))*(imag(-y(end,1).
*conj(y(end,3)) + y(end,3).*conj(y(end,1))))*xx_i(end,:)' +...
              exp(-xx_r(end,:)*Beta_1(:,3))*2*abs(y(end,1)).^2*(xx_i(end,:)
*Beta_2(:,2))*xx_i(end,:)'+...
              sqrt(-1)*exp(-xx_r(end,:)*Beta_1(:,3))*(-y(end,1).*conj(y(end,2)).
*(b(end).') + conj(y(end,1).*conj(y(end,2)).*(b(end).')))*xx_i(end,:)');
        temp_mat_91 = bsxfun(@times, ik9, xx_i(2:nfreq,:));
        temp_mat_92 = bsxfun(@times, ck9, bsxfun(@times,xx_i(2:nfreq,:),xx_i(2:
nfreq,:)*Beta_2(:,3)));
        temp_mat_93 = bsxfun(@times, xx_i(2:nfreq,:),dk9);
        gr9 = gr9 + sum(bsxfun(@times, (temp_mat_91 + temp_mat_92 + temp_mat_93),
exp(-xx_r(2:nfreq,:)*Beta_1(:,3))))' + ...
              0.5 * (exp(-xx_r(1,:)*Beta_1(:,3))*(imag(-y(1,2).*conj(y(1,3))
+ y(1,3).*conj(y(1,2))))*xx_i(1,:)' +...
              exp(-xx_r(1,:)*Beta_1(:,3))*2*abs(y(1,2)).^2*(xx_i(1,:)
*Beta_2(:,3))*xx_i(1,:)'+...
              sqrt(-1)*exp(-xx_r(1,:)*Beta_1(:,3))*(-y(1,1).*conj(y(1,2)).
*(a(1).') + conj(y(1,1).*conj(y(1,2)).*(a(1).')))*xx_i(1,:)')+...
              0.5 * (exp(-xx_r(end,:)*Beta_1(:,3))*(imag(-y(end,2).*conj(y(end,3))
+ y(end,3).*conj(y(end,2))))*xx_i(end,:)' +...
              exp(-xx_r(end,:)*Beta_1(:,3))*2*abs(y(end,2)).^2*(xx_i(end,:)
*Beta_2(:,3))*xx_i(end,:)'+...
              sqrt(-1)*exp(-xx_r(end,:)*Beta_1(:,3))*(-y(end,1).*conj(y(end,2)).
*(a(end).') + conj(y(end,1).*conj(y(end,2)).*(a(end).')))*xx_i(end,:)');
    end
    gr = [gr1;gr2;gr3;gr4;gr5;gr6;gr7;gr8;gr9];
    gr_index = (1:(dimen^2*nBeta)).*kron(chol_index(Phi_temp,:),ones(nBeta,1)');
    gr_index = gr_index(find(gr_index~=0));
    gr = gr(gr_index);
end

gr=-gr;

function [gr] = Gradient2(yobs_tmp, chol_index, Phi_temp, tau_temp_1,...
              tau_temp_2, Beta_temp_1, Beta_temp_2, sigmasqalpha, nbasis, nseg)

%%%%%%%%%%%%%%%%%%%%%%%%%%%%%%%%%%%%%%%%%%%%%%%%%%%%%%%%%%%%%%%%%%%%%%%%%%%%%%
% Calculate log gradients for coefficients selected to be then same
%
% Input:
%     1) x - initial values for coefficient of basis functions need to
%     be optimized
%     2) yobs_tmp - time series data within the segment
%     3) chol_index - index matrix
%     4) Phi_temp - which component changed
%     5) tau_temp_1 - smoothing parameters for the first segment
%     6) tau_temp_2 - smoothing parameters for the second segment
%     7) Beta_temp_1 - current coefficients for the first segment
%     8) Beta_temp_2 - current coefficients for the second segment
%     9) sigmasqalpha - smoothing parameters for the constant in real
%     components
```

```
%     10) nbasis - number of basis functions used
%     11) nseg - number of observations in the first segment
%   Main Outputs:
%     2) gr - gradients for optimization process
%
%   Required programs: lin_basis_func, Beta_derive1
%%%%%%%%%%%%%%%%%%%%%%%%%%%%%%%%%%%%%%%%%%%%%%%%%%%%%%%%%%%%%%%%%%%%%%%%%%%%%%%%

yobs_tmp_1 = yobs_tmp(1:nseg,:);
yobs_tmp_2 = yobs_tmp(nseg+1:end,:);
[grad1] = Gradient1(yobs_tmp_1, chol_index, Phi_temp, tau_temp_1, Beta_temp_1,
sigmasqalpha, nbasis);
[grad2] = Gradient1(yobs_tmp_2, chol_index, Phi_temp, tau_temp_2, Beta_temp_2,
sigmasqalpha, nbasis);

gr = grad1 + grad2;


function [Beta_out, m_out, m, yobs_tmp]=Hamilt1(chol_index, Phi_temp,...
                      j, yobs, tau_temp, Beta_temp, xi_temp)
%%%%%%%%%%%%%%%%%%%%%%%%%%%%%%%%%%%%%%%%%%%%%%%%%%%%%%%%%%%%%%%%%%%%%%%%%%%%%%%%
% Does HMC updates for the coefficents that are different
%
%   Input:
%       1) chol_index - index matrix
%       2) Phi_temp - indicate variable that determine which coefficient
%       should be different
%       3) j - one of two segments
%       4) yobs - time series observations
%       5) tau_temp - current smoothing parameters
%       6) Beta_temp - current coefficients
%       7) xi__temp - current partitions
%   Main Outputs:
%       1) Beta_out - vector of coefficients
%       2) m_out - updated momentum variables
%       3) m - initial momentum variables
%       4) yobs_tmp - time series observations in selected segment
%
%   Required programs:  Gradient1
%%%%%%%%%%%%%%%%%%%%%%%%%%%%%%%%%%%%%%%%%%%%%
  global nbasis sigmasqalpha dimen nBeta M ee

  %pick right portion of the data
  if j>1
      yobs_tmp = yobs((xi_temp(j-1)+1):xi_temp(j),:);
  else
      yobs_tmp = yobs(1:xi_temp(j),:);
  end

  select = chol_index(Phi_temp,:).*(1:dimen^2);
  select = select(select~=0);


  ll = nBeta*length(select);
  [gr]=Gradient1(yobs_tmp, chol_index, Phi_temp, tau_temp,...
                      Beta_temp, sigmasqalpha, nbasis);
  Beta_old = reshape(Beta_temp(:,select),ll,1);
  Beta_out = Beta_old;
```

```
% generate momentum variable
m = mvnrnd(zeros(ll,1),M*eye(ll))';
% determine leap number and step
stepsize = unifrnd(0,2*ee);
leap = randsample(1:2*ceil(1/ee),1);
%leap = randsample(1:(1/stepsize),1);

m_out = m + 0.5*gr*stepsize;
for i=1:leap
    Beta_out = Beta_out + stepsize*(1/M)*eye(ll)*m_out;
    Beta_temp(:,select) = reshape(Beta out,nBeta,length(select));
    if i==leap
        [gr] = Gradient1(yobs_tmp, chol_index, Phi_temp, tau_temp,...
                        Beta_temp, sigmasqalpha, nbasis);
        m_out = m_out + 0.5*gr*stepsize;
    else
        [gr] = Gradient1(yobs_tmp, chol_index, Phi_temp, tau_temp,...
                        Beta_temp, sigmasqalpha, nbasis);
        m_out = m_out + 1*gr*stepsize;
    end
end
m_out = -m_out;


function[Beta_out, m_out, m, yobs_tmp] = Hamilt2(chol_index, Phi_temp, j,
yobs,...
            tau_temp_1, tau_temp_2, Beta_temp_1, Beta_temp_2, xi_temp, nseg)

%%%%%%%%%%%%%%%%%%%%%%%%%%%%%%%%%%%%%%%%%%%%%%%%%%%%%%%%%%%%%%%%%%%%%%%%%%%%%%%%
% Does HMC updates for the coefficents that are the same
%
%  Input:
%      1) chol_index - index matrix
%      2) Phi_temp - indicate variable that determine which coefficient
%      should be different
%      3) j - the first segment
%      4) yobs - time series observations
%      5) tau_temp_1 - current smoothing parameters for the first segment
%      6) tau_temp_2 - current smoothing parameters for the second
%      segment
%      7) Beta_temp_1 - current coefficients for the first segment
%      8) Beta_temp_2 - current coefficients for the second segment
%      9) xi__temp - current partitions
%      10) nseg - number of observation in the first segment
%  Main Outputs:
%      1) Beta_out - vector of coefficients
%      2) m_out - updated momentum variables
%      3) m - initial momentum variables
%      4) yobs_tmp - time series observations in selected segment
%
%  Required programs: Gradient2
%%%%%%%%%%%%%%%%%%%%%%%%%%%%%%%%%%%%%%%%%%%%%%%%%%%%%%%%%%%%%%%%%%%%%%%%%%%%%%%%
global dimen nbasis sigmasqalpha nBeta M ee


%pick right portion of the data
if j==1
    yobs_tmp = yobs(1:xi_temp(j+1),:);
```

```
    elseif j==length(xi_temp)-1
        yobs_tmp = yobs(xi_temp(j-1)+1:end,:);
    else
        yobs_tmp = yobs(xi_temp(j-1)+1:xi_temp(j+1),:);
    end

    aa = zeros(2^(dimen^2),1);
    for i=1:2^(dimen^2)
        aa(i)=sum(chol_index(i,:)~=chol_index(Phi_temp,:));
    end
    Phi_inv = find(aa==dimen^2);

    select = chol_index(Phi_inv,:).*(1:dimen^2);
    select = select(select~=0);

    ll = nBeta*length(select);
    [gr] = Gradient2(yobs_tmp, chol_index, Phi_inv, tau_temp_1,...
            tau_temp_2, Beta_temp_1, Beta_temp_2, sigmasqalpha, nbasis, nseg);
    Beta_old = reshape(Beta_temp_1(:,select),ll,1);
    Beta_out = Beta_old;
    % generate momentum variable
    m = mvnrnd(zeros(ll,1),M*eye(ll))';
    % determine leap number and stepsize
    stepsize = unifrnd(0,2*ee);
    leap = randsample(1:2*ceil(1/ee),1);

    m_out = m + 0.5*gr*stepsize;
    for i=1:leap
        Beta_out = Beta_out + stepsize*(1/M)*eye(ll)*m_out;
        Beta_temp_1(:,select) = reshape(Beta_out,nBeta,length(select));
        Beta_temp_2(:,select) = reshape(Beta_out,nBeta,length(select));
        if i==leap
            [gr] = Gradient2(yobs_tmp, chol_index, Phi_inv, tau_temp_1,...
                    tau_temp_2, Beta_temp_1, Beta_temp_2, sigmasqalpha, nbasis, nseg);
            m_out = m_out + 0.5*gr*stepsize;
        else
            [gr] = Gradient2(yobs_tmp, chol_index, Phi_inv, tau_temp_1,...
                    tau_temp_2, Beta_temp_1, Beta_temp_2, sigmasqalpha, nbasis,
nseg);
            m_out = m_out + 1*gr*stepsize;
        end
    end
    m_out = -m_out;

function [xx_r,xx_i]= lin_basis_func(freq_hat)% freq are the frequencies

%%%%%%%%%%%%%%%%%%%%%%%%%%%%%%%%%%%%%%%%%%%%%%%%%%%%%%%%%%%%%%%%%%%%%%%%%%%%%%%
% Produces linear basis functions
%
%   Input:
%       1) xx_r - linear basis function for real Cholesky components
%       2) ts - linear basis function for imaginary Cholesky components
%   Main Outputs:
%       1) freq_hat - frequencies used
%
%%%%%%%%%%%%%%%%%%%%%%%%%%%%%%%%%%%%%%%%%%%%%%%%%%%%%%%%%%%%%%%%%%%%%%%%%%%%%%%
    global nBeta dimen
    nfreq_hat=length(freq_hat);
    xx_r = ones((nfreq_hat),nBeta);
    xx_i = ones((nfreq_hat),nBeta);
    for j=2:nBeta
```

```
        xx_r(:,j) = sqrt(2)* cos(2*pi*(j-1)*freq_hat)/(2*pi*(j-1));
    end
    for j =1:nBeta
        xx_i(:,j) = sqrt(2)*sin(2*pi*j*freq_hat)/(2*pi*j);
    end

function[ai] = matpower(a,alpha)
%%%%%%%%%%%%%%%%%%%%%%%%%%%%%%%%%%%%%%%%%%%%%%%%%%%%%%%%%%%%%%%%%%%%%%%%%%%%%%%%
% Calculate matrix power
%
%   Input:
%       1) a - a matrix
%       2) alpha - desired power
%   Main Outputs:
%       1) ai - matrix with desired power
%
%%%%%%%%%%%%%%%%%%%%%%%%%%%%%%%%%%%%%%%%%
       small = .000001;
       if numel(a) == 1
           ai=a^alpha;
       else
          [p1, ~] = size(a);
          [eve, eva] = eig(a);
           eva = diag(eva);
           eve = eve./(repmat((diag((eve)'*eve).^0.5),1,p1));
           index = 1:p1;
           index = index(eva>small);
           evai = eva;
           evai(index) = (eva(index)).^(alpha);
           ai = eve*diag(evai)*(eve)';
       end
end

function[Beta_mean,Beta_var,yobs_tmp] = postBeta1(chol_index, Phi_temp,...
                        j, yobs, tau_temp, Beta_temp, xi_temp)
%%%%%%%%%%%%%%%%%%%%%%%%%%%%%%%%%%%%%%%%%%%%%%%%%%%%%%%%%%%%%%%%%%%%%%%%%%%%%%%%
% Calculate the mean and variance of normal approximation for coefficient
% of basis functions that are different across segments
%
%   Input:
%       1) chol_index - index matrix
%       2) Phi_temp - indicate variable that determine which coefficient
%       should be different
%       3) j - one of two segments
%       4) yobs - time series observations
%       5) tau_temp - current smoothing parameters
%       6) Beta_temp - current coefficients
%       7) xi__temp - current partitions
%   Main Outputs:
%       1) Beta_mean - mean of approximated distribution
%       2) Beta_var - variance of approximated distribution
%       3) yobs_tmp - time series observations in selected segment
%
%   Required programs:  Beta_derive1
%%%%%%%%%%%%%%%%%%%%%%%%%%%%%%%%%%%%%%%%%%%%%%%%%%%%%%%%%%%%%%%%%%%%%%%%%%%%%%%%

  global nbasis nBeta sigmasqalpha dimen options
```

```
  %var_inflate_1 is for variance inflation for first real part of Cholesky components;
  %var_inflate_1 is for variance inflation for rest of components

  %pick right portion of the data
  if j>1
       yobs_tmp = yobs((xi_temp(j-1)+1):xi_temp(j),:);
  else
       yobs_tmp = yobs(1:xi_temp(j),:);
  end

  %provide initial values
  x = zeros(chol_index(Phi_temp,:)*repmat(nBeta,dimen^2,1),1);

  %optimization process
  [Beta_mean,~,~,~,~,Beta_inv_var] = fminunc(@Beta_derive1, x, options, ...
           yobs_tmp, chol_index, Phi_temp, tau_temp, Beta_temp, sigmasqalpha, nbasis);
  %Beta_var = Beta_inv_var\eye(size(Beta_inv_var));
  Beta_var = matpower(Beta_inv_var,-1);

function[Beta_mean,Beta_var,yobs_tmp] = postBeta2(chol_index, Phi_temp, j, yobs,...
           tau_temp_1, tau_temp_2, Beta_temp_1, Beta_temp_2, xi_temp, nseg)
  global dimen options nbasis nBeta sigmasqalpha

%%%%%%%%%%%%%%%%%%%%%%%%%%%%%%%%%%%%%%%%%%%%%%%%%%%%%%%%%%%%%%%%%%%%%%%%%%%%%%%
% Calculate the mean and variance of normal approximation for coefficient
% of basis functions that are the same across segments
%
%   Input:
%       1) chol_index - index matrix
%       2) Phi_temp - indicate variable that determine which coefficient
%       should be different
%       3) j - the first segment
%       4) yobs - time series observations
%       5) tau_temp_1 - current smoothing parameters for the first segment
%       6) tau_temp_2 - current smoothing parameters for the second
%        segment
%       7) Beta_temp_1 - current coefficients for the first segment
%       8) Beta_temp_2 - current coefficients for the second segment
%       9) xi__temp - current partitions
%       10) nseg - number of observation in the first segment
%   Main Outputs:
%       1) Beta_mean - mean of approximated distribution
%       2) Beta_var - variance of approximated distribution
%       3) yobs_tmp - time series observations in selected segment
%
%   Required programs:  Beta_derive2
%%%%%%%%%%%%%%%%%%%%%%%%%%%%%%%%%%%%%%%%%%%%%%%%%%%%%%%%%%%%%%%%%%%%%%%%%%%%%%%

  %pick right portion of the data
  if j==1
     yobs_tmp = yobs(1:xi_temp(j+1),:);
  elseif j==length(xi_temp)-1
     yobs_tmp = yobs(xi_temp(j-1)+1:end,:);
  else
     yobs_tmp = yobs(xi_temp(j-1)+1:xi_temp(j+1),:);
  end
```

```
    aa = zeros(2^(dimen^2),1);
    for i=1:2^(dimen^2)
        aa(i)=sum(chol_index(i,:)~=chol_index(Phi_temp,:));
    end
    Phi_inv = find(aa==dimen^2);

    %provide initial values
    x = zeros(chol_index(Phi_inv,:)*repmat(nBeta,dimen^2,1),1);

    %optimization process
    [Beta_mean,~,~,~,~,Beta_inv_var] = fminunc(@Beta_derive2, x, options,
yobs_tmp, chol_index, Phi_inv,...
                                    tau_temp_1, tau_temp_2, Beta_temp_1, Beta_temp_2,
sigmasqalpha, nbasis, nseg);
    %Beta_var = Beta_inv_var\eye(size(Beta_inv_var));
    Beta_var = matpower(Beta_inv_var,-1);

function [log_whittle] = whittle_like(yobs_tmp, Beta)

%%%%%%%%%%%%%%%%%%%%%%%%%%%%%%%%%%%%%%%%%%%%%%%%%%%%%%%%%%%%%%%%%%%%%%%%%%%%%%%%
% Calculate local Whittle likelihood
%
%   Input:
%       1) yobs_tmp - time series in the segment
%       2) Beta - coefficient of basis functions
%   Main Outputs:
%       1) log_white - log local whittle likelihood
%   Require programs: lin_basis_function
%%%%%%%%%%%%%%%%%%%%%%%%%%%%%%%%%%%%%%%%%%%%%%%%%%%%%%%%%%%%%%%%%%%%%%%%%%%%%%%%

global dimen
Beta_1 = Beta(:,1:(dimen + dimen*(dimen-1)/2));
Beta_2 = Beta(:,(dimen + dimen*(dimen-1)/2 + 1): end);

dim = size(yobs_tmp);
n = dim(1);
nfreq = floor(n/2);
tt = (0:nfreq)/(2*nfreq);
yy = fft(yobs_tmp)/sqrt(n);
y = yy(1:(nfreq+1),:);
nf = length(y);

[xx_r, xx_i]=lin_basis_func(tt);

%theta's
theta = zeros(dimen*(dimen-1)/2,nf );
for i=1:(dimen*(dimen-1)/2)
  theta_real = xx_r * Beta_1(:,i+dimen);
  theta_imag = xx_i * Beta_2(:,i);
  theta(i,:) = theta_real + sqrt(-1)*theta_imag;
end

%delta's
delta_sq = zeros(dimen,nf );
for i=1:dimen
  delta_sq(i,:) = exp(xx_r * Beta_1(:,i));
end
```

```
if dimen==2
   if (mod(n,2)==1) %odd n
         log_whittle = -sum(log(delta_sq(1,2:end))' + log(delta_sq(2,2:end))' + ...
                     conj(y(2:end,1)).*y(2:end,1).*exp(-xx_r(2:end,:)*Beta_1(:,1)) + ...
                     conj(y(2:end,2) - theta(2:end).'.*y(2:end,1)).*(y(2:end,2) - theta
(2:end).'.*y(2:end,1)).*exp(-xx_r(2:end,:)*Beta_1(:,2))) - ...
                     0.5*(log(delta_sq(1,1))' + log(delta_sq(2,1))' + conj(y(1,1)).
*y(1,1).*exp(-xx_r(1,:)*Beta_1(:,1)) - ...
                     conj(y(1,2) - theta(1).'.*y(1,1)).*(y(1,2) - theta(1).'.*y(1,1)).
*exp(-xx_r(1,:)*Beta_1(:,2)));
  else
      log_whittle = -sum(log(delta_sq(1,2:nfreq))' + log(delta_sq(2,2:nfreq))' + ...
                   conj(y(2:nfreq,1)).*y(2:nfreq,1).*exp(-xx_r(2:nfreq,:)
*Beta_1(:,1)) + ...
                   conj(y(2:nfreq,2) - theta(2:nfreq).'.*y(2:nfreq,1)).*(y(2:nfreq,2)
- theta(2:nfreq).'.*y(2:nfreq,1)).*exp(-xx_r(2:nfreq,:)*Beta_1(:,2))) - ...
                     0.5*(log(delta_sq(1,1)) + log(delta_sq(2,1)) + conj(y(1,1)).
*y(1,1).*exp(-xx_r(1,:)*Beta_1(:,1)) + ...
                     conj(y(1,2) - theta(1).'.*y(1,1)).*(y(1,2) - theta(1).'.*y(1,1)).
*exp(-xx_r(1,:)*Beta_1(:,2))) - ...
                     0.5*(log(delta_sq(1,end)) + log(delta_sq(2,end)) + conj(y(end,1)).
*y((nfreq+1),1).*exp(-xx_r(end,:)*Beta_1(:,1)) + ...
                     conj(y(end,2) - theta(end).'.*y(end,1)).*(y(end,2) - theta
(end).'.*y(end,1)).*exp(-xx_r(end,:)*Beta_1(:,2)));
  end
elseif dimen==3
  if (mod(n,2)==1) %odd n
      log_whittle = -sum(log(delta_sq(1,2:end))' + log(delta_sq(2,2:end))' + log
(delta_sq(3,2:end))' + ...
                     conj(y(2:end,1)).*y(2:end,1).*exp(-xx_r(2:end,:)*Beta_1(:,1)) + ...
                     conj(y(2:end,2) - theta(1,2:end).'.*y(2:end,1)).*(y(2:end,2) -
theta(1,2:end).'.*y(2:end,1)).*exp(-xx_r(2:end,:)*Beta_1(:,2))+...
                     conj(y(2:end,3) -(theta(2,2:end).'.*y(2:end,1)+theta(3,2:end).'.
*y(2:end,2))).*(y(2:end,3) -(theta(2,2:end).'.*y(2:end,1)+theta(3,2:end).'.
*y(2:end,2))).*...
                     exp(-xx_r(2:end,:)*Beta_1(:,3))) - ...
                     0.5*(log(delta_sq(1,1))' + log(delta_sq(2,1))' + log(delta_sq
(3,1))' + ...
                     conj(y(1,1)).*y(1,1).*exp(-xx_r(1,:)*Beta_1(:,1)) + ...
                     conj(y(1,2) - theta(1,1).'.*y(1,1)).*(y(1,2) - theta(1,1).'.
*y(1,1)).*exp(-xx_r(1,:)*Beta_1(:,2))+...
                     conj(y(1,3) -(theta(2,1).'.*y(1,1)+ theta(3,1).'.*y(1,2))).
*(y(1,3) -(theta(2,1).'.*y(1,1)+ theta(3,1).'.*y(1,2))).*...
                     exp(-xx_r(1,:)*Beta_1(:,3)));

  else
      log_whittle = -sum(log(delta_sq(1,2:nfreq))' + log(delta_sq(2,2:nfreq))' +
log(delta_sq(3,2:nfreq))' + ...
                     conj(y(2:nfreq,1)).*y(2:nfreq,1).*exp(-xx_r(2:nfreq,:)
*Beta_1(:,1)) + ...
                     conj(y(2:nfreq,2) - theta(1,2:nfreq).'.*y(2:nfreq,1)).
*(y(2:nfreq,2) - theta(1,2:nfreq).'.*y(2:nfreq,1)).*exp(-xx_r(2:nfreq,:)
*Beta_1(:,2))+...
                     conj(y(2:nfreq,3) -(theta(2,2:nfreq).'.*y(2:nfreq,1)+ theta(3,2:
nfreq).'.*y(2:nfreq,2))).*(y(2:nfreq,3) -(theta(2,2:nfreq).'.*y(2:nfreq,1)+ theta
(3,2:nfreq).'.*y(2:nfreq,2))).*...
                     exp(-xx_r(2:nfreq,:)*Beta_1(:,3))) - ...
                     0.5*(log(delta_sq(1,1))' + log(delta_sq(2,1))' + log(delta_sq
(3,1))' + ...
                     conj(y(1,1)).*y(1,1).*exp(-xx_r(1,:)*Beta_1(:,1)) + ...
                     conj(y(1,2) - theta(1,1).'.*y(1,1)).*(y(1,2) - theta(1,1).'.
*y(1,1)).*exp(-xx_r(1,:)*Beta_1(:,2))+...
```

```
                conj(y(1,3) -(theta(2,1).'.*y(1,1)+ theta(3,1).'.*y(1,2))).
*(y(1,3) -(theta(2,1).'.*y(1,1)+ theta(3,1).'.*y(1,2))).*...
                exp(-xx_r(1,:)*Beta_1(:,3)))-...
                0.5*(log(delta_sq(1,end))' + log(delta_sq(2,end))'
+ log(delta_sq(3,end))' + ...
                conj(y(end,1)).*y(end,1).*exp(-xx_r(end,:)*Beta_1(:,1)) + ...
                conj(y(end,2) - theta(1,end).'.*y(end,1)).*(y(end,2) - theta
(1,end).'.*y(end,1)).*exp(-xx_r(end,:)*Beta_1(:,2))+...
                conj(y(end,3) -(theta(2,end).'.*y(end,1)+theta(3,end).'.
*y(end,2))).*(y(end,3) -(theta(2,end).'.*y(end,1)+theta(3,end).'.*y(end,2))).*...
                exp(-xx_r(end,:)*Beta 1(:,3)));
    end
end

function[A,nseg_new,xi_prop,tau_prop,Beta_prop,Phi_prop,seg_temp]=...

within(chol_index,ts,nexp_temp,tau_temp,xi_curr_temp,nseg_curr_temp,
Beta_curr_temp,Phi_temp)
global nobs dimen nBeta M prob_mm1 tmin

%%%%%%%%%%%%%%%%%%%%%%%%%%%%%%%%%%%%%%%%%%%%%%%%%%%%%%%%%%%%%%%%%%%%%%%%%%%%%%
%  Does the within-model move in the paper
%
%   Input:
%       1) chol_index - index matrix
%       2) ts - TxN matrix of time series data
%       3) nexp_temp - number of segments
%       5) tau_temp -  vector of smoothing parameters
%       6) xi_curr_temp - current partitions
%       7) nseg_curr_temp - current number of observations in each segment
%       8) Beta_curr_temp - current vector of coefficients
%       9) Phi_temp - which component changed
%       10) Z - together with Phi_temp, indicating how components should
%       be update
%   Main Outputs:
%       1) PI - acceptance probability
%       2) nseg_new - new number of observations in each segment
%       3) xi_prop - proposed partitions
%       4) tau_prop - proposed smoothing parameters
%       5) Beta_prop - proposed coefficients
%       6) Phi_prop - proposed indicator variable
%
%   Required programs: postBeta1, postBeta2, Beta_derive1, Beta_derive2,
%   whittle_like
%%%%%%%%%%%%%%%%%%%%%%%%%%%%%%%%%%%%%%%%%%%%%%%%%%%%%%%%%%%%%%%%%%%%%%%%

xi_prop = xi_curr_temp;
Beta_prop = Beta_curr_temp;
nseg_new = nseg_curr_temp;
tau_prop = tau_temp;
Phi_prop = Phi_temp;

if nexp_temp>1
    %*********************************************************
    % If contains more than one segments
    %*********************************************************
```

```
  seg_temp = unidrnd(nexp_temp-1);  %Drawing Segment to cut
  u = rand;
  cut_poss_curr = xi_curr_temp(seg_temp);
  nposs_prior = nseg_curr_temp(seg_temp) + nseg_curr_temp(seg_temp+1) - 2*tmin+1;

  %Determing if the relocation is a big jump or small jump
  if u<prob_mm1
     if nseg_curr_temp(seg_temp)==tmin && nseg_curr_temp(seg_temp+1)==tmin
        nposs=1; %Number of possible locations for new cutpoint
                  new_index=unidrnd(nposs);%Drawing index of new cutpoint
                  cut_poss_new = xi_curr_temp(seg temp)- 1 + new_index;
     elseif nseg_curr_temp(seg_temp)==tmin
                  nposs=2; %Number of possible locations for new cutpoint
                  new_index=unidrnd(nposs);%Drawing index of new cutpoint
                  cut_poss_new = xi_curr_temp(seg_temp)- 1 + new_index;
     elseif nseg_curr_temp(seg_temp+1)==tmin
                  nposs=2; %Number of possible locations for new cutpoint
                  new_index = unidrnd(nposs); %Drawing index of new cutpoint
                  cut_poss_new = xi_curr_temp(seg_temp) + 1 - new_index;
     else
                  nposs=3;% Number of possible locations for new cutpoint
                  new_index = unidrnd(nposs);%Drawing index of new cutpoint
                  cut_poss_new = xi_curr_temp(seg_temp) - 2 + new_index;
      end
  else
             new_index=unidrnd(nposs_prior);%
             cut_poss_new = sum(nseg_curr_temp(1:seg_temp-1)) - 1 + tmin+new_index;
  end

  xi_prop(seg_temp)=cut_poss_new;
  if seg_temp>1
      %Number of observations in lower part of new cutpoin
             nseg_new(seg_temp) = xi_prop(seg_temp) - xi_curr_temp(seg_temp-1);
  else
             nseg_new(seg_temp) = xi_prop(seg_temp);
  end
  %Number of observations in upper part of new cutpoint
     nseg_new(seg_temp+1) = nseg_curr_temp(seg_temp) + nseg_curr_temp(seg_temp+1) -
nseg_new(seg_temp);

%=============================================================
%Evaluating the cut Proposal density for the cut-point at the cureent
%and proposed values
%=============================================================
  if(abs(cut_poss_new-cut_poss_curr)>1)
     log_prop_cut_prop=log(1-prob_mm1)-log(nposs_prior);
     log_prop_cut_curr=log(1-prob_mm1)-log(nposs_prior);
  elseif nseg_curr_temp(seg_temp)==tmin  nseg_curr_temp(seg_temp+1)==tmin
     log_prop_cut_prop=0;
     log_prop_cut_curr=0;
  else
     if (nseg_curr_temp(seg_temp)==tmin || nseg_curr_temp(seg_temp+1)==tmin)
        %log_prop_cut_prop=log(1-prob_mm1)-log(nposs_prior)+log(1/2)+log
(prob_mm1);
        log_prop_cut_prop=log(1/2)+log(prob_mm1);
      else
        %log_prop_cut_prop=log(1-prob_mm1)-log(nposs_prior)+log(1/3)+log
(prob_mm1);
        log_prop_cut_prop=log(1/3)+log(prob_mm1);
      end
```

```
        if(nseg_new(seg_temp)==tmin || nseg_new(seg_temp+1)==tmin)
           %log_prop_cut_curr=log(1-prob_mm1)-log(nposs_prior)+log(1/2)+log
 (prob_mm1);
           log_prop_cut_curr=log(1/2)+log(prob_mm1);
        else
           %log_prop_cut_curr=log(1-prob_mm1)-log(nposs_prior)+log(1/3)+log
 (prob_mm1);
           log_prop_cut_curr=log(1/3)+log(prob_mm1);
        end
    end


    need = sum(Beta_curr_temp(:,:,seg_temp) - Beta_curr_temp(:,:,seg_temp+1));
    need = need./need;
    need(isnan(need))=0;
    aa = zeros(2^(dimen^2),1);
    for i=1:2^(dimen^2)
        aa(i)=sum(chol_index(i,:)==need);
    end
    Phi_need = find(aa==dimen^2);
    select = find(chol_index(Phi_need,:)~=0);
    select_inv = find(chol_index(Phi_need,:)==0);

%=============================================================================
%Evaluating the Loglikelihood, Priors and Proposals at the current values
%=============================================================================

    loglike_curr = 0;
    for j=seg_temp:seg_temp+1
          if j>1
             yobs_tmp = ts((xi_curr_temp(j-1)+1):xi_curr_temp(j),:);
           else
             yobs_tmp = ts(1:xi_curr_temp(j),:);
           end
         %Loglikelihood at current values
         [log_curr_spec_dens] = whittle_like(yobs_tmp,Beta_curr_temp(:,:,j));
         loglike_curr = loglike_curr + log_curr_spec_dens;
    end

%=============================================================================
%Evaluating the Loglikelihood, Priors and Proposals at the proposed values
%=============================================================================

%%%%%%%%%%%%%%%%%%%%%%%%%%%%%%%%%%%%%%%%%%%%%%%%%%%%%%%%%%%%%%%%%%%%%%%%%%%%%%%
%For coefficient of basis functions that are the same across two segments
%%%%%%%%%%%%%%%%%%%%%%%%%%%%%%%%%%%%%%%%%%%%%%%%%%%%%%%%%%%%%%%%%%%%%%%%%%%%%%%

    if Phi_need ~=2^(dimen^2)
        [Beta_out, m_out, m, ~] = Hamilt2(chol_index, Phi_need, seg_temp,...
                        ts, tau_prop(:,seg_temp), tau_prop(:,seg_temp+1),...
                        Beta_prop(:,:,seg_temp), Beta_prop(:,:,seg_temp+1), xi_prop,
nseg_new(seg_temp));

        Beta_prop(:,select_inv,seg_temp) = reshape(Beta_out,nBeta,length
(select_inv));
        Beta_prop(:,select_inv,seg_temp+1) = Beta_prop(:,select_inv,seg_temp);
        m_curr_1 = -0.5*m'*((1/M)*eye(length(m)))*m;
        m_prop_1 = -0.5*m_out'*((1/M)*eye(length(m_out)))*m_out;
    else
```

```
        m_curr_1 = 0;
        m_prop_1 = 0;
    end
%%%%%%%%%%%%%%%%%%%%%%%%%%%%%%%%%%%%%%%%%%%%%%%%%%%%%%%%%%%%%%%%%%%%%%%%%%%%%%%%
% For coefficient of basis functions that are different across two segments
%%%%%%%%%%%%%%%%%%%%%%%%%%%%%%%%%%%%%%%%%%%%%%%%%%%%%%%%%%%%%%%%%%%%%%%%%%%%%%%%
    loglike_prop = 0;
    yobs_tmp_2 = cell(2,1);
    m_prop_2 = 0;
    m_curr_2 = 0;
    for j=seg_temp:seg_temp+1

        [Beta_out, m_out, m, yobs_tmp]=Hamilt1(chol_index, Phi_need,...
                            j, ts, tau_prop(:,j), Beta_prop(:,:,j), xi_prop);
        if j==seg_temp
            yobs_tmp_2{1} = yobs_tmp;
        else
            yobs_tmp_2{2} = yobs_tmp;
        end

        Beta_prop(:,select,j) = reshape(Beta_out,nBeta,length(select));
        m_curr_2 = m_curr_2 - 0.5*m'*((1/M)*eye(length(m)))*m;
        m_prop_2 = m_prop_2 - 0.5*m_out'*((1/M)*eye(length(m_out)))*m_out;
    end

    %Loglikelihood at proposed values
    for j=seg_temp:seg_temp+1
        if j==seg_temp
            [log_curr_spec_dens] = whittle_like(yobs_tmp_2{1},Beta_prop(:,:,j));
        else
            [log_curr_spec_dens] = whittle_like(yobs_tmp_2{2},Beta_prop(:,:,j));
        end
        loglike_prop = loglike_prop+log_curr_spec_dens;
    end

    %proposal density
    log_proposal_curr =  log_prop_cut_curr;
    log_proposal_prop =  log_prop_cut_prop;

    %target density
    log_prior_cut_prop=0;
        log_prior_cut_curr=0;
    for k=1:nexp_temp-1
        if k==1
            log_prior_cut_prop=-log(nobs-(nexp_temp-k+1)*tmin+1);
                        log_prior_cut_curr=-log(nobs-(nexp_temp-k+1)*tmin+1);
                else
                        log_prior_cut_prop=log_prior_cut_prop - log(nobs-xi_prop
(k-1)-(nexp_temp-k+1)*tmin+1);
                        log_prior_cut_curr=log_prior_cut_curr - log
(nobs-xi_curr_temp(k-1)-(nexp_temp-k+1)*tmin+1);
        end
    end

    log_target_prop = loglike_prop + log_prior_cut_prop + m_prop_1 + m_prop_2;
    log_target_curr = loglike_curr + log_prior_cut_curr + m_curr_1 + m_curr_2;

else
```

```
    %**********************************************************
    % If contains only one segment
    %**********************************************************
    nseg_new = nobs;
    seg_temp = 1;
    %==========================================================================
    %Evaluating the Loglikelihood, Priors and Proposals at the proposed values
    %==========================================================================
    Phi_need = 2^(dimen^2);

    [Beta_out, m_out, m]=Hamilt1(chol_index, Phi_need,
                        1, ts, tau_temp, Beta_curr_temp, xi_curr_temp);
    Beta_prop(:,:,1) = reshape(Beta_out,nBeta,dimen^2);
    m_curr =  - 0.5*m'*((1/M)*eye(length(m)))*m;
    m_prop =  - 0.5*m_out'*((1/M)*eye(length(m_out)))*m_out;

    %Loglike at proposed values
    [loglike_prop] = whittle_like(ts,Beta_prop(:,:,1));

    %Loglike at current values
    [loglike_curr] = whittle_like(ts,Beta_curr_temp(:,:,1));

    log_target_prop = loglike_prop + m_prop;
    log_target_curr = loglike_curr + m_curr;
    log_proposal_curr =  0;
    log_proposal_prop =  0;
end

%************************************************************
%Calculations acceptance probability
%************************************************************
A = min(1,exp(log_target_prop-log_target_curr+log_proposal_curr-
log_proposal_prop));

function [out, fitparams] = MultiSpect_diag(ts,varargin)

%%%%%%%%%%%%%%%%%%%%%%%%%%%%%%%%%%%%%%%%%%%%%%%%%%%%%%%%%%%%%%%%%%%%%%%%%%
% Program for the MCMC nonstationary multivariate spectrum analysis paper
% 07/02/2016
% Does the MCMC iterations for bivariate time series
%%%%%%%%%%%%%%%%%%%%%%%%%%%%%%%%%%%%%%%%%%%%%%%%%%%%%%%%%%%%%%%%%%%%%%%%%%

%%%%%%%%%%%%%%%%%%%%%%%%%%%%%%%%%%%%%%%%%%%%%%%%%%%%%%%%%%%%%%%%%%%%%%%%%%
% Extract information from the option parameters
%%%%%%%%%%%%%%%%%%%%%%%%%%%%%%%%%%%%%%%%%%%%%%%%%%%%%%%%%%%%%%%%%%%%%%%%%%
% If params is empty, then use the default parameters
if nargin==1
    params = OptsMultiSpect();
else
    params = varargin{1};
end;

%%%%%%%%%%%%%%%%%%%%%%%%%%%%%%%%%%%%%%%%%%%%%%%%%%%%%%%%%%%%%%%%%%%%%%%%%%
%II) Run the estimation proceedure
%%%%%%%%%%%%%%%%%%%%%%%%%%%%%%%%%%%%%%%%%%%%%%%%%%%%%%%%%%%%%%%%%%%%%%%%%%

warning('off','MATLAB:nearlySingularMatrix')
nloop = params.nloop;
```

```
%number of total MCMC iterations
nwarmup = params.nwarmup;
%number of warmup period
nexp_max = params.nexp_max;
%Input maximum number of segments

global dimen nobs nbasis nBeta sigmasqalpha tau_up_limit ...
        prob_mm1 tmin var_inflate_1 var_inflate_2 options
options = optimset('Display','off','GradObj','on','Hessian','on','MaxIter',10000,...
   'MaxFunEvals',10000,'TolFun',1e-5,'TolX',1e-5);

nbasis = params.nbasis;
%number of linear smoothing spline basis functions
nBeta = nbasis + 1;
%number of coefficients for real part of components
sigmasqalpha = params.sigmasqalpha;
%smoothing parameter for alpha
tau_up_limit = params.tau_up_limit;
%the prior for smoothing parameters
prob_mm1 = params.prob_mm1;
%the probability of small jump, 1-prob of big jump
tmin = params.tmin;
%minimum number of observation in each segment
var_inflate_1 = params.v1(1);
var_inflate_2 = params.v1(1);
nfreq_hat = params.nfreq;
freq_hat=(0:nfreq_hat)'/(2*nfreq_hat);

%%%%%%%%%%%%%%%%%%%%%%%%%%%%%%%%%%%%%%%%%%%%%%%%%%%%%%%%%%%%%%%%%%%%%%%%%%%%%%%%
% II) Run the estimation proceedure
%%%%%%%%%%%%%%%%%%%%%%%%%%%%%%%%%%%%%%%%%%%%%%%%%%%%%%%%%%%%%%%%%%%%%%%%%%%%%%%%

dim = size(ts);
nobs = dim(1);
dimen = dim(2);

spect_hat = cell(nexp_max,1);
for j=1:nexp_max
   spect_hat{j} = zeros(2,2,nfreq_hat+1,j,nloop+1);
end

nexp_curr = params.init;  %initialize the number of segments

%initialize tausq
for j=1:nexp_curr
        tausq_curr(:,j,1)=rand(4,1)*tau_up_limit;
end

%initilize the location of the changepoints for j=1:nexp_curr(1)
xi_curr = zeros(nexp_curr,1);
nseg_curr = zeros(nexp_curr,1);
for j=1:nexp_curr
   if nexp_curr==1
      xi_curr = nobs;
      nseg_curr = nobs;
   else
      if j==1
                        nposs = nobs-nexp_curr*tmin+1;
                        xi_curr(j) = tmin + unidrnd(nposs)-1;
                        nseg_curr(j) = xi_curr(j);
```

```
        elseif j>1 && j<nexp_curr
                        nposs=nobs-xi_curr(j-1)-tmin*(nexp_curr-j+1)+1;
                        xi_curr(j)=tmin+unidrnd(nposs)+xi_curr(j-1)-1;
                        nseg_curr(j)=xi_curr(j)-xi_curr(j-1);
        else
                        xi_curr(j)=nobs;
                        nseg_curr(j)=xi_curr(j)-xi_curr(j-1);
        end
    end
end

%index matrix for which components of cholesky decomposition changed
chol_index = [0 0 0 0; 1 0 0 0; 0 1 0 0; 0 0 1 0; 0 0 0 1;...
            1 1 0 0; 1 0 1 0; 1 0 0 1; 0 0 1 1; 0 1 0 1;...
            0 1 1 0; 1 1 1 0; 1 1 0 1; 0 1 1 1; 1 0 1 1;...
            1 1 1 1];
% 1: no change; 2-5: one changes; 6-11: two changes; 12-15: three changes
% 16: all changed

Beta_curr = zeros(nBeta,4,nexp_curr);
Phi_temp = repmat(16,nexp_curr,1);
for j=1:nexp_curr
        [Beta_mean, Beta_var,~] = postBeta1(chol_index, Phi_temp(j), j, ts,
tausq_curr, Beta_curr, xi_curr);
        Beta_curr(:,:,j)=reshape(mvnrnd(Beta_mean,0.5*(Beta_var+Beta_var')),nBeta,4);
end
Phi_curr=[repmat(16,nexp_curr-1,1);0];
%jumping probabilities
epsilon=zeros(nloop,1);
met_rat=zeros(nloop,1);
rat_birth=[];
rat_death=[];


%%%initialize .mat file
S.('ts') = ts;
S.('varargin') = varargin;
S.('nexp') = [];
S.('nseg') = [];
S.('xi') = {};
S.('Beta') = {};
S.('tausq') = {};
S.('spect_hat') = {};
S.('Spec_1') = {};
S.('Spec_2') = {};
S.('Coh') = {};
S.('Phi') = {};
if params.convdiag==1
   S.('conv_diag') = [];
end

save(params.fname, '-struct', 'S', '-v7.3');

%create matfile object to use in MCMC sampler
out = matfile(params.fname, 'Writable', true);

%create data structures to store temporary results in between ouputs
Beta_tmp = cell(params.batchsize,1);
spect_hat_tmp = cell(params.batchsize,1);
nexp_tmp = zeros(params.batchsize,1);
```

```
nseg_tmp = cell(params.batchsize,1);
tausq_tmp = cell(params.batchsize,1);
xi_tmp = cell(params.batchsize,1);
Phi_tmp = cell(params.batchsize,1);
Spect_1_tmp = cell(params.batchsize,1);
Spect_2_tmp = cell(params.batchsize,1);
Coh_tmp = cell(params.batchsize,1);
if params.convdiag==1
   convdiag_tmp = zeros(params.batchsize,4*nBeta+4);
end

% preallocate for the worst case memory use
for i=1:params.batchsize
   Beta_tmp{i} = zeros(nBeta, 4, params.nexp_max);
   spect_hat_tmp{i} = zeros(2,2,nfreq_hat+1,params.nexp_max);
   nexp_tmp(i) = params.nexp_max;
   nseg_tmp{i} = zeros(params.nexp_max,1);
   tausq_tmp{i} = zeros(4,params.nexp_max);
   xi_tmp{i} = zeros(params.nexp_max,1);
   Phi_tmp{i} = zeros(params.nexp_max,1);
   Spect_1_tmp{i} = zeros(nfreq_hat+1,nobs);
   Spect_2_tmp{i} = zeros(nfreq_hat+1,nobs);
   Coh_tmp{i} = zeros(nfreq_hat+1,nobs);
   if params.convdiag==1
      convdiag_tmp(i,:)=zeros(1,4*nBeta+4);
   end
end

%set batch index
batch_idx = 1;


%%%%%%%%%%%%%%%%%%%%%%%%%%%%%%%%%%%%%%
% run the loop
%%%%%%%%%%%%%%%%%%%%%%%%%%%%%%%%%%%%%%
for p=1:nloop
  tic;
  if(mod(p,100)==0)
     fprintf('iter: %g of %g \n' ,p, nloop)
  end
  if p<nwarmup
      var_inflate_1=1;
      var_inflate_2=1;
  else
      var_inflate_1=1;
      var_inflate_2=1;
  end

  %========================
  %BETWEEN MODEL MOVE
  %========================

  kk = length(find(nseg_curr>2*tmin)); %Number of available segments

  %===========================
  %Deciding on birth or death
  if kk==0 %Stay where you (if nexp_curr=1) or join segments if there are no available
segments to cut
```

```
    if nexp_curr==1
        nexp_prop = nexp_curr; %Stay
        log_move_prop = 0;
        log_move_curr = 0;
    else
        nexp_prop = nexp_curr-1; %death
        log_move_prop = 0;
        if nexp_prop==1
            log_move_curr = 1;
        else
            log_move_curr = log(0.5);
        end
    end
else
    if nexp_curr==1
        nexp_prop = nexp_curr + 1; %birth
        log_move_prop = 0;
        if nexp_prop==nexp_max
            log_move_curr = 0;
        else
            log_move_curr = log(0.5);
        end
    elseif nexp_curr==nexp_max
        nexp_prop = nexp_curr-1;   %death
        log_move_prop = 0;
        if nexp_prop==1
            log_move_curr = 0;
        else
            log_move_curr = log(0.5);
        end
    else
        u = rand;
        if u<0.5
            nexp_prop = nexp_curr+1; %birth
            if nexp_prop==nexp_max;
                log_move_curr = 0;
                log_move_prop = log(0.5);
            else
                log_move_curr=log(0.5);
                log_move_prop=log(0.5);
            end
        else
            nexp_prop = nexp_curr-1; %death
            if nexp_prop==1
                log_move_curr = 0;
                log_move_prop = log(0.5);
            else
                log_move_curr = log(0.5);
                log_move_prop = log(0.5);
            end
        end
    end
end


if nexp_prop<nexp_curr
    %Do Death step
```

```
      [met_rat(p),nseg_prop,xi_prop,tausq_prop,Beta_prop, Phi_prop]=
death(chol_index,ts,nexp_curr,nexp_prop,...
               tausq_curr,xi_curr,nseg_curr,Beta_curr,Phi_curr,
log_move_curr,log_move_prop);
      rat_death=[rat_death met_rat(p)];
   elseif nexp_prop>nexp_curr
      %Do Birth step
      [met_rat(p),nseg_prop,xi_prop,tausq_prop,Beta_prop, Phi_prop]= birth
(chol_index,ts,nexp_curr,nexp_prop,...
               tausq_curr,xi_curr,nseg_curr,Beta_curr,Phi_curr,log_move_curr,
log_move_prop);
       rat_birth=[rat_birth met_rat(p)];
   else
       xi_prop=xi_curr;
       nseg_prop=nseg_curr;
       tausq_prop=tausq_curr;
       Beta_prop=Beta_curr;
       Phi_prop=Phi_curr;
       met_rat(p) = 1;
   end
   u = rand;
   if u<met_rat(p)
                nexp_curr=nexp_prop;
                xi_curr=xi_prop;
                nseg_curr=nseg_prop;
                tausq_curr=tausq_prop;
                Beta_curr=Beta_prop;
      Phi_curr=Phi_prop;
   end
   %========================
   %WITHIN MODEL MOVE
   %========================
      %Drawing a new cut point and Beta simultaneously
   %update coeffiecient of linear basis function

   [epsilon(p),nseg_new,xi_prop,~,Beta_prop,Phi_prop,seg_prop]= ...
   within(chol_index,ts,nexp_curr,tausq_curr, xi_curr,...
                                   nseg_curr, Beta_curr, Phi_curr);
   u = rand;
   if (u<epsilon(p)|| p==1)
      if nexp_curr>1
         for j=seg_prop:seg_prop+1
            Beta_curr=Beta_prop;
            xi_curr=xi_prop;
            nseg_curr=nseg_new;
            Phi_curr=Phi_prop;
         end
      else
         Beta_curr=Beta_prop;
      end
   end

   %Drawing tau
   for j=1:nexp_curr
      for i=1:3
         tau_a = nbasis/2;
         tau_b = sum(Beta_curr(2:end,i,j).^2)/2;
         u=rand;
```

```
            const1 = gamcdf(1/tau_up_limit,tau_a,1/tau_b);
            const2 = 1-u*(1-const1);
            tausq_curr(i,j) = 1/gaminv(const2,tau_a,1/tau_b);
        end
        tau_a = nbasis/2;
        tau_b = sum(Beta_curr(1:nBeta,4,j).^2)/2;
        u=rand;
        const1 = gamcdf(1/tau_up_limit,tau_a,1/tau_b);
        const2 = 1-u*(1-const1);
        tausq_curr(4,j) = 1/gaminv(const2,tau_a,1/tau_b);
    end

    %===================================
    %Estimating Spectral Density
    %===================================
    [xx_r, xx_i] = lin_basis_func(freq_hat); %produce linear basis functions

    for j =1:nexp_curr
        %getting the coefficients of linear basis functions
        g1 = Beta_curr(:,1:3,j);
        g2 = Beta_curr(1:nBeta,4,j);

        theta_real = xx_r * g1(:,3);
        theta_imag = xx_i * g2;
        theta = theta_real+sqrt(-1)*theta_imag;

        delta_sq_hat = zeros(2,nfreq_hat+1);
        for q=1:2
            delta_sq_hat(q,:) = exp(xx_r * g1(:,q))';
        end
        %produce the spectral density matrix
        for k=1:(nfreq_hat+1)
            TT = eye(2);
            TT(2,1) = -theta(k);
            spect_hat{nexp_curr}(:,:,k,j,p+1) = ...
            inv(TT)*diag(delta_sq_hat(:,k))*inv(TT');
        end
    end

    %=============================================
    % Get time-varying spectra and coherence
    %=============================================
    Spec_1_est = zeros(nfreq_hat+1,nobs);
    Spec_2_est = zeros(nfreq_hat+1,nobs);
    Coh_est = zeros(nfreq_hat+1,nobs);

    % first spectrum
    if((p+1)>nwarmup)
        spec_hat_curr=squeeze(spect_hat{nexp_curr}(1,1,:,:,p+1));
        for j=1:nexp_curr
            if(j==1)
              Spect_1_est(:,1:xi_curr(j))=repmat(spec_hat_curr(:,j),1,xi_curr(j));
            else
               Spect_1_est(:,xi_curr(j-1)+1:xi_curr(j))= ...
               repmat(spec_hat_curr(:,j),1,xi_curr(j)-xi_curr(j-1));
            end
        end
    end
```

```
% second spectrum
if((p+1)>nwarmup)
    spec_hat_curr=squeeze(spect_hat{nexp_curr}(2,2,:,:,p+1));
    for j=1:nexp_curr
        if(j==1)
            Spect_2_est(:,1:xi_curr(j))=repmat(spec_hat_curr(:,j),1,xi_curr(j));
        else
            Spect_2_est(:,xi_curr(j-1)+1:xi_curr(j))=...
            repmat(spec_hat_curr(:,j),1,xi_curr(j)-xi_curr(j-1));
        end
    end
end

% coherence
if((p+1)>nwarmup)
    spec_hat_curr=abs(squeeze(spect_hat{nexp_curr}(2,1,:,:,p+1))).^2./...
      (squeeze(spect_hat{nexp_curr}(1,1,:,:,p+1)).*squeeze(spect_hat
{nexp_curr}(2,2,:,:,p+1)));
    for j=1:nexp_curr
       if(j==1)
          Coh_est(:,1:xi_curr(j))=repmat(spec_hat_curr(:,j),1,xi_curr(j));
       else
          Coh_est(:,xi_curr(j-1)+1:xi_curr(j))=repmat(spec_hat_curr(:,j),1,
xi_curr(j)-xi_curr(j-1));
       end
    end
end

%create convergence diagnostics
if params.convdiag==1
    convdiag_curr=convdiag(nobs, nBeta, nseg_curr, nexp_curr, Beta_curr,
tausq_curr);
end

%output to temporary container
Beta_tmp{batch_idx} = Beta_curr;
spect_hat_tmp{batch_idx} = spect_hat;
nexp_tmp(batch_idx) = nexp_curr;
nseg_tmp{batch_idx} = nseg_curr;
tausq_tmp{batch_idx} = tausq_curr;
xi_tmp{batch_idx} = xi_curr;
Phi_tmp{batch_idx} = Phi_curr;
if((p+1)>nwarmup)
    Spect_1_tmp{batch_idx} = Spect_1_est;
    Spect_2_tmp{batch_idx} = Spect_2_est;
    Coh_tmp{batch_idx} = Coh_est;
end
if params.convdiag==1
    convdiag_tmp=convdiag_curr;
end

if(mod(p,params.batchsize)==0)
    %output data to .mat file
    out.Beta((p-params.batchsize+1):p,1) = Beta_tmp;
    out.spect_hat((p-params.batchsize+1):p,1) = spect_hat_tmp;
    out.nexp((p-params.batchsize+1):p,1)=nexp_tmp;
    out.nseg((p-params.batchsize+1):p,1)=nseg_tmp;
```

```
        out.tausq((p-params.batchsize+1):p,1) = tausq_tmp;
        out.xi((p-params.batchsize+1):p,1) = xi_tmp;
        out.Phi((p-params.batchsize+1):p,1) = Phi_tmp;
        out.Spect_1((p-params.batchsize+1):p,1) = Spect_1_tmp;
        out.Spect_2((p-params.batchsize+1):p,1) = Spect_2_tmp;
        out.Coh((p-params.batchsize+1):p,1) = Coh_tmp;
        if params.convdiag==1
            out.convdiag((p-params.batchsize+1):p,1:nbeta+1)=convdiag_tmp;
        end

        %reset temporary data containers
        Beta_tmp = cell(params.batchsize,1);
        spect_hat_tmp = cell(params.batchsize,1);
        nexp_tmp = zeros(params.batchsize,1);
        nseg_tmp = cell(params.batchsize,1);
        tausq_tmp = cell(params.batchsize,1);
        xi_tmp = cell(params.batchsize,1);
        Phi_tmp = cell(params.batchsize,1);
        Spect_1_tmp = cell(params.batchsize,1);
        Spect_2_tmp = cell(params.batchsize,1);
        Coh_tmp = cell(params.batchsize,1);
        if params.convdiag==1
            convdiag_tmp = zeros(params.batchsize,4*nBeta+4);
        end

        %reset batch index
        batch_idx=0;

    end

    %increment batch index
    batch_idx = batch_idx + 1;

    tms(p) = toc;
    if params.verb ==1
        fprintf('sec / min hr: %g %g %g \n' ,[tms(p),sum(tms(1:p))/60, sum(tms(1:
p))/(60*60)]')
    end

end

fitparams = struct('nloop', nloop, ...
             'nwarmup', nwarmup, ...
             'timeMean', mean(tms), ...
             'timeMax', max(tms(2:end)), ...
             'timeMin', min(tms),...
             'timeStd', std(tms(2:end)));
end

function[spectra, coh] = MultiSpect_interval(zt, spect_matrices, fit, params,
alphalevel)

dim = size(zt); nobs = dim(1); dimen=dim(2);
spectra = cell(dimen,1);
coh = cell(dimen,1);
if dimen==3
```

```
temp1 = zeros(params.nfreq+1, (params.nloop-params.nwarmup));
temp2 = zeros(params.nfreq+1, (params.nloop-params.nwarmup));
temp3 = zeros(params.nfreq+1, (params.nloop-params.nwarmup));
temp4 = zeros(params.nfreq+1, (params.nloop-params.nwarmup));
temp5 = zeros(params.nfreq+1, (params.nloop-params.nwarmup));
temp6 = zeros(params.nfreq+1, (params.nloop-params.nwarmup));
spect11 = zeros(params.nfreq+1,nobs,3);
spect22 = zeros(params.nfreq+1,nobs,3);
spect33 = zeros(params.nfreq+1,nobs,3);
coh21 = zeros(params.nfreq+1,nobs,3);
coh31 = zeros(params.nfreq+1,nobs,3);
coh32 = zeros(params.nfreq+1,nobs,3);
for j=1:nobs
   if(mod(j,10)==0)
     fprintf('Completed: %g of %g observations \n' ,j, nobs)
   end
   for p=1:params.nloop
     if(p>params.nwarmup)
       if length(fit(fit(1).nexp_curr(p)).xi(:,p))==1
          k=1;
       else
          k = min(find(j<=fit(fit(1).nexp_curr(p)).xi(:,p)));
       end
       temp1(:,p-params.nwarmup) = squeeze(spect_matrices{fit(1).nexp_curr(p)}
(1,1,:,k,p));
       temp2(:,p-params.nwarmup) = squeeze(spect_matrices{fit(1).nexp_curr(p)}
(2,2,:,k,p));
       temp3(:,p-params.nwarmup) = squeeze(spect_matrices{fit(1).nexp_curr(p)}
(3,3,:,k,p));
      temp4(:,p-params.nwarmup) = abs(squeeze(spect_matrices{fit(1).nexp_curr
(p)}(2,1,:,k,p))).^2./...
(squeeze(spect_matrices{fit(1).nexp_curr(p)}(1,1,:,k,p)).*squeeze
(spect_matrices{fit(1).nexp_curr(p)}(2,2,:,k,p)));
      temp5(:,p-params.nwarmup) = abs(squeeze(spect_matrices{fit(1).nexp_curr
(p)}(3,1,:,k,p))).^2./...
(squeeze(spect_matrices{fit(1).nexp_curr(p)}(1,1,:,k,p)).*squeeze
(spect_matrices{fit(1).nexp_curr(p)}(3,3,:,k,p)));
      temp6(:,p-params.nwarmup) = abs(squeeze(spect_matrices{fit(1).nexp_curr
(p)}(3,2,:,k,p))).^2./...
(squeeze(spect_matrices{fit(1).nexp_curr(p)}(2,2,:,k,p)).*squeeze
(spect_matrices{fit(1).nexp_curr(p)}(3,3,:,k,p)));
     end
   end
   spect11(:,j,1) = mean(temp1,2); spect11(:,j,2:3) =  quantile
(temp1,[alphalevel/2, 1-alphalevel/2],2);
   spect22(:,j,1) = mean(temp2,2); spect22(:,j,2:3) =  quantile
(temp2,[alphalevel/2, 1-alphalevel/2],2);
   spect33(:,j,1) = mean(temp3,2); spect33(:,j,2:3) =  quantile
(temp3,[alphalevel/2, 1-alphalevel/2],2);
   coh21(:,j,1) = mean(temp4,2); coh21(:,j,2:3) =  quantile(temp4,[alphalevel/2,
1-alphalevel/2],2);
   coh31(:,j,1) = mean(temp5,2); coh31(:,j,2:3) =  quantile(temp5,[alphalevel/2,
1-alphalevel/2],2);
   coh32(:,j,1) = mean(temp6,2); coh32(:,j,2:3) =  quantile(temp6,[alphalevel/2,
1-alphalevel/2],2);
end

spectra{1}=spect11;spectra{2}=spect22;spectra{3}=spect33;
coh{1}=coh21;coh{2}=coh31; coh{3}=coh32;
```

```
elseif dimen==2

   temp1 = zeros(params.nfreq+1, (params.nloop-params.nwarmup));
   temp2 = zeros(params.nfreq+1, (params.nloop-params.nwarmup));
   temp3 = zeros(params.nfreq+1, (params.nloop-params.nwarmup));
   spect11 = zeros(params.nfreq+1,nobs,3);
   spect22 = zeros(params.nfreq+1,nobs,3);
   coh21 = zeros(params.nfreq+1,nobs,3);
   for j=1:nobs
       if(mod(j,10)==0)
         fprintf('Completed: %g of %g observations \n' ,j, nobs)
       end
       for p=1:params.nloop
           if(p>params.nwarmup)
             if length(fit(fit(1).nexp_curr(p)).xi(:,p))==1
               k=1;
              else
               k = min(find(j<=fit(fit(1).nexp_curr(p)).xi(:,p)));
              end
             temp1(:,p-params.nwarmup) = squeeze(spect_matrices{fit(1).nexp_curr
(p)}(1,1,:,k,p));
             temp2(:,p-params.nwarmup) = squeeze(spect_matrices{fit(1).nexp_curr
(p)}(2,2,:,k,p));
              temp3(:,p-params.nwarmup) = abs(squeeze(spect_matrices{fit(1).
nexp_curr(p)}(2,1,:,k,p))).^2./...
(squeeze(spect_matrices{fit(1).nexp_curr(p)}(1,1,:,k,p)).*squeeze
(spect_matrices{fit(1).nexp_curr(p)}(2,2,:,k,p)));
         end
       end
       spect11(:,j,1) = mean(temp1,2); spect11(:,j,2:3) =  quantile
(temp1,[alphalevel/2, 1-alphalevel/2],2);
       spect22(:,j,1) = mean(temp2,2); spect22(:,j,2:3) =  quantile
(temp2,[alphalevel/2, 1-alphalevel/2],2);
       coh21(:,j,1) = mean(temp3,2); coh21(:,j,2:3) =  quantile(temp3,
[alphalevel/2, 1-alphalevel/2],2);
   end

   spectra{1}=spect11;spectra{2}=spect22;
   coh{1}=coh21;

end

function[post_partitions]=MultiSpect_partition(zt, fit, params)

dim = size(zt);nobs = dim(1);
post_partitions = (histc(fit(1).nexp_curr(params.nwarmup+1:params.nloop),1:
params.nexp_max)'/length(params.nwarmup+1:params.nloop))';

figure
histogram(fit(1).nexp_curr(params.nwarmup+1:params.nloop),'FaceAlpha',1,...

'Normalization','probability','BinMethod','integers','BinWidth',.5,'BinLimits',
[1,10])
title('Histogram of the Number of Partitions')
xlabel('The number of partitions')
```

```
ylabel('Probability')
for j=1:params.nexp_max
        kk=find(fit(1).nexp_curr(params.nwarmup+1:params.nloop)==j);
  if ~isempty(kk) && j>1
      figure
      hold
      title(['Plot of Partition Points Given ',int2str(j), ' Segments'])
      for k=1:j-1
          plot(fit(j).xi(k,kk+params.nwarmup))
      end
      for k=1:j-1
          figure
          hold
          title(['Histogram of Location of Partition ', int2str(k), ',' ' Given
',int2str(j), ' Segments'])
          histogram(fit(j).xi(k,kk+params.nwarmup),'FaceAlpha',1,...
             'Normalization','probability','BinMethod','integers','BinWidth',20,
'BinLimits',[1,nobs])
    end
  end
end

end

function[spectra, coh] = MultiSpect_surface(zt, spect_matrices, fit, params)
dim = size(zt); nobs = dim(1); dimen=dim(2);
spectra = cell(dimen,1);
coh = cell(dimen,1);
if dimen==3
   s_11=zeros(params.nfreq+1,nobs);
   s_22=zeros(params.nfreq+1,nobs);
   s_33=zeros(params.nfreq+1,nobs);
   for p=1:params.nloop
      if(p>params.nwarmup)
        xi_curr=fit(fit(1).nexp_curr(p)).xi(:,p);
        spec_hat_curr_11=squeeze(spect_matrices{fit(1).nexp_curr(p)}(1,1,:,:,p));
        spec_hat_curr_22=squeeze(spect_matrices{fit(1).nexp_curr(p)}(2,2,:,:,p));
        spec_hat_curr_33=squeeze(spect_matrices{fit(1).nexp_curr(p)}(3,3,:,:,p));
        for j=1:fit(1).nexp_curr(p)
            if(j==1)
s_11(:,1:xi_curr)=s_11(:,1:xi_curr(j))+repmat(spec_hat_curr_11(:,j),1,xi_curr
(j))/(params.nloop-params.nwarmup);
s_22(:,1:xi_curr)=s_22(:,1:xi_curr(j))+repmat(spec_hat_curr_22(:,j),1,xi_curr
(j))/(params.nloop-params.nwarmup);
s_33(:,1:xi_curr)=s_33(:,1:xi_curr(j))+repmat(spec_hat_curr_33(:,j),1,xi_curr
(j))/(params.nloop-params.nwarmup);
            else
               s_11(:,xi_curr(j-1)+1:xi_curr(j))=s_11(:,xi_curr(j-1)
+1:xi_curr(j))+...
                  repmat(spec_hat_curr_11(:,j),1,xi_curr(j)-xi_curr(j-1))/
(params.nloop-params.nwarmup);
               s_22(:,xi_curr(j-1)+1:xi_curr(j))=s_22(:,xi_curr(j-1)
+1:xi_curr(j))+...
                  repmat(spec_hat_curr_22(:,j),1,xi_curr(j)-xi_curr(j-1))/
(params.nloop-params.nwarmup);
               s_33(:,xi_curr(j-1)+1:xi_curr(j))=s_33(:,xi_curr(j-1)
+1:xi_curr(j))+...
                  repmat(spec_hat_curr_33(:,j),1,xi_curr(j)-xi_curr(j-1))/
(params.nloop-params.nwarmup);
```

```
          end
        end
      end
    end

    coh_21=zeros(params.nfreq+1,nobs);
    coh_31=zeros(params.nfreq+1,nobs);
    coh_32=zeros(params.nfreq+1,nobs);
    for p=1:params.nloop
        if(p>params.nwarmup)
           xi_curr=fit(fit(1).nexp_curr(p)).xi(:,p);
           spec_hat_curr_21=abs(squeeze(spect_matrices{fit(1).nexp_curr(p)}
(2,1,:,:,p))).^2./...
(squeeze(spect_matrices{fit(1).nexp_curr(p)}(1,1,:,:,p)).*squeeze
(spect_matrices{fit(1).nexp_curr(p)}(2,2,:,:,p)));
           spec_hat_curr_31=abs(squeeze(spect_matrices{fit(1).nexp_curr(p)}
(3,1,:,:,p))).^2./...
(squeeze(spect_matrices{fit(1).nexp_curr(p)}(1,1,:,:,p)).
*squeeze(spect_matrices{fit(1).nexp_curr(p)}(3,3,:,:,p)));
           spec_hat_curr_32=abs(squeeze(spect_matrices{fit(1).nexp_curr(p)}
(3,2,:,:,p))).^2./...
(squeeze(spect_matrices{fit(1).nexp_curr(p)}(3,3,:,:,p)).*squeeze
(spect_matrices{fit(1).nexp_curr(p)}(2,2,:,:,p)));
           for j=1:fit(1).nexp_curr(p)
               if(j==1)
coh_21(:,1:xi_curr(j))=coh_21(:,1:xi_curr(j))+repmat(spec_hat_curr_21(:,j),1,
xi_curr(j))/(params.nloop-params.nwarmup);
  coh_31(:,1:xi_curr(j))=coh_31(:,1:xi_curr(j))+repmat(spec_hat_curr_31(:,j),1,
xi_curr(j))/(params.nloop-params.nwarmup);
  coh_32(:,1:xi_curr(j))=coh_32(:,1:xi_curr(j))+repmat(spec_hat_curr_32(:,j),1,
xi_curr(j))/(params.nloop-params.nwarmup);
               else
                  coh_21(:,xi_curr(j-1)+1:xi_curr(j))=coh_21(:,xi_curr(j-1)+1:
xi_curr(j))+...
                    repmat(spec_hat_curr_21(:,j),1,xi_curr(j)-xi_curr(j-1))/
(params.nloop-params.nwarmup);
                  coh_31(:,xi_curr(j-1)+1:xi_curr(j))=coh_31(:,xi_curr(j-1)+1:
xi_curr(j))+...
                     repmat(spec_hat_curr_31(:,j),1,xi_curr(j)-xi_curr(j-1))/
(params.nloop-params.nwarmup);
                  coh_32(:,xi_curr(j-1)+1:xi_curr(j))=coh_32(:,xi_curr(j-1)+1:
xi_curr(j))+...
                     repmat(spec_hat_curr_32(:,j),1,xi_curr(j)-xi_curr(j-1))/
(params.nloop-params.nwarmup);
               end
           end
        end
    end

    spectra{1}=s_11;spectra{2}=s_22;spectra{3}=s_33;
    coh{1}=coh_21;coh{2}=coh_31; coh{3}=coh_32;
elseif dimen==2
    s_11=zeros(params.nfreq+1,nobs);
    s_22=zeros(params.nfreq+1,nobs);
    for p=1:params.nloop
        if(p>params.nwarmup)
          xi_curr=fit(fit(1).nexp_curr(p)).xi(:,p);
          spec_hat_curr_11=squeeze(spect_matrices{fit(1).nexp_curr(p)}(1,1,:,:,p));
```

```
        spec_hat_curr_22=squeeze(spect_matrices{fit(1).nexp_curr(p)}(2,2,:,:,p));
        for j=1:fit(1).nexp_curr(p)
          if(j==1)
s_11(:,1:xi_curr)=s_11(:,1:xi_curr(j))+repmat(spec_hat_curr_11(:,j),1,xi_curr
(j))/(params.nloop-params.nwarmup);
s_22(:,1:xi_curr)=s_22(:,1:xi_curr(j))+repmat(spec_hat_curr_22(:,j),1,xi_curr
(j))/(params.nloop-params.nwarmup);
          else
             s_11(:,xi_curr(j-1)+1:xi_curr(j))=s_11(:,xi_curr(j-1)+1:xi_curr(j))+...
               repmat(spec_hat_curr_11(:,j),1,xi_curr(j)-xi_curr(j-1))/(params.
nloop-params.nwarmup);
             s_22(:,xi_curr(j-1)+1:xi_curr(j))=s_22(:,xi_curr(j-1)+1:xi_curr(j))+...
               repmat(spec_hat_curr_22(:,j),1,xi_curr(j)-xi_curr(j-1))/(params.
nloop-params.nwarmup);
              end
          end
      end
   end

   coh_21=zeros(params.nfreq+1,nobs);
   for p=1:params.nloop
     if(p>params.nwarmup)
        xi_curr=fit(fit(1).nexp_curr(p)).xi(:,p);
        spec_hat_curr_21=abs(squeeze(spect_matrices{fit(1).nexp_curr(p)}(2,1,:,:,
p))).^2./...

(squeeze(spect_matrices{fit(1).nexp_curr(p)}(1,1,:,:,p)).*squeeze
(spect_matrices{fit(1).nexp_curr(p)}(2,2,:,:,p)));
        for j=1:fit(1).nexp_curr(p)
          if(j==1)

coh_21(:,1:xi_curr(j))=coh_21(:,1:xi_curr(j))+repmat(spec_hat_curr_21(:,j),1,
xi_curr(j))/(params.nloop-params.nwarmup);
          else
             coh_21(:,xi_curr(j-1)+1:xi_curr(j))=coh_21(:,xi_curr(j-1)+1:
xi_curr(j))+...
             repmat(spec_hat_curr_21(:,j),1,xi_curr(j)-xi_curr(j-1))/(params.
nloop-params.nwarmup);
              end
          end
      end
   end

   spectra{1}=s_11;spectra{2}=s_22;
   coh{1}=coh_21;
end

function [spect_hat, freq_hat, diagparams, fitparams] = MultiSpect(zt,varargin)

%%%%%%%%%%%%%%%%%%%%%%%%%%%%%%%%%%%%%%%%%%%%%%%%%%%%%%%%%%%%%%%%%%%%%%%%%%%%
% Program for the MCMC nonstationary multivariate spectrum % analysis paper
% Does the MCMC iterations for bivariate or trivariate time series
%  Input:
%    1)zt - time series data. It should be T by N matrix. T is
%    the length of the time series; N is the dimension (2 or 3)
%    2)varargin - Tunning parameters, which can be set %  by the program
%    OptsMultiSpec.m
%         See the documnetation in that program for details and default
%         values.
```

```
%   Main Outputs:
%     1) spec_hat - a cell contains spectral matrix
%     2) freq_hat - vector of frequencies considered.
%     3) diagparams - contains draws of parameters, including:
%        diagparams.tausq - smoothing parameters,
%        diagparams.Beta - coefficients,
%        diagparams.xi - locations of partitions
%        diagparams.nseg - number of observations in each block,
%        diagparams.nexp_curr - number of partitions,
%        diagparams.Phi - set of coefficients that changed,
%        diagparams.epsilon - acceptance rate of within-model move,
%        diagparams.bet_birth - number of accepted birth step,
%        diagparams.bet_death - number of accepted death step,
%        diagparams.bet_within - number of accepted within step,
%        diagparams.tms - tum time in second,
%        diagparams.convdiag_out - an optional convergence diagnostic
%        output
%     4) fitparams - 6 dimensional structural array
%        fitparams.nloop - number of iterations run
%        fitparams.nwarmup - length of the burn-in
%        fitparams.timeMean - average iteration run time time in
%        seconds
%        fitparams.timeMax - maximum iteration run time in seconds
%        fitparams.timeMin - minimum iteration run time in seconds
%        fitparams.timeStd - standard deviation of iteration run times
%                     in seconds
%%%%%%%%%%%%%%%%%%%%%%%%%%%%%%%%%%%%%%%%%%%%%%%%%%%%%%%%%%%%%%%%%%%%%%%%%%%%%%
% I) Extract information from the option parameters
%%%%%%%%%%%%%%%%%%%%%%%%%%%%%%%%%%%%%%%%%%%%%%%%%%%%%%%%%%%%%%%%%%%%%%%%%%%%%%
% If params is empty, then use the default parameters
if nargin==1
   params = OptsMultiSpect();
else
   params = varargin{1};
end

nloop = params.nloop;
%number of total MCMC iterations
nwarmup = params.nwarmup;
%number of warmup period
nexp_max = params.nexp_max;
%Input maximum number of segments

global dimen nobs nbasis nBeta sigmasqalpha tau_up_limit ...
       prob_mm1 tmin M ee options
if verLessThan('matlab','9.2')
   options = optimset('Display','off','GradObj','on','Hessian','on','MaxIter',
10000,...
   'MaxFunEvals',10000,'TolFun',1e-6,'TolX',1e-6);
else
 options = optimoptions(@fminunc,'Display','off','GradObj','on','Hessian','on','
MaxIter',10000,...
   'MaxFunEvals',10000,'TolFun',1e-6,'TolX',1e-6,'Algorithm','trust-region');
end
nbasis = params.nbasis;
%number of linear smoothing spline basis functions
nBeta = nbasis + 1;
%number of coefficients
sigmasqalpha = params.sigmasqalpha;
```

```
%smoothing parameter for alpha
tau_up_limit = params.tau_up_limit;
%the prior for smoothing parameters
prob_mml = params.prob_mml;
%the probability of small jump, 1-prob of big jump
tmin = params.tmin;
%minimum number of observation in each segment
M = tau_up_limit;
%scale value for momentum variable
ee = params.ee;
%step size in HMC
nfreq_hat = params.nfreq;
freq_hat=(0:nfreq_hat)'/(2*nfreq_hat);

%%%%%%%%%%%%%%%%%%%%%%%%%%%%%%%%%%%%%%
% II) Run the estimation proceedure
%%%%%%%%%%%%%%%%%%%%%%%%%%%%%%%%%%%%%%

dim = size(zt);
nobs = dim(1);
dimen = dim(2);
ts = zt;
% ts = zeros(nobs,dimen);
% for i=1:dimen
%   x = zt(:,i);
%   xmat = [ones(length(x),1) linspace(1,length(x),%     length(x))'];
%   linfit=inv(xmat'*xmat)*xmat'*x;
%   ts(:,i)=zt(:,i)- xmat*linfit;
% end

tausq = cell(nexp_max,1);
%big array for smoothing parameters
Beta = cell(nexp_max,1);
%big array for coefficients
spect_hat = cell(nexp_max,1);
xi = cell(nexp_max,1);
%Cutpoint locations xi_1 is first cutpoint, xi_) is beginning of timeseries
nseg = cell(nexp_max,1);
%Number of observations in each segment
Phi = cell(nexp_max,1);
%index for component change
tms = zeros(1,nloop);
for j=1:nexp_max
  tausq{j}=ones(dimen^2,j,nloop+1);
  Beta{j} = zeros(nBeta,dimen^2,j,nloop+1);
  spect_hat{j} = zeros(dimen,dimen,nfreq_hat+1,j,nloop+1);
  xi{j}=ones(j,nloop+1);
       nseg{j}=ones(j,nloop+1);
  Phi{j}=zeros(j,nloop+1);
end

%%%%%%%%%%%%%%%%%%%%%%%%%%%%%%
% initilize the MCMC iteration
%%%%%%%%%%%%%%%%%%%%%%%%%%%%%%

nexp_curr=nexp_max*ones(nloop+1,1);
%big array for number of segment
```

```
nexp_curr(1) = params.init;
%initialize the number of segments

%initialize tausq
for j=1:nexp_curr(1)
        tausq{nexp_curr(1)}(:,j,1)=rand(dimen^2,1)*tau_up_limit;
end

%initilize the location of the changepoints for j=1:nexp_curr(1)
for j=1:nexp_curr(1)
  if nexp_curr(1)==1
     xi{nexp_curr(1)}(j,1) = nobs;
     nseg{nexp_curr(1)}(j,1) = nobs;
  else
     if j==1
                    nposs = nobs-nexp_curr(1)*tmin+1;
                    xi{nexp_curr(1)}(j,1) = tmin + unidrnd(nposs)-1;
                    nseg{nexp_curr(1)}(j,1) = xi{nexp_curr(1)}(j,1);
    elseif j>1 && j<nexp_curr(1)
                    nposs=nobs-xi{nexp_curr(1)}(j-1,1)-tmin*(nexp_curr(1)-j+1)+1;
                    xi{nexp_curr(1)}(j,1)=tmin+unidrnd(nposs)+xi{nexp_curr(1)}
(j-1,1)-1;
                    nseg{nexp_curr(1)}(j,1)=xi{nexp_curr(1)}(j,1)-xi
{nexp_curr(1)}(j-1,1);
    else
                    xi{nexp_curr(1)}(j,1)=nobs;
                    nseg{nexp_curr(1)}(j,1)=xi{nexp_curr(1)}(j,1)-xi
{nexp_curr(1)}(j-1,1);
    end
  end
end

%index matrix for which components of cholesky decomposition changed
chol_index = chol_ind(dimen);

xi_temp = xi{nexp_curr(1)}(:,1);
tau_temp = tausq{nexp_curr(1)}(:,:,1);
Beta_temp = Beta{nexp_curr(1)}(:,:,:,1);
Phi{nexp_curr(1)}(1:(end-1),1)=2^(dimen^2);
Phi_temp = repmat(2^(dimen^2),nexp_curr(1));
for j=1:nexp_curr(1)
       [Beta_mean, Beta_var,~] = postBeta1(chol_index, Phi_temp(j), j, ts, tau_temp
(:,j), Beta_temp(:,:,j,1), xi_temp);
  if min(eig(0.5*(Beta_var+Beta_var')))<0
     Beta{nexp_curr(1)}(:,:,j,1) = reshape(mvnrnd(Beta_mean,0.5*
(eye(nBeta*dimen^2))),nBeta,dimen^2);
  else
     Beta{nexp_curr(1)}(:,:,j,1) = reshape(mvnrnd(Beta_mean,0.5*(Beta_var
+Beta_var')),nBeta,dimen^2);
  end
end

%jumping probabilities
epsilon=zeros(nloop,1);
met_rat=zeros(nloop,1);
```

```
bet_death = 0;
bet_birth = 0;
with = 0;

%create convergence diagnostics
if params.convdiag==1
   convdiag_out = zeros((dimen^2)*nBeta+dimen^2,nloop);
   convdiag_out(:,1)=conv_diag(nobs, nBeta,  nseg{nexp_curr(1)}(:,1),...
             nexp_curr(1), Beta{nexp_curr(1)}(:,:,:,1), tausq{nexp_curr(1)}
(:,:,1),dimen);
end

%%%%%%%%%%%%%%%%%%%%%%%%%%%%%%%%%%%%%%%%%%%%
% run the loop
%%%%%%%%%%%%%%%%%%%%%%%%%%%%%%%%%%%%%%%%%%%%
for p=1:nloop
  tic;
  if(mod(p,50)==0)
     fprintf('iter: %g of %g \n' ,p, nloop)
  end

  %========================
  %BETWEEN MODEL MOVE
  %========================
  kk = length(find(nseg{nexp_curr(p)}(:,p)>2*tmin)); %Number of available segments

  %%%%%%%%%%%%%%%%%%%%%%%%%%%%%%%%
  %Deciding on birth or death
  %%%%%%%%%%%%%%%%%%%%%%%%%%%%%%%%

  if kk==0 %Stay where you (if nexp_curr=1) or join segments if there are no
available segments to cut
     if nexp_curr(p)==1
        nexp_prop = nexp_curr(p); %Stay
        log_move_prop = 0;
        log_move_curr = 0;
     else
        nexp_prop = nexp_curr(p)-1; %death
        log_move_prop = 0;
        if nexp_prop==1
           log_move_curr = 1;
        else
           log_move_curr = log(0.5);
        end
     end
  else
     if nexp_curr(p)==1
        nexp_prop = nexp_curr(p) + 1; %birth
        log_move_prop = 0;
        if nexp_prop==nexp_max
           log_move_curr = 0;
        else
           log_move_curr = log(0.5);
        end
     elseif nexp_curr(p)==nexp_max
        nexp_prop = nexp_curr(p)-1; %death
        log_move_prop = 0;
        if nexp_prop==1
           log_move_curr = 0;
        else
          log_move_curr = log(0.5);
        end
```

```
    else
        u = rand;
        if u<0.5
            nexp_prop = nexp_curr(p)+1; %birth
            if nexp_prop==nexp_max;
                log_move_curr = 0;
                log_move_prop = log(0.5);
            else
                log_move_curr=log(0.5);
                log_move_prop=log(0.5);
            end
        else
            nexp_prop = nexp_curr(p)-1; %death
            if nexp_prop==1
                log_move_curr = 0;
                log_move_prop = log(0.5);
            else
                log_move_curr = log(0.5);
                log_move_prop = log(0.5);
              end
          end
    end
end

xi_curr_temp = xi{nexp_curr(p)}(:,p);
Beta_curr_temp = Beta{nexp_curr(p)}(:,:,:,p);
nseg_curr_temp = nseg{nexp_curr(p)}(:,p);
tau_curr_temp = tausq{nexp_curr(p)}(:,:,p);
Phi_curr_temp = Phi{nexp_curr(p)}(:,p);

if nexp_prop<nexp_curr(p)
    %Death step
    [met_rat(p),nseg_prop,xi_prop,tausq_prop,Beta_prop, Phi_prop]= death
(chol_index,ts,nexp_curr(p),nexp_prop,...
            tau_curr_temp,xi_curr_temp,nseg_curr_temp,Beta_curr_temp,
Phi_curr_temp, log_move_curr,log_move_prop);
elseif nexp_prop>nexp_curr(p)
     %Birth step
     [met_rat(p),nseg_prop,xi_prop,tausq_prop,Beta_prop, Phi_prop]= birth
(chol_index,ts,nexp_curr(p),nexp_prop,...
        tau_curr_temp,xi_curr_temp,nseg_curr_temp,Beta_curr_temp,Phi_curr_temp,
log_move_curr,log_move_prop);
else
     xi_prop=xi{nexp_curr(p)}(:,p);
     nseg_prop=nseg{nexp_curr(p)}(:,p);
     tausq_prop=tausq{nexp_curr(p)}(:,:,p);
     Beta_prop=Beta{nexp_curr(p)}(:,:,p);
     Phi_prop=Phi{nexp_curr(p)}(:,p);
     met_rat(p) = 1;
end
u = rand;
if u<met_rat(p)
    if nexp_prop<nexp_curr(p)
        bet_death = bet_death + 1;
    elseif nexp_prop>nexp_curr(p)
        bet_birth = bet_birth + 1;
    end
```

```
                nexp_curr(p+1)=nexp_prop;
                xi{nexp_curr(p+1)}(:,p+1)=xi_prop;
                nseg{nexp_curr(p+1)}(:,p+1)=nseg_prop;
                tausq{nexp_curr(p+1)}(:,:,p+1)=tausq_prop;
                Beta{nexp_curr(p+1)}(:,:,:,p+1)=Beta_prop;
    Phi{nexp_curr(p+1)}(:,p+1)=Phi_prop;
else
                nexp_curr(p+1)=nexp_curr(p);
                xi{nexp_curr(p+1)}(:,p+1)=xi{nexp_curr(p+1)}(:,p);
                nseg{nexp_curr(p+1)}(:,p+1)=nseg{nexp_curr(p+1)}(:,p);
                tausq{nexp_curr(p+1)}(:,:,p+1)=tausq{nexp_curr(p+1)}(:,:,p);
                Beta{nexp_curr(p+1)}(:,:,:,p+1)=Beta{nexp_curr(p+1)}(:,:,:,p);
    Phi{nexp_curr(p+1)}(:,p+1)=Phi{nexp_curr(p+1)}(:,p);
end

%========================
%WITHIN MODEL MOVE
%========================

     %Drawing a new cut point and Beta simultaneously
%update coeffiecient of linear basis function
xi_curr_temp=xi{nexp_curr(p+1)}(:,p+1);
Beta_curr_temp=Beta{nexp_curr(p+1)}(:,:,:,p+1);
tau_temp=tausq{nexp_curr(p+1)}(:,:,p+1);
nseg_curr_temp=nseg{nexp_curr(p+1)}(:,p+1);
Phi_temp = Phi{nexp_curr(p+1)}(:,p+1);
[epsilon(p),nseg_new,xi_prop,tausq_prop,Beta_prop,Phi_prop,seg_temp]= ...
within(chol_index,ts,nexp_curr(p+1),tau_temp, xi_curr_temp,...
                          nseg_curr_temp, Beta_curr_temp,Phi_temp);
u = rand;
if (u<epsilon(p)|| p==1)
     with = with + 1;
     if nexp_curr(p+1)>1
        for j=seg_temp:seg_temp+1
          Beta{nexp_curr(p+1)}(:,:,j,p+1)=Beta_prop(:,:,j);
          xi{nexp_curr(p+1)}(j,p+1)=xi_prop(j);
          nseg{nexp_curr(p+1)}(j,p+1)=nseg_new(j);
          Phi{nexp_curr(p+1)}(j,p+1)=Phi_prop(j);
        end
    else
        Beta{nexp_curr(p+1)}(:,:,p+1)=Beta_prop;
    end
else
     Beta{nexp_curr(p+1)}(:,:,:,p+1)=Beta_curr_temp;
     xi{nexp_curr(p+1)}(:,p+1)=xi_curr_temp;
     nseg{nexp_curr(p+1)}(:,p+1)=nseg_curr_temp;
     Phi{nexp_curr(p+1)}(:,p+1)=Phi_temp;
end

%Drawing tau
for j=1:nexp_curr(p+1)
   for i=1:dimen^2
     if ismember(i,1:dimen + dimen*(dimen-1)/2)
        tau_a = nbasis/2;
        tau_b = sum(Beta{nexp_curr(p+1)}(2:end,i,j,p+1).^2)/2;
        u=rand;
        const1 = gamcdf(1/tau_up_limit,tau_a,1/tau_b);
        const2 = 1-u*(1-const1);
        tausq{nexp_curr(p+1)}(i,j,p+1) = 1/gaminv(const2,tau_a,1/tau_b);
```

```
        else
           tau_a = nBeta/2;
           tau_b = sum(Beta{nexp_curr(p+1)}(1:nBeta,i,j,p+1).^2)/2;
           u=rand;
           const1 = gamcdf(1/tau_up_limit,tau_a,1/tau_b);
           const2 = 1-u*(1-const1);
           tausq{nexp_curr(p+1)}(i,j,p+1) = 1/gaminv(const2,tau_a,1/tau_b);
         end
      end
   end

   tms(p) = toc;
   if params.verb ==1
      fprintf('sec / min hr: %g %g %g \n' ,[tms(p),sum(tms(1:p))/60, sum(tms
(1:p))/(60*60)]')
   end
   %====================================
   %Estimating Spectral Density
   %====================================
   [xx_r, xx_i] = lin_basis_func(freq_hat); %produce linear basis functions

   for j =1:nexp_curr(p+1)
       %getting the coefficients of linear basis functions
       g1 = Beta{nexp_curr(p+1)}(:,1:(dimen + dimen*(dimen-1)/2),j,p+1);
       g2 = Beta{nexp_curr(p+1)}(1:nBeta,(dimen + dimen*(dimen-1)/2 + 1):end,j,p+1);

       theta = zeros(dimen*(dimen-1)/2,(nfreq_hat+1));
       for i=1:dimen*(dimen-1)/2
         theta_real = xx_r * g1(:,i+dimen);
         theta_imag = xx_i * g2(:,i);
         theta(i,:) = (theta_real + sqrt(-1)*theta_imag).';
       end

       delta_sq_hat = zeros(2,nfreq_hat+1);
       for q=1:dimen
         delta_sq_hat(q,:) = exp(xx_r * g1(:,q)).';
       end

       %produce the spectral density matrix
       for k=1:(nfreq_hat+1)
         TT = eye(dimen);
         TT(2,1) = -theta(1,k);
         if dimen==3
            TT(3,1) = -theta(2,k);
            TT(3,2) = -theta(3,k);
         end
         spect_hat{nexp_curr(p+1)}(:,:,k,j,p+1) = ...
         inv(TT)*diag(delta_sq_hat(:,k))*inv(TT');
       end

   end

   %create convergence diagnostics
   if params.convdiag==1
      convdiag_out(:,p+1)=conv_diag(nobs, nBeta,  nseg{nexp_curr(p+1)}(:,p+1),...
         nexp_curr(p+1), Beta{nexp_curr(p+1)}(:,:,:,p+1), tausq{nexp_curr(p+1)}
(:,:,p+1),dimen);
   end
```

```
end
%%%%%%%%%%%%%%%%%%%%%%%%%%%%%%%%%%%%%%%%%%%%%%%%%%%%%%%%%%%%%%%%%%%%%%%%%%%
% III) Collect outputs
%%%%%%%%%%%%%%%%%%%%%%%%%%%%%%%%%%%%%%%%%%%%%%%%%%%%%%%%%%%%%%%%%%%%%%%%%%%
fitparams = struct('nloop', nloop, ...
            'nwarmup', nwarmup, ...
            'timeMean', mean(tms), ...
            'timeMax', max(tms(2:end)), ...
            'timeMin', min(tms),...
            'timeStd', std(tms(2:end)));
if params.convdiag==1
   diagparams = struct('tausq', tausq,...
            'Beta', Beta,...
            'xi', xi,...
            'nseg', nseg,...
            'nexp_curr', nexp_curr,...
            'Phi', Phi,...
            'epsilon', epsilon,...
            'bet_birth', bet_birth,...
            'bet_death', bet_death,...
            'with', with,...
            'time', tms,...
            'convdiag_out', convdiag_out);
else
   diagparams = struct('tausq', tausq,...
            'Beta', Beta,...
            'xi', xi,...
            'nseg', nseg,...
            'nexp_curr', nexp_curr,...
            'Phi', Phi,...
            'epsilon', epsilon,...
            'bet_birth', bet_birth,...
            'bet_death', bet_death,...
            'with', with,...
            'time', tms);
end

end

function [param] = OptsMultiSpect(varargin)
% This function sets the optional input arguments for the function
% MCBSpec().
%
%  (I) ONLY DEFAULT PARAMETERS
%      If only defult values for the parameters are desired, then either:
%
%         a) the SECOND argument in MultiSpec() can be left missing, or
%
%         b) params=setOptions() can be defined and used as the second
%         argument of MultiSpec().
%
% (II) USING NONDEFAULT PARAMETERS
%      If some options other than the default are desired:
%
%         1) Set all default parameters using (Ia) above.
%
%         2) Change desired parameters.
%
```

```
%
% PARAMETERS
%
%   nloop    -  The number of iterations run.
%               Default: 6000.
%   nwarmup -  The length of the burn-in.
%               Default: 200.
%   nexp_max-  The maximum number of segments allowed, adjust % when
               time series get longer.
%               Default: 10
%   tmin     -  The minimum number of observation in each segment.
%               Default: 100
%   prob_mml-  The probability of small jump, 1-prob of big
%   jump.
%               Default: 0.8
%   nbasis -    The number of linear smoothing spline basis
% functions.
%               Default: 10
%   tau_up_limit - The normal variance prior for smoothing
% parameters.
%               Default:  10^4.
%   sigmasqalpha -   The smoothing parameter for alpha.
%               Default: 10^4
%   init   -    Initial number of partitions
%               Default: 3
%   nfreq -     The number of frequencies for spectrm
%               Default: 50
%   verb   -    An indicator if the iteration number should be printed
%               at the completion of each iteration.   1 is
% yes, 0 is
%             no.
%   covdiag -    An indicator if the diagnostic should be
%   ee      -    Step size in Hamiltonian Monte Carlo
%                Default is 0.1


param = struct('nloop',10000, 'nwarmup',2000, ...
      'nexp_max',10, 'tmin',60, 'prob_mml',0.8 , 'nbasis',10, ...
      'tau_up_limit',10^5 , 'sigmasqalpha',10^5, 'init',3,...
      'nfreq',50, 'verb',1, 'convdiag',0, 'ee',0.1);

param.bands = {};

end


%%%%%%%%%%%%%%%%%%%%%%%%%%%%%%%%%%%%%%%%%%%%%%%%%%%%%%%
% read in data and implement the analysis
%%%%%%%%%%%%%%%%%%%%%%%%%%%%%%%%%%%%%%%%%%%%%%%%%%%%%%%

fdata = csvread('C:/Bookdata/WW9.csv',1,1);
dimen = size(fdata);
nobs = dimen(1);

startDate = datenum('04-01-1990');
endDate = datenum('12-31-2011');
xData = linspace(startDate,endDate,nobs);
figure
```

```
subplot(1,3,1)
plot(xData,fdata(:,1))
title('DJIA')
datetick('x','yyyy','keeplimits')
subplot(1,3,2)
plot(xData,fdata(:,2))
title('NASDAQ')
datetick('x','yyyy','keeplimits')
subplot(1,3,3)
plot(xData,fdata(:,3))
title('S&P500')
datetick('x','yyyy','keeplimits')

params = struct('nloop',12000, 'nwarmup',4000, ...
         'nexp_max',10, 'tmin',60, 'prob_mml',0.8 , 'nbasis',10, ...
         'tau_up_limit',10^5, 'sigmasqalpha',10^5, 'init',2,...
         'nfreq',50, 'verb',1, 'convdiag',0, 'ee', 0.05);

%%% Note: he code can be run using default parameters
%%% [spect_matrices, freq_hat, fit, fit_diag] = MultiSpect(zt)

rng(201701204);
[spect_matrices, freq_hat, fit, fit_diag] = MultiSpect(fdata,params);

[spect, coh] = MultiSpect_surface(fdata, spect_matrices, fit, params);

figure
subplot(1,3,1); meshc( xData, freq_hat,real(spect{1})); set
(gca,'YDir','reverse');  title('$ f_{11}$','Interpreter','LaTex'); xlabel
('time'); ylabel('freq'); datetick('x','yyyy','keeplimits')
subplot(1,3,2); meshc( xData, freq_hat,real(spect{2})); set
(gca,'YDir','reverse');  title('$ f_{22}$','Interpreter','LaTex'); xlabel
('time'); ylabel('freq'); datetick('x','yyyy','keeplimits')
subplot(1,3,3); meshc( xData, freq_hat,real(spect{3})); set
(gca,'YDir','reverse');  title('$ f_{33}$','Interpreter','LaTex'); xlabel
('time'); ylabel('freq'); datetick('x','yyyy','keeplimits')
figure
subplot(1,3,1); meshc( xData, freq_hat,real(coh{1})); set(gca,'YDir','reverse');
title('$ \rho_{21}^2$','Interpreter','LaTex'); xlabel('time'); ylabel('freq');
datetick('x','yyyy','keeplimits')
subplot(1,3,2); meshc( xData, freq_hat,real(coh{2})); set(gca,'YDir','reverse');
title('$ \rho_{31}^2$','Interpreter','LaTex'); xlabel('time'); ylabel('freq');
datetick('x','yyyy','keeplimits')
subplot(1,3,3); meshc( xData, freq_hat,real(coh{3})); set(gca,'YDir','reverse');
title('$ \rho_{32}^2$','Interpreter','LaTex'); xlabel('time'); ylabel('freq');
datetick('x','yyyy','keeplimits')
```

Projects

1. Find a m-dimensional stationary vector time series data set of your interest with m no less than 5. Complete your detailed nonparametric analysis of multivariate spectrum with a written report and analysis software code.

2. Find a m-dimensional stationary vector time series data set of your interest with m no less than 5. Complete your detailed VARMA spectral analysis with a written report and analysis software code.

3. Use the data set from Project 1 to perform a VARMA spectral analysis and compare the results from the two methods.
4. Use the data set from Project 2 to perform a nonparametric analysis of multivariate spectrum, and compare the results from the two methods.
5. Find a m-dimensional nonstationary vector time series data set of your interest and complete your detailed analysis of multivariate spectral analysis with a written report and analysis software code.

References

Bandyopadhyay, S., Jentsch, C., and Rao, S.S. (2017). A spectral domain test for stationarity of spatio-temporal data. *Journal of Time Series Analysis* **38**: 326–351.

Bartlett, M.S. (1950). Periodogram analysis and continuous spectra. *Biometrika* **37**: 1–16.

Blackman, R.B. and Tukey, J.W. (1959). *The Measurements of Power Spectrum from the Point of View of Communications Engineering*. Dover Publications.

Brillinger, D.R. (2002). *Time Series: Data Analysis and Theory*. Philadelphia: SIAM.

Chang, J., Hall, P., and Tang, C.Y. (2017). A frequency domain analysis of the error distribution from noisy high-frequency data. *Biometrika* **103**: 1–16.

Dahlhaus, R. (1996). Fitting time series model to nonstationary process. *Annals of Statistics* **25**: 1–37.

Dahlhaus, R. (2000). A likelihood approximation for locally stationary process. *Annals of Statistics* **28**: 1762–1794.

Dai, M. and Guo, W. (2004). Multivariate spectral analysis using Cholesky decomposition. *Biometrika* **91**: 629–643.

Daniell, P.J. (1946). Discussion on symposium on autocorrelation in time series. *Journal of the Royal Statistical Society* (Suppl. 8): 88–90.

Davis, R.A., Lee, T.C.M., and Rodriguez-Yam, G.A. (2006). Structural break estimation for nonstationary time series models. *Journal of the American Statistical Association* **101**: 223–239.

Eubank, R.L. and Hsing, T. (2008). Canonical correlation for stochastic processes. *Stochastic Processes and their Applications* **118**: 1634–1661.

Goodman, N. (1963). Statistical analysis based on a certain multivariate complex Gaussian distribution. *Annals of Mathematical Statistics* **34**: 152–177.

Grant, A.J. and Quinn, B.G. (2017). Parametric spectral discrimination. *Journal of Time Series Analysis* **38**: 838–864.

Gu, C. and Wahba, G. (1993). Semiparametric analysis of variance with tensor product thin plate splines. *Journal of the Royal Statistical Society, Series B* **55**: 353–368.

Guo, W. and Dai, M. (2006). Multivariate time-dependent spectral analysis using Cholesky decomposition. *Statistica Sinica* **16**: 825–845.

Hannan, E.J. (1970). *Multiple Time Series*. Wiley.

Hansen, M.H. and Yu, B. (2001). Model selection and the principle of minimum description length. *Journal of the American Statistical Association* **96**: 746–774.

Krafty, R.T. and Collinge, W.O. (2013). Penalized multivariate Whittle likelihood for power spectrum estimation. *Biometrika* **100**: 447–458.

Li, Z. and Krafty, R.T. (2018). Adaptive Bayesian time-frequency analysis of multivariate time series. To appear in *Journal of the American Statistical Association*.

Ombao, H., von Sachs, R., and Guo, W. (2005). SLEX analysis of multivariate nonstationary time series. *Journal of the American Statistical Association* **100**: 519–531.

Parzen, E. (1961). Mathematical considerations in the estimation of spectra. *Technometrics* **3**: 167–190.

Parzen, E. (1963). Notes on Fourier analysis and spectral windows. Technical report No. 48, Office of Naval Research.

Pawitan, Y. (1996). Automatic estimation of the cross-spectrum of a bivariate time series. *Biometrika* **83**: 419–432.

Pawitan, Y. and O'Sullivan, F. (1994). Nonparametric spectral density estimation using penalized whittle likelihood. *Journal of the American Statistical Association* **89**: 600–610.

Priestley, M.B. (1965). Evolutionary spectral and non-stationary process. *Journal of the Royal Statistical Society, Series B* **27**: 204–237.

Priestley, M.B. (1966). Design relations for non-stationary processes. *Journal of the Royal Statistical Society, Series B* **28**: 228–240.

Priestley, M.B. (1967). Power spectral analyses of non-stationary random processes. *Journal of Sound and Vibration* **6**: 86–97.

Priestley, M.B. (1981). *Spectral Analysis and Time Series*, vol. 1 and 2. Academic Press.

Qin, L. and Wang, Y. (2008). Nonparametric spectral analysis with applications to seizure characterization using EEG time series. *Annals of Applied Statistics* **2**: 1432–1451.

Rao, T.S. and Terdick, G. (2017). On the frequency variogram and on frequency domain methods for the analysis of spatio-temporal data. *Journal of Time Series Analysis* **38**: 308–325.

Ray, E.L., Sakrejda, K., Lauer, S.A., Johansson, M.A., and Reich, N.G. (2017). Infectious disease prediction with kernel conditional density estimation. *Statistics in Medicine* **36**: 4908–4929.

Riedel, K. and Sidorenko, A. (1995). Minimum bias multiple taper spectral estimation. *IEEE Transaction on Signal Processing* **43**: 188–195.

Rissanen, J. (1989). *Stochastic Complexity in Statistical Inquiry*. Singapore: Word Scientific.

Rosen, O. and Stoffer, D.S. (2007). Automatic estimation of multivariate spectra via smoothing splines. *Biometrika* **94**: 335–345.

Schwarz, K. and Krivobokova, T. (2016). A unified framework for spline estimators. *Biometrika* **103**: 103–120.

Slepian, D. (1978). Prolate spheroidal wave functions, Fourier analysis, and uncertainty – V: the discrete case. *Bell System Technical Journal* **57**: 1371–1430.

Thompson, D.J. (1982). Spectrum estimation and harmonic analysis. *Proceedings of the IEEE* **70**: 1055–1096.

Wahba, G. (1990). *Spline Models for Observational Data, CBMS-NSF Regional Conference Series in Applied Mathematics*. Philadelphia: SIAM.

Walden, A.T. (2000). A unified view of multitaper multivariate spectral estimation. *Biometrika* **87**: 767–788.

Wei, W.W.S. (2006). *Time Series Analysis Univariate & Multivariate Methods*, 2e. Pearson Addison-Wesley.

Wei, W.W.S. (2008). Spectral analysis. In: *Handbook of Longitudinal Research, Design, Measurement, and Analysis* (ed. S. Menard), 601–620. Academic Press.

Whittle, P. (1953). Estimation and information in stationary time series. *Arkiv för Matematik* **2**: 423–434.

Whittle, P. (1954). Some recent contributions to the theory of stationary processes. In: *A Study in the Analysis of Stationary Time Series*, 2e (ed. H.O. Wold), 196–228. Stockhol: Almqristand Witsett.

Wilson, G.T. (2017). Spectral estimation of the multivariate impulse response. *Journal of Time Series Analysis* **38**: 381–391.

Wood, S.N. (2011). Fast stable restricted maximum likelihood and marginal likelihood estimation of semiparametric generalized linear models. *Journal of the Royal Statistical Society, Series B* **73**: 3–36.

Zhang, S. (2016). Adaptive spectral estimation for nonstationary multivariate time series. *Computational Statistics & Data Analysis* **103**: 330–349.

10

Dimension reduction in high-dimensional multivariate time series analysis

The vector autoregressive (VAR) and vector autoregressive moving average (VARMA) models have been widely used to model multivariate time series, because of their ability to represent the dynamic relationships among variables in a system and their usefulness in forecasting unknown future values. However, when the number of dimensions is very large, the number of parameters often exceeds the number of available observations, and it is impossible to estimate the parameters. A suitable solution is clearly needed. In this chapter, after introducing some existing methods, we will suggest the use of contemporal aggregation as a dimension reduction method, which is very natural and simple to use. We will compare our proposed method with other existing methods in terms of forecast accuracy through both simulations and empirical examples.

10.1 Introduction

Multivariate time series are of interest in many fields such as economics, business, education, psychology, epidemiology, physical science, geoscience, and many others. When modeling multivariate time series, the VAR and VARMA models are possibly the most widely used models, because of their capability to represent the dynamic relationships among variables in a system and their usefulness in forecasting unknown future values. These models are described in many time series textbooks including Hannan (1970), Hamilton (1994), Reinsel (1997), Wei (2006), Lütkepohl (2007), Tsay (2013), Box et al. (2015), and many others.

Let $\mathbf{Z}_t = [Z_{1,t}, Z_{2,t}, \ldots, Z_{m,t}]'$, $t = 0, \pm 1, \pm 2, \ldots$, be a m-dimensional jointly stationary real-valued vector process so that $E(Z_{i,t}) = \mu_i$ is constant for each $i = 1, 2, \ldots, m$ and the cross-covariances between $Z_{i,t}$ and $Z_{j,s}$, $E(Z_{i,t} - \mu_i)(Z_{j,s} - \mu_j)$, for all $i = 1, 2, \ldots, m$ and

Multivariate Time Series Analysis and Applications, First Edition. William W.S. Wei.
© 2019 John Wiley & Sons Ltd. Published 2019 by John Wiley & Sons Ltd.
Companion website: www.wiley.com/go/wei/datasets

$j = 1, 2, \ldots, m$, are functions only of the time difference $(s - t)$. A useful class of vector time series models is the following vector autoregressive moving average model of order p and q, shortened to VARMA(p,q),

$$\mathbf{\Phi}_p(B)\mathbf{Z}_t = \mathbf{\theta}_0 + \mathbf{\Theta}_q(B)\mathbf{a}_t, \tag{10.1}$$

where $\mathbf{\Phi}_p(B) = \mathbf{\Phi}_0 - \mathbf{\Phi}_1 B - \cdots - \mathbf{\Phi}_p B^p$ and $\mathbf{\Theta}_q(B) = \mathbf{\Theta}_0 - \mathbf{\Theta}_1 B - \cdots - \mathbf{\Theta}_q B^q$ are autoregressive and moving average matrix polynomials of order p and q, respectively, $\mathbf{\Phi}_i$ and $\mathbf{\Theta}_j$ are nonsingular $m \times m$ matrices, and $\mathbf{a}_t$ is a sequence of m-dimensional white noise processes with mean zero vector and positive definite variance–covariance matrix $\boldsymbol{\varepsilon}$. Since one can always invert $\mathbf{\Phi}_0$ and $\mathbf{\Theta}_0$, and combine them into $\boldsymbol{\varepsilon}$, with no loss of generality, we will assume in the following discussion that $\mathbf{\Phi}_0 = \mathbf{\Theta}_0 = \mathbf{I}$, the $m \times m$ identity matrix. Furthermore, any VARMA model can be approximated by a vector AR model, in practice, one often uses the following vector autoregressive model of order p, shortened to VAR(p),

$$\left(\mathbf{I} - \mathbf{\Phi}_1 B - \cdots - \mathbf{\Phi}_p B^p\right)\mathbf{Z}_t = \mathbf{\theta}_0 + \mathbf{a}_t, \tag{10.2}$$

or

$$\mathbf{Z}_t = \mathbf{\theta}_0 + \mathbf{\Phi}_1 \mathbf{Z}_{t-1} + \cdots + \mathbf{\Phi}_p \mathbf{Z}_{t-p} + \mathbf{a}_t,$$

where the zeros of $|\mathbf{I} - \mathbf{\Phi}_1 B - \cdots - \mathbf{\Phi}_p B^p|$ lie outside of the unit circle or, equivalently, the roots of $|\lambda^p \mathbf{I} - \lambda^{p-1}\mathbf{\Phi}_1 - \cdots - \mathbf{\Phi}_p| = 0$ are all inside of the unit circle. With no loss of generality, in the following discussion, we will mostly use the VAR(p) model with zero mean for our illustrations.

With the development of computers and the internet, we have had a data explosion. For a m-dimensional multivariate time series, m being numbered in the hundreds or thousands is very common. If we simply consider $m = 100$, a simple VAR(2) model has a minimum of $2(100 \times 100) = 20{,}000$ parameters. For observations obtained yearly, we cannot estimate the model parameters, even with a hundred years of data.

To solve this problem, after introducing some existing methods, we will suggest the use of aggregation as a dimension reduction method, which is very natural and simple to use. We will compare our proposed method with other existing methods in terms of forecast accuracy through both simulations and empirical examples. The chapter will be organized as follows. Section 10.2 introduces and discusses several existing methods to handle high-dimensional time series. The proposed procedure is introduced in Section 10.3. Monte Carlo simulations and empirical data analysis are presented in Sections 10.4 and 10.5, respectively. Lastly, further discussions and remarks are given in Section 10.6.

A condensed version for the first five sections of this chapter was presented in an invited talk at the 2017 ICSA (International Chinese Statistical Association) Applied Statistical Symposium in Chicago when the organization celebrated its 30th year anniversary. For a formal paper, please see Wei (2018). The supplementary appendix at the end of the chapter is from Li and Wei (2017).

10.2 Existing methods

In this section, we briefly review the existing methods that handle time series modeling in high-dimensional settings, including various regularization methods (Section 10.2.1), the space–time STAR model if the data is collected from different locations (Section 10.2.2), model-based clustering (Section 10.2.3), and the factor model (Section 10.2.4).

10.2.1 Regularization methods

Let $\mathbf{Z}_t = [Z_{1,t}, Z_{2,t}, \ldots, Z_{m,t}]'$, $t = 1, 2, \ldots, n$, be a zero-mean m-dimensional time series with n observations. It is well known that the least squares method can be used to fit the VAR(p) model by minimizing

$$\sum_{t=1}^{n} \left\| \mathbf{Z}_t - \sum_{k=1}^{p} \mathbf{\Phi}_k \mathbf{Z}_{t-k} \right\|_2, \tag{10.3}$$

where $\|\|_2$ is Euclidean (L^2) norm of a vector. In practice, more compactly, with data $\mathbf{Z}_t = [Z_{1,t}, Z_{2,t}, \ldots, Z_{m,t}]'$, $t = 1, 2, \ldots, n$, we can present the VAR(p) model in Eq. (10.2) in the matrix form,

$$\underset{n\times m}{\mathbf{Y}} = \underset{(n\times mp)}{\mathbf{X}} \underset{(mp\times m)}{\mathbf{\Phi}} + \underset{(n\times m)}{\boldsymbol{\xi}}, \tag{10.4}$$

where

$$\mathbf{Y} = \begin{bmatrix} \mathbf{Z}_1' \\ \mathbf{Z}_2' \\ \vdots \\ \mathbf{Z}_n' \end{bmatrix}, \mathbf{X} = \begin{bmatrix} \mathbf{X}_1' \\ \mathbf{X}_2' \\ \vdots \\ \mathbf{X}_n' \end{bmatrix}, \mathbf{\Phi} = \begin{bmatrix} \mathbf{\Phi}_1' \\ \mathbf{\Phi}_2' \\ \vdots \\ \mathbf{\Phi}_p' \end{bmatrix}, \boldsymbol{\xi} = \begin{bmatrix} \mathbf{a}_1' \\ \mathbf{a}_2' \\ \vdots \\ \mathbf{a}_n' \end{bmatrix},$$

and

$$\mathbf{X}_t' = \left[\mathbf{Z}_{t-1}', \mathbf{Z}_{t-2}', \ldots, \mathbf{Z}_{t-p}'\right].$$

So, minimizing Eq. (10.3) is equivalent to

$$\underset{\mathbf{\Phi}}{\operatorname{argmin}} \|\mathbf{Y} - \mathbf{X}\mathbf{\Phi}\|_F, \tag{10.5}$$

where $\|\|_F$ is the Frobenius norm of the matrix.

For a VAR model in high-dimensional setting, many regularization methods have been developed, which assume sparse structures on coefficient matrices $\mathbf{\Phi}_k$ and use regularization procedure to estimate parameters. These methods include the Lasso (Least Absolute Shrinkage and Selection Operator) method, the lag-weighted lasso method, and the hierarchical vector autoregression method, among others.

10.2.1.1 The lasso method

One of the most commonly used regularization methods is the lasso method proposed by Tibshirani (1996) and extended to the vector time series setting by Hsu et al. (2008). Formally, the estimation procedure for the VAR model is through

$$\underset{\mathbf{\Phi}}{\operatorname{argmin}} \left\{ \|\mathbf{Y} - \mathbf{X}\mathbf{\Phi}\|_F + \lambda \|vec(\mathbf{\Phi})\|_1 \right\}, \tag{10.6}$$

where the second term is the regularization through L_1 penalty with λ being its control parameter. λ can be determined by cross-validation. The lasso method does not impose any special assumption on the relationship of lag orders and tends to over select the lag order p of the VAR model. This leads us to the development of some modified methods.

10.2.1.2 The lag-weighted lasso method

Song and Bickel (2011) proposed a method that incorporates the lag-weighted lasso (lasso and group lasso structures) approach for the high-dimensional VAR model. They placed group lasso penalties introduced by Yuan and Lin (2006) on the off-diagonal terms and lasso penalties on the diagonal terms. More specifically, if we denote $\mathbf{\Phi}(j, -j)$ as the vector composed of off-diagonal elements $\{\phi_{j,i}\}_{i \neq j}$, and $\mathbf{\Phi}_k(j, j)$ as the $j$thdiagonal element of $\mathbf{\Phi}_k$, then the regularization for $\mathbf{\Phi}_k$ is

$$\sum_{j=1}^{m} \|\mathbf{\Phi}_k(j,-j)\mathbf{W}(-j)\|_2 + \lambda \sum_{j=1}^{m} w_j \,|\, \mathbf{\Phi}_k(j,j) \,|, \tag{10.7}$$

where $\mathbf{W}(-j) = \text{diag}(w_1, \ldots, w_{j-1}, w_{j+1}, \ldots, w_m)$, an $(m-1) \times (m-1)$ diagonal matrix with w_j being the positive real-valued weight associated with the jth variable for $1 \leq j \leq m$, which is chosen to be the standard deviation of $Z_{j,t}$. λ is the control parameter that controls the extent to which other lags are less informative than its own lags. The first term of Eq. (10.7) is the group lasso penalty, the second term is the lasso penalty, and they impose regularization on other lags and its own lags, respectively. Let $0 < \alpha < 1$ and $(k)^\alpha$ be the other control parameter for different regularization for different lags; the estimation procedure is based on

$$\underset{\mathbf{\Phi}_1,\ldots,\mathbf{\Phi}_p}{\arg\min} \left\{ \|\mathbf{Y} - \mathbf{X\Phi}\|_F + \sum_{k=1}^{p} k^\alpha \left[\sum_{j=1}^{m} \|\mathbf{\Phi}_k(j,-j)\mathbf{W}(-j)\|_2 + \lambda \sum_{j=1}^{m} w_j \|\mathbf{\Phi}_k(j,j)\|_1 \right] \right\}. \tag{10.8}$$

10.2.1.3 The hierarchical vector autoregression (HVAR) method

More recently, Nicholson et al. (2016) proposed the HVAR method for high-dimensional time series. Particularly, they assume various predefined sparse assumptions on the coefficient matrices of the VAR model. Let $\mathbf{\Phi}_k(i)$ be the ith row of the coefficient matrix $\mathbf{\Phi}_k$ and $\mathbf{\Phi}_k(i, j)$ be the (i, j)th element of the coefficient matrix $\mathbf{\Phi}_k$. To express their model, we denote

$$\mathbf{\Phi}_{k:p} = [\mathbf{\Phi}_k, \ldots, \mathbf{\Phi}_p]' \in \mathbf{R}^{m(p-k+1) \times m},$$

$$\mathbf{\Phi}_{k:p}(i) = [\mathbf{\Phi}_k(i), \ldots, \mathbf{\Phi}_p(i)]' \in \mathbf{R}^{m(p-k+1) \times 1},$$

and

$$\mathbf{\Phi}_{k:p}(i,j) = [\mathbf{\Phi}_k(i,j), \ldots, \mathbf{\Phi}_p(i,j)]' \in \mathbf{R}^{(p-k+1) \times 1}.$$

Consider the $m \times m$ matrix of elementwise coefficient lags L as

$$L_{i,j} = \max\{k : \mathbf{\Phi}_k(i,j) \neq 0\}, \tag{10.9}$$

where we let $L_{i,j} = 0$ if $\mathbf{\Phi}_k(i,j) = 0$ for all $k = 1, \ldots, p$. Thus, each $L_{i,j}$ denotes the maximal coefficient lag for the (i, j)th component, meaning $L_{i,j}$ is the smallest k such that $\mathbf{\Phi}_{k+1:p}(i, j) = 0$.

The method includes three types of sparse structures for coefficient matrices of the VAR model. They are (i) the componentwise structure, which allows each of the m marginal equations from Eq. (10.2) to have its own maximal of lag orders, but requires all components within each equation to share the same maximal lag orders, such that $L_{i,j} = L_i$ for $i = 1, \ldots, m$; (ii) the own-other structure, which assumes a series' own lags are more informative than lags from other series and emphasizes the importance of diagonal elements of the coefficient matrices $\mathbf{\Phi}_k$, such that $L_{i,j} = L_i^{other}$ for $i \neq j$ and $L_{i,i} \in \{L_i^{other}, L_i^{other} + 1\}$, for $i = 1, \ldots, m$, and (iii) the elementwise structure, which places no stipulated relationship.

The parameter estimation is based on a convex optimization algorithm. For the componentwise structure, the parameters are estimated through

$$\underset{\mathbf{\Phi}}{\operatorname{argmin}}\left\{\frac{1}{2}\|\mathbf{Y}-\mathbf{X\Phi}\|_F+\lambda\sum_{i=1}^{m}\sum_{k=1}^{p}\|\mathbf{\Phi}_{k:p}(i)\|_2\right\}, \tag{10.10}$$

where again λ is the control parameter controlling sparsity such that bigger λ means $\hat{\mathbf{\Phi}}_{k:p}(i) = 0$ for more i and for smaller k. This means that if $\hat{\mathbf{\Phi}}_{k:p}(i) = 0$, then $\hat{\mathbf{\Phi}}_{k':p}(i) = 0$, for all $k' > k$. For the own-other structure, the objective function is

$$\underset{\mathbf{\Phi}}{\operatorname{argmin}}\left\{\frac{1}{2}\|\mathbf{Y}-\mathbf{X\Phi}\|_F+\lambda\sum_{i=1}^{m}\sum_{k=1}^{p}\left[\|\mathbf{\Phi}_{k:p}(i)\|_2+\|\mathbf{D}\|_2\right]\right\}, \tag{10.11}$$

where $\mathbf{D}$ is a vector concatenating $\mathbf{\Phi}_k(i, -i) = \{\mathbf{\Phi}_k(i, j) : j \neq i\}_{(m-1)\times 1}$ and $\mathbf{\Phi}_{(k+1):p}(i)$. The additional second penalty allows coefficient matrices to be sparse such that the influence of component i itself may be nonzero at lag k even though the influence of other components is zero at that lag. This indicates that for all $k' > k$., $\hat{\mathbf{\Phi}}_k(i) = 0$ implies $\hat{\mathbf{\Phi}}_{k'}(i) = 0$, and $\hat{\mathbf{\Phi}}_k(i, i) = 0$ implies $\hat{\mathbf{\Phi}}_{k+1}(i, -i) = 0$. Finally, for (iii) the elementwise structure, the objective function is given by

$$\underset{\mathbf{\Phi}}{\operatorname{argmin}}\left\{\frac{1}{2}\|\mathbf{Y}-\mathbf{X\Phi}\|_F+\lambda\sum_{i=1}^{m}\sum_{j=1}^{m}\sum_{k=1}^{p}\|\mathbf{\Phi}_{k:p}(i,j)\|_2\right\}. \tag{10.12}$$

This structure in Eq. (10.12) indicates that each of the components of coefficient matrix can have its own maximum lags. Thus, this is the most flexible structure proposed by Nicholson et al. (2016), which performs well if $L_{i,j}$ differs for all i and j, but would be suboptimal if $L_{i,j} = L_i$. The HVAR Method can be fitted by using R package *BigVAR* in the CRAN, which is a network of FTP and web servers around the world that store identical, up-to-date, versions of code and documentation for R.

10.2.2 The space–time AR (STAR) model

Similar to the regularization methods that control the values of parameters, when modeling time series associated with spaces or locations, it is very likely that many elements of $\mathbf{\Phi}_k$ are not significantly different from zero for pairs of locations that are spatially far away and uncorrelated given information from other locations. Thus, a model incorporating spatial information is not only helpful for parameter estimation, but also for dimension reduction and forecasting.

For a zero-mean stationary spatial time series, the space–time autoregressive moving average $\mathbf{STARMA}\left(p_{a_1,\ldots,a_p}, q_{m_1,\ldots,m_q}\right)$ model is defined by

$$\mathbf{Z}_t = \sum_{k=1}^{p}\sum_{\ell=0}^{a_k}\phi_{k,\ell}\mathbf{W}^{(\ell)}\mathbf{Z}_{t-k} + \mathbf{a}_t - \sum_{k=1}^{q}\sum_{\ell=0}^{m_k}\theta_{k,\ell}\mathbf{W}^{(\ell)}\mathbf{a}_{t-k}, \tag{10.13}$$

where the zeros of $\det\left(\mathbf{I}-\sum_{k=1}^{p}\sum_{\ell=0}^{a_k}\phi_{k,\ell}\mathbf{W}^{(\ell)}B^k\right)=0$ lie outside the unit circle, $\mathbf{a}_t$ is a Gaussian vector white noise process with zero-mean vector $\mathbf{0}$, and covariance matrix structure

$$E\left[\mathbf{a}_t\mathbf{a}_{t+k}'\right] = \begin{cases} \boldsymbol{\varepsilon}, & \text{if } k=0, \\ \mathbf{0}, & \text{if } k\neq 0, \end{cases} \tag{10.14}$$

and $\boldsymbol{\varepsilon}$ is an $m \times m$ symmetric positive definite matrix. The $\mathbf{STARMA}\left(p_{a_1,\ldots,a_p}, q_{m_1,\ldots,m_q}\right)$ model becomes a space–time autoregressive $\mathbf{STAR}\left(p_{a_1,\ldots,a_p}\right)$ model when $q = 0$. The STAR models were first introduced by Cliff and Ord (1975) and further extended to STARMA models by Pfeifer and Deutsch (1980a, b, c). Since a stationary model can be approximated by an autoregressive model, because of its easier interpretation, the most widely used STARMA models in practice are $\mathbf{STAR}\left(p_{a_1,\ldots,a_p}\right)$ models,

$$\mathbf{Z}_t = \sum_{k=1}^{p}\sum_{\ell=0}^{a_k}\phi_{k,\ell}\mathbf{W}^{(\ell)}\mathbf{Z}_{t-k} + \mathbf{a}_t, \tag{10.15}$$

where $\mathbf{Z}_t$ is a zero-mean stationary spatial time series or a proper differenced and transformed series of a nonstationary spatial time series.

The spatial information is introduced to the model by weighting matrices $\mathbf{W}^{(\ell)} = \left[w_{(i,j)}^{(\ell)}\right]$. Suppose that there are a total of m locations and we let $\mathbf{Z}_t = [Z_{1,t}, Z_{2,t}, \ldots, Z_{m,t}]'$ be the vector of times series of these m locations. Based on the spatial orders, with respect to the time series at location i, we will assign weights related to this location, $w_{i,j}^{(\ell)}$, such that they are nonzero only when the location j is the ℓth order neighbor of location i, and the sum of these weights is equal to 1. In other words, with respect to location i, we have $\sum_{j=1}^{m} w_{i,j}^{(\ell)} = 1$, where

$$w_{i,j}^{(\ell)} = \begin{cases} (0,1], & \text{if location } j \text{ is the } \ell th \text{ order neighbor of location } i, \\ 0, & \text{otherwise.} \end{cases}$$

Combining these weights, $w_{i,j}^{(\ell)}$, for all m locations, we have the spatial weight matrix for the neighborhood, $\mathbf{W}^{(\ell)} = \left[w_{(i,j)}^{(\ell)}\right]$, which is an $m \times m$ matrix with $w_{(i,j)}^{(\ell)}$ being nonzero if and only if locations i and j are in the same ℓth order neighbor and each row sums to 1. The weight can be chosen to reflect physical properties such as border length or distance of neighboring locations. One can also assign equal weights to all of the locations of the same spatial order. Clearly, $\mathbf{W}^{(0)} = \mathbf{I}$, an identity matrix, because each location is its own zeroth order neighbor.

It should be noted that the space–time autoregressive moving average (STARMA) model is a special case of the VARMA model,

$$\mathbf{\Phi}_p(B)\mathbf{Z}_t = \mathbf{\Theta}_q(B)\mathbf{a}_t, \tag{10.16}$$

where $\mathbf{\Phi}_p(B) = \mathbf{I} - \sum_{k=1}^{p}\sum_{\ell=0}^{a_k}\phi_{k,\ell}\mathbf{W}^{(\ell)}B^k$, and $\mathbf{\Theta}_q(B) = \mathbf{I} - \sum_{k=1}^{q}\sum_{\ell=0}^{m_k}\theta_{k,\ell}\mathbf{W}^{(\ell)}B^k$. For more detailed discussion of models related to both time and space, we refer readers to Chapter 8.

10.2.3 The model-based cluster method

Clustering or cluster analysis is a methodology that has been used by researchers to group data into homogeneous groups for a long time and may have originated in the fields of anthropology and psychology. There are many methods of clustering, including subjective observation and various distance methods for similarity. Earlier works include Tryon (1939), Cattell (1943), Ward (1963), Macqueen (1967), McLachlan and Basford (1988), among others. We will discuss these further in the last section. These methods were extended to the model-based cluster approach with an associated probability distribution by researchers including Banfield and Raftery (1993), Fraley and Raftery (2002), Wang and Zhou (2008), Scrucca (2010), and others. More recently, Wang et al. (2013) introduced a robust model-based clustering method for forecasting high-dimensional time series, and in this section, we will use their approach as an illustration. Let p_h be the probability a time series belongs to cluster h. The method first groups multiple time series into H mutually exclusive clusters, $\sum_{h=1}^{H} p_h = 1$, and assumes that each mean adjusted time series in a given cluster follows the same AR(p) model. Thus, for the ith time series that is in cluster h, we have,

$$Z_{i,t} = \sum_{k=1}^{p} \phi_k^{(h)} Z_{i,t-k} + \sigma_h \varepsilon_{i,t}, \text{ for } t = p+1, \ldots, n, \tag{10.17}$$

where $h = 1, 2, \ldots, H$, the $\varepsilon_{i,t}$ are $i.\ i.\ d.\ N(0, 1)$ random variables, independent across time and series. Let $\mathbf{\theta}_h = \left(\phi_1^{(h)}, \phi_2^{(h)}, \ldots, \phi_p^{(h)}, \sigma^{(h)}\right)$ be the vector of all parameters in cluster h, and $\mathbf{\Theta} = (\mathbf{\theta}_1, \ldots, \mathbf{\theta}_H, \mathbf{\eta})$, where $\mathbf{\eta} = (p_1, \ldots, p_H)$. The estimation procedure is accomplished through the Bayesian Markov Chain and Monte Carlo method.

10.2.4 The factor analysis

In previous sections, we have mainly reviewed methods that are based on the VAR model but with different model constraints and estimation procedures. However, there exist many other models for multivariate time series analysis, such as the transfer function model (Box et al. 2015), the state space model (Kalman 1960), and canonical correlation analysis (Box and Tiao 1977). More recently, Stock and Watson (2002a, b) introduced the factor model for dimension reduction and forecasting. The dynamic orthogonal component analysis was proposed by Matteson and Tsay (2011). In this review section, we will concentrate on the factor model.

The factor model is also called the diffusion index approach and can be written as

$$\mathbf{Z}_t = \mathbf{L}\mathbf{F}_t + \boldsymbol{\varepsilon}_t, \tag{10.18}$$

where $\mathbf{F}_t = (F_{1,t}, F_{2,t}, \ldots, F_{k,t})'$ is a $(k \times 1)$ vector of factors at time t, $\mathbf{L} = [\ell_{i,j}]$ is a $(m \times k)$ loading matrix, $\ell_{i,j}$ is the loading of the ith variable on the jth factor, $i = 1, 2, \ldots, m$, $j = 1, 2, \ldots, k$, and $\boldsymbol{\varepsilon}_t = (\varepsilon_{1,t}, \ldots, \varepsilon_{m,t})'$ is a $(m \times 1)$ vector of noises with $E(\boldsymbol{\varepsilon}_t) = \mathbf{0}$, and $\mathrm{Cov}(\boldsymbol{\varepsilon}_t) = \boldsymbol{\Sigma}$. Let $Z_{i,t+\ell}$ be ith component of $\mathbf{Z}_{t+\ell}$, once values of factors are obtained, we can build a forecast equation for the ℓ – step ahead forecast, such that

$$Z_{i,t+\ell} = \boldsymbol{\beta}'\mathbf{F}_t + \varepsilon_{i,t+\ell}, \tag{10.19}$$

where $\boldsymbol{\beta} = (\beta_1, \ldots, \beta_k)'$ denotes the coefficient vector and $\varepsilon_{i,t+\ell}$ is a sequence of uncorrelated zero-mean random variables. Note that the Eq. (10.19) can be further extended to:

$$Z_{i,t+\ell} = \boldsymbol{\beta}'\mathbf{F}_t + \boldsymbol{\alpha}'\mathbf{X}_{i,t} + \varepsilon_{i,t+\ell}, \tag{10.20}$$

where $\mathbf{X}_{i,t}$ is a $m \times 1$ vector of lagged values of $Z_{i,t+\ell}$ and/or other observed variables. We follow the approach proposed by Bai and Ng (2002) plus the penalty term $k[(m+n)/mn] \log[mn/(m+n)]$ to select the number of factors in our simulation studies and empirical examples. Other methods or penalties as described in Bai and Ng (2002) can also be used although this is beyond the scope of this chapter.

10.3 The proposed method for high-dimension reduction

In many applications, a large number of individual time series may follow a similar pattern so that we can aggregate them together. By doing so, we can reduce the dimension of the multivariate time series to a manageable and meaningful size. Specifically, we will concentrate on the VAR model described in Section 10.2 and propose aggregation as our method of dimension reduction.

Given a vector time series, assume that after model identification, it follows the VAR (p) model,

$$\mathbf{Z}_t = \sum_{k=1}^{p} \boldsymbol{\Phi}_k \mathbf{Z}_{t-k} + \mathbf{a}_t, \tag{10.21}$$

where $\mathbf{Z}_t$ is mean adjusted stationary m-dimensional original time series. Let

$$\mathbf{Y}_t = \mathbf{A}\mathbf{Z}_t, \tag{10.22}$$

where $\mathbf{A}$ is a $s \times m$ aggregation matrix with $s < m$, and $\mathbf{Y}_t = [Y_{1,t}, \ldots, Y_{s,t}]'$. Presently, the elements in $\mathbf{A}$ are assumed to be binary, such that its (i, j) element is 1 when $Z_{j,t}$ is included in the aggregate $Y_{i,t}$, and is 0 otherwise. In other words, the elements of row i in $\mathbf{A}$ construct $Y_{i,t}$ as the sum of designated elements of $\mathbf{Z}_t$. We will call $\mathbf{Y}_t$ the aggregate series and $\mathbf{Z}_t$ the non-aggregate series. It can be shown that the aggregate series $\mathbf{Y}_t$ will also follow a VAR(p) model. However, in practice, we will normally use the same model identification procedure to fit a VAR(P) model for some P such that

$$\mathbf{Y}_t = \sum_{k=1}^{P} \mathbf{\Phi}_k^{(a)} \mathbf{Y}_{t-k} + \boldsymbol{\xi}_t, \tag{10.23}$$

where $\mathbf{\Phi}_k^{(a)}$ for $k = 1, \ldots, P$ are $s \times s$ coefficient matrices, and $\boldsymbol{\xi}_t$ follows s-dimensional i.i.d. normal distribution with mean vector zero and covariance $\mathbf{\Sigma}^{(a)}$. The order P can be selected by existing methods such as AIC, BIC, and sequential likelihood ratio test (a detailed review of order selection methods can be found in Lütkepohl, 2007).

By using the aggregation, we reduce the dimension of the time series from m to s. Suppose we are interested in the $\ell-$step ahead forecast $\hat{\mathbf{Y}}_t(\ell)$ for the aggregate variable $\mathbf{Y}_{t+\ell}$. There are two ways to forecast: (i) forecasting from the non-aggregate data first and then aggregating its forecasts. Mathematically, this can be represented as

$$\hat{\mathbf{Y}}_t(\ell) = \mathbf{A}\hat{\mathbf{Z}}_t(\ell), \tag{10.24}$$

where $\hat{\mathbf{Z}}_t(\ell)$ is the $\ell-$step ahead forecasts from the model in Eq. (10.21); (ii) modeling and forecasting directly from the aggregates from the aggregate model in Eq. (10.23). Our proposed method takes the second procedure to reduce the dimension when modeling the data.

For the VARMA and VAR models, many results (Rose, 1977, Tiao and Guttman, 1980, Wei and Abraham, 1981, Kohn, 1982, and Lütkepohl, 1984) have shown that it is preferable to forecast the original time series first and then aggregate the forecasts, rather than forecast the aggregate time series directly. They also established the conditions for those two methods to be equivalent, which can be summarized here:

Theorem 10.1 Consider a m-dimensional non-aggregate VARMA (p,q) model, $\mathbf{Z}_t = \sum_{i=1}^{p} \mathbf{\Phi}_i \mathbf{Z}_{t-i} + \mathbf{a}_t - \sum_{j=1}^{q} \mathbf{\Theta}_j \mathbf{a}_t$, and its s-dimensional aggregate, $\mathbf{Y}_t = \mathbf{A}\mathbf{Z}_t$, modeled with a VARMA($p,q$) model, $\mathbf{Y}_t = \sum_{i=1}^{p} \mathbf{\Phi}_i^{(a)} \mathbf{Y}_{t-i} + \boldsymbol{\xi}_t - \sum_{j=1}^{q} \mathbf{\Theta}_j^{(a)} \boldsymbol{\xi}_t$. The condition for the forecasts of $\mathbf{Y}_t$ from an aggregate model to be equivalent to the aggregate of forecasts from a non-aggregate model are: $\mathbf{A}\mathbf{\Phi}_i = \mathbf{\Phi}_i^{(a)}\mathbf{A}$, for $i = 1, \ldots, p$ and $\mathbf{A}\mathbf{\Theta}_j = \mathbf{\Theta}_j^{(a)}\mathbf{A}$, for $j = 1, \ldots, q$, where $\boldsymbol{\xi}_t = \mathbf{A}\mathbf{a}_t$.

For the STARMA and STAR models, Gehman (2016) proved similar results. Given a non-aggregate data that follows a STARMA model and modeling its aggregate data as the same order as the non-aggregate data, the mean squared forecast error is always larger when using the aggregate model under the assumption that parameters are known.

Results shown here are based on the assumptions that all parameters are known. When parameters are unknown, Lütkepohl (1984) showed that forecasts from the aggregate data may outperform forecasts from the non-aggregate data, since parameter estimates could be noisy. This argument is more obvious in the high-dimension setting since so many parameters need to be estimated, which supports the reasoning that forecasting from the aggregate data could be better in some situations.

10.4 Simulation studies

In this section, we evaluate the performance of different methods in forecasting aggregates via Monte Carlo simulations. We consider three scenarios that were all simulated from the $m = 50$-dimensional VAR(1) model

$$\mathbf{Z}_t = \mathbf{\Phi}_1 \mathbf{Z}_{t-1} + \mathbf{a}_t, \tag{10.25}$$

where $\mathbf{\Phi}_1$ is coefficient matrix and $\mathbf{a}_t$ is vector white noise, which is simulated from a 50-dimensional normal random variable with zero-mean vector and identity covariance matrix. The number of observations used for in-sample modeling and estimation are set to be $n = 100, 500$. An additional five out-of-sample observations are used to compute the mean squared forecast error. To compare the performances, we simulate 200 realizations for each scenario. We consider two aggregation schemes. First, two-region aggregation, indicating that we aggregate the first 25 time series and the last 25 time series. Thus, the resulting aggregated time series is bivariate. Second, total aggregate, meaning that the aggregation matrix A is a row vector with all elements equal to one. We choose mean squared forecast error (*MSFE*) as the evaluation metric and define *MSFE*(ℓ) as the one-step-ahead forecast mean squared error of forecasts, such that

$$MSFE(\ell) = \frac{1}{200s} \sum_{k=1}^{S} \sum_{i=1}^{200} \left[Y_{k,t+\ell}^{(i)} - \hat{Y}_{k,t}^{(i)}(\ell) \right]^2 .$$

Methods compared in this section include: (i) the VAR model based on non-aggregate data and estimated through the least square; (ii) the univariate AR model for each time series with the lag orders selected by AIC, denoted by AR; (iii) the lasso method; (iv) the lag-weighted lasso method; (v) the HVAR method with componentwise structure, denoted by HVAR-C; (vi) the HVAR method with own-other structure, denoted by HVAR-OO; (vii) the HVAR method with elementwise structure, denoted by HVAR-E; (viii) the factor model with one lag; (ix) the model-based cluster method with maximum four clusters; and (x) the proposed method.

10.4.1 Scenario 1

In scenario 1, we assume $\mathbf{\Phi}_1$ to be a diagonal matrix with the diagonal elements generated from uniform distribution $U(0.2, 0.4)$. This is a very simple case in which there is no interdependence between each individual time series, and the AR coefficients for each series are similar. Thus, a simple model based on the univariate AR model for each time series would possibly produce fairly reasonable fitting and forecasts.

Table 10.1 displays the MSFEs and corresponding standard deviations of two-region aggregation. The smallest MSFE in each category are in boldface to facilitate presentation. It appears that the VAR method has much larger MSFE compared to all other methods when $n = 100$. This is due to large parameter estimation errors when n is relatively small. As the series length n increases to 500, the MSFE of the VAR method approaches other methods. Although all methods except the VAR method perform similarly in terms of MSFE, the proposed method produces the smallest MSFE in most cases. Further, it seems that all regularization methods produce similar MSFEs. Table 10.2 presents the MSFEs and their standard deviations of total

Table 10.1 Scenario 1 (based on 200 repetitions) mean square forecast errors (MSFE) and standard deviations of two-region aggregation with the smallest MSFEs in boldface.

Sample	ℓ	VAR	AR	Lasso	Lag-weighted	HVAR-C	HVAR-OO	HVAR-E	Factor	Cluster	Proposed
$n = 100$	1	43.5	22.98	24.19	24.02	25.05	23.4	24.22	26.97	23.31	**22.25**
		(61.25)	(35.27)	(39.41)	(36.41)	(35.76)	(33.92)	(33.93)	(36.53)	(36.17)	(36.32)
	2	52.64	34.46	35.29	35.31	35.23	35.45	34.85	34.95	34.78	**34.05**
		(81.01)	(56.19)	(55.94)	(56.12)	(56.98)	(57.13)	(57.13)	(56.53)	(56.03)	(53.47)
	3	24.76	**22.91**	23.28	23.19	22.96	23.04	23.08	29.78	23.31	23.07
		(36.65)	(34.63)	(35.04)	(34.87)	(34.8)	(34.88)	(34.89)	(39.8)	(34.54)	(35.01)
$n = 500$	1	29.49	28.18	28.8	28.45	28.15	27.76	27.87	28.55	28.75	**27.68**
		(43.46)	(43.36)	(43.06)	(42.15)	(41.95)	(42.1)	(42.1)	(41.5)	(40.96)	(40.96)
	2	28.18	27.16	27.49	27.22	28.19	27.36	27.28	26.21	27.67	**26.08**
		(42.25)	(41.48)	(41.3)	(41.21)	(43.11)	(42.04)	(41.9)	(39.14)	(41.75)	(41.1)
	3	**27.79**	28.11	27.94	27.95	27.97	27.89	27.95	28.32	28.38	27.93
		(33.64)	(34.78)	(34.05)	(34.25)	(34.24)	(34.18)	(34.31)	(38.01)	(34.9)	(34.36)

Table 10.2 Scenario 1 (based on 200 repetitions) MSFEs and standard deviations of total aggregation with the smallest MSFEs in boldface.

Sample	ℓ	VAR	AR	Lasso	Lag-weighted	HVAR-C	HVAR-OO	HVAR-E	Factor	Cluster	Proposed
n = *100*	1	98.65	49.63	50.2	50.1	50.17	48.81	49.33	55.29	49.68	**48.67**
		(147.22)	(70.68)	(68.91)	(69.23)	(73.43)	(71.76)	(68.28)	(76.65)	(70.37)	(70.09)
	2	94.76	**61.17**	63.42	62.31	62.28	61.35	61.28	66.84	61.2	61.63
		(129.28)	(84.39)	(86.69)	(84.23)	(87.52)	(84.53)	(84.32)	(97.42)	(84.12)	(86.52)
	3	43.98	40.98	42.13	41.89	41.42	41.76	41.66	45.23	**40.72**	41.31
		(63.44)	(59.97)	(61.32)	(60.93)	(60.17)	(60.61)	(60.59)	(63.84)	(56.53)	(60.42)
n = *500*	1	57.89	54.79	57.24	57.32	57.89	57.5	55.57	59.07	54.25	**53.98**
		(−82.42)	(74.66)	(80.69)	(80.14)	(82.14)	(76.02)	(75.32)	(81.88)	(75.17)	(73.77)
	2	57.07	56.77	57.06	56.92	57.58	56.96	56.99	58.42	56.28	**56.16**
		(86.77)	(86.31)	(87.73)	(87.58)	(89.71)	(87.19)	(87.07)	(87.97)	(85.9)	(86.77)
	3	58.33	58.79	58.36	58.21	58.41	58.44	58.46	**54.07**	58.72	58.97
		(70.67)	(70.76)	(70.66)	(70.37)	(70.79)	(70.49)	(70.47)	(63.81)	(70.65)	(71.59)

aggregation. The results in Table 10.2 are nearly consistent with results in Table 10.1 and the proposed method is still among one of the best methods.

10.4.2 Scenario 2

In scenario 2, the coefficient matrix $\mathbf{\Phi}_1$ is generated from a "band" matrix pattern shown in Figure 10.1 where purple points correspond to nonzero entries and white areas correspond to zero entries. The nonzero diagonal entries of $\mathbf{\Phi}_1$ are fixed to be 0.3 and the nonzero off-diagonal elements are fixed at 0.1. This coefficient structure indicates that each time series depends largely on its own past, and weakly depends on other series that are close. Tables 10.3 and 10.4 present the MSFEs of two different aggregation schemes. Again, the MSFEs and standard deviations of VAR method are much larger than all other methods when series length $n = 100$. For $n = 100$, the proposed method outperforms all other methods. For $n = 500$, the proposed method outperforms all other methods when $\ell = 1$ and 2, and the factor model performs best when $\ell = 3$. Among all regularization methods, the HVAR-OO produces relative smaller MSFEs when $n = 100$. This is because HVAR-OO assumes the diagonal elements to be more informative that are close to the true coefficient matrix structure. HVAR-E has smaller MSFEs than other regularization methods when $n = 500$. This is due to its flexible structure assumption.

10.4.3 Scenario 3

In the scenario 3, the coefficient matrix $\mathbf{\Phi}_1$ is generated from a "cluster" matrix pattern. We set the diagonal elements of $\mathbf{\Phi}_1$ to be 0.3. Then, we randomly select $2m$ elements from off-diagonal elements of $\mathbf{\Phi}_1$ and assign each of them with the value 0.1 (see Figure 10.2). Tables 10.5 and 10.6 show the MSFEs and the corresponding standard deviations. In this more complicated simulation setting, the AR model and the model-based cluster method have very large MSFEs for both of the sample sizes we consider. This is because they largely ignore the interdependences between each time series. For both $n = 100$ and 500, the proposed method outperforms other methods in terms of MSFEs.

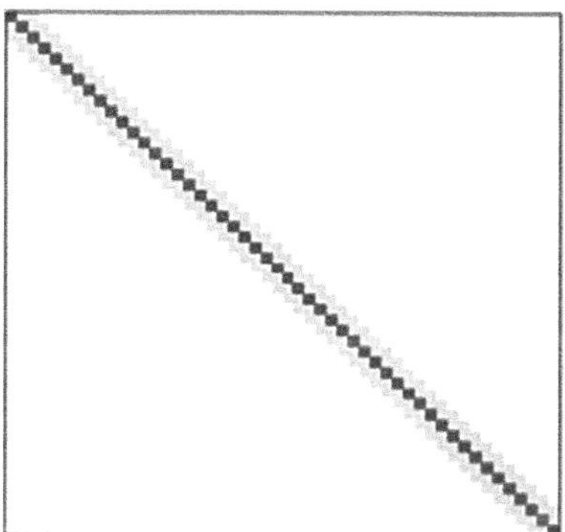

Figure 10.1 Pattern of $\mathbf{\Phi}_1$ in scenario 2.

Table 10.3 Scenario 2 (based on 200 repetitions) MSFEs and standard deviations of two-region aggregation with the smallest MSFEs in boldface.

Sample	ℓ	VAR	AR	Lasso	Lag-weighted	HVAR-C	HVAR-OO	HVAR-E	Factor	Cluster	Proposed
$n = 100$	1	60.07	35.38	40.67	40.53	35.09	31.61	33.03	33.49	33.29	**25.68**
		(88.08)	(47.16)	(55.97)	(54.37)	(49.4)	(42.04)	(44.24)	(41.79)	(47.6)	(34.05)
	2	96.52	64.89	67.18	65.01	61.28	55.88	56.71	57.22	61.19	**40.89**
		(131.14)	(83.75)	(92.72)	(89.13)	(83.62)	(75.92)	(79.64)	(66.61)	(80.1)	(62.56)
	3	139.34	88.5	91.18	89.18	93.26	85.64	86.97	94.67	85.4	**75.77**
		(203.14)	(130.39)	(127.07)	(121.25)	(124.12)	(126.97)	(124.33)	(126.11)	(119.4)	(114.31)
$n = 500$	1	26.62	37.17	28.61	27.02	29.31	28.36	26.67	33.87	38.27	**23.07**
		(43.12)	(41.48)	(45.08)	(41.1)	(46.53)	(45.12)	(44.63)	(42.08)	(55.61)	(39.34)
	2	47.43	66.17	52.04	51.9	51.31	50.57	48.34	51.86	67.27	**42.4**
		(86.33)	(100.05)	(89.29)	(82.21)	(95.19)	(90.69)	(84.9)	(73.49)	(104.11)	(73.76)
	3	82.48	100.91	84.78	84.21	87.72	86.46	85.38	**74.07**	102.03	78.53
		(108.93)	(123.56)	(107.83)	(101.3)	(112.24)	(109.56)	(106.01)	(104.64)	(121.23)	(97.67)

Table 10.4 Scenario 2 (based on 200 repetitions) MSFEs and standard deviations of total aggregation with the smallest MSFEs in boldface.

Sample	ℓ	VAR	AR	Lasso	Lag-weighted	HVAR-C	HVAR-OO	HVAR-E	Factor	Cluster	Proposed
$n = 100$	1	121.27	67.72	79.5	76.31	73.5	63.52	66.45	78.79	77.03	**53.19**
		(166.05)	(88.77)	(105.12)	(102.18)	(107.03)	(89.22)	(91.97)	(100.34)	(102.87)	(70.24)
	2	214.58	132.14	128.11	118.21	129.32	117.72	121.51	118.62	130.98	**95.11**
		(286.97)	(173.65)	(172.1)	(161.25)	(176.69)	(161.75)	(170.92)	(161.4)	(163.21)	(121.5)
	3	297.62	182.33	172.08	169.94	185.36	170.15	166.59	192.95	188.52	**150.99**
		(426.84)	(266.46)	(233.07)	(233.18)	(256.29)	(245.52)	(234.41)	(247.35)	(231.23)	(202.51)
$n = 500$	1	54	69.94	54.99	52.28	55.17	54.05	52.72	60.59	62.12	**49.16**
		(69.08)	(90.14)	(74.65)	(70.53)	(74.23)	(70.61)	(91.97)	(91.83)	(99.46)	(63.31)
	2	94.76	116.69	93.82	91.82	95.19	98.82	89.16	99.45	118.62	**82.04**
		(125.59)	(157.2)	(130.48)	(122.12)	(131.1)	(125.08)	(122.00)	(139.41)	(158.12)	(118.87)
	3	186.63	214.26	187.51	186.21	192.41	189.95	187.12	**166.28**	216.6	172.06
		(254.26	(302.01)	(247.53)	(232.72)	(252.89)	(253.2)	(248.16)	(215.91)	(312.1)	(232.37)

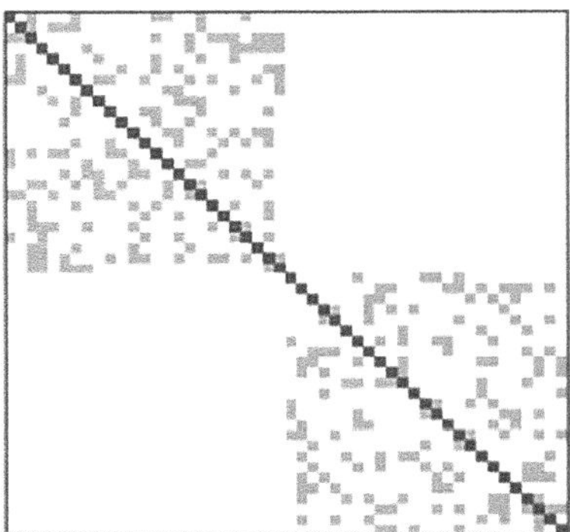

Figure 10.2 Pattern of $\mathbf{\Phi}_1$ in scenario 3.

10.5 Empirical examples

In this section, two real data examples are considered, one using the macroeconomic time series data and the other using sexually transmitted disease (STD) time series data.

10.5.1 The macroeconomic time series

We compare MSFEs of different methods and assess the effectiveness of the proposed method through the collection of time series of U.S. macroeconomic indicators. The data is collected from Stock and Watson (2009) and Koop (2013). The full data list contains 168 quarterly macroeconomic variables from Quarter 1 of 1959 to Quarter 4 of 2007, representing information about many aspects of the US economy. We retrieve 61 time series from the full dataset. Readers who are interested in this data can view their papers for further details. Variables which are originally at a monthly frequency are transformed to quarterly by taking average of three months in a quarter. Seasonally adjustments are made if necessary, leading to a dataset with $m = 61$ and $n = 196$. Time series are transformed to stationary using the suggestion of Stock and Watson (2009) with $n = 195$ remaining, and listed as WW10 in the Data Appendix. Table 10.7 contains brief descriptions of each variable, and the aggregation group they belong to, along with a transformation code, where 1 = first differencing of log of variables, and 2 = second differencing of log of variables.

Those 61 time series can be aggregated into three main macroeconomic measures by their nature: gross domestic product (GDP), industrial production index (IPS), and constant elasticity of substitution (CES).

The main interest of this section is on accurately forecasting three aggregate variables: GDP, IPS, and CES, since they are important measurements of the US economy. In this application, we used data from Quarter 1 of 1959 to Quarter 3 of 1992 for model fitting, and then compute the rolling out-of-sample one-step-ahead forecasts, starting from Quarter 4 of 1992 to Quarter 4 of 2007. The MSFEs of the univariate AR, VAR, lasso, lag-weighted lasso, HVAR-C, HVAR-OO, HVAR-E, the factor model, the model-based cluster, and the proposed method are compared in this application and the results given in Table 10.8.

The three regularization methods, including HVAR-C, HVAR-OO, and HVAR-E, perform the best as their MSFEs are below 0.7. The proposed method performs close to those three methods and has smaller MSFEs than all other methods. The benchmark univariate AR method

Table 10.5 Scenario 3 (based on 200 repetitions) MSFEs and standard deviations of two-region aggregation with the smallest MSFEs in boldface.

Sample	ℓ	VAR	AR	Lasso	Lag-weighted	HVAR-C	HVAR-OO	HVAR-E	Factor	Cluster	Proposed
$n = 100$	1	58.59	105.66	59.34	58.13	42.43	41.65	47.79	83.27	102.13	**26.91**
		(79.16)	(152.43)	(93.34)	(92.61)	(62.31)	(62.71)	(69.91)	(81.52)	(131.09)	(36.94)
	2	90.44	172.88	92.25	89.76	71.34	72.69	87.25	122.54	168.1	**50.61**
		(107.71)	(266.76)	(147.69)	(151.00)	(102.08)	(106.57)	(126.1)	(196.01)	(231.22)	(71.77)
	3	200.81	369.7	221.68	198.28	196.56	198.79	231.06	233.49	310.36	**151.58**
		(312.03)	(539.85)	(332.73)	(293.89)	(279.96)	(280.07)	(331.01)	(376.56)	(423.91)	(211.32)
$n = 500$	1	21.98	63.4	25.65	25.39	24.09	23.28	23.31	31.04	64.82	**21.4**
		(31.26)	(86.55)	(32.79)	(31.08)	(32.72)	(30.24)	(30.23)	(44.46)	(83.28)	(26.74)
	2	49.08	110.83	56.89	56.89	56.19	52.42	51.88	61.46	112.21	**47.2**
		(61.87)	(161.49)	(79.29)	(79.29)	(78.34)	(70.28)	(69.55)	(90.62)	(167.51)	(67.96)
	3	114.33	202.02	125.73	125.24	120.38	121.67	123.55	**97.45**	205.16	116.06
		(160.87)	(314.87)	(178.88)	(175.12)	(175.33)	(179.78)	(177.49)	(132.56)	(310.29)	(157.49)

Table 10.6 Scenario 3 (based on 200 repetitions) MSFEs and standard deviations of total aggregation with the smallest MSFEs in boldface.

Sample	ℓ	VAR	AR	Lasso	Lag-weighted	HVAR-C	HVAR-OO	HVAR-E	Factor	Cluster	Proposed
$n = 100$	1	107.19	182.26	116.37	103.15	84.45	80.48	93.58	133.16	188.28	**54.95**
		(149.76)	(257.29)	(170.56)	(149.18)	(122.7)	(118.01)	(127.87)	(222.4)	(217.39)	(74.07)
	2	186.46	408.39	179.01	170.24	140.49	140.87	166.53	196.96	292.90	**112.13**
		(260.73)	(294.77)	(246.78)	(250.17)	(175.49)	(174.82)	(210.67)	(333.81)	(402.15)	(146.59)
	3	346.11	544.34	356.22	341.03	319.08	326.22	369.65	338.71	751.73	**308.98**
		(595.51)	(761.25)	(532.16)	(541.36)	(440.95)	(452.28)	(522.6)	(556.76)	(548.12)	(453.83)
$n = 500$	1	49.95	208.07	54.48	53.28	53.19	53.09	53.52	86.99	133.21	**48.03**
		(67.17)	(132.74)	(84.02)	(80.97)	(70.37)	(72.87)	(73.3)	(124.47)	(202.15)	(64.02)
	2	96.22	217.57	111.54	110.27	101.45	101.99	103.41	113.79	216.27	**93.04**
		(137.05)	(357.48)	(174.73)	(173.67)	(149.36)	(155.19)	(157.41)	(155.56)	(357.20)	(136.73)
	3	206.39	368.43	234.19	230.59	221.89	226.44	224.87	**171.55**	364.12	219.96
		(267.12)	(522.56)	(312.03)	(309.1)	(290.89)	(308.09)	(296.66)	(275.44)	(502.16)	(285.35)

Table 10.7 Variables used in Section 10.5.1.

Variable	Description	Code	Group
GDP252	Real personal consumption exp: quantity index	1	GDP
GDP253	Real personal consumption exp: durable goods	1	GDP
GDP254	Real personal consumption exp: nondurable goods	1	GDP
GDP255	Real personal consumption exp: services	1	GDP
GDP256	Real gross private domestic inv: quantity index	1	GDP
GDP257	Real gross private domestic inv: xed inv	1	GDP
GDP258	Real gross private domestic inv: nonresidential	1	GDP
GDP259	Real gross private domestic inv: nonres structure	1	GDP
GDP260	Real gross private domestic inv: nonres equipment	1	GDP
GDP261	Real gross private domestic inv: residential	1	GDP
GDP266	Real gov consumption exp, gross inv: federal	2	GDP
GDP267	Real gov consumption exp, gross inv: state and local	2	GDP
GDP268	Real final sales of domestic product	2	GDP
GDP269	Real gross domestic purchases	2	GDP
GDP271	Real gross national product	2	GDP
GDP272	Gross domestic product: price index	2	GDP
GDP274	Personal cons exp: durable goods, price index	2	GDP
GDP275	Personal cons exp: nondurable goods, price index	2	GDP
GDP276	Personal cons exp: services, price index	2	GDP
GDP277	Gross private domestic investment, price index	2	GDP
GDP278	Gross priv dom inv: fixed inv, price index	2	GDP
GDP279	Gross priv dom inv: nonresidential, price index	2	GDP
GDP280	Gross priv dom inv: nonres structures, price index	2	GDP
GDP281	Gross priv dom inv: nonres equipment, price index	2	GDP
GDP282	Gross priv dom inv: residential, price index	2	GDP
GDP284	Exports, price index	2	GDP
GDP285	Imports, price index	2	GDP
GDP286	Government cons exp and gross inv, price index	2	GDP
GDP287	Gov cons exp and gross inv: federal, price index	2	GDP
GDP288	Gov cons exp and gross inv: state and local, price index	2	GDP
GDP289	Final sales of domestic product, price index	2	GDP
GDP290	Gross domestic purchases, price index	2	GDP
GDP291	Final sales to domestic purchasers, price index	2	GDP
GDP292	Gross national products, price index	2	GDP
IPS11	Industrial production index: products total	1	IPS
IPS299	Industrial production index: final products	1	IPS
IPS12	Industrial production index: consumer goods	1	IPS
IPS13	Industrial production index: consumer durable	1	IPS
IPS18	Industrial production index: consumer nondurable	1	IPS
IPS25	Industrial production index: business equipment	1	IPS
IPS32	Industrial production index: materials	1	IPS

(*continued overleaf*)

Table 10.7 (*continued*)

Variable	Description	Code	Group
IPS34	Industrial production index: durable goods materials	1	IPS
IPS38	Industrial production index: nondurable goods material	1	IPS
IPS43	Industrial production index: manufacturing	1	IPS
IPS307	Industrial production index: residential utilities	1	IPS
IPS306	Industrial production index: consumer fuels	1	IPS
CES275	Avg hrly earnings, prod wrkrs, nonfarm-goods prod	2	CES
CES277	Avg hrly earnings, prod wrkrs, nonfarm-construction	2	CES
CES278	Avg hrly earnings, prod wrkrs, nonfarm-manufacturing	2	CES
CES003	Employees, nonfarm: goods-producing	1	CES
CES006	Employees, nonfarm: mining	1	CES
CES011	Employees, nonfarm: construction	1	CES
CES015	Employees, nonfarm: manufacturing	1	CES
CES017	Employees, nonfarm: durablc goods	1	CES
CES033	Employees, nonfarm: nondurable goods	1	CES
CES046	Employees, nonfarm: service providing	1	CES
CES048	Employees, nonfarm: trade, transportation, and utilities	1	CES
CES049	Employees, nonfarm: wholesale trade	1	CES
CES053	Employees, nonfarm: retail trade	1	CES
CES088	Employees, nonfarm: financial activities	1	CES
CES140	Employees, nonfarm: government	1	CES

Table 10.8 MSFEs of forecasting three aggregate macroeconomics variables (GDP, IPS, CES).

	MSFE
Univariate AR	0.838
VAR	1.537
Lasso	0.744
Lag-weighted lasso	0.752
HVAR-C	0.683
HVAR-OO	0.667
HVAR-E	0.699
Factor model	0.883
Model-based cluster clustering	1.465
The proposed method	0.715

outperforms the VAR, the factor model, and the model-based cluster, but does not perform as well as the proposed aggregation method.

10.5.2 The yearly U.S. STD data

In this section, we provide an illustration using a spatial time series introduced in Chapter 8. Recall that the data set contains yearly sexually transmitted disease STD morbidity rates reported to the National Center for HIV/AIDS, viral Hepatitis, STD, and TB Prevention (NCHHSTP), Center for HIV, and Centers for Disease Control and Prevention (CDC) from 1984 to 2014. The dataset was retrieved from the CDC's website and includes 50 states plus DC. The rates per 100,000 persons are calculated as the incidence of STD reports, divided by the population, and multiplied by 100,000.

For the analysis, we standardized each time series and removed data from the following states, Montana, North Dakota, South Dakota, Vermont, Wyoming, Alaska, and Hawaii, due to missing data. Hence, the dimension of data is $m = 44$ and $n = 29$. The data is listed as WW8c in the Data Appendix and shown in Figure 10.3. In modeling STD data, researchers are interested in forecasting aggregate data based on nine Morbidity and Mortality Weekly Report (MMWR) regions or four STD regions as shown in Figure 10.4.

We used the first 24 observations for model fitting, and the rest of the observations for evaluating the forecasting performance. The MSFEs averaged across the lags are reported. Methods considered include: the univariate AR, VAR, lasso, lag-weighted lasso, HVAR-C, HVAR-OO, HVAR-E, the factor model, the model-based cluster, and the proposed method. In addition, we also add the STAR model for comparison in this application, as it is one of most naturally

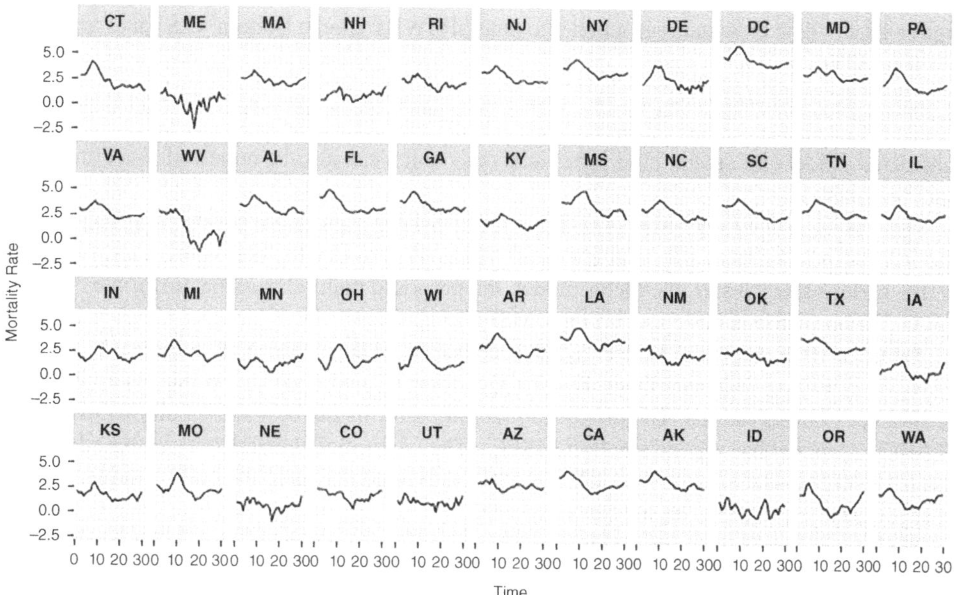

Figure 10.3 The U.S. yearly sexually transmitted disease (STD) time series for each state.

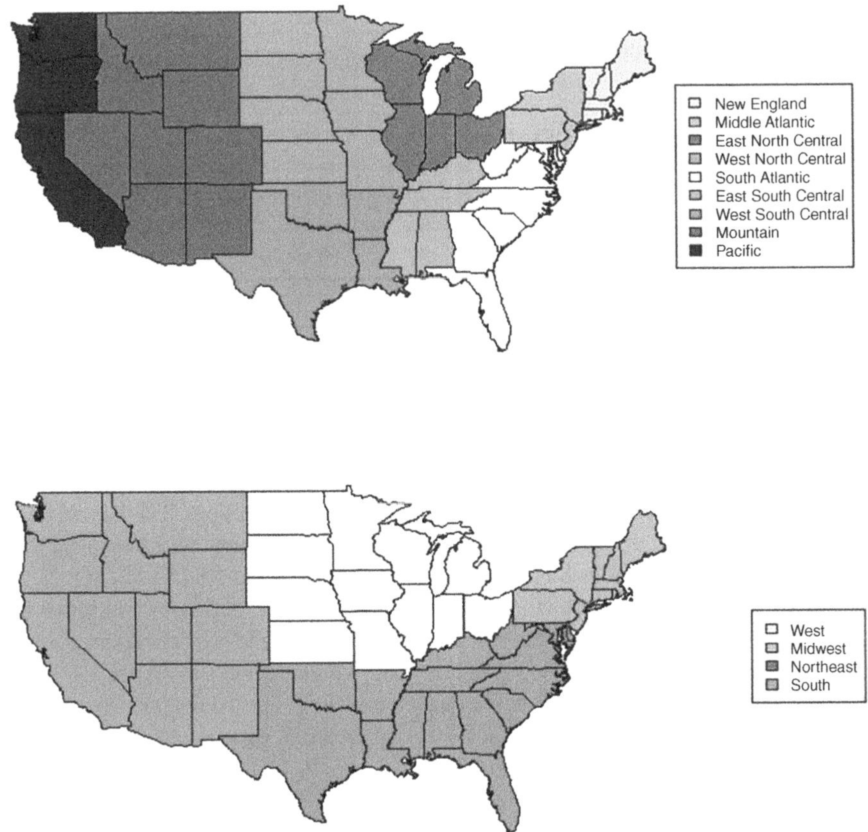

Figure 10.4 Top: U.S. states grouped into nine MMWR regions; Bottom: U.S. states grouped into four STD regions.

considered models for spatial time series analysis, which can also be reviewed as a dimension reduction method.

The MSFEs when forecasting the STD at nine MMWR regions are presented in Table 10.9. The VAR fails to estimate the parameters. STAR($2_{2,\,1}$) performs the best in this case, followed by the model-based cluster, the factor model, the proposed method, and the univariate AR. All regularization methods have much larger MSFEs.

Table 10.10 displays the MSFEs when forecasting STDs in four STD regions. Again, the VAR fails to estimate the parameters due to the number of parameters to estimate is bigger than the number of observations. The proposed method performs the best among all methods. The STAR($2_{2,\,1}$) model has second smallest MSFEs. The factor model and the model-based cluster also perform reasonably well. The MSFE for the univariate AR is in the middle range. Again, all regularization methods do not perform well.

We have analyzed many procedures in this section and summarized their estimation results in Tables 10.8–10.10. In supporting the results in these tables and for readers' reference, the outputs of the analysis and estimation from various procedures are provided in this chapter's appendix.

Table 10.9 MSFEs in forecasting STD rate in the nine MMWR regions.

	MSFE
Univariate AR	4.54
VAR	NA
Lasso	5.63
Lag-weighted lasso	5.05
HVAR-C	5.79
HVAR-OO	5.65
HVAR-E	5.56
STAR($2_{2,1}$)	**3.73**
Factor model	4.51
Model-based cluster clustering	3.97
Proposed method	4.53

Table 10.10 MSFEs in forecasting STD rate in the four STD regions.

	MSFE
Univariate AR	14.10
VAR	NA
Lasso	19.10
Lag-weighted lasso	19.27
HVAR-C	19.81
HVAR-OO	19.16
HVAR-E	18.75
STAR($2_{2,1}$)	10.91
Factor model	13.96
Model-based cluster clustering	12.56
Proposed method	**10.65**

10.6 Further discussions and remarks

10.6.1 More on clustering

Clustering is a method of grouping a set of elements or a complicated data set into some homogeneous groups. Other than subjective observation, a very nature objective method of clustering is based on a distance and similarity measure. In time domain, given two r-dimensional observations, $\mathbf{X} = (x_1, x_2, \ldots, x_r)$ and $\mathbf{Y} = (y_1, y_2, \ldots, y_r)$. Some commonly used distance and similarity measures include the following.

1. Euclidian distance:

$$D_E(\mathbf{X},\mathbf{Y}) = \sqrt{(x_1-y_1)^2 + (x_2-y_2)^2 + \cdots + (x_r-y_r)^2}.$$

2. Minkowski distance:

$$D_M(\mathbf{X},\mathbf{Y}) = \left[\sum_{i=1}^{r} |x_i - y_i|^r\right]^{1/r}.$$

3. Canberra distance:

$$D_C(\mathbf{X},\mathbf{Y}) = \sum_{i-1}^{r} \frac{|x_i - y_i|}{|x_i + y_i|}.$$

4. Kullback–Leibler (KL) distance:

   Statistically, we see that a good distance and similarity measure is a distribution-based or a model-based approach briefly introduced in Section 10.2.3, where an associated probability distribution is used in clustering. In terms of frequency domain spectral density distributions, given two spectra, $f_i(\omega_k)$ and $f_j(\omega_k)$, $0 < \omega_k < 1/2$, the most commonly used similarity measure of spectrum is the KL distance, defined as

$$KL(f_i, f_j) = \sum\nolimits_{0<\omega_k<1/2} \left[\frac{f_i(\omega_k)}{f_j(\omega_k)} - \log\frac{f_j(\omega_k)}{f_i(\omega_k)} - 1\right], \tag{10.26}$$

   and its symmetric form as the quasi-distance $D(f_i, f_j) = KL(f_i, f_j) + KL(f_j, f_i)$. It turns out that

$$D(f_i, f_j) = \sum\nolimits_{0<\omega_k<1/2} \left[\frac{f_i(\omega_k)}{f_j(\omega_k)} + \frac{f_j(\omega_k)}{f_i(\omega_k)} - 2\right], \tag{10.27}$$

   which is the distance we will use in the frequency domain.

Based on a distance measure, we have hierarchical and nonhierarchical clustering methods. Given a set of N elements, the hierarchical clustering method includes the following steps:

Step 1. Begin with N clusters by assuming each of the elements is an individual cluster. The distances between clusters are the same as the distances between elements.

Step 2. Combine the two elements that are most similar with minimum distance and form a new cluster. This leads to one less cluster.

Step 3. Compute distances between the newly formed cluster and other remaining clusters.

Step 4. Repeat steps 2 and 3 a total $(N - 1)$ times so that all become a single cluster after the algorithm terminates, with the results shown in a dendrogram.

It should be noted that step 3 can be done in different ways or algorithms. The first way is "single linkage" in which the minimum distance between each element of newly formed cluster and other clusters is computed. The second way is "complete linkage" in which the maximum distance between each element of newly formed cluster and other clusters is computed. The third way is "average linkage" in which the average distance between each element of newly formed cluster and other clusters is computed. The fourth way is "median linkage" in which the median distance between each element is used as introduced by Tibshirani et al. (2001).

In terms of nonhierarchical methods, the most popular approach is the K-means method, which assigns each item to the cluster having the nearest centroid (mean) or assigned center point.

In terms of deciding an optimal number of clusters, we will use the total within sum of squares (TWSS), which measures total intra-cluster variation and we would like it to be as small as possible. We will illustrate its use through an empirical example in the next section.

10.6.2 Forming aggregate data through both time domain and frequency domain clustering

In modeling the U.S. yearly STD data, we created aggregate data based on MMWR regions and four STD regions. We can also form the aggregate groups using clustering as illustrated in the following.

10.6.2.1 Example of time domain clustering

Recall the data set of U.S. yearly STD morbidity rates from 1984 to 2013, which we have analyzed earlier. It contains 43 states plus Washington D.C. So, we have a set of $m = 44$ elements, each with a length of $n = 29$. Based on the model-based clustering method, the 44 elements are classified into three clusters with the detail given in Table 10.11.

10.6.2.2 Example of frequency domain clustering

Next, we will show that the clustering of the STD data can also be done through frequency domain spectral matrices. Specifically, we will use the kernel smoothing method discussed in Chapter 9 to obtain the estimates of spectral matrices. We choose the Daniell smoothing window with bandwidth $m = 5$. The resulting individual spectra for 44 elements are presented in Figure 10.5.

Table 10.11 The model-based clustering of the STD data.

Cluster	Number of states	States
1	16	CT, NY, DC, PA, FL, GA, MS, IL, MI, OH, WI, AR, LA, MO, CA, WA
2	15	MA, NJ, DE, MD, VA, AL, NC, SC, TN, IN, TX, IA, KS, ID, OR
3	13	ME, NH, RI, WV, KY, MN, NM, OK, NE, CO, UT, AZ, NV

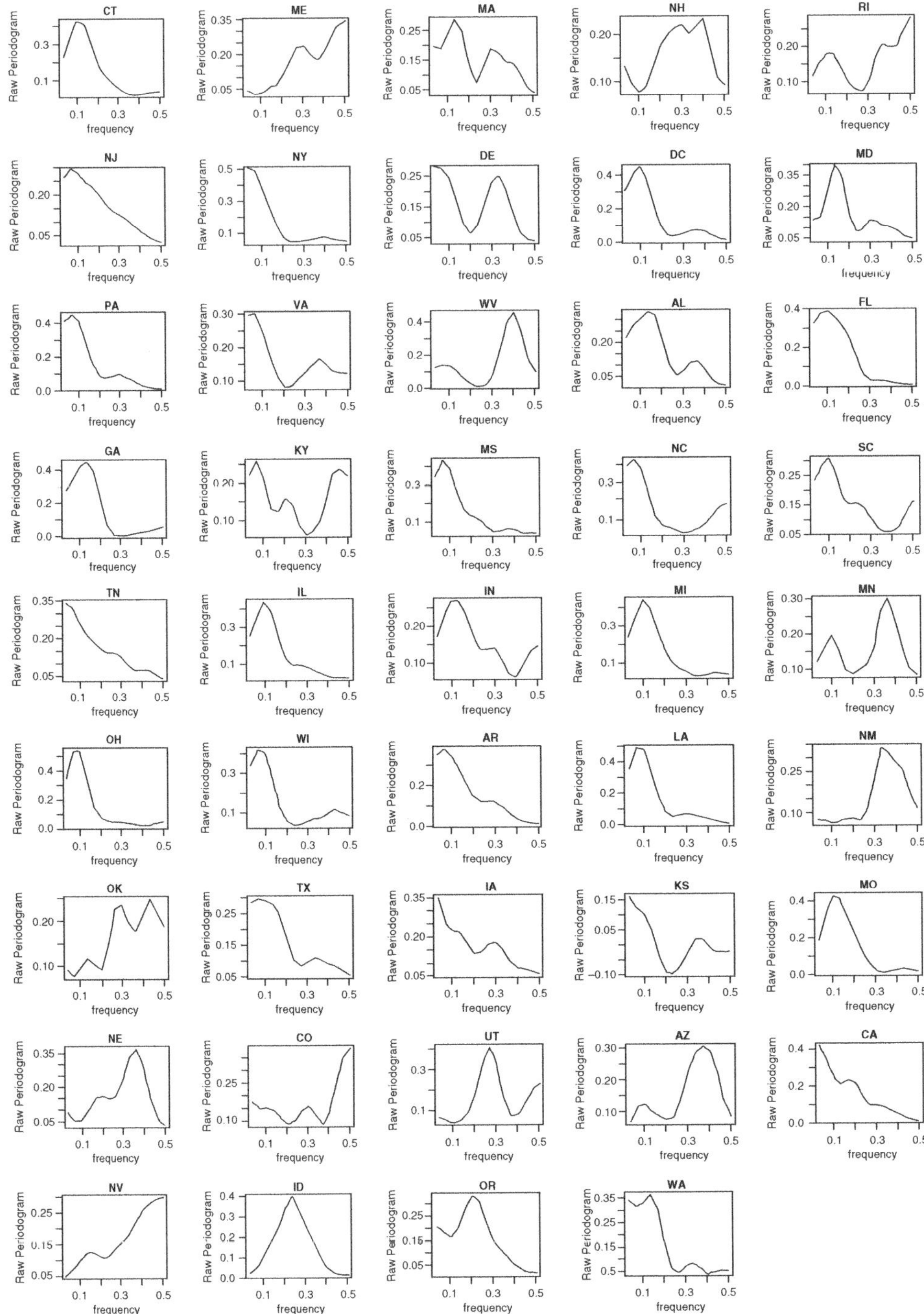

Figure 10.5 The spectra of the STD data for 43 states plus D.C.

10.6.2.2.1 Clustering using similarity measures

Based on the *KL* distance and its quasi-distance form, $D(f_i, f_j)$, we compute TWSS, which measures total intra-cluster variation. Figure 10.6 is the plot for TWSS across different numbers of clusters from the STD data set. The location of a bend (knee) in the plot is generally considered to be an indicator of the appropriate number of clusters. It shows that five or six clusters seem to be a good choice.

10.6.2.2.2 Clustering by subjective observation

The spectrum matrices in Figure 10.5 clearly indicate a group or cluster pattern. Intuitively, the spectrum of each state can be roughly categorized into the following clusters, which are highly consistent with the MMWR regions. Based on the similarity of spectrums, we can group them in the following six clusters given in Table 10.12.

10.6.2.2.3 Hierarchical clustering

The dendrograms of the four algorithms for the STD data are shown in Figures 10.7–10.10.

10.6.2.2.4 Nonhierarchical clustering using the K*-means method*

The Partitioning Around Medoids (PAM) is a robust version of the *K*-means clustering algorithm. It begins with a selection *K* objects as medoids, and then we minimize the sum of the dissimilarities of the observations to their closest cluster. We apply PAM to the spectrums of STD data set with $K = 4$, and obtain the following result in Table 10.13 and its cluster plot in Figure 10.11.

It turns out that the clustering patterns found by the model-based clustering and the PAM method are very similar. Specifically, the cluster 1 of the model-based clustering method is almost identical to the cluster 1 of PAM; the cluster 2 of the model-based clustering method corresponds to the clusters 3 and 4 of PAM method; and cluster 3 of the model-based clustering method is close to the cluster 2 of the PAM method.

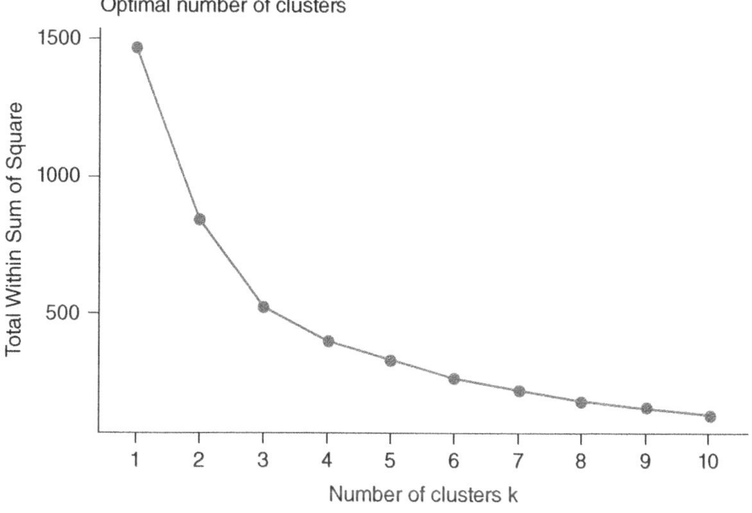

Figure 10.6 Total within sum of squares plot.

Table 10.12 Clustering by subjective observation.

Cluster	Number of states	States
1	3	ME, NH, RI These states are in the North East region.
2	4	CT, NJ, NY, PA These states are in the Middle Atlantic region.
3	12	MA, DE, DC, MD, VA, AL, FL, GA, MS, SC, NC, TN Most of these states are in the South Atlantic and East South Central regions.
4	14	IL, IN, KY, MI, MN, OH, WV, WI, KS, MO, IA, TX, LA, AR These states are in the East North Central, Western North Central, and West South Central regions.
5	8	OK, NE, AZ, NM, NV, ID, UT, CO These states are in the West South Central and Mountain regions.
6	3	CA, OR, WA These states are in the Pacific region.

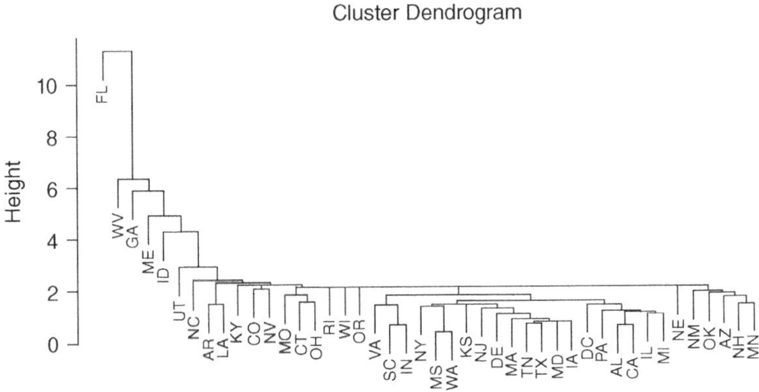

Figure 10.7 Single linkage dendrogram for STD spectra.

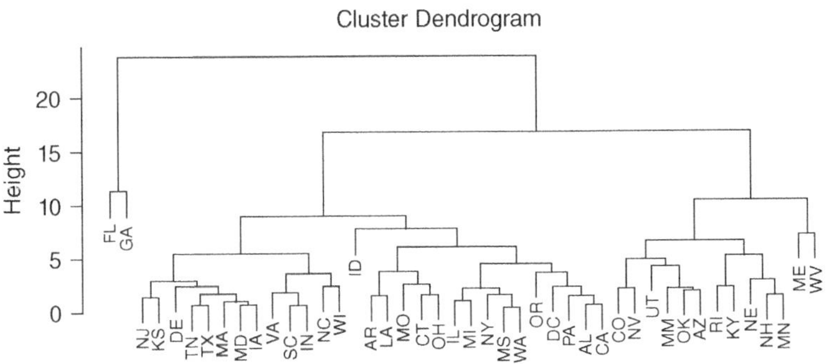

Figure 10.8 Complete linkage dendrogram for STD spectra.

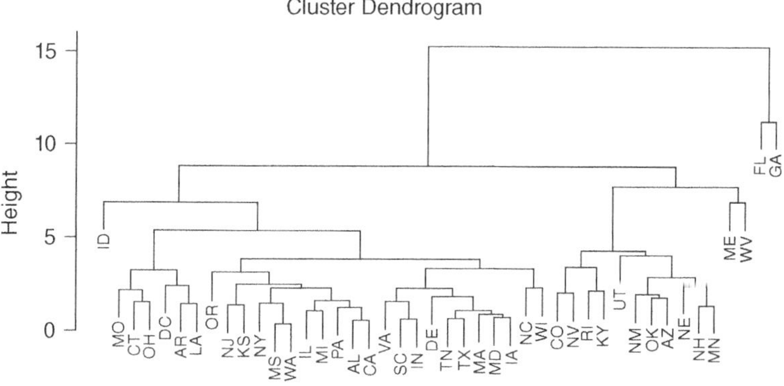

Figure 10.9 Average linkage dendrogram for STD spectra.

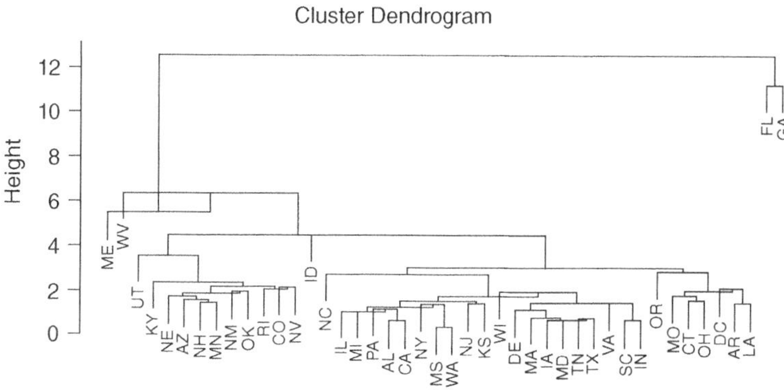

Figure 10.10 Median linkage dendrogram for STD spectra.

Table 10.13 PAM clustering based on STD spectrums with $K = 4$.

Cluster	Number of states	States
1	17	CT, NY, DC, PA, AL, FL, GA, MS, TN, IL, MI, OH, AR, LA, MO, CA, WA
2	9	ME, NH, NM, OK, NE, CO, UT, AZ, NV
3	13	MA, RI, DE, VA, WV, KY, NC, SC, IN, MN, WI, TX, IA
4	5	NJ, MD, KS, ID, OR

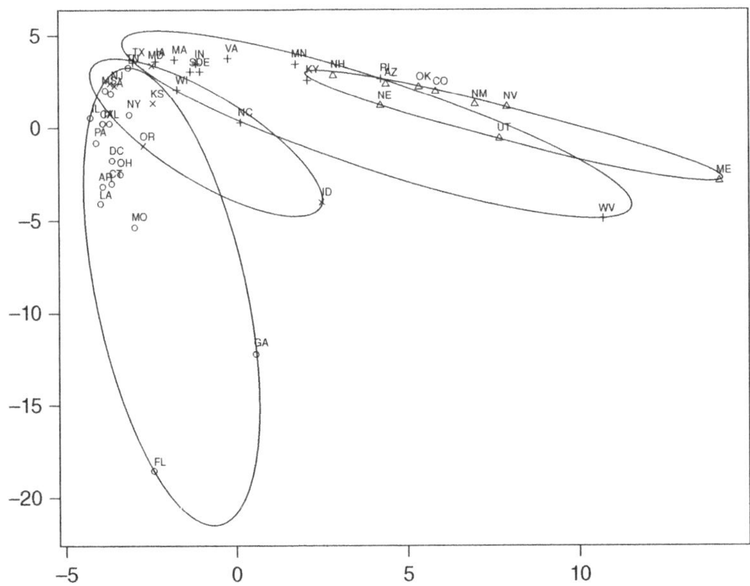

Figure 10.11 PAM cluster plot of the STD data set.

For more discussion and examples of clustering, we refer readers to Tibshirani et al. (2001), Johnson and Wichern (2007), and Izenman (2008), among others.

10.6.3 The specification of aggregate matrix and its associated aggregate dimension

Big data and high-dimensional problems are everywhere in the age of fast computers and the internet. To address this, we propose aggregation as a dimension reduction method. It is very natural and simple to use, and its performance in forecasting is supported by both simulation and empirical examples as a good method for dimension reduction.

The aggregation matrix $\mathbf{A}$ and its associated s can occur in many different specifications. Even based on practical considerations, we can specify different forms of $\mathbf{A}$ and s. By choosing $s = 1$, $\mathbf{A}$ becomes a row vector, and any m-dimensional multivariate time series $\mathbf{Z}_t$ will aggregate to become a univariate time series Y_t. Most of the time, the result of aggregation is meaningful. For example, sales data can be specified in terms of regions or kinds (categories). With regard to housing sales of the 3144 US counties, this data can be aggregated into housing sales for each of the 50 states, into the housing sales of the four regions (East, West, North, and South), and further into the total housing sales of the entire country. We can also specify $\mathbf{A}$ and s based on data-driven considerations, which we will leave to our readers to try.

10.6.4 Be aware of other forms of aggregation

It is important to note that aggregation as a tool for high dimension reduction that we have suggested is contemporal aggregation. Other forms of aggregation include temporal aggregation for a flow variable such as industrial production and a systematic sampling for a stock variable such as the price of a given commodity. However, these techniques are not useful for high dimension reduction, in fact, they will aggravate the problem, since temporal aggregation and systematic sampling lead to very serious information loss as shown in Wei (2006, chapter 20). In fact, limitations of temporal aggregation and systematic sampling have been studied by many researchers including Amemiya and Wu (1972), Brewer (1973), Tiao and Wei (1976), Weiss (1984), Stram and Wei (1986), Lütkepohl (1987), Teles and Wei (2000, 2002), Breitung and Swanson (2002), Teles et al. (1999, 2008), Lee and Wei (2017), and many more. An interesting related topic is how to recover the information loss of temporal aggregation and systematic sampling. One can use either temporal disaggregation or bootstrap, which we will not discuss in this book and refer readers to Wei and Stram (1990), Meyer and Kreiss (2015), Kim and Wei (2018), and Tewes (2018), among others.

Before closing this chapter, we want to point out that high dimension is an important issue in multivariate time series analysis. We introduce a simple and useful method to reduce dimension. For more discussion and applications on high-dimensional problems, we refer readers to Chen and Qin (2010), Huang et al. (2010), Flamm et al. (2013), Cho and Fryzlewicz (2015), Giraud (2015), Hung et al. (2016), Lee et al. (2016), Gao et al. (2017), Huang and Chiang (2017), and Wood et al. (2017), among others.

10.A Appendix: Parameter estimation results of various procedures

The details of parameter estimation results for the two empirical examples in Section 10.5 are given next.

10.A.1 Further details of the macroeconomic time series

Since we used out-of-sample, the estimation results shown in this section is based on in-sample data from quarter 1 of 1959 to Quarter 4 of 2007.

10.A.1.1 VAR(1)

Based on the AIC criteria, the VAR(1) model was fitted to the data. The least square estimation was used and the estimated coefficient matrix is

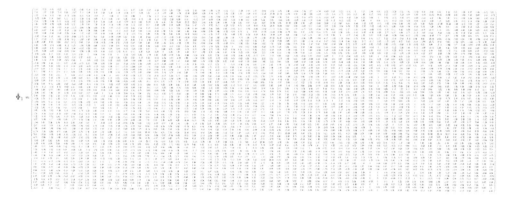

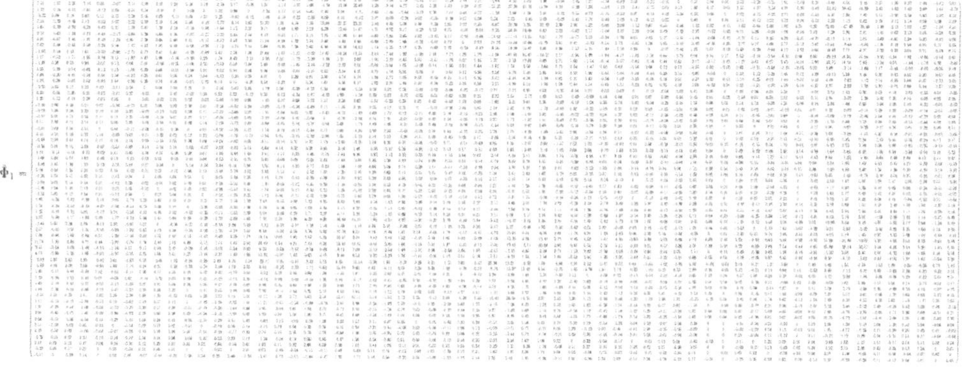

It appears that the matrix is not sparse and is noisy. We will compare the estimated coefficient matrix with estimates from other regularization methods in following sections.

10.A.1.2 Lasso

The maximum allowed lag order is set to be 4. The estimates of coefficient matrices are

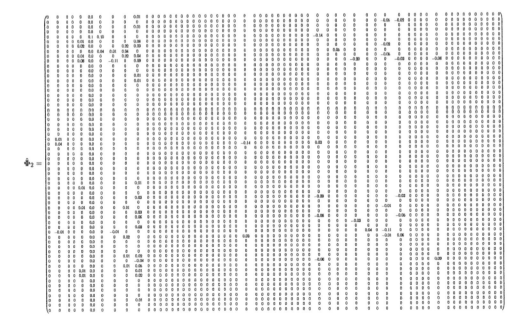

$$\hat{\Phi}_1 = \text{[illegible]}$$

$$\hat{\Phi}_2 = \text{[illegible]}$$

$\hat{\Phi}_3 =$ [illegible]

$\hat{\Phi}_4 =$ [illegible]

The estimates of those matrices are much sparser compared to the estimates from the VAR, using least squares.

10.A.1.3 Componentwise

The maximum allowed lag order is set to be 4. The estimates of coefficient matrices are

$$\hat{\Phi}_1 = \text{[illegible]}$$

$$\hat{\Phi}_2 = \text{[illegible]}$$

$$\hat{\Phi}_3 = \text{[illegible]}$$

$\hat{\Phi}_4 =$ [illegible]

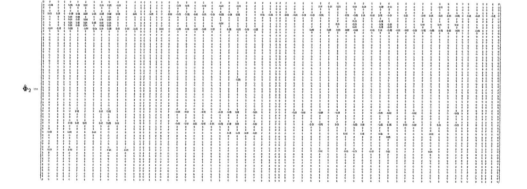

10.A.1.4 Own-other

The maximum allowed lag order is set to be 4. The estimates of coefficient matrices are

$\hat{\Phi}_1 =$ [illegible]

$\hat{\Phi}_2 =$ [illegible]

$\hat{\Phi}_3 =$ [illegible]

$\hat{\Phi}_4 =$ [illegible]

10.A.1.5 Elementwise

The maximum allowed lag order is set to be 4. The estimates of coefficient matrices are

$$\hat{\boldsymbol{\Phi}}_1 = \text{[illegible]}$$

$$\hat{\boldsymbol{\Phi}}_2 = \text{[illegible]}$$

$\hat{\Phi}_3 =$ [illegible]

$\hat{\Phi}_4 =$

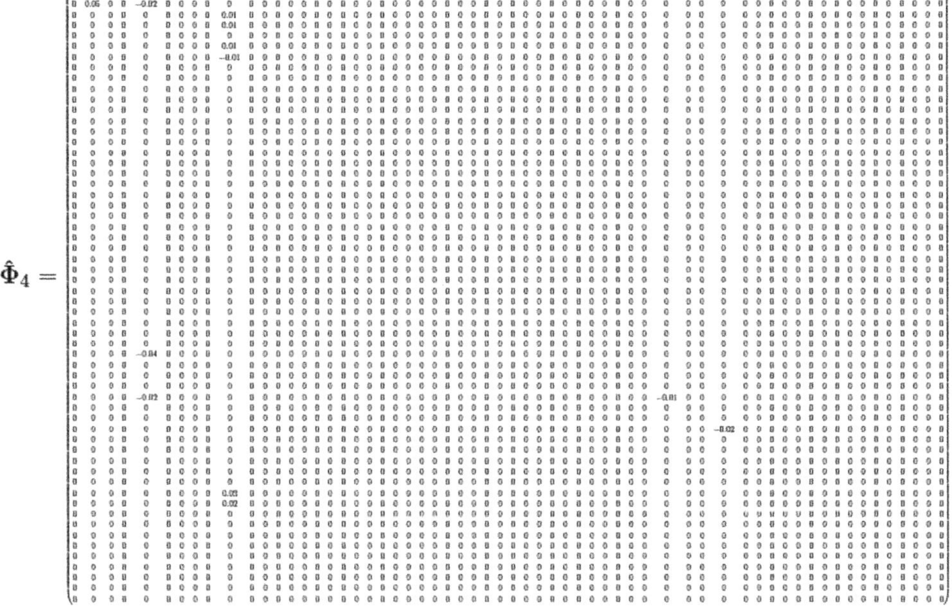

10.A.1.6 The factor model

In this section, we show the equation of forecasting the first component $Z_{1,t}$, which is GDP252. Based on Bai and Ng (2002), two factors are selected. The first factors $F_{1,t}$ and the second factor $F_{2,t}$ are shown in Figure 10.A.1.

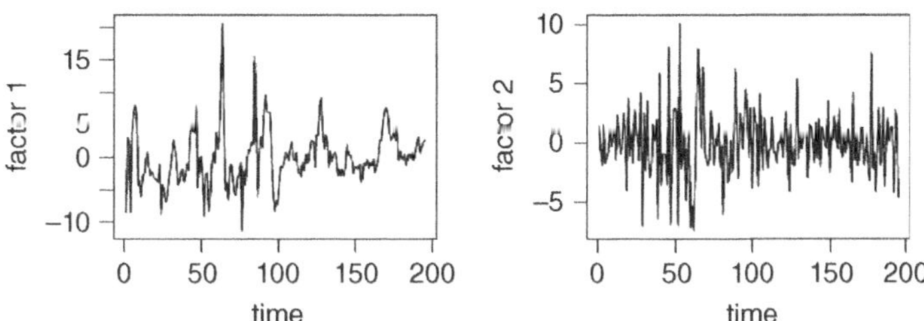

Figure 10.A.1 The two factors used in factor modeling for the first component $Z_{1,t}$ GDP252.

Suppose we want to obtain ℓ – step ahead forecast for the first component, $Z_{1,t}$, of those time series by using factors $\mathbf{F}_t$. The equation is:

$$Z_{1,t+\ell} = \boldsymbol{\beta}'\mathbf{F}_t + \alpha X_{1,t} + \varepsilon_{1,t},$$

where $X_{1,t}$ is the first lag of $Z_{1,t}$. The estimated parameters are presented in Table 10.A.1.

Table 10.A.1 The estimated parameters of the factor model for the first component $Z_{1,t}$, GDP252.

Estimate	$\hat{\beta}_1$	$\hat{\beta}_2$	$\hat{\alpha}$
	0.0005	−0.0001	0.7599

10.A.1.7 The model-based cluster

The maximum number of clusters allowed is set to be 4 and the AR(1) model is fitted for each cluster. The individual time series is assigned to cluster i if its probability of being in cluster i is the highest. Table 10.A.2 presents probabilities of each time series being in specific clusters (first four columns) and the finally assigned cluster for each time series (the last column). Table 10.A.2 shows that no time series is within cluster 4 based on the estimated probability. Moreover, most of "GDPXXX" time series are assigned to cluster 1, most of "CESXXX" time series are in cluster 2, and most of "IPSXXX" time series are in cluster 2 and 3. The models of cluster 1, 2, and 3 can be written as

$$Z_{i,t} = \phi_1^{(i)} Z_{i,t-1} + \sigma^{(i)} \varepsilon_{i,t},\ i \in 1,2,3$$

Table 10.A.2 Macroeconomic time series with probabilities of series being in four different clusters and the finally assigned cluster for each series.

	Cluster 1	Cluster 2	Cluster 3	Cluster 4	Cluster Assigned
GDP252	0.505	0.441	0.053	0.002	1
GDP253	0.001	0.000	0.704	0.295	3
GDP254	0.499	0.501	0.000	0.000	2
GDP255	0.559	0.098	0.328	0.015	1
GDP256	0.016	0.000	0.626	0.357	3
GDP257	0.540	0.093	0.352	0.015	1
GDP258	0.554	0.097	0.342	0.006	1
GDP259	0.214	0.042	0.677	0.067	3
GDP260	0.162	0.039	0.727	0.072	3
GDP261	0.629	0.208	0.161	0.002	1
GDP266	0.095	0.000	0.653	0.251	3
GDP267	0.558	0.299	0.143	0.000	1
GDP268	0.500	0.300	0.186	0.015	1
GDP269	0.556	0.302	0.142	0.000	1
GDP271	0.690	0.310	0.000	0.000	1
GDP272	0.502	0.297	0.186	0.015	1
GDP274	0.559	0.098	0.328	0.015	1
GDP275	0.559	0.098	0.328	0.015	1
GDP276	0.559	0.098	0.328	0.015	1
GDP277	0.557	0.098	0.328	0.017	1
GDP278	0.558	0.299	0.143	0.000	1
GDP279	0.502	0.298	0.186	0.015	1
GDP280	0.557	0.219	0.215	0.009	1
GDP281	0.511	0.349	0.128	0.013	1
GDP282	0.484	0.038	0.373	0.104	1
GDP284	0.559	0.098	0.328	0.015	1
GDP285	0.136	0.000	0.634	0.230	3
GDP286	0.745	0.110	0.143	0.002	1
GDP287	0.838	0.049	0.036	0.077	1
GDP288	0.860	0.135	0.002	0.002	1
GDP289	0.860	0.135	0.002	0.002	1
GDP290	0.860	0.135	0.002	0.002	1
GDP291	0.499	0.501	0.000	0.000	2
GDP292	0.499	0.501	0.000	0.000	2
IPS11	0.499	0.501	0.000	0.000	2
IPS299	0.499	0.501	0.000	0.000	2
IPS12	0.860	0.135	0.002	0.002	1
IPS13	0.042	0.000	0.684	0.273	3
IPS18	0.860	0.135	0.002	0.002	1
IPS25	0.499	0.501	0.000	0.000	1
IPS32	0.061	0.000	0.694	0.245	3

Table 10.A.2 *(continued)*

	Cluster 1	Cluster 2	Cluster 3	Cluster 4	Cluster Assigned
IPS34	0.068	0.002	0.708	0.222	3
IPS38	0.499	0.501	0.000	0.000	2
IPS43	0.499	0.501	0.000	0.000	2
IPS307	0.000	0.000	0.680	0.320	3
IPS306	0.010	0.000	0.730	0.260	3
CES275	0.499	0.501	0.000	0.000	2
CES277	0.499	0.501	0.000	0.000	2
CES278	0.499	0.501	0.000	0.000	2
CES003	0.499	0.501	0.000	0.000	2
CES006	0.196	0.016	0.627	0.161	3
CES011	0.499	0.501	0.000	0.000	2
CES015	0.499	0.501	0.000	0.000	2
CES017	0.499	0.501	0.000	0.000	2
CES033	0.499	0.501	0.000	0.000	2
CES046	0.499	0.501	0.000	0.000	2
CES048	0.499	0.501	0.000	0.000	2
CES049	0.499	0.501	0.000	0.000	2
CES053	0.499	0.501	0.000	0.000	2
CES088	0.499	0.501	0.000	0.000	2
CES140	0.499	0.501	0.000	0.000	2

where the estimated parameters are shown in Table 10.A.3.

Table 10.A.3 The estimated parameters of the model-based clustering for the macroeconomic time series.

Estimates	$\hat{\phi}_1^{(1)}$	$\hat{\sigma}^{(1)}$	$\hat{\phi}_1^{(2)}$	$\hat{\sigma}^{(2)}$	$\hat{\phi}_1^{(3)}$	$\hat{\sigma}^{(3)}$
	0.398	0.033	0.541	0.032	0.155	0.039

10.A.1.8 The proposed method

The 61 time series are aggregated into trivariate time series, GDP, IPS, and CES. We fit a VAR (p) model,

$$\mathbf{Y}_t = \sum_{k=1}^{p} \mathbf{\Phi}_k^{(a)} \mathbf{Y}_{t-k} + \boldsymbol{\varepsilon}_t .$$

The best fitted model is $p = 2$ based on AIC. Estimates of coefficient matrices along with their standard errors (in parenthesis) are shown next:

$$\hat{\mathbf{\Phi}}_1^{(a)} = \begin{bmatrix} \underset{(0.12)}{0.50} & \underset{(0.10)}{0.27} & \underset{(0.04)}{-0.56} \\ \underset{(0.13)}{0.35} & \underset{(0.12)}{0.27} & \underset{(0.04)}{-0.39} \\ \underset{(0.31)}{0.08} & \underset{(0.25)}{0.10} & \underset{(0.10)}{0.24} \end{bmatrix}, \quad \hat{\mathbf{\Phi}}_2^{(a)} = \begin{bmatrix} \underset{(0.12)}{0.33} & \underset{(0.10)}{-0.17} & \underset{(0.04)}{-0.09} \\ \underset{(0.13)}{0.28} & \underset{(0.11)}{-0.24} & \underset{(0.04)}{-0.07} \\ \underset{(0.29)}{0.05} & \underset{(0.25)}{-0.09} & \underset{(0.10)}{0.24} \end{bmatrix}$$

10.A.2 Further details of the STD time series

10.A.2.1 VAR

Since the number of dimension $m = 44$ is much greater than the number of observations $n = 29$, the ordinary least square estimation without any restrictions failed for this model.

10.A.2.2 Lasso

$$\hat{\mathbf{\Phi}}_1 = [illegible]$$

$$\hat{\mathbf{\Phi}}_2 = [illegible]$$

$$\hat{\Phi}_3 = \begin{pmatrix} \text{[illegible]} \end{pmatrix}$$

$$\hat{\Phi}_4 = \begin{pmatrix} \text{[illegible]} \end{pmatrix}$$

10.A.2.3 Componentwise

$$\hat{\Phi}_1 = \begin{pmatrix} \text{[illegible]} \end{pmatrix}$$

$$\hat{\Phi}_2 = \text{[illegible]}$$

$$\hat{\Phi}_3 = \text{[illegible]}$$

$$\hat{\Phi}_4 = \begin{pmatrix}
0&0&0&0&0&0&0&0&0&0&0&0&0&0&0&0&0&0&0&0&0&0&0&0&0&0&0&0&0&0&0&0&0&0&0&0&0&0&0&0&0&0&0&0\\
0&0&0&0&0&0&0&0&0&0&0&0&0&0&0&0&0&0&0&0&0&0&0&0&0&0&0&0&0&0&0&0&0&0&0&0&0&0&0&0&0&0&0&0\\
0&0&0&0&0&0&0&0&0&0&0&0&0&0&0&0&0&0&0&0&0&0&0&0&0&0&0&0&0&0&0&0&0&0&0&0&0&0&0&0&0&0&0&0\\
0&0&0&0&0&0&0&0&0&0&0&0&0&0&0&0&0&0&0&0&0&0&0&0&0&0&0&0&0&0&0&0&0&0&0&0&0&0&0&0&0&0&0&0\\
0&0&0&0&0&0&0&0&0&0&0&0&0&0&0&0&0&0&0&0&0&0&0&0&0&0&0&0&0&0&0&0&0&0&0&0&0&0&0&0&0&0&0&0\\
0&0&0&0&0&0&0&0&0&0&0&0&0&0&0&0&0&0&0&0&0&0&0&0&0&0&0&0&0&0&0&0&0&0&0&0&0&0&0&0&0&0&0&0\\
0&0&0&0&0&0&0&0&0&0&0&0&0&0&0&0&0&0&0&0&0&0&0&0&0&0&0&0&0&0&0&0&0&0&0&0&0&0&0&0&0&0&0&0\\
0&0&0&0&0&0&0&0&0&0&0&0&0&0&0&0&0&0&0&0&0&0&0&0&0&0&0&0&0&0&0&0&0&0&0&0&0&0&0&0&0&0&0&0\\
0&0&0&0&0&0&0&0&0&0&0&0&0&0&0&0&0&0&0&0&0&0&0&0&0&0&0&0&0&0&0&0&0&0&0&0&0&0&0&0&0&0&0&0\\
0&0&0&0&0&0&0&0&0&0&0&0&0&0&0&0&0&0&0&0&0&0&0&0&0&0&0&0&0&0&0.01&0&0&0&0&0&0&0&0&0&0&0&0&0\\
0&0&0&0&0&0&0&0&0&0&0&0&0&0&0&0&0&0&0&0&0&0&0&0&0&0&0&0&0&0&0&0&0&0&0&0&0&0&0&0&0&0&0&0\\
0&0&0&0&0&0&0&0&0&0&0&0&0&0&0&0&0&0&0&0&0&0&0&0&0&0&0&0&0&0&0&0&0&0&0&0&0&0&0&0&0&0&0&0\\
0&0&0&0&0&0&0&0&0&0&0&0&0&0&0&0&0&0&0&0&0&0&0&0&0&0&0&0&0&0&0&0&0&0&0&0&0&0&0&0&0&0&0&0\\
0&0&0&0&0&0&0&0&0&0&0&0&0&0&0&0&0&0&0&0&0&0&0&0&0&0&0&0&0&0&0&0&0&0&0&0&0&0&0&0&0&0&0&0\\
0&0&0&0&0&0&0&0&0&0&0&0&0&0&0&0&0&0&0&0&0&0&0&0&0&0&0&0&0&0&0&0&0&0&0&0&0&0&0&0&0&0&0&0\\
0&0&0&0&0&0&0&0&0&0&0&0&0&0&0&0&0&0&0&0&0&0&0&0&0&0&0&0&0&0&0&0&0&0&0&0&0&0&0&0&0&0&0&0\\
0&0&0&0&0&0&0&0&0&0&0&0&0&0&0&0&0&0&0&0&0&0&0&0&0&0&0&0&0&0&0&0&0&0&0&0&0&0&0&0&0&0&0&0\\
0&0&0&0&0&0&0&0&0&0&0&0&0&0&0&0&0&0&0&0&0&0&0&0&0&0&0&0&0&0&0&0&0&0&0&0&0&0&0&0&0&0&0&0\\
0&0&0&0&0&0&0&0&0&0&0&0&0&0&0&0&0&0&0&0&0&0&0&0&0&0&0&0&0&0&0&0&0&0&0&0&0&0&0&0&0&0&0&0\\
0&0&0&0&0&0&0&0&0&0&0&0&0&0&0&0&0&0&0&0&0&0&0&0&0&0&0&0&0&0&0&0&0&0&0&0&0&0&0&0&0&0&0&0\\
0&0&0&0&0&0&0&0&0&0&0&0&0&0&0&0&0&0&0&0&0&0&0&0&0&0&0&0&0&0&0&0&0&0&0&0&0&0&0&0&0&0&0&0\\
0&0&0&0&0&0&0&0&0&0&0&0&0&0&0&0&0&0&0&0&0&0&0&0&0&0&0&0&0&0&0&0&0&0&0&0&0&0&0&0&0&0&0&0\\
0&0&0&0&0&0&0&0&0&0&0&0&0&0&0&0&0&0&0&0&0&0&0&0&0&0&0&0&0&0&0&0&0&0&0&0&0&0&0&0&0&0&0&0\\
0&0&0&0&0&0&0&0&0&0&0&0&0&0&0&0&0&0&0&0&0&0&0&0&0&0&0&0&0&0&0&0&0&0&0&0&0&0&0&0&0&0&0&0\\
0&0&0&0&0&0&0&0&0&0&0&0&0&0&0&0&0&0&0&0&0&0&0&0&0&0&0&0&0&0&0&0&0&0&0&0&0&0&0&0&0&0&0&0\\
0&0&0&0&0&0&0&0&0&0&0&0&0&0&0&0&0&0&0&0&0&0&0&0&0&0&0&0&0&0&0&0&0&0&0&0&0&0&0&0&0&0&0&0\\
0&0&0&0&0&0&0&0&0&0&0&0&0&0&0&0&0&0&0&0&0&0&0&0&0&0&0&0&0&0&0&0&0&0&0&0&0&0&0&0&0&0&0&0\\
0&0&0&0&0&0&0&0&0&0&0&0&0&0&0&0&0&0&0&0&0&0&0&0&0&0&0&0&0&0&0&0&0&0&0&0&0&0&0&0&0&0&0&0\\
0&0&0&0&0&0&0&0&0&0&0&0&0&0&0&0&0&0&0&0&0&0&0&0&0&0&0&0&0&0&0&0&0&0&0&0&0&0&0&0&0&0&0&0\\
0&0&0&0&0&0&0&0&0&0&0&0&0&0&0&0&0&0&0&0&0&0&0&0&0&0&0&0&0&0&0&0&0&0&0&0&0&0&0&0&0&0&0&0\\
0&0&0&0&0&0&0&0&0&0&0&0&0&0&0&0&0&0&0&0&0&0&0&0&0&0&0&0&0&0&0&0&0&0&0&0&0&0&0&0&0&0&0&0\\
0&0&0&0&0&0&0&0&0&0&0&0&0&0&0&0&0&0&0&0&0&0&0&0&0&0&0&0&0&0&0&0&0&0&0&0&0&0&0&0&0&0&0&0\\
0&0&0&0&0&0&0&0&0&0&0&0&0&0&0&0&0&0&0&0&0&0&0&0&0&0&0&0&0&0&0&0&0&0&0&0&0&0&0&0&0&0&0&0\\
0&0&0&0&0&0&0&0&0&0&0&0&0&0&0&0&0&0&0&0&0&0&0&0&0&0&0&0&0&0&0&0&0&0&0&0&0&0&0&0&0&0&0&0\\
0&0&0&0&0&0&0&0&0&0&0&0&0&0&0&0&0&0&0&0&0&0&0&0&0&0&0&0&0&0&0&0&0&0&0&0&0&0&0&0&0&0&0&0\\
0&0&0&0&0&0&0&0&0&0&0&0&0&0&0&0&0&0&0&0&0&0&0&0&0&0&0&0&0&0&0&0&0&0&0&0&0&0&0&0&0&0&0&0\\
0&0&0&0&0&0&0&0&0&0&0&0&0&0&0&0&0&0&0&0&0&0&0&0&0&0&0&0&0&0&0&0&0&0&0&0&0&0&0&0&0&0&0&0\\
0&0&0&0&0&0&0&0&0&0&0&0&0&0&0&0&0&0&0&0&0&0&0&0&0&0&0&0&0&0&0&0&0&0&0&0&0&0&0&0&0&0&0&0\\
0&0&0&0&0&0&0&0&0&0&0&0&0&0&0&0&0&0&0&0&0&0&0&0&0&0&0&0&0&0&0&0&0&0&0&0&0&0&0&0&0&0&0&0\\
0&0&0&0&0&0&0&0&0&0&0&0&0&0&0&0&0&0&0&0&0&0&0&0&0&0&0&0&0&0&0&0&0&0&0&0&0&0&0&0&0&0&0&0\\
0&0&0&0&0&0&0&0&0&0&0&0&0&0&0&0&0&0&0&0&0&0&0&0&0&0&0&0&0&0&0&0&0&0&0&0&0&0&0&0&0&0&0&0\\
0&0&0&0&0&0&0&0&0&0&0&0&0&0&0&0&0&0&0&0&0&0&0&0&0&0&0&0&0&0&0&0&0&0&0&0&0&0&0&0&0&0&0&0\\
0&0&0&0&0&0&0&0&0&0&0&0&0&0&0&0&0&0&0&0&0&0&0&0&0&0&0&0&0&0&0&0&0&0&0&0&0&0&0&0&0&0&0&0\\
0&0&0&0&0&0&0&0&0&0&0&0&0&0&0&0&0&0&0&0&0&0&0&0&0&0&0&0&0&0&0&0&0&0&0&0&0&0&0&0&0&0&0&0
\end{pmatrix}$$

10.A.2.4 Own-other

The maximum allowed lag order is set to be 4.

$$\hat{\Phi}_1 =$$ [illegible]

$$\hat{\Phi}_2 =$$ [illegible]

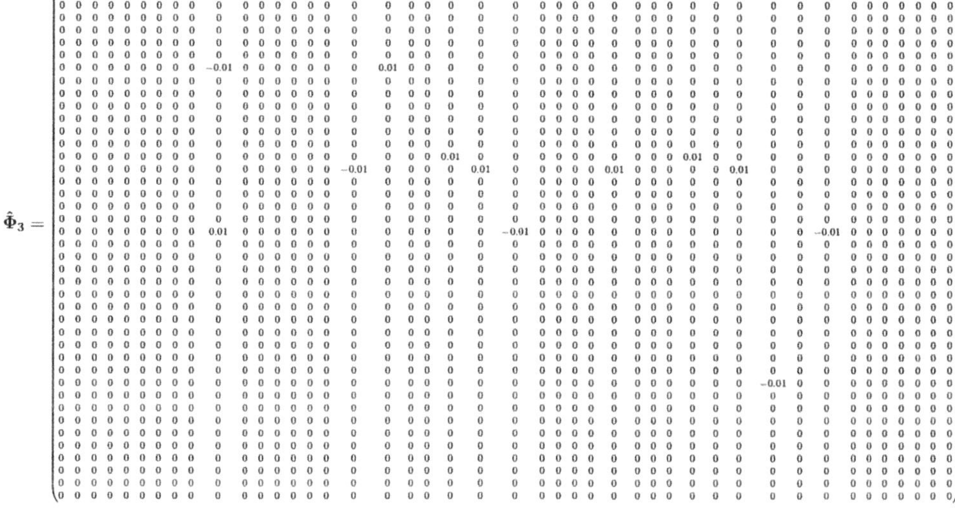

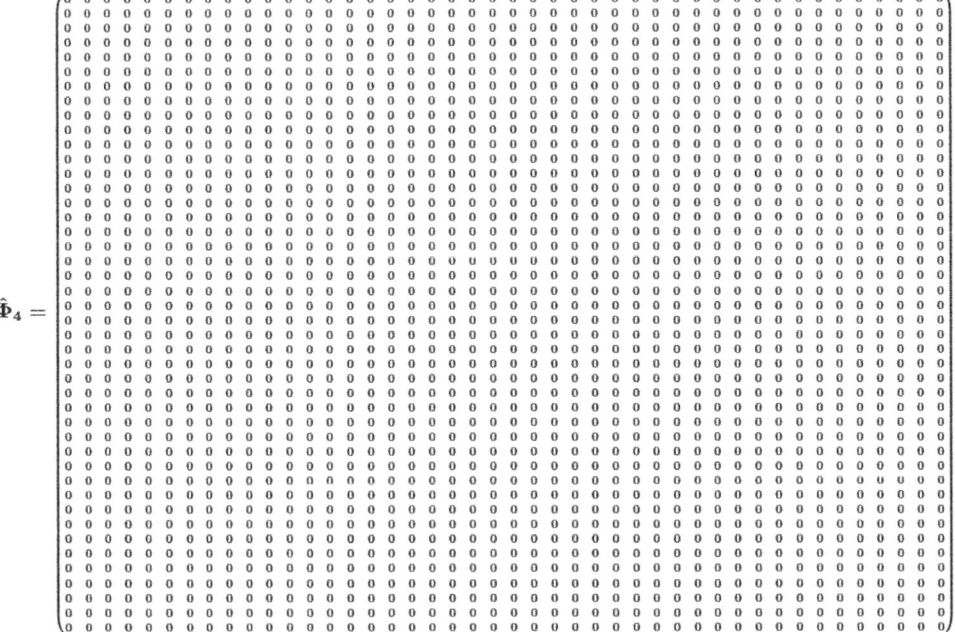

10.A.2.5 Elementwise

The maximum allowed lag order is set to be 4.

$$\hat{\Phi}_1 = \begin{pmatrix}
0&0&0&0&0&0&0&0&0&0&0&0&0&0&0.42&0&0&0&0&0&0&0&0&0&0&0&0&0&-0.01&0&0&0&0&0&0&0&0&0&0&0&0.21&0&0&0 \\
0&0&0&0&0&0&0&0&0&0&0&0&0&0&0&0&0&0&0&0&0&0&0&0&0&0&0&0&0&0&0&0&0&0&0&0&0&0&0&0&0&0&0&0 \\
0.47&0&0&0&0&0&0&0.08&0&0&0&0&0&0&0&0&0&0&0&0&0&0&0&0&0&0&0&0&0&-0.14&0&0&0&0&0&0&0&0&0&0&0&0&0&0 \\
0&-0.03&0&0&0&0&0&0&0&0&0&0&0.18&0&0&0&-0.04&0&0&0.01&0&0&-0.01&0.06&0&0&0&0.05&0&0&0&0&0&0&0&0.09&0&0&0&0&0&0&0&0 \\
0&0&0.28&0&0&0.15&0&0&0&0&0&0&0&0&0&0&0&0&0&0&0&0&0&0&0&0&0&0&0&0&0&0&0&0&0&0&0&0&0&0&0.31&0&0&0 \\
0.42&0&0&0&0&0&0&0.17&0&0&0&0&0&0&0.11&0&0&0&0&0&0&0&0&0&0&0&0&0&0&0&0&0&0&0&0&0&0&0&0&0&0&0&0&0 \\
0.48&0&0&0&0&0.01&0&0&0&0&0&0&0&0&0&0&0&0&0&0&0&0&0&0&0&0&0&0&0&0&0&0&0&0&0&0&0&0&0&0&0&0&0&0 \\
0&0&0&0&0&0&0&0&0&0&0&0.12&0.16&0&0&0.02&0&0&0&0&0&0&0&0.08&0&-0.06&0&0&0&0&0&0&0&0&0&0.04&0&0&0&0&0.35&0&0&0 \\
0.46&0&0&0&0&0&0&0&0&0&0&0&0&0&0&0.21&0&0&0&0&0&0&0&0&0&0&0&0&0&0&0&0&0&0&0&0&0&0&0&0&0&0&0&0 \\
0.45&0&0&0&0&0&0&0&0&0&0&0&0&0&0&0.01&0&0&0&0&0&0&0&0&0&0&0&0&0&-0.09&0&0&0&0&-0.01&0&0&0&0&0&0&0&0&0 \\
0.50&0&0&0&0&0&0&0&0&0&0&0&0.01&0&0.03&0&0&0&0&0&0&0&0&0&0&0&0&0&0&0&0&0&0&0&0&0.16&0&0&0&0&0&0&0&0 \\
0&0&0&0&0&0.19&0&0&0&0&0.40&0&0&0&0&0&0&0&0&0&0&0&0&0&0&0&0&0&0&0&0&0&0&0&0&0&0&0&0&0&0&0&0&0 \\
0&0&0.07&0&0&0&0&0&0.16&0&0&0&0&0&0&0&0&0&0&0&0&0&0&0&0&0&0&0&0&0&0&0&0&0&0&0&0&0&0.21&0&0.34&0&0&0 \\
0.14&0&0&0&0&0&0&0&0&0&0.30&0&0&0&0&0.25&0&0&0&0&0&0&0&0&0&0&0&0&0&0&0&0&0&0&0&0&0&0&0&0&0&0&0&0 \\
0&0&0&0&0&0&0&0&0&0&0&0&0&0&0.04&0&0&-0.01&0&0&-0.13&0&0&0&0&0&-0.06&0&0&0&0&0&0&0&0&0&0&0&0&0&0&0&0&0 \\
0.31&0&0&0&0&0&0&0&0&0&0&0&0&0&0.21&0&0&0&0&0&0&0&0&0&0&0&0&0&0&0&-0.03&0&0&0&0&0&0&0&0&0&0.02&0&0&0 \\
0&0&0&0&0&0&0&0&0&0&0&0&0&0&0&0&0&0&0.55&0&0&0&0&0&0&0&0&0&0&0&0&-0.02&0&0&0&0&0&0&0&0&0&0&0&0 \\
0&0&0&0&0&0&0&0&0&0&0&0&0.18&0&0&0&0&0&0.15&0&0&0&0&0&0&0.31&0&0&0&0&0&0&0.04&0&0.16&0&0&0&0&0&0&0&0&0 \\
0&0&0&0&0&0&0&0.03&0&0&0&0.18&0.01&0&0&0&0&0&0&0.27&0&0&0&0.13&0&0&0.10&0&0&0&0&0&0&0&0&0&0&0&0&0&0&0&0&0 \\
0&0&0&0&0&0.18&0&0&0.27&0&0&0&0&0&0&0&0&0&0&0&0.28&0&0&0&0&0&0&0&0&0&0&0&0&0&0&0&0&0&0&0&0&0&0&0 \\
0.41&0&0&0&0&0&0&0.23&0&0&0&0&0&0&0&0&0&0&0&0&0&0&0&0&0&0&0&0&0&0&0&0&0&0&0&0&0&0&0&0&0&0&0&0 \\
0&0&0&0&0.09&0.27&0&0&0.23&0&0&0.12&0&0.03&0&0&0&0&0&0&0&0&0&0&0&0&0&0&0&0&0&0&0&0&0&0&0&0&0&0&0&0&0&0 \\
0&-0.05&-0.03&0&0&0&0&0&0&0&0&0&0&0&0&0&0&0&0.31&0&0&0.06&0&0&0&0&0&0.07&0&0&0&0&0&0&0&0&0&0&0&0&0&0&0&0 \\
0&0&0&0&0&0.34&0&0&0&0&0.11&0&0&0&0&0.25&0&0&0&0&0&0&0&0&0&0&0&0&0&0&0&0&0&0&0&0&0&0&0&0&0&0&0&0 \\
0&0&0&0&0&0&0&0&0&0&0&0&0.18&0&0&0&0&0&0&0&0&0&0&0.03&0&0&0&0&0&0&0&0&0&0&0&0&0&0&0&0&0&0&0&0 \\
0&0&0&0&0&0&0&0.13&0&0&0&0&0&0&0&0&0&0&0.08&0&0&0&0&0&0&0&0.05&0.17&0.14&0&0&0&0&0&0.24&0&0&0&0&0&0&0&0&0 \\
0&0&0&0&0&0&0&0&0&0&0.46&0&0&0&0&0&0&0&0.20&0&0&0.13&0&0&0&0&0&0&0&0&0&0&0&0&0&0&0&0&0&0&0&0&0&0 \\
0&0&0&0&0&0.16&0&0&0&0&0&0.20&0&0.19&0&0&0&0&0&0&0&0.02&0&0.01&0&0&0&0&0&0&0&0.08&0&0&0&0&0&0&0&0&0&0&0&0 \\
0&0&0&0&0&0&0&0.20&0&0&0.34&0&0&0&0&0&0&0&0.08&0&0&0.05&0&0&0&0&0&0&0.11&0&0&0&0&0&0&0&0&0&0&0&0&0&0&0 \\
0&0&0&0&0&0&0&0&0&0&0&-0.02&0&0&0&0&0&0&0&0&0&0&0&-0.03&0&0&0&0&0&0&0&0&0&0&0&0&0&0&0&0&-0.29&0&0&0 \\
0&-0.04&0&0&0&0&0&0&0&0&0&0&0.21&0&0&0&0&0&0&0&0&0.16&-0.13&0&0&0&0&0&0&-0.19&0&0&0&0&0&0&0&0&0&0&0&0&0&0 \\
0.23&0&0&0&0&0.01&0&0&0&0&0.24&0&0&0&0&0&0&0&0&0&0&0&0&0&0&0&0&0&0&0&0&0&0&0&0&0&0&0&0&0&0&0&0&0 \\
0&0&0&0&0&0.10&0&0&0&0&0&0&0&0&0&0&0&0&0&0&0&0&0&0&0&0.05&0.05&0&0&-0.02&0&0&0&0&0.24&0&0&0&0&0&0&0&0&0 \\
0&0&0&0&0&0&0&0&0&0.22&0&0&0&0.25&0&0&0&0&0&0&0&0&0&0.22&0&0&0&0&0&0&0&0&0&0&0&0&0&0&0&0&0&0&0&0 \\
0&0&0&0&0&0&0&0&0&0&0&0&0&0&0&0&0&0&0.19&0&0&0&0&0&0&0&0.19&0&0&0&0&0&0&0.42&0&0&0&0&0&0&0&0&0&0 \\
0&0&0&0&0&0&0&-0.01&0&0.16&0&0&0&0&0&0&0&0&0&0&0&0&0&0&0&0&0&0&0&0&0&0&0&0&0&0&0&0&0&0&0.13&0&0&0 \\
0&0&0&0&0&0&0&0&0&0&0&0&0&0&0&0&0&0&0&0&0&0&0&0&0&0.12&0&0&0&0&0&0&0&0&0&0&0&0&0&0&-0.13&0&0&0 \\
0&0&0&0&0&0&0&0&0&0&0&0&0&0&0&0&0&0&0&0&0&0&0&0&0&0&0&0&0&0&0.06&0&0&0&0&0&0&0&0&0&0&0&0&0 \\
0.12&0&0&0&0&0&0&0&0&0&0&-0.03&0&0&0&0&0&0&0&0&0&0&0&0&0&0&-0.05&0&-0.06&-0.17&0&0&0&0&0&0&0&0&0&0&-0.14&-0.09&0&0 \\
0&0&0&0&0&0&0&0&0&0&0&0&0&0&0&0&0&0&0&0&-0.10&0&0&0&0&0&0&0&0&0&0&0&0&0&0&0&0&0&0&0&0&0&0&0 \\
0&0&0&0&0&0&0&0&0&0&0&0&0&0&0&0&0&0&0&0&0&0&0&-0.07&0&0&0&0&0&0&0&0&0&0&0&0&0&0&0&0&-0.24&0&0&0 \\
0&0&0&0&0&0&0&0&0&0&0&0&0&0&0&0&0&0&0&0&0&0&0&0&0&0&0&0&0&0&0&0&0&0&0&0&0&0&0.03&0&-0.14&0&0&0 \\
-0.07&0&0&0&0&0&0&0&0&0&0&0&0&0&0&0&0&0&0&0&0&0&0&0&0&0&0&0&0&0&0&0&0&0&0&0&0&0&0&0&0&0&0&0 \\
0&0&0&0&0&0&0&0&0&0&0&0&0&0&0.36&0&0&0&0&0&-0.06&0&0&0&0&0&-0.01&0&0&0&0&0&0&0&0&0&0&0&0&0&0&0&0&0
\end{pmatrix}$$

$\hat{\mathbf{\Phi}}_2 =$

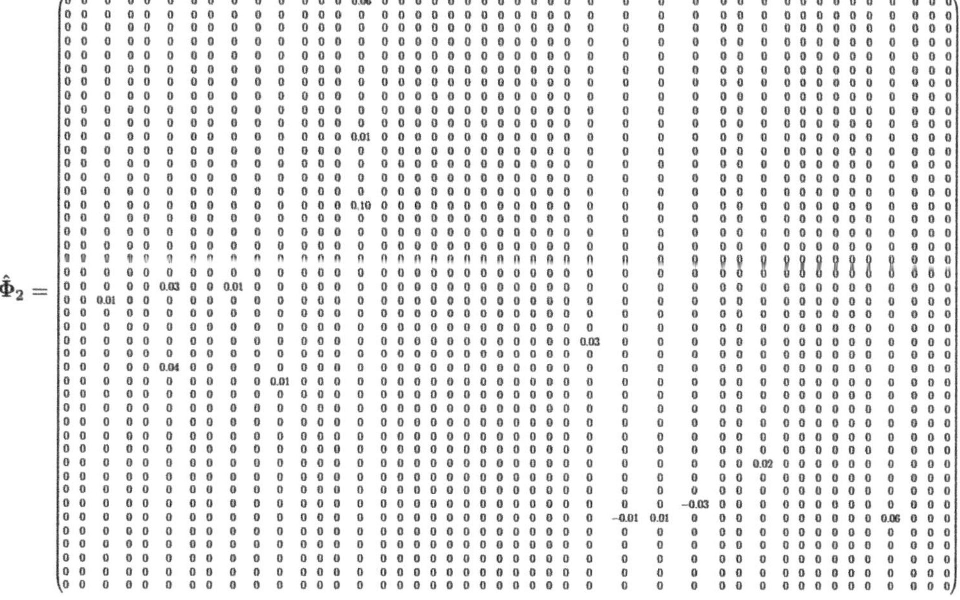

$\hat{\mathbf{\Phi}}_3 =$

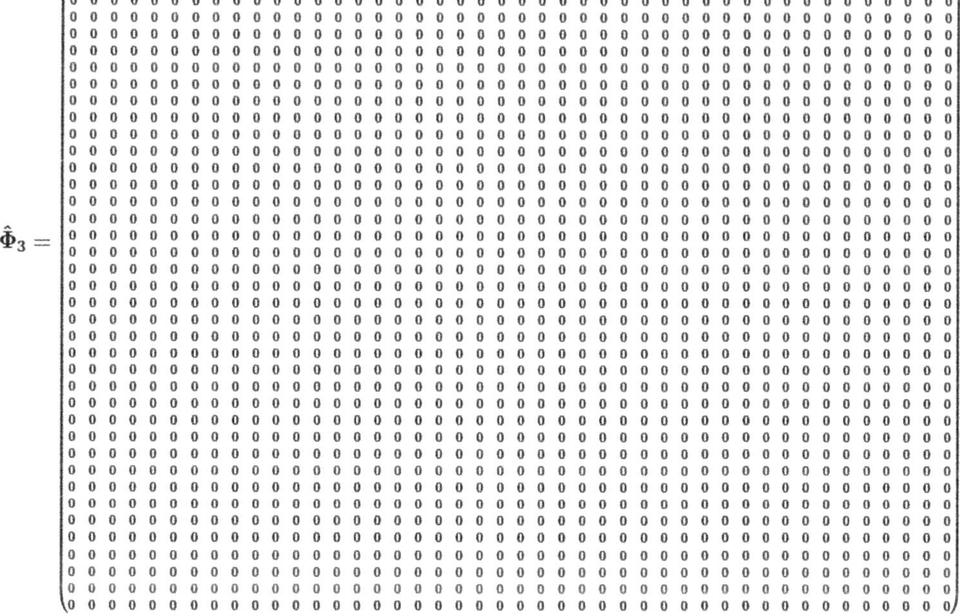

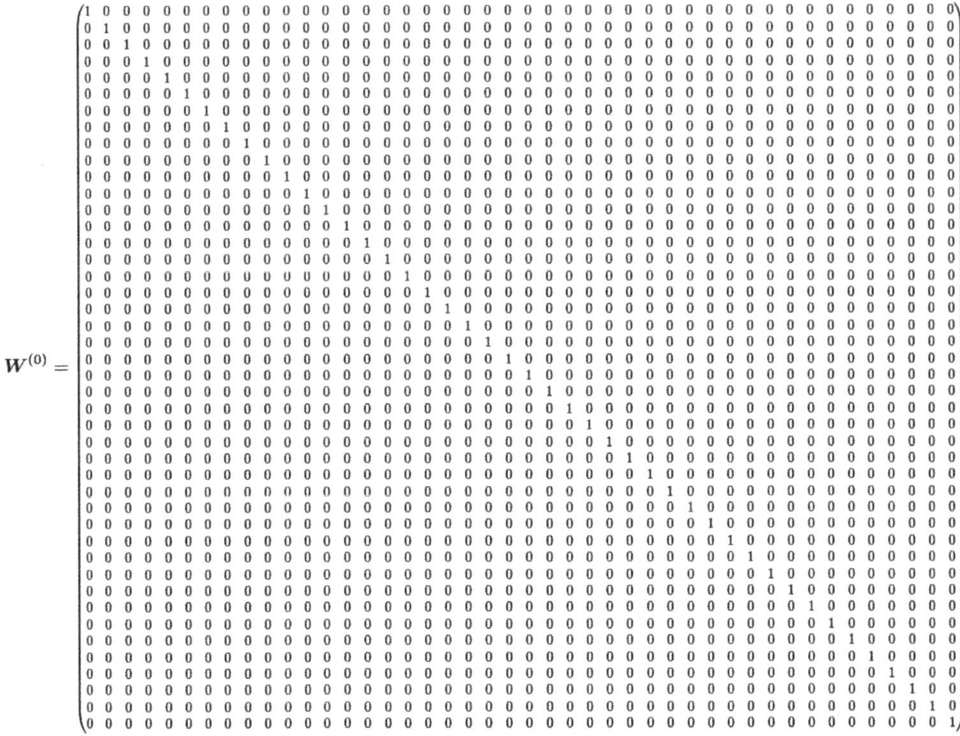

10.A.2.6 The STAR model

Since the STD data are collected from different states, which are spatially related, the space–time modeling can be used. Based on AIC criteria, the STAR($2_{2,\,1}$) model is used and can be written as

$$\mathbf{Z}_t = \sum_{k=1}^{2}\sum_{\ell=0}^{a_k}\phi_{k,\ell}\mathbf{W}^{(\ell)}\mathbf{Z}_{t-k} + \mathbf{a}_t, a_1 = 2, a_2 = 1.$$

The estimates are presented in Table 10.A.4.

Table 10.A.4 Estimated parameters for the STAR model.

Estimates	$\hat{\phi}_{1,0}$	$\hat{\phi}_{1,1}$	$\hat{\phi}_{1,2}$	$\hat{\phi}_{2,0}$	$\hat{\phi}_{2,1}$
	0.074	0.286	0.239	−0.126	0.065

Moreover, the weighting matrices, $\mathbf{W}^{(0)}$, $\mathbf{W}^{(1)}$, and $\mathbf{W}^{(2)}$, are specified by using spatial relationship between states and are shown next:

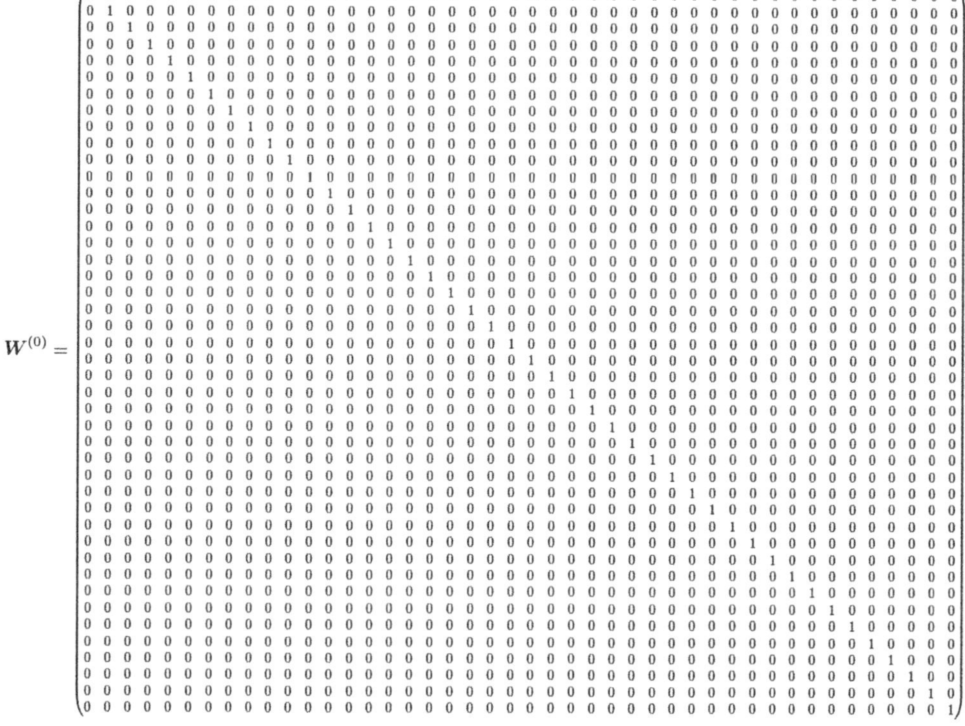

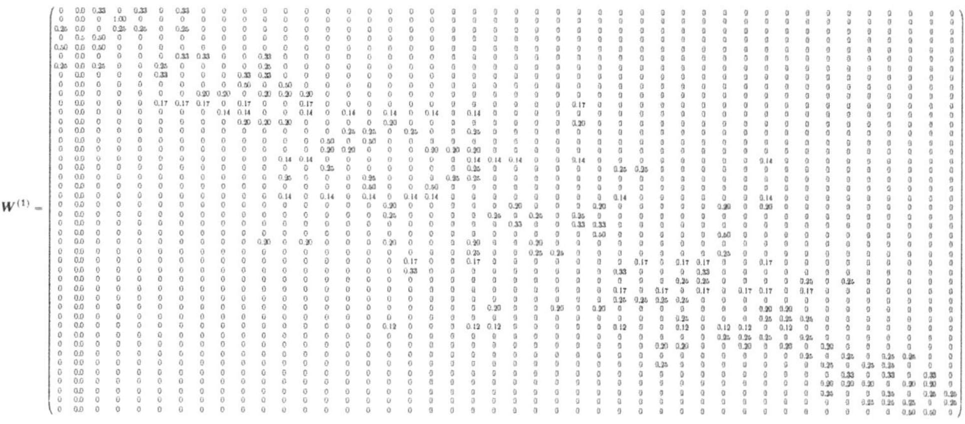

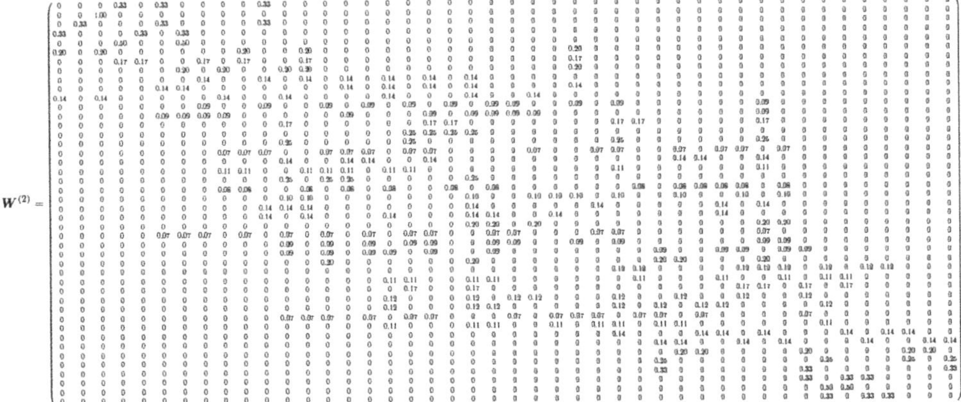

10.A.2.7 The factor model

In this section, we show the equation for forecasting the first component $Z_{1,t}$, which is the STD rate in Connecticut. Six factors are selected, and the estimated factor scores are shown in Figure 10.A.2.

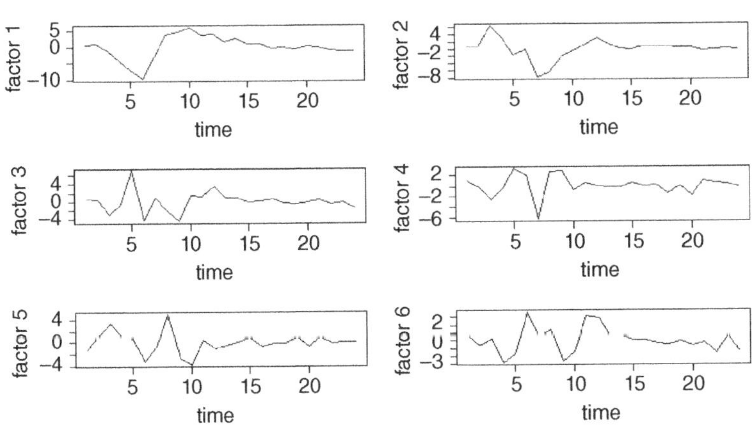

Figure 10.A.2 The plot of estimated six common factor scores for the STD data set.

To compute the ℓ – step ahead forecast for the first component, $Z_{1,t+\ell}$, we will use the following model equation, which is:

$$Z_{1,t+\ell} = \boldsymbol{\beta}'\mathbf{F}_t + \alpha X_{1,t} + \varepsilon_{1,t},$$

where $X_{1,t}$ is the first lag of $Z_{1,t}$. The estimated parameters are presented in Table 10.A.5.

Table 10.A.5 The estimated parameters of the factor model for the first component, STD rate in Connecticut.

Estimate	$\hat{\beta}_1$	$\hat{\beta}_2$	$\hat{\beta}_3$	$\hat{\beta}_4$	$\hat{\beta}_5$	$\hat{\beta}_6$	$\hat{\alpha}$
	0.054	0.232	−0.056	−0.100	−0.208	−0.203	0.234

10.A.2.8 The model-based cluster

The maximum number of clusters allowed is set to be 4 and the AR(1) model is fitted for each cluster. Each time series is assigned to cluster i if its probability of being in cluster i is the highest. Probabilities of each time series being in specific clusters (first four columns) and the finally assigned cluster for each time series (the last column) are presented in Table 10.A.6. It shows

Table 10.A.6 STD time series with probabilities of the series being in four different clusters and the finally assigned cluster for each series.

	Cluster 1	Cluster 2	Cluster 3	Cluster 4	Assigned Cluster
CT	0.02	0.62	0.00	0.36	2
ME	0.05	0.00	0.94	0.01	3
MA	0.23	0.12	0.14	0.51	4
NH	0.10	0.00	0.85	0.05	3
RI	0.13	0.00	0.83	0.04	3
NJ	0.10	0.36	0.01	0.53	4
NY	0.02	0.66	0.00	0.32	2
DE	0.21	0.16	0.08	0.55	4
DC	0.02	0.61	0.00	0.36	2
MD	0.18	0.22	0.03	0.57	4
PA	0.02	0.63	0.00	0.35	2
VA	0.23	0.12	0.13	0.52	4
WV	0.04	0.00	0.96	0.00	3
AL	0.07	0.42	0.00	0.50	4
FL	0.01	0.70	0.00	0.29	2
GA	0.01	0.68	0.00	0.31	2
KY	0.23	0.03	0.48	0.26	3
MS	0.03	0.55	0.00	0.42	2
NC	0.07	0.43	0.00	0.49	4
SC	0.17	0.23	0.03	0.57	4
TN	0.11	0.34	0.01	0.54	4
IL	0.03	0.56	0.00	0.40	2
IN	0.19	0.19	0.06	0.56	4
MI	0.03	0.59	0.00	0.38	2
MN	0.13	0.22	0.34	0.31	3

(continued overleaf)

Table 10.A.6 *(continued)*

	Cluster 1	Cluster 2	Cluster 3	Cluster 4	Assigned Cluster
OH	0.01	0.76	0.00	0.23	2
WI	0.07	0.43	0.00	0.50	4
AR	0.03	0.56	0.00	0.41	2
LA	0.01	0.74	0.00	0.26	2
NM	0.04	0.00	0.96	0.00	3
OK	0.07	0.00	0.92	0.01	3
TX	0.12	0.31	0.01	0.55	4
IA	0.17	0.22	0.14	0.47	4
KS	0.18	0.20	0.06	0.56	4
MO	0.02	0.64	0.00	0.34	2
NE	0.06	0.00	0.92	0.02	3
CO	0.16	0.02	0.68	0.15	3
UT	0.12	0.01	0.78	0.09	3
AZ	0.07	0.00	0.92	0.01	3
CA	0.05	0.49	0.00	0.45	2
NV	0.04	0.00	0.95	0.00	3
ID	0.24	0.09	0.22	0.45	4
OR	0.17	0.23	0.04	0.56	4
WA	0.02	0.68	0.00	0.31	2

that no time series was assigned to cluster 1. The models of cluster 1, 2, 3, and 4 can be written as

$$Z_{i,t} = \phi_1^{(i)} Z_{i,t-1} + \sigma^{(i)} \varepsilon_{i,t}, i \in 1,2,3,4$$

where the estimated parameters are shown in Table 10.A.7.

Table 10.A.7 The estimated parameters of the model-based clustering for the STD data set.

Estimates	$\hat{\phi}_1^{(1)}$	$\hat{\sigma}^{(1)}$	$\hat{\phi}_1^{(2)}$	$\hat{\sigma}^{(2)}$	$\hat{\phi}_1^{(3)}$	$\hat{\sigma}^{(3)}$	$\hat{\phi}_1^{(4)}$	$\hat{\sigma}^{(4)}$
	0.121	1.032	0.518	0.929	−0.246	0.989	0.349	0.999

10.A.2.9 The proposed method

The 44 time series are aggregated into MMWR region (9 time series) and fit a VAR(p) model, such that

$$\mathbf{Y}_t = \sum_{k=1}^{p} \mathbf{\Phi}_k^{(a)} \mathbf{Y}_{t-k} + \boldsymbol{\varepsilon}_t.$$

The best fitted model is $p = 1$ based AIC. The estimate of the coefficient matrix and the associated standard error (in parenthesis) are shown next.

$$\hat{\boldsymbol{\Phi}}_1^{(a)} =$$

$$\begin{pmatrix}
-0.14 & 0.47 & -0.32 & 0.26 & 0.21 & -0.26 & 0.01 & -0.04 & 0.23 \\
(0.3) & (0.17) & (0.17) & (0.26) & (0.29) & (0.18) & (0.19) & (0.56) & (0.33) \\
0.12 & 0.46 & -0.04 & 0.28 & 0.17 & -0.06 & -0.48 & -0.2 & 0.5 \\
(0.59) & (0.33) & (0.34) & (0.53) & (0.59) & (0.35) & (0.37) & (1.12) & (0.66) \\
-0.54 & 1.06 & 0.10 & 0.42 & 0.12 & -0.38 & 0.15 & -0.16 & -0.2 \\
(0.36) & (0.2) & (0.21) & (0.32) & (0.36) & (0.21) & (0.23) & (0.68) & (0.4) \\
-0.09 & 0.17 & 0.46 & -0.02 & 0.25 & -0.17 & -0.11 & 0.18 & -0.09 \\
(0.28) & (0.16) & (0.16) & (0.25) & (0.27) & (0.16) & (0.17) & (0.52) & (0.31) \\
0.02 & 0.8 & -0.32 & -0.21 & 0.5 & 0.25 & -0.1 & -0.05 & 0.63 \\
(0.19) & (0.11) & (0.11) & (0.17) & (0.19) & (0.11) & (0.12) & (0.36) & (0.21) \\
-0.31 & 0.45 & -0.07 & 0.5 & 0.03 & 0.23 & -0.01 & -0.56 & 0.69 \\
(0.37) & (0.21) & (0.21) & (0.33) & (0.37) & (0.22) & (0.23) & (0.7) & (0.41) \\
-0.39 & 0.89 & -0.02 & 0.3 & 0.27 & -0.1 & -0.14 & -0.22 & -0.1 \\
(0.44) & (0.25) & (0.25) & (0.39) & (0.44) & (0.26) & (0.28) & (0.83) & (0.49) \\
0.13 & 0.41 & 0.41 & -0.44 & -0.48 & -0.66 & 0.51 & -0.72 & 1.08 \\
(0.21) & (0.12) & (0.12) & (0.19) & (0.21) & (0.12) & (0.13) & (0.4) & (0.23) \\
0.16 & -0.05 & 0.23 & -0.14 & -0.12 & -0.53 & 0.13 & -0.17 & 0.76 \\
(0.34) & (0.19) & (0.2) & (0.3) & (0.34) & (0.2) & (0.21) & (0.64) & (0.38)
\end{pmatrix}$$

If we aggregate the 44 time series into STD region (4 time series) and fit a VAR(p) model, the best fitted model is $p = 4$ with the following estimates of the associated coefficient matrices.

$$\hat{\boldsymbol{\Phi}}_1^{(a)} = \begin{pmatrix}
0.65 & -0.08 & 0.8 & -1.3 \\
(0.37) & (0.12) & (0.24) & (0.19) \\
0.16 & -0.01 & 0.1 & -0.24 \\
(0.9) & (0.29) & (0.58) & (0.46) \\
0.54 & 0.66 & 0.32 & -0.5 \\
(0.81) & (0.26) & (0.53) & (0.41) \\
0.46 & 0.17 & 0.31 & -0.27 \\
(0.62) & (0.2) & (0.41) & (0.32)
\end{pmatrix}
\qquad
\hat{\boldsymbol{\Phi}}_2^{(a)} = \begin{pmatrix}
0.65 & -0.08 & 0.8 & -1.3 \\
(0.37) & (0.12) & (0.24) & (0.19) \\
0.16 & -0.01 & 0.1 & -0.24 \\
(0.9) & (0.29) & (0.58) & (0.46) \\
0.54 & 0.66 & 0.32 & -0.5 \\
(0.81) & (0.26) & (0.53) & (0.41) \\
0.46 & 0.17 & 0.31 & -0.27 \\
(0.62) & (0.2) & (0.41) & (0.32)
\end{pmatrix}$$

$$\hat{\boldsymbol{\Phi}}_3^{(a)} = \begin{pmatrix}
0.61 & -0.58 & 0.56 & 0.38 \\
(0.46) & (0.15) & (0.3) & (0.23) \\
0.43 & 0.06 & 0.06 & -0.17 \\
(0.61) & (0.2) & (0.4) & (0.31) \\
0.78 & 0.01 & -0.21 & -0.55 \\
(0.9) & (0.29) & (0.58) & (0.46) \\
1.34 & -0.21 & -0.04 & -0.48 \\
(0.73) & (0.23) & (0.47) & (0.37)
\end{pmatrix}
\qquad
\hat{\boldsymbol{\Phi}}_4^{(a)} = \begin{pmatrix}
1 & -0.09 & -1.79 & 0.03 \\
(0.41) & (0.13) & (0.27) & (0.21) \\
0.19 & -0.64 & -0.61 & 0.46 \\
(0.46) & (0.15) & (0.3) & (0.24) \\
0.16 & -0.13 & -0.68 & 0.28 \\
(0.78) & (0.25) & (0.51) & (0.4) \\
0.35 & -0.63 & -1.32 & 0.94 \\
(0.59) & (0.19) & (0.38) & (0.3)
\end{pmatrix}$$

Software code

R code for the macroeconomic time series

```
> library(MTS)

> setwd("C:/Bookdata/")
> koopact <- read.csv("WW10.csv",head=TRUE)
> attach(koopact)
> dat <- as.matrix(cbind(GDP252, GDP253, GDP254, GDP255,
GDP256,
                     GDP257, GDP258, GDP259, GDP260, GDP261,
                     GDP266, GDP267, GDP268, GDP269, GDP271,
                     GDP272, GDP274, GDP275, GDP276, GDP277,
                     GDP278, GDP279, GDP280, GDP281, GDP282,
                     GDP284, GDP285, GDP286, GDP287, GDP288,
                     GDP289, GDP290, GDP291, GDP292, IPS11,
                     IPS299, IPS12, IPS13, IPS18, IPS25,
                     IPS32, IPS34, IPS38, IPS43, IPS307,
IPS306,
                     CES275, CES277, CES278, CES003, CES006,
                     CES011, CES015, CES017, CES033, CES046,
                     CES048, CES049, CES053, CES088, CES140))

> dat_fit <- dat[1:131,]
> A <- as.matrix(rbind(c(rep(1,34),rep(0,27)),
          c(rep(0,34),rep(1,12),rep(0,15)),
          c(rep(0,46),rep(1,15))))
> A_dat_fit <- t(A%*%t(dat_fit))
> A_dat <- t(A%*%t(dat))
> A_dat_out <- A_dat[132:195,]

#=====================================================
# VAR
#=====================================================
> mse_var <- matrix(0,64,3)
> for (j in 1:64){
    m1 <- VAR(dat[1:(131+j-1),],p=1,output=FALSE,
include.mean=FALSE)
    Xf <- VARpred(m1,1)
    Xf <- Xf$pred
    error1 <- (A%*%Xf - A_dat_out[j,])^2
    mse_var[j,] <- error1
}
> m1 <- VAR(dat[1:(131+64-1),],p=1,output=FALSE,
include.mean=FALSE)
```

```
# estimation
> m1$coef # estimator
> m1$secoef # standard error
# forecast
> f <- VARpred(m1,1)$pred
> A%*%f

#==================================================
# The proposed
#==================================================
> mse_var <- matrix(0,64,3)
> for (j in 1:64){
    oo <- VARorder(A_dat[1:(131+j-1),],5)$aicor
    m1 <- VAR(A_dat[1:(131+j-1),],p=1,output=FALSE,
include.mean=TRUE)
    Xf <- VARpred(m1,1)
    Xf <- Xf$pred
    error1 <- (Xf - A_dat_out[j,])^2
    mse_var[j,] <- error1
}

> m1 <- VAR(A_dat[1:(131+64-1),],p=2,output=FALSE,
include.mean=FALSE)
# estimation
> M1 <- matrix(as.vector(rbind(as.character
(round(m1$Phi[1:3,1:3],2)),
            paste("(",round(m1$secoef[1:3,1:3],2),")",
sep=""))), nrow=6,byrow=FALSE)
> M1 # lag 1
> M2 <- matrix(as.vector(rbind(as.character
(round(m1$Phi[1:3,4:6],2)),
           paste("(",round(m1$secoef[4:6,1:3],2),")",
sep=""))), nrow=6,byrow=FALSE)
> M2 # lag 2
# forecast
> f2 <- VARpred(m1,1)$pred
> f2
```

R Code for the STD data

```
> Library(MTS)

> SexDise <- read.csv("C:/Bookdata/WW8c.csv", header=TRUE)
> A_MMWR <- as.matrix(read.csv("C:/Bookdata/WW8c-R9.csv"))
> A_STD <- as.matrix(read.csv("C:/Bookdata/WW8c-R4.csv"))
```

```
> d1 <- as.matrix(SexDise[,-1])
> rownames(d1) <- SexDise[,1]

# Take difference and mean-center the data
> d2 <- d1[-1,]-d1[-nrow(d1),]
> d3 <- (d2 - t(matrix(colMeans(d2),ncol(d2),nrow(d2))))/t
(matrix(apply(d2,2,sd),ncol(d2),nrow(d2)))

# Final data matrix
> Z <- as.matrix(d3)
# Identify # of locations (r) and # of time points (T)
> Z.fit <- Z[1:24,]
> r <- ncol(Z.fit)
> T <- nrow(Z.fit)

> Z.MMWR <- Z%*%t(A_MMWR)
> Z.MMWR.fit <- Z.MMWR[1:24,]

> Z.STD <- Z%*%t(A_STD)
> Z.STD.fit <- Z.STD[1:24,]

#=====================================================
#  MMWR
#=====================================================

> fit.VAR.MMWR = VAR(Z.MMWR.fit,p=1, include.mean = FALSE)

# estimation
> m2 <- VAR(Z.MMWR.fit,p=1,output=FALSE,include.mean=FALSE)
> M <- matrix(as.vector(rbind(as.character(round(m2
$Phi,2)),
                  paste("(",round(m2$secoef,2),")",
sep=""))), nrow=18,byrow=FALSE)
> M

# forecast
> f <- VARpred(fit.VAR.MMWR, h = 5, orig = 0,
Out.level = F)$pred
> f

#=====================================================
#  SDT
#=====================================================
> VARorder(Z.STD.fit, maxp = 4, output = T)
> fit.VAR.STD = VAR(Z.STD.fit,p=4, include.mean = FALSE)
```

```
# estimation
> m2 <- VAR(Z.STD.fit,p=4,output=FALSE, include.mean=FALSE)
> M1 <- matrix(as.vector(rbind(as.character(round(m2$Phi
[,1:4],2)),
                paste("(",round(m2$secoef[1:4,],2),")",
sep=""))), nrow=8,byrow=FALSE)
> M1 # lag 1
> M2 <- matrix(as.vector(rbind(as.character(round
(m2$Phi[,5:8],2)),
                paste("(",round(m2$secoef[5:8,],2),")",
sep=""))), nrow=8,byrow=FALSE)
> M2 # lag 2
> M3 <- matrix(as.vector(rbind(as.character(round
(m2$Phi[,9:12],2)),
                paste("(",round(m2$secoef[9:12,],2),")",
sep=""))), nrow=8,byrow=FALSE)
> M3 # lag 3
> M4 <- matrix(as.vector(rbind(as.character
(round(m2$Phi[,13:16],2)),
                paste("(",round(m2$secoef[13:16,],2),")",
sep=""))), nrow=8,byrow=FALSE)
> M4 # lag 4

# forecast
> f2 <- VARpred(fit.VAR.STD, h = 5, orig = 0,
Out.level = F)$pred
> f2
```

R Code for clustering

```
if("cluster" %in% rownames(installed.packages()) == FALSE)
{install.packages("cluster")}
if("pscl" %in% rownames(installed.packages()) == FALSE)
{install.packages("pscl")}
if("factoextra" %in% rownames(installed.packages()) ==
FALSE)
{install.packages("factoextra")}
if("MCMCpack" %in% rownames(installed.packages()) == FALSE)
{install.packages("MCMCpack")}
if("fBasics" %in% rownames(installed.packages()) == FALSE)
{install.packages("fBasics")}
if("mvtnorm" %in% rownames(installed.packages()) == FALSE)
{install.packages("mvtnorm")}

library(cluster)
library(astsa)
```

```
library(factoextra)
library(pscl)
library(MCMCpack)
library(mvtnorm)
library(fBasics)
source("MBcluster.R")

"MBcluster" <- function(data,p,K,mcmc,prior=NULL,
differ=FALSE,start.values=NULL){
# Perform Model-Based clustering analysis on the given data
# K              - number of clusters fitted for the data
# p              - number of lags of AR(p) model fitted for
each cluster
# prior          - prior=list(phi0,Sigma0,n0,s02,e0)
#                  the priors of AR parameters are normal
distribution
#                  with mean phi0 and covariance Sigma0; priors
of sigma2 are
#                  inverse-gamma distribution with alpha=n0/2
and beta=n0*s02/2;
#                  the priors of group indicator S are Dirichlet
distribution with
#                  parameter e0
# mcmc           - mcmc=list(burnin,rep)
#                  burnin is the burn-in period of mcmc;
#                  rep is the number of runs of mcmc
# start.values - start.values=list(phi,sigma2)
#                  where phi is a K*(p+1) matrix and sigma2
is a K*1 vector
data=as.matrix(data)
N=length(data[1,])  # Number of time series
colname=rep(0,N)
if (length(colnames(data))==0){
  for (n in 1:N){
    colname[n]=paste('Series ',n)
  }
}
else {colname=colnames(data)}

if (differ==TRUE){
  data=apply(data,2,diff )
}
T=length(data[,1])
## setup default prior
if(is.null(prior)){
```

```
  cat("Use default priors","\n")
  phi0=rep(0,p+1)
  Sigma0_phi = 10^4*diag(p+1)
  n0=2
  sig02=0.5
  e0=rep(3,K)
  prior=list(phi0=phi0,Sigma0_phi=Sigma0_phi,n0=n0,
sig02=sig02,e0=e0)
}
#
# Initialization
#
s=rep(0,N)
S=matrix(0,(mcmc$burnin+mcmc$rep),N)
phi=matrix(0,K*(mcmc$burnin+mcmc$rep),(p+1))
sigma2=rep(0,K*(mcmc$burnin+mcmc$rep))
pr=matrix(0,(mcmc$burnin+mcmc$rep),K)
Sigma_inv=NULL
Sigma1_inv_phi=NULL
eta=rdirichlet(1,prior$e0)
p0=log(eta)
n0=prior$n0
sig02=prior$sig02
phi0=prior$phi0
Sigma0_inv=kron(diag(K),solve(prior$Sigma0_phi))
n0sig02=n0*sig02
if (is.null(start.values)){
# use k-means to define initial values of ar-parameters
  init=kmeans(t(data),K)
  center=init$center
  for (k in 1:K){
   mfit=arima(center[k,],order=c(p,0,0),method='CSS')
# 1st element of phi is unconditional mean,
# 2nd to (p+1)th elements are ar-parameters
   phi[k,1]=mfit$coef[p+1]
   phi[k,2:(p+1)]=mfit$coef[1:p]
  }
  sigma2[1:K]=rigamma(1,n0/2,n0sig02/2)
}
else {
  phi[1:K,]=start.values$phi
  sigma2[1:K]=start.values$sigma2
}
# Use Gibbs sampler to draw group indicator S,
sigma2, and phi
```

```
for (run in 2:(mcmc$burnin+mcmc$rep)){
# Draw S
  lik=matrix(0,K,N)
  post=matrix(0,K,N)
  for (n in 1:N){
    y=data[T:(p+1),n]
    X=rep(1,T-p)
    for (q in 1:p){
      X=cbind(X,data[(T-q):(p+1-q),n])
    }
    #compute log-likelihood and posterior probability
    for (k in 1:K){
      lik[k,n]=sum(dnorm(y,X%*%phi[(run-2)*K+k,],
sqrt(sigma2[(run-2)*K+k]),log=TRUE))
      post[k,n]=p0[k]+lik[k,n]
    }
  post[,n]=post[,n]-max(post[,n])
  s[n]=sample(1:K,size=1,prob=exp(post[,n]),replace=TRUE)
  }
  ss=unique(s)
  # if there is no individual drawn to the cluster,
  # assign those with top [N/K] posterior probabilities
to that cluster
  a=rep(0,N)
  k=1
  while ((length(ss)<K)=K)){
    if (sum(ss==k)==0){
      s1=sort.list(post[k,a==0],decreasing=TRUE)
[1:floor(N/K)]
      s[s1]=k
      a[s1]=1
    }
    ss=unique(s)
    k=sample(1:K,1)
  }
  phi1=rep(0,(p+1)*K)
  Sigma_inv[[run]]=matrix(0,(p+1)*K,(p+1)*K)
  Sigma1_inv=matrix(0,(p+1)*K,(p+1)*K)
  for (k in 1:K){
    # Draw sigma2
    data_pool=data[,s==k]
    n1=n0+(T-p)
    n1sig12=n0sig02
    pooldim=sum(s==k)
    data_pool=matrix(data_pool,T,pooldim)
```

```
    y=c(data_pool[T:(p+1),])
    X=rep(1,T-p)
    for (q in 1:p){
      X=cbind(X,data_pool[(T-q):(p+1-q),1])
    }
    if (pooldim>1){
      for (m in 2:pooldim){
        Xm=rep(1,T-p)
        for (q in 1:p){
          Xm=cbind(Xm,data_pool[(T-q):(p+1-q),m])
        }
      X=rbind(X,Xm)
      n1=n1+(T-p)
      }
    }
    n1sig12=n1sig12+t(y-X%*%phi[((run-2)*K+k),])
%*%(y-X%*%phi[((run-2)*K+k),])
    sigma2[(run-1)*K+k]=rigamma(1,n1/2,n1sig12/2)
    # Draw phi
    # define phi1 and Sigma1_inv as the mean vector and
covariance matrix
    # of the posterior distribution of phi,
Sigma1_inv_phi=Sigma1_inv%*%phi1
    Sigma1_inv[((k-1)*(p+1)+1):(k*(p+1)),
((k-1)*(p+1)+1):(k*(p+1))]=
    Sigma0_inv[((k-1)*(p+1)+1):(k*(p+1)),
((k-1)*(p+1)+1):(k*(p+1))]+
    t(X)%*%X/sigma2[(run-1)*K+k]
    Sigma1_inv_phi[[k]]=
    Sigma0_inv[((k-1)*(p+1)+1):(k*(p+1)),((k-1)*(p+1)+1):
(k*(p+1))]%*%phi0+
    t(X)%*%y/sigma2[(run-1)*K+k]
    phi1[((k-1)*(p+1)+1):(k*(p+1))]=
    solve(Sigma1_inv[((k-1)*(p+1)+1):(k*(p+1)),
((k-1)*(p+1)+1):(k*(p+1))])%*%
    Sigma1_inv_phi[[k]]
    phi[((run-1)*K+k),]=rmvnorm(1,phi1[((k-1)*(p+1)+1):
(k*(p+1))],
    solve(Sigma1_inv[((k-1)*(p+1)+1):(k*(p+1)),
((k-1)*(p+1)+1):(k*(p+1))]))
    Sigma_inv[[run]][((k-1)*(p+1)+1):(k*(p+1)),
((k-1)*(p+1)+1):(k*(p+1))]=
    Sigma1_inv[((k-1)*(p+1)+1):(k*(p+1)),((k-1)
*(p+1)+1):(k*(p+1))]
  }
```

```
  S[run,]=s
  # print(run)
  # print(s)
  # print(phi[((run-1)*K+1):(run*K),])
  # print(sigma2[((run-1)*K+1):(run*K)])
}
phiGibbs=phi[-c(1:(K*mcmc$burnin)),]
sigma2Gibbs=sigma2[-c(1:(K*mcmc$burnin))]
SGibbs=S[-c(1:mcmc$burnin),]
km=kmeans(cbind(phiGibbs,sqrt(sigma2Gibbs)),K,nstart=25)
estim=km$centers
cluster=matrix(0,mcmc$rep,N)
pr=matrix(0,mcmc$rep,K)
SigmaGibbs=NULL
# label switching
for (i in 1:mcmc$rep){
  switch=sort.list(km$cluster[((i-1)*K+1):(i*K)],
decreasing=FALSE)
  phitemp=phiGibbs[((i-1)*K+1):(i*K),]
  phiGibbs[((i-1)*K+1):(i*K),]=phitemp[switch,]
  sigma2temp=sigma2Gibbs[((i-1)*K+1):(i*K)]
  sigma2Gibbs[((i-1)*K+1):(i*K)]=sigma2temp[switch]
  SigmaGibbs[[i]]=solve(Sigma_inv[[i+mcmc$burnin]])
  for (j in 1:N){
    cluster[i,j]=km$cluster[(i-1)*K+SGibbs[i,j]]
  }
}
Sigmahat=matrix(0,(p+1)*K,(p+1)*K)
for (i in 1:mcmc$rep){
  Sigmatemp=SigmaGibbs[[i]]
  for (k in 1:K){
    pr[i,k]=sum(cluster[i,]==k)/N
    SigmaGibbs[[i]][((k-1)*(p+1)+1):(k*(p+1)),
((k-1)*(p+1)+1):(k*(p+1))]=
    Sigmatemp[(switch[k]*(p+1)-p):(switch[k]*(p+1)),
          (switch[k]*(p+1)-p):(switch[k]*(p+1))]
  }
  Sigmahat=Sigmahat+SigmaGibbs[[i]]
}
prhat=apply(pr,2,mean)
Sigmahat=Sigmahat/mcmc$rep
#
prob=matrix(0,K,N)
for (k in 1:K){
  prob[k,]=apply(cluster==k,2,mean)
```

```
}
#
# Calculate Marginal Likelihood with Generalized Harmonic
Mean
#
logp=rep(0,mcmc$rep)
for (i in 1:mcmc$rep){
  if (min(pr[i,])>0){
  L=matrix(0,N,K)
  for (j in 1:N){
    y=data[T:(p+1),j]
    X=rep(1,T-p)
    for (q in 1:p){
       X=cbind(X,data[(T-q):(p+1-q),j])
    }
    for (k in 1:K){
      L[j,k]=log(pr[i,k])+sum(dnorm(y,X%
*%phiGibbs[(i-1)*K+k,],sqrt(sigma2Gibbs[(i-1)*K+k]),
log=TRUE))
    }
  }
  Lmax=max(L)
  Ld=L-Lmax
  logp[i]=logp[i]+N*Lmax+sum(log(apply(exp(Ld),1,sum)))
  }
}
logp=logp[logp>0]
A=max(logp)
logphat=log(1/mean(1/exp(logp-A)))+A
#
# Output
#
rn=rep(0,K)
for (i in 1:K){
  rn[i]=paste('Cluster',i)
}
rownames(estim)=rn
cn=rep(0,p+2)
cn[p+2]=paste('sigma')
for (i in 1:(p+1)){
  cn[i]=paste('phi',i-1)
}
colnames(estim)=cn

rn=rep(0,K)
for (i in 1:K){
```

```
  rn[i]=paste('Cluster',i)
}
rownames(prob)=rn
cn=rep(0,N)
for (i in 1:N){
  cn[i]=colname[i]
}
colnames(prob)=cn
cls=vector('list',K)
for (k in 1:K){
  if (length(which(prob[k,]==apply(prob,2,max)))>0){
    cls[[k]]=colname[which(prob[k,]==apply(prob,2,max))]
  }
}
clsn=rep(0,N)
for (i in 1:N){
  clsn[i]=which(prob[,i]==apply(prob,2,max)[i])[1]
}
#
cat('\n Estimation for Cluster Parameters:\n')
cat('Number of Clusters: K=',K,'\n')
cat('Number of Lags in AR model: p=',p,'\n')
print(estim,digits=5)
cat('\n Classification Probabilities:\n')
print(t(prob),digits=3)
cat('\n Classification:\n')
for (k in 1:K){
  if (length(cls[[k]])>0){
    cat('Cluster',k,':\n')
    print(cls[[k]])
  }
}
cat('\n Marginal LogLikelihood:',logphat,'\n')
#
#
#
result=list(estim=estim,prob=prob,pr=pr,
cluster=cls,clsindex=clsn,phiGibbs=phiGibbs,
sigma2Gibbs=sigma2Gibbs,
SigmaInv_Gibbs=Sigma_inv,S_draw=SGibbs,MarginalLik=logphat)
return(result)
}

SexDise <- read.csv("C:/Bookdata/WW8C.csv",header=TRUE)
d1 <- as.matrix(SexDise[,-1])
```

```
rownames(d1) <- SexDise[,1]
d2 <- diff(d1)
d3 <- (d2 - t(matrix(colMeans(d2),ncol(d2),nrow(d2))))
/t(matrix(apply(d2,2,sd),ncol(d2),nrow(d2)))

#=============================================
L = c(3,3)    # degree of smoothing
spec = mvspec(d3, spans=L, kernel="daniel", detrend=FALSE,
taper=0, plot=FALSE)
f = spec$fxx/(2*pi)  # estimated spectral matrix
freq = spec$freq
par(mfrow=c(4,5),mar=c(4.2, 3, 1, 1))
plot(spec$freq,Re(f[1,1,]),mgp=c(2,1,0),type='l',
lty=1,ylab="Spectrum",xlab="Frequency",
main=colnames(d3)[1])
plot(spec$freq,Re(f[2,2,]),mgp=c(2,1,0),type='l',
lty=1,ylab="Spectrum",xlab="Frequency",
main=colnames(d3)[2])
plot(spec$freq,Re(f[3,3,]),mgp=c(2,1,0),type='l',
lty=1,ylab="Spectrum",xlab="Frequency",
main=colnames(d3)[3])
plot(spec$freq,Re(f[4,4,]),mgp=c(2,1,0),type='l',
lty=1,ylab="Spectrum",xlab="Frequency",
main=colnames(d3)[4])
plot(spec$freq,Re(f[5,5,]),mgp=c(2,1,0),type='l',
lty=1,ylab="Spectrum",xlab="Frequency",
main=colnames(d3)[5])
plot(spec$freq,Re(f[6,6,]),mgp=c(2,1,0),type='l',
lty=1,ylab="Spectrum",xlab="Frequency",
main=colnames(d3)[6])
plot(spec$freq,Re(f[7,7,]),mgp=c(2,1,0),type='l',
lty=1,ylab="Spectrum",xlab="Frequency",
main=colnames(d3)[7])
plot(spec$freq,Re(f[8,8,]),mgp=c(2,1,0),type='l',
lty=1,ylab="Spectrum",xlab="Frequency",
main=colnames(d3)[8])
plot(spec$freq,Re(f[9,9,]),mgp=c(2,1,0),type='l',
lty=1,ylab="Spectrum",xlab="Frequency",
main=colnames(d3)[9])
plot(spec$freq,Re(f[10,10,]),mgp=c(2,1,0),type='l',
lty=1,ylab="Spectrum",xlab="Frequency",
main=colnames(d3)[10])
plot(spec$freq,Re(f[11,11,]),mgp=c(2,1,0),type='l',
lty=1,ylab="Spectrum",xlab="Frequency",
main=colnames(d3)[11])
```

```
plot(spec$freq,Re(f[12,12,]),mgp=c(2,1,0),type='l',
lty=1,ylab="Spectrum",xlab="Frequency",
main=colnames(d3)[12])
plot(spec$freq,Re(f[13,13,]),mgp=c(2,1,0),type='l',
lty=1,ylab="Spectrum",xlab="Frequency",
main=colnames(d3)[13])
plot(spec$freq,Re(f[14,14,]),mgp=c(2,1,0),type='l',
lty=1,ylab="Spectrum",xlab="Frequency",
main=colnames(d3)[14])
plot(spec$freq,Re(f[15,15,]),mgp=c(2,1,0),type='l',
lty=1,ylab="Spectrum",xlab="Frequency",
main=colnames(d3)[15])
plot(spec$freq,Re(f[16,16,]),mgp=c(2,1,0),type='l',
lty=1,ylab="Spectrum",xlab="Frequency",
main=colnames(d3)[16])
plot(spec$freq,Re(f[17,17,]),mgp=c(2,1,0),type='l',
lty=1,ylab="Spectrum",xlab="Frequency",
main=colnames(d3)[17])
plot(spec$freq,Re(f[18,18,]),mgp=c(2,1,0),type='l',
lty=1,ylab="Spectrum",xlab="Frequency",
main=colnames(d3)[18])
plot(spec$freq,Re(f[19,19,]),mgp=c(2,1,0),type='l',
lty=1,ylab="Spectrum",xlab="Frequency",
main=colnames(d3)[19])
plot(spec$freq,Re(f[20,20,]),mgp=c(2,1,0),type='l',
lty=1,ylab="Spectrum",xlab="Frequency",
main=colnames(d3)[20])

dev.new()
par(mfrow=c(4,5),mar=c(4.2, 3, 1, 1))
plot(spec$freq,Re(f[21,21,]),mgp=c(2,1,0),type='l',
lty=1,ylab="Spectrum",xlab="Frequency",
main=colnames(d3)[21])
plot(spec$freq,Re(f[22,22,]),mgp=c(2,1,0),type='l',
lty=1,ylab="Spectrum",xlab="Frequency",
main=colnames(d3)[22])
plot(spec$freq,Re(f[23,23,]),mgp=c(2,1,0),type='l',
lty=1,ylab="Spectrum",xlab="Frequency",
main=colnames(d3)[23])
plot(spec$freq,Re(f[24,24,]),mgp=c(2,1,0),type='l',
lty=1,ylab="Spectrum",xlab="Frequency",
main=colnames(d3)[24])
plot(spec$freq,Re(f[25,25,]),mgp=c(2,1,0),type='l',
lty=1,ylab="Spectrum",xlab="Frequency",
main=colnames(d3)[25])
```

```
plot(spec$freq,Re(f[26,26,]),mgp=c(2,1,0),type='l',
lty=1,ylab="Spectrum",xlab="Frequency",
main=colnames(d3)[26])
plot(spec$freq,Re(f[27,27,]),mgp=c(2,1,0),type='l',
lty=1,ylab="Spectrum",xlab="Frequency",
main=colnames(d3)[27])
plot(spec$freq,Re(f[28,28,]),mgp=c(2,1,0),type='l',
lty=1,ylab="Spectrum",xlab="Frequency",
main=colnames(d3)[28])
plot(spec$freq,Re(f[29,29,]),mgp=c(2,1,0),type='l',
lty=1,ylab="Spectrum",xlab="Frequency",
main=colnames(d3)[29])
plot(spec$freq,Re(f[30,30,]),mgp=c(2,1,0),type='l',
lty=1,ylab="Spectrum",xlab="Frequency",
main=colnames(d3)[30])
plot(spec$freq,Re(f[31,31,]),mgp=c(2,1,0),type='l',
lty=1,ylab="Spectrum",xlab="Frequency",
main=colnames(d3)[31])
plot(spec$freq,Re(f[32,32,]),mgp=c(2,1,0),type='l',
lty=1,ylab="Spectrum",xlab="Frequency",
main=colnames(d3)[32])
plot(spec$freq,Re(f[33,33,]),mgp=c(2,1,0),type='l',
lty=1,ylab="Spectrum",xlab="Frequency",
main=colnames(d3)[33])
plot(spec$freq,Re(f[34,34,]),mgp=c(2,1,0),type='l',
lty=1,ylab="Spectrum",xlab="Frequency",
main=colnames(d3)[34])
plot(spec$freq,Re(f[35,35,]),mgp=c(2,1,0),type='l',
lty=1,ylab="Spectrum",xlab="Frequency",
main=colnames(d3)[35])
plot(spec$freq,Re(f[36,36,]),mgp=c(2,1,0),type='l',
lty=1,ylab="Spectrum",xlab="Frequency",
main=colnames(d3)[36])
plot(spec$freq,Re(f[37,37,]),mgp=c(2,1,0),type='l',
lty=1,ylab="Spectrum",xlab="Frequency",
main=colnames(d3)[37])
plot(spec$freq,Re(f[38,38,]),mgp=c(2,1,0),type='l',
lty=1,ylab="Spectrum",xlab="Frequency",
main=colnames(d3)[38])
plot(spec$freq,Re(f[39,39,]),mgp=c(2,1,0),type='l',
lty=1,ylab="Spectrum",xlab="Frequency",
main=colnames(d3)[39])
plot(spec$freq,Re(f[40,40,]),mgp=c(2,1,0),type='l',
lty=1,ylab="Spectrum",xlab="Frequency",
main=colnames(d3)[40])
```

```
dev.new()
par(mfrow=c(1,4),mar=c(4.2, 3, 1, 1))
plot(spec$freq,Re(f[41,41,]),mgp=c(2,1,0),type='l',
lty=1,ylab="Spectrum",xlab="Frequency",
main=colnames(d3)[41])
plot(spec$freq,Re(f[42,42,]),mgp=c(2,1,0),type='l',
lty=1,ylab="Spectrum",xlab="Frequency",
main=colnames(d3)[42])
plot(spec$freq,Re(f[43,43,]),mgp=c(2,1,0),type='l',
lty=1,ylab="Spectrum",xlab="Frequency",
main=colnames(d3)[43])
plot(spec$freq,Re(f[44,44,]),mgp=c(2,1,0),type='l',
lty=1,ylab="Spectrum",xlab="Frequency",
main=colnames(d3)[44])

#===============================================
L = c(3,3)    # degree of smoothing
f = mvspec(d3, spans=L, kernel="daniel",
detrend=FALSE,
taper=0, plot=FALSE)$fxx/(2*pi)
l.eig <- c()
for(k in 1:dim(f)[3]){
    u <- eigen(f[,,k],symmetric= TRUE, only.values = TRUE)
    l.eig[k] <- u$values[1]
}
plot(spec$freq,l.eig,type='l',xlab="Frequency",
ylab="First Principal Component")

#===============================================
L = c(3,3)    # degree of smoothing
DD <- matrix(0,44,44)
f = mvspec(d3, spans=L, kernel="daniel", detrend=FALSE,
taper=0, plot=FALSE)$fxx/(2*pi)
for (i in 1:43){
    for(j in (i+1):44){
        for(k in 1:15){
            tr1 = Re(sum(solve(f[i,i,k], f[j,j,k])))
            tr2 = Re(sum(solve(f[j,j,k], f[i,i,k])))
            DD[i,j] = DD[i,j] + (tr1 + tr2 - 2*1)
        }
    }
}
DD = (DD + t(DD))/29
colnames(DD) = colnames(d3)
rownames(DD) = colnames(d3)
```

```
fviz_nbclust(DD, FUN = hcut, method = "wss")

#  partitioning around medoids
clu = pam(DD,4,diss=TRUE,stand=TRUE)
summary(clu)
clu$cluster==1
clusplot(DD, clu$cluster,col.clus=1, label=3, lines=0,
col.p=1,cex=0.6)
clu$cluster[clu$cluster==1]
clu$cluster[clu$cluster==2]
clu$cluster[clu$cluster==3]
clu$cluster[clu$cluster==4]

# hierarchical clustering
clu1=hclust(dist(DD),method = "single")
clu2=hclust(dist(DD),method = "complete")
clu3=hclust(dist(DD),method = "average")
clu4=hclust(dist(DD),method = "median")
fviz_nbclust(DD, FUN = hcut, method = "wss")
fviz_nbclust(DD, FUN = hcut, method = "silhouette")
par(mfrow=c(1,1))
plot(clu1, cex = 0.7)
plot(clu2, cex = 0.7)
plot(clu3, cex = 0.7)
plot(clu4, cex = 0.7)

###################################################
mcmc <- list(burnin=10000,rep=50000)
set.seed(19880901)
m1 <- MBcluster(d3,1,4,mcmc=mcmc)
```

Projects

1. Find a high-dimensional time series data set in a social science field, perform its principal component and factor analyses, and complete your analysis with a written report and attached software code.

2. Use the aggregation method to build an aggregate model for the data set from Project 1 and complete your analysis with a written report and attached software code.

3. Perform your analyses in Projects 1 and 2 with $(n - 5)$ observations, then compare the results in terms of their next five-step forecasts.

4. Find a high-dimensional time series data set in a natural science-related field, and perform its analysis with a written report and associated software code.

5. Find a real high-dimensional vector time series data set of your interest, use various dimension reduction methods to analyze your data set, make comparisons, and complete your analysis with a written report and associated software code.

References

Amemiya, T. and Wu, R.Y. (1972). The effect of aggregation on prediction in the autoregressive model. *Journal of American Statistical Association* **339**: 628–632.

Bai, J. and Ng, S. (2002). Determining the number of factors in approximate factor models. *Econometrica* **70**: 191–221.

Banfield, J. and Raftery, A. (1993). Model-based cluster Gaussian and non-Gaussian clustering. *Biometrics* **49**: 803–821.

Box, G.E.P., Jenkins, G.M., Reinsel, G.C., and Ljung, G.M. (2015). *Time Series Analysis, Forecasting and Control*, 5e. Wiley.

Box, G.E.P. and Tiao, G.C. (1977). A canonical analysis of multiple times series. *Biometrika* **64**: 355–370.

Breitung, J. and Swanson, N.R. (2002). Temporal aggregation and spurious instantaneous causality in multiple time series models. *Journal of Time Series Analysis* **23**: 651–666.

Brewer, K.R.W. (1973). Some consequences of temporal aggregation and systematic sampling for ARMA and ARMAX models. *Journal of Econometrics* **1**: 133–154.

Cattell, R.B. (1943). The description of personality: basic traits resolved into clusters. *Journal of Abnormal and Social Psychology* **38**: 476–506.

Chen, S.X. and Qin, Y.L. (2010). A two-sample test for high-dimensional data with applications to gene-set testing. *The Annals of Statistics* **38**: 808–835.

Cho, H. and Fryzlewicz, P. (2015). Multiple-change-point detection for high dimensional time series via sparsified binary segmentation. *Journal of the Royal Statistical Society, Series B* **77**: 475–507.

Cliff, A.D. and Ord, J. (1975). Model building and the analysis of spatial pattern in human geography. *Journal of the Royal Statistical Society, Series B* **37**: 297–348.

Flamm, C., Graef, A., Pirker, S., Baumgartner, C. and Deistler, M. (2013). Influence analysis for high-dimensional time series with an application to epileptic seizure onset zone detection. *Journal of Neuroscience Methods* **214**: 80–90.

Fraley, C. and Raftery, A. (2002). Model-based cluster clustering, discriminant analysis, and density estimation. *Journal of the American Statistical Association* **97**: 458–470.

Gao, J., Han, X., Pan, G., and Yang, Y. (2017). High dimensional correlation matrices: the central limit theorem and its applications. *Journal of Royal Statistical Society, Series B* **79**: 677–693.

Gehman, A. (2016). *The Effects of Spatial Aggregation on Spatial Time Series Modeling and Forecasting*, PhD dissertation, Temple University.

Giraud, C. (2015). *Introduction to High-Dimensional Statistics*. Chapman and Hall/CRC.

Hamilton, J.D. (1994). *Time Series Analysis*. Princeton University Press.

Hannon, E.J. (1970). *Multiple Time Series*. Wiley.

Hsu, N., Hung, H., and Chang, Y. (2008). Subset selection for vector autoregressive processes using lasso. *Computational Statistics and Data Analysis* **52**: 3645–3657.

Huang, M.Y. and Chiang, C.T. (2017). An efficient semiparametric estimation approach for the sufficient dimension reduction model. *Journal of American Statistical Association* **112**: 1296–1310.

Huang, H.C., Hsu, N.J., Theobald, D.M., and Breidt, F.J. (2010). Spatial lasso with applications to GIS model selection. *Journal of Computational and Graphical Statistics* **19** (4): 963–983.

Hung, H., Liu, C.Y., and Lu, H.H.S. (2016). Sufficient dimension reduction with additional information. *Biostatistics* **17**: 405–421.

Izenman, A.J. (2008). *Modern Multivariate Statistical Techniques*. Springer.

Johnson, R.A. and Wichern, D.W. (2007). *Applied Multivariate Statistical Analysis*, 6e. Pearson Prentice Hall.

Kalman, R.E. (1960). A new approach to linear filtering and prediction problems. *Transactions of the ASME – Journal of Basic Engineering, Series D* **82**: 35–45.

Kim, H.C., and Wei, W.W.S. (2018). *Block bootstrap and effect of aggregation and systematic sampling*. A research manuscript.

Kohn, R. (1982). When is an aggregate of a time series efficiently forecast by its past? *Journal of Econometrics* **18**: 337–349.

Koop, G.M. (2013). Forecasting with medium and large Bayesian VARs. *Journal of Applied Econometrics* **28**: 177–203.

Lee, B.H. and Wei, W.W.S. (2017). The use of temporally aggregation data on detecting a mean change of a time series process. *Communications in Statistics-Theory and Methods* **46** (12): 5851–5871.

Lee, N., Choi, H., and Kim, S.H. (2016). Bayes shrinkage estimation for high-dimensional VAR models with scale mixture of normal distributions for noise. *Computational Statistics and Data Analysis* **101**: 250–276.

Li, Z., and Wei, W.W.S. (2017). *The use of contemporal aggregation for high-dimensional time Series*. A research manuscript.

Lütkepohl, H. (1984). Forecasting contemporaneously aggregated vector ARMA processes. *Journal of Business and Economic Statistics* **2**: 201–214.

Lütkepohl, H. (1987). *Forecasting Aggregated Vector ARMA Processes*. Springer.

Lütkepohl, H. (2007). *New Introduction to Multiple Time Series Analysis*. Springer.

MacQueen, J. (1967). Some methods for classification and analysis of multivariate observations. In: *Proceedings of the 5th Berkeley Symposium on Mathematical Statistics and Probability*, vol. **1** (ed. L.M. Cam and J. Neyman), 281–297. University of California Press.

Matteson, D.S. and Tsay, R.S. (2011). Dynamic orthogonal components for multivariate time series. *Journal of the American Statistical Association* **106**: 1450–1463.

McLachlan, G. and Basford, K.E. (1988). *Mixture Models: Inference and Applications to Clustering*. Marcel Dekker.

Meyer, M. and Kreiss, J. (2015). On the vector autoregressive sieve bootstrap. *Journal of Time Series Analysis* **36**: 377–397.

Nicholson, W.B., Bien, J., and Matteson, D.S. (2016). High dimensional forecasting via interpretable vector autoregression, arXiv: 1412.525v2.

Pfeifer, P.E. and Deutsch, S.J. (1980a). A three-stage iterative procedure for space-time modeling. *Technometrics* **22** (1): 35–47.

Pfeifer, P.E. and Deutsch, S.J. (1980b). Identification and interpretation of the first order space-time ARMA models. *Technometrics* **22**: 397–408.

Pfeifer, P.E. and Deutsch, S.J. (1980c). Stationary and inevitability regions for low order STARMA models. *Communications in Statistics Part B: Simulation and Computation* **9**: 551–562.

Reinsel, G.C. (1997). *Elements of Multivariate Time Series Analysis*, 2e. Springer-Verlag.

Rose, D. (1977). Forecasting aggregates of independent ARIMA process. *Journal of Econometrics* **5**: 323–345.

Scrucca, L. (2010). Dimension reduction for model-based cluster clustering. *Statistics and Computing* **20** (4): 471–484.

Song, S., and Bickel, P. (2011). Large vector autoregressions, arXiv: 1106.3915.

Stock, J.H. and Watson, M.W. (2002a). Forecasting using principal components from a large number of predictors. *Journal of the American Statistical Association* **97**: 1167–1179.

Stock, J.H. and Watson, M.W. (2002b). Macroeconomic forecasting using diffusion index. *Journal of Business and Economics Statistics* **20**: 1147–1162.

Stock, J.H. and Watson, M.W. (2009). Forecasting in dynamic factor models subject to structural instability. In: *The Methodology and Practice of Econometrics* (ed. N. Shephard and J. Castle), 173–206. Oxford: Oxford University Press.

Stram, D.O. and Wei, W.W.S. (1986). Temporal aggregation in the ARIMA process. *Journal of Time Series Analysis* **7** (4): 279–292.

Teles, P., Hodgess, E., and Wei, W.W.S. (2008). Testing a unit root based on aggregate time series. *Communications in Statistics – Theory and Methods* **37**: 565–590.

Teles, P. and Wei, W.W.S. (2000). The effects of temporal aggregation on tests of linearity of a time series. *Computational Statistics and Data Analysis* **34**: 91–103.

Teles, P. and Wei, W.W.S. (2002). The use of aggregate time series in testing for Gaussianity. *Journal of Time Series Analysis* **23**: 95–116.

Teles, P., Wei, W.W.S., and Crato, N. (1999). The use of aggregate time series in testing for long memory. *Bulletin of International Statistical Institute* **3**: 341–342.

Tewes, J. (2018). Block bootstrap for the empirical process of long-range dependent data. *Journal of Time Series Analysis* **39**: 28–53.

Tiao, G.C. and Guttman, I. (1980). Forecasting contemporal aggregate of multiple time series. *Journal of Econometrics* **12**: 219–230.

Tiao, G.C. and Wei, W.W.S. (1976). Effect of temporal aggregation on the dynamic relationship of two time series variables. *Biometrika* **63**: 513–523.

Tibshirani, R. (1996). Regression shrinkage and selection via the lasso. *Journal of Royal Statistical Society, Series B* **58**: 267–288.

Tibshirani, R., Walther, G., and Hastie, T. (2001). Estimating the number of clusters in a data set via the gap statistics. *Journal of Royal Statistical Association, Series B* **63**: 411–423.

Tryon, R.C. (1939). *Cluster Analysis: Correlation Profile and Orthometric (Factor) Analysis for the Isolation of Unities in Mind and Personality*. Edwards Brothers.

Tsay, R.S. (2013). *Multivariate Time Series Analysis with R and Financial Applications*. Wiley.

Wang, S. and Zhou, J. (2008). Variable selection for model-based high-dimensional clustering and its application to microarray data. *Biometrics* **64**: 440–448.

Wang, Y., Tsay, R.S., Ledolter, J., and Shrestha, K.M. (2013). Forecasting simultaneously high-dimensional time series: a robust model-based clustering approach. *Journal of Forecasting* **32**: 673–684.

Ward, J.H. (1963). Hierarchical grouping to optimize an objective function. *Journal of the American Statistical Association* **58**: 234–244.

Wei, W.W.S. (2006). *Time Series Analysis – Univariate and Multivariate Methods*, 2e. Boston: Pearson Addison-Wesley.

Wei, W.W.S. and Abraham, B. (1981). Forecasting contemporal time series aggregates. *Communications in Statistics – Theory and Methods* **10**: 1335–1344.

Wei, W.W.S. and Stram, D.D.O. (1990). Disaggregation of time series models. *Journal of Royal Statistical Society, Series B* **52**: 453–467.

Wei, W.W.S. (2018). Dimension reduction in high dimensional multivariate time series analysis. To appear in *Contemporary Biostatistics with Biopharmaceutical Application*, ICSA Book Series in Statistics.

Weiss, A.A. (1984). Systematic sampling and temporal aggregation in time series models. *Journal of Econometrics* **26**: 255–278.

Wood, S.N., Li, Z., Shaddick, G., and Augustin, N.H. (2017). Generalized additive models for gigadata: modeling the U.K. black smoke network daily data. *Journal of the American Statistical Association* **112**: 1199–1210.

Yuan, M. and Lin, Y. (2006). Model selection and estimation in regression with grouped variables. *Journal of the Royal Statistical Society, Series B* **68**: 49–67.

Author index

Multivariate Time Series Analysis and Applications, First Edition. William W.S. Wei.
© 2019 John Wiley & Sons Ltd. Published 2019 by John Wiley & Sons Ltd.
Companion website: www.wiley.com/go/wei/datasets

Subject index

Multivariate Time Series Analysis and Applications, First Edition. William W.S. Wei.

© 2019 John Wiley & Sons Ltd. Published 2019 by John Wiley & Sons Ltd.

Companion website: www.wiley.com/go/wei/datasets

Printed and bound by CPI Group (UK) Ltd, Croydon, CR0 4YY

27/10/2024